Marine Particles: Analysis and Characterization

Geophysical Monograph Series

Including

IUGG Volumes
Maurice Ewing Volumes
Mineral Physics Volumes

GEOPHYSICAL MONOGRAPH SERIES

Geophysical Monograph Volumes

1 **Antarctica in the International Geophysical Year** *A. P. Crary, L. M. Gould, E. O. Hulburt, Hugh Odishaw, and Waldo E. Smith (Eds.)*
2 **Geophysics and the IGY** *Hugh Odishaw and Stanley Ruttenberg (Eds.)*
3 **Atmospheric Chemistry of Chlorine and Sulfur Compounds** *James P. Lodge, Jr. (Ed.)*
4 **Contemporary Geodesy** *Charles A. Whitten and Kenneth H. Drummond (Eds.)*
5 **Physics of Precipitation** *Helmut Weickmann (Ed.)*
6 **The Crust of the Pacific Basin** *Gordon A. Macdonald and Hisashi Kuno (Eds.)*
7 **Antarctica Research: The Matthew Fontaine Maury Memorial Symposium** *H. Wexler, M. J. Rubin, and J. E. Caskey, Jr. (Eds.)*
8 **Terrestrial Heat Flow** *William H. K. Lee (Ed.)*
9 **Gravity Anomalies: Unsurveyed Areas** *Hyman Orlin (Ed.)*
10 **The Earth Beneath the Continents: A Volume of Geophysical Studies in Honor of Merle A. Tuve** *John S. Steinhart and T. Jefferson Smith (Eds.)*
11 **Isotope Techniques in the Hydrologic Cycle** *Glenn E. Stout (Ed.)*
12 **The Crust and Upper Mantle of the Pacific Area** *Leon Knopoff, Charles L. Drake, and Pembroke J. Hart (Eds.)*
13 **The Earth's Crust and Upper Mantle** *Pembroke J. Hart (Ed.)*
14 **The Structure and Physical Properties of the Earth's Crust** *John G. Heacock (Ed.)*
15 **The Use of Artificial Satellites for Geodesy** *Soren W. Henricksen, Armando Mancini, and Bernard H. Chovitz (Eds.)*
16 **Flow and Fracture of Rocks** *H. C. Heard, I. Y. Borg, N. L. Carter, and C. B. Raleigh (Eds.)*
17 **Man-Made Lakes: Their Problems and Environmental Effects** *William C. Ackermann, Gilbert F. White, and E. B. Worthington (Eds.)*
18 **The Upper Atmosphere in Motion: A Selection of Papers With Annotation** *C. O. Hines and Colleagues*
19 **The Geophysics of the Pacific Ocean Basin and Its Margin: A Volume in Honor of George P. Woollard** *George H. Sutton, Murli H. Manghnani, and Ralph Moberly (Eds.)*
20 **The Earth's Crust: Its Nature and Physical Properties** *John G. Heacock (Ed.)*
21 **Quantitative Modeling of Magnetospheric Processes** *W. P. Olson (Ed.)*
22 **Derivation, Meaning, and Use of Geomagnetic Indices** *P. N. Mayaud*
23 **The Tectonic and Geologic Evolution of Southeast Asian Seas and Islands** *Dennis E. Hayes (Ed.)*
24 **Mechanical Behavior of Crustal Rocks: The Handin Volume** *N. L. Carter, M. Friedman, J. M. Logan, and D. W. Stearns (Eds.)*
25 **Physics of Auroral Arc Formation** *S.-I. Akasofu and J. R. Kan (Eds.)*
26 **Heterogeneous Atmospheric Chemistry** *David R. Schryer (Ed.)*
27 **The Tectonic and Geologic Evolution of Southeast Asian Seas and Islands: Part 2** *Dennis E. Hayes (Ed.)*
28 **Magnetospheric Currents** *Thomas A. Potemra (Ed.)*
29 **Climate Processes and Climate Sensitivity (Maurice Ewing Volume 5)** *James E. Hansen and Taro Takahashi (Eds.)*
30 **Magnetic Reconnection in Space and Laboratory Plasmas** *Edward W. Hones, Jr. (Ed.)*
31 **Point Defects in Minerals (Mineral Physics Volume 1)** *Robert N. Schock (Ed.)*
32 **The Carbon Cycle and Atmospheric CO_2: Natural Variations Archean to Present** *E. T. Sundquist and W. S. Broecker (Eds.)*
33 **Greenland Ice Core: Geophysics, Geochemistry, and the Environment** *C. C. Langway, Jr., H. Oeschger, and W. Dansgaard (Eds.)*
34 **Collisionless Shocks in the Heliosphere: A Tutorial Review** *Robert G. Stone and Bruce T. Tsurutani (Eds.)*
35 **Collisionless Shocks in the Heliosphere: Reviews of Current Research** *Bruce T. Tsurutani and Robert G. Stone (Eds.)*
36 **Mineral and Rock Deformation: Laboratory Studies—The Paterson Volume** *B. E. Hobbs and H. C. Heard (Eds.)*
37 **Earthquake Source Mechanics (Maurice Ewing Volume 6)** *Shamita Das, John Boatwright, and Christopher H. Scholz (Eds.)*
38 **Ion Acceleration in the Magnetosphere and Ionosphere** *Tom Chang (Ed.)*
39 **High Pressure Research in Mineral Physics (Mineral Physics Volume 2)** *Murli H. Manghnani and Yasuhiko Syono (Eds.)*
40 **Gondwana Six: Structure, Tectonics, and Geophysics** *Gary D. McKenzie (Ed.)*

41 **Gondwana Six: Stratigraphy, Sedimentology, and Paleontology** *Garry D. McKenzie (Ed.)*

42 **Flow and Transport Through Unsaturated Fractured Rock** *Daniel D. Evans and Thomas J. Nicholson (Eds.)*

43 **Seamounts, Islands, and Atolls** *Barbara H. Keating, Patricia Fryer, Rodey Batiza, and George W. Boehlert (Eds.)*

44 **Modeling Magnetospheric Plasma** *T. E. Moore and J. H. Waite, Jr. (Eds.)*

45 **Perovskite: A Structure of Great Interest to Geophysics and Materials Science** *Alexandra Navrotsky and Donald J. Weidner (Eds.)*

46 **Structure and Dynamics of Earth's Deep Interior (IUGG Volume 1)** *D. E. Smylie and Raymond Hide (Eds.)*

47 **Hydrological Regimes and Their Subsurface Thermal Effects (IUGG Volume 2)** *Alan E. Beck, Grant Garven, and Lajos Stegena (Eds.)*

48 **Origin and Evolution of Sedimentary Basins and Their Energy and Mineral Resources (IUGG Volume 3)** *Raymond A. Price (Ed.)*

49 **Slow Deformation and Transmission of Stress in the Earth (IUGG Volume 4)** *Steven C. Cohen and Petr Vaníček (Eds.)*

50 **Deep Structure and Past Kinematics of Accreted Terranes (IUGG Volume 5)** *John W. Hillhouse (Ed.)*

51 **Properties and Processes of Earth's Lower Crust (IUGG Volume 6)** *Robert F. Mereu, Stephan Mueller, and David M. Fountain (Eds.)*

52 **Understanding Climate Change (IUGG Volume 7)** *Andre L. Berger, Robert E. Dickinson, and J. Kidson (Eds.)*

53 **Plasma Waves and Istabilities at Comets and in Magnetospheres** *Bruce T. Tsurutani and Hiroshi Oya (Eds.)*

54 **Solar System Plasma Physics** *J. H. Waite, Jr., J. L. Burch, and R. L. Moore (Eds.)*

55 **Aspects of Climate Variability in the Pacific and Western Americas** *David H. Peterson (Ed.)*

56 **The Brittle-Ductile Transition in Rocks** *A. G. Duba, W. B. Durham, J. W. Handin, and H. F. Wang (Eds.)*

57 **Evolution of Mid Ocean Ridges (IUGG Volume 8)** *John M. Sinton (Ed.)*

58 **Physics of Magnetic Flux Ropes** *C. T. Russell, E. R. Priest, and L. C. Lee (Eds.)*

59 **Variations in Earth Rotation (IUGG Volume 6)** *Dennis D. McCarthy and William E. Carter (Eds.)*

60 **Quo Vadimus *Geophysics for the Next Generation* (IUGG Volume 10)** *George D. Garland and John R. Apel (Eds.)*

61 **Cometary Plasma Processes** *Alan D. Johnstone (Ed.)*

62 **Modeling Magnetospheric Plasma Processes** *Gordon R. Wilson (Ed.)*

Maurice Ewing Volumes

1 **Island Arcs, Deep Sea Trenches, and Back-Arc Basins** *Manik Talwani and Walter C. Pitman III (Eds.)*

2 **Deep Drilling Results in the Atlantic Ocean: Ocean Crust** *Manik Talwani, Christopher G. Harrison, and Dennis E. Hayes (Eds.)*

3 **Deep Drilling Results in the Atlantic Ocean: Continental Margins and Paleoenvironment** *Manik Talwani, William Hay, and William B. F. Ryan (Eds.)*

4 **Earthquake Prediction—An International Review** *David W. Simpson and Paul G. Richards (Eds.)*

5 **Climate Processes and Climate Sensitivity** *James E. Hansen and Taro Takahashi (Eds.)*

6 **Earthquake Source Mechanics** *Shamita Das, John Boatwright, and Christopher H. Scholz (Eds.)*

IUGG Volumes

1 **Structure and Dynamics of Earth's Deep Interior** *D. E. Smylie and Raymond Hide (Eds.)*

2 **Hydrological Regimes and Their Subsurface Thermal Effects** *Alan E. Beck, Grant Garven, and Lajos Stegena (Eds.)*

3 **Origin and Evolution of Sedimentary Basins and Their Energy and Mineral Resources** *Raymond A. Price (Ed.)*

4 **Slow Deformation and Transmission of Stress in the Earth** *Steven C. Cohen and Petr Vaníček (Eds.)*

5 **Deep Structure and Past Kinematics of Accreted Terranes** *John W. Hillhouse (Ed.)*

6 **Properties and Processes of Earth's Lower Crust** *Robert F. Mereu, Stephan Mueller, and David M. Fountain (Eds.)*

7 **Understanding Climate Change** *Andre L. Berger, Robert E. Dickinson, and J. Kidson (Eds.)*

8 **Evolution of Mid Ocean Ridges** *John M. Sinton (Ed.)*

9 **Variations in Earth Rotation** *Dennis D. McCarthy and William E. Carter (Eds.)*

10 **Quo Vadimus *Geophysics for the Next Generation*** *George D. Garland and John R. Apel (Eds.)*

Mineral Physics Volumes

1 **Point Defects in Minerals** *Robert N. Schock (Ed.)*

2 **High Pressure Research in Mineral Physics** *Murli H. Manghnani and Yasuhiko Syono (Eds.)*

Geophysical Monograph 63

Marine Particles: Analysis and Characterization

David C. Hurd
Derek W. Spencer
Editors

American Geophysical Union

Geophysical Monograph Series

Library of Congress Cataloging in Publication Data

Marine particles : analysis and characterization / David C. Hurd,
Derek W. Spencer, editors.
p. cm. — (Geophysical monograph : 63)
ISBN 0-87590-028-3
1. Marine sediments. I. Hurd, David C. II. Spencer, Derek.
III. American Geophysical Union. IV. Series.
GC380.15.M38 1991
551.46'083—dc20 91-30337
CIP

ISBN 0-87590-028-3

Printed in the United States of America.

TABLE OF CONTENTS

INTRODUCTION AND RATIONALE FOR THE WORKSHOP

David C. Hurd[1] and Derek W. Spencer[2]

[1] National Science Foundation, [2] Woods Hole Oceanographic Institution

In the past twenty or so years, there has been an increasing amount of interest in suspended particulates in seawater. This interest has concentrated on defining the distribution and variance of particle mass, composition, and flux. It is now well recognized that biogenic and other particle fluxes in the ocean define the principle mechanisms of non-conservative behavior that influence all reactive material. There is, however, concern that the analytical and sampling variability associated with our methods must be sufficiently within our control in order to discern important temporal and spatial variability of particles and fluxes. A review of papers in the recent literature which deal with analyzing the chemical and physical properties of particles collected by traps, pumps and other samplers, reveals that there is a considerable diversity of approaches and techniques being used by the community of active researchers in this field.

Many of the techniques have been in existence for some time yet there is no single source which provides a review and catalogue of existing methods for the characterization of marine particles. With the growing interest in marine particles a volume(s) that would bring together the collective experiences of many researchers would be a valuable resource for the community, as well as a primary teaching aid.

At this time it appears that a number of the techniques employed in marine particle research are quite robust. Others, to say the least, are less so, with some approaches entirely experimental and others pragmatic and/or arbitrary. The workshop and this volume try to obtain some consensus regarding the spectrum of techniques so that new persons entering the field would at least have a common ground for discussion with experienced workers. If several methods are equally valid, all have been included; if others have been shown to be clearly flawed, these have been noted with an explanation; if controversies have yet to be resolved, the issues have been clearly stated so that the general direction of progress is clear.

The range of disciplines and topics that are intrinsic to the 'Analysis and Characterization of Marine Particles' is large. To mention a few; the engineering of sampling devices, the chemical analysis of complex phases, the characterization of particle size, morphology and hydrodynamic properties, the determination of surface properties are all of concern for researchers interested in marine particles be they biologists, chemists, geologists or physicists. The size and diversity of the subject is such that some focussing of effort was deemed both advisable and necessary. This first workshop concentrated on the sampling and chemical analysis of marine particles with primary emphasis on the analytical techniques. Future workshops may address the optical, acoustic and hydrodynamic properties and the morphological, microbiological and biological aspects.

The workshop on sampling and chemical analysis of marine particles brought together leading researchers in the field of marine particle analysis. The object of this meeting has been to produce a monograph which would

1) be a comprehensive overview of the present practices,
2) provide an effective teaching resource for graduate level oceanography courses and
3) be a reference for experienced researchers.

It is important to note that although we have convened a representative cross-section of the most active workers in the field of particle analysis and characterization, not all who could have participated were able to be present for reasons of practical group dynamics. As a result, we are sure that the references listed, although representative of many important researchers' work are by no means complete. Students and others relatively new to the field are encouraged to use the general subject headings and such services as Citations Index and Subject Index to round out their portfolio of critical references for their areas of interest.

Workshop Format

In 1989, discussions with several of our colleagues outlined the need for a document that would bring together the experiences of researchers and provide a comprehensive review of current practices. Subsequently an organizational meeting was held in Washington D.C. to outline a procedure by which such a document could be produced. Participants at this meeting, subsequently to become the steering committee, defined a workshop as the principal focus of a community effort that would

Marine Particles: Analysis and Characterization
Geophysical Monograph 63

collect and review contributions from leading researchers in the analysis of marine particles. The workshop was held at the East-West Center in Hawaii on Jan 20-25, 1991. The workshop activities focused on four groups that cover the principal classes of analyses that are performed on marine particles. The groups, which concentrated on methods, accuracy, precision, intercalibration and reference standards, were:

Major Bioelements (C, H, O, N, P, S, Si)

Chair D. Karl, Rapporteur G. Harrison; Techniques for determinations of total mass flux and the principal nutrient chemical species.

Organic Properties

Chair C. Lee, Rapporteur S. Henrichs; Techniques for the major classes of organic compounds.

Inorganic and Radioactive properties

Co-chairs R. Anderson and G. Luther, Rapporteur R. Collier; Techniques for major and trace elements as well as stable and radioactive isotopes.

Mineralogical and Surface Properties

Chair P. Biscaye, Rapporteur D. DeMaster; Techniques for quantitative and qualitative mineralogical determinations and surface chemical properties.

In addition, a significant part of the workshop was devoted to sampling and sample manipulation techniques, topics that cross-cut the groups. Cross-cutting discussion sessions, which focussed on the interactions between sample handling techniques and property determinations, are:

Sampling and Sample Preservation

Chair R. Feely, Rapporteur J. Trefry; Sampling techniques, bottle and in situ filtration, sediment traps. Wet and dry sample preservation and archiving.

Sample Manipulations and Speciation

Chair W. Landing, Rapporteur P. Yeats; Sample manipulations including washing, species separation techniques, leaching and sample splitting.

The chairs of the above discussion groups and sessions together with the conference conveners, D. Hurd and D.W. Spencer, formed the Steering Committee.

The Steering Committee identified leading contributors to each of the above areas and letters of invitation were sent to potential participants. In order to have the workshop operate as efficiently as possible, participants were required to provide, by October 1, 1990, prior contributions as requested in the following excerpt from the invitation letter:

a) A description of sampling and analytical methods employed by your laboratory together with appropriate (and correct) references. In this regard, please do not restrict yourself to analytical or sampling methods that apply only to the group to which you have been assigned. The assignments are to ensure reasonable coverage of the subject matter and most of you perform analyses in more than one of the groups. Please provide descriptions, comments and references to methods of any type that you use.

b) Anecdotal information on technique -- the kind that may not get published yet contributes to your artistry.

c) Comments and suggestions for important workshop topics of discussion.

These contributions were initially edited by the chairs and then collated and distributed to all participants at the workshop. These contributions formed the basis for most of the final monograph. The workshop focussed on refining and revising the initial document and on the cross-cutting discussions attempting to resolve some of the more contentious procedures and techniques.

The ease with which corrections were made and new text was generated at the workshop was made possible by a number of the participants bringing portable computers; these added to the eight which had been rented specifically for the workshop. Also there were 'breakout rooms' available for cross-cutting discussions and small group writing activities. As a result of the hard work of the participants and such computer word processor availability, by the end of the workshop week, approximately 90% of the finished text was on diskette and in the hands of the editors. The editing and layout of the final manuscript was performed using Microsoft "Word for Windows".

In organizing this volume we have elected to present the results of the cross-cutting discussion sessions first. The material covered in chapters 1 and 2 was generated largely at the workshop and the majority of the more contentious issues concerning sampling and sample manipulations of various kinds are presented here. These issues must be faced by all researchers involved in the analysis of marine particles and they are presented first.

The remaining chapters expand on techniques that are currently prevalent for the measurement of the total suite of chemical and mineralogical properties.

A subject and author index are included as appendix I and II. In addition we include a glossary of abbreviations as appendix III and, rather than identify trade names throughout the volume, we include them together with a list of manufacturers that are referenced in the text in appendix IV.

Acknowledgements. The workshop and this volume were supported by the National Science Foundation and the Department of Energy under grant OCE20/823025 to the Woods Hole Oceanographic Institution. We are grateful to all the participants for their dedication and efforts, without which the production of this volume would not have been possible. We wish to thank, particularly, the session and group chairs for their excellent work in organizing and running the workshop break-out sessions and their attention to timely production of manuscripts. We thank Mr. James McMahon, Director of the East-West Center, and his staff for their services, cheerfully provided, during the workshop. The facilities of the East-West Center were ideal for our purposes. Both the workshop and this volume have benefitted greatly from the skills and perseverance of Mary Zawoysky and we appreciate the many hours spent by Helen Hurd in checking references.

Table I-1. WORKSHOP PARTICIPANTS

Conference Conveners:	
David C. Hurd	National Science Foundation
Derek W. Spencer	Woods Hole Oceanographic Institution
CROSS-CUTTING SESSION -- *Sampling and Preservation Techniques*	
Richard A. Feely *Chair*	NOAA/Pacific Marine Env. Laboratory
John H. Trefry *Rapporteur*	Florida Institute of Technology
Bruce Monger *Assist. Rapporteur*	University of Hawaii, Manoa
CROSS-CUTTING SESSION -- *Sample Manipulations and Speciation*	
William M. Landing *Chair*	Florida State University
Philip A. Yeats *Rapporteur*	Bedford Institute of Oceanography
Christopher Sabine *Assist. Rapporteur*	University of Hawaii, Manoa
Major Bioelements GROUP	
David M. Karl *Chair*	University of Hawaii, Manoa
W. Glen Harrison *Rapporteur*	Bedford Institute of Oceanography
John Dore *Assist. Rapporteur*	University of Hawaii, Manoa
Alice Alldredge	University of California, Santa Barbara
Gregory A. Cutter	Old Dominion University
Lewis E. Fox	Harvard University
George A. Knauer	University of Southern Mississippi
Patricia A. Matrai	University of Miami
Jonathan H. Sharp	University of Delaware
Stephen V. Smith	University of Hawaii, Manoa
Bodo von Bodungen	Universitat Kiel
Paul Wassmann	University of Tromsø
Organic Properties GROUP	
Cindy Lee *Chair*	State University of N. Y.
Susan M. Henrichs *Rapporteur*	University of Alaska
Ricardo Letelier *Assist. Rapporteur*	University of Hawaii
Ronald Benner	University of Texas at Austin

Table I-1. (continued)

Robert R. Bidigare	University of Hawaii, Manoa
Bruce Brownawell	State University of N. Y.
David Carlson	Oregon State University
Jan W. de Leeuw	Technical University of Delft
John Hedges	University of Washington
Kenneth Mopper	University of Miami
Brian N. Popp	University of Hawaii, Manoa
Daniel D. Repeta	Woods Hole Oceanographic Institution
Eiichiro Tanoue	Meteorological Research Institute, Tokyo
John K. Volkman	CSIRO Marine Laboratories
Stuart G. Wakeham	University of Georgia
Inorganic and Radioactive Properties Group	
Robert Anderson *Co-Chair*	Lamont-Doherty G.O., Columbia University
George Luther *Co-Chair*	University of Delaware
Robert W. Collier *Rapporteur*	Oregon State University
Peter Sedwick *Assist. Rapporteur*	University of Hawaii, Manoa
Michael P. Bacon	Woods Hole Oceanographic Institution
James Bishop	Lamont-Doherty G.O., Columbia University
James Cowen	University of Hawaii, Manoa
Keith Hunter	University of Otago
Kelly Kenison-Falkner	Massachusetts Institute of Technology
S. Bradley Moran	Dalhousie University
Michael J. Mottl	University of Hawaii, Manoa
Robert M. Sherrell	Massachusetts Institute of Technology
Edward R. Sholkovitz	Woods Hole Oceanographic Institution
Jens Skei	Norwegian Institute for Water Research
Herbert L. Windom	University of Georgia
Mineralogy and Surface Properties Group	
Pierre E. Biscaye *Chair*	Lamont-Doherty G.O., Columbia University
David DeMaster *Rapporteur*	North Carolina State University
Ursula Ginster *Assist. Rapporteur*	University of Hawaii, Manoa
Peter Betzer	University of South Florida
Andrew C. Campbell	Massachusetts Institute of Technology
Eric H. De Carlo	University of Hawaii, Manoa
Gerhard Fischer	Universitat Bremen
Bruce Honeyman	Texas A. & M. University
Jacques Jedwab[1]	Universite Libre de Bruxelles
Sjerry van der Gaast	Netherlands Institute for Sea Res.
Steering Committee	
Robert Anderson	Pierre E. Biscaye
Richard A. Feely	David C. Hurd
David M. Karl	William M. Landing
Cindy Lee	George Luther

[1] Jaques Jedwab was unable to attend the workshop. His written contribution is included in chapter 6.

CHAPTER 1. PARTICLE SAMPLING AND PRESERVATION

WORKSHOP REPORT

Introduction by

Chair: Richard A. Feely, Rapporteur: John H. Trefry, Assistant Rapporteur: Bruce Monger

1. An Introduction to Particle Sampling and Preservation

Suspended particles play a major role in the geochemical cycles of many elements and compounds in the oceans. Furthermore, chemical transformations and fluxes of particles in the water column have a large impact on the distributions of most dissolved substances. Recognition of this role has motivated oceanographers to develop innovative methods for studying particles in the sea over the past thirty years. The size spectrum, mineralogy and chemical composition of particles in the oceans can vary significantly from one environmental setting to another (Figure 1-1). Thus, marine scientists have employed a variety of different sampling techniques including shipboard filtration procedures, in situ filtration and sediment traps. In this overview chapter, we will outline present sample collection and preservation techniques while also identifying problems to be addressed in future research. The reader will be directed to specific chapters in this monograph for detailed descriptions of current problems and potential solutions.

1.1. Bottle Filtration Methods

At present, the largest body of information about the chemical composition of marine particles has been derived with samples collected by discrete bottles and filtered aboard ship. Samples are collected either by attaching Teflon-coated Niskin or Go-Flo bottles at various levels to a Kevlar line suspended from a ship, or by attaching Niskin bottles to a rosette which is lowered to specified depths and triggered. The samples are returned to the ship and processed by a variety of techniques that are described in subsequent chapters.

During the past thirty years, oceanographers have addressed many of the technical questions for collection of reliable samples. This includes procedures for collection and extraction as well as techniques for separation of suspended particles from seawater. Studies conducted in the late 1970's demonstrated that extraction of subsamples from Niskin bottles often resulted in data that were erroneously low in suspended mass determinations and elemental concentrations [Gardner, 1977; Calvert and McCartney, 1979]. The magnitude of the errors became obvious when suspended mass determinations by shipboard filtrations were compared with data collected by in situ filtration [Bishop and Edmond, 1976] (Figure 1-2). Suspended mass determinations of the same water mass (0-400 m) by in situ filtration were often several tens of percent to 100% higher than what was obtained by shipboard filtration using GEOSECS in-line vacuum filtration methodology. Clearly, some of the fast settling particles were not being removed from the Niskin bottles (Bishop, this volume, page 229). Periodic shaking of the samples and modification of the Niskin bottles so that all of the sample in the bottle could be filtered minimized this problem (Feely et al., this volume, page 251 ; Windom and Smith, this volume, page 317).

Chemists and oceanographers in the 1970's gathered a considerable amount of information on the physical dynamics and efficiency of filtration by various types of filters. The results are summarized in a number of papers [Sheldon and Sutcliffe, 1969; Sheldon, 1972; Bishop and Edmond, 1976; Stumm, 1977; Fan et al., 1978; Cranston and Buckley, unpublished, 1972]. Several experiments with glass beads and natural particulate material were carried out to determine particle loading, size distributions and particle retention characteristics of Millipore, Nuclepore, glass fiber and silver filters. Some filters lost mass during the experiments (Table 1-1). Studies with controls indicated that Nuclepore filters provided the best combination of attributes including low initial mass, low weight loss during filtration, high particle retention efficiency and high stability over time. One drawback of Nuclepore filters is that they tend to adsorb dissolved organic acids and possibly other organic substances on their surfaces and, consequently, analyses for particulate organic compounds using these filters are extremely difficult. Fortunately, glass fiber filters (e.g., Gelman AE, Whatman GF/C and GF/F) or microquartz fiber filters (Whatman QMA) are suitable for collection and analysis of particulate organic substances [Conte, 1989; Henrichs, this volume, page 139].

Nuclepore filters are true membrane filters with discrete holes of relatively specific pore size ranges. In contrast, glass fiber filters are made from overlapping

Marine Particles: Analysis and Characterization
Geophysical Monograph 63

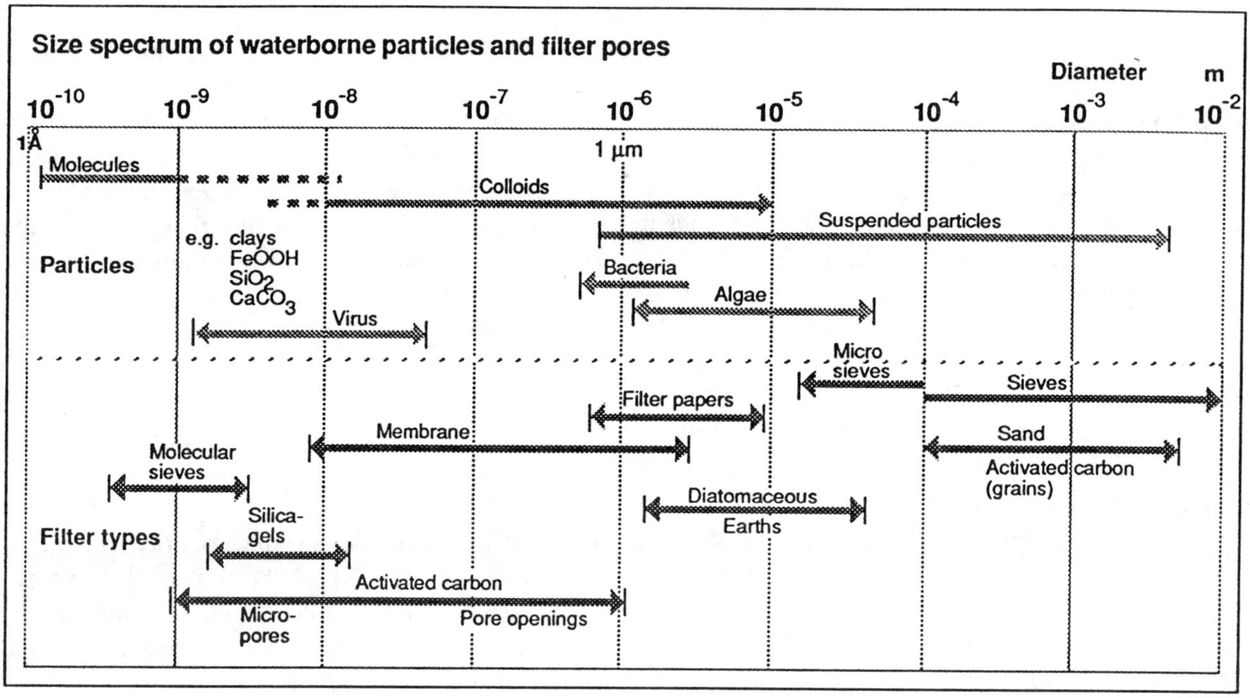

Fig. 1-1. Particle size spectrum of particulate materials in seawater and pore sizes of filters [after Stumm, 1977].

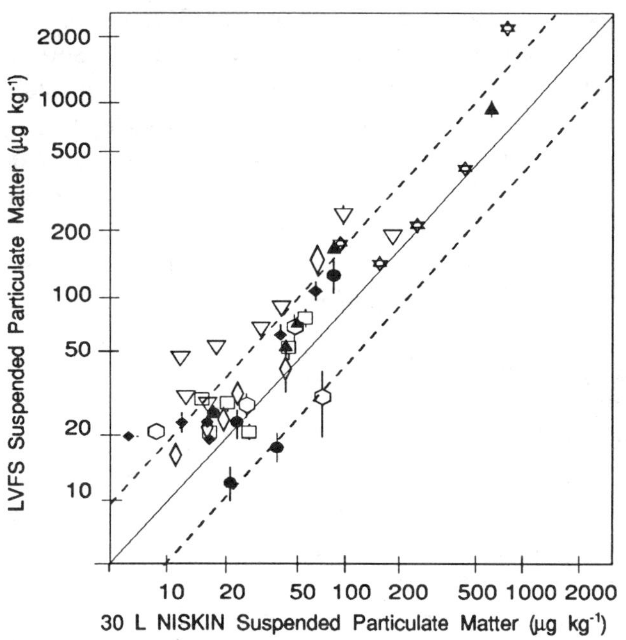

Fig. 1-2. Dry weight concentration intercalibration, LVFS and 30 L Niskin [from Bishop and Edmond, 1976]. The LVFS data have been corrected for a 20% error in flow meter calibration discovered in 1981 [reported in Bishop et al. 1987].

glass fibers. However, in terms of the size distributions of particles collected by these vastly different filter types, both retain about the same particles. There are several possible reasons for this (see Henrichs, this volume, page 139, for details), but it appears that both Nuclepore filters and glass fiber filters collect particles by adsorption to their surfaces rather than acting as true screens. Consequently, both filter types retain particles that are smaller than their nominal pore sizes.

Recently, several researchers have used other types of filter materials in their work. Sholkovitz (this volume, page 295) describes the use of hydrophilic Teflon-like, Durapore filters for studies of rare earth elements in fine-grained particles from estuaries. These filters have more mass than Nuclepore filters, but still provide good stability during filtration. They have the added advantage of reducing the incidence of sample leakage by providing a better seal against the filter holders. Cross-flow filtration, which has been used in microbiological research, has recently been applied by oceanographers to study trace metals in colloidal-sized particles [Whitehouse et al., 1986; Moran and Moore, 1989; Moran, this volume, page 275]. Anopore filters have been used for collection and analysis of marine bacteria. As these and other new filter types are added to our

TABLE 1-1.

Weight Changes in Grams in Control Filters During Particle Retention Experiments*, Data from Buckley and Cranston (unpublished manuscript, 1972)

Filter Type	Millipore MF	Nuclepore	Glass Fiber	Silver
Mean Weight Before	0.09444	0.01765	0.14050	0.48184
Mean Weight After	0.09419	0.01767	0.14034	0.48150
Mean Weight Change	-0.00025	+0.00002	-0.00016	-0.00034
Std. Dev. of Weight Change	0.00037	0.00006	0.00021	0.00013

* These values are based on 19 control filters used during various experiments. They were pre-weighed, exposed to a 1 L deionized water filtration during the respective experiment, and weighed at the end of the experiment along with the experimental filters.

arsenal for collection of marine particles, we must continue to be vigilant in our efforts to determine their relative particle size retention characteristics, weighing capabilities, and adsorption characteristics.

1.2. In Situ Filtration Methods

During the early 1970's, in situ filtration systems became popular for collection of large amounts of suspended particles. The in situ filtration systems generally collect much larger samples than conventional Niskin bottles. Powered by batteries or ship's power, in situ filtration systems can process as much as 25,000 L per sample (see summary table in Bishop, this volume, page 229). These large volume samples facilitate a variety of chemical, mineralogical, morphological, and biological measurements. When careful compositional studies are coupled with accurate sedimentation models, the data can be used to provide an estimate of material fluxes [Bishop et al., 1987]. In situ systems also provide a statistically better probability for capturing rare large particles often missed by conventional Niskin samples. Furthermore, because the samples are concentrated on the filters at depth, many of the potential sources of shipboard contamination are eliminated. The paper by Sherrell (this volume, page 285) provides a discussion of many shipboard procedures and sample handling techniques associated with in situ filtration. Possible drawbacks of in situ filtration systems are their complexity and cost ($2,000-$25,000 per unit).

The paper by Bishop (this volume, page 229) provides an excellent summary of comparisons between samples collected with Niskin bottles and samples collected with in situ filtration systems. As stated earlier, if proper procedures are not followed, subsampling from Niskin bottles can lead to underestimates of suspended mass and elemental concentrations. However, as Bishop points out, reasonably good agreement, within about 10 percent, can be obtained when proper procedures are followed. This is especially true for deep waters.

Figure 1-3 shows the results of our comparisons of suspended mass determinations using 30 L Niskin bottles suspended from a rosette and two different in situ filtration systems: the FIT in situ pump deployed 10 m above the rosette, and a McLane Water Transfer System deployed on the same rosette as the Niskin bottles. The samples were collected in the northeast Pacific in hydrothermal plumes over the Juan de Fuca Ridge. Even though the samples were collected in the hydrothermal plumes over different time scales (almost instantaneous for the Niskin samples and up to two hours of filtration for the in situ pump samples) there was remarkably good agreement between the three sets of data. The samples are currently being analyzed to identify any compositional differences. This type of experiment needs to be repeated for a variety of environmental settings before we can assume that such techniques are interchangeable. Although we may obtain good results for suspended mass determinations, we will not necessarily achieve consistent results for chemical composition. As an example, Bishop (this volume, page 229) reports tenfold differences in particulate organic nitrogen concentrations between bottle samples and in situ filtration samples during the Warm Core Rings Experiment in the Atlantic Ocean. In this case, the tenfold higher nitrogen values from bottle samples appeared to be an artifact of handling the water sample prior to filtration. The exact reasons for these differences are still uncertain; however, it is clear that more comparisons of this kind are required before we can say that we fully understand how differences in sample handling procedures, particle loadings on the filters, and filter pore sizes affect the final results.

1.3. Sample Preservation Methods

Once marine particles are collected on filters, screens, or in sediment traps, they can continue to undergo physical and chemical changes which alter their composition. This is due to a number of factors including changes that the particles were undergoing in the ocean and changes resulting from the physical and chemical conditions of the particles "new environment". This is particularly true for biogenic components in the samples which can undergo cell lysis, bacterial decomposition, dissolution, and disruption with only mild changes in their environmental setting. The trick for the analyst is to develop an analytical scheme which preserves the samples as they existed in nature. Of course, this is somewhat dependent upon the suite of measurements being planned for the samples. What may be an appropriate preservation step for one kind of analysis could be detrimental for another.

Particularly good examples of the preservation dilemma are described in the papers by Trefry and Trocine, Jedwab, and Windom and Smith, pages 311, 399, 317, in this monograph. In order to preserve filtered samples for the determination of mass loadings many investigators wash their filters with distilled water adjusted to pH 8.0 ± 0.3 with Ultrex ammonium hydroxide to remove sea salts from the filter. They try to keep the volume of rinse water to a minimum to prevent extensive breakdown of the biogenic components in the sample. Although this procedure is very important for suspended mass determinations, the paper of Collier (this volume, page 235) suggests that even under these mild conditions some cell breakdown occurs and some of the biogenic components of the samples are lost to the filtrate. To prevent this from occurring, Windom and Smith (this volume, page 317) recommend storing the samples saturated with a small volume of ambient seawater for shipment to the laboratory. However, Jedwab (this volume, page 399) has suggested that storing samples with some amount of seawater on the filters enhances the possibility of formation of unwanted gypsum particles on the filters which can lead to erroneously high mass loadings. Collier (this volume, page 235) points out that the release of biogenic components to the rinse water is time dependent and recommends that: 1) the rinses are few in number and volume; 2) the rinsing process is conducted quickly; and 3) the rinse water is saved and analyzed for biogenic components.

Similar dilemmas exist for other procedures as well. For many organic analyses, an important step in the preservation process calls for freezing the samples at -30° to -20° C for periods of up to several months prior to analysis. Freezing can cause some biological materials to break down and release cellular material (e.g., lipids). Some procedures call for quick freezing of the samples to minimize this effect.

In-situ Pump vs Niskin Bottle Comparison of Total Suspended Matter Concentrations in µg/l

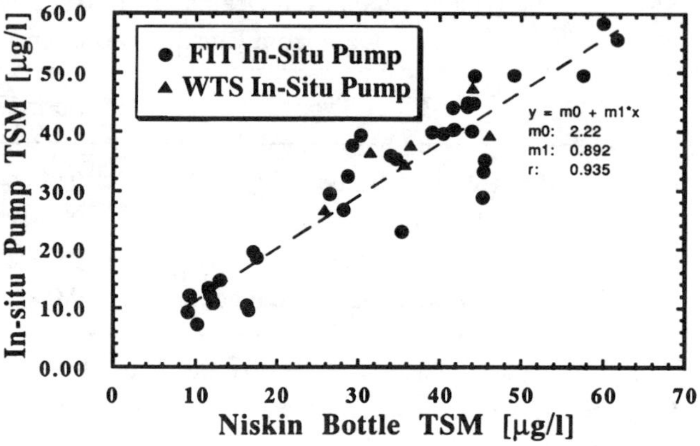

Fig. 1-3. Plot of the comparison of suspended mass determinations by in situ filtration (FIT in situ pump (Sherrell, this volume, page 285) and McLane Water Transfer System) versus 30 L Niskin bottle samples using pressure filtration.

1.4. Preservation Procedures For Sediment Trap Samples

In recent years, many oceanographers have extended their research on marine particles by developing sediment trap procedures for determining fluxes of chemical, mineralogical, and biogenic components through the water column. These studies have great potential for elucidating geochemical cycles for many elements and compounds in the oceans. However, the results are only useful if the integrity of samples remains intact during collection (see discussion in Knauer, this volume, page 79). To be most accurate, the samples collected with sediment traps must retain their original composition even though they may spend up to several months in the traps. Consequently, many investigators have elected to use various kinds of "preservatives" to ensure the integrity of the samples. In a recent review of the use of preservatives, the U.S. GOFS Working Group on sediment trap technology and sampling [U.S. GOFS Working Group, 1989] made the following recommendations:

1. For most applications, a poison or preservative should be used in sediment trap deployments, regardless of duration of deployment. If short-term experiments

are conducted without a poison or preservative, then replicate traps with poison/preservative should be used.

2. The specific poison/preservative to be used (or lack thereof) should be determined by the scientific objectives of the trap experiment.
3. The concentration of the poison/preservative should be determined both before and after deployment to ensure an effective concentration level during deployment.
4. Poisons, preservatives or other additives should be prepared in a seawater-based medium of a density greater than the seawater surrounding the sediment traps.
5. For carbon flux (soluble + particulate) measurements, there is no universally adequate poison/preservative; all of the treatments listed either add soluble carbon for which adequate corrections are not feasible at present or introduce potentially serious artifacts from swimmers which cannot be removed completely. Particulate organic carbon flux determined after picking of macro-zooplankton swimmers should be regarded as a best estimate until adequate prevention of swimmers can be developed. Picking out the swimmers can result in either an underestimation or overestimation of flux depending on the community present.
6. Future work should be directed at eliminating the swimmer problems and in determining whether soluble carbon measurements can be made in the presence of the large C background resulting from the use of formalin.

Table 1-2 provides a list of some of the commonly used poisons and preservatives along with their advantages and disadvantages [U.S. GOFS Working Group, 1989]. No combination of poisons/preservatives is suitable for all kinds of sediment trap experiments. Consequently, more research is required to determine the best combination of poisons/preservatives for the various kinds of experiments that would be suitable for sediment traps.

Another problem with sediment traps is that many of the constituents of interest are solubilized via chemical and microbiological processes during the deployment or after sediment trap recovery. This is particularly true for several of the biogenic components (e.g., C, N and P). The loss of components to the soluble pool can be significant, up to 50 to 90% in some cases, and can have serious impacts on the flux calculations. The soluble fraction of the particulate flux can also be lost due to diffusion and mixing with the overlying water while the sample cup is open for collection.

Although some of these losses can be minimized by using density solutions in the traps and by analyzing the soluble fraction in the cups for the elements of interest, further research on preservatives which prevent dissolution and/or degradation from occurring needs to be implemented.

1.5. Procedures For Marine Snow

Marine snow particles are delicate aggregates of algae, bacteria, fecal matter and biogenic detritus larger than 500 μm in diameter. Because of their large size and sometimes high abundances, particularly in the upper regions of the oceans, they play an important role in the biogeochemical cycles of many elements and compounds. Because these particles are large, rare and extremely delicate relative to microscopic particles, oceanographers have developed specialized techniques to collect the particles individually using SCUBA and submersible techniques. Alldredge (this volume, page 46) describes several procedures for collecting intact individual aggregates underwater using various sizes of syringe barrels. For trace element studies, precleaned Teflon tubing is attached to the syringes and aggregates are sucked into the tubing along with the surrounding water. The samples are processed in much the same way that sediment trap samples are processed with all of the same precautions regarding dissolution and degradation losses.

1.6. Procedures For Samples From Reducing Environments

Collection and preservation of particles from anoxic waters require special handling procedures. As discussed in Skei (this volume, page 303), several important geochemical reactions occur at the interface between oxic and anoxic waters and, therefore, special care must be taken to avoid any exposure to oxygen during filtration and sample handling. If samples are filtered aboard ship, they should be kept in an inert atmosphere, such as nitrogen, until the samples are dried in a desiccator. Similar precautions are required for reduced minerals in hydrothermal vent particles. The paper by Campbell (this volume, page 375) recommends filtration and storage of hydrothermal particles under an argon atmosphere.

TABLE 1-2.
Common Poisons and Preservatives, Advantages and Disadvantages [From U.S. GOFS Working Group, 1989]

Additive	Advantages	Disadvantages/Precautions	Concentration
None	• live in situ experiments	• transient swimmers can alter apparent flux • sample may degrade during collection	
Preservatives			
Formalin	• hardens tissue (advantage for physical removal of swimmers) • good for SEM • effective biocide	• binds proteins • may interfere with total (soluble) C determination • may be titrated during deployment as it is incorporated into tissue • interferes with stable isotope measurements • pH sensitive, needs careful buffering to prevent carbonate dissolution • may be contaminated with N and interfere with soluble N determination	2-5% buffered
Glutaraldehyde	• as for formalin • good for SEM and TEM	• may precipitate in seawater • needs careful buffering with arsenic (which interferes with the measurement of P)	
Salt (brine)	• used as gradient to retain poison/ preservative in trap • high density keeps it in the bottom of trap	• poor biocide • may attract swimmers • enhances carbonate dissolution	30-300 gL^{-1}
Poisons			
Chloroform	• may prevent extra-cellular enzyme activity	• dissolves pigments and lipids • causes difficulty in splitting of samples • may attract swimmers	neat or saturated seawater
Sodium azide	• introduces no particulate carbon • good buffer for calcium carbonate	• interferes with N determination • may interfere with particulate N determination	10-15 $g L^{-1}$
Mercuric chloride	• good biocide	• may introduce heavy metal contaminants • precipitation in sulfidic waters may reduce effective concentration • may alter redox state in trap (e.g., reduction of Hg in metal cod end in sub-oxic waters)	10 $mL L^{-1}$ saturated solution
Antibiotics	• useful for specialized biocide applications	• generally ineffective	200 $mg L^{-1}$

2. The Workshop Report by Members of the Working Group "Particle Sampling and Preservation"

The following report from the Working Group on Sampling and Preservation and the subsequent chapters in this monograph provide a comprehensive summary of our present state of knowledge concerning sampling and preservation procedures for marine particles. They are intended to be used as a technical aid for students and scientists in designing future research studies. These reports show that we have made many technical advancements in our understanding of sample collection and preservation procedures; however, there are still many unanswered questions. We hope that this monograph will serve to stimulate development of new techniques and approaches for studying marine particles.

The cross-cutting section on sampling and preservation techniques concentrated their deliberations on identifying and describing the best approaches for collection and preservation of open ocean particles. The discussion focused on how materials and techniques vary as a function of the analyses planned for the samples. Four general groups of analyses were considered as follows: inorganic components, organic components, minerals and surface properties, and bioelements.

In addition to discussing specific techniques, considerable emphasis was given to a comparison of the collection of particles with bottle samplers versus in situ pumps. A common theme throughout the discussion was that the choice of specific techniques and experimental design are directly related to the problem at hand or the hypothesis being tested. The reader is encouraged to read the workshop report and the accumulated papers in this context.

2.1. Contamination

Contamination concerns during the sampling of marine particles are summarized in Table 1-3.

2.1.1. Inorganics.

Sources of contamination during shipboard collection and handling of marine suspended particles include the following: hydrowires, rosette frames and sacrificial anodes, natural surface slicks and surface slicks associated with ship's bilge, ships garbage and the ship's laboratory atmosphere. Contamination from hydrowires can be minimized by the use of clean stainless steel or, perhaps preferably, by use of Kevlar line. Often a length of Kevlar line is attached to the end of the ship's hydrowire and separated on the hydrowinch drum from metallic line using plastic sheeting. Sample bottles deployed on a rosette are still exposed to potential contamination from the metal frame and sacrificial zinc anodes. During deployment, definitive isolation of the sample bottles from these sources of contamination is difficult; however, use of plastic coatings and stainless steel anodizing on rosettes and associated equipment will help minimize potential sources. While on deck, sample bottles should be covered with plastic to isolate them from the rosette and other sources of contamination.

The discharge of bilge and garbage during deployment of sampling gear should not be allowed. This is a simple management problem. Avoiding surface slicks is more difficult. The use of Go-Flo sample bottles minimizes contamination by surface slicks because the bottles open only after submerging to depths of about five meters. Nevertheless, the outside of the bottle can still be contaminated during passage through the slick. "Soaking" for a period of time after deployment should minimize contamination of bottles and pumps due to passage through surface slicks.

A major opportunity for contamination is during sample handling and processing in the ship's laboratory after collection. To minimize this problem, transfer from sample bottle to filtering apparatus should be carried out in a closed system as much as possible. Any air entering the sample bottle as water is withdrawn should be filtered. All filtration should be conducted in clean benches or in modular clean vans. All manipulation of filters should also be carried out in similar clean environments.

2.1.2. Organics.

There are very real concerns about potential contamination during collection and shipboard handling of particle samples to be subjected to organic analyses.

TABLE 1-3.

Contamination Concerns when Sampling Marine Particles

Particle Analyses	Potential Contaminants
Inorganic	Hydrowire, rosette (Zn anodes), sample bottles, ship discharges, surface slick, filtration area aboard ship, ship sources of dust, stack gases, paint chips
Organic	Same general concerns as for the inorganic group plus additional concerns as listed below: For lipids: oil, sewage discharges, suntan lotion For amino acids: hair, fingerprints, smoking, ammonia from cleaning solutions, powdered gloves For lignins: paper towels, kimwipes, pencil shavings
Minerals	Gloves with talc, sand/litter used for spills, ship's air filter
Bioelements	Similar to organics plus stack discharge

The specific sources of contamination are different depending on the organic compounds to be determined as described below. A major source of contamination for one set of compounds may present no problem for another. In general, however, shipboard operations tend to be relatively "dirty" with respect to most organic compounds, and scientific operations used for one type of sampling may present conflicting requirements with those used for other types (i.e. sampling for organic materials entails markedly different constraints than those used for trace metal work). A common set of precautions may apply for universal organic geochemical work. For example, oil and sewage discharges from the ship must be avoided, as should slicks at the air-sea interface, unless that is the phenomenon being studied. Airborne dust and stack exhaust particles must be avoided as samples are filtered, i.e. as air enters pressure or vacuum filtration systems, and as filters are handled. Powdered gloves should not be used. Suntan lotion can be a source of contamination for several compound classes. Blank filters must be rigorously carried through all sampling and handling procedures. Compound-specific concerns are listed below.

Lipids: The fuel and lubricating oils, paints, stack condensates, and sewage discharges inherent in ship operations and the environments around ships are all potential sources of catastrophic contamination. This is especially true for hydrocarbons, but other lipid classes may be adversely affected as well. Sampling gear, such as hydrowire, water bottles, pumps, particle traps, etc., must be closely protected from this type of ubiquitous contamination, and all may need to be cleaned extensively prior to use.

Amino acids: The main sources of amino acid contamination are biological products, both human and otherwise. Caution must be taken to prevent the introduction of hair, eyelashes, fingerprints, and residual organic matter from previous samples into samples. Many cleaning solutions and all cigarette smoke contain ammonia which will interfere with amino acid analyses.

Sugars: Cellulose-based products, such as paper and starch-powdered gloves are the major potential sources of contamination of samples to be analyzed for sugars.

Lignins: Since lignins are uniquely derived from wood, the sources of lignin contamination obviously are wood-based products, such as pencils, paper, and wood dust and chips.

Pigments: The high molecular-specificity of pigments to nearly exclusively phytoplankton sources means that contamination of pigment samples is a relatively minor problem. None of the contamination sources listed above is relevant to pigment analyses.

2.1.3. Minerals and surfaces.

Potential contaminants include any mineral phases present aboard ship. A common concern here is for talc. Thus, talc-free gloves should be used. It should be noted that talc is abundant globally and is used as a carrier for pesticides in aerial spraying. When talc is a mineral of interest, precaution must be heightened.

Other shipboard contaminants include stack effluents, paint chips and the sand or "kitty litter" used as an absorbent for oil spills on deck. Concerns for contamination on deck prompt use of a filter on the air inlet to 30 liter Niskin bottles so that the air drawn in as water is removed is clean.

2.1.4. Bioelements.

Most contamination concerns previously described for sampling marine particles apply when samples are to be analyzed for bioelements. During sample collection, the most important potential sources of contaminants are the sampling devices themselves. Thorough pre-cleaning (usually with HCl and double distilled water) is necessary to minimize this problem. Other concerns, particularly for near-surface samples, are ship discharges and particles adhered to the sea surface microlayer.

When a sampler is brought on deck, care must be taken to avoid introduction of dust, hair, oil and other shipboard materials such as stack effluents and paint chips which may contain the elements of interest. Sediment traps should be capped immediately upon retrieval to prevent these types of contamination. When using Niskin or Go-Flo bottles, direct in-line filtration can be used to remove the possibility of sample/contaminant contact during transfer.

Filtration equipment may also be a source of contamination, and also must be pre-cleaned. If pressure filtration is utilized, the source gas should be passed through an air filter. Filters themselves may contaminate for particular bioelements and thus proper choice of filter and treatment can be critical (see section on filter types below).

Contamination for bioelements during sample storage is unlikely, provided that filters are stored in clean, sealed containers. If freezing is required, it would be prudent to keep samples well away from other sources of organic matter such as fish.

2.2. Filter Types

Filtration media fall into two broad categories, sieves and depth filters, based on the physical morphology and mechanism of particle separation. Sieves have uniform pores which prevent passage of larger particles and have a relatively strict cut-off dimension whereas depth filters have pores with non-uniform, tortuous pathways which separate particles by a combination of size restriction, impaction and sticking. Table 1-4 summarizes the characteristics of several kinds of filters used in the sampling and analysis of marine particles.

2.2.1. Sieve filters.

Polycarbonate membrane filters. Polycarbonate, e.g. Nuclepore, filters are the primary sieve type filters which have been broadly used for collecting marine particles. They have the advantages of low tare weight, low water retention and low blanks for many elements. Principal disadvantages include (1) incompatibility with organic carbon analysis because of the polycarbonate matrix and because of possible adsorption of DOC; (2) low porosity which results in low flow rates and low loading capacity.

Because this is a sieve filter, particles are retained on the surface of the filter plane. This makes polycarbonate filters suitable for direct optical and chemical analysis of particles on the filter. These techniques include light and electron microscopy, x-ray fluorescence, x-ray diffraction, scanning electron microscopy, optical image analysis, and other solid phase spectroscopic methods. The flat smooth surface also allows particles to be removed by scraping and/or sonication before analysis. This property also allows lateral movement of particles during sampling and handling, and could result in inhomogeneous distribution across the filter face. These filters have a smoother side which is preferred as the sample surface for optical techniques. However, it has been reported that using the other side may increase flow rates.

Significant blanks have been reported for Cr and Br on these filters. Chromium can be removed by acid leaching with concentrated HCl and low Br batches of the filters are available from Nuclepore. The filter matrix can be digested in strong base (such as ammonium hydroxide) and certain organic solvents (see manufacturers' specifications). It has been reported that oxidation of the filters with nitric acid is relatively easy after the filters have first been digested in ammonium hydroxide.

Nuclepore filters are often considered incompatible for analysis of bioelements because destructive analysis methods are used. It is possible to remove material from filters for CNP analysis and for isotopic analysis of C and N (Knauer and Sharp, this volume, pages 79 and 87). Nuclepore polyester filters have been used to collect particles for pigment analysis, but polycarbonate filters should not be used as they contain a yellow pigment (Bidigare, this volume, page 119).

Alumina filters. Anopore filters are high porosity aluminum oxide sieve filters with a uniform honeycomb structure which results in relatively high initial flow rates compared to polycarbonate sieves. They have been used for collection of particles for total organic analysis and stable N isotope determination [Altabet, 1990]. Anopore filters are untested for trace elements, but the metallic matrix probably precludes use for analysis of some trace metals. The Anopore filter appears desirable for analysis of bioelements because of its superior retention of submicron particles. At present, major problems exist for handling of filters for automated CN analysis.

Silver filters. Silver filters are microporous metallic membranes which have been used for organic matter analysis and some mineralogical applications. The membranes are relatively fragile, difficult to handle, and are expensive. As discussed for Anopore filters, the silver matrix is inappropriate for many trace metal applications and is not widely used.

TABLE 1-4. Summary of Filter Characteristics

Filter Types	Polycarbonate	Silver	Alumina*	Cellulose	Micro-quartz	Glass fiber	Fluorocarbon**
Manufacturers	m, n, p	f, m	i	g, m, n, w	w	w	g, m
TSM	good	good	ok	ok	ok	ok	ok
Depth vs. Sieve	sieve	sieve	sieve	depth	depth	depth	depth
Recommended for:							
organic carbon	no	yes	?	no	yes	yes	?
bioelements	Si, S	Si, S, N, P	probably	Si	S, N, P	S, N, P	?
mineralogy	yes	yes	no	yes	limited	limited	?
trace metals	yes	no	no	?	yes	no	yes
Comments:							
advantages	good for SEM, XRF, XRD, optical microscopy						withstand harsh treatment
disadvantages	Cr, Br blanks	expensive heavy tare	heavy tare	mass loss	fiber loss heavy tare	fiber loss heavy tare high inorganic blanks	heavy tare

* trade name Anopore.

** includes Teflon which requires wetting solvent and PVDF (Durapore) in both hydrophilic and hydrophobic forms.

? unknown, further testing required.

Manufacturer key:

f = Fischer; g = Gelman; i = Isotech; m = Millipore; n = Nuclepore; p = Poretics; w = Whatman

2.2.2. Depth filters.

Cellulose filters. Cellulose filters, available as acetate and nitrate esters, are depth filters which have been used for radionuclides and some other inorganic analytes.Their principal advantages are low cost, availability and ease of filter matrix digestion. Disadvantages are mass loss during filtration, heavier tare weights and water retention. Although the cellulose (Millipore) filters have had limited use for minor element analyses in marine particles, they have been widely used for atmospheric aerosol sampling and respectable trace element blanks are well documented. Cellulosic filters are not acceptable for analysis of bioelements.

Quartz fiber filters. Quartz fiber filters are depth filters made of a disordered bed of individual fibers arrayed with a nominal mean pore size. As a result of this structure, they have high flow rates and total particle capacity, little post-sampling particle redistribution and the ability to be cut for subsampling with plastic punches. They can be used for optical examination of particles because the filter can be rendered transparent with immersion oil, although this examination may be complicated because particles do not lie in a single plane. Evidence to date suggests little DOC adsorption on quartz filters, which have been used successfully for bulk POC analyses as well as for specific classes such as lipids. Quartz filters survive CHN analysis without fusing in the furnace. Disadvantages include high tare weights and fragility when wet, requiring high loading for precise TSM determination and additional physical support to maintain integrity when wet.

Applications of quartz filters have been limited to large volume in situ filtration, where high loading results in large sample to blank ratios. Inorganic blanks can be greatly reduced by rigorous acid cleaning, although trace metal blanks are not fully documented. Use for organic analysis requires overnight combustion at about 500° C because shorter combustion times may not result in complete removal of the carbon blank. Solvent cleaning may also remove organic blanks. Obviously, quartz filters cannot be used for opal and silicate mineral analysis. Water absorption has also been observed to interfere with H analysis.

Quartz filters have been used for particulate S analysis. Their features of high melting point and ease of handling make them desirable for sulfur as well as particulate CN analyses, although they have not been used extensively for C and N at this time.

Quartz microfiber filters should be suitable for organic analyses where glass fiber filters are presently used. Conte and Bishop [1988] used them to collect particles for lipid analysis. Quartz filters are more expensive than glass fiber filters of equivalent size, and have a similar pore size to Gelman AE filters. This filter type may well prove to be useful for providing particles for both organic and inorganic analyses on the same material.

Glass fiber filters. Glass fiber filters are similar in structure to quartz filters, but are made of much less pure silica. They have been widely used to analyze organic C, N and P and major inorganic elements at high particle loading. In general, other than cost, glass fiber filters offer no advantages over quartz media. For many applications, however, glass fiber filters are inferior to quartz. Blanks for minor elements and natural radionuclides are much higher and cannot be greatly reduced by acid cleaning. Zinc stearate is used as a releasing agent during manufacture and poses a substantial blank source.

In most studies of organic components, particles are collected on glass fiber filters. Whatman GF/F (nominal pore size 0.7 μm) and Gelman AE (nominal pore size 1 μm) are commonly used. Whatman GF/C filters are rarely used because they retain fewer of the less than 1 μm sized particles. Glass fiber filters are readily cleaned of organic contamination by solvent rinsing, or more commonly by heating in a muffle furnace at 450 to 500° C overnight, although this procedure weakens the filters and makes them more brittle. Filters without binder should be used. Some investigators use 2 or 3 filters in the same filter holder. This system is more robust, and some particles which pass through the first filter are retained by the second. The third filter can be used as a quasi-blank. A disadvantage of glass fiber filters is that broken fibers accumulate in the filtrate and may interfere with colloid studies.

Fluorocarbon filters. Teflon filters are hydrophobic depth filters made of expanded fluorocarbon polymers with very high porosity. These filters are very hydrophobic and require wetting with a water miscible organic solvent such as methanol prior to use with aqueous media. Although these filters have not been widely used for marine particles, they have some potentially positive characteristics. Teflon can be rigorously cleaned with both inorganic acids and bases as well as organic solvents so that filter blanks can be very low. High porosity allows higher flow rates than polycarbonate filters. Complete digestion of particles embedded in the matrix may be hampered by the hydrophobicity of the polymer.

Teflon filters are usually bonded to a support screen of polyethylene or polypropylene. Although they are available without the support, stretching and dimensional changes under heating are problems. One variety is laminated to a coarse Teflon support, but these tend to delaminate and curl on heating. Durapore filters, like the Teflon filters, can withstand strong acids and are hydrophilic (Sholkovitz, this volume, page 295).

2.2.3. Filter screens and support media.

A variety of filters and screens have been used to exclude larger particles from the sample or to provide size-fractionated particles [Bishop et al., 1977; Wakeham, 1990; Wakeham and Beier, 1991]. Prefiltering is particularly important for organic analyses, because inclusion of zooplankton in the sample may dramatically bias the organic composition. Nitex and steel screens are most commonly used.

Support media for filters include stainless steel, Teflon, porous polyethylene and a variety of other plastics. The choice of support is largely dependent on the analyses to be carried out on the filter.

2.3. Rinsing

Suggested rinses for marine particles on filters are summarized in Table 1-5.

TABLE 1-5.
Suggested Rinses For Marine Particles on Filters

Inorganic	Q-water. Q-water at pH 8.
Organic	Do not rinse. Do not dry.
Minerals	Deionized water. Some concern for carbonate solution in long rinses.
Bioelements	Generally do not rinse. For sulfur, rinse with isotonic solutions.

2.3.1. Inorganics.

Selective removal of sea salt from filters may be difficult due to potential losses of other constituents. It is therefore generally recommended that rinsing of filters should be avoided if not critical for the analyses to be performed.

Rinsing of sea salt from membrane filters may be desirable in the following cases:

1) Gravimetric analyses of total suspended matter to calculate the elemental composition on a salt-free basis. This makes it possible to compare results with those of bottom sediments and sediment trap material.
2) Analyses of elements like Ca, Mg, K, Cl and S on a salt-free basis.
3) Analyses of Nuclepore filters by SEM/EDXA without the influence of salt crystals.

If the filters are to be analyzed for organic compounds, the filters are normally not rinsed. In this case one has to rely on subsamples to be able to express the results on a salt-free, dry weight basis. Salt corrections may be done by analyzing for some of the salt elements (i.e., Cl, Na) and recalculating the dry weight on a salt-free basis.

The ease and efficiency of totally removing the salt will depend on the filter types. Nuclepore filters retain the smallest amount of salt. In contrast, rinsing a glass fiber filter essentially free of salt will require a larger amount of solution, and artifacts of rinsing may be more significant with these filters.

The most commonly used rinsing fluid is distilled water, with and without pH adjustments (usually to 8 to match seawater). By raising the pH the loss of carbonates and manganese is reduced.

Isotonic solutions, e.g., ammonium chloride, ammonium formate, are frequently used to rinse filters loaded with biogenic particles. This minimizes the problem of cell rupturing. The wash solutions can and should be retained so that analyses can be made to quantify the loss since ammonium is an extremely effective ion-exchanger of cations.

The effect of rinsing will depend on the time the sample is exposed to the rinsing solution and the volume of fluid. A quick rinse to reduce time and volume is recommended (e.g., 10 mL for 47 mm diameter membrane filters). Suction of air or pressure filtering of nitrogen through the filter prior to rinsing will reduce the amount of rinsing solution required. However, tests have shown that prior drying of the filter causes severe losses of elements [Price and Skei, 1975].

Tests on the effects of rinsing filters with distilled water without adjusting the pH to 8 have shown a 60 to 70% loss of P and 50 to 80% loss of Mn from particulate matter collected in the productive layer of a fjord [Price and Skei, 1975]. At depth these losses decrease to about 20%, indicating that such losses are related to elements associated with planktonic organic matter. In contrast, Al, Si and Ti, associated predominantly with detrital material, show no losses. Collier and Edmond [1984] have also documented the loss of elements from the planktonic organic matter. Analyses of rinse solutions after washing suspended particles with pH-adjusted distilled water have shown losses of generally less than 10% for Zn and 20 to 50% for Cd [Sherrell, 1989].

Readers are referred to the following papers in chapter 5 of this volume for more details related to the rinsing of particle samples: (1) Collier, page 235, (2) Feely et al., page 251, (3) Landing and Lewis, page 263, (4) Sherrell, page 285, (5) Skei, page 303, (6) Trefry and Trocine, page 311 and (7) Yeats, page 321.

2.3.2. Organics.

The general practice for organic analyses is not to rinse the filter after sample collection. Furthermore, in most instances the particles collected on the filter are not dried. One allowable exception to this practice is for lignins.

2.3.3. Minerals and surface properties.

The preferred technique for rinsing filters in the field is to use a small volume of deionized water and then to draw air (filtered in a laminar hood if possible) through the filter for several minutes to remove the majority of the water. Special care should be taken to keep the volume of rinse solution to a minimum to preserve fragile or more soluble phases such as carbonate and celestite.

2.3.4. Bioelements.

No rinse is necessary or routinely used for C, N or P. For S, rinsing is required to remove sulfate introduced from sea salt. An isotonic solution is used for the S rinse to prevent cell rupture and loss of potentially volatile sulfur. Solutions of ammonium formate and sodium chloride have been used for this purpose.

2.4. In Situ Filtration Versus Bottle Samples

There are three approaches to particle sampling in the marine environment. The first uses 30 L (or smaller) Niskin or Go-Flo bottles which isolate a sample of water at depth and return it to the ship for filtration. The major difference between Niskin and Go-Flo bottles is that the latter are designed to open several meters below the surface to avoid surface slick and airborne contamination and have an external closing mechanism. A second approach employs submersible pumping systems to collect particles at depth. Volumes of seawater up to 25,000 L have been sampled by in situ filtration. The purpose of this section is to present the pros and cons of each approach. Further discussion comparing bottle and pump sampling is given in the papers by Bishop (this volume, page 229) and Sherrell (this volume, page 285). The third approach involves direct in situ collection of particles by SCUBA or submersible and is discussed in detail by Alldredge (this volume, page 46).

There are several basic questions to ask when considering either pumps or bottles for suspended matter sampling:

1) What volume do you need to sample?
2) What samplers are available?
3) What is the best manner to get your sample out of the water?
4) Are there any contamination "prohibitions" of either technique?
5) Do you need to compromise and share samples with other studies?
6) What is the impact of a chosen approach on shiptime? Is shiptime limited?
7) How much does the chosen approach cost in terms of equipment and labor?

With bottle sampling, the most serious problem is caused by sedimentation of particulate material to the bottom of the sampler and incomplete recovery of this material during filtration [Bishop and Edmond, 1976; Gardner, 1977; Trefry and Trocine, this volume, page 311]. It is best to filter the entire contents of the bottle. In some coastal situations and always in the euphotic zone, this requires the use of smaller bottle sizes, multiple filters, or filters with a larger surface area. When samples are shared from the same bottle, it is impossible to predict the impact of the sedimentation bias on the results. Swirling or shaking of the bottle to overcome the sedimentation bias is sometimes used but can alter partitioning between dissolved, colloidal, and particulate size fractions and therefore is not recommended.

Extreme care also must be taken to avoid contamination, particularly when small quantities (sometimes less than 300 μg) of SPM are recovered. Direct in-line pressure (preferred because one maintains a closed system) or vacuum (with appropriate in-line air filtration) filtration of seawater from the bottle produces the most consistent results. Water should be processed as soon as possible to minimize changes within the sample. Transfer of the sample to interim holding containers prior to filtration should be avoided. Pressure differentials of 5 to 15 psi are routinely used, although there is no consensus on the optimum pressure differential to be used. There is some feeling that this issue needs further study in the case of organic matter. Excessive pressure differential or flow velocity may disrupt fragile cells with loss of cell fluids. In no case is the use of open filter funnels recommended. Bottles are not recommended for investigations of large aggregate particles due to their relatively rare occurrence and the greater probability of destroying their morphology. On the other hand, properly conducted bottle sampling is considered to be quite suitable for investigating dissolved/colloid/particle partitioning.

Pump systems offer many advantages over bottles but are presently not as widely available. The clear advantages of pumps are the large volume of water that can be filtered and the lack of sedimentation bias. Thirty to 1000 times more particulate material can be obtained by pumps compared to bottles, and pumping separates particles from the ambient solution without delay. In some cases, size-fractionation of samples (e.g. 1 to 53 μm and greater than 53 μm or any other fractionation scheme of choice) is also accomplished in situ during sampling, thereby enabling studies of the composition of various particle size classes.

The large quantity of material obtainable using pumps gives the advantage of a greatly expanded suite of determinations possible on a single sample and minimizes blank problems. For many radionuclide and organic analyses, in situ filtration is the only practical method for obtaining a sample of sufficient size. The lack of sample handling results in more efficient sampling of large particles and better preserved aggregate morphology. As for bottles, samples should be processed as soon as possible after a station in order to avoid biases due to sea salt precipitation, e.g. $CaSO_4$, and organic matter degradation. Some pump samples are large enough to permit a variety of subsampling and preservation schemes at sea.

Pump systems are also capable of providing a temporally integrated sample. They are easily deployed from moorings and can be deployed to collect samples over a variety of different time-scales depending on the experimental requirements.

Some concern still exists about the hydrodynamical aspects of in situ pumping. For example, turbulent flow fields create inhomogeneity (which can be seen in particle rich environments) and particle manipulation during pumping can distort samples.

One conclusion of the workshop is that more rigorous comparisons of sampling modes are needed. These include comparisons of the different in situ filtration devices presently in use to evaluate the differences between samples collected with pump systems relative to water bottles.

A deep ocean CTD/rosette/bottle cast takes about 4 hours. A full depth profile taken using an equivalent number of pumps requires 2 to 12 hours of extra ship time. The return of orders of magnitude more material per unit of shiptime and the resulting simplification of analyses (inorganic, organic, mineral and bioelement) can offset the additional shiptime required for pumping. Presently, individual investigators are limited by the number of pumps at their disposal, necessitating multiple casts to obtain the vertical sampling resolution possible with standard rosette/bottle systems.

In considering the question of which sampling method is preferred for collection of suspended matter, the group was not able to identify any specific advantage of bottles over in situ filtration apart from the widespread availability of bottles. However, in situ techniques for collection of filtrate have yet to be perfected, and therefore bottle sampling is a necessity for most studies of dissolved constituents. Exceptions to this are various adsorber cartridges that have been developed for some dissolved species such as radionuclides. There is a trend within the oceanographic community toward greater use of in situ filtration techniques for SPM sampling.

For the investigation of rarer large particles, attention must be given to the statistics of their distribution in deciding on the necessary volume to filter. Adequate representation of larger size classes of particles requires that large volumes of water be filtered to achieve good sampling statistics.

Large, relatively rare aggregates of marine snow and fecal pellets exist in all ocean waters. In coastal areas, these particles range from about 0.1 to 100 particles per liter [Alldredge and Silver, 1988]. These abundances are usually adequate to insure that aggregates are sampled representatively by a 30 L Niskin bottle. If the bottle is to be subsampled, it should be agitated or shaken to break larger aggregates apart and distribute them through the bottle. But shaking may also produce new POM from the dissolved pool, introducing additional error. The magnitude of this error relative to that introduced by non-representative sampling of large aggregates is unknown, but may be significant.

In the open ocean and deep sea, large particles in the marine snow range are considerably rarer. In extreme cases, 30 L bottles may miss them entirely. For bioelements, undersampling of rare, large particles may introduce relatively little error. For example, moderately sized aggregates of 30 μg mass each, occurring at an abundance of 1 aggregate per 30 L in seawater containing 100 $\mu g\ L^{-1}$ would contribute less than 10% to the total suspended matter per liter. The concentrations of specific inorganic and organic compounds in marine snow are little known. However, undersampling of large particles may have greater effects on these measurements. For example, Hebel et al. [1986] found 100% of the particulate Al, Fe and Mn was associated with larger aggregates.

Inclusion of rare aggregates may also produce anomalously high measurements, especially if the bottle is subsampled and the subsample contains an aggregate. The occurrence of outlying samples is a clue that the volume of seawater sampled may be too small, and replicate or larger samples should be considered.

Bottles sample instantaneously at the time of tripping whereas pumps tend to average over space and time. Both techniques can be applied to meter-scale vertical variability in the SPM field. Prior to sampling, small-scale SPM variability is best investigated optically using either nephelometers or transmissometers as these instruments sample volumes of water much smaller and faster than either method. Considerations of spatial variability must be included in interpretation of results by either method.

Soluble phases, such as $SrSO_4$, may dissolve during filtration due to exposure to large volumes of water. This problem affects bottle and pump sampling equally because the volume of water filtered per gram of suspended matter recovered is constant. Existing data [Bishop et al., 1977] indicated that this is not a major problem. Another larger unexplored problem is the potential loss of "interstitial water" in aggregates during filtration. SCUBA-based methods may minimize this problem but no comparisons of these two methodologies have been made.

No comparative studies of the differences in abundance, morphology, and chemistry of macroscopic aggregate particles collected by in situ filtration and SCUBA techniques have been made. Some studies have been performed to investigate the effects of filtration speed on aggregate recovery [Bishop, 1982] which suggest near quantitative recovery of most phases but a cross check against SCUBA collected particles would be most valuable.

TABLE 1-6.
Storage Practices for Marine Particles

Inorganic	Dry in petri dish at 60° C. Air dry in laminar flow hood. Dry by vacuum desiccation. Freeze in vial.
Organic	Freeze quickly (a few minutes). Do not let samples dry.
Minerals	Air dry or by vacuum desiccation. Save reduced minerals (e.g. sulfides) in bag with argon.
Bioelements	Freeze dry.

2.5. Storage of Particle Samples

Storage practices for filters are summarized in Table 1-6.

2.5.1. Inorganics.

Storage techniques for particle samples intended for inorganic analysis include variations on two basic methods: freezing and drying. Some groups leave filters wet and immediately freeze them. This procedure prevents effective rinsing of sea salt because rinse-resistant evaporates may form, eliminating accurate mass determination and complicating analysis of the particles for major sea salt ions.

Drying immediately after sampling is more commonly employed. Techniques include immediate oven drying at 60° C, air-drying on ship followed by desiccation on shore, immediate vacuum desiccation, and air-drying only. All of these methods apparently work well for most inorganic analyses, although there is some (largely undocumented) concern about loss of volatile components (e.g. methylated metalloids such as Hg) during any drying procedure. Careful control to prevent atmospheric particle contamination during drying is important; drying in open lab/ship environment is discouraged.

Drying procedures do have important implications for obtaining accurate mass determination. Absorption or adsorption of water vapor on the filter matrix, on residual sea salt crystals, or on the particulate matter itself can significantly bias observed mass. As with all chemical determinations, appropriate procedural blanks for mass determination should be carried through all handling and storage steps.

2.5.2. Organics.

Organic matter can be rapidly degraded by microorganisms during storage. Although the total carbon degradation may be quantitatively small, the impact on specific classes of labile organic compounds may be significant. A number of studies detailing the effects of storage on specific suites of organic compounds have been reported, and it is generally recognized that different classes of organic compounds respond differently to storage. Rapid freezing of a sample to temperatures less than -20° C is the most widely used method of preservation, but there is no general consensus on the specific temperature required for prolonged storage. Temperatures to -72° C (liquid N_2) have been reported. Low temperatures are generally considered to be safer. Desiccation and freeze drying are not considered adequate for most classes of labile organic compounds (proteins, amino acids, pigments and lipids), but are adequate for resistant biopolymers, such as lignins and carbohydrates.

There have been relatively few studies of sample shelf life. For relatively labile organic compounds, analyses must be made less than 1 year from collection. For lignins, carbohydrates and other resistant biopolymers, little degradation is observed after 5 to 10 years of storage. Shelf life is somewhat dependant on the sample matrix. Highly purified samples or standards are the most susceptible to degradation followed by filtered particulate matter, sediment trap and sediment samples. Purified pigment standards preserved by rapid freezing in liquid N_2 show slight but quantitatively significant degradation over several months, however pigment analyses of sediment subsamples analyzed four years apart showed no measurable degradation. It is generally felt that unless the stability with respect to storage is known, analyses should be made as soon after sample collection as possible.

2.5.3. Minerals and surface properties.

The preferred technique for drying filters in the field is to rinse the sample with a small volume of deionized water and then draw air, filtered in a laminar hood if possible, through the filter for several minutes to remove the majority of the water. Subsequently, the filter is transferred to a vacuum desiccator for removal of any remaining water. Storage of the filters in vacuum is preferred to avoid the potential for bacterial growth, but dried filters stored in plastic vials generally are free from growth for periods of months to years. When a large number of filters are processed, some investigators dry their filters at 60° C. Some volatile organics may be lost by this procedure, but the effect on mineralogy is minimal. When mineralogical samples are collected from low oxygen environments, filtration should be carried out in a glove bag under an inert gas atmosphere. Filter storage in plastic bags (Seal-A-Meal technique) filled with argon maintains samples in a low oxygen environment for a period of months (see Campbell, this volume, page 379).

2.5.4. Bioelements.

Particle samples for the bioelements C, N, P, S, and Si should be stored by either deep-freezing or drying if the analysis cannot be carried out in appropriate time after filtration. For all bioelements, samples may be shock-frozen and the storage temperature should be at maximum -20° C. Drying of the samples can be done in an oven (60° to 80° C) or in a vacuum desiccator, except for sulfur samples. For the latter, drying temperature should not exceed 60° C and vacuum drying is not recommended because of the possible loss of volatile compounds. Samples for suspended and sediment trap material can also be stored freeze-dried, except for sulfur.

2.6. Sample Preservation: Sediment Traps

Poisons/preservatives have been added to sediment traps prior to deployment to decrease bacterial degradation of trapped matter and also to prevent grazing by "swimmers" inside the traps [U.S. GOFS Working Group, 1989]. We note that although both poisons and preservatives are effective biocides, their modes of action often differ. For example, preservatives tend to "harden" the tissues of organisms making the separation and/or removal from trap material easier. Conversely, organisms associated with poisoned trap materials generally become soft, making intact removal difficult. Table 1-2 gives an overview of many of the various poisons/preservatives which have been commonly used. Here we discuss some of the caveats involved with the use of these additives.

At this time, there are no ideal poisons/preservatives (additives). For example, regardless of the particular additive used, various compounds or elements associated with the particular phases will partition into the surrounding soluble phase. Nonetheless, the choice of additive is strongly parameter dependent. For example, if bulk bioelements (e.g. C and N) are to be measured, preservatives such as formaldehyde and glutaraldehyde are to be preferred because of their hardening characteristics which tend to limit losses from the particulate to the dissolved phases. Preservatives are effective biocides even at low concentrations [Lee et al., 1991]. Preservatives must be used if contamination of "swimmers", i.e. those organisms which swim into the traps and die and are not considered as part of the particle flux [Knauer et al. 1979, 1984], is anticipated. The hardening characteristics of preservatives allow efficient removal of "swimmers" whereas poisons render these organisms "too soft" to be completely removed.

TABLE 1-7.

Approximate Concentrations of Treatment Which Reduce Bacterial Uptake of Glutamic Acid to 1% of the Activity in the Unpoisoned Material. Data from Lee et al. [1991].

Treatment	1% Concentration
Sodium chloride	N.E. (100 gL^{-1})
Streptomycin	N.E. (2 gL^{-1})
Ampicillin	N.E. (> 3 gL^{-1})
Penicillin	N.E. (> 3 gL^{-1})
Sodium azide	N.E. (10 gL^{-1})
Formaldehyde	0.11% (0.037 M)
Chloroform	50% aqueous saturation
Mercuric chloride	<50 mg L^{-1} (180 μM)

N.E. = not effective at the highest concentration tested (shown in parenthesis)

Some disadvantages include bias of isotopic analysis of C, analytical interferences with respect to C/N in the soluble phases, and dissolution of carbonates, if not adequately buffered.

Other less commonly employed biocides include sodium azide, chloroform, antibiotics and high concentrations of salts, e.g. NaCl (Table 1-7). Sodium azide has been widely used in sediment traps. It has the advantage of not introducing carbon contamination, but interferes with the determination of dissolved N [Knauer et al., 1984]. Mercuric chloride is a good biocide, but may introduce heavy metal contamination as well as altering the redox state in the traps. The use of this poison may be preferable when stable isotopes are to be analyzed. Chloroform is a high density poison and therefore requires no additional salts. It is an efficient biocide and may prevent extra-cellular enzyme activity [U.S. GOFS Working Group, 1989; Lee et al., 1991]. However, it significantly dissolves pigments and lipids in even small concentration [Gundersen and Wassmann, 1990]. As with brine solutions, chloroform also appears to attract "swimmers" and causes difficulty in sample splitting. Salt brine alone is a poor biocide and may increase "swimmer" contamination through positive attraction [Lee et al., 1991]. Antibiotics may be useful for specialized biocide application, but are generally ineffective with respect to the preservation of sediment trap materials.

Although the above comments have been primarily directed towards the measurements of bulk elements, it should be understood that many of the caveats discussed above also apply to trap-collected inorganic material. Knauer et al. [1984] have discussed the effects of various poisons and preservatives on trace metal concentrations associated with sediment trap materials. They found differential partitioning between the particulate and soluble phases. For example, greater than 70% of Cd and Mn were lost to the trap solutions whereas most of the Fe tended to remain in the particulate phase regardless of poisons/preservatives used.

Clearly, the use of additives is a complex aspect of the overall aspect of sediment trap technology; and each parameter of interest, e.g. organics, inorganics, isotopic composition, requires its own unique experimental protocol with respect to the choice of poison or preservative. These aspects have been discussed in some detail in the literature and the reader is referred to Knauer et al. [1984] and Table 1-2.

2.6.1. Preservation of organic matter in sediment traps.

Organic matter on particles collected in sediment traps is subject to microbial degradation and, when present within an organism, autolysis. Loss of organic matter in traps deployed with no poison or preservative treatment can be rapid [Gardner et al., 1983]. This degradation can greatly alter the molecular-level organic-compound composition of the particles, even when the total elemental composition does not change greatly. In order to prevent microbial transformations, and loss of organic matter, various poisons, preservatives, or antibiotic treatments have been used. For organic analyses, several treatments appear to be effective. Formalin and $HgCl_2$ are preferred and can be used depending on the particular compounds being analyzed. Enzymatic activity has not been as extensively investigated and whether this has a significant influence on organic matter decomposition remains to be addressed. In addition, use of formalin and $HgCl_2$ do not prevent leaching of soluble materials from particulate matter although formalin may be slightly better at preventing dissolution than $HgCl_2$ [Lee et al., 1991; Table 1-7].

One limitation to the use of various treatments to prevent microbial degradation in sediment traps is the resultant collection of "swimmers".

2.6.2. The swimmer problem.

Swimmers represent a particular problem for the accurate measurement of the compositions and fluxes of organic compounds collected in sediment traps [U.S. GOFS Working Group, 1989]. This is especially true for biochemicals such as amino acids and wax esters which are both sensitive to microbial degradation and concentrated in zooplankton. Swimmers also leak organic compounds that can interfere with accurate measurements of the corresponding total, particulate plus dissolved, fluxes. Unfortunately, the swimmer problem is particularly acute in the surface ocean where most organic matter recycling occurs.

A common response to the swimmer problem is to physically remove all recognizable zooplankton remains by sieving or picking (see previous discussion) from treated particle collections. This technique, however, is time consuming and sometimes is not sufficient for biochemicals enriched in zooplankton (Peterson et al., in preparation). Another strategy is to physically bar randomly swimming zooplankton from poisoned sample accumulators while passing directly sinking particles.

One example of this approach is a separator module containing a labyrinth of concentric funnels through which naturally raining particles are sequentially focused into a treated collector [Coale, 1990]. A second design (Peterson et al., in preparation) involves an indented rotating sphere (IRS) valve mounted flush beneath a collar located on the inner wall of a cylindrical trap. A microprocessor-controlled electric motor rotates the sphere at a predetermined rate and periodicity, at which time a small notch in the ball selectively transfers particles collected along the collar/ball contact to the lower portion of the trap where they accumulate in a poisoned brine solution that is not susceptible to washout. Although the two previous barrier traps greatly minimize swimmers, neither are 100% effective and removal of some swimmers by picking or sieving is still necessary. Both designs also should minimize sample reprocessing by living zooplankton, but do not guarantee against elevated fluxes resulting from unnaturally high concentrations of living zooplankton congregated within the upper section of the trap. Efficient means of physically or chemically repelling living zooplankton from sediment traps are still needed and should be further investigated.

Acknowledgements. The authors wish to express their thanks to Karen Conlan and Ryan Whitney of the Pacific Marine Environmental Laboratory for providing graphics and typing support, respectively. This work was supported by the National Oceanic and Atmospheric Administration with funds from the NOAA VENTS Program and by the National Science Foundation (OCE-9012825).

References

Altabet, M. A., Organic C, N and stable isotope composition of particulate matter collected on glass fiber and aluminum oxide filters, Limnol. Oceanogr., 35, 902-909, 1990.

Alldredge, A. L., and M. W. Silver, Characteristics, dynamics and significance of marine snow, Prog. Oceanogr., 20, 41-82, 1988.

Bishop, J. K. B., The effect of filtration speed on large particle catch during large volume in situ filtration, EOS Trans. Am. Geophys. Union, 63, 46, 1982.

Bishop, J. K. B., and J. M. Edmond, A new large volume filtration system for the sampling of oceanic particulate matter, J. Mar. Res., 34, 181-198, 1976.

Bishop, J. K. B., J. M. Edmond, D. R. Ketten, M. P. Bacon, and W. B. Silker, The chemistry, biology and vertical flux of particulate matter from the upper 400 m of the equatorial Atlantic Ocean, Deep-Sea Res., 24, 511-548, 1977.

Bishop, J. K. B., J. C. Stepien, and P. H. Wiebe, Particulate matter distributions, chemistry and flux in the Panama Basin: response to environmental forcing, Progr. Oceanogr., 17, 1-59, 1987.

Calvert, S. E., and M. J. McCartney, A note on the effect of incomplete recovery of large particles from water samplers on the chemical composition of oceanic particulate matter, Limnol. Oceanogr., 24, 532-536, 1979.

Coale, K. H., Labyrinth of doom: A device to minimize the "swimmer" component in sediment trap collections, Limnol. Oceanogr., 35, 1376-1381, 1990.

Collier, R. W., and J. Edmond, The trace element geochemistry of marine biogenic particulate matter, Prog. Oceanogr., 13, 113-199, 1984.

Conte, M. H., and J. K. B. Bishop, Nanogram quantification of non-polar lipid classes in environmental samples by high performance thin layer chromatography, Lipids, 23, 493-500, 1988.

Fan, K. C., C. Leasburge, Y. Hyun, and J. Gentry, Clogging in Nuclepore filters: cap formation models, Atmos. Environ., 12, 1797-1802, 1978.

Gardner, W. D., Incomplete extraction of rapidly settling particles from water samples, Limnol. Oceanogr., 22, 764-768, 1977.

Gardner, W. D., H. D. Hinga, and J. Marra, Observations of the degradation of biogenic material in the deep ocean with implications on the accuracy of sediment trap fluxes, J. Mar. Res., 41, 195-214, 1983.

Gundersen, K., and P. Wassmann, Use of chloroform in sediment traps: caution advised, Mar. Ecol Prog. Ser., 64, 187-195, 1990.

Hebel, D., G. A. Knauer, and J. H. Martin, Trace metals in large agglomerates (marine snow), J. Plankton Res., 8(4), 819-824, 1986.

Knauer, G. A., J. M. Martin, and K. W. Bruland, Fluxes of particulate carbon, nitrogen, and phosphorous in the upper water column of the northeast Pacific, Deep Sea Res., 26, 97-108, 1979.

Knauer, G. A., D. M. Karl, J. M. Martin, and C. N. Hunter, In situ effects of selected preservatives on total carbon, nitrogen and metals collected in sediment traps, J. Mar. Res., 42, 445-462, 1984.

Lee, C., J. I. Hedges, S. G. Wakeham, and N. Zhu, Effectiveness of poisons and preservatives in retarding bacterial activity in sediment trap material and their effect on the collection of swimmers, Limnol. Oceanogr., in press, 1991.

Moran, S. B., and R. M. Moore, The distribution of colloidal aluminum and organic carbon in coastal and open ocean waters off Nova Scotia, Geochim. Cosmochim. Acta, 53, 2519-2527, 1989.

Price, N. B., and J. M. Skei, Areal and seasonal variations in the chemistry of suspended particulate matter in a deep water fjord, Estuarine Coastal Mar. Sci., 3, 349-369, 1975.

Sheldon, R. W., Size separation of marine seston by membrane and glass fiber filters, Limnol. Oceanogr., 17(3), 494-498, 1972.

Sheldon, R. W., and W. H. Sutcliffe Jr., Retention of marine particles by screens and filters, Limnol. Oceanogr., 14, 441-444, 1969.

Sherrell, R. M., The trace metal geochemistry of suspended oceanic particulate matter, Ph.D. thesis, 211 pp., MIT-WHOI Joint Program in Oceanography, 1989.

Skei, J. M., Formation of framboidal iron sulfide in the water of a permanently anoxic fjord-Framvaren, South Norway, Mar. Chem., 23, 345-352, 1988.

Stumm, W., Chemical interaction in particle separation, Environ. Sci. Technol., 11(12), 1066-1070, 1977.

U.S. GOFS Working Group, G. A. Knauer and V. Asper, Co-chairs, Sediment trap technology and sampling, Report No. 10, November 1988, 94 pp., Woods Hole, Massachusetts, 1989.

Wakeham, S. G., Algal and bacterial hydrocarbons in particulate matter and interfacial sediment of the Cariaco Trench, Geochim. Cosmochim. Acta, 54, 1325-1336, 1990.

Wakeham, S. G., and J. A. Beier, Fatty acid and sterol biomarkers as indicators of particulate organic matter source and alteration processes in the water column of the Black Sea, Deep-Sea Res., in press, 1991.

CHAPTER 2. SAMPLE MANIPULATION AND SPECIATION

WORKSHOP REPORT

by
Chair: William M. Landing, Rapporteur: Philip A. Yeats, Assistant Rapporteur: Chris Sabine
and Members of the Working Group;

1. Introduction

The discussion in the cross-cutting session on sample manipulations and speciation covered a number of important topics and is summarized in this chapter under the following headings:

1. Physical separation schemes
2. Chemical and analytical speciation schemes
3. Surface properties.

A summary of the discussion, cross-referenced against the individual contributions and other chapters in this volume, appears below. Throughout the chapter, reference to specific manufacturers or sources of equipment and supplies does not imply endorsement by any individual or governmental body. The information is offered as examples suggested by the workshop participants.

As noted above, the discussion in this session covered many topics. However, if there was one overriding theme to the tone of this discussion, it was the issue of colloidal particles, especially colloidal organic matter, and the biogeochemical consequences of this rather poorly understood pool of material lying between the truly "dissolved" and the filterable "particulate" categories.

Increasing attention has been paid to marine colloids in recent years for a number of reasons. Paramount among these for the inorganic chemist has been the development of scavenging models that incorporate the partitioning of metals between dissolved and particulate phases. It has been suggested in some of these studies [e.g. Morel and Gschwend, 1987] that discrepancies between experimental results and theory could be explained if the effects of colloidal sized particles were included. Another major impetus for interest in colloids has been generated by the dissolved organic matter (DOM) controversy [Sugimura and Suzuki, 1988]. They report that greater than 40% of the extra DOM is in the greater than 20,000 MW colloidal size fraction, while Benner (this volume, page 181), Brownawell (this volume, page 187), Carlson et al. [1985], and Whitehouse et al. [1990] all report that the "colloidal" organic material collected using regular or tangential flow ultrafiltration represents no more than 10% of the total DOM pool. Whether these discrepancies simply represent spatial and temporal variability or are primarily due to differences in analytical methodologies is unresolved at this time. Finally, studies of marine colloidal material may provide insight into problems that bridge the gap between inorganic and organic marine chemistry, including metal-organic complexation, particle surface properties, and metal bioavailability.

Colloids are traditionally defined as particles that reside in the 1 nm to 1 μm size range. Because of their very small size, they have large surface areas per unit volume. Thus surface effects are magnified and properties such as surface tension and zeta potential can be very important. Because of the large surface area, colloids can have an important effect on dissolved-particle interactions even if the concentrations of colloids are quite small.

Marine scientists have often used 0.4 μm filtration to separate "dissolved" from "particulate" material. It is now clear that there are very many discrete, identifiable particles smaller than this. Marine bacteria generally range in size down to 0.1 μm, while viruses and "Koike" particles [Koike et al., 1990] can be collected using 0.02 μm filtration. Membrane filters are available down to 15 nm (and perhaps smaller) but the flow rates and the volumes that can be filtered are very small. The size range for which there is less information is in the 1 to 20 nm range, roughly between 1,000 and 750,000 nominal molecular weight, as calculated from diffusion coefficients for globular proteins assuming frictional coefficients for spherical particles. Expressions such as "colloidal DOM or DOC", while appearing as literal contradictions in terms, are therefore referring to this relatively poorly understood category of material.

Several elements, including particle reactive elements like Mn, Fe and Al and others such as Se, are extensively associated with colloids in river waters. In offshore waters, the amount of Al, Mn, and Fe in the colloidal size range falls to only a few percent or less of

Marine Particles: Analysis and Characterization
Geophysical Monograph 63

the less than 0.4 μm "dissolved" fraction [Moran and Moore, 1989; Whitehouse et al., 1990]. Colloidal copper concentrations in marine waters can be considerably higher [Hasle and Abdullah, 1981; Whitehouse et al., 1990]. Recent publications by Coale and Bruland [1988] and Donat and Bruland [1990] have demonstrated that Cu and Zn in oceanic surface waters are strongly complexed by unknown, but presumably organic, ligands. These ligands appear to be somewhat element specific, are found in low (nM) concentrations, and have high complexation constants. Understanding the nature of the organic colloids should help in understanding the marine chemistry of elements which may also be complexed by organic material in the sea.

Colloids seem to play a much bigger role in marine organic chemistry. Of current concern is the accuracy of high temperature DOC catalytic oxidation techniques and the report that a considerable quantity of this "new" DOC is colloidal [Sugimura and Suzuki, 1988]. Consideration of the nature and reactivity of this colloidal DOC may help to explain the apparently anomalous behavior of the "new" DOC. Separation of colloids is also very important in the study of small particles such as bacteria, viruses, and "Koike" particles; certain types of organic matter such as biopolymers (proteins, polysaccharides, microbial exudates); "humic" substances of terrestrial and marine origin; and aggregates of smaller compounds produced by turbulent, and other, physical processes.

2. Physical Separation Schemes

Concerns regarding alteration of particle morphology during collection and handling apply across the spectrum of collection and analytical techniques. Aggregates may be disrupted by turbulence, pressure changes, etc. Agitation and turbulence during sample collection and handling can also create larger particles from smaller ones, thereby biasing the analytical results. Studies which require the retention of particle morphology should be designed appropriately.

Use of physical separation schemes in studies of trace metal environmental chemistry has been reviewed by de Mora and Harrison [1983], and a paper dealing with such methods in river water was published by Moore et al. [1979]. Additional papers dealing with colloidal Fe and DOM have been published by Lewis Fox (this volume, page 65). A review of the use of ultrafiltration applied to DOM studies has been presented by Carlson et al. [1985]. Papers in this volume that describe or comment on separation procedures are listed in Table 2-1.

2.1. Pore Size Separations

The de-facto "standard" use of 0.4 μmpore-size filters in the study of the inorganic properties of marine particles comes primarily from a matter of convenience, in that a reasonably large sample can be processed in a reasonable amount of time. Glass fiber filters (0.7 μm) are generally used when studying organic material. A discussion of the various filter types, matrix materials, and manufacturers can be found in Chapter 1.

One should recognize that filtration using any filter of a nominal pore size should not imply a distinct particle size cutoff. Sheldon [1972] has reported retention vs. particle size for a variety of filter types and pore sizes and the information is summarized again in de Mora and Harrison [1983]. This type of information is also usually available from the filter manufacturers. Johnson and Wangersky [1985] have also addressed this issue. Generally speaking, tortuous-path "depth" filters with large nominal pore sizes (8 μm Millipore cellulose ester filters, for example) can efficiently trap much smaller particles. Screen-type filters (Nuclepore, Millipore, or Poretics track-etched polycarbonate or Anopore aluminum oxide) generally have a sharper particle size cutoff. Unfortunately, as any filter becomes loaded and the flow rate drops, the filtration characteristics change, generally leading to increased trapping of small particles. Impaction of particles smaller than the nominal pore size using polycarbonate filters is a recognized problem. Particles on heavily loaded filters may also experience the surface regulation phenomenon, leading to potential alterations in chemistry due to localized changes in surface properties of the trapped particles.

Filtration using very small pore sizes (less than 0.05 μm) is problematic due to the low flow rates and small sample volumes. For colloidal particles, other techniques such as ultra-filtration may be more appropriate.

Attempts to perform sequential filtration to obtain size distribution information may yield artifacts. Turbulence during filtration can lead to aggregation of small particles. Cavitation during vacuum filtration can also lead to aggregation. Organic-rich waters may re-equilibrate following filtration, leading to aggregation of new filterable particles rich in organic carbon, iron, and aluminum.

The common practice in POC determinations of using a primary and secondary glass fiber filter to correct for organic carbon adsorption on the primary filter may be confounded by these effects.

Table 2-1.
Physical Separation Techniques Described or Cited by the Contributing Authors

Chapter 3. Major Bioelements	
Alldredge	Individual aggregates by SCUBA diver
Karl et al.	Trap swimmers by 335 μm screen
Knauer	Trap swimmers removed manually
Chapter 4. Organic Properties	
Henrichs	Flow cytometry (bacteria)
Benner	Tangential-flow ultrafiltration (bacteria, viruses, colloidal DOM)
Brownawell	Tangential-flow ultrafiltration (colloidal DOM)
Tanoue	Electrophoresis of proteins
Chapter 5. Inorganic and Radioactive Properties	
Fleer and Bacon	1 mm sieving of trap samples
Hunter	Electrophoresis (surface coatings); Ultracentrifugation/TEM (size distribution analysis.)
Yeats	Tangential-flow ultrafiltration (inorganics)
Moran	Tangential-flow ultrafiltration (inorganics, colloidal DOM)
Skei	Pore-size effects (FeS)
Sherrell	Pore-size effects (pumps, TSM)
Landing and Lewis	Chromatography resins (FeS, $Fe(OH)_3$)
Chapter 6. Mineralogy and Surface Properties	
De Carlo	Pore-size separation (clays)
Fischer	Sieving trap and net samples
Van der Gaast	Sieving trap samples

While recognizing the potential artifacts associated with any filtration technique, simple size fractionation of marine particulate matter can be carried out without resorting to elaborate schemes or instruments. For example, track-etched polycarbonate membrane filters can be purchased with pore sizes down to 0.015 μm, much smaller than the pore sizes used for routine marine work, typically 0.2 to 1.0 μm. This offers the opportunity for collection and size fractionation of particles extending into the colloidal range. The simplest apparatus for size fractionation would require only a syringe and a filter holder adapted to the syringe with Luer fittings. The smallest pore diameter through which significant volumes of water can be forced using the syringe technique is 0.03 to 0.05 μm. Thus, useful information can be obtained by filtration over the 0.03 to 0.4 μm size range using this simple technology. The most obvious limitation would be that the analytical method must be sufficiently sensitive to detect the analyte of interest on the relatively small sample sizes one is limited to with these filters. Use of larger diameter filters obviously increases the filterable volume (as a function of the radius squared) but may also involve an increase in the filter blank contamination levels.

There are several examples of the use of different pore-size filters to size fractionate suspended material. Henrichs and Williams [1985] have fractionated particulate combined amino acids using pore sizes as low as 0.1 μm. Skei (this volume, page 303) has used 0.4 μm and 0.2 μm filters to study particulate FeS in Framvaren Fjord. Lewis and Landing [1991] tried a similar experiment in the Black Sea, observing that a significant fraction of the less than 0.4 μm "dissolved" Fe(II) in the anoxic zone appeared to behave as a small colloidal-sized species between 7 and 50 nm in diameter. K. Falkner (personal communication, 1991) also reports the successful use of 0.1 μm Teflon filters for analysis of particulate platinum group elements and gold.

R. Anderson (Lamont-Doherty Geological Observatory, personal communication, 1991) reports that filtration of anoxic Black Sea bottom water through a sequence of 0.4, 0.1, and 0.03 μm Nuclepore polycarbonate membrane filters produced no detectable change in the concentration of dissolved uranium. In contrast, applying the same technique to Black Sea sediment pore water showed that 40% of the less than 0.4 μm "dissolved" U was removed by refiltering through a 0.1 μm filter. Furthermore, not even one drop of the pore water could be forced through a 0.03 μm filter. These results indicated that greater than 0.03 μm colloids were sufficiently abundant in the pore waters to immediately clog the filter, and that a significant amount of the "dissolved" uranium is contained in a colloidal phase. While these results do not pertain directly to particles from the water column, they do demonstrate the ability of relatively simple sampling techniques to provide information about the abundance and geochemical importance of colloids.

2.2. Ultrafiltration

Use of standard pressurized ultra-filtration systems for the study of marine colloidal dissolved organic matter (DOM) has been recently reviewed by Carlson et al. [1985]. They concluded that polarization artifacts caused by colloid build-up on the membrane may lead to serious artifacts. Recently, ultrafiltration methods employing high shear across the membrane surface have come into use by marine scientists. The high shear reduces the polarization effect, and the large filtration areas allow for processing of very large volume samples, tens to hundreds of liters.

These techniques are most commonly known as tangential-flow ultrafiltration (TFF), and use either flat plate or cartridge membranes (cross-flow ultrafiltration; CFF) or bundles of hollow fibers (including spiral-wound cartridges; HFF). Four contributions in this volume refer to the use of these methods (Benner, Brownawell, Moran, and Yeats, pages 181, 187, 275 and 321) and describe applications in both organic and inorganic marine chemistry. Brownawell makes special mention of the use of TFF as opposed to adsorptive extraction of "humic" materials at low pH, emphasizing that the nature of the material collected with these two disparate methods may be quite different.

As a filtration scheme, this technique is subject to most of the concerns, discussed above and in Chapter 1, regarding filtration. The nominal pore-size of any membrane is thus only approximate, and specific particle-membrane interactions can strongly affect the retention of a given colloidal particle. Pre-filtration may also be required, with the attendant potential artifacts. In addition, at very high concentration factors, formerly dilute colloidal suspensions become more concentrated with the potential for particle-particle interaction artifacts. There is a great need to standardize these techniques in seawater, using a wide variety of organic and inorganic colloidal materials.

There are several manufacturers of TFF equipment. Flat and cartridge membrane systems are sold by Millipore (Pelicon system), Osmonics, and Filtron, while hollow fiber equipment is sold by Amicon and Osmonics. The nominal pore-size cutoff for the various systems ranges from 1 nm (500-1,000 MW) to 0.65 μm. Several membrane types are available for the flat plate and cartridge systems including polysulfone and regenerated cellulose. The regenerated cellulose (Millipore) and modified polysulfone (Filtron) are reported by the manufacturers to be "low protein binding". The hollow fibers are composed of polysulfone. Refer to the filter matrix table in Chapter 1 for warnings regarding the behavior of various membrane matrices. Both systems cost about $5,000 (in 1990) for a complete pump, reservoir, and membrane setup.

Flow rates are a function of pore-size, back-pressure, filtration area, ionic strength, and temperature. Seawater cations (Ca and Mg) can seriously retard the flow rate, presumably due to specific cation-membrane interactions. Because of the high cross flow shear, initial particle load and the extent of concentration limit the flow rate only slightly. It has been reported that for a given pore size, membrane area, and system back pressure, the hollow fiber systems offer higher flow rates. The Pelicon system has been reported to routinely allow 60 L hr^{-1} filtration rates. Removal of adsorbed trapped particles from the membranes can be accomplished by rapid back flushing, chemical leaching, or physical removal (flat plate membrane design).

Potential sources of contamination (in addition to concerns, discussed in Chapter 1, are the shedding of silicone rubber tubing in peristaltic pumps, or the exposure of metallic surfaces when using Teflon diaphragm pumps. Glass fiber pre-filtration can release small glass fibers into the sample. Finally, delivering a large-volume sample to a TFF system generally requires pumping and/or sample mixing, with the potential aggregation artifacts noted above.

2.3. Ultracentrifugation

Ultracentrifugation is useful for the concentration of particles for morphological study and individual particle analysis. A limitation is the small sample volume which can be conveniently processed. Because of the small number of particles generally studied by electron microscopic techniques, environmental contamination by dust, fibers from clothing and laboratory wipes can be a problem. Adoption of clean room techniques, especially the use of laminar-flow hoods, would be prudent for work of this type.

Reports on the use of ultracentrifugation of marine water samples to collect colloidal particles on SEM or TEM stubs are appearing in the literature [Reid et al., 1990; Hunter, this volume, page 251]. Microanalysis of marine particles using STEM is facilitated by ultracentrifugation to concentrate material prior to fixing and thin sectioning (Cowen, this volume, page 433). Recent work by Mark Wells and Ed Goldberg on this technique is described in Longhurst et al. (submitted, 1991).

2.4. Size-Exclusion Chromatography

There are many potential applications of SEC to the size-fractionation of small particles, especially colloidal DOM. The method has been discussed recently by Reid et al. [1990]. Brownawell (this volume, page 187) also discusses this technique and reports some information on resin materials and manufacturers. Careful calibration and standardization of the method as applied to the marine field is still required. Caution is necessary when choosing an appropriate gel or resin type such that it is compatible with the particles being treated. One must also consider potential specific particle-resin adsorption or affinity interactions. Generally, pre-concentration is required for samples from open ocean environments. Use of developing techniques such as tangential-flow ultrafiltration for the concentration of colloidal material, followed by SEC treatment seems appropriate.

2.5. Density Gradient Centrifugation

The technique does not appear to be widely used in the study of marine particles at this time. The method has been used to separate cell wall fragments and internal organelles of marine diatoms following mild cell disruption (Landing, unpublished work, 1990), and is also used in geological research for mineral separations and in biochemical research for macromolecule separations and characterization.

2.6. Dialysis

Dialysis membranes (generally cellulose-based) and dialysis equipment are commonly available from most scientific suppliers including VWR, Fisher Scientific, etc.

Fox [1988] discusses the use of passive dialysis to isolate dissolved Fe (less than 1,000 MW) from river and estuarine samples. He reports good agreement with tangential-flow ultrafiltration (Lewis Fox, Harvard University, personal communication, 1991).

2.7. Flow Cytometry

Flow cytometry is being used extensively by microbiologists and other marine scientists studying small organisms. The technique offers the capability of sorting particles into, generally, three "bins" or categories based on features such as size, charge or spectral properties. They are generally limited to particles larger than 100 nm, and particle concentrations greater than 10^6 mL^{-1}. Sample sizes are relatively small, processing 1 mL in 60-90 minutes (1,000 particles sec^{-1}). Sorting particles by size requires roughly a factor of two difference in size. Instruments are relatively expensive ($100k to $250k) and available from several sources (Cytofluorograph 2S, EPICS-753). Some possible applications of this equipment are described by Henrichs (this volume, page 139).

2.8. Field-Flow Fractionation

Use of FFF has been described by Beckett et al. [1987] and is used extensively by J. Calvin Giddings of the University of Utah. The method uses small volume samples, may require pre-concentration, and has been applied to the study of freshwater humic substances [Beckett et al., 1987]. Interfacing the instrument to other equipment such as ICP-MS is being pursued. It does not appear to have been applied to the study of marine colloidal DOM.

2.9. Particle Counters

Limited to particles larger than 300 nm (using a 12 μm aperture), the newest instruments (Elzone 80XY) can be used to obtain detailed size distribution information. Samples should be pre-filtered to remove particles larger than the aperture which is used. One must be aware of potential artifacts associated with this intermediate handling. Koike et al. [1990] have reported the existence of very small particles which are counted using the Elzone instrument. A recent NATO-sponsored workshop at the Bedford Institute of Oceanography on sub-micron particles compared the use of such particle counting techniques with other techniques including ultracentrifugation. A report from this workshop has been submitted for publication (Longhurst et al., 1991).

2.10. Electrophoresis

A widely used technique in the biomedical and biochemical fields, classical gel electrophoresis has been used to study proteins in seawater (Tanoue, this volume, page 163). The principles of capillary electrophoresis have been recently reviewed by Olechno et al. [1990].

Individual particle electrophoresis has been described by Hunter (this volume, page 251). This application is primarily oriented towards study of particle surface chemistry via investigation of the surface charge. With laser detection, particles as small as 100 nm can be studied, however natural particle populations are much more difficult to study than are laboratory preparations of uniform size. Sophisticated computer imaging may expand the capabilities of this technique in the future. Individual particle interactions can be studied by altering ionic strength, pH, and solution composition. Instruments are available from Rank Brothers (UK), Malvern Instruments (UK) and Zeta Corp. (USA) and range in cost from $5,000-$10,000 (in 1990).

3. "Selective" Chemical and Analytical Methods

Wet chemical leaching schemes have been widely used to study marine particles and sediments and a classic early paper in the field was written by Chester and Hughes [1967]. Many people have developed leaching schemes to study a variety of solid phases, and the methods used by the participants in this workshop are outlined in Table 2.2. A review by Chao [1984] and an article by Tessier et al. [1979] describe "selective" chemical leaching methods for rock and sediment samples.

While there are many examples in the literature on the use of various "selective" chemical treatments, and while many of these treatments can be adjusted to extract or dissolve predominantly one phase or another, there are few reports which conclusively demonstrate the selective, quantitative extraction of a given particulate phase. Acidic treatments attack many phases, at differing rates, while neutral or basic treatments may result in re-precipitation of extracted elements. For example, alteration of clay minerals by acidic, basic, or oxidizing treatments may affect the resultant XRD patterns. In addition, some schemes are sequential while others use a fresh sub-sample for each different treatment step, and neither approach has any obvious advantage under all conditions.

The critical issue regarding methods such as these is that they are, by definition, "operational". Conditions including reagent concentrations, reagent/solids ratios, duration of treatment, temperature, etc. are all important to the analytical results. The best advice regarding "selective" chemical treatments is probably "let the analyst beware", meaning that the claims of any operational scheme must be thoroughly investigated.

Despite these caveats, such methods are in wide use, and the application of a given scheme to a set of similar samples in a standardized fashion can yield interesting geochemical information. The following summarizes a few particular examples of techniques or species for which problems were noted in the workshop discussion.

Oxidizing treatments intended to selectively remove organic matter (H_2O_2, Br_2, HCl-HNO_3, etc.) may result in the oxidative precipitation of Mn and Fe with concurrent scavenging of other species (Al, Cu, etc.).

Selective dissolution of opal is reviewed by DeMaster (this volume, page 363) and discussed in Chapter 6. Corrections for alumino-silicate dissolution using Al measurements on the leach solutions may be inadequate for a variety of reasons. The Si/Al ratio in the mineral detritus may not be known accurately, and the dissolution of very reactive amorphous Al-Si phases may confound the determinations.

Selective removal of particulate inorganic carbon from samples prior to high-temperature combustion (CHN analysis) for particulate carbon measurements is discussed in Chapters 3 and 6. Samples are generally leached with dilute HCl or exposed to fumes of concentrated HCl. Calculations of carbonate content using excess Ca in acidic leaching treatments may be inaccurate due to Ca exchange from clays and other particles. Wet chemical techniques for particulate carbon analysis exist in the literature, may be approved for use on EPA-sponsored research, and should be cross-compared with CHN analysis.

While total particulate phosphorus can be determined by several methods (Sharp, this volume, page 87; Walsh et al., this volume page 91; Chapter 3), there seems to be no appropriate method for distinguishing between organic and inorganic particulate phosphorus. A method involving the use of sodium dodecyl sulfate (SDS) to preferentially extract organic material from sediments and sediment trap samples has been investigated (Sue Vink, University of Hawaii, personal communication, 1991) and is briefly mentioned by Collier (this volume, page 235).

Techniques for particulate sulfur measurements are described in papers by Cutter, de Leeuw, Luther, and Matrai (this volume, pages 57, 207, 273 and 83). While it is possible to selectively determine inorganic sulfur speciation, methods for organic sulfur compounds are not as well-developed. Generally, total organic sulfur is determined by the difference between total particulate

Table 2-2. "Selective" Chemical and Analytical Techniques Described or Cited by the Contributing Authors.

Chapter 3. Major Bioelements	
von Bodungen	0.1 M HCl leach for carbonates; Korreleff digestion for POP; NaOH for opal
Cutter	1 M HCl for carbonates; S speciation by GC
Harrison	30% H_2O_2/UV oxidation for POP; 0.5% Na_2CO3 for opal
Knauer	Concentrated HCl fumes for carbonates
Smith (Walsh et al.)	6 M HCl leach for carbonates
Chapter 4. Organic Properties	
Bidigare	Chlorophylls/carotenoids
Repeta	Chlorophylls/carotenoids
Hedges	Lignin, amino acids, carbohydrates
Henrichs	Amino acids, carbohydrates
Mopper and Furton	Polysaccharides, amino acids, lipids
Wakeham and Volkman	Lipids
Chapter 5. Inorganic and Radioactive Properties	
Collier	1 M NaAc/HAc (pH 5) for carbonate and sorbed; 1% Na dodecyl sulfate/0.2 M $NaHCO_3$ for organics; 1 M NH_2OH-HCl for oxyhydroxides; HF/HNO_3 for refractory
Feely et al.	0.04 N NH_2OH-HCl/25% HAc; refractory by XRF
Landing and Lewis	25% HAc; 2 M HCl-1 M HNO_3; HF/HCl/HNO_3 for refractory
Yeats	25% HAc; HF/HCl/HNO_3 for refractory
Chapter 6. Mineralogy and Surface Properties	
De Carlo	HAc/NaAc; NH_2OH-HCl/HNO_3; NH_2OH-HCl/HAc; Citrate-Bicarbonate-Dithionite; HF/HCl/HNO_3 for refractory
DeMaster	Opal dissolution review
Fischer	Extraction of various organic classes; 6 N HCl for carbonates; 1 M NaOH for opal; 15% H_2O_2 and 20% HAc leaching of clays for XRF; HF/$HClO_4$ for refractory
Van der Gaast	NaAc/HAc (pH 5) for carbonate; Na_2CO_3 or NaOH for opal; Br_2 or H_2O_2 for organics; Citrate-dithionite for FeO_x

sulfur and total inorganic sulfur. A summary on particulate sulfur can be found in Chapter 3.

Metalloid elements (Se, As, Ge) are generally analyzed using chemical treatments such as hydride generation which allow for speciation determinations. Total Se, Se(IV), Se(V), and Se(0) can be selectively determined, and organic Se calculated by difference [Cutter, 1985; Velinsky and Cutter, 1990]. Total As and Sb, As(III) and Sb(III), and As(V) and Sb(V) can also be measured [Andreae and Froelich, 1984]. Samples for redox speciation measurements must be protected from atmospheric oxygen, and are generally stored frozen.

The selective extraction and analysis of specific organic compound classes is discussed in Chapter 4, and the potential loss of organic compounds during filter washing is discussed in Chapter 1. Wet chemical treatments designed to selectively solubilize or oxidize organic matter (H_2O_2, SDS) have been used but must be carefully investigated.

Selective extraction of metals from various solid phases using chelating agents such as EDTA and crown ethers may be possible, and the ideal ligand for such treatments would neither adsorb to surfaces nor enhance particle dissolution. This approach has been tried for rare earth element extractions without success to date, apparently because rare earth elements are particularly prone to re-exchange onto exposed surfaces [Sholkovitz, 1989].

4. Surface Chemistry, Coatings, and Properties

It is clear that particles in seawater are coated to some extent with organic compounds [Balistrieri et al., 1981], attached either by non-specific adsorption or by specific molecule-surface interactions. Several papers in this volume discuss this issue (Honeyman, page 438, Hunter, page 251). Hunter describes the use of electrophoretic mobility to study surface charge. An important contrast between model adsorption studies and natural organic-particle interactions is that natural systems do not often exhibit adsorption/desorption reversibility.

Large quantities of colloidal marine organic material collected using TFF would provide interesting natural material for adsorption and surface chemistry studies. As discussed above, the chemical properties of colloidal DOM collected using TFF are significantly different from compounds collected using hydrophobic adsorption at low pH (XAD-2 technique). Jan de Leeuw (personal communication, 1991) has suggested that the exudate from cyanobacteria (a common, easily cultured marine microbe) might also serve as a conveniently-obtained "natural" model compound for investigations on the chemistry and properties of surface films and coatings. The need for more research on the interactions between particles and colloidal DOM (including microbial exudates) is clearly indicated.

Microbial extracellular polymer materials (EPM) have long been known to be important in microbial attachment and surface films [Fletcher and Floodgate, 1973; Costerton et al., 1974; Paerl, 1975; Geesy et al., 1977; Marshall, 1988] and in bacterial flocculation [Friedman et al., 1969; Deinema and Zevenhuizen, 1971]. Recently, microbial EPM have been directly implicated in the formation of marine aggregates [Biddanda, 1985, 1986]. Aggregates collected by submersible from mid-depths possessed significant numbers of bacteria that were encapsulated in extensive extracellular polymers [Cowen and Silver, 1984]. Sediment trap studies reported by Cowen and Bruland [1985] as well as those associated with the hydrothermal plume VENTS program (Cowen, unpublished data, 1990) revealed large amounts of such polymer secretions associated with general trap material but not with distinct fecal pellets. EPM also have important roles in metal oxide formation (see Chapter 6) and in carbon and energy cycling [Wrangstadh et al., 1990].

Methods used to study the properties of EPM include direct observation using transmission electron microscopy (TEM) (Figures 6.48 and 6.49, Cowen, this volume, page 429) and chemical analysis of isolated exudates of cultured cells [Friedman et al., 1969; Deinema and Zevenhuizen, 1971; Massalski and Leppard, 1979; Ghiorse and Hirsch, 1979; Biddanda, 1986; Cowen and Li, in press; Cowen, this volume, page 429]. Future development of radiolabeled molecular probes and specific heavy metal stains in conjunction with TEM may provide a more detailed bridge between laboratory studies on the properties of EPM and bulk organic-chemical analysis (see Chapter 4) of the polymer content of field samples. For example, Ruthenium Red is thought to specifically stain acidic polysaccharides [Luft, 1966]; TEM and elemental mapping (X-ray fluorescence) can be employed to reveal the physical associations of stained polymers in a heterogeneous particle field.

5. Summary

Many physical separation and chemical speciation techniques have been employed for the study of the chemistry and properties of marine particles, however all may be subject to potential artifacts. For any application, a careful assessment of the method using detailed calibration, including the use of standard reference materials, must be performed. Intercomparison with other independent methods demonstrating good agreement for a given separation helps to provide confidence in any procedure.

There is also a growing potential for combining physical or chemical separation techniques with modern or more sophisticated analytical instrumentation. A particularly good example is the use of tangential flow ultrafiltration to pre-concentrate colloidal marine material for characterization using a variety of spectroscopic methods and other analytical tools.

Acknowledgements. We are appreciative of the contributions made by all participants to the discussion in our session, and particularly for the written contributions from Bob Anderson and Jim Cowen.

References

Andreae, M. O., and P. N. Froelich, Arsenic, antimony, and germanium in the Baltic Sea, Tellus, 36, 101-117, 1984.

Balistrieri, L., P. G. Brewer, and J. W. Murray, Scavenging residence times of trace metals and surface chemistry of sinking particles in the deep ocean, Deep-Sea Res., 28, 101-121, 1981.

Beckett, R., Z. Jue, and J. C. Giddings, Determination of molecular weight distributions of fulvic and humic acids using flow field-flow fractionation, Environ. Sci. Technol., 21, 289-295, 1987.

Biddanda, B. A., Microbial synthesis of macroparticulate matter, Mar. Ecol. Prog. Ser., 20, 241-251, 1985.

Biddanda, B. A., Structure and function of marine microbial aggregates, Oceanol. Acta, 9, 209-211, 1986.

Carlson, D., M. Brann, T. Mague, and L. Mayer, Molecular weight distribution of dissolved organic materials in seawater determined by ultrafiltration: a re-examination, Mar. Chem., 16, 155-171, 1985.

Chao, T. T., Use of partial dissolution techniques in geochemical exploration, J. Geochem. Explor., 20, 101-135, 1984.

Chester, R., and M. J. Hughes, A chemical technique for the separation of ferro-manganese minerals, carbonate minerals and adsorbed trace elements from pelagic sediments, Chem. Geol., 2, 249-262, 1967.

Coale, K. H., and K. W. Bruland, Copper complexation in the northeast Pacific, Limnol. Oceanogr., 33, 1084-1101, 1988.

Costerton, J. W., J. M. Ingram, and K.-J. Cheng, Structure and function of the cell envelope of gram-negative bacteria, Bacteriol. Rev., 38, 87-110, 1974.

Cowen, J. P., and K. W. Bruland, Metal deposits associated with bacteria: implications for Fe and Mn marine biogeochemistry, Deep-Sea Res., 32, 253-272, 1985.

Cowen, J. P., and Y.-H. Li, The influence of a changing bacterial community on trace metal scavenging in a deep-sea particle plume, J. Mar. Res., in press, 1991.

Cowen, J. P., and M. W. Silver, The association of iron and manganese with bacteria on marine macroparticulate material, Science, 224, 1340-1342, 1984.

Cutter, G. A., Determination of selenium speciation in biogenic particles and sediments, Anal. Chem., 57, 2951-2955, 1985.

de Mora, S. J., and R. M. Harrison, The use of physical separation techniques in trace metal speciation studies, Water Res., 17, 723-733, 1983.

Deinema, M. H., and L. P. T. M. Zevenhuizen, Formation of cellulose fibrils by gram-negative bacteria and their role in bacterial flocculation, Arch. Mikrobiol., 78, 42-57, 1971.

Donat, J. R., and K. W. Bruland, Complexation of zinc by natural organic ligands in Sargasso Sea surface waters: comparison to the North Pacific, EOS Trans. Am. Geophys. Union, 71(2), 153, 1990.

Fletcher, M., and G. D. Floodgate, An electron-microscopic demonstration of an acidic polysaccharide involved in the adhesion of a marine bacterium to solid surfaces, J. General Microbiol., 74, 325-334, 1973.

Fox, L. E., The solubility of colloidal ferric hydroxide and its relevance to iron concentrations in river water, Geochim. Cosmochim. Acta, 52, 771-777, 1988.

Friedman, B. A., P. R. Dugan, R. M. Pfister, and C. C. Remsen, Structure of exocellular polymers and their relationship to bacterial flocculation, J. Bacteriol., 98, 1328-1334, 1969.

Geesy, G. G., W. T. Richardson, Y. G. Yeomans, R. T. Irvin, and J. W. Costerton, Microscopic examination of natural sessile bacterial populations from an alpine stream, Can. J. Microbiol., 23, 1733-1736, 1977.

Ghiorse, W. C., and P. Hirsch, An ultrastructural study of Fe and Mn deposition associated with extracellular polymers of Pedomicrobium-like budding bacteria, Arch. Microbiol., 123, 213-226, 1979.

Hasle, J. R., and M. I. Abdullah, Analytical fractionation of dissolved copper, lead and cadmium in coastal seawater, Mar. Chem., 10, 487-503, 1981.

Henrichs, S. M., and P. M. Williams, Dissolved and particulate amino acids and carbohydrates in the sea surface microlayer, Mar. Chem., 17, 141-163, 1985.

Johnson, B. D., and P. J. Wangersky, Seawater filtration: particle flow and impact considerations, Limnol. Oceanogr., 30, 966-971, 1985.

Koike, I., S. Hara, K. Terauchi, and K. Kogure, Role of sub-micrometre particles in the ocean, Nature, 345, 242-244, 1990.

Lewis, B. L., and W. M. Landing, The biogeochemistry of Mn and Fe in the Black Sea, Deep-Sea Res., in press, 1991.

Luft, J. H., Ruthenium red staining of the striated muscle cell membrane and the myotendinal junction, in Sixth Int. Congr. for Electron Micros. Proc., 794 pp., Kyoto, 1966.

Marshall, K. C., Adhesion and growth of bacteria at surfaces in oligotrophic habitats, Can. J. Microbiol., 34, 503-506, 1988.

Massalski, A., and G. G. Leppard, Morphological examination of fibrillar colloids associated with algae and bacteria in lakes, J. Fish. Res. Board Can., 36, 922-938, 1979.

Moore, R., J. Burton, P. Williams, and M. Young, The behavior of dissolved organic material, iron and manganese in estuarine mixing, Geochim. Cosmochim. Acta, 43, 919-926, 1979.

Moran, S., and R. Moore, The distribution of colloidal aluminum and organic carbon in coastal and open ocean waters off Nova Scotia, Geochim. Cosmoschim. Acta, 53, 2519-2527, 1989.

Morel, F. M. M., and P. M. Gschwend, The role of colloids in the partitioning of solutes in natural waters, in Aquatic Surface Chemistry, edited by W. Stumm, pp. 405-422, J. Wiley and Sons, New York, 1987.

Olechno, J. D., J. M. Y. Tso, J. Thayer, and A. Wainwright, Capillary electrophoresis: a multifaceted technique for analytical chemistry. Parts 1, 2, and 3, Am. Lab., 22, 1990.

Paerl, H. W., Microbial attachment to particles in marine and freshwater ecosystems, Microbial Ecol., 2, 73-83, 1975.

Reid, P. M., A. E. Wilkinson, E. Tipping, and M. N. Jones, Determination of molecular weights of humic substances by analytical (UV scanning) ultracentrifugation, Geochim. Cosmochim. Acta, 54, 131-138, 1990.

Sheldon, R. W., Size separation of marine seston by membrane and glass fiber filters, Limnol. Oceanogr., 17, 494-498, 1972.

Sholkovitz, E. R., Artifacts associated with the chemical leaching of sediments for rare-earth elements, Chem. Geol., 77, 47-51, 1989.

Sugimura, Y., and Y. Suzuki, A high-temperature catalytic oxidation method of non-volatile dissolved organic carbon in seawater by direct injection of liquid sample, Mar. Chem., 24, 105-131, 1988.

Tessier, A., P. G. C. Campbell, and M. Bisson, Sequential extraction procedure for the speciation of particulate trace metals, Anal. Chem., 51, 844-851, 1979.

Velinsky, D. J., and G. A. Cutter, Determination of elemental selenium and pyrite-selenium in sediments, Anal. Chim. Acta, 235, 419-425, 1990.

Whitehouse, B. G., P. A. Yeats, and P. M. Strain, Cross-flow filtration of colloids from aquatic environments, Limnol. Oceanogr., 35, 1368-1375, 1990.

Wrangstadh, M., U. Szewzyk, J. Ostling, and S. Kjelleberg, Applied Environ. Microbiol., 56, 2065-2072, 1990.

CHAPTER 3. MAJOR BIOELEMENTS

WORKSHOP REPORT

by

Chair: David M. Karl, Rapporteur: W. G. Harrison, Assistant Rapporteur: John Dore
and Members of the Working Group.

1. Introduction

The particulate matter pool in the ocean is comprised of both living and non-living components; the latter includes organic and inorganic fractions. However, most of the particulate matter in the ocean is of biotic origin. Thus biotic processes and transformations largely determine the types, size distributions, abundances and chemical compositions of particles in seawater. Carbon (C), nitrogen (N), oxygen (O), hydrogen (H), phosphorus (P) and sulfur (S) are the major components of living organisms and organic detritus. Accurate measurement of these "bioelements" is particularly important in assessing the chemistry and dynamics of particulate matter in the ocean.

During the discussion of the Bioelements Working Group, it became clear that a consensus exists within the community regarding the best analytical methods for measuring bioelements. Obtaining samples representative of nature and handling them so as not to alter their chemical composition was far more problematic. The following report begins by clarifying the nature of the diverse particle pools in the ocean. Sampling and handling problems unique to bioelements are then discussed. The report concludes with recommendations for the accurate chemical analyses of C, N, P and S. Details of these methods are discussed in the individual contributions. More detailed information can be obtained from articles provided in the reference list.

2. Origins and Pools of Marine Particles

Organic matter in the ocean exists along a continuum from small molecular weight compounds, e.g., hexoses, fatty acids, amino acids, nucleotides, to whale-sized particles [Sheldon and Parsons, 1967; Sharp, 1973]. Typically, an operational distinction is made between "dissolved" and "particulate" pools using either 0.2, 0.45, 0.7 or 1.2 μm pore size filters, depending upon the specific analytical measurements. It is important to note that by this operational definition, colloids and small particles including most bacteria and all viruses are sometimes included in the "dissolved" matter pool.

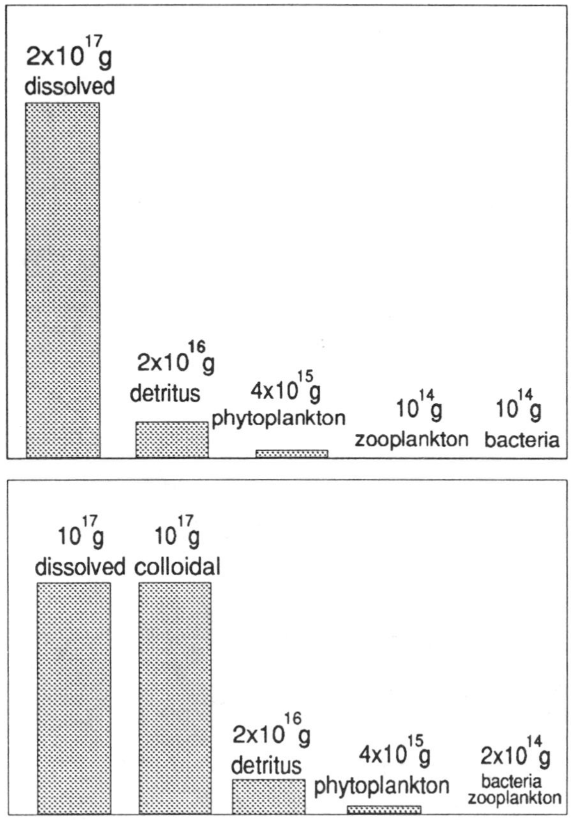

Fig. 3-1. Top: Distribution of organic carbon in the oceans [after Cauwet, 1978]. Bottom: Importance of organic colloids in seawater [after Cauwet, 1978].

In most oceanic environments, the inventory of organic matter, usually expressed as organic carbon, is dominated by dissolved materials (Figure 3-1). Consequently, one cannot ignore the presence of dissolved organic matter (DOM) when analyzing and interpreting particulate matter distributions in the sea. Recently, the accuracy of historical measurements of DOM in seawater has been called into question [Sugimura and Suzuki, 1988; Williams and Druffel, 1988]. Apparently, there exists a large pool of unidentified carbon compounds that are difficult to oxidize by conventional analytical procedures. It is conceivable that an analogous

Marine Particles: Analysis and Characterization
Geophysical Monograph 63

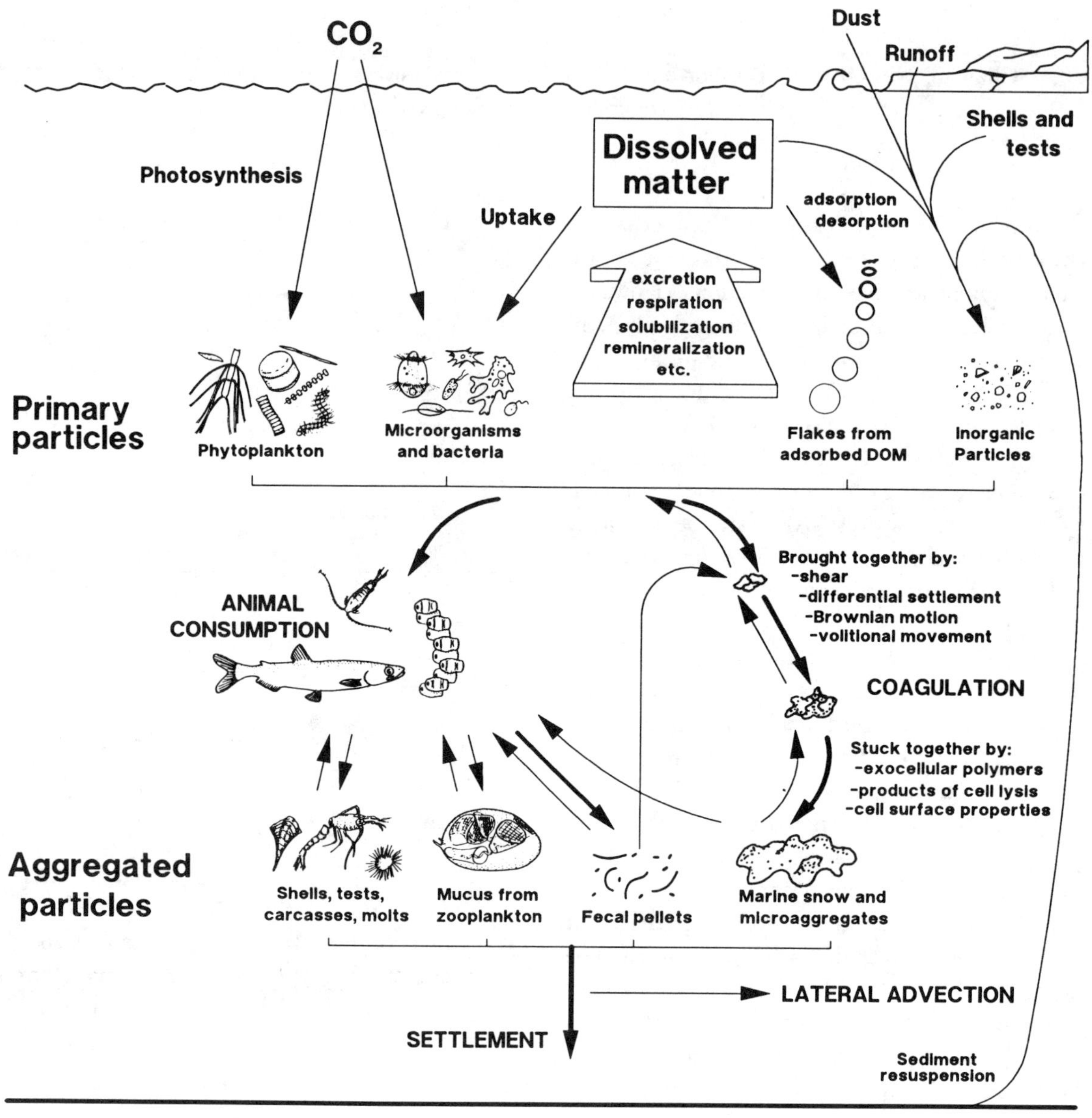

Fig. 3-2. Major sources of particles in seawater. Primary particles are inorganic particles or organic particles formed directly from inorganic or dissolved organic sources via carbon fixation, microbial uptake or adsorption. Aggregated particles are formed by the coagulation of primary particles or by animal consumption.

component of chemically-resistant particulate organic matter (POM) also exists in seawater.

The diverse origins of particles in seawater are summarized in Figure 3-2. Primary particles are those materials that arise directly from either inorganic or dissolved organic sources. While some primary particles in the ocean are inorganic, originating from windblown dust, terrigenous run-off and resuspension of bottom sediments, the majority are organic particles. The contribution of inorganic sources to the primary particle pool depends upon location and is particularly sensitive to proximity to the shore or bottom.

Organic primary particles arise from a variety of sources but most are produced through photosynthesis by eucaryotic and procaryotic algae and cyanobacteria. Transformation of dissolved organic matter into particles can also occur through processes that may be quite significant but poorly quantified. Monomeric organic compounds are known to bind abiotically to dissolved macromolecules [Carlson et al., 1985], thereby affecting the molecular weight distribution of organic matter in seawater. High molecular weight DOM adsorbed to bubble surfaces becomes flakes of particulate matter ranging from a few to tens of microns as the bubbles burst or dissolve [Johnson and Cooke, 1980; Kepkay and Johnson, 1988, 1989]. Humic acids and other dissolved organic and colloidal materials form particles via salt flocculation as river water mixes with seawater [Sholkovitz, 1976; Fox, 1983]. It is also well known that bacteria can transform DOM into organic particles [Parsons and Strickland, 1962; Azam et al., 1983] and there is some evidence that DOM can also chemically precipitate in standing seawater [Riley, 1970]. Finally, the role of photochemical processes in the formation and transformations of organic matter in the marine environment also needs to be considered [Mopper and Stahovec, 1986; Kieber et al., 1989].

Once formed, primary particles in seawater undergo numerous transformations, most of which involve alteration and aggregation into new particles with substantially different properties. Some living primary particles die and thereby directly contribute to the formation of detritus. Most, however, become aggregated through two major pathways: consumption by organisms and physical coagulation [Alldredge and Silver, 1988]. Primary particles and other aggregated particles are consumed by organisms, especially protozoans, zooplankton and nekton, and converted to new particulate phases including animal biomass such as living animal bodies, calcareous or siliceous tests, molts, eggs, carcasses, fecal pellets and feeding webs and to dissolved and inorganic phases through respiration, excretion and bacterial decomposition. Primary particles may also aggregate into larger particles via physical coagulation processes. Physical transport mechanisms such as Brownian motion, turbulent shear and differential settlement join smaller particles together. These components stick together at rates dependent upon their surface properties and the nature of the "sticky" substances produced by microbes living on their surfaces. Coagulation processes may be particularly important in aggregating phytoplankton blooms and in producing larger sinking particles.

The multiple origins of particles in seawater serve to graphically illustrate the diverse nature of the types of particle pools which exist simultaneously in any parcel of seawater. Moreover, transformations between these pools are bidirectional and dynamic. The diverse and changing nature of these pools, both in terms of chemical composition and size, must be kept in mind when designing any program for the chemical analysis of marine particulate matter. It is particularly important when considering analyses of bioelements. Interpretation of the results of chemical analyses may ultimately depend upon the investigator having a quantitative understanding of the types of particles making up a sample of filtered seawater or a sediment trap sample.

3. Problems of Sample Collection and Handling Unique to Bioelements

Particulate material in the water column is segregated into suspended particles and sinking particles, which can be characterized by their respective residence times and sinking speeds in the water column [Riley, 1970]. Traditionally, suspended particles are collected by either water bottles or by in situ filtration. Although most of the problems associated with the collection and sample handling of bioelements are common to marine particles in general, the following aspects deserve particular attention:

1.) The cleaning of equipment appropriate to the particular analysis by removing the surface films, adsorbed trace elements etc.;
2.) Homogeneity of the sample must be maintained. This can be achieved by either filtration of the total contents of the water bottle or by gentle agitation of the water sample prior to taking a sufficient subsample. Care must be taken to avoid de novo formation of particles via shaking.
3.) The presence of high concentrations of macrozooplankton may bias the suspended sample. Prescreening of the sample may reduce this problem. However, care must be taken that mucilaginous algae (e.g., *Phaeocystis*) or diatom chains are not excluded by this process.
4.) The choice of filter must be compatible with the size spectrum of particles to be sampled.
5.) The application of vacuum/pressure filtration should be conducted so as to avoid sample damage such as cell lysis.
6.) If samples cannot be analyzed in reasonable time, they can be stored either dry or deep frozen.

The sinking particulate pool can be sampled by pumps (see sampling section), by sediment traps or directly using SCUBA or submersibles. The sampling of macroaggregates (e.g., marine snow) is specifically discussed by Alldredge (this chapter, page 46).

Since sediment traps were first introduced in the 1970's, a number of problems have arisen with respect to their use. Examples of such concerns include trap design, various hydrodynamic considerations, appropriate sample preservation, mode of deployment (e.g., moored vs. free floating) and potential contamination from the inclusion of organisms not considered as part of the passively sinking flux [i.e., swimmers; Knauer et al., 1979, 1984]. All of these issues have been discussed in detail in the U.S. GOFS report #10 [1989] and some are reiterated in this chapter (see reports by von Bodungen et al., Karl et al., Knauer, and Wassmann, pages 47, 73, 81 and 99). The latter reports include the use of traps in both eutrophic and oligotrophic ocean waters, appropriate blank preparation and the use of various types of density solutions.

4. Chemical Analyses of Specific Bioelements

The major bioelements are routinely analyzed in many oceanographic laboratories. In most cases similar standard methods of analysis are employed. The working group recommends the following procedures for the chemical analysis of each bioelement. Details of the methods used in individual laboratories may be found in the individual contributions.

4.1. Analysis of Particulate Carbon and Nitrogen

Particulate carbon (PC) and nitrogen (PN) are the most commonly measured bioelements in the ocean, partially because of the convenient simultaneous determination of these two elements with commercial elemental analyzers. Consequently, of the three elements commonly reported as components of the "Redfield ratio" (i.e. C, N, P), particulate phosphorus (PP) is the least frequently determined.

Both organic and inorganic forms of PC are found in seawater. Relatively easy separation protocols (e.g., acidification) have been established so that particulate organic carbon (POC) can be directly distinguished from particulate inorganic carbon (PIC). There are no common marine nitrogen minerals, but ammonium can adsorb to or form in particles, creating an inorganic pool of PN. This pool is not usually considered to be significant. Phosphate can adsorb to particles, and phosphate minerals do exist in seawater. Sequential extractions have been used to separate particulate phosphorus (PP) phases (see individual papers in this chapter). Since much of the phosphorus bound in plankton can be lost with inorganic extraction protocols, such separations are not satisfactory for determining particulate organic phosphorus (POP).

4.1.1. Method.

Particulate matter on a filter or in powdered form, from sediment traps or sedimentary material, is analyzed by high temperature oxidation usually followed by detection of carbon as CO_2 and nitrogen as N_2 by thermal conductivity detection (TCD) gas chromatography using a commercial elemental analyzer. The advantage of the TCD is that both elements can be determined in a single sample.

4.1.2. Sample acquisition.

With bottle sampling, care should be taken to assure that quantitative recovery of a representative aliquot of the full sample or the full sample itself is used since suspended material may be unevenly distributed [Wangersky, 1974]. Although metal-free cleaning procedures [Fitzwater et al., 1982] are not necessary, care should be taken to assure that C and N contamination does not occur during sample collection. Sample volume should be dictated by the sensitivity of the analyzer used and the concentration and distribution of the suspended material in the environment being sampled. If infrequent large particles are encountered on filters, prescreening of water samples or removal of such particles may be desired and, if done, should be reported.

There are unique problems with sediment trap sample preparation (e.g., contamination, "swimmers", sample splitting) that are discussed below. If samples are collected by in situ pumps, considerations about particle destruction and/or production should be made.

4.1.3. Collection on filters.

Commonly, glass fiber filters (e.g., Whatman GF/F or GF/C) are used to collect particulate matter, although silver membrane and quartz fiber filters are sometimes used. Recent applications of aluminum oxide filters, i.e., Anopore, have been made with apparent good retention of submicron particles, although the filters are expensive and difficult to handle [Altabet, 1990].

A relatively low pressure differential (0.3 to 0.5 atmospheres) should be used in vacuum or pressure filtration to minimize cell disruption. Care should be taken to prevent the filters from running dry until the end of the filtration. A common problem for glass fiber filters is a change in the nominal pore size with loading of particulate matter on the filter because of accumulation of particles on the surface and in the interstices of the filter. Therefore, attempts should be made to maintain consistent filter loading.

Sea salt residues in the filters are not ordinarily a problem. Rinsing, for example with deionized water, might compromise the integrity of the sample (e.g., cell lysis) and is not required for CNP analysis and therefore is not ordinarily recommended, however see Knauer (this chapter, page 81) for exceptions with trap collected POC and PON. In certain environments (e.g., coccolithophorid blooms, coral reef environments), carbonate minerals can contribute a significant portion of PC. Under such circumstances, removal of PIC by direct acidification can be done to separate between total PC and POC. Fuming is not recommended. The acid used must be strong enough to dissolve carbonate particles embedded in organic matrices. Total PC can be analyzed on an untreated filter while POC can be measured after calcium carbonate removal, with PIC determined by difference. Differential temperature techniques for separation between PIC and POC [Telek and Marshall, 1974; Hirota and Szyper, 1975] are not recommended because there can be significant portions of the PIC (particularly high-Mg calcite) that decompose below the recommended 500° C and significant portions of the POC that survive temperatures above that.

Several of us have seen evidence of losses (up to 40%) of PN and POC with acid treatment. If acid rinsing is used, the rinse should be collected and analyzed for dissolved organic carbon and nitrogen (DOC, DON; Walsh et al., this volume, page 93). The recommended method for acidification in other than high carbonate environments is addition of acid in a modified sample analyzer cup (Cutter, this chapter, page 57).

4.1.4. Sample storage.

It is desirable to minimize sample handling in order to avoid contamination. Therefore, it is best to store samples in a vacuum desiccator directly in sample cups for the instrumental analysis. As an alternative, samples can be frozen with care to minimize contamination in the freezer and subsequently desiccated before analysis. There is no indication of time limits for storage.

4.1.5. Instrumentation.

Although much of the PC and PN analyses reported in the literature were performed on manual instruments, primarily Hewlett-Packard [Sharp, 1974] and Perkin-Elmer [Gordon and Sutcliffe, 1974], instruments available today are automated full gas chromatographic instruments (Carlo Erba, Europa, Perkin-Elmer, Leeman - see individual papers in this chapter). Operation should follow manufacturer recommendations with modifications determined by empirical evidence and from revisions from manufacturer and other users.

4.1.6. Standards and blanks.

The primary standard should be stable and not hygroscopic; it should have a C/N ratio appropriate to the material being analyzed (commonly used compounds are EDTA, acetanilide, and sulfanilamide). Because these pure compounds are easily analyzed and may not adequately represent instrument performance with more recalcitrant compounds under all conditions, it has been recommended that an appropriate natural material reference standard be used. For example, National Institute of Standards and Testing (NIST) "Orchard Leaves" and "Estuarine Sediment" have been used. Broader use of such reference standards is strongly recommended and it is suggested that other appropriate NIST standards be established for use by the oceanographic community.

Appropriate blank correction is necessary because of contamination of samples from handling and storage, and contamination from dissolved organic matter retained on filter pads. The instrument blank can usually be accounted for during calibration. The best method to determine the handling and storage contamination is to treat the filter blank exactly as a sample. A filter can retain seawater which includes a small but significant amount of dissolved carbon; there is no comparable nitrogen blank problem because of much lower concentrations of dissolved nitrogen. If a filter blank is wetted with filtered seawater before treatment like a sample, this will account for the retained carbon. If there is concern about dissolved organic matter (DOM) contributing a blank problem (e.g., organic compounds adhering to the matrix of the filter and aggregation of colloidal particles from turbulence during filtration), appropriate blank corrections should be made. One method to approach this problem is to use an additional filter (or filters) below the sample collection filter in the filtration step [Menzel, 1967; Holm-Hansen et al., 1966]. It has been suggested that filtering large volumes of water in low particle environments could decrease the influence of the DOM adhesion by increasing the amount of particulate matter on the filter. However, this could also indicate that the low, relatively invariant particulate content in the deep ocean might be only dissolved contamination.

4.1.7. Accuracy, precision and detection limits.

Due to the unknown nature of organic matter in seawater, we consider accuracy difficult to assess. If the instrument for CN analysis is properly calibrated, the accuracy must be 100%.

Analytical precision (replicability of a homogeneous standard material) should be within ±5-10%. The coefficient of variation of replicate sample filters of suspended matter has been shown to be approximately ±10%.

The analytical detection limits of the various instruments vary somewhat, and are about 0.1 to 0.4 μM C or N. Samples containing greater than 0.5 μM N and 1 μM C can be run with confidence on most of the instruments.

4.2. Analysis of Phosphorus

4.2.1. Method.

Particulate matter on a filter or in powdered form such as from sedimentary material of sediment traps is usually analyzed by high temperature combustion and acid hydrolysis followed by detection by standard colorimetric analysis of phosphate. Various other oxidation methods (e.g., UV irradiation, persulfate oxidation) have been used previously; these have been shown to suffer from variable and incomplete conversion of bound phosphorus to orthophosphate (see individual contributions in this chapter). Moreover, the PP methods discussed here undoubtedly miss some of the mineral phosphate pools (e.g., some of the apatite minerals).

4.2.2. Sample acquisition.

Sample acquisition procedures used for PC and PN are also appropriate for PP. No unique precautions are noted.

4.2.3. Collection on filters.

If a desired objective is direct comparison of P data to C and N data, it is important to follow the same sample filtration procedures (e.g., filter type, filter loading) used for carbon and nitrogen. There is not a sea salt problem with phosphorus comparable to that with carbon due to the comparatively low dissolved phosphate concentration in seawater.

4.2.4. Sample storage.

It is desirable to minimize sample handling to avoid contamination. It is convenient to place wet filters directly in analysis vials or tubes and dry in an oven immediately after collection, before storing the samples in a desiccator at room temperature. As an alternative, samples can be frozen with care to minimize contamination in the freezer. There is no indication of time limits for storage.

4.2.5. Analysis.

The high temperature method for destruction of organic phosphorus is commonly used and convenient [see individual papers in this chapter]. The UV oxidation method [Armstrong et al., 1966] was suspect for incomplete recovery of phosphorus from organic phosphorus compounds (Solorzano, personal communication, 1980). However, a UV method with higher output lamp appears to compare favorably with high temperature combustion (Harrison, this chapter, page 71). Persulfate digestion for PP analysis has been found by several laboratories (e.g., Smith and Walsh, unpublished, 1981) to yield non-quantitative recovery.

4.2.6. Standards and blanks.

Because complete recovery of phosphate from organic compounds has been well established [Solorzano and Sharp, 1980], a number of laboratories do routine standardization with orthophosphate only. Other laboratories do daily standardization with organic phosphorus compounds such as NIST Orchard Leaves or glucose-6-phosphate.

Blanks can be run by treating an unused filter in the same fashion as a sample. When this is done carefully, the final colorimetric absorption is similar to an orthophosphate analysis deionized water blank. For routine analysis, it is advisable to run a filter blank with a batch of samples to verify that there is no contamination problem.

4.2.7. Accuracy, precision and limits of detection.

With complete recovery of organic standards, the accuracy of the method is 100%. The analytical precision of the method is nominally the same as that of the orthophosphate method [Strickland and Parsons, 1972]. The precision for particulate phosphorus on filters has been determined in several laboratories at about ± 10%. Strickland and Parsons [1972] report the limits of detection for the orthophosphate method as 0.03 μM (μM L^{-1}). The detection limits for the particulate phosphorus method will depend upon the concentration of phosphorus in the final product water after all sample handling (influenced by volume of water filtered, dilution, etc.).

4.3. H, O and Total Caloric Content

The measurement of carbon-specific H, O and caloric content of suspended and sediment trap-collected organic matter is of intrinsic interest in the assessment of the available energy content. These measurements are fundamental to studies of trophodynamics and energy flux [Paine, 1971], but are rarely conducted in oceanographic studies.

In theory, quantitative biochemical composition (i.e., total protein, lipid and carbohydrate) of bulk organic matter should be stoichiometrically related to H elemental analyses and to the bomb caloric value. Of these determinations, CHN elemental composition appears to be the most straightforward analysis. It requires only small amounts of particulate organic matter (≤ 250 μg dry weight) and can be measured simultaneously with C and N using commercially-available instruments. Gnaiger and Bitterlich [1984] found excellent agreement between proximate biochemical composition and CHN determinations for a variety of organic materials, including detritus and algae.

Because quantitative measurements of H and O are not routinely conducted, we have decided against making any recommendation regarding detailed analytical procedures. The fact that several investigators have published data on CHN content of particulate matter [Hallegraeff, 1978; Gnaiger and Bitterlich, 1984; Martin et al., 1987] indicates that these measurements are possible. A major source of error in H measurements is the presence of residual water in the sample materials. Consequently, it is imperative that all samples are thoroughly dried and retained in an absolutely dehydrated condition until final analysis. Gnaiger and Bitterlich [1984] provide a comprehensive discussion of the "residual water" problem and provide analytical procedures designed for the Carlo Erba Model #1160 Elemental Analyzer to satisfy these objectives.

4.4. Sulfur

4.4.1. Method description.

The basis of determining sulfur in marine particles is the high temperature combustion of filters or powdered material (e.g., from sediment traps) to produce sulfur dioxide. This SO_2 is then quantified using sulfur-specific detectors (Matrai, this chapter, page 83) or gas chromatography with thermal conductivity detection (GC-TCD; Cutter, this chapter, page 57). Although the GC-TCD method has higher detection limits, it is also capable of simultaneously determining carbon and nitrogen. It is important to recognize that particulate sulfur can exist in many chemical forms, e.g., organic compounds such as methionine and inorganic compounds such as pyrite. Therefore, the combustion efficiency must be evaluated for a wide class of compounds and materials in order to assure accurate determinations of total sulfur. In this respect, standard reference materials must always be analyzed.

4.4.2. Instrumentation.

Two commercially available systems are capable of determining sulfur in marine particulate matter: the Antek 771 pyroreactor with 701C sulfur detector [Matrai, 1989], and the Carlo Erba NA-1500 elemental analyzer (Cutter, this volume, page 57). Generally, these instruments can be used with minimal modifications.

4.4.3. Sample acquisition and storage.

Although there are many techniques for acquiring suspended particles in the water column, three important considerations are:

1) the method should collect samples that are representative of marine suspended matter;
2) the method must not compromise the chemical composition or speciation of a sample;
3) the method should be logistically acceptable, e.g., relatively low cost, simple to use at sea.

The procedures recommended here and in the Carbon and Nitrogen section generally meet these criteria.

The choice of filter type and size is a difficult one and requires certain compromises. The simultaneous determination of carbon, nitrogen and sulfur excludes the use of most filters with sub-micron pore-sizes (e.g., Nuclepore polycarbonate filters). Aluminum oxide and silver filters are too costly for most laboratories and too fragile to be folded into the tin sample cups used with the Carlo Erba or quartz spoon used by the Antek instrument. Thus, it is necessary to use quartz or glass fiber filters (nominal pore size ≥ 0.7 μm). In the case of the Carlo Erba analyzer, the size of the sample cups and the capability of the instrument also limit the filter diameter to 13 millimeters.

It is recommended that samples be filtered at least in duplicate, and preferably in triplicate, in order to assess overall precision. After filtration, the only sample pretreatment for sulfur that is performed on board ship is to remove inorganic sulfate which might remain on the filter after the seawater evaporates; we wish to determine the sulfur in the particles and not that from evaporated sea salts. Any rinse must be of a similar ionic strength in order to avoid cell rupture and the potential loss of sulfur. For the specific determination of particulate sulfur, Matrai [1989] used ammonium formate, but for simultaneous CNS determinations (i.e., Carlo Erba) this reagent may contaminate the sample with excess nitrogen and carbon. Thus, the Carlo Erba procedure utilizes a simple sodium chloride solution. With either desalting method the filter is immediately frozen in order to minimize artifacts due to bacterial growth and degradation of the organic matter.

Sample processing at the shore-based laboratory includes drying the filters or sediment trap materials. This must be done at low temperatures (40 to 60° C) to avoid the loss of volatile sulfur compounds such as dimethylsulfide. Furthermore, lyophilization should be avoided since very high losses of total sulfur have been observed in lacustrine sediments [Amaral et al., 1989].

4.4.4. Standards and blanks.

An important consideration in the determination of particulate sulfur is the choice of standards. Because pure sulfanilamide can be obtained commercially, is stable during storage, and is not hygroscopic, it makes an excellent primary standard. Similarly, methionine is a stable sulfur standard. However, neither is very representative of marine particles. In this respect, standard reference materials such as NIST "Estuarine Sediment" may be useful for evaluating accuracy. Estuarine Sediment has an uncertified sulfur value of 0.96%, and using data over the last 5 years from the Cutter lab, has a sulfur concentration of 0.96 ± 0.04% (n=114).

Filter blanks should be processed in a manner similar to that of samples. This can include soaking the filter in filtered seawater, and then processing as a sample (i.e., wash, freeze, and dry).

4.4.5. Accuracy, precision and detection limits.

The absolute detection limit (3 times the standard deviation of blanks) for the Carlo Erba instrument is 0.50 μg S. Assuming that 2 L of seawater are filtered, this translates to a relative detection limit of 7.8 nM S L^{-1}. For solid samples (i.e., sediment trap material), an average of 10 mg is analyzed, giving a relative detection limit of 1.6 μM S g^{-1}. The precision of the method for suspended (filtered) particles, expressed as the relative standard deviation, is 12% for sulfur at the 0.06 μM S L^{-1} level. For solid phase samples (i.e., sediments), precision (as RSD) is 2% at 0.3 mM S g^{-1}. Accuracy using the Carlo Erba is assured by the analysis of Estuarine Sediment for every 12 samples or one standard for each tray on the autosampler.

The absolute limit of detection for the Antek instrument is 0.01 μg (11% CV) at the highest sensitivity setting available, corresponding to a relative detection limit of 0.3 nM S L^{-1} for a 1 L sample. The precision of the method is approximately 4% (CV) for sulfur signals greater than 2 μg and 11% for the smallest amount routinely tested (0.5 μg S). The accuracy of the Antek has been tested by analyzing known amounts of sulfur in several compounds chosen to represent various sulfur-containing metabolite pools present in phytoplankton, as well as directly analyzing a SO_2 gas standard. No significant differences have been observed among the standard curves of S-methionine and other organic sulfur-containing compounds tested [Matrai, 1989]. Inorganic sulfur, such as that present in sodium sulfate and the ester sulfate carrageenan, has shown a significantly lower signal. This reduces the possibility of sea salt-sulfur contamination.

5. Analysis of Total Mass

Methods and problems associated with total mass are discussed in detail in the inorganic section and in the contributions of Bishop, Karl et al. and Sherrell (this volume, pages 229, 73 and 289). Certain aspects of the measurement of total mass by the methods described there are particularly important for bioelements. We often wish to know the mass of a particular bioelement as a function of total mass. Two problems arise here. First, where possible the sample for mass should be taken from the same collecting bottle as the sample for C, N, P and S. This will reduce variability due to real variability in nature expressed by two different bottle samples. Second, mass is most accurately determined on Nuclepore filters while C, N, P and S are usually measured using material collected on glass fiber filters. These filters have different retention characteristics and care must be exercised in selecting a membrane filter of appropriate pore size to match the retention characteristics of the glass fiber filters. Sheldon [1972] reports a nominal pore size of 0.7 μm for GF/F glass fiber filters. Recommendations for comparable Nuclepore pore sizes range from 0.2 to 0.8 μm depending upon the volume of sample to be filtered and the nature of its contained particles.

References

Alldredge, A. L., and M. W. Silver, Characteristics, dynamics and significance of marine snow, Prog. Oceanogr., 20, 41-82, 1988.

Altabet, M. A., Organic C, N, and stable isotopic composition of particulate matter collected on glass fiber and aluminum oxide filters, Limnol. Oceanogr., 35, 902-909, 1990.

Amaral, J. A., R. H. Hesslein, J. W. M. Rudd, and D. E. Fox, Loss of total sulfur and changes in sulfur isotopic ratios due to drying of lacustrine sediments, Limnol. Oceanogr., 34, 1351-1358, 1989.

Armstrong, F. A. J., P. M. Williams, and J. D. H. Strickland, Photo-oxidation of organic matter in sea water by ultra-violet radiation, analytical and other applications, Nature, 211, 481-483, 1966.

Azam, F., T. Fenchel, J. G. Field, J. S. Gray, L. A. Meyer-Reil, and F. Thingstad, The ecological role of water-column microbes in the sea, Mar. Ecol. Prog. Ser., 10, 257-263, 1983.

Carlson, D. J., L. M. Mayer, M. L. Brann, and T. H. Mague, Binding of monomeric organic compounds to macromolecular dissolved organic matter in seawater, Mar. Chem., 16, 141-153, 1985.

Cauwet, G., Organic chemistry of sea water particulates; concepts and developments, Oceanol. Acta, 1, 99-105, 1978.

Fitzwater, S. E., G. A. Knauer, and J. H. Martin, Metal contamination and its effect on primary production measurements, Limnol. Oceanogr., 27, 544-551, 1982.

Fox, L. E., Geochemistry of humic acid during estuarine mixing, in Aquatic and Terrestrial Humic Materials, edited by R. F. Christman, and E. T. Gjessing, Ann Arbor Science Publ., pp. 407-427, 1983.

Fox, L. E., The removal of dissolved humic acid during estuarine mixing, Estuarine Coastal Shelf Sci., 16, 431-440, 1983.

Gnaiger, E., and G. Bitterlich, Proximate biochemical composition and caloric content calculated from elemental CHN analysis: a stoichiometric concept, Oecologia, 62, 289-298, 1984.

Gordon, D. C., Jr., and W. H. Sutcliffe, Jr., Filtration of seawater using silver filters for particulate nitrogen and carbon analysis, Limnol. Oceanogr., 19, 989-993, 1974.

Hallegraeff, G. M., Caloric content and elementary composition of seston of three Dutch freshwater lakes, Arch. Hydrobiol., 83, 80-98, 1978.

Hirota, J., and J. P. Szyper, Separation of total particulate carbon into inorganic and organic components, Limnol. Oceanogr., 20, 896-900, 1975.

Holm-Hansen, O., J. D. H. Strickland, and P. M. Williams, A detailed analysis of biologically important substances in a profile off Southern California, Limnol. Oceanogr., 11, 548-561, 1966.

Johnson, B. D., and R. C. Cooke, Organic particle and aggregate formation from bubble dissolution, Limnol. Oceanogr., 25, 653-661, 1980.

Kepkay, P. E., and B. D. Johnson, Microbial response to organic particle generation by surface coagulation in seawater, Mar. Ecol. Prog. Ser., 48, 193-198, 1988.

Kepkay, P. E., and B. D. Johnson, Coagulation of bubbles allows microbial respiration of oceanic dissolved organic carbon, Nature, 338, 63-65, 1989.

Kieber, D. J., J. McDaniel, and K. Mopper, Photochemical source of biological substrates in sea water: implications for carbon cycling, Nature, 341, 637-639, 1989.

Knauer, G. A., J. H. Martin, and K. W. Bruland, Fluxes of particulate carbon, nitrogen, and phosphorus in the upper water column of the northeast Pacific, Deep-Sea Res., 26, 97-108, 1979.

Knauer, G. A., D. M. Karl, J. H. Martin, and C. N. Hunter, In situ effects of selected preservatives on total carbon, nitrogen and metals collected in sediment traps, J. Mar. Res., 42, 445-462, 1984.

Martin, J. H., G. A. Knauer, D. M. Karl, and W. W. Broenkow, VERTEX: carbon cycling in the northeast Pacific, Deep-Sea Res., 34, 267-285, 1987.

Matrai, P. A., Determination of sulfur in ocean particulates by combustion-florescence, Mar. Chem., 26, 227-238, 1989.

Menzel, D. W., Particulate organic carbon in the deep sea, Deep-Sea Res., 14, 229-238, 1967.

Mopper, K., and W. L. Stahovec, Sources and sinks of low molecular weight organic carbonyl compounds in seawater, Mar. Chem., 19, 305-321, 1986.

Paine, R. T., The measurement and application of the calorie to ecological problems, Ann. Rev. Ecol. Sys., 2, 145-164, 1971.

Parsons, T. R., and J. D. H. Strickland, On the production of particulate organic carbon by heterotrophic processes in sea water, Deep-Sea Res., 8, 211-222, 1962.

Riley, G. A., Particulate organic matter in sea water, Adv. Mar. Biol., 8, 1-118, 1970.

Sharp, J. H., Size classes of organic carbon in seawater, Limnol. Oceanogr., 18, 441-447, 1973.

Sharp, J. H., Improved analysis for "particulate" carbon and nitrogen, Limnol. Oceanogr., 19, 984-989, 1974.

Sheldon, R. W., Size separation of marine seston by membrane and glass fiber filters, Limnol. Oceanogr., 17, 494-498, 1972.

Sheldon, R. W., and T. R. Parsons, A continuous size spectrum for particulate matter in the sea, J. Fish. Res. Board Can., 24, 909-915, 1967.

Sholkovitz, E. R., Flocculation of dissolved organic and inorganic matter during the mixing of river water and seawater, Geochim. Cosmochim. Acta, 40, 831-845, 1976.

Solorzano, L., and J. H. Sharp, Determination of total dissolved phosphorus and particulate phosphorus in natural waters, Limnol. Oceanogr., 25, 756-760, 1980.

Strickland, J. D. H., and T. R. Parsons, A Practical Handbook of Seawater Analysis, second ed., 310 pp., Bull. Fish. Res. Board Can., vol. 167, 1972.

Sugimura, Y., and Y. Suzuki, A high-temperature catalytic oxidation method for the determination of non-volatile dissolved organic carbon in seawater by direct injection of a liquid sample, Mar. Chem., 24, 105-131, 1988.

Telek, G., and N. Marshall, Using a CHN analyzer to reduce carbonate interference in particulate organic carbon analyses, Mar. Biol., 24, 219-221, 1974.

U.S. GOFS Working Group, G. A. Knauer and V. Asper, Co-chairs, Sediment trap technology and sampling, Report No. 10, November 1988, 94 pp., Woods Hole, Massachusetts, 1989.

Wangersky, P. J., Particulate organic carbon: sampling variability, Limnol. Oceanogr., 19, 980-984, 1974.

Williams, P. M., and E. Druffel, Dissolved organic matter in the ocean: comments on a controversy, Oceanogr. Mag., 1, 14-17, 1988.

IN SITU COLLECTION AND LABORATORY ANALYSIS OF MARINE SNOW AND LARGE FECAL PELLETS

Alice L. Alldredge
Department of Biological Sciences, University of California, Santa Barbara, California 93106

1. Introduction

Many of the settling particles in the oceanic water column are sufficiently large to be visible to the human eye and, thus, are amenable to individual collection directly in situ. These particles include marine snow (aggregates of algae, bacteria, fecal matter and detritus larger than 500 μm in diameter) and the visible fecal pellets of macrozooplankton and nekton. Collection of large, individual particles by scuba divers or submersibles can provide information on the characteristics of settling particulate matter unattainable by other sampling methods. For example, the biological and chemical composition of settling particles collected individually can be determined with minimal contamination from smaller, suspended particles. The microscale distributions of many elements, compounds, and organisms associated with large settling particles can also be assessed. Bacteria, protozoans, phytoplankton, ammonium, nitrate, phosphate, and many trace metals are usually enriched by one or more orders of magnitude within large particles [Alldredge and Silver, 1988]. Moreover, collection of large particles intact allows quantitative investigations of the rates of biological and chemical processes occurring within settling particles. Oxygen production and depletion, photosynthesis, decomposition, nutrient regeneration, remineralization, adsorption/desorption, and numerous microbial processes and trophic interactions all occur actively within the microenvironment of large particles. Investigations of these processes within undisturbed marine snow and fecal pellets allows quantification of the significance of large settling particles as sites for biological and chemical transformations in the water column.

2. Techniques of Direct, In Situ Collection

We use blue-water SCUBA techniques [Hamner, 1975] to collect large suspended particles in surface waters and submersibles for collections at deeper depths [Silver and Alldredge, 1981; Alldredge and Youngbluth, 1985]. SCUBA techniques allow hundreds of individual particles of known type to be visually identified underwater and selectively collected quickly with minimal expense and equipment. Any investigator with access to SCUBA equipment and a small boat can utilize these techniques. Aggregates can be collected by divers at very precise depths with minimal inclusion of surrounding seawater so as to maintain the integrity of their unique microenvironments. Safe SCUBA collection is limited to the upper 50 m of the water column, however, and can be difficult or dangerous in rough seas.

Submersibles allow direct collection of individual large particles to unlimited depths. However, submersibles are expensive and usually only eight to twenty individual particles can be collected on any single dive. Only very large particles (centimeter-sized) can be collected easily and the large sampling vessels required collect large volumes (liters) of surrounding seawater with each aggregate. Despite the limitations of small sample size and expense, carefully planned studies using submersibles can yield measurements unattainable by other means.

2.1. Methods Using SCUBA

We routinely collect three types of samples using SCUBA: 1) individual or pooled aggregates of marine snow or fecal pellets in small volumes of surrounding seawater, 2) background seawater samples which contain no visible aggregates, and 3) bulk seawater samples which contain large aggregates at ambient concentrations. Background seawater samples provide "blanks" with which to correct aggregate samples.

Intact individual aggregates. Intact, individual aggregates are collected in 1 mL, 6 mL or 20 mL syringes whose tip ends have been cut off. The syringe barrels are stoppered with a syringe plunger at each end and strung together (through a hole in one plunger) with electrical wire into circles of 20 to 50 syringes each for easy handling underwater [Alldredge, 1979]. A single, experienced diver can collect over 300 individual particles in one 50 minute dive if marine snow is abundant. Tests of the toxicity of the syringe plungers reveal that they do not alter ^{14}C or ^{3}H-thymidine uptake [Gotschalk, 1988].

Marine Particles: Analysis and Characterization
Geophysical Monograph 63

Aggregates may also be collected in larger reaction vessels such as 500 mL sartorius jars, for manipulative studies underwater. Radioisotopes, acetylene gas, or other substances can be injected underwater and the samples incubated without ever being removed from the ocean [Alldredge and Silver, 1982; Martinez et al., 1983].

Large, pooled aggregate samples. The large multi-aggregate samples needed for most chemical analysis may be collected using regular or ultra-clean techniques. We collect samples for standard chemical analyses (nutrients, POC, PON, pigments, dry weight etc.) using unaltered 25 mL syringes. Ten to forty aggregates are sucked through the syringe tip (no needle attached) and the syringe tip is then capped with the plastic base of a syringe needle (needles are cut off, the bases are ground smooth for safety and plugged with glue). Aggregates may be disrupted slightly as they are drawn into the syringe. The syringes are strung together for easy handling underwater. An experienced diver can collect 300-400 particles during a 50 minute dive if marine snow is abundant.

Ultra-clean pooled aggregate samples. We collect large quantities of undisrupted aggregates or samples requiring clean or metal-free conditions in pre-cleaned Teflon tubing filled with distilled water. The inside diameter of the tubing is selected to insure that aggregates are not disrupted. One end of the clean tubing is attached to a one-way valve on a spring-loaded syringe. Particles are sucked into the free end of the tubing along with a consistent preset volume of surrounding seawater (usually 0.5 or 1 mL) which progressively displaces the distilled water. When the tubing is full of seawater and aggregates the syringe is disconnected and the tubing ends are plugged. The diver keeps track of the number of particles collected or calculates them from the volume of the tube. Several tens of meters of tubing may be coiled into a dive bag allowing a single diver to collect several hundred particles per dive with this method.

Background seawater. We collect background seawater underwater in 60 mL syringes by carefully drawing in seawater so as to exclude visible particles.

Bulk seawater. We collect bulk seawater in 4 L polyethylene bottles. These are filled with seawater (or distilled water if clean conditions are required) just below the surface, purged with SCUBA air at depth, and inverted to fill with seawater.

Collection using submersibles. Large particles may be collected from submersibles in individual vessels or sucked through a funnel and tubing onto an in-line filter. We have collected centimeter-sized aggregates in 5 L transparent mechanically triggered Van Dorn bottle-type samplers or in hydraulically operated cylindrical samplers [Silver and Alldredge, 1981; Alldredge and Youngbluth, 1985]. The submersible is maneuvered so as to place a targeted particle into the collecting vessel and the vessel is then closed. Considerable skill is required of the pilot. Background seawater is usually taken from the aggregate vessels. Bulk seawater may be collected separately.

2.2. *Assessment of Large Particle Abundance In Situ*

Determination of the abundance of large particles in situ is essential in order to assess the contribution of aggregates to water column biology and chemistry. We use two methods.

Visual counts by divers. A scuba diver swims multiple transects counting aggregates passing though a hoop attached to a hand-held flow meter [Trent et al., 1978]. A ruler on the hoop helps assess approximate particle size.

In situ camera systems. Computerized image analysis of the film or tape from in situ camera systems yields quantitative data on particle size distributions and abundances with depth. Two types have been built. Still camera systems [Honjo et al., 1984; Wells and Shanks, 1987; Gardner and Walsh, 1990; Kranck and Milligan, 1989] take a photograph of a lighted rectangular slab of water about once every eight seconds. The volume photographed is usually kept large in order to obtain an adequate sample size. However, the slab volume may be adjusted so as to measure particles down to about 50 μm in size. The images are not prohibitively affected by background illumination. Purchase of the initial system and film processing can be expensive.

Video cameras (A. Alldredge, unpublished, 1990) also produce images of a slab of water but give continuous images allowing for very large sample size and accurate, continuous profiles of particle size distributions with depth. Resolution is lower than that of still cameras. Video cameras are best deployed at night or below 100 m because of their sensitivity to background illumination.

3. Laboratory Handling and Chemical Analysis of Marine Snow and Fecal Pellets

Slurries of marine snow are handled and analyzed in the laboratory in a manner similar to those employed for sediment trap samples. We use a similar initial procedure for almost all types of measurements. If several types of measurements are to be made on particles from one station, the aggregate samples are combined in a laboratory vessel to produce an aggregate slurry of a known volume containing a known particle abundance. The slurry is mixed gently and replicate subsamples containing ten to twenty aggregates each are removed with a large bore pipette for various analyses such as pigments, dry weight, ^{14}C uptake, CHN, bacterial abundance etc. This allows intercomparison of various measurements from the same sample.

In some cases, multiple measurements of single particles may be required. We have successfully disrupted and subsampled individual particles to obtain measurements of bacterial abundance, dry weight, and tritiated thymidine uptake or of Chlorophyll *a*, dry weight, and ^{14}C uptake on the same particle [Alldredge and Gotschalk, 1990; Simon et al., 1990]. Gentle agitation with a tissue grinder effectively disrupts most aggregates without breaking incorporated cells or reducing bicarbonate uptake [Gotschalk and Alldredge, 1989].

Background seawater blanks are subtracted from aggregate samples, generally assuming that the background seawater makes up 90% to 100% of the sample by volume. In most cases this assumption introduces little error since the seawater blanks tend to be very low relative to the aggregate values. In those cases where the aggregates are of large volume, the aggregate slurry is allowed to settle in a graduated cylinder for about 1 hour and the volume of the background seawater is estimated. The volume of the background seawater in individual particle syringes may also be quantified directly in the syringe using the mL gradations on the syringe barrels.

The following methods are used in analyzing marine snow samples.

3.1. Dry Weight (Mass)

Individual aggregates or 1 to 10 mL subsamples of an aggregate slurry are filtered onto pre-weighed Nuclepore filters, rinsed very quickly with distilled water, dried overnight in a desiccator and reweighed on a Cahn Electrobalance. Individual particles as small as $3\,\mu g$ can be weighed to an accuracy of $1\,\mu g$.

3.2. Ash-free Dry Weight

Samples are filtered onto pre-washed, pre-weighed Whatman GF/F glass fiber filters, rinsed, dried, reweighed, and ashed at 450° C [Strickland and Parsons, 1972]. A sample size of 10 to 50 aggregates per filter is required.

3.3. Aggregate Volume

Aggregate volumes can be obtained by photographing or video recording aggregates in the laboratory [Alldredge et al., 1990; Alldredge and McGillivary, in press, 1991] or in the field [Alldredge and Gotschalk, 1988] and analyzing the images via image analysis. Aggregates may also be drawn into pipettes and the volume estimated. This latter technique is not as accurate and may result in aggregate disruption if the bore of the pipette is small. Approximate estimates of aggregate volume may also be obtained from dry weight using the equations in Figures 2a and 2b in Alldredge and Gotschalk [1988].

3.4. POC and PON

Ten to fifty aggregates are filtered onto each Whatman GF/F pre-ashed glass fiber filter and analyzed with a Perkin Elmer CHN Analyzer according to Sharp [1974].

3.5. Interstitial Dissolved Oxygen

Particles should be collected in glass syringes unless they are to be fixed immediately underwater. The plastics used in disposable syringes allow oxygen exchange [Evers et al., 1972]. We perform Winkler analysis on individual aggregate samples [Strickland and Parsons, 1972] of 3-5 mL by scaling down the reagent additions and titrating with a microburette. Interstitial oxygen concentrations within each particle can be calculated if aggregate volume relative to background seawater volume is known. For immediate dissolved oxygen determinations we collect marine snow in vials stoppered with serum stoppers. The diver adds the manganous sulfate and alkaline iodide reagents to the vial underwater (dispensed via syringe) immediately after each particle is collected. Oxygen gradients within individual aggregates may also be determined with a microelectrode in the laboratory [Alldredge and Cohen, 1987].

3.6. Interstitial Nutrients

We routinely measure inorganic nutrient content of the interstitial water within aggregates using standard techniques modified for small sample size [Strickland and Parsons, 1972] or flow injection techniques [Johnson et al., 1985]. Samples containing 8 to 10 aggregates collected in situ in clean syringes are filtered through an Acrodisc syringe filter and the filtrate analyzed. By subtraction of background seawater values (seawater and aggregate volumes need to be known) one can calculate the nM of each nutrient within the interstitial fluid of the aggregates themselves. Estimates of nutrient enrichment within each aggregate may be obtained without knowing aggregate volume.

3.7. Chlorophyll

We measure Chlorophyll *a* and phaeopigments in individual aggregates or aggregate slurry subsamples using standard fluorometric methods described in Strickland and Parsons [1972].

References

Alldredge, A. L., The chemical composition of macroscopic aggregates in two neritic seas, Limnol. Oceanogr., 24, 855-866, 1979.

Alldredge, A. L., and Y. Cohen, Can microscale chemical patches persist in the sea? Microelectrode study of marine snow, fecal pellets, Science, 235, 689-691, 1987.

Alldredge, A. L., and C. Gotschalk, In situ settling behavior of marine snow, Limnol.Oceanogr., 33, 339-351, 1988.

Alldredge, A. L., and C. Gotschalk, The relative contribution of marine snow of different origins to biological processes in coastal waters, Cont. Shelf Res., 10, 41-58, 1990.

Alldredge, A. L., and P. McGillivary, The attachment probabilities of marine snow and their implications for particle coagulation processes in the ocean, Deep-Sea Res., in press, 1991.

Alldredge, A. L., and M. W. Silver, Abundance and production rates of floating diatom mats (*Rhizisolenia castracanei* and *R. imbricata var. shrubsolei*) in the eastern Pacific Ocean, Mar. Biol., 66, 83-88, 1982.

Alldredge, A. L., and M. W. Silver, Characteristics, dynamics and significance of marine snow, Prog. Oceanogr., 20, 41-82, 1988.

Alldredge, A. L., and M. J. Youngbluth, The significance of macroscopic aggregates (marine snow) as sites of heterotrophic bacterial production in the mesopelagic zone of the subtropical Atlantic, Deep-Sea Res., 32, 1445-1456, 1985.

Alldredge, A., T. C. Granata, C. C. Gotschalk, and T. D. Dickey, The physical strength of marine snow and its implications for particle disaggregation in the ocean, Limnol. Oceanogr., 35, 1415-1428, 1990.

Evers, W., G. B. Racz, and A. A. Levy, A comparative study of plastic (polypropylene) and glass syringes in blood-gas analysis, Anesth. Analg., 51, 92-97, 1972.

Gardner, W. D., and I. D. Walsh, The role of aggregates in horizontal and vertical flux across a continental margin, Deep-Sea Res., 37, 401-412, 1990.

Gotschalk, C. C, Diatom aggregates: Potential sites of enhanced primary production and nutrient regeneration in upwelling systems, Masters thesis, 53 pp., Univ. of California, Santa Barbara, 1988.

Gotschalk, C. C., and A. L. Alldredge, Enhanced primary production and nutrient regeneration within aggregated marine diatoms, Mar. Biol., 103, 119-129, 1989.

Hamner, W. M., Underwater observations of blue-water plankton: logistics, techniques and safety procedures for divers at sea, Limnol. Oceanogr., 20, 1045-1051, 1975.

Honjo, S., K. W. Doherty, Y. C. Agrawal, and V. L. Asper, Direct optical assessment of large amorphous aggregates (marine snow) in the deep ocean, Deep-Sea Res., 31, 67-76, 1984.

Johnson, K. S., R. L. Petty, and J. Thomsen, Flow injection analysis of seawater and micronutrients, in Mapping Strategies in Chemical Oceanography, edited by A. Zirino, pp. 7-30, American Chemical Society, Washington D.C., 1985.

Kranck, K., and T. G. Milligan, Macroflocs from diatoms: in situ photography of particles in Bedford Basin, Nova Scotia, Mar. Ecol. Prog. Ser., 44, 183-189, 1989.

Martinez, L. A., M. W. Silver, J. M. King, and A. L. Alldredge, Nitrogen fixation by floating diatom mats: a source of new nitrogen to oligotrophic oceanic waters, Science, 221, 152-154, 1983.

Sharp, J. H., Improved analysis for "particulate" organic carbon and nitrogen from seawater, Limnol. Oceanogr, 19, 984-989, 1974.

Silver, M. W., and A. L. Alldredge, Bathypelagic marine snow: vertical transport system and deep-sea algal and detrital community, J. Mar. Res., 39, 501-530, 1981.

Simon, M., A. L. Alldredge, and F. Azam, Bacterial carbon dynamics on marine snow, Mar. Ecol. Prog. Ser., 65, 205-211, 1990.

Strickland, J. D. H., and T. R. Parsons, A Practical Handbook of Seawater Analysis, second edition, 310 pp., Bull. Fish. Res. Board Can., vol. 167, 1972.

Trent, J. D., A. L. Shanks, and M. W. Silver, In situ laboratory measurements on macroscopic aggregates in Monterey Bay, California, Limnol. Oceanogr., 23, 626-635, 1978.

Wells, J. T., and A. L. Shanks, Observations and geologic significance of marine snow in a shallow-water, partially enclosed marine embayment, J. Geophys. Res., 92, 13,185-13,190, 1987.

SAMPLING AND ANALYSIS OF SUSPENDED AND SINKING PARTICLES IN THE NORTHERN NORTH ATLANTIC

Bodo v. Bodungen, Marita Wunsch and Heike Fürderer

Sonderforschungsbereich 313, Kiel University, Heinrich-Hecht-Platz 10, D2300 Kiel 1, FRG

1. Introduction

Comparative studies of pelagic system structure and related particle flux have been conducted in the Norwegian Basin and in the Greenland Sea since 1986/1987. These studies are part of a so-called SONDERFORSCHUNGSBEREICH 313 (Special Research Project), which is entitled: Changes In The Environment: The Northern North Atlantic. The program also encompasses investigations of the benthal, near-bottom transport, sediment accumulation, and biogeochemistry of the sediments as well as geophysical, paleoecological and oceanographic studies and modelling efforts. The overall goal of this program is to establish fluxes of biogenic elements and the related sedimentary record in the contemporary northern North Atlantic and to interpret glacial/interglacial variations in circulation, ventilation and sedimentation in this area.

Within the framework of this program, vertical export of particulate biogenic matter is recorded continuously at two time-series stations (70° N 00° W; and 73° N 07° W) with sediment traps, which are moored in three depth horizons with weekly to monthly sampling intervals. In different seasons, selected according to recurring patterns in quantity and quality of vertical flux, process-oriented studies are carried out with high spatial and temporal resolution [Peinert et al., 1987; Bathmann et al., 1990].

In this program, sampling and analysis of bioelements are done in accordance with widely used standard procedures. The investigations in the Greenland Sea are part of the JGOFS program; thus, core measurement protocols given for the NORTH ATLANTIC SPRING BLOOM STUDY are followed in general [JGOFS Report No. 6, 1990]. However, sample handling and methods are slightly altered with regard to requirements of sampling sites and particular problems addressed. Prior to such adaptations improvements versus caveats are evaluated carefully.

Marine Particles: Analysis and Characterization
Geophysical Monograph 63

2. Water Column Sampling and Sample Handling

Preferably Niskin bottles mounted on a rosette sampler combined with CTD and fluorescence probes are used. When two or more water bottles are required for a discrete depth horizon, the individual water bottles are closed with as little delay as possible on the same cast. Continuous in situ recording of fluorescence has shown that internal waves can alter the vertical distribution of chlorophyll *a*-containing particles on time scales of minutes to days (R. Peinert and S. Podewski, Institut Meereskunde, Kiel, FRG, unpublished data, Figure 3-3, 1990). Sampling of the surface layer, therefore, is carried out as rapidly as possible to avoid biases in the vertical profile due to temporal inhomogeneities. At times and locations where strong internal wave action is expected, sampling at density horizons may be more appropriate than sampling from standard depths.

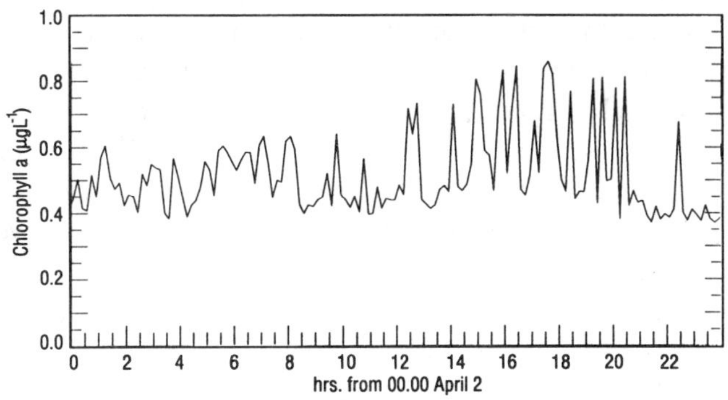

Fig. 3-3. Changes in chlorophyll *a* (CHLA) over the course of one day in 60 m water depth measured with a self-contained in situ fluorimeter. The recordings were taken at 18° N; 20° W on April 2, 1989 during METEOR-Cruise M10/1 and data were recorded in 10 min intervals (Courtesy of Peinert and Podewski). The time of day is given on the horizontal axis.

After recovery of the water bottles, the entire contents are drained into a large carboy. Immediately thereafter the respective subsamples for the various measurements are taken. Agitation of the carboy to guarantee an even distribution of particles is gentle to avoid de novo formation of particles from dissolved and colloidal material. In general samples are not prescreened as sieves may retain long diatom chains and mucilaginous aggregates, e.g. *Phaeocystis* colonies. Zooplankton organisms, which are visible on filters, are removed carefully with forceps. Only when large gelatinous zooplankton, such as salps, are abundant, coarse sieves are applied for gentle prescreening.

2.1. Choice of Filters and Filtration

Particulate and dissolved matter is operationally defined by the choice of a particular filter (see workshop report, Major Bioelements, this volume, page 33). In general the question as to which filter type should be preferred cannot be answered unequivocally. Compromises must be made with regard to filter material and pore size. As particulate organic carbon (POC) is one of the key elements in biological oceanography the analytical requirements for POC measurements mostly set the stage for the selection of a filter. These are met by glass fiber filters (GF/F), which can also be used for a variety of other bioelements. These filters retain particles of 0.7-0.8 μm in size [Sheldon, 1972]. Retention of smaller particles, e.g. submicron particles, [Koike et al., 1990] increases with filtration volume. Aluminum oxide filters (Anopore), which retain more submicron particles [Altabet, 1990] and can be used for POC measurements, have been tried in our lab. These filters are very delicate and handling easily causes considerable damage to the filters. For total mass, POC, particulate organic nitrogen (PON), particulate organic phosphorus (POP) and carbonate we use GF/F filters, thus accepting an unaccountable loss of small particles. For biogenic silicate Nuclepore filters with a pore size of 0.8 μm are used.

The volume for filtration is chosen in accordance with the sensibility of the respective analysis. In surface waters of less than 100 m of water depth, mostly 1 L is sufficient. From deeper water 3 to 5 L are filtered in triplicate to guarantee reliable numbers. Filtration volume is kept as small as possible to avoid long filtration times and particle damage. The filtration is carried out at 0.2 to 0.4 atmospheres in vacuum. After filtration the cone of the device is rinsed carefully with filtered seawater as often mucilaginous particles stick to the filtration cone.

2.2. Total Mass (synonym: Seston, Dry Weight)

Pretreatment of filters. GF/F filters are rinsed with filtered (0.2 μm membrane filters) deionized water and precombusted at 500° C for 4 hours. Thereafter they are rinsed again from both sides, dried and cooled in a desiccator and weighed. Filters are stored dry in plates each holding 40 filters.

After the sample has been filtered it is rinsed three times with a small squirt of deionized water to remove the salt. Samples are stored deep frozen (-18° C) or dried. Out of 40 filters, four are taken as blanks and processed in exactly the same way as those with the samples.

Before reweighing, filters are oven dried and cooled in a desiccator. First, the four blank filters are weighed and then ten filters are weighed with sample. Blank filters are repeatedly reweighed after weighing of 10 filters with sample. Weights for total mass are corrected by the average weight change of the blanks. The procedure is described in detail by Lenz [1971]. Analytical precision, replicability in triplicate samples, can be within ± 10% if filters are handled carefully and loss of brittle glass fibers is minimized.

2.3. Combustible, Ash Free Dry Weight

After evaluation of total mass the filters are used in some cases for estimates of the combustible fraction in sediment trap samples. Filters are combusted at 500° C for 4 hours and reweighed after cooling following the same procedure as described for total mass.

2.4. Carbonate

In most cases seston filters are further processed for carbonate. After weighing for total mass, filters are acidified by fuming with concentrated hydrochloric acid for 48 hours in a box. When the samples contain large amounts of foraminifera and pteropods, as is often the case in sediment trap samples, the filters are fumed for another 24 or 48 hours.

Filters are rinsed several times with deionized water and oven dried. After reweighing, the weight difference to total mass is related to carbonate. Loss of organic matter due to the rinsing procedure is comparatively small when expressed as absolute weight.

The weight loss method was compared to estimates of carbonate by measuring calcium and magnesium in the collected rinse water (D. Schulz, Institut Meereskunde, Kiel, FRG, unpublished data, 1991). The latter yielded mostly lower numbers than those obtained by the weight loss calculation. The sum of all weights contributing to total mass accounted for only 80% of total mass when carbonate was included as calcium/magnesium equivalent, but was close to 100% when "weight loss carbonate" was used. The reasons for this discrepancy are not as yet clear (D. Schulz, Institut Meereskunde, Kiel, FRG, personal communication, 1991).

Better results than with the weight loss method are obtained by combusting acidified and non-acidified samples in a carbon analyzer. This method, however, gives better results only for small particle loads on the filter.

2.5. POC and PON

Pretreatment of GF/F filters follows the same procedure as for total mass. Filtration is done at lowest possible vacuum to avoid lysis and leaching of organic substances from the cells. Filters are stored deep frozen or oven dried.

Prior to the measurements, filters are acidified in a fume box for 48 hours. The measurements are conducted in a Heraeus CHN-analyzer at a combustion temperature of 1000° C. The analytical precision is within ± 4% with acetanilide standard. With samples containing greater than 1 μM C and N the most reliable results are achieved.

Blank filters are handled in the same way as the samples. Four blanks are run per 40 samples. An instrument blank is established once a day during calibration. Standards are prepared with acetanilide containing C and N over the range which is expected in the samples. Standard runs are made several times a day dependent on the correct performance of the instrument. In general the analyses are performed according to the outline given by Sharp [1974].

2.6. Particulate Phosphorus

Pretreatment of GF/F filters is the same as described above. It has been noted occasionally that untreated filters yielded high blanks. Although phosphorus contamination on the filters cannot be removed by furnace combustion the oxidation of organic contaminants may further the removal of phosphorus compounds during the rinsing procedure. Filtration procedure and sample storage are the same as for POC.

The acidified samples are wet-chemically oxidized with potassium peroxydisulfate by autoclaving the filter in 35 mL of deionized water in a closed bottle for 30 minutes, followed by centrifugation and colorimetric measurement of orthophosphate. The procedure is that of the Koroleff method for total phosphorus [Grasshoff, 1976]. The detection limit is given with 0.08 μM and a precision of less than ± 10% can be obtained from replicate measurements of a homogeneous water sample.

For blank corrections, empty filters are treated in exactly the same way as the samples. For standards ordinary orthophosphate solutions are used in routine measurements. Synthetic organic phosphorus compounds have been used occasionally. A 100% recovery rate of both standard solutions does not necessarily indicate complete destruction of bound phosphorus to orthophosphate. There is no direct method to differentiate inorganic from organic phosphorus in the particulate pool. In oceanic waters, particulate phosphorus may be more or less organically bound.

The persulfate oxidation method for particulate phosphorus is well suited for routine measurements on board ship. However, incomplete oxidation of bound phosphorus to orthophosphate may occur. High temperature combustion combined with acid hydrolysis results in better accuracy (Sharp, this volume page 87, Major Bioelements Workshop Report, this volume, page 33).

2.7. Biogenic Particulate Silicate (Opal)

Polycarbonate filters with a pore size of 0.8 μm are used for the filtration. For compatibility with the particle loads on GF/F filters in general, a pore size of 0.4 μm may be more realistic (Major Bioelements Workshop Report, this volume, page 33). However, there are most likely no opal particles less than 0.8 μm in the ocean and this pore size may be advantageous for not retaining smaller mineral particles. Cellulose acetate filters have also been tried. These filters, however, yield high blanks due to turbidity caused by fine filter particles. Glass sinters in the filtration device have been found to cause considerable contamination. They should be replaced by Teflon or PVC sinters or plates. Filters are stored deep frozen or oven dried.

Wet-alkaline digestion of the samples basically follows the description of Krause et al. [1983] and Paasche [1973]. The filters are digested in 25 mL of prewarmed 0.1 N sodium hydroxide (85° C). The bottles are placed in an agitated warm water bath held at the same temperature. Time-course digestions as described by Eggimann et al. [1980] were carried out to determine the appropriate digestion time and to evaluate biases introduced by the digestion of mineral particles. In cultures of *Thallassiosira fluviatilis* and *Phaeodactylum tricornutum*, a digestion time of 30 minutes was sufficient. After a 1 hour digestion time for suspended particles, the bias introduced by mineral particles was less than 2% [Puch, 1990]. This bias is higher in high turbidity coastal waters. Therefore, suspended particles from open ocean environments are routinely digested for 1 hour.

The detection limit is 0.03 μM. Precision for replicate measurements of a homogeneous water sample is ± 20%. The accuracy, defined by the recovery in the particulate pool in relation to the uptake of reactive silicate in various batch cultures with natural Antarctic phytoplankton populations, including silicoflagellates, varies between 94 and 100% [Puch, 1990].

The highly varying nature of siliceous particles in moored sediment traps due to clay minerals on organic particles, resuspension and advective transport requires time course digestions on each individual trap sample. For samples from drifting traps in surface layers of open ocean environments a 1 hour standard digestion time is justified. An overview of alternative measurements of opal and a critical evaluation is given by DeMaster (this volume, page 367).

3. Sampling of Sinking Particles With Traps

Unless otherwise specified, the analysis for the major bioelements from trap samples is carried out in the same way as described above . Problems concerning trap design, trapping efficiency and poisons vs. preservation agents will not be discussed in detail. An overview of these important aspects of trapping is given in U.S. GOFS Report No. 10 [1989]. Along with the following description of particle flux measurements, some results from experiments related to these aspects are also presented.

3.1. Particle Trap Design and Deployment Depth

In our studies funnel- (cone-)shaped traps with a baffle, of aspect ratio 5:1, have been used since 1975. The basic design is described by Zeitzschel et al. [1978]. Funnel shaped traps with rotating collection cup holders, up to 20 cups, were chosen because a large part of our program is based on annual moorings. Besides this long-term record of particle flux drifting sediment traps are deployed during process-oriented studies. Short-term moorings with up to six traps are also deployed to fill the gap between recovery and redeployment of the annual moorings and to follow flux patterns to the seafloor in relation to the concomitant pelagic and benthic investigations as well as for experimental purposes.

Depths of deployment for the different operation modes are chosen according to physical structure and biological zonation of the water column. In the annual moorings the uppermost trap is placed just below the winter mixed layer and well above overwintering depth of zooplankton. Further traps are deployed below the overwintering depth and close to the seafloor. The latter were deployed previously 300 m off the seafloor to record the vertical supply to the benthal. In both investigation areas this distance from the seabottom was not sufficient to exclude resuspended particles from trap collections. Throughout the year measured flux rates for total mass in the bottom near traps were up to two times higher than those recorded in mid-water depths. The appropriate distance from the seafloor in which the "true" vertical input can be recorded will have to be established in moorings with higher vertical resolution in bottom near waters.

Drifting traps and those in short-term moorings are used to obtain higher spatial resolution in surface and bottom near water. Depth of euphotic zone, vertical distribution of grazers, reproductive behavior, daily vertical migration of zooplankton, as well as the extent of the bottom-near nepheloid layer are the most important criteria in selection of deployment depths.

3.2. Preparation of the Collector Cups

For each trap, water from the deployment depth is filtered and 10% of the seawater is evaporated, to reduce loss of solutes due to diffusion by a slight salinity gradient between the cups and the funnel. A hypersaline brine solution is not used as this may attract swimmers [Peterson and Dam, 1990]. The water is analysed for "blanks" for later measurements of dissolved properties in the supernatant water. One milliliter of slightly undersaturated mercuric chloride per 100 mL seawater in the cup is added as poison.

The choice of an optimal poison, e.g. mercuric chloride, sodium acid, or preserving agent, e.g. formaldehyde, glutaraldehyde, is dependent on specific problems addressed [Knauer et al., 1984]. In long-term moorings, losses and changes of unpoisoned or unpreserved trap material can be considerable due to feeding activities of zooplankton (swimmers) and bacterial degradation [Gardner et al., 1983; Silver et al., 1984; Lee et al., 1988].

In our studies we use mercuric chloride as this agent apparently does not interfere with measurements of stable carbon and nitrogen isotopes. Lee et al. [in press, 1991] report that mercuric chloride suppresses microbial activity quite efficiently. However, decay of organic matter in mercuric chloride poisoned samples is not completely prevented in long-term deployments (see below). A disadvantage is that swimmers become mushy and are difficult to remove in contrast with the use of formaldehyde, which is recommended in the JGOFS protocols.

3.3. Comparison of Poisoned and Unpoisoned Daily Trap Samples

During process-oriented studies our drifting traps are equipped with a single collector cup. The cups are recovered daily and the samples are split, filtered and stored deep frozen immediately thereafter. In a twin drifter, where two traps are deployed side by side, we compared poisoned and unpoisoned samples. The traps were deployed 100 m below the sea surface in August/September 1990 in the Greenland Sea.

In Figure 3-4 quantity and composition of swimmers in the two traps are depicted. Generally more swimmers of the four taxons depicted were found in the poisoned trap. This indicates that a considerable number of swimmers may enter and leave the unpoisoned trap. Copepods apparently do more so than ostracods (*Conchoecia borealis*). Among the copepod swimmers *Metrida longa* was most abundant in the traps, greater than 60%, and also dominated the copepod population in the water column above the traps. *Calanus hyperboreus*, which contributed only less than 5% to the copepod population, made up for more than 30% of the copepod swimmers in the traps. These species apparently exhibit a different "affinity" to traps.

The daily POC-flux measured in the poisoned and unpoisoned trap agreed within ±25%, except for two occasions where flux rates in the poisoned trap were considerably higher (Figure 3-5a). Differences in PON flux were smaller (Figure 3-5b), but showed the same general tendency of higher flux rates measured in the poisoned traps. The C:N ratio ranged from 6.3 to 7.0 and 6.4 to 7.9 in the unpoisoned and poisoned trap, respectively. Besides one occasion, the numbers for the 28.08 and 29.08 represent the 24 hour average of a 48 hour sampling interval; the measured flux rates for chlorophyll *a* equivalents (chl *a* + phaeopigments) show a conformity to within ±15% and were generally somewhat higher in the poisoned trap (Figure 3-5c).

The observed pattern in the two traps is not likely to have been caused by the presence or absence of poison alone. Twin traps used in other investigations, where both traps contained poison, showed differences in measured flux rates of up to ±15% due to trap-flow interactions [Peinert et al., 1987; cf. Butman, 1986]. During splitting procedures, biases dependent on the degree of homogeneity of the sample may be introduced.

In this experiment, no differences in the amount of fecal pellets were observed in the two traps, indicating that defecation by swimmers in the unpoisoned trap was negligible. The larger amounts of dead animals in the

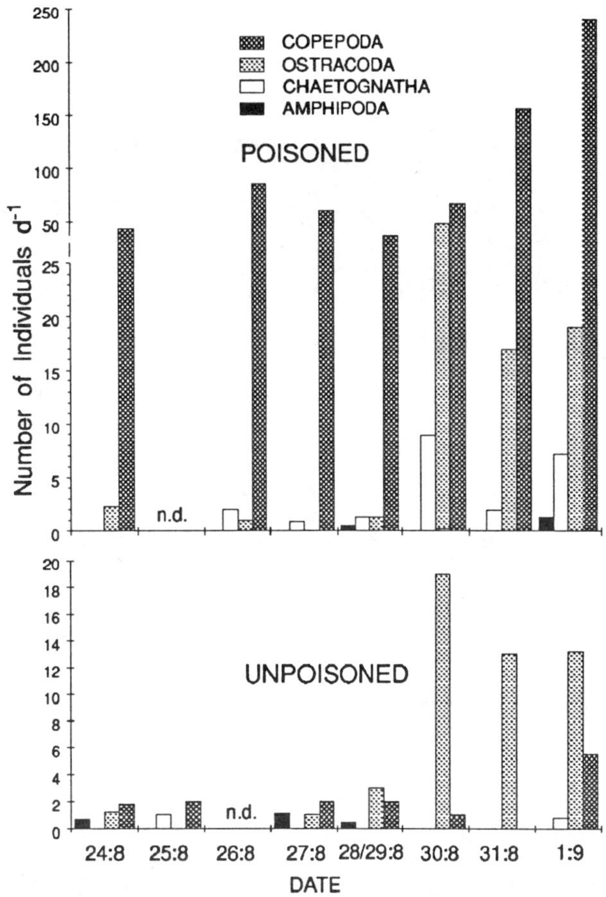

Fig. 3-4. Numbers and composition of swimmers in a poisoned (upper panel) and unpoisoned trap (lower panel). Twin traps were deployed as drifters in 100 m water depth from 24:08:91 to 01:09:91 at 73° N; 05° W in the Greenland Sea.

poisoned traps introduce more biases due to leaching, expectoration of guts and demolition during picking procedure than swimmers entering and leaving the trap. Moreover, poison-free trap samples are of considerable advantage for sample handling and analysis. In short-term deployments (ca. 1 day) we prefer an omission of poisons and preservatives, when samples are processed immediately after recovery. To minimize swimmer caused biases, results from unpreserved traps with screens were combined with those from preserved traps without screens [Karl and Knauer, 1989]. There have also been attempts with particular trap designs to separate swimmers from the sinking particles [Coale, 1990].

3.4. Sample Handling Prior to Splitting

Directly prior to splitting, 50 mL of supernatant water is withdrawn with a syringe equipped with a filter holder. A part of the sample is then transferred carefully

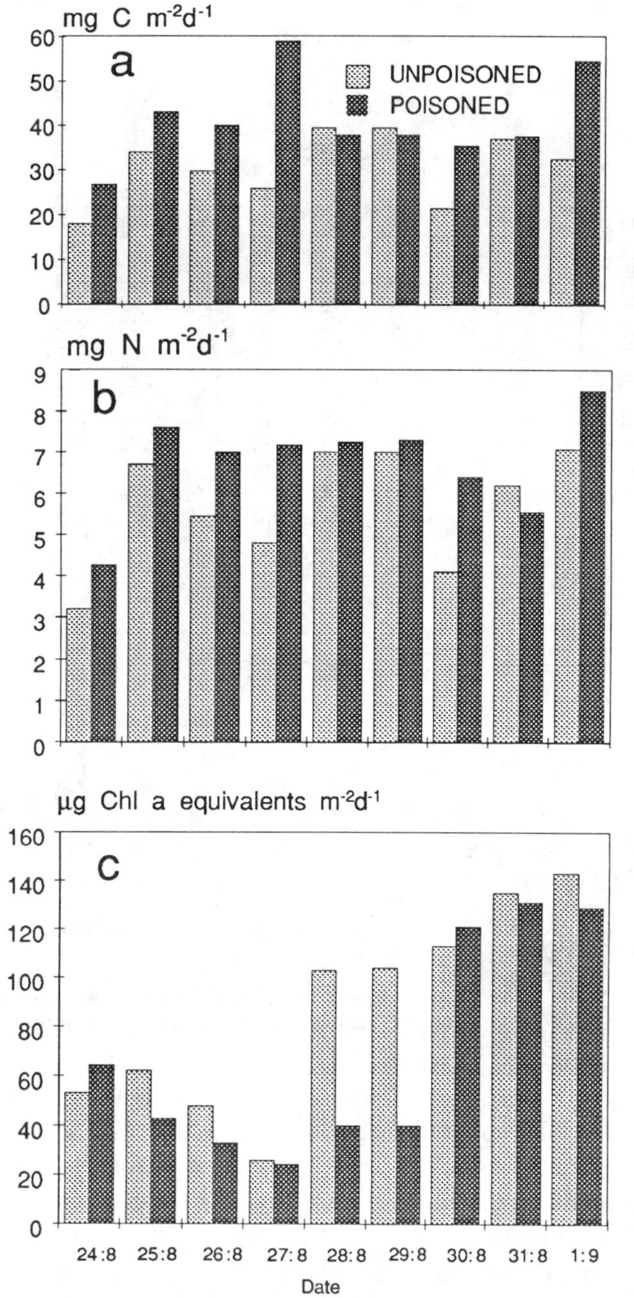

Fig. 3-5. Daily flux rates of particulate organic carbon (POC, a), nitrogen (PON, b) and chlorophyll *a* equivalents (Chl *a* + phaeopigments, c) in a poisoned and unpoisoned trap (see Figure 3-4).

into a petri dish for microscopical inspection and photography of macroscopic properties, such as large fecal strings and aggregates. A minute amount of the sample is removed for later SEM-analysis, when fragile particles, bearing valuable information, are abundant and destruction of these particles during splitting is expected. From the first cursory inspection of the sample, decisions are made as to whether the sample should be size fractionated and appropriate size classes are chosen.

Very large swimmers are picked directly from the sample and are thoroughly cleaned from adhering detrital particles. After this, samples are screened carefully through a set of sieves with different mesh width according to the composition of the swimmers. Additionally, after splitting and filtration, subsamples and filters are controlled once more for swimmers. In samples, for which detailed information about the vertical distribution of zooplankton are available, animals which did live during the sampling interval exclusively far above the deployment depth are either not picked or analyzed separately after picking. For the latter case the results are included in the respective flux rates [Bathmann et al., in press, 1991].

3.5. Splitting of Samples

Trap samples are poured on a 1 m long groove, which is slightly inclined towards a rotation splitter with eight compartments. The sample is rinsed into the rapidly turning splitter under a steady flow of filtered seawater. Care is taken that the sample is evenly dispersed in the groove. The subsamples located vis a vis on the splitter are poured together. The quarters obtained are then used for further splitting into subsamples. Large subsamples of slimy material are less reproducible than smaller subsamples. In such cases triplicates for total mass of different subsample sizes are made to evaluate the most reliable size of the subsamples. Very sticky samples are homogenized prior to splitting. With this wet-splitting procedure followed by filtration of the subsamples a precision within ±6% is achieved in triplicates. Precision may be improved by subsampling out of a large freeze dried portion of the sample (Fischer and Wefer, this volume, page 391).

3.6. Losses of Particulate Matter During Deployment

Losses of particulate matter in the collector cups during deployments may occur due to physico-chemical dissolution (e.g. silicate, carbonate, celestite) and by decay and/or leaching of organic material. Loss by dissolution may be prevented by filling the cups with saturated solution of the respective compounds. However, such a procedure most likely will cause other unexpected and undesired problems. In the following measurements of dissolved inorganic nitrogen, silicate and phosphorus in the supernatant water of a long-term mooring are presented.

In Figure 3-6, flux rates for PON, particulate biogenic silicate (PSi) and particulate organic phosphorus (POP) in 500 m water depth from an annual mooring in the Greenland Sea are depicted as well as the respective inorganic species in the supernatant water. In the case of nitrogen, ammonia made up for 99% of DIN (ammonia + nitrite + nitrate) in the supernatant water. Concentrations rose to levels far above 0.5 mM dm^{-3} (Figure 3-6a). Assuming that DIN is derived from the initial PON flux 7.8 to 39% of the PON flux was lost to the dissolved phase. This is most likely an underestimation as some diffusion out of the cups may occur. Concentrations of DIN are correlated neither to the amount of PON in the cups nor to the length and season of the sampling intervals nor the time lag between sampling and analysis. There is also no correlation between DIN concentrations and amount of swimmers in the cups. On average the loss to the dissolved inorganic phase was 25 ± 10%.

The clear seasonal trend in PSi flux is also seen in the concentrations of reactive silicate in the cups (Figure 3-6b). Concentrations in spring rose to almost 1 mM dm^{-3} approaching saturation level, which is 1.024 mM dm^{-3} according to Wollast [1974]. In spring, however, only 4% of the original PSi-flux underwent dissolution. During winter, dissolution amounted to 45 to 55% of the sedimented particles. The average annual loss to the dissolved phase was 33 ± 17%. Dissolution of sinking biogenic PSi may vary seasonally and regionally as interspecific differences in the dissolution of diatom valves and radiolarians can be expected [Hurd, 1972].

Furthermore, dissolution may be dependent on whether diatoms sank in phytodetritus aggregates or fecal pellets. Experiments addressing this problem are presently being carried out. For phosphorus, between 0.2 and 1.4 mM dm^{-3} were found in the supernatant water (Figure 3-6c). Ninety four to ninety eight percent of the original POP flux was in the dissolved inorganic phase. The N:P ratios (molar) of around 1 in the supernatant water indicate high leaching of phosphate from the particulate pool. Maximum numbers of swimmers were found from the end of October through February. Dissolved phosphorus in the cups can be considerably contaminated by leaching of swimmer tissues (G. Knauer, personal communication, 1991), although there is no clearly related pattern.

The trapped material, thus, is altered considerably during deployment or storage in the collector cups despite the use of a poison by which microbial activity is stifled effectively [Lee et al., in press, 1991]. Minimum corrections for particle flux are straightforward for PSi-flux by estimates of reactive silicate in the supernatant water and for carbonate by calcium and magnesium

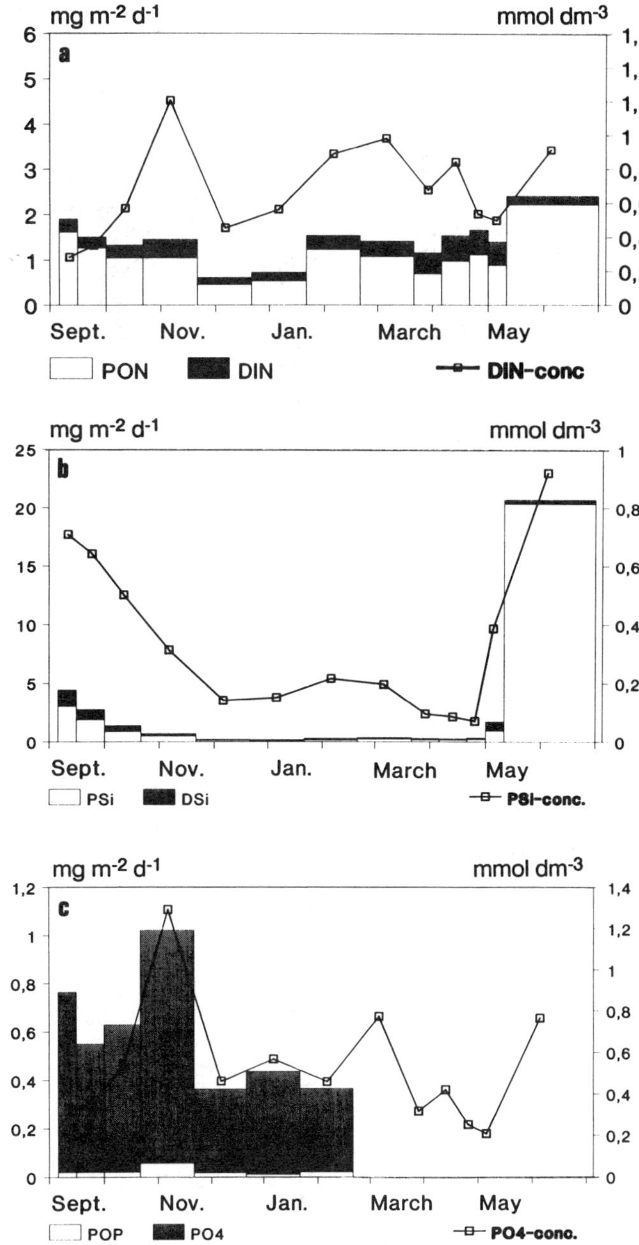

Fig. 3-6. Flux rates of particulate organic nitrogen (PON, a), particulate biogenic silicate (PSi, b) and particulate organic phosphorus (POP, c) in 500 m water depth from September 1988 to June 1989 in the Greenland Sea at 72° N 10° W. Concentrations of dissolved inorganic nitrogen (DIN = ammonia + nitrite + nitrate), silicate (DSi) and phosphate (PO_4) are also depicted. Shaded areas indicate dissolved inorganic concentrations expressed as daily flux rates on a weight basis.

measurements. In the case of strontium, valuable new information may be derived with regard to sinking fluxes of acantharians which are rarely reported from trap

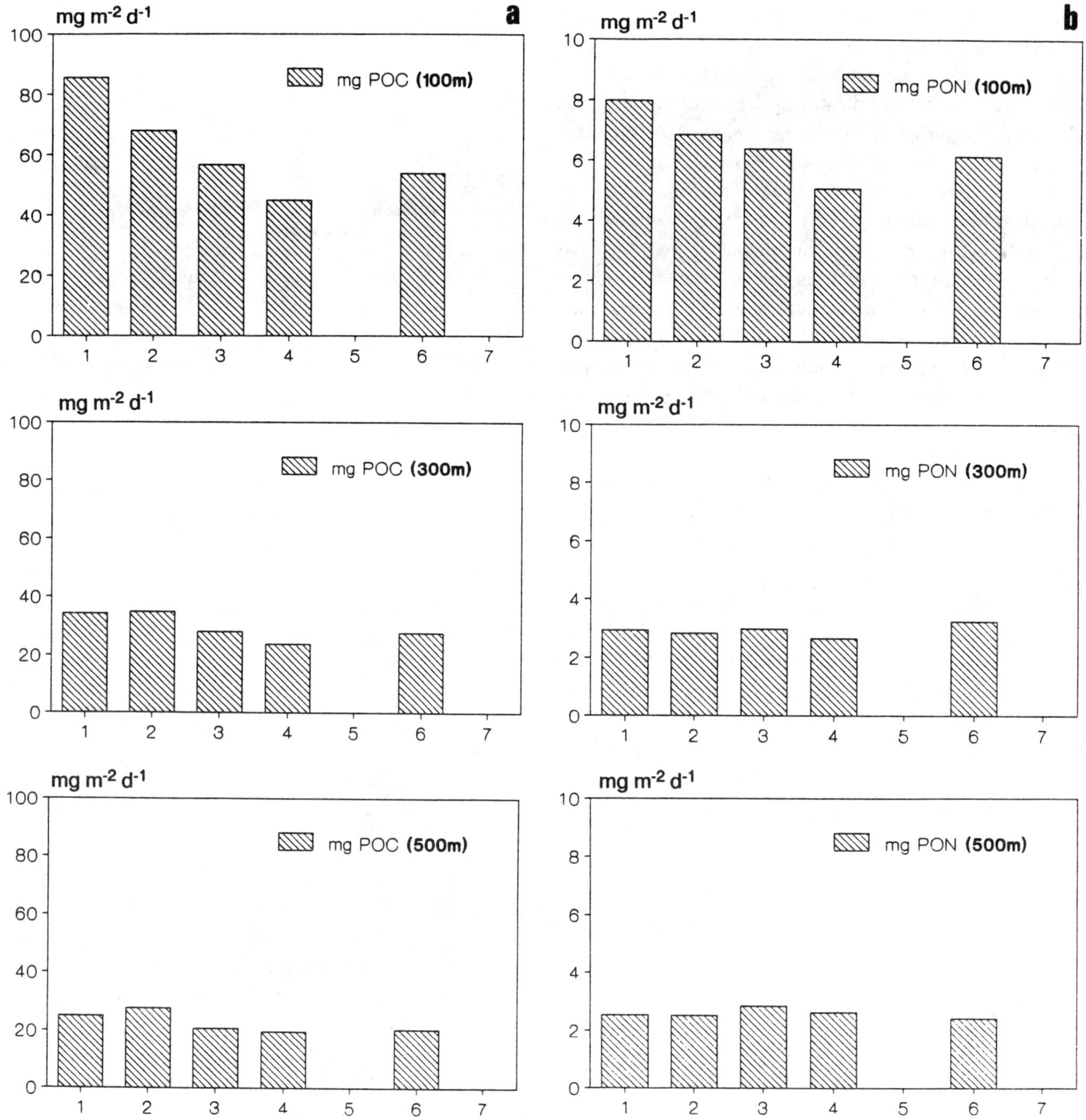

Fig. 3-7. Decrease of particulate organic carbon (POC, a) and nitrogen (PON, b) during a time course experiment with sediment trap material collected in 100 m, 300 m and 500 m between 29:06:89 and 09:07:89 in the Greenland Sea (72° N 07° W). Numbers on the horizontal axis indicate the number of months elapsed after recovery of the traps.

collections as their celestite skeletons (strontium sulfate) dissolve readily in sea water [Bernstein et al., 1987].

For POP, PON and POC fluxes, in addition to the inorganic dissolved species the dissolved organic fraction must be measured in the supernatant water (Karl et al.; Knauer, this volume, pages 73 and 81). Furthermore, for proper correction of PON and POC flux the abundance of nitrogen gases and carbon dioxide in the collector cup would have to be examined. Such estimates will be very problematic as upon recovery and sample handling unaccountable losses will occur due to degassing.

To learn about the loss of POC and PON the

decrease over time was measured directly in dissolution experiments with natural sediment trap samples. For this purpose a short-term mooring with traps in 100, 300, and 500 m water depth was deployed in June/July 1989 in the Greenland Sea. Traps were equipped with one poisoned collector cup and the sampling interval was 10 days. The trap samples were processed, including swimmer picking, 4 weeks after recovery of the traps. One subsample was measured immediately, four more subsamples were kept at 0° C to simulate longer deployment periods and were analyzed in month and two-month intervals (Figures 3-7a, b). Unfortunately the loss of particulate matter for the first 4 weeks could not be followed in this experiment. At present, a second experiment with natural trap samples from autumn 1990, in which the first measurements were made immediately after recovery, is under way.

The decrease over time in POC and PON is most apparent in the sample from the 100 m trap. The decrease in PON was tenfold higher in this sample than the increase in DIN. Apparently a large portion of the PON is moved to the dissolved organic pool. POC decrease in the 300 m and 500 m trap is less pronounced and PON remains remarkably constant over 6 months. In all samples the C:N ratios (molar) decrease over time, from 12.5 to 10.4, from 13.6 to 10.3 and from 11.5 to 10.0 in the 100 m, 300 m and 500 m sample, respectively. The low initial C:N ratio in the 500 m sample can be related to phytodetritus from an ice-edge bloom which had settled out of the surface waters prior to the deployment of the mooring. The upper two traps contained more detrital matter and calanoid fecal pellets. We can only speculate about the apparent higher loss rates in POC as compared to PON. Assuming microbial and/or exoenzymatic degradation as reasons for losses from the particulate pool, POC in the cups may have been a more attractive C source than the dissolved organic pool, whereas for nitrogen the opposite could have been the case.

The results briefly outlined here point towards the complexity of loss rates of particulate matter during deployment time of sediment traps. It appears that no general correction to the initial flux rates can be applied, rather, for each individual sampling interval and compound an individual correction has to be made to improve the accuracy of particle flux measurements. The consequent measurements of various properties in the supernatant water impose more work than the analysis of the particles, which, however, is worthwhile, considering the benefits from long-term deployments.

Acknowledgements. We would like to thank Dr. Derek Spencer, Dr. Dave Hurd, and Mary Zawoysky for organizing the stimulating workshop and dealing with the manuscript. A. Antia, E. Bauerfeind, W. Koewe and M. Voss gave valuable support in the fieldwork and completion of the manuscript. This work was supported by the Deutsche Forschungsgemeinschaft in the frame work of the Sonderforschungsbereich 313. This is publication No. 141 of the Sonderforschungsbereich 313.

References

Altabet, M. A., Organic C, N, and stable isotopic composition of particulate matter collected on glass fiber and aluminum oxide filters, Limnol. Oceanogr., 35, 902-909, 1990.

Bathmann, U. V., R. Peinert, T. T. Noji, and B. v. Bodungen, Pelagic origin and fate of sedimenting particles in the Norwegian Sea, Prog. Oceanogr., 24, 117-125, 1990.

Bathmann, U. V., T. T. Noji, and B. v. Bodungen, Sedimentation of pteropods in the Norwegian Sea in autumn, Deep-Sea Res., in press, 1991.

Bernstein, R. E., P. R. Betzer, R. A. Feely, R. H. Byrne, M. F. Lamb, and A. F. Michaels, Acantharian fluxes and strontium to chlorinity ratios in the North Pacific Ocean, Science, 237, 1490-1494, 1987.

Butman, C. A., Sediment trap biases in turbulent flows: results from a laboratory flume study, J. Mar. Res., 44, 645-693, 1986.

Coale, K. H., Labyrinth of doom: a device to minimize the "swimmer" component in sediment trap collections, Limnol. Oceanogr., 35, 1376-1381, 1990.

Eggimann, D. W., F. T. Manheim, and P. R. Betzer, Dissolution and analysis of amorphous silica in marine sediments, J. Sediment. Petrol., 50, 215-225, 1980.

Gardner, W. D., K. R. Hinga, and J. Marra, Observations on the degradation of biogenic material in the deep ocean with implications on accuracy of sediment trap fluxes, J. Mar. Res., 41, 195-214, 1983.

Grasshoff, K., Methods of Seawater Analysis, 419 pp., Verlag Chemie, New York and Weinheim, 1976.

Hurd, D. C., Factors affecting solution rate of biogenic opal in seawater, Earth Planet. Sci. Lett., 15, 411-417, 1972.

JGOFS Report No. 6, Core measurement protocols, reports of the core measurement working groups, 40 pp., 1990.

Karl, D. M., and G. A. Knauer, Swimmers: a recapitulation of the problem and a potential solution, Oceanography, 2, 32-35, 1989.

Knauer, G. A., D. M. Karl, J. H. Martin, and C. N. Hunter, In situ effects of selected preservations of total carbon, nitrogen and metals collected in sediment traps, J. Mar. Res., 42, 445-462, 1984.

Koike, I., H. Shigemitsu, T. Kasuki, and K. Kazuhiro, Role of sub-micrometre particles in the ocean, Nature, 345, 242-244, 1990.

Krause, G. L., C. L. Schelske, and C. O. Davies, Comparison of three wet-alkaline methods of digestion of biogenic silica in water, Freshwater Biol., 13, 73-81, 1983.

Lee, C., S. G. Wakeham, and J. I. Hedges, The measurement of oceanic particle flux - are swimmers a problem?, Oceanography, 1, 34-36, 1988.

Lee, C., J. I. Hedges, S. G. Wakeham, and N. Zhu, Effectiveness of various treatments in retarding microbial activity in sediment trap material and their effect on the collection of swimmers, Limnol. Oceanogr., in press, 1991.

Lenz, J., Zur Methode der Sestonbestimmung, Kieler Meeresforsch., 27, 180-193, 1971.

Paasche, E., Silicon and the ecology of marine plankton diatoms. I. *Thalassiosira pseudonana (Cyclotella nana)* grown in a chemostat with silicate as limiting nutrient, Mar. Biol., 19, 117-126, 1973.

Peinert, R., U. Bathmann, B. v. Bodungen, and T. Noji, The impact of grazing on spring phytoplankton growth and sedimentation in the Norwegian Current, in Particle Flux in the Ocean, edited by E. T. Degens, E. Izdar, and S. Honjo, pp. 149-164, SCOPE/UNEP Sdb. 62, 1987.

Peterson, W., and H. G. Dam, The influence of copepod "swimmers" on pigment fluxes in brine-filled vs. ambient seawater-filled sediment traps, Limnol. Oceanogr., 35, 448-455, 1990.

Puch, M., Zum Silikathaushalt des Pelagials im Europäischen Nordmeer, M.S. thesis, 66 pp., Univ. Kiel, 1990.

Sharp, J. H., Improved analysis for "particulate" carbon and nitrogen, Limnol. Oceanogr., 19, 984-989, 1974.

Sheldon, R. W., Size separation of marine seston by membrane and glass fiber filters, Limnol. Oceanogr., 17, 494-498, 1972.

Silver, M. W., M. M. Gowing, D. C. Brownlee, and J. O. Corliss, Ciliated protozoa associated with oceanic sinking debris, Nature, 309, 246-248, 1984.

U.S. GOFS Working Group, G. A. Knauer and V. Asper, Co-chairs, Sediment trap technology and sampling, Report No. 10, November 1988, 94 pp., Woods Hole, Massachusetts, 1989.

Wollast, R., The silica problem, in The Sea, edited by E. D. Goldberg, Vol. 5, pp. 359-392, Wiley & Sons, New York, 1974.

Zeitzschel, B., P. Diekmann, and L. Uhlmann, A new multisample sediment trap, Mar. Biol., 45, 285-288, 1978.

DETERMINATION OF CARBON, NITROGEN, SULFUR, AND INORGANIC SULFUR SPECIES IN MARINE PARTICLES

Gregory A. Cutter and Joel Radford-Knoery

Department of Oceanography, Old Dominion University, Norfolk, Virginia 23529-0276

1. Overview

Numerous papers have presented data on the carbon and nitrogen content of marine particles [e.g., Knauer et al., 1979; Feely et al., 1981; Karl and Knauer, 1984] with only brief descriptions of the methodology employed. Furthermore, a few analytical papers have described the methods in more detail [e.g., Sharp, 1974; Simpson, 1982; Verardo et al., 1990]. However, only a few papers have described the distributions of particulate sulfur in the oceanic water column [Matrai and Eppley, 1989] and the methods to determine it [Matrai, 1989]. In light of recent attention to the global sulfur cycle and its role in global change, data on the distributions, concentrations, and fluxes of particulate sulfur are needed. Correspondingly, accurate and precise methods that can detect sulfur at low concentrations must be utilized. While the need for accuracy is obvious, precision must be excellent in order to resolve sulfur's biogeochemical behavior. From a mechanistic standpoint, determinations of particulate sulfur speciation can further reveal the processes that affect the element's behavior. This approach has proven quite beneficial in the examination of sulfur in marine sediments (e.g., mechanisms of pyritization, etc.), and applying speciation techniques to the water column may be similarly rewarding. As an example, the data in Figure 3-8 show that acid volatile sulfides (e.g., FeS, ZnS, CuS) comprise a small portion of the suspended sulfur in the Sargasso Sea.

While specialized instruments such as the flame photometric detector allow the unique determination of sulfur, several commercially available instruments can determine carbon, nitrogen, and sulfur simultaneously. This paper focuses on the use of such an instrument, the Carlo Erba NA-1500, for the determination of particulate carbon, nitrogen, and sulfur. In addition, methods for the characterization of inorganic sulfur speciation will be briefly described as a means of stimulating more efforts in this regard. The discussion includes methods for the collection of suspended particulate samples and their preparation prior to instrumental analysis.

Marine Particles: Analysis and Characterization
Geophysical Monograph 63

2. Experimental Section

2.1. Apparatus

Sampling. Water samples should be taken with Go-Flo bottles (greater than or equal to 5 L), or alternatively a pumping system for shallow waters. The Go-Flo's should be modified by replacing the upper air bleed valve with a Nylon O-Seal Straight Thread Connector (Swagelok; 3/8" tube; Part # NY-600-1-OR) in order to pressurize the bottle with nitrogen. In addition, the bottom sample valve should be removed, and the hole threaded to accommodate a Teflon plug valve [Berghof/America; 3/8" plug valve; Part # TPV-S0604-F(K)]; the thread size is determined by the plug valve used, e.g., the nut on one side of the plug valve is removed, and this side is screwed into the Go-Flo. A length of polyethylene or Teflon tubing, depending on the other samples being taken from the Go-Flo, is attached to the Go-Flo side of the valve so that it extends to the bottom of the bottle. The end is bevelled so that it is flush with the bottom ball valve. In this configuration, the Go-Flo can be pressurized to a maximum of 4 to 8 psi nitrogen and the water directed through a filter, or series of filters on a manifold.

Sample processing. Filters for CNS are held in 13 mm, Gelman Swinney Filter Holders (Gelman Part # 4317), while filters for sulfur speciation are held in 25 mm Swin-Lok devices (Nuclepore Part # 420200). Very little pressure differential develops across the glass fiber filters used for CNS, and thus either vacuum or pressure filtration can be used. For vacuum filtration, a manifold which can hold six Swinney holders is used. A sample from the Go-Flo is placed in a polyethylene bottle (up to 5 L) with "tubulation" (tubing port at the bottom) that is then connected to the filter holder on the manifold using polyethylene tubing. Samples for sulfur speciation are pressure filtered by connecting a manifold that can hold three Swin-Lok devices to the Go-Flo using polyethylene tubing.

Instrumentation. Samples are weighed using a Cahn Model 29 microbalance, capable of weighing to 0.1 μg. The balance is calibrated daily using a 200 mg NIST Class M calibration weight.

The simultaneous determination of CNS uses a Carlo Erba NA-1500 with the following settings and parameters:

GC column: 2 m x 16 mm (o.d.) stainless steel, packed with 50/80 mesh Porapak QS	
Combustion furnace temperature	1010° C
GC column oven and detector block temp.	80° C
Filament temperature	190° C
He (ultra high purity) carrier gas flow rate	95 mL min^{-1}
Helium reference flow rate	80 mL min^{-1}
Helium purge flow rate	60 mL min^{-1}
Oxygen (ultra high purity) flow rate	40 mL min^{-1}
oxygen gauge pressure	1 kg cm^{-2}
oxygen loop volume	10 mL
Run time	550 sec
Sample start time	20-25 sec
Sample stop time	60 sec
Oxygen inject time	80 sec
Retention times (approximate)	
N_2	70 sec
CO_2	95 sec
SO_2	425 sec

The determination of inorganic sulfur speciation utilizes the apparatus described in Cutter and Oatts [1987] with no modifications required for the analysis of suspended matter.

2.2. Reagents and Standards

An ultrapure sulfanilamide standard (16.27% N, 41.84% C, 18.62% S) is obtained from Carlo Erba, dried at 40° C to constant weight, and stored in a desiccator. NIST Estuarine Sediment (SRM 1646) is used as a secondary standard that is more representative of marine particulate matter. This homogenized material has been analyzed in our laboratory over the last 5 years and contains 0.18 ± 0.01% N, 1.79 ± 0.05 % C, and 0.96 ± 0.04% S (n=114).

Reagents for the elemental analyzer include the oxidation catalyst (tungstic trioxide on alumina), a desiccant (magnesium perchlorate), and reduced copper wire. While the catalyst must be obtained from Carlo Erba, the copper can be obtained from other sources. However, the copper must be of high purity and free of oxidation in order to obtain satisfactory results for sulfur. In this respect, copper wire which has some oxidation can be cleaned by only filling the combustion tube with copper as described below, placing it in the cooled furnace, and passing hydrogen through the helium carrier line at 10 mL min^{-1}. Place a length of Tygon tubing on the end of the combustion tube and vent the hydrogen into a fume hood. After thirty minutes of hydrogen purging, turn on the instrument and raise the furnace temperature to 600° C. Eight hours of heating in the presence of hydrogen will remove any surface oxidation, and after cooling the tube can be filled as below. However, extreme caution should be exercised with this procedure, and the furnace must be not be heated until a pure hydrogen atmosphere is in the combustion tube.

A 1 M HCl solution is prepared by dilution of redistilled, concentrated hydrochloric acid with distilled water. The other reagents are reagent grade, and include vanadium pentoxide, dichloromethane, methanol, and sodium chloride. A saline solution for rinsing filters which has a salinity near that of the samples (e.g., 35 ppt) is made using sodium chloride and distilled water. Before use, the vanadium pentoxide is placed in a porcelain crucible and heated to 500° C in a muffle furnace for 24 hours.

In addition to these reagents, household aluminum foil, nickel foil (15 cm wide x 0.13 mm thick), and large tin sample cups (Carlo Erba) are required. For CNS determinations, 13 mm Gelman A/E glass fiber filters (binder-free) are used. For the determination of particulate sulfur speciation, 25 mm Nuclepore 0.4 μm polycarbonate membranes are used.

2.3. Procedures

Preparation of filters and sample cups. Filters for CNS are cleaned by placing fifty to one hundred in a clean 100 mL beaker, adding 30 mL of methanol, covering with a watch glass, and refluxing on a warm hot plate for 10 minutes. The methanol is decanted and the procedure is repeated once more. After the second methanol rinse, 30 mL of dichloromethane is added, the beaker covered, and placed on a warm hot plate to reflux for 5 minutes. After decanting the dichloromethane, the procedure is repeated. After removing the dichloromethane, the covered beaker is placed in a 100° C oven for 2-4 hours to remove any remaining solvent. Filters are stored in a solvent-cleaned glass jar with lid. Filters for sulfur speciation require no cleaning prior to use.

Aluminum sample cups are prepared by cutting out about 2 cm diameter discs from aluminum foil using a sharpened cork boring tool. The sample cup is then made by molding the disc around the end of a 6 mm diameter glass or metal rod (edges rounded slightly to prevent tearing the foil); gloves should be worn during handling to minimize contamination. These aluminum cups are then cleaned (as above) and placed inside large tin sample cups using forceps to produce the sample cups used for organic carbon determinations. The tin+aluminum sample cups and tin (only) cups used for total CNS determinations are solvent cleaned using the procedure given above. Clean sample cups are stored in either glass jars or petri dishes.

Combustion tube preparation. Quartz combustion furnace tubes (opaque, Carlo Erba) are packed in the following sequence (bottom to top): 1 cm quartz wool; 8 cm copper wire; 1 cm quartz wool; 10 cm oxidation catalyst; and a nickel foil tube with 0.5 cm quartz wool at the bottom (against the top of the oxidation catalyst). The nickel foil tube is made by rolling a 5.5 cm wide x 15 cm long piece of nickel foil into a tube that has an outer diameter slightly smaller than the inner diameter of the combustion tube, so that it can easily slide inside the combustion tube, even when heated. After rolling, a bottom corner of the foil is bent inward so that the quartz wool is kept from sliding out. The prepared nickel tube is placed in the combustion tube using a pair of hemostats until it rests on top of the oxidation catalyst. Depending on the quality of the copper, a combustion tube packed in this manner should be capable of 175-200 analyses.

Ship-board sample processing. For oceanic samples, 2-5 L of seawater should be filtered for CNS; coastal samples (i.e., higher particle loads) typically clog the filters after 0.5-1 liter. After the sample, 100 mL of the saline solution should be passed through the filter (vacuum filtration works best for this) and then air is drawn through the filter to dry it. Using forceps remove the filter from the holder, fold it in half so that the particles are inside, place in a 1 mL polyethylene snap-cap vial, and freeze immediately. Process blanks include rinsing filters with saline solution and then treating them as samples. It is recommended that samples be filtered at least in duplicate; if total CNS (no acidification) and organic carbon (after acidification) are being determined, then at least four filters per sample must be collected.

The collection of suspended matter for sulfur speciation entails filtering 1-2 liters of water via pressure filtration, passing nitrogen across the filter to remove most of the water, immediately transferring the filter to a 5 mL snap-cap vial (fold so that the particles are inside the filter), and freezing. No pretreatment is necessary.

Laboratory processing. To facilitate manipulations in the laboratory, a sample tray that holds up to 50 tin sample cups is constructed of 1.5 cm thick acrylic plastic. Holes that are 7 mm in diameter and 7 mm deep are drilled to hold the sample cups. For the determination of total CNS, tin sample cups are placed in the sample tray, and sample filters are transferred from the vials into the cups using two pairs of forceps (some folding is required). The first 15 positions of the sample tray (i.e., positions 0-14) should be left open for standards and blanks. In addition to the sample filters, three process blank filters should be placed in cups in positions 12-14. The tray is then placed in an oven set at 40° C for approximately 4 hours to dry the filters. After drying, about 10 mg of vanadium pentoxide is placed in each cup, and the cup gently sealed and rolled into a compact ball using two pairs of forceps (cover the lab bench with aluminum foil in case a sample cup is dropped). The top of the cup should first be pinched closed, and then rolled over several times. The cup can then be compacted and formed into a ball, taking care not to tear the tin or leave any edges of tin foil protruding (these can jam the autosampler).

For the determination of organic carbon, sample filters are transferred into Sn+Al sample cups as above (also three blank filters). Using an Eppendorf pipette, 25 μL of distilled water are dispensed in the cup to wet the filter, followed by 25 μL of 1 M HCl. After reacting for 1 hour, an additional 25 μL of 1 M HCl is added. The sample tray is placed in a vacuum desiccator set at 40° C for 2 hours.

This procedure should treat at least 2 mg of calcium carbonate (i.e., to remove the carbonate carbon), but if any effervescence is noted on the second HCl addition or if high carbonate levels are expected, the above procedure should be repeated once more. After the final vacuum drying, 10 mg vanadium pentoxide should be added, the aluminum cup should be folded over inside the tin cup using forceps, and then the cup rolled and formed as before. These procedures for total CNS and organic carbon can be applied to other marine particles such as materials from sediment traps and net tows, or sediments. The only caution is that the acidification procedures must be done very carefully to avoid sample loss; it is very important to completely wet the sample before adding acid.

Sample analysis - CNS. The sample tray should be filled in the following order:

Positions 0-1:	sulfanilamide conditioners (ca. 5 mg sulfanilamide, exact weight not required)
Positions 2-4:	tin cup + V_2O_5 blanks
Positions 5-8:	sulfanilamide standards (exact weight, but keep it close to 1 mg, tin cup) + V_2O_5
Positions 9-11:	NIST Estuarine Sediment + V_2O_5 (tin cups)
Positions 12-14:	filter blanks
Positions 15-49:	samples

In the parameter table given above, the sample start time is listed as 20-25 seconds. When the conditioning runs are analyzed, the flash combustion should be monitored through the viewing port. The bright flash combustion which lasts for 10-20 seconds should begin within 5-10 seconds after the sample is dropped. If it does not, then the sample time should be adjusted appropriately. After it is assured that flash combustion occurs properly, continue until the last Estuarine Sediment is analyzed. Calculate the calibration factors and the CNS values for the sediment standard. If the Estuarine Sediment results agree with those listed above, then continue with the analyses of the remaining samples.

After one tray is analyzed, the combustion residues should be removed as follows:

1. Prepare another nickel foil tube and quartz wool plug as above.
2. Turn off the helium carrier gas at the tank.
3. Disconnect the autosampler from the top of the combustion tube.
4. Using a long pair of hemostats, pull the nickel foil tube out of the combustion tube (CAUTION: IT'S VERY HOT).
5. Insert the new tube until it rests on the top of the oxidation catalyst.
6. Reassemble the autosampler and check for leaks. Continue analyses, but run two additional conditioners.

When running unacidified samples with high (greater than 30%) carbonate concentrations, it is important to monitor the chromatograms. It has been observed with calcareous sediments that the sulfur dioxide peak begins to broaden severely after the analysis of 15-20 samples. If this occurs, remove the debris as above and continue the analyses.

The three calibration factors for N, C, and S are calculated as follows:

$$K = (\%T \times WS)/(PA\text{-}B)$$

where,

%T = theoretical % of N, C, or S in the standard
WS = weight of standard in milligrams
PA = peak area of N, C, or S
B = average of three tin + V_2O_5 blanks

K is determined for each element on each of the four standards analyzed, and the averages of the four values used in later calculations.

The CNS content of a solid sample (e.g., sediment) is calculated as follows:

$$\%\ C, N, \text{or } S = (K \times (PA\text{-}B))/WS$$

where,

K = the average calibration factor calculated above
PA = peak area of N, C, or S
B = average of three tin + V_2O_5 blanks
WS = weight of sample in milligrams

The CNS content of a filtered sample is calculated as follows:

$$\mu g\ C, N, \text{or } S/L = (K \times 1000 \times (PA\text{-}FB))/VF$$

where,

K = the average calibration factor calculated above
PA = peak area of N, C, or S
FB = average of three filter + V_2O_5 + Sn-Al cup blanks
VF = volume filtered in liters

Sample analysis - inorganic sulfur speciation. The procedures for sediment analysis described by Cutter and Oatts [1987] should be followed. The only exception is that for filters, no extraction of elemental sulfur with carbon tetrachloride is used. The filter is placed in the stripping vessel and the determinations of FeS, greigite, and pyrite are performed sequentially on the same filter. Samples should be analyzed in triplicate.

3. Discussion

3.1. Sample Acquisition and Preparation

Although there are many techniques for acquiring suspended particles in the water column, three important considerations are: (1) the method should collect samples that are representative of marine suspended matter; (2) the method must not compromise the chemical composition or speciation of a sample; and (3) the method should be logistically acceptable (e.g., relatively low cost, simple to use at sea). The procedures recommended here generally meet these three criteria. Go-Flo bottles are commercially available, and with simple modifications can be used for pressure filtration. Thus, sample handling is minimized and for certain forms of particulate sulfur (e.g., the metal monosulfides), the inert nitrogen atmosphere preserves their chemical speciation. Furthermore, by extending a tube to the bottom of Go-Flo, the entire contents of the bottle (including any particles which might settle during processing) are transferred to the filter(s). For the determination of total carbon, nitrogen, and sulfur, and organic carbon, vacuum filtration of the Go-Flo contents is an acceptable alternative (see discussion to follow).

The choice of filter type and size is a difficult one and requires certain compromises. The simultaneous determination of carbon, nitrogen, and sulfur excludes the use of most filters with sub-micron pore-sizes (e.g., Nuclepore polycarbonate filters). Aluminum oxide and silver filters are too costly for most laboratories and too fragile to be folded into the tin sample cups. Thus, it is necessary to use glass fiber filters with nominal pore size greater than or equal to 0.8 μm. The size of the sample cups and the capability of the instrument also limit the filter diameter to 13 millimeters. For the determination of inorganic sulfur speciation, contamination is not a problem and filters such as 0.4 μm polycarbonate membranes can be used.

Most methods for the determination of carbon and nitrogen [e.g., Verardo et al., 1990] recommend the cleaning of sample cups and filters using heat, e.g., heating cups and filters in a muffle furnace at 550° C. However, we have found that sequential solvent rinses, with low temperature heating, minimize CNS blank values in cups and filters, without causing these materials to become brittle as is often the case with muffle furnace heating.

After filtration, the only sample pretreatment for CNS filters that is performed on board ship is to remove inorganic sulfate which might remain on the filter after the seawater evaporates. Any rinse must be of a similar ionic strength in order to avoid cell rupture and the potential loss of CNS. For the determination of particulate sulfur Matrai [1989] used ammonium formate, but this might contaminate the sample with excess nitrogen and carbon. Thus, we utilize a sodium chloride solution. Afterwards the filter is immediately frozen in order to minimize artifacts due to bacterial growth and degradation of the organic matter.

Prior to analysis the only laboratory pretreatment for total CNS involves placing the filters, or other particulate matter such as that from sediment traps, in the tin sample cups and drying them at 40° C. In this manner the potential loss of volatile compounds is minimized. In order to determine organic carbon, it is necessary to first remove any carbonate in the filtered or solid sample. Acidification of each sample using sulfurous acid to drive off carbon dioxide has been described by Verardo et al. [1990]. The advantages of this "in situ" method include elimination of losses of acid-soluble organic carbon and, for high carbonate samples, the correction for weight loss after acidification. The method described here is exactly the same as that of Verardo et al. [1990] except that it uses hydrochloric acid. Since metal sulfides such as FeS and ZnS might be lost during acidification, any organic carbon method should not be used to determine total sulfur. In addition, we have observed up to 20% losses in nitrogen for several sediment samples after acidification. For these reasons, two separate determinations should be made; one for total CNS and one for organic carbon. Thus, the acidification method, whether it is that described here or the Verardo et al. [1990] procedure, is the choice of the user.

3.2 Instrumental Procedures.

The basis of the determination used here is the formation of carbon dioxide, molecular nitrogen, and sulfur dioxide, and their gas chromatographic detection. Verardo et al. [1990] thoroughly describe the determination of carbon and nitrogen using the Carlo Erba NA-1500. However, there are two potential problems encountered with this approach for sulfur. The first is the formation of SO_2 from such refractory compounds as pyrite (FeS_2) and gypsum ($CaSO_4$). To overcome this, combustion should occur in an oxidizing atmosphere at greater than 1500° C. The furnace in the NA-1500 is held at 1010° C and the sample is dropped into a pure oxygen atmosphere; 10 mL of O_2 is injected into the furnace just before the sample is introduced. Under these conditions, the tin cup undergoes an exothermic reaction which produces temperatures approaching 1700° C (the "flash combustion"). Vanadium pentoxide is added to each sample to further ensure complete oxidation. Analyses of calcium sulfate and pyrite using the Carlo Erba NA-1500 showed that quantitative recoveries for sulfur could only be achieved if vanadium pentoxide was added.

The second problem with the determination of sulfur is the reactivity of sulfur dioxide. The major difficulty here is the copper reductant at the bottom of the combustion tube. The function of this copper is to reduce oxides of nitrogen to N_2, to reduce any SO_3 to SO_2, and to remove oxygen which co-elutes with N_2 in the chromatogram. Impurities in the copper affect the recovery of sulfur, but this must be tested empirically. By running standards in combustion tubes filled with different brands of copper wire, one can select the copper that gives the best recovery.

Another consideration in the analysis of particulate CNS is the choice of standards. Since pure sulfanilamide can be obtained commercially, is stable during storage, and not hygroscopic, it makes an excellent primary standard. However, it is not very representative of marine particles. To overcome this, we have used NIST Estuarine Sediment which has an uncertified sulfur value of 0.96%. Having analyzed this material for the last five years using sulfanilamide as the primary standard, we have obtained the results given above. These results not only confirm the uncertified value, but also provide carbon and nitrogen concentrations which are relatively easy to obtain in comparison to sulfur. Thus, Estuarine Sediment can be used as a secondary standard to check the performance and accuracy of the instrument prior to analyzing actual samples.

Using the procedures described above, we have determined the absolute detection limits (3 x std. dev. of blanks) for the instrument to be 0.48 μg N, 0.92 μg C, and 0.50 μg S. Assuming that 2 L are filtered, this translates to relative detection limits of 17 nM N L^{-1}, 38 nM C L^{-1}, and 7.8 nM S L^{-1}. For solid samples, an average of 10 mg are analyzed, giving relative detection limits of 3.4 μM N g^{-1}, 7.7 μM C g^{-1}, and 1.6 μM S g^1. The precisions of the method for suspended (filtered) particles, expressed as relative standard deviation, are 9% for carbon at 8.3 μM C L^{-1}, 8% for nitrogen at 0.7 μM N L^{-1}, and 12% at 0.06 μM S L^{-1}. For solid phase samples, such as sediments or sediment trap material, precisions for the three elements (as RSD) are 1.1% for nitrogen at 0.1 mM N g^{-1}, 0.7% for carbon at 1.7 mM C g^{-1}, and 2% for sulfur at 0.3 mM S g^{-1}.

Cutter and Oatts [1987] describe a method for the determination of FeS, greigite (Fe_3S_4), and pyrite in sediments. Briefly, this technique utilizes the selective formation of hydrogen sulfide from these sulfur species. Hydrogen sulfide is then determined using gas chromatography-photo-ionization detection, a method that has an absolute detection limit of 12 ng S. Because of this low detection limit, the method can be easily applied to the analysis of suspended particles. However, certain modifications to the procedures, and thus redefinitions of the species that are determined, are necessary. First, since the Cutter and Oatts [1987] methods were designed for sediments, the "acid volatile sulfide", which is the H_2S liberated by 0.5 M HCl, is considered to be primarily FeS. However in the water column, acid volatile sulfide likely includes other metal monosulfides such as ZnS and CuS. Secondly, prior to the determination of pyrite (via Cr(II) reduction), the Cutter and Oatts [1987] protocol extracts elemental sulfur with carbon tetrachloride; thus, pyrite is determined uniquely. With polycarbonate filters, this elemental sulfur extraction would be difficult and might affect the precision. For the analysis of suspended particles, the extraction step is omitted, and the Cr(II) reduction step produces hydrogen sulfide from pyrite and elemental sulfur. In this fashion, the determinations of acid volatile sulfide (AVS), greigite, and pyrite + elemental sulfur can be performed on one filter sequentially. Additionally, the concentration of particulate organic sulfur can be estimated as the difference between total sulfur (using the Carlo Erba) and AVS + greigite + pyrite + elemental sulfur.

Using a volume of 1 L, the relative detection limit for suspended AVS, greigite, and pyrite is 0.38 nM S L^{-1}. However, by using large (142 mm diameter) Nuclepore

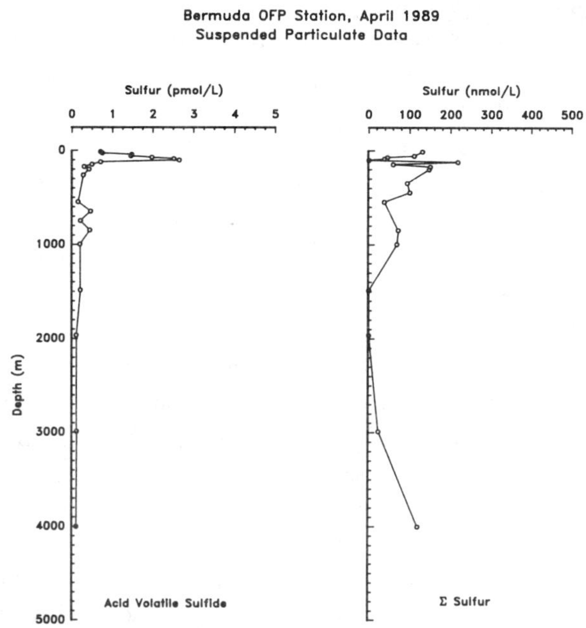

Fig. 3-8. Depth profiles of total particulate sulfur and acid volatile sulfide on suspended (filtered) particles at the Ocean Flux Program Station near Bermuda in April 1989.

membranes to filter 30 L, the detection limits can be lowered even further (e.g., results shown in Figure 3-8). The precision for these species at concentrations of 3 nM S L^{-1} is better than 10% (RSD). Using sulfur speciation methods such as these, in combination with total sulfur data, the biogeochemical processes affecting sulfur can begin to be elucidated.

Acknowledgements. This work was supported by NSF Grant OCE-8800371.

References

Cutter, G. A., and T. J. Oatts, Determination of dissolved sulfide and sedimentary sulfur speciation using gas chromatography-photo-ionization detection, Anal. Chem., 59, 717-721, 1987.

Feely, R. A., G. J. Massoth, and W. M. Landing, Major- and trace-element composition of suspended matter in the northeast Gulf of Alaska: relationships with major sources, Mar. Chem., 10, 431-453, 1981.

Karl, D. M., and G. A. Knauer, Vertical distribution, transport, and exchange of carbon in the northeast Pacific Ocean: evidence for multiple zones of biological activity, Deep-Sea Res., 31, 221-143, 1984.

Knauer, G. A., J. H. Martin, and K. W. Bruland, Fluxes of particulate carbon, nitrogen, and phosphorus in the upper water column of the northeast Pacific, Deep-Sea Res., 26, 97-108, 1979.

Matrai, P. A., Determination of sulfur in ocean particulates by combustion-florescence, Mar. Chem., 26, 227-238, 1989.

Matrai, P. A., and R. W. Eppley, Particulate sulfur in the waters of the southern California Bight, Global Biogeochem. Cycles, 3, 89-103, 1989.

Sharp, J. H, Improved analysis for "particulate" organic carbon and nitrogen from seawater, Limnol. Oceanogr., 19, 984-989, 1974.

Simpson, W. R., Particulate matter in the oceans - sampling methods, concentration, size distribution, and particle dynamics, Oceanogr. Mar. Biol., Ann. Rev., 20, 119-172, 1982.

Verardo, D. J., P. N. Froelich, and A. McIntyre, Determination of organic carbon and nitrogen in marine sediments using the Carlo Erba NA-1500 Analyzer, Deep-Sea Res., 37, 157-165, 1990.

DETERMINATION OF AQUATIC HUMIC ACID CARBON AND NITROGEN BY HIGH TEMPERATURE COMBUSTION

Lewis E. Fox

108 Pierce Hall, Harvard University, 29 Oxford St., Cambridge, Massachusetts 02138

1. Introduction

Humic substances refer to an enormously complex class of organic compounds generally attributed to breakdown products of biogenic organic matter on land or planktonic excretions in seawaters [see, for example, Schnitzer and Khan, 1972, for terrestrial humics; and Stuermer and Harvey, 1974; Harvey et al., 1984 for seawater humics]. Although these substances have been thoroughly investigated, their chemical structures are variable causing their identification to be dependent on general chemical characteristics. They are defined as yellow to black colored, moderately high molecular weight, alkali soluble, polyelectrolytes [Schnitzer and Khan, 1972; Stevenson and Butler, 1969; Kalle, 1966]. Modern extraction of humic substances from water is done with XAD sorbant resins [Thurman and Malcolm, 1981; Stuermer, 1975; Stuermer and Harvey, 1974; Gagosian and Stuermer, 1977]. Humic substances are subdivided according to their solubility in strong acid solution. Under these conditions, the soluble portion is defined as fulvic acid, the insoluble portion, humic acid. The method described below is specifically used for humic acid [see Fox, 1983], but is of equal value for measuring fulvic acid after appropriate modification of the initial extraction procedure.

2. Principle of the Method

Humic acid is extracted according to classical definition, desiccated, and oxidized over copper oxide to CO_2 and NO_x at 1200° C. NO_x is subsequently reduced to N_2 over copper metal at 500° C. Water is removed with a dry ice cold trap. Both CO_2 and N_2 are separated and quantified by gas chromatography using Porapack Q as column packing and a thermal conductivity detector (Figure 3-9). Sample output is standardized against ethylene diamine tetracetic acid (EDTA).

2.1. Sample Storage

Humic acids should be extracted from water samples as soon as possible after collection. All standard precautions against organic contamination must be observed. Samples cannot be frozen prior to extraction. After extraction, dried samples may be stored indefinitely.

2.2. Sample Preparation and Isolation

Analysis of aquatic humic acid carbon and nitrogen is basically a modified version of a standard particulate organic carbon and nitrogen analysis [Sharp, 1974] after modified extraction. Aquatic humics are extracted according to the classic definition, solidified, and oxidized to CO_2 and N_2 at high temperature. Aquatic humic acid is measured using the following procedure [see Fox, 1981, 1983]. Duplicate 100 mL water samples are filtered through pre-baked (450-500° C for 2 hours) Whatman GF/C filters. To 100 mL of filtrate, 0.8 mL of 1 M H_2SO_4 is added to aggregate humic acid. This lowers the sample pH to between 1.9 and 2.0. The acidified sample is immediately re-filtered through pre-baked, Whatman GF/C filters. There is a linear decrease in recovery with time amounting to 10% every 15 minutes for the first half hour after acidification. The recovery loss maximizes at 40% after 90 minutes. After filtration, the filters are rolled and placed in pre-baked, aluminum foil cylinders. The foil wrappers are constructed by wrapping 3-4 cm wide slices of aluminum foil around a glass stirring rod with a 6 mm outside diameter. The rod is carefully "slid-out" of the foil leaving a hollow aluminum foil cylinder. The cylinders are stored and baked in a glass petri dish [Sharp, 1974]. The filters are dried in a vacuum desiccator until combustion.

2.3. Equipment and Reagents

Equipment. The prepared samples may be analyzed in a C-H-N analyzer modified for this analysis [see Sharp, 1974]. A combustion gas chromatograph which affords better control over sensitivity can be constructed much more cheaply. A schematic for such a system is given in Figure 3-9. The oxidation and reduction furnaces can be purchased from Hewlett-Packard as replacements for their commercially available instrument.

Gas supply. Helium is required as carrier gas. High purity He is necessary and the addition of a molecular sieve trap is recommended. Contaminants in the carrier

Marine Particles: Analysis and Characterization
Geophysical Monograph 63

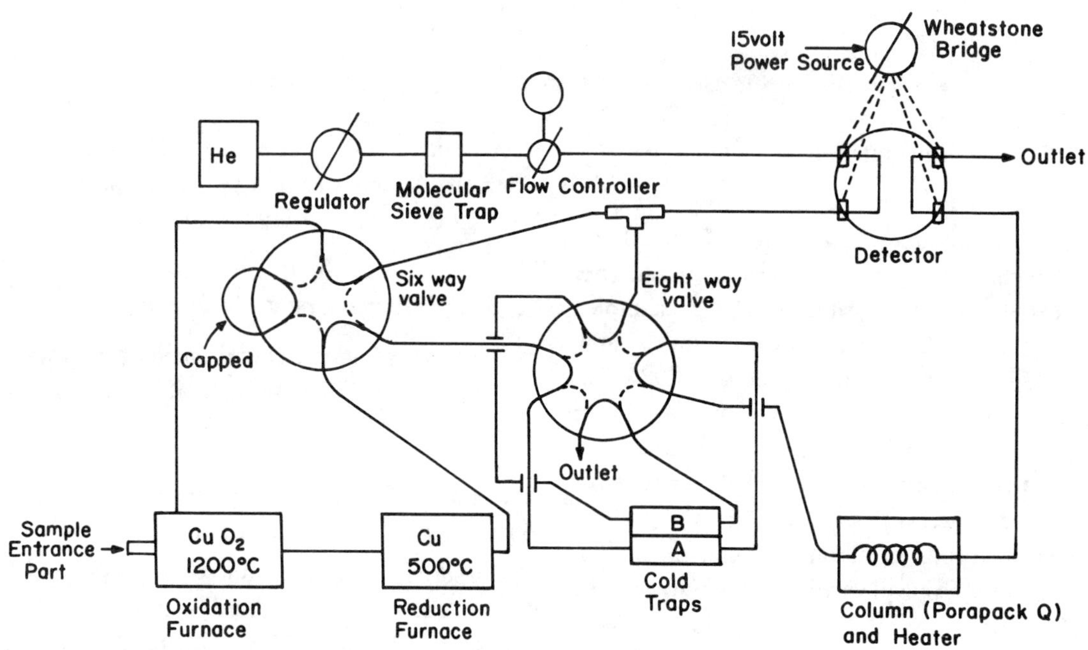

Fig. 3-9. Schematic diagram for combustion analysis of particulate carbon and nitrogen.

gas significantly shorten the column life-time. The tank is fitted with a dual-stage regulator. A high precision flow controller is installed down stream for flow control. The flow rate is adjusted for peak separation and definition. Our system generally runs at 35 psi.

Cold trap. A cold trap which consists of a 6 mm, O.D., Pyrex U-tube, about 30 cm long, is placed in line, downstream of the oxidation-reduction furnaces. The trap is designed to remove H_2O from the gas stream. The trap is immersed in a dewar containing methanol-dry ice slurry when in use. To prevent clogging and associated flow variations, another trap is mounted adjacent to the one in use. This trap is flushed with He and valved so it can be put in-line to replace the first trap without interruption of sample processing. When the second trap is engaged, the first is flushed with He. Trapping out water significantly reduces sample processing time.

2.4. Standards

EDTA is used to standard the humic C-N analysis. The choice of EDTA is discussed by Sharp [1974]. Appropriate concentrations of EDTA solution are prepared and μL quantities are pipetted onto Whatman GF/C filters. Typically a 20 mM EDTA solution is prepared and 10-150 μL is pipetted on to filters. The filters will absorb up to 500 μL of solution.

2.5. Blank Determination

Blanks (background carbon) are determined from the y-intercept of a graph of humic acid carbon as a function of five different volumes of the same sample. Background carbon will vary depending on the DOC of the aquatic environment under study. Estuarine blanks are observed to range between 5-20 μM C [Fox, 1983]. Nitrogen background is typically insignificant, but the same blank determination as used for C should be applied.

2.6. Sensitivity, Precision, Recovery Efficiency

Recovery of humic acid from solutions of humic acids extracted from solids was 98.5% [Fox, 1983].

Detection limits for the method depend on the size of the blank. Typically 0.5 μM C and 0.1 μM N are resolvable. The precision of the analysis is dependent on the uniformity of humic acid aggregation. This is observed to be somewhat variable depending on the organic composition of a particular environment. Variations of from 1 to 10% are observed. We typically claim 10% precision.

References

Fox, L. E., The geochemistry of humic acid and iron during estuarine mixing, Ph.D. dissertation, 219 pp., University of Delaware, Newark, DE, 1981.

Fox, L. E., The removal of dissolved humic acid during estuarine mixing, Estuarine Coastal Shelf Sci., 16, 431-440, 1983.

Gagosian, R. B., and D. H. Stuermer, The cycling of biogenic compounds and their diagenetically transformed products in seawater, Mar. Chem., 5, 605-632, 1977.

Harvey, G. R., D. A. Boran, S. R. Piotrowicz, and C. P. Weisel, Synthesis of marine humic substances from unsaturated lipids, Nature, 309, 244-246, 1984.

Kalle, K., The problem of Gelbstoff in the sea, Oceanogr. Mar. Biol. Ann. Rev., 4, 91-104, 1966.

Schnitzer, M., and S. U. Khan, Humic Substances in the Environment, 327 pp., Marcel Dekker, Inc., New York, 1972.

Sharp, J. H., Improved analysis of "particulate" organic carbon and nitrogen from sea water, Limnol. Oceanogr., 19, 984-989, 1974.

Stevenson, F. J., and J. H. A. Butler, Chemistry of humic acids and related pigments, in Organic Chemistry, edited by G. Eglinton, and M. T. J. Murphy, pp. 534-557, Springer-Verlag, New York, 1969.

Stuermer, D. H., The characterization of humic substances in seawater, Ph.D. dissertation, Woods Hole Oceanographic Institution, Woods Hole, MA, 1975.

Stuermer, D. H., and G. R. Harvey, Humic substances in seawater, Nature, 50, 480-481, 1974.

Thurman, E. M., and R. L. Malcolm, Preparative isolation of aquatic humic substances, Environ. Sci. Technol., 15, 463-466, 1981.

PARTICULATE C, N, P AND SI ANALYSIS AT THE BEDFORD INSTITUTE OF OCEANOGRAPHY, CANADA

W. G. Harrison and E. J. H. Head

Biological Oceanography Division, Bedford Institute of Oceanography, Dept. of Fisheries & Oceans
Box 1006, Dartmouth, Nova Scotia, Canada B2Y 4A2

1. Introduction

The Biological Oceanography Division of the Bedford Institute routinely measures bioelements (C, N, P, Si) in suspended and sinking biogenic particles in support of a number of research programs focusing on primary and secondary production processes in coastal and oceanic waters of the NW Atlantic.

2. Particulate Organic Carbon and Nitrogen

Analytical principle. Automated elemental analysis is conducted using commercial equipment.

Sampling. We collect suspended particles by Niskin bottle or use a shallow water pump system [Herman et al., 1984]. Generally no pretreatment for screening for large particles is done unless warranted by filter inspection. Sedimenting material is collected using the MULTITRAP particle interceptor device and processed after "swimmers" are removed [Knauer et al., 1979]. Subsamples are then filtered under vacuum, at less than 200 mm Hg, onto glass fiber filters (Whatman GF/F, 24 mm) which have been precombusted for 2 hrs at ~ 475° C; higher temperatures make filters brittle and difficult to "pelletize". Filters are then transferred to clean plastic petri dishes and frozen at -20° C and later dried before analysis or dried immediately at 80° C for 2 hrs and stored at room temperature. Samples are run in duplicate or triplicate. Precombusted filters serve as "blanks".

Procedure. Except for "pelletizing" the sample, for accommodation in an instrument autosampler, all steps are automated. Our "operational precision", or mean coefficient of variation of triplicate samples analyses, is ~ 10% for carbon and nitrogen (Figure 3-10). Primary standards are Cyclohexanone-2, 4-Dinitrophenyl hydrazone and/or Acetanilide, weighed on a Cahn 31 ultramicrobalance. Details of standardization and computations are given in Karl et al. (this chapter, page 73).

Marine Particles: Analysis and Characterization
Geophysical Monograph 63

3. Particulate Organic Phosphorus

Analytical principle. This procedure is based on the oxidation of organic-P linkages to orthophosphate by exposure to high energy ultraviolet (UV) radiation. Orthophosphate is subsequently determined by conventional colorimetry. This is a modification of the method of Armstrong et al. [Strickland and Parsons, 1972] as described by Perry [1976]. NOTE: this procedure does not break polyphosphate linkages which requires an additional acid hydrolysis step [Strickland and Parsons, 1972].

Sampling. Sample collection and treatment are essentially the same as for organic carbon and nitrogen except that the glass fiber filters are rinsed with distilled water after combustion and just before sample filtration to remove any residual phosphate. Storage is by freezing at -20° C; there is no drying step in this procedure.

Procedure. Filters are placed in quartz tubes with ~ 50 mL DW and 2 drops of 30% H_2O_2 as an oxygen source. Tubes are then placed in the UV apparatus (with care taken to orient filters with sample material facing UV lamp) and exposed for ~ 1.5 hr. Filtrates are then analyzed for orthophosphate colorimetrically [Strickland and Parsons, 1972]. The commercial UV-oxidation unit (La Jolla Scientific Co.) we use is comprised of a sample chamber accommodating 12-25 x 300 mm quartz tubes, a UV lamp of 1200W, and a power supply/control unit. Analytical range is 0.03 to 2.5 μM but could be extended on the lower end to ~ 0.006 μM using solvent extraction procedures [Strickland and Parsons, 1972]. Precision is ± 0.03 μM at the 0.5 μM level [Strickland and Parsons, 1972]; our mean "operational precision" for duplicate samples is < 10% (Figure 3-10). Glucose-6-phosphate is used for standardization. NOTE: We have found organic-P recoveries with this procedure to be comparable (~ 100%) with those of high temperature combustion methods [Solorzano and Sharp, 1980] over the concentration range we have tested (0.2 to 4.5 μM) although Solorzano (personal communication, 1990) cautions that this may not always be the case; her

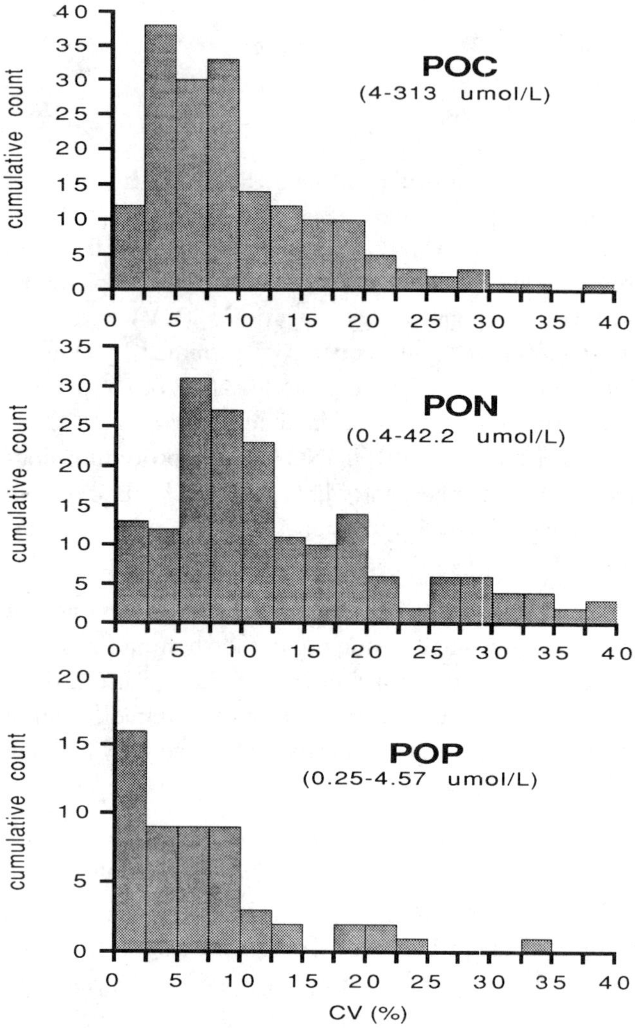

Fig. 3-10. Precision (coefficient of variation) of replicate analyses of particulate organic C, N and P from field samples collected in coastal and oceanic waters of the NW Atlantic.

studies revealed poor recoveries with adenylates (ATP, ADP and AMP, for example).

4. Biogenic (Particulate) Silicon

Analytical principle. Polymeric-Si is chemically converted to soluble-Si which is followed by conventional colorimetric analysis. This is a modification [Conover et al., 1986] of the procedure originally described by Paasche [1980].

Sampling. Sample collection, treatment and storage are essentially the same as for organic C, N, and P except that particles are concentrated on Nuclepore membrane filters (0.2 μm) instead of on glass fiber filters.

Procedure. Filters are heated in polycarbonate or Teflon tubes with 8 mL of 0.5% sodium carbonate at 80-90° C for 2 h. After cooling, samples are neutralized with 10% HCl, added dropwise, using methyl orange, at pH 3-4, as an indicator. Samples are then made up to 10 mL volume with distilled water for colorimetric analysis [Strickland and Parsons, 1972]. Detection limit is ~ 0.03 μM and our "operational precision" for duplicate analyses is ~ 10-20%. Sodium hexafluorsilicate is used for standardization.

References

Conover, R. J., R. Durvasula, S. Roy, and R. Wang, Probable loss of chlorophyll-derived pigments during passage through the gut of zooplankton and the question of superfluous feeding, Limnol. Oceanogr., 31, 878-887, 1986.

Herman, A. W., M. R. Mitchell, and S. W. Young, A continuous pump sampler for profiling copepods and chlorophyll in the upper oceanic layers, Deep-Sea Res., 31, 439-450, 1984.

Knauer, G. A., J. H. Martin, and K. W. Bruland, Fluxes of particulate carbon, nitrogen and phosphorus in the upper water column of the northeast Pacific, Deep-Sea Res., 26, 97-108, 1979.

Paasche, E., Silicon content of five marine plankton diatom species measured with a rapid filter method, Limnol. Oceanogr., 25, 474-480, 1980.

Perry, M. J., Phosphate utilization by an oceanic diatom in phosphorus-limited chemostat culture and in the oligotrophic waters of the central north Pacific, Limnol. Oceanogr., 21, 88-107, 1976.

Solorzano, L., and J. H. Sharp, Determination of total dissolved phosphorus and particulate phosphorus in natural waters, Limnol. Oceanogr., 25, 756-760, 1980.

Strickland, J. D. H., and T. R. Parsons, A Practical Handbook of Seawater Analysis, second ed., 310 pp., Bull. Fish. Res. Board Can., vol. 167, 1972.

PROCEDURES FOR PARTICULATE CARBON, NITROGEN, PHOSPHORUS AND TOTAL MASS ANALYSES USED IN THE US-JGOFS HAWAII OCEAN TIME-SERIES PROGRAM

D. M. Karl[1,2], J. E. Dore[1], D. V. Hebel[2] and C. Winn[1,2]

[1]Department of Oceanography and [2]Division of Biological Oceanography
School of Ocean and Earth Science and Technology, University of Hawaii, Honolulu, Hawaii 96822

1. Summary

A deep-water hydrostation has been established near Oahu, Hawaii as one component of the U.S. Joint Global Ocean Flux Study (U.S.-JGOFS) research program. Since October 1988, oceanographic cruises have been conducted for the purpose of studying the time-varying rates of several biogeochemical processes, including primary production of organic matter and particulate matter export from the euphotic zone. This contribution describes briefly our sampling and analysis protocols with an emphasis on the measurements of the carbon (PC), nitrogen (PN) and phosphorus (PP) concentrations of particulate matter suspended in the water column and the downward vertical flux of PC, PN, PP and total mass (TM) as measured using free-drifting sediment traps.

2. Introduction

In June 1988, a team of scientists at the University of Hawaii received funding from the NSF Global Geosciences Program to establish and maintain a deep-water hydrographic station at a site near the island of Oahu, Hawaii (22° 45′N, 158° W). The Hawaii Ocean Time-Series (HOT) program was funded initially for a five-year period but is expected to continue for at least 10-15 years, and possibly longer. This hydrostation will serve as the open ocean benchmark for combined JGOFS- and WOCE-related research activities, and will comprise the single most extensive investigation of its kind in the Pacific Ocean. The HOT site has been named Station ALOHA (A Long-term Oligotrophic Habitat Assessment) to reflect our primary focus on low-nutrient environments.

Data collected during the first 12 cruises of HOT, from October 1988 to November 1989 have been published [Chiswell et al., 1990] along with a detailed field and laboratory protocols manual which outlines, in addition to the analytical methods employed, the time-series sampling strategy and the HOT program quality control/quality assurance programs [Karl et al., 1990].

Marine Particles: Analysis and Characterization
Geophysical Monograph 63

This contribution discusses our sampling procedures for suspended and rapidly sinking particles and outlines the post-collection treatments and analytical detection systems used for the measurements of particulate carbon (PC), nitrogen (PN), phosphorus (PP) and total mass (TM) in the HOT program. Blanks become very important when conducting low-level measurements of PC, PN and PP.

3. Definitions and Caveats

We define "particles," operationally for all but total mass analysis, as those materials which are retained by a Whatman micro-fine glass fiber filter, catalog #GF/F 0.7 μm nominal retention [Sheldon, 1972], acknowledging the fact that some particulate material is known to pass through this filter. The particle retention characteristics are also known to vary with filter loading and, hence, may not be constant even during the filtration of a single water sample. We also know that certain dissolved compounds are likely to adsorb to the glass fiber filter matrix, or bind to or react with the collected particulate matter, thereby contributing to the total "particle" concentration. While sample blank treatments can be designed to evaluate the approximate magnitude of these effects, no approach is universally acceptable, or accepted.

In presenting our methods, we have focused on total elemental masses and mass fluxes without consideration as to the chemical form of the given element. Consequently, we derive estimates of PC, PN and PP which are not taken to be equivalent to the organically-bound fraction. Methods currently in use to distinguish inorganic from organic C, N and P cannot rigorously distinguish these important particle classes. Additional research will be required in order to resolve these outstanding issues.

In recent years, an important scientific controversy regarding the accuracy of dissolved organic carbon determinations has emerged [Williams and Druffel, 1988]. Historical measurements of C [Sugimura and Suzuki, 1988] and N [Suzuki et al., 1985] dissolved in seawater may be in error due to the presence of as yet unidentified compounds that are not fully oxidized except in the presence of a platinum catalyst. By analogy to these recalcitrant dissolved materials, it is conceivable that similar compounds are present in the particulate or colloidal fractions of seawater. If so, it may be necessary to re-evaluate the "standard methods" of catalysis which have been used historically for the measurements of PC, PN and PP. However, until this present controversy is resolved, or at least until the higher DOC and DON estimates are confirmed by independent laboratories, it may be inappropriate to call into question, a priori, the entire historical literature.

4. General Precautions

Because suspended particle concentrations and particle fluxes in oligotrophic habitats are expected to be low, special attention must be paid to the preparation and processing of water samples and sediment trap collections in order to reduce potentially contaminating particles. For example, sediment traps should be capped immediately after filling and immediately after retrieval. Particular attention should be paid to airborne and/or shipboard particulate contamination sources. In addition, the time interval between trap retrieval and subsample filtration should be minimized in order to limit the inclusion of extraneous abiotic particles and the post-collection solubilization of particles. For water sample analysis, it is important to process large volumes of water in order to increase the signal-to-noise ratio and precision estimates of the determinations.

5. Collection and Processing of Suspended Particles

5.1. Collection

Suspended particulate matter is collected from discrete water depths using a SeaBird CTD-rosette system with external temperature, conductivity, dissolved oxygen and fluorescence sensors and an internal pressure sensor. A General Oceanics 24-place pylon and an aluminum rosette containing twenty four 12 L sample bottles are used to obtain water from predetermined depths. The CTD and rosette are deployed on a 3-conductor cable allowing for the real-time display of data and enabling the collection of water samples from areas of interest relative to hydrographic and biochemical features. The CTD system takes 24 samples per second and the raw data are stored both in the computer and, for redundancy, on VHS-format video tapes. CTD data manipulation and sensor calibration are presented elsewhere [Chiswell et al., 1990; Karl et al., 1990].

Generally, 10-12 consecutive casts to 1000 m are made over a 36 hr period, at approximately 3 hr intervals, to span the local inertial period (~ 31 hr) and three semi-diurnal tidal cycles. This sampling is performed so that tidal or near-inertial variability can be estimated and removed, thereby minimizing aliasing by these components. During this 36 hr intensive sampling period, most of the water samples for the GOFS program objectives are collected. To the extent possible, depth profiles of specific chemical constituents are collected from a single or from contiguous casts. This is done in order to produce profiles which minimize the effects of time-dependent density structure variation within the water column.

The HOT Program 24-place aluminum rosette was manufactured by Scripps Institution of Oceanography's Oceanographic Data Facility (ODF). The 12 L polyvinylchloride (PVC) sample bottles, also manufactured by SIO-ODF, are equipped with Buna-N rubber O-rings, Teflon-coated steel springs and standard General Oceanics sampling valves.

5.2. Processing

Seawater is collected in 12 L sample bottles and transferred directly to acid-cleaned filtration bottles. The samples are transferred via Tygon tubing which incorporates an in-line 202 μm Nitex screen prefilter to remove zooplankton or other rare particles which might otherwise affect the precision of the estimate. The filtration bottles are 4 and 12 L polyethylene aspirator bottles fitted with a valve assembly and tubing connection.

Once the rosette sampler is on deck, the vent valve of each sample bottle is opened and one end of the drawing tube is attached to the outflow spigot of the sampling bottle and the other end to the tubing connector on the cap of the filtration bottle. The filtration bottle valve is opened and 100-200 mL is run through the transfer tube and valve assembly to rinse the filtration bottle. The bottle and cap are rinsed three times. During the rinsing and filling operation, the filtration bottle cap valve is used to control and direct the sample flow. After rinsing, the cap is placed over the bottle mouth (without tightening), the valve opened and the polyethylene bottle filled to the calibration mark.

After filling, the filtration bottles are inverted and placed into the filtration rack. The contents are then pressure filtered (4-7 psi nitrogen gas) through combusted in-line 25 mm GF/F filters. Filters used for the determination of PP are combusted and acid-washed. Following filtration, clean forceps are used to transfer each filter to a clean plastic petri dish containing a ~ 6 cm^2 piece of combusted foil or, in the case of PP, into combusted 12 x 75 mm glass test tubes covered with foil. The sample is labeled and stored frozen, at -20° C, in preparation for the final measurement of PC, PN and PP. Combusted filters and, in the case of PP, combusted/acid-washed filters, are used as procedural blanks.

6. Collection and Processing of Sinking Particles

6.1. Collection

Sinking particles are collected using a free-drifting "Knauer-design" sediment trap array [Knauer et al., 1979; Martin et al., 1987]. Twelve individual 0.0039 m^2 mouth opening polycarbonate particle interceptor traps (PITs) with honeycomb baffles are attached to polyvinylchloride (PVC) cross members and are positioned on the 1/2" polypropylene line at depths of 150, 300 and 500 m. A spar buoy is equipped with a VHF radio transmitter, an Argos satellite transmitter and two xenon strobe lights. Generally, deployments are made for a 72 hr period on each monthly cruise.

Prior to deployment, each trap is rinsed thoroughly with tap water then with deionized distilled water (DDW) before filling with a high density solution to prevent advective-diffusive loss of extractants and preservatives during the deployment period and to eliminate flushing of the traps during recovery [Knauer et al., 1979]. The trap solution is prepared by dissolving 50 g of NaCl in each liter of surface seawater. This brine solution is pressure filtered through a 0.5 μm filter cartridge prior to the addition of 10 mL of 100% formalin per liter of solution. The individual traps are filled and a portion of the trap solution is saved for analysis of the solution blanks.

6.2. Processing: PC, PN and PP

Upon recovery, individual traps are capped and transported to the shipboard portable laboratory for analysis. Care is taken not to mix the higher density trap solutions with the overlying seawater. Trap samples are processed from deep to shallow in order to minimize potential cross contamination. The depth of the interface between the high density solution and the overlying seawater is marked on each trap. The overlying seawater is then aspirated with a plastic tube attached to a vacuum system in order to avoid disturbing the high density solution. Because some sinking particulate material collects near the interface between the high density solution and the overlying seawater, the seawater is removed only to a depth that is 5 cm above the previously identified interface. After the overlying seawater has been removed from all the traps at a given depth, the contents of each trap is passed through a 335 μm Nitex screen to remove the so-called "swimmers" (i.e., contaminating zooplankton and micronekton which entered the traps in a living state and are not truly part of the passive flux). Immediately before this sieving process, the contents of each trap are mixed to disrupt large amorphous particles. The traps are rinsed with a portion of the less than 335 μm sample in order to recover all particulate matter, and the 335 μm Nitex screen is examined to determine whether residual material, in addition to the "swimmers", is present. If so, the screens are rinsed again with a portion of the 335 μm filtrate. After all traps from a given depth have been processed, the 335 μm screen is removed and placed into a vial containing 20 mL of formalin-seawater solution, and stored at 4° C for subsequent microscopic examination and organism identification and enumeration.

6.3. Processing: Total Mass

Three of the 12 traps at each water depth are used for the determination of mass flux. At our shore-based laboratory, triplicate 250 mL subsamples of the time-zero high density trap solution (procedural blank) and equivalent volumes from individual traps (starting with the deepest depth and working up), are vacuum filtered through tared 25 mm 0.2 μm Nuclepore membrane filters prepared as follows: (1) rinse filters three times with distilled water and place rinsed filter on a ~ 6 cm^2 foil square (to reduce static electricity) in a plastic 47 mm petri dish, (2) fold the foil in half over the filter and place the petri dish in a drying oven with the lid ajar for 2 hours at 55° C, remove and cool in desiccator for 30 minutes, (3) weigh filter to constant weight (i.e., repeat oven drying, cooling and weighing until a relative standard deviation of less than 0.01% is achieved), on a microbalance capable of 0.1 μg resolution, record weights (to the nearest 0.1 μg) on label tape placed on top of the petri dish.

After the last of the sample has passed through the tared filter, the walls of the filter funnel are washed with three consecutive 5 mL rinses of an approximately isotonic (1 M) ammonium formate solution to remove seawater salts (note: during each rinse, allow the ammonium formate solution to completely cover the filter). Return the processed filter to its petri dish, record sample number (on the dish and data sheet), and place in a drying oven at 55° C for 8 hours (or alternatively, store in a desiccator, if an oven is not

Table 3-1.
Equipment, Supplies and Reagents for Particulate C and N

Water sampling bottles and optional rosette/ CTD system	Low pressure filtration apparatus (4-7 psi nitrogen gas), or equivalent
Perkin-Elmer Model #2400 (or equivalent)	Carbon/Nitrogen analyzer with integrator
Cahn (or equivalent) electronic balance	O_2 and He gases
Muffle furnace (for 450° C)	Drying oven
Combusted 2.5 cm GF/F filters (450° C, 4 hours)	Ethanol-cleaned forceps for handling filters
Combusted foil (450° C, 4 hours)	Aluminum foil for covering work area
Vacuum filtration system	Clean petri dishes
Acetanilide standard	Formalin (100%)
NaCl	

immediately available), and dry to constant weight (as above).

7. Particulate Carbon (PC) and Nitrogen (PN)

7.1. Principle

Both PC and PN can be measured using commercially-available instruments which detect the gaseous products of high-temperature combustion. The Perkin-Elmer model 2400 CHN analyzer used in the GOFS project, combines the classical Pregal and Dumas methods for the determination of carbon and nitrogen, respectively. The samples are combusted in a pure oxygen environment, the gases mixed and separated by frontal chromatography. The separated gases are determined stepwise by a thermal conductivity detector.

7.2. Analysis

Standard procedures for instrument warm-up are followed. Primary PC/PN standards are prepared using acetanilide (C_8H_9NO; mol. wt. = 135.16). Standards are made to encompass the range ~ 30-600 μg per sample. Instrument conditions include: combustion temperature, 925° C; reduction temperature, 640° C; O_2 gas, 15 psig; He gas, 20 psig. The O_2 used is ultrapure grade (99.995 mole % minimum). The equipment, supplies and reagents are listed in Table 3-1.

8. Particulate Phosphorus (PP)

8.1. Principle

The procedure presented here is a modification of one used by the Hawaii Institute of Marine Biology Analytical Services laboratory at the University of Hawaii. The method relies on the release of organically-bound phosphorus compounds as orthophosphate, by high temperature combustion at 450-500° C. The orthophosphate is then extracted with 0.5 N HCl at 90° C and neutralized with 6 M sodium hydroxide. The liberated orthophosphate is reacted with a mixed reagent of molybdic acid, ascorbic acid and trivalent antimony to form phosphomolybdic acid. This heteropoly acid is then reduced to the colored molybdenum blue complex by ascorbic acid and the solution is measured spectrophotometrically. This procedure measures all forms of reactive phosphorus which can be released by combustion and acid hydrolysis.

8.2. Analysis

Samples are combusted in precombusted, acid-rinsed 12 x 75 mm test tubes at 475-500° C for 3 hours in a muffle furnace. The samples are then allowed to cool and are immersed in 3 mL of 0.5 M HCl. The test tube is then heated for 90 minutes at 90° C in a water bath.

The samples are allowed to cool, vortexed for 20 seconds and vacuum filtered through a combusted acid-washed GF/F filter attached to an acid cleaned syringe. Vortexing and filtration is repeated with three additional 3 mL aliquots of 0.5 M HCl and washes are combined in special acid washed test tubes marked volumetrically at the 20 mL level. After processing all samples, 1 drop of phenolphthalein solution and approximately 1 mL of 6 M NaOH is added to each test tube. While mixing, 0.5 M NaOH is added dropwise to produce a light pink color. The tubes are then filled to the 20 mL mark with DDW.

Two mL of mixed reagent are added to samples and standards. These are mixed thoroughly. Color is developed for 10 minutes and absorbance is read at 860 nm against a DDW reference. Standards are corrected for absorbance of reagent blanks, and samples are corrected for absorbance of filter or procedural blanks. Equipment, supplies and reagents for particulate P are listed in Table 3-2.

Table 3-2.
Equipment, Supplies and Reagents for Particulate Phosphorus

Water sampling bottles and optional rosette /CTD system	Varian DMS-100S spectrophotometer and 1-cm cuvette, or equivalent
Low pressure filtration apparatus (4-7 psi nitrogen gas), or equivalent	Acid-washed 12 x 75 mm glass tubes, combusted (450° C, 3 hr)
Water bath (for 90° C)	Combusted, acid-rinsed GF/F filters
Acid-rinsed glassware	Acid-rinsed vacuum filtering assembly
Glass distilled deionized water (DDW)	Muffle furnace (for 450° C)

Reagents

HCl (0.5 M and 1 M)	
NaOH (0.5 M and 6 M)	
H_2SO_4 (2.5 M)	
ammonium molybdate solution:	Dissolve 15 g of ACS grade ammonium paramolybdate $[(NH_4)_6 Mo_7O_{24} \cdot 4H_2O]$, in 500 mL DDW. Store in plastic bottle in the dark.
ascorbic acid solution:	Dissolve 0.5 μg of ACS ascorbic acid in 10 mL DDW (5.4% wt/vol). Prepare fresh.
potassium antimonyl-tartrate solution:	Dissolve 0.34 g of ACS potassium antimonyl-tartrate (tartaremetic), in 250 mL DDW. Solution is stable for many months.
mixed reagent:	Mix together 10 mL ammonium molybdate, 25 mL 2.5 M H_2SO_4, 10 mL 5.4% ascorbic acid and 5 mL potassium antimony tartrate. Prepare fresh.
phosphate standard solution (1000 μM):	Dissolve 0.1361 g of dry (110° C for 2 hrs) KH_2PO_4, in 1000 mL of DDW. NOTE: Store in a dark bottle with 1 mL of chloroform.
working phosphate standard (100 μM):	Dilute 10 mL of the stock standard to 100 mL, using a volumetric flask. Dilute the working standard to prepare a series of standards to cover the range from 0.05-10 μM.
phenolphthalein indicator	

9. Data Reduction and Calculations

9.1. *Suspended Particles: PC and PN*

Calculate μg per sample of PC or PN from standard curve using:

$$\mu g = (x-b)/m$$

where:

x = (sample counts - blank counts)
b = y intercept of regression line
m = slope of regression line

Calculate $\mu g\ C\ L^{-1}$ or $\mu g\ N\ L^{-1}$ using:

$$\mu g\ L^{-1} = (\mu g\ sample^{-1}) / (volume\ filtered\ in\ liters)$$

9.2. *Suspended Particles: PP*

Calculate $\mu M\ L^{-1}$ of PP from standard curve using:

$$\mu M\ L^{-1} = (x - b)/m$$

where:

x = corrected absorbance of sample
b = y intercept of regression line
m = slope of regression line

Calculate $\mu g\ P\text{-}P0_4\ L^{-1}$ using:

$\mu g\ P\ L^{-1}$ in sample extract = $\mu M\ L^{-1}$ sample x 30.97376 $\mu g\ \mu M^{-1}$

$\mu g\ P\ mL^{-1}$ = ($\mu g\ P\ L^{-1}$ in extract) /(1000 $mL\ L^{-1}$)

$\mu g\ P\text{-}PO_4\ L^{-1}$ = ($\mu g\ P\ mL^{-1}$ x 20.0 mL of extract) / (vol in L filtered)

9.3. *Sinking Particles: PC, PN and PP*

The quantities of particulate C, N and P in the screened trap solutions are determined using methods described above. Six replicate traps are used for C/N determinations and three additional traps for P. Typically, 1.5-2 L are used for a single C/N or P measurement. Equivalent volumes of the time-zero sediment trap solution, filtered through the appropriate filters are used as C, N or P blanks.

C, N and P flux is calculated as follows:

$$\text{mg C (or N, P) m}^{-2}\,\text{d}^{-1} = \frac{[(C_s - C_b)] * V_t}{V_f * A_t * t}$$

where:

C_s = carbon (mg) in sample
C_b = carbon (mg) in blank
V_t = volume of trap (L)
V_f = volume filtered (L)
A_t = cross-sectional area of trap (m^2)
t = deployment period (d)

9.4. Sinking Particles: Mass

Mass flux is calculated as follows:

$$\text{mg (dry wt.) m}^{-2}\,\text{d}^{-1} = \frac{[(W_a - W_b) - W_{bl}] * V_t}{V_f * A_t * 1000 * t}$$

where:

W_a = filter weight after filtration (μg)
W_b = filter weight before filtration (μg)
W_{bl} = net weight of blank solution (μg)
V_t = volume of trap (L)
V_f = volume filtered (L)
A_t = cross-sectional area of trap (m^2)
1000 = conversion factor ($\mu g\ mg^{-1}$)
t = deployment period (d)

10. Precision and Accuracy

Estimates of the precision of our suspended PC, PN and PP measurements are based on a number of replicate samples taken during the first year of the HOT program. The standard deviation of replicate determinations (or the difference if n=2) is divided by the mean to yield a coefficient of variation (cv; expressed as percent); the mean cv for the determinations of PC, PN and PP are 12%, 10% and 18%, respectively. In addition, the accuracy obtained with the PE-2400 CHN analyzer has been estimated from determinations of the C and N contents of reference standards analyzed along with samples during each analytical run. This accuracy, expressed as the mean difference between the analytical determination and the known C and N contents of the standards, is found to be 6% for PC and 13% for PN. Note that standards containing less than 10 μg N and/or less than 70 μg C are not included in this latter calculation; the performance of the PE-2400 is suspect at such low levels, thus sample volumes and trap deployment times must be kept large enough to avoid entering this lower range.

The precision of our PC, PN, PP and TM flux determinations is estimated from replicate traps deployed at each given depth during year one of the HOT program. Again, an overall mean cv is calculated for each analysis. The resulting precisions are 23%, 28%, 32% and 18% for fluxes of PC, PN, PP and TM, respectively. Note that at the lowest fluxes, the analytical blank produces a sizeable fraction of the total signal; similarly, the blank can be large in the analyses of suspended particles from deep in the water column where concentrations are very low. Tabulated below are the typical ranges in analytical blanks encountered during year one, expressed as percentages of the total signal produced by a sample in the course of a given analysis:

Suspended PC:	5-17%
Suspended PN:	5-17%
Suspended PP:	6-29%
PC Flux:	17-54%
PN Flux:	13-55%
PP Flux:	12-43%
TM Flux:	18-41%

Acknowledgements. We thank our HOT program colleagues, and especially U. Magaard, for their assistance and support, L. Lum for her assistance in the preparation of this manuscript and the National Science Foundation Division of Ocean Sciences for financial support.

References

Chiswell, S., E. Firing, D. Karl, R. Lukas, and C. Winn, Hawaii Ocean Time-series Data Report 1, 1988-1989, SOEST Technical Report #1, University of Hawaii, 1990.

Karl, D. M., C. D. Winn, D. V. W. Hebel, and R. Letelier, Hawaii Ocean Time-series Program field and laboratory protocols, University of Hawaii, 1990.

Knauer, G. A., J. H. Martin, and K. W. Bruland, Fluxes of particulate carbon, nitrogen, and phosphorus in the upper water column of the northeast Pacific, Deep-Sea Res., 26, 97-108, 1979.

Martin, J. H., G. A. Knauer, D. M. Karl, and W. W. Broenkow, VERTEX: carbon cycling in the northeast Pacific, Deep-Sea Res., 34, 267-285, 1987.

Sheldon, R. W., Size separation of marine seston by membrane and glass fiber filters, Limnol. Oceanogr., 17, 494-498, 1972.

Sugimura, Y., and Y. Suzuki, A high temperature catalytic oxidation method of non-volatile dissolved organic carbon in seawater by direct injection of liquid samples, Mar. Chem., 24, 105-131, 1988.

Suzuki, Y., Y. Sugimura, and T. Ito, A catalytic oxidation method for the determination of total nitrogen dissolved in seawater, Mar. Chem., 16, 83-97, 1985.

Williams, P. M., and E. R. M. Druffel, Dissolved organic matter in the ocean: comments on a controversy, Oceanogr. Mag., 1, 14-17, 1988.

THE ANALYTICAL DETERMINATION OF MASS FLUX, INORGANIC AND ORGANIC CARBON AND NITROGEN FLUX IN RAPIDLY SINKING PARTICLES COLLECTED IN SEDIMENT TRAPS

George Knauer

University of Southern Mississippi, Center for Marine Sciences, Stennis Space Center, Mississippi 39529

1. Introduction

While many of the following analytical trap-related procedures can be applied in some aspects to a number of trap designs, these particular procedures were developed primarily for used with the small (i.e. mouth opening = 0.0039 m^2) Multitrap design described in Knauer et al. [1979]. This design was chosen to (a) allow for increased replication of up to 16 individual traps/depth, and/or (b) multiple treatments (e.g. can prepare eight traps for organic material analysis, 8 for trace metal analysis each with it's own set of analytical characteristics), (c) to facilitate the removal of contaminating "swimmers" [Knauer et al, 1984] and (d) to facilitate ease in cleaning due to their small size. For additional information on various aspects of sediment trap technology, see U.S. GOFS Working Group [1989].

2. Trap Materials and Preparation

2.1. Materials

Individual trap collectors, baffles and retaining collars are normally constructed of either cast acrylic or polycarbonate. It should be emphasized that all of the materials associated with the collection of particulate matter should be given special consideration. For example, we have found that acrylic-type plastics are extremely difficult to properly clean with respect to metal analysis. On the other hand, if strong organic in-trap solutions such as chloroform are to be used, plastic materials should be avoided since the plastic may be damaged or the samples contaminated with various binders etc. In the case of the use of in-trap organic solvents, we have had success using specially constructed all-glass traps. Typically, acrylic traps are used for the collection of various organic materials; polycarbonate traps are used to collect materials that will be used for trace metal analysis since they are more easily cleaned for metals. Individual collectors have an inside diameter of 7.3 cm and without baffles (see below) are 54 cm in length. The bottom of each individual collector is fitted with a cod-end and polyethylene frit which is used to support a 90 mm, 0.8 μm polycarbonate Nuclepore filter. The cod-ends are machined from ultra-high molecular weight polyethylene rod (rod from U. S. Plastics, valves from Cole-Parmer); the filter frits are cut from 1.6 mm porous polyethylene sheet (35 μm pore size, Scientific Products). The top of each individual collector is fitted with a baffle system designed to reduce turbulent flow across the trap mouth [Soutar et al., 1977]. The baffle system consists of 15 smaller acrylic tubes (1.27 cm I.D.) cemented into a 7.6 cm length of acrylic trap tubing. The tops of each of the smaller acrylic tubes are milled to a wall thickness of 0.06 mm to reduce surface area.

2.2. Preparation

We have found the following cleaning procedures to be adequate in preparing the various trap components used for the collection of bulk mass, inorganic and organic materials. The trap tubes, baffles and cod-ends should be soaked at least overnight in a strong detergent in order to remove associated surface films and other particulate materials. We have found the detergent "Micro" to be sufficient for this purpose. After soaking in detergent, it is important to rinse with tap water in order to completely remove all traces of the detergent; soft waters (e.g. Milli-Q) do not efficiently remove detergents. Following the detergent treatment, all parts must be thoroughly rinsed in tap water, soaked at least 1 hr in 0.5 N HCl and rinsed in Milli-Q water.

The filter frits are not soaked in detergent since their porous nature makes the complete removal of the detergent difficult. Instead, they are soaked overnight in 0.5 N HCl followed by repeated rinsing with Milli-Q water. The Nuclepore membrane filters are soaked overnight in 5.0 N HCl, rinsed once with 5 N HCl followed by thorough rinsing with Milli-Q water. When the cleaning procedure is complete, the trap components are allowed to dry in a clean-air environment and the membrane filters are tared to constant weight in a vacuum desiccator. The trap parts are then assembled. Both the baffle and cod-end assembly (i.e. cod-end, frit and filter) are attached to the trap tube proper with polyethylene tape. Each assembled trap is stored in a

Marine Particles: Analysis and Characterization
Geophysical Monograph 63

plastic bag to avoid contamination by ambient particulate matter.

3. Preparation of Density Solution and Trap Blanks

3.1. Density Solution

While not all trap systems employ the use of density solutions, we recommend their use since they facilitate the retention of trapped materials as well as the maintenance of poisons/preservatives or other solutes in the cod-end. It has been our policy to use seawater collected from an open ocean environment (i.e. oligotrophic waters) as our starter solution since such waters are (a) generally low in DOC and therefore DOC to POC conversions due to agitation/bubbles are minimized and (b) they are naturally buffered and unless gram amounts of material are collected the dissolution of carbonate materials is avoided.

Immediately after seawater collection, using Go-Flo's cleaned as above, 50 g L^{-1} of reagent grade NaCl and 20 mL of 37% formalin L^{-1} are added. This is followed by gravity filtration through a 0.5 μm 10 inch cartridge filter. It is important to filter the density solution after adding the salt and formalin since reagent grade NaCl often contains particulate contaminants and formalin may contain unspecified quantities of insoluble particulate paraformaldehyde. It should also be pointed out that the actual concentration of a given poison/preservative will depend on the length of deployment and the amount of material collected.

3.2. Trap Blanks

It has been our experience that in general, some particulate material is produced de novo during handling either through agitation or bubble formation during the filtration/filling process or during deployment. In the case of the system described here (e.g. it would be a function of the density solution volume), the blank amounts to approximately 100-200 μg which in most cases will present no problem.

Trap blanks are prepared in the standard manner and filled completely to the top with density solution to prevent bubble formation. The traps are then sealed shut using a polyethylene cap and deployed along with the other individual collectors.

4. Post Recovery Preparation, Shipboard

Sediment trap collected materials require extensive microscopic examination to remove contaminating swimmers (see below). In general, this removal process is best performed at a shore-based laboratory. For this reason, samples are only processed to the point of microscopic examination on board ship.

In the case of the system described here (as well as some others), the upper portion of the density solution is often flushed away and replaced by ambient seawater. The degree of flushing depends on such factors as length of deployment, whether the traps were deployed or recovered open or closed, sea state etc. As soon as possible after recovery, this layer of seawater should be removed to prevent possible particulate contamination from this allochthonous source. The ambient seawater-density solution interface can be easily located by gently rocking the collector and looking for changes in optical density caused by turbulence at the interface. The overlying seawater can then be removed by siphon with a length of acid-cleaned Teflon tubing.

After the overlying seawater has been removed, the remaining density solution is drained through the cod-end valve and discarded. In some instances, it may be desirable to retain the density solution since it has been shown that some phases of the trapped particulate material may partition into the surrounding cod-end solutions. This has been shown to be the case for metals for example [Knauer et al., 1984]. Karl and Knauer [1989] have shown this to be the case for POC/PON. However, when certain poisons/preservatives are used, it is difficult to analyze these solutions for organic carbon due, for example, to the overwhelming concentration of formaldehyde.

After the density solution has been removed, the cod-end is removed and any particulate material adhering to the inside walls of the collector are rinsed into the cod-end using a squeeze bottle with prepared density solution. The filter is allowed to drain to a "moist" condition, transferred to a cleaned (90 mm) petri dish, covered and sealed with parafilm to maintain moistness. It is important to rinse the collector with the same density solution used in the initial preparation especially if there are plans to analyze the density solution. Use of alternative solutions, such as distilled water, may alter the concentration of poison/preservative and result in decomposition of the particulate material. In addition, it is important to maintain moistness; if the filter is allowed to go dry, the particles will adhere to both the filter and swimmers making accurate sample preparation impossible. On the other hand, if too much liquid is left in the petri dish, the free liquid can spill over and under the filter edge. The petri dishes should be maintained face up during transport.

5. Laboratory Analysis

5.1. Swimmer Removal

As early as 1979, Knauer et al. reported that sediment trap samples could be grossly contaminated by organisms that were not part of the passive flux and which actively swam into the traps and were killed by the poison/preservative. A partial solution to this problem has been to remove obvious swimmers according to the criteria described in Wheeler [1967] which involves microscopic examination of the preserved samples followed by removal using jewelers forceps. While this method has met with some success [Knauer et al., 1984] some investigators have suggested that this approach is not always possible especially in the case of gelatinous zooplankton or other microzooplankton such as various naupliar stages [Michaels et al., 1990]. Some investigators have tried to solve this problem by using a series of screens [Karl and Knauer, 1989]. More recently, Hedges et al. (Univ. of Washington, personal communication, 1990) have described a system designed to separate active swimmers from the passive material before it reaches the cod-end. Such an approach has been employed by Coale [1990] with some success although it is questionable if these latter approaches can adequately deal with the gelatinous or microzooplankton aspects. In any case, it is our opinion that no single technique has yet been shown to be entirely adequate and sediment trap practitioners must decide on how best to solve this problem based on their own particular needs. At present, it seems that the most efficacious and safest procedure is the microscopic method. This is accomplished using a dissecting scope at 10-50x. While removing swimmers, special care should be taken to remove any particles adhering to individual organisms. It has also been our experience, that in the picking process obvious differences with respect to the use of poisons (e.g. $HgCl_2$, NaN_3) or preservatives (e.g. formalin, glutaraldehyde) are apparent. With formalin, the swimmers are hardened and relatively intact and therefore easy to remove. With poisons, the swimmers are often soft, disintegrate easily and therefore tend to fragment during the removal process.

5.2. Salt Removal

After microscopic examination and swimmer removal, the filter is carefully transferred to a lint-free wipe (Scott Paper Co.) and the sample gently scraped with a scalpel into the center of the filter to form a bolus. During this process, the filter is frequently rinsed using pH adjusted Milli-Q water to remove salts. (The Milli-Q water is prepared by the addition of one drop of 24% NH_4OH solution to 500 mL of Milli-Q water.) We have found this rinse/blot procedure to be quite efficient in salt removal. Because even a small amount of retained salt can greatly influence the gravimetrically determined flux calculation (especially when using small traps deployed for short periods in oligotrophic waters), it is important to remove as much salt as possible. We have found it best to transfer the filter to a new wipe a number of times during the rinsing process.

5.3. Determination of Mass, Total Carbon, Organic Carbon and Organic Nitrogen

Mass flux. The rinsed filter is returned to a clean petri dish and the bolus dried to a constant weight in a vacuum desiccator. If the filter is dried on the petri dish, static electricity charges can be generated when the filter is peeled off the dish. These charges can be strong enough to result in the loss of the sample by repelling it off the filter. To avoid this, an alpha emitting ionizing unit can be used in the balance (Staticmaster). An alternative approach is to cut a small circle from a lint-free wipe and place it between the filter and the petri dish.

The dry weight minus the filter tare weight divided by the days deployed and the trap mouth cross-sectional area equals the total mass flux in (mg m^{-2} d^{-1}). We note that this method accounts for any material trapped in the filter pores and therefore not included in the bolus. This fact becomes important when dealing with small sample sizes.

Total carbon and organic carbon and nitrogen. Following the determination of total mass flux, the dried bolus is scraped from the filter with a scalpel, homogenized in an acetone cleaned agate mortar and pestle, and transferred to a clean shell vial for storage. For organic carbon and nitrogen analysis, a 100-200 μg subsample is weighed into an acetone cleaned silver boat. The boat is placed into a well drilled into a Teflon block and fumed with 100 mL of 12N HCl in a clean desiccator for 36-48 h to remove carbonates [Hedges and Stern, 1984]. To determine total carbon (i.e. organic plus inorganic) a second subsample is weighed into a tin boat but is not fumed. Both samples are then run on a suitable carbon/nitrogen analyzer and organic carbon is calculated by difference. Because of the small sample sizes, we use a Carlo-Erba Nitrogen/Carbon analyzer calibrated with acetanilide as the primary standard. As this analyzer yields results in percent, particulate carbon or nitrogen flux is simply calculated by multiplying the percent times total mass flux which gives results in mg C or N.

References

Coale, K. H., Labyrinth of doom: a device to minimize the "swimmer" component in sediment trap collections, Limnol. Oceanogr., 35, 1376-1381, 1990.

Hedges, J. I., and J. H. Stern, Carbon and nitrogen determination of carbonate-containing solids, Limnol. Oceanogr., 29, 657-663, 1984.

Karl, D. M., and G. A. Knauer, Swimmers: a recapitulation of the problem and a potential solution, Oceanogr. Mag., 2, 32-35, 1989.

Knauer, G. A., J. H. Martin, and K. W. Bruland, Fluxes of particulate carbon, nitrogen and phosphorus in the upper water column of the northeast Pacific, Deep-Sea Res., 26, 97-108, 1979.

Knauer, G. A., D. M. Karl, J. H. Martin, and C. N. Hunter, In situ effects of selected preservatives on total carbon, nitrogen, and metals collected in sediment traps, J. Mar. Res., 42, 445-462, 1984.

Michaels, A. F., M. W. Silver, M. W. Gowing, and G. A. Knauer, Cryptic zooplankton "swimmers" in upper ocean sediment traps, Deep-Sea Res., 37, 1285-1296, 1990.

Soutar, A., S. A. Kling, P. A. Crill, E. Duffrin, and K. W. Bruland, Monitoring the marine environment through sedimentation, Nature, 255, 136-139, 1977.

U.S. GOFS Working Group, G. A. Knauer and V. Asper, Co-chairs, Sediment trap technology and sampling, Report No. 10, November 1988, 94 pp., Woods Hole, Massachusetts, 1989.

Wheeler, E. H., Copepod detritus in the deep sea, Limnol. Oceanogr., 12, 697-701, 1967.

DETERMINATION OF SULFUR IN MARINE PARTICLES

Patricia A. Matrai

Rosenstiel School of Marine and Atmospheric Science
Division of Marine and Atmospheric Chemistry, 4600 Rickenbacker Causeway, Miami, Florida 33149-1098

1. Introduction

An analytical methodology is presented which permits the reliable determination of sub-microgram per liter levels of organic sulfur present in particulate matter (POS) occurring in natural waters. First, particulate matter is collected on filters and combusted at high temperature in an argon/oxygen atmosphere. Second, sulfur is detected by exposure of the reaction product to ultraviolet radiation. This produces a fluorescent emission which is then quantified.

2. Materials and Methods

2.1. Standards

All chemicals used are of reagent grade. Solutions are prepared using deionized distilled water. Standards are stored refrigerated and checked prior to analysis. *DL*-methionine is used as the working standard. Other standards tested include *L*-cysteine, ferredoxin, iota-carrageenan, sodium sulfate (all from Sigma), and sulfur dioxide gas (99.98% anhydrous; MG Industries) [Matrai, 1989].

2.2. Analytical Procedure

Analyses are carried out with a commercial Antek 771 pyroreactor unit coupled with an Antek 701C sulfur detector. Standards are placed into the pyroreactor using quartz spoons. Antek boat carriers have a small, top-loading opening; larger quartz spoons were made to accommodate filters. Prior to analysis, traces of sulfur are removed from the quartz boat carriers by combusting them until no sulfur is detected. Liquid standards are spotted on pre-combusted Whatman QM/A quartz fiber filters, at 500° C during 4 hr, subsequently dried at 60° C for 3 minutes. Analytical blanks are determined by measuring sulfur levels prior to the addition of standards: no sulfur is detected from empty quartz boats or pre-combusted blank filters. A less expensive avenue is to use GF/F glass fiber filters wrapped with pre-combusted aluminum foil to prevent any buildup of residue in the quartz boats after combustion; a lining of quartz wool will also protect the sample carrier from glass buildup. Gaseous standards are injected directly into the pyroreactor.

The standard, carried in a quartz boat, is mechanically introduced into the pyroreactor unit. The most advanced and least costly mechanical injector has consisted of a variable speed drill attached to a positioning clamp which holds the end of the injecting rod; a syringe injector has also been used. Slower and constant sample introduction, e.g., time ~ 15 secs, 10-20 secs, results in better reproducibility. Injection rate and reproducibility have a strong bearing on precision and accuracy. Manual introduction of the quartz boat results in variable operator-related replication.

The standards are combusted in a stream of O_2 and Ar at high temperature (1100° C) within the pyroreactor unit, inside a quartz tube partially filled with quartz chips. This treatment converts all sulfur in the standard to sulfur dioxide (SO_2). The SO_2 produced is carried into the sulfur detector where it is exposed to UV radiation inside the fluorescence chamber. The resulting light emission, proportional to the amount of sulfur present, is sensed by a photomultiplier tube with output to a Hewlett Packard 3390A integrator.

The instrumental conditions used are an argon flow of about 70 mL min^{-1} (setting of 3 on the Antek flowmeter), an inlet-oxygen flow of about 60 mL min^{-1} (setting of 2.5 on flowmeter) and a pyro-oxygen flow of about 350 mL min^{-1} (setting of 3.5 on flowmeter). The Antek sulfur detector has eight sensitivity levels (1- to 2000-fold), of which the 100-fold is used for these standards and most samples of suspended particulate matter; sediment trap samples, generally with higher levels of organic sulfur, need a less sensitive setting. Under these conditions, the time of analysis is about 2 minutes from the moment of injection.

2.3. Sample Collection and Handling

Particulate matter (or cultured phytoplankton cells) is collected by gently (less than 5 in Hg) filtering up to 2 L of seawater (or approx. 5 mL of culture) through a pre-combusted QM/A filter. Gravity filtering of

Marine Particles: Analysis and Characterization
Geophysical Monograph 63

cultured cells decreases the potential loss of intracellular, potentially volatile, sulfur-containing compounds. Samples are desalted by washing with isotonic ammonium formate 3% solution (AF) to prevent cell breakage [Parsons et al., 1961; Omori and Ikeda, 1984]. Care is also taken to rinse the edge of the filter. Once filtration is over, any additional suction and exposure to air of the samples should be minimized to prevent cell breakage. Samples are stored at -20° C in prebaked glass vials.

Prior to analysis filters are first dried at 60° C for 10 min. In the case of filters with cells from phytoplankton cultures, drying time affects the retention on the filter of sulfur compounds potentially volatile such as dimethylsulfoniopropionate (DMSP) (lost as dimethylsulfide, DMS). In the case of DMSP, up to 20% of its intracellular levels may be volatilized with drying times of 1 hour. Lyophilization has been reported to yield an even higher loss of up to 86%, in relation to oven-drying, in the case of sulfur-containing organic compounds present in sediments [Amaral et al., 1989]. Each sample is combusted in the pyroreactor following the same procedure previously described for the standards.

2.4. *Effects of Pyro-Oxygen Flow Rates and Combustion Temperatures*

Oxygen flow rate and combustion conditions have been examined in order to determine optimal operating conditions. A four-and-a-half fold increase in the pyro-oxygen flow rate from about 186 mL min^{-1} to about 830 mL min^{-1} resulted in a decrease of about 60% in the detector response. Inlet-oxygen and argon flow rates were held constant at 60 and 70 mL min^{-1}, respectively. Increasing the pyro-oxygen flow rate quenches the UV-induced fluorescence of SO_2 [Greenough and Duncan, 1961]. On combustion sulfur-containing compounds can yield some sulfur trioxide (SO_3); higher levels of pyro-oxygen may affect the reaction equilibrium $SO_2 + 1/2\ O_2 = SO_3$, favoring the formation of SO_3. This would be at the expense of SO_2 which is the basis of the fluorescence measurement. It has been reported that if the gases leaving the combustion chamber are cooled before detection, as in this case, the SO_2/SO_3 equilibrium is not established which favors the formation of SO_2 [van Grondelle et al., 1978]. However, reducing the pyrolysis temperature from 1100° C to 950° C produced only a small increase (6%) in the detector response. The use of catalysts such as tungsten oxide [Kirsten, 1979] or the presence of reduced copper [Pella and Colombo, 1978] could minimize the formation of SO_3 and enhance the yield of inorganic sulfur-containing compounds (i.e., sulfate) if such determination is desired. A pyro-oxygen flow rate of about 360 mL min^{-1} or lower and a pyrolysis temperature of 1100° C provide maximum response, complete combustion and good separation of the organic and inorganic sulfur fractions. The yield of SO_2 under these conditions is higher than 92%.

2.5. *Rinsing Procedures*

Desalting, which removes the sea salt sulfate, is necessary to enhance the signal of the organic sulfur present in the sample of particles. Sulfate can also be measured by this method; however, its reaction with O_2 is sufficiently slower to allow its signal to be detected as a separate, consecutive peak, if present, during the integration process. Separate detection of each peak requires the use of a separate integrator such as the HP 3390-A since the Antek built-in integrator does not have the capability to separate peaks (i.e., it detects total counts). To insure that there is no interaction between the seawater salt, AF solution, and any sulfur-containing compounds, standard methionine was spotted on filters previously soaked in filtered seawater and rinsed with AF. These were compared to untreated replicates of the standard. It is recommended, however, that field blanks be collected by soaking filters with filtered seawater and rinsing with AF in a manner similar to the samples. No significant differences have been observed with the results from filters treated only with AF and subsequently spotted with the standard, or filters spotted only with the standard.

2.6. *Calibration Procedures*

Standardization of the system is conducted with QM/A filters spotted with increasing amounts of S-methionine. Polynomial standard curves are prepared at each of the routinely used attenuation levels of the detector. The precision of the method is approximately 4% (CV) for sulfur signals greater than 2 μg and 11% for the smallest amount routinely tested (0.5 μg S). The limit of detection is 0.01 μg (11% CV) at the highest sensitivity setting available.

The accuracy of the method has been tested by analyzing known and equal amounts of sulfur in several compounds chosen to represent various sulfur-containing metabolite pools present in phytoplankton as well as gaseous SO_2. No significant differences have been observed among the standard curves of S-methionine and other organic sulfur-containing compounds tested [Matrai, 1989]. Inorganic sulfur, such as that present in sodium sulfate and the ester sulfate carrageenan, has shown a significantly lower signal. This reduces the possibility of sea salt-sulfur contamination.

References

Amaral, J. A., R. H. Hesslein, J. W. M. Rudd, and D. E. Fox, Loss of total sulfur and changes in sulfur isotopic ratios due to drying of lacustrine sediments, Limnol. Oceanogr., 34, 1351-1358, 1989.

Greenough, K. F., and A. B. F. Duncan, The fluorescence of sulfur dioxide, J. Am. Chem. Soc., 83, 555-560, 1961.

Kirsten, W. J., Automatic methods for the simultaneous determination of carbon, hydrogen, nitrogen, and sulfur alone in organic and inorganic materials, Anal. Chem., 51, 1173-1179, 1979.

Matrai, P. A., Determination of sulfur in ocean particulates by combustion-florescence, Mar. Chem., 26, 227-238, 1989.

Omori, M., and T. Ikeda, Methods in Marine Zooplankton, 332 pp., Wiley-Interscience, New York, 1984.

Parsons, T. R., K. Stephens, and J. D. H. Strickland, On the chemical composition of eleven species of phytoplankters, J. Fish Res. Board Can., 18, 1001-1006, 1961.

Pella, E., and B. Colombo, Simultaneous carbon-hydrogen-nitrogen and sulfur microdetermination by combustion and gas chromatography, Mikrochim. Acta, 1, 271-286, 1978.

van Grondelle, M. C., P. J. Zeen, and F. Van de Craats, The combustion-coulometric determination of sulphur in gaseous and liquid organic products, Anal. Chim. Acta, 100, 439-450, 1978.

TOTAL MASS AND PARTICULATE CARBON, NITROGEN AND PHOSPHORUS

Jonathan H. Sharp

College of Marine Studies, University of Delaware, Lewes, Delaware 19958

1. Introduction

Most of my research effort in the past decade has been in estuarine and coastal waters. The routine analyses, although mostly usable for oceanic waters as well, have been aimed at use in these more particle-rich environments. Since estuarine waters are much richer in non-organic particles that cause problems unique to these environments, I will discuss filtration as a separate issue first. In addition to routine mass analyses, I have been involved in natural abundance stable isotope research with Marilyn Fogel at the Carnegie Institute of Washington Geophysical Laboratory for several years. Thus, our procedure for carbon and nitrogen isotope analyses of seston is also described. Our routine procedures are described below.

2. Filtration Prior to Particle Analyses

There has been a great deal of discussion on particle size and choice of filters prior to both particle analyses and analyses of the filtrate. I tried to contribute to a simplification of the subject a number of years ago [Sharp, 1973] for oceanic waters; there have been numerous studies done since. I will restrict my description here to the solutions that we have made for routine sampling in areas ranging from tidal fresh waters with large urban influences to coastal waters. We have tried to use routine procedures for these diverse environments since we often sample this extreme gradient in the period of 2 to 3 days.

Our current procedure does not include any regular screening of samples prior to filtration; any large objects are removed if seen. Samples are generally collected in Niskin bottles on a rosette sampler and aliquots drawn directly into poly or glass bottles to transport to the laboratory for filtration. As much as is possible, we attempt to take all samples (for salinity, dissolved oxygen, dissolved inorganic carbon, pH, seston, chlorophyll, particulate CNP, primary productivity, phytoplankton enumeration, bacterial biomass, bacterial production, nutrient utilization, nutrients, dissolved organic CNP, and humic acid CN) from a single Niskin bottle. We use 1 μm as our nominal particle cutoff by employing 1.0 μm Nuclepore or GF/C filters. I am convinced from years of work in muddy estuary through clear coastal waters that all membrane, including glass fiber, filters have an effective pore size moderated by the amount of material filtered. Only the Nuclepore is truly a sieve with anything approximating an exact pore size. I think that the GF/C filters have a particle retention characteristic that varies with the nature and loading of particulate matter being filtered and probably also varies with "dissolved" organic matter that adheres to the surface. I use the GF/C filters purely for pragmatic reasons and assume that they perform as though they have a pore size of about 1 μm (± 100 to 200%).

Depending upon the analysis, we use 25 mm, 47 mm, or 144 mm diameter filters. The volume of water filtered varies with the attempt to avoid excessive clogging and prolonged filtration. Depending upon analysis, filtration is done with mild vacuum, mild pressure, or by gravity. Generally, Nuclepore filters, due to very low porosity, force us into more prolonged filtration and clogging. Since we are often interested in retaining microorganisms in good physiological condition, we keep mild filtration conditions and as short of duration of filtering as possible.

3. Seston Analysis

3.1. Method

Particulate matter from a water sample is collected on an individually tared Nuclepore filter. Final and tare weights are made on dry filters with a microelectrobalance.

3.2. Procedure

We use 47 mm diameter 1.0 μm pore size Nuclepore filters and gravimetric analysis between final and tare weight. Weights are measured on a Cahn Model 29 Automatic Electrobalance. In earlier work, we used 0.4 μm Nuclepore filters instead of the 1.0 μm. I do not think that there is much difference in seston retained in the estuarine environment between these two filters. I have made a cursory comparison of seston on 0.2, 0.4, and 1.0 μm Nuclepore filters with samples from the tidal

Marine Particles: Analysis and Characterization
Geophysical Monograph 63

river portion of the Delaware Estuary. There was no significant difference between the three filters with this water that is dominated by clays with a carbon content on the order of 4 to 6%.

3.3. Comments

Along the estuarine gradient, we generally find 2 to 100 mg L^{-1} seston content with rare samples close to 300 mg/L. Samples closer to coastal waters with sestons of 2-3 mg L^{-1} usually have 10 to 30% C content; those from the turbidity maximum with sestons of greater than 50 mg L^{-1} are generally in the 2 to 3% C content. We do not measure seston except in estuarine waters. We have made a direct comparison of seston estimates on our Nuclepore filters to those done on GF/C filters with the water quality Standard Methods approach [APHA, 1989]. For samples above 10 mg L^{-1}, the comparison was good; for smaller samples, the Standard Methods approach cannot resolve the numbers due to very high tare weights caused by weighing in a crucible.

4. Particulate Carbon and Nitrogen

4.1. Method

Particulate matter from a water sample is collected on a glass fiber filter immediately upon collection of the water sample. The filter is stored at room temperature in a vacuum desiccator until analysis later on land. Simultaneous C and N analysis is obtained by high temperature pyrolysis of the sample filter with detection of carbon as CO_2 and nitrogen as N_2 by thermal conductivity gas chromatography using a commercial elemental analyzer.

4.2. Procedure

We have been using 2.4 (now 2.5) cm GF/C filters for sample collection and analysis on a Hewlett Packard 185 B CHN analyzer for the past two decades [described in Sharp, 1974]. The procedure was developed for oceanic samples with filtration of 500 to 5000 mL. We now use it in estuarine and coastal waters with filtration of 50 to 2000 mL. The procedure involves rolling of the filter immediately after filtration and storage at room temperature in aluminum tubes in a desiccator under vacuum between collection and analysis.

4.3. Standards and Blanks

The Hewlett Packard CHN analyzer is standardized periodically with disodium EDTA and the standard curve is usually stable for several years. A standard curve of 50 to 100 standards is prepared with distribution of weights evenly over the expected analysis range; usually around 10 to 100 μg C and 3 to 20 μg N. Examples of standard curves are given in Figure 3-11. A small positive intercept is shown in both curves; however, these are not significantly different from the regression lines going through the zero intercept. Regular analysis of samples is preceded by 2 to 3 blank sample rods and 2 to 3 standards to assure that instrument behavior is correct. These are not used in calculating data for samples. Calculations are made using only the slopes of the standard curves with appropriate blank corrections.

The CHN analyzer has a small consistent carbon blank and no nitrogen blank, determined by running a sample boat with "catalyst" in the sample rod. An appropriate sampling blank can be determined by treating a filter that has had no water filtered through it like a sample filter, carrying it through all the handling and storage steps. When this is done, the filter blank has a significant increase in C above the machine blank (about 1 to 5 μg C) and contains no N, like the machine blank.

The limits of detection for the analysis of particulate C and N are determined by a combination of the machine detection limits and blanks. Functionally, the limits are approximately 0.5 μM N and 1 μM C.

4.4. Comments

Routinely, we have not made any attempt to eliminate inorganic C or N in the samples. I know of no serious problems with contamination of particulate samples by inorganic nitrogen in any marine environment. In carbonate-rich oceanic environments, especially those with dominance of the phytoplankton by coccolithophorids, it is routine to try to eliminate inorganic carbon influences. In temperate nearshore environments, there is little concern with inorganic carbon and usually "particulate carbon" and "particulate organic carbon" are used interchangeably.

4.5. Alternate Method

We have recently purchased a Europa GC-MS system for tracer addition C and N uptake work with ^{13}C and ^{15}N. With quantitative filtration, it is possible to obtain amounts of C and N as well as isotopic composition of each element from single samples. We are in the process of an intercalibration of the Hewlett Packard, the Europa, and a Carlo Erba CNS analyzer for particulate C and N analysis.

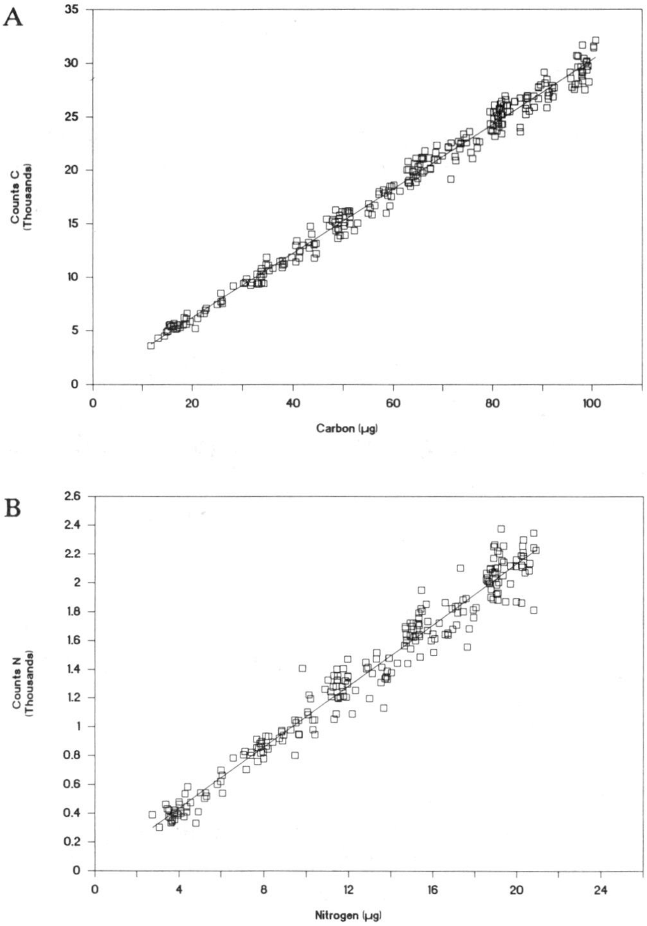

Fig. 3-11. Example of standard curves for the Hewlett Packard CHN analyzer. A. Carbon standardization curve: Y = 301.1X + 219.7 (r^2=0.988). B. Nitrogen standardization curve: Y = 106.3X + 9.54 (r^2=0.964).

5. Particulate Phosphorus

5.1. Method

Particulate matter from a water sample is collected on a glass fiber filter immediately upon collection of the water sample. The filter is treated and dried at sea and stored at room temperature for later analysis on land. Analysis is by high temperature combustion with recovery of phosphorus as phosphate and colorimetric analysis.

5.2. Procedure

We have been using 2.5 cm GF/C filters for sample collection with analysis by the method of Solorzano and Sharp [1980]. The method was developed with attention to 100% recovery of dissolved organic phosphorus compounds in both distilled water and seawater. After filtration, the sample is rinsed twice with 2 mL aliquots of 0.17 M Na_2SO_4 and then placed in a baked glass scintillation vial. It is soaked with 2 mL of 0.017 M $MgSO_4$ (failure to add the salt in fresh and low salinity waters gives poor recovery due to phosphate losses at high temperature) and then dried in a drying oven at 95° C. After storage at room temperature, the vial is baked at 450 to 500° C for 2 hr. After cooling, 5 mL of 0.2 M HCl is added and the vial is heated to 80° C for 30 min to hydrolyse any polyphosphate present or formed at high temperature. After this hydrolysis step, the sample is cooled, the supernatant is transferred to a centrifuge tube along with 5 mL distilled water, to rinse the vial; 1 mL of phosphate mixed reagent is added and the sample is centrifuged for 5 minutes followed by the routine phosphate colorimetric analysis.

5.3. Standards and Blanks

We have analyzed a variety of organic phosphorus compounds and inorganic phosphate salts and have found complete recovery of phosphate [Solorzano and Sharp, 1980]. We find no difference between standardizing by running an organic phosphorus compound through the full procedure and using only the colorimetric phosphate standard. Therefore we routinely standardize only at the final orthophosphate step.

An appropriate blank for particulate phosphorus analysis is made by treating a filter that has had no seawater filtered through it like a regular sample. Unless there is contamination in sample handling or storage, this type of blank gives the same colorimetric absorbance as a deionized water blank for the orthophosphate analysis. We routinely prepare at least one filter blank and carry it through the processing with every batch of samples to verify that there is no contamination problem.

5.4. Comments

In oceanic waters, the particulate phosphorus should be almost exclusively "organic". In estuarine waters, an appreciable portion of the measured particulate matter is probably in inorganic forms, especially phosphate coadsorbed on iron oxides and hydroxides. We have not been able to develop any direct method to segregate between inorganic and organic P. We have used modeling with standard CP ratios from particulate carbon to determine how much of the measured matter is organic [Lebo and Sharp, in press, 1991]; we have also estimated particulate phosphorus pools with sequential leaching [Lebo, in press, 1991].

6. Carbon and Nitrogen Isotopes in Seston

6.1. Method

Particulate matter from a water sample is filtered on a Nuclepore filter and backwashed into a centrifuge bottle. The slurry is then transferred to an aluminum drying dish, dried, and stored in a small vial for later analysis on land. Mass spectral analyses are performed after Dumas combustion and cryogenic distillation. Carbon, as CO_2, and nitrogen, as N_2, are analyzed separately on triple collector mass spectrometers for isotopic composition.

6.2. Procedure

The procedure described in Cifuentes et al. [1988] is used with slight modification. Water samples are collected in Niskin bottles and 10 to 30 L transferred to large PVC reservoirs from which they are forced by nitrogen pressure through Teflon tubing to 144 mm Nuclepore filters of 1 μm pore size. The reservoirs and filter setups are used in a modular manifold fashion so that multiple samples can be processed at one time. After filtration, seston from the filters is washed, with gentle scraping by rubber policemen, into centrifuge bottles and centrifuged immediately on board ship. The slurry from centrifugation is transferred to an aluminum drying dish and dried in a 60° C oven with a nitrogen atmosphere. The dried seston is then transferred to small vials and stored in a desiccator until mass spectral analysis at a later date.

6.3. Comments

It has not been possible to quantitatively collect the seston from the large volume filtration so that this analysis is only for isotopic ratios. The isotope analyses have been done through cooperative effort with Marilyn Fogel at the Geophysical Laboratory of Carnegie Institute of Washington.

References

Cifuentes, L. A., J. H. Sharp, and M. L. Fogel, Stable carbon and nitrogen isotope biogeochemistry in the Delaware Estuary, Limnol. Oceanogr., 33, 1102-1115, 1988.

Lebo, M. E., Particle-bound phosphorus along an urbanized coastal plain estuary, Mar. Chem., in press, 1991.

Lebo, M. E., and J. H. Sharp, Phosphorus distributions along the Delaware Estuary, Estuaries, in revision, 1991.

Sharp, J. H., Size classes of organic carbon in seawater, Limnol. Oceanogr., 18, 441-447, 1973.

Sharp, J. H., Improved analysis for "particulate" carbon and nitrogen, Limnol. Oceanogr., 19, 984-989, 1974.

Solorzano, L., and J. H. Sharp, Determination of total dissolved phosphorus and particulate phosphorus in natural waters, Limnol. Oceanogr., 25, 756-760, 1980.

Standard Methods for the Examination of Water and Wastewater, vol. 16, 1268 pp., American Public Health Association (APHA), 1988.

REVIEW OF PARTICULATE C:N:P METHODS USED BY UH/HIMB ANALYTICAL SERVICES

T. W. Walsh, S. V. Smith, S. Vink

School of Ocean and Earth Science and Technology, and Department of Oceanography,
University of Hawaii, Honolulu, Hawaii 96822

1. Introduction

The primary focus of our laboratory's C:N:P methods was initially geared toward high carbonate (80-99%) sediments. Working with materials containing only a small amount of organic matter in a matrix containing a large amount of inorganic carbon has forced us to develop high-sensitivity analytical methods which minimize interference of inorganic carbon in the determination of organic carbon. These methods can be adapted for materials with lower carbonate percentages. As presented here, the method is separated into two parts: one for total particulate carbon (TPC), particulate organic carbon (POC), and total particulate nitrogen (TPN); the other for particulate phosphorus (PP). Particulate inorganic carbon (PIC) is the difference between TPC and POC.

2. Particulate Carbon and Nitrogen

Particulate carbon in marine sediments and suspended materials includes both organic matter and calcareous materials. Samples with the C fraction dominated by PIC present analytical problems in determining POC and TPN accurately. Hirota and Szyper [1975] suggested that CHN analyses on replicate samples with and without ashing at 500° C could be used to measure PIC and TPC. They then obtained POC by difference. However, we have found that ashing (500° C, 4 hr) to remove POC before analysis of PIC can also remove small amounts of PIC (apparently high-Mg calcite); this loss occurs at temperatures in excess of about 375° C.

This loss is generally minor with respect to estimates of PIC, but can cause very large proportional errors in the determination of POC in samples with a small fraction of organic matter and a large fraction of carbonate minerals. For example in a sample with 0.2% POC and 10% PIC, a release of 5% of the PIC as CO_2 during thermal oxidation of organic matter would increase the estimated POC to 0.7%.

Marine Particles: Analysis and Characterization
Geophysical Monograph 63

It has also been established that organic oxidation of refractory compounds in sediments may not be complete at temperatures below about 1050° C [Gibbs, 1977]. Because of the relative thermal lability of some carbonate phases and refractivity of some organic matter, any thermal oxidation within this temperature range will release incompletely both POC and PIC to CO_2.

Direct analysis of PN in carbonate-rich sediments is also problematical, for the following reason. The POC content of calcareous sediments is typically well below 1%, and the N content of such samples is typically well below 0.1% which is below nitrogen detection limits for samples of sufficiently small mass to analyze with an elemental CHN analyzer. We have also found that acidification to remove carbonates before CHN determinations of POC and PN will convert significant fractions of the sediment organic matter to dissolved organic carbon (DOC), and total dissolved nitrogen (TDN).

We have therefore modified methods developed by previous researchers [e.g., Weliky et al., 1983; Sandstrom et al., 1986]. In general, CHN analysis (1100° C with catalyst) is used on an aliquot of the sample to measure TPC; this standard technique requires no further explanation. Aliquots of the samples are then acidified to remove carbonate minerals. CHN analysis is then performed on the filterable acid residue to measure sample POC and PN, uncorrected for organic matter released during acidification. Dissolved organic carbon (DOC) and total dissolved nitrogen (TDN) analyses are performed on an aliquot of the acid filtrate. The DOC and TDN, adjusted back to the total volume of that filtrate and corrected for acid and reagent blanks, are added to the uncorrected POC and PN, to give corrected sample POC and PN. PIC is determined as the difference between TPC and POC.

Apparatus Required:

1) Autoanalyzer (or spectrophotometer) for the measurement of NO_3 and NO_2, or a total N analyzer capable of measuring TDN in the 10-100 μM concentration range.
2) Total Carbon Analyzer or other analytical technique for DOC or TOC.

3) Photoxidation unit (or screw capped test tubes, rack, and autoclave; or other analytical technique) for TDN digestion.
4) Elemental CHN analyzer for POC and PN measurements.
5) Grinder or mortar and pestle for homogenizing sediment samples.
6) Precombusted (500° C, 4 hours) glass fiber filters (Whatman GF/F 2.1 cm) and Al foil squares.
7) Precleaned (methanol wiped) plastic CHN petri dishes.
8) Precleaned (10% HCl rinsed) 125 mL LPE or PP bottles.
9) Millipore precleaned (10% HCl) glass filtration setup.

Reagents Required:

1) Distilled deionized water (DDW).
2) 6 N HCl solution.
3) 1 N $NaHCO_3$ solution.
4) Sulfanilamide, NED, H_3PO_4 (85%), NH_4Cl for NO_3 & NO_2 analysis.
5) Sodium persulfate for DOC analysis.

2.1. Procedure (For Dried, Ground Carbonate Sediments)

1) Preclean (10% HCl rinse and 4-5 DDW rinses) 125 mL LPE bottles (~ 2 times the number of samples).
2) Weigh out ~ 250 mg of ground homogeneous sample sediment into each bottle. Also add ~ 250 mg of reagent grade $CaCO_3$ into 2 or 3 "blank" bottles.
3) Add 2 mL of 6 N HCl and swirl acid around to contact sediment; allow to bubble and dissolve ~ 10 minutes.
4) Add 20 mL of DDW, cap and swirl and shake vigorously.
5) Add 60 mL of DDW, cap, swirl and shake vigorously, let stand for at least 3-4 hours; repeat shaking once each hour; leaving overnight is satisfactory, but not recommended.
6) Filter each solution (~ 82 mL) through a precombusted GF/F filter into a clean (10% HCl) 250 mL filter flask. Use two ~ 10 mL DDW aliquots as final rinses; first rinse the sample bottle and pour into filter funnel and then rinse only the funnel to lower the residual acid on the filter). One or two 5 mL additional rinses are preferred.
7) Save filter for CHN analysis by storing on the Al foil square into a petri dish and dry at 65° C for ~ 48-72 hours; note initial sediment weight on dish.
8) Partially neutralize the filtrate by adding ~ 8 mL of 1 N $NaHCO_3$; if a white precipitate forms, add 6 N HCl dropwise until the precipitate dissolves.
9) Pour filtrate into a clean tared sample bottle (from step 1), weigh and record total volume (x mL) of sample digest. Pour out some sample to allow ~ 1/3-1/4 head space, cap, and store frozen until TDN and DOC analyses.
10) Treat the "blank" bottles similarly and store with each sample batch.

2.2. Analytical Measurements

1) Analyze filters on a CHN analyzer for particulate carbon (POC) and nitrogen (PN) content (in μg-C and N).
2) Analyze filtrates for TDN concentration using the photoxidation (uv-irradiation) digestion method followed by autoanalyzer NO_3 and NO_2 analysis, or alternative TDN technique. Convert TDN to μg-N (in x mL of filtrate).
3) Also measure filtrates for DOC concentration using the O. I. Model 700 carbon analyzer (or alternative technique). Convert DOC to μg-C (in x mL of filtrate).
4) Calculate final results:

$$\%POC = (\mu g\ POC + \mu g\ DOC / \mu g\ sample) \times 100$$
$$\%PN = (\mu g\ PN + \mu g\ TDN / \mu g\ sample) \times 100$$

In 1, 2 and 3 above each sample must be corrected for the average batch blank.

2.3. Analytical Notes

1) A persulfate oxidation may be used in place of the UV technique for TDN analyses.
2) The pH of the final filtrate should be between 3 and 7. Adjustments will vary with sample type.
3) The PP analysis, as described below, should be on a subsample of the same sediment used for C and N.
4) Typical reagent $CaCO_3$ blanks give values less than 0.01% POC, < 0.002% PN, and < 0.003% PP.
5) This method has been applied to a variety of carbonate sediments. There is no reason that this method could not be applied to suspended material or glass fiber filters.
6) Attempts to estimate PP by this method were unsuccessful, primarily due to interference problems when analyzing the acid filtrate via UV oxidation for TDP.

TABLE 3-3.
C:N:P Method, Hawaii Institute of Marine Biology Analytical Services; Typical Results

	---------Filter----------			------Filtrate------		Ash	-----------Particle-----------		
SAMPLE I.D.	Wgt (mg)	μg POC	μg PN	μg DOC	μg TDN	μg PP	%OC	%N	%P
BLANKS:									
Reag A	0	46	0.1	25	7	0.4			
Reag B	0	49	0.1	29	8	0.3			
$CaCO_3$									
Blank Corrected									
$CaCO_3$ A	250	8	0.1	3	5	6	0.005	0.002	0.002
$CaCO_3$ B	250	31	0.1	7	4	8	0.011	0.002	0.003
Carbonate Sediments									
FN-30	250	367	54	135	25	42	0.201	0.032	0.017
FN-26	250	398	52	227	30	43	0.250	0.033	0.017
FN-26	250	347	39	233	33	39	0.232	0.029	0.016
FN-22	250	1003	216	245	46	69	0.521	0.105	0.028
FN-22	250	1107	243	263	46	71	0.548	0.116	0.028
Other Sediments									
10-1	247	211	24	130	21	**	0.138	0.018	0.070
10-2	271	285	31	188	35	**	0.175	0.024	0.071
3	247	309	42	156	36	**	0.189	0.032	0.051
6	255	123	12	156	26	**	0.109	0.015	0.076
18	260	450	56	146	21	**	0.229	0.030	0.077
22-1*	139	761	152	124	45	**	0.636	0.142	0.045
22-2*	136	538	111	127	49	**	0.489	0.117	0.046
15	137	74	12	92	28	**	0.121	0.021	0.007
33	247	191	29	85	17	**	0.112	0.019	0.011
26	250	400	53	143	33	**	0.217	0.034	0.019
30	146	205	36	69	22	**	0.188	0.039	0.027

Notes: * This sample not homogeneous enough (fine powder preferred).
** These PP's done on various weights.

3. Particulate Phosphorus

The method presented here is similar to methods employed by soil scientists [Aspila et al., 1976; Greweling, 1976] and was modified for application to samples including marine carbonates, powdered organic tissues, and seawater particulate material collected on glass fiber filters. Additionally, the method has been used on Hawaiian soils, and coral skeletons/tissue/mucus. The daily PP calibration standard used is the NBS standard orchard leaves (0.21% P). We have found this method to be superior to other wet chemical digestions, in terms of oxidation completeness, range of applicable sample types, and analytical simplicity. We have also found that acidification alone will extract >90% of P in sediment samples, similar fractions of P in shallow water suspended material, and 30-50% of P in various organic material.

Apparatus Required:

1) Muffle Furnace - capable of 500-550° C.
2) Porcelain crucibles - 2 gram size is good; can fit 20-25 of these into a Pyrex dish. Precombust crucibles at 500° C for 4 hours.
3) Small glass test tubes for filters.
4) 50 mL graduated (marks at 12.5, 25, 35 and 50 mL) test tubes (20-40 or so).
5) Vortex mixer for 50 mL tubes.
6) Glass or Oxford pipettes (2 mL).
7) Autoanalyzer or Spectrophotometer set up for PO_4 colorimetry.
8) Grinder or mortar and pestle.
9) Boiling water bath (optional).

TABLE 3-4.
PP Method, Hawaii Institute of Marine Biology Analytical Services; Typical Results

SAMPLE Type / ID	PIC (%)	Wgt (mg)	[PW] (μM)	Ash + Acid % PP	Acid only % PP	Acid Soluble% of total P
$CaCO_3$	12	144	0.07	0.0007	0.0007	100
Carbonate Sediment						
X-10	~ 11	311	4.97	0.0216		
X-19	~ 11	219	3.89	0.0240		
X-20	~ 11	190	3.32	0.0237	0.0219	92
X-24	~ 11	191	2.90	0.0206		
X-23	~ 11	220	3.61	0.0222		
X-29	~ 11	198	4.67	0.0320		
X-30	~ 11	217	4.88	0.0305	0.0301	99
X-30	~ 11	214	5.10	0.0324	0.0306	94
X-31	~ 11	210	5.67	0.0365		
Miscellaneous						
Kahana Bay	6	*	0.0677		0.0617	91
S. D. Bay		0.4	*	0.0533	0.0456	86
Pel. Clay	0.1	*	0.2081		0.2011	97
Grasses						
#62 Ratail	~ 0	92	4.03	0.0593		
#62 Ratail	~ 0	90	3.96	0.0598		
#63 Ratail	~ 0	115	4.53	0.0534	0.0197	37
Algae						
X-30 ALG	~ 0	82	5.38	0.0891		
X-29 ALG	~ 0	130	8.51	0.0886		
Sargassum	~ 0	46	4.96	0.1448	0.0436	30
Miscellaneous Suspended Material						
Plankton	~ 0	*		0.1220	0.0606	50
Sed Trap	~ 0	*		0.0828	0.0790	95
CG susp.	~ 0	*		0.0414	0.0415	100
TP susp.	~ 0	*		0.0796	0.0673	85
HI susp.	~ 0	*		0.0234	0.0208	89
Stream	~ 0	*		0.0429	0.0328	76
NBS Standard Orchard Leaves (0.21% P)						
Std 1	0	34	5.38	0.2100	0.1091	52
Std 1	0	72	11.13	0.2079	0.1055	51

NOTES: 1) * Analyses done on various sample weights.
2) X sediments from Christmas Island.
3) Clays from Hawaii; Suspended from Tomales Bay, California.

Reagents Required:

1) Distilled deionized water (DDW), or least low-PO_4, distilled water ($<0.2 \mu M\ PO_4$).
2) 10% (v/v) reagent HCl, for cleaning.
3) 6.0 Normal reagent HCl.
4) 0.5 Normal $NaHCO_3$.
5) PO_4 reagents (for whatever method is used).

3.1. Procedure (For Dried Ground Materials/Filters)

1) After precombusting, preclean ashing crucibles by soaking in 10% HCl overnight, rinsing with DDW, and drying, at 105° C, upside down in oven.
2) Weigh out the ground samples into pre-weighed crucibles according to expected % P concentrations; record weights.

Sample Type	% P levels µg-at/L	Weight to use (mg)
Carbonate Sediment	0.01 - 0.1	100 - 50
Other Sediments	0.05 - 0.5	50 - 15
Algae/Grasses	0.1 - 0.5	30 - 15
SW Filters (glass)*	0.01 - 0.5	Record volume.

Notes: Number crucibles in pencil on bottom before weighing. *Glass filters with water column particles should be folded with material facing inside and can be cooked in small glass test tubes (in metal rack - SEM sample order in rack) or wrapped in Al foil or in crucible, if it fits.

3) Place crucibles in a Pyrex dish, cover with Al foil and bake at 500-550° C in a furnace from 4-5 hours to overnight. Remove and cool. The samples should be run within a day of ashing.
4) During ashing, preclean 50 mL test tubes with 10% HCl and DDW. Start up Autoanalyzer for PO_4 analyses with range of 0.1-10 μM PO_4.
5) Acidification of Ash: Dump the ash material or filter out of each crucible into numbered 50 mL test tubes. Note: if there is only a small amount of ash left over you can add a drop of 6 N HCl into crucible with a little DDW and pour into tube - watch contamination from crucible sides/fingers. Add 2.0 mL of 6 N HCl to each tube and bring volume to the 25 mL mark with DDW, rinsing down any ash on the side of the tube. Mix ~ 20 seconds on a vortex mixer.
6) Acid hydrolysis, optional: Place 50 mL covered test tubes into a hot water bath (90° C) for ~ 1 hour then cool. We rarely use this step; there is little change during hydrolysis.
7) Neutralization and dilution: Pipette 2.0 mL of the 25 mL digest solution into another clean 50 mL test tube. Add 2.0 mL of 0.5 N $NaHCO_3$ and dilute to the 35 mL mark with DDW. Mix for ~ 20 seconds on a vortex mixer.
8) Analyze the 35 mL product (PW) waters for PO_4 concentration by your preferred method. PO_4 will probably be in the 2 to 10 μM PO_4 range for sediments, etc., while filters will be probably much lower, usually less than 1 μM PO_4. Note: the blank, all but no ash added, will be very low - near 0.1 μM PO_4.
9) Calculate the final PP - sample concentration as follows:
   A) To get concentration of 25 mL digest water (DW) in μM/liter:

   $$([PW]-[Blank]) \times 17.5 = [DW]$$

   B) To get total μg-P in sample:

   $$[DW] \times 0.025 \times 31 = \mu g\text{-}P/sample$$

   C) And therefore:

   $$\%PP = \frac{mg\text{-}P}{mg\ sample} \times 100$$

3.2. Analytical Notes

1) One of the few problems with this method is too low or too high pH in the final product water being measured. This pH problem will vary a little according to the sample type and colorimetric method used. We use an AAII-Technicon seawater PO_4 method which is slightly modified; a pH range between 3 and 8 seems fine. Carbonate sediments are usually in the good range while organics and filters tend to be lower, sometimes too low (pH<2). Watch AA peaks; if they are sharp and/or jagged, adjust pH up with a little more bicarbonate. We have also used filtered seawater (<0.2 μM PO_4) in place of DDW as a diluent and the peaks were excellent; however, this uses a lot of seawater.
2) The biggest problem with filter seawater particles is usually the low mass of P collected on a given glass filter relative to the blank. Filtering large volumes is obviously desirable; however, when clogging starts, time becomes a factor. The best notes for filters are:
   a) If you can see an obvious colored spot, PP will be measurable.
   b) Take at least three to five blank filter samples.
   c) Like a CHN sample, the sample bottle must be constantly mixed while filtering; otherwise, duplicate sample filters can be surprisingly very different.
   d) With filters, all levels are "very low" i.e. - blank filters of ~ 0.1-0.2 μM-PO_4 with samples of ~ 0.3-1.0 μM. Everything in contact with these filters must be micro-clean (precombusted and HCl/DDW cleaned filters and HCl/DDW cleaned glassware. Store folded inside a HCl/DDW cleaned small glass test tube - cover with Al-foil and freeze.
3) The ashed material, especially for high-organic samples, can be extremely light and flaky, hence easily blown out of a small crucible. Each crucible

containing these samples should therefore be covered with Al foil.

References

Aspila, K. I., Agemian, H., and A. S. Y. Chau, A semi-automated method for the determination of inorganic, organic, and total phosphate in sediment, Analyst, 101, 187-197, 1976.

Gibbs, R. J., Effect of combustion temperature and time, and of the oxidation agent used in organic carbon and nitrogen analyses of sediments and dissolved organic material, J. Sediment. Petrol., 47, 547-550, 1977.

Greweling, T., Chemical analysis of plant tissue, Search Agriculture, Agronomy, 6, Cornell University Ag. Exp. Sta., N. Y. State College of Ag. and Life Sciences, Ithaca, 1976.

Hirota, J., and J. P. Szyper, Separation of total particulate carbon into inorganic and organic components, Limnol. Oceanogr., 20, 896-900, 1975.

Sandstrom, M. W., F. Tirendi, and A. Nott, Direct determination of organic carbon in modern reef sediments and calcareous organisms after dissolution of carbonate, Mar. Geol., 70, 321-329, 1986.

Weliky, K., E. Suess, C. A. Ungerer, P. J. Müller, and K. Fischer, Problems with accurate carbon measurements in marine sediments and particulate matter in seawater: a new approach, Limnol. Oceanogr., 28, 1252-1259, 1983.

SAMPLING AND ANALYSIS OF MARINE PARTICLES WITH PEBENOCO (PELAGIC-BENTHIC COUPLING IN THE NORWEGIAN COASTAL ZONE), UNIVERSITY OF TROMSØ, NORWAY

Paul Wassmann

Norwegian College of Fishery Science, University of Tromsø, N-9001 Tromsø, Norway

1. Introduction

The sampling and analysis of bioelements at the University of Tromsø are more or less carried out according to standard methods. These can be studied in the literature cited in this contribution and are discussed in the workshop report of this chapter. Below, more emphasis will be given to deviations from these standard methods.

2. Sampling

2.1. Water Column

Water column sampling is done with PVC-Niskin bottles (singular or on rosette) at fixed depths throughout the water column or at depths which have been chosen following the examination of water column structure such as density and in vivo fluorescence.

2.2. Sediment Traps

Three types of sediment traps are used: single, cylindrical traps (H = 500 mm, H/D ratio = 5), double, cylindrical, gimbaled traps (H = 1800 mm and 600 mm, H/D = 10 and 6, respectively) and funnel-shaped, automatic Honjo-type (Mark-7). While the simple traps are used at various depths at anchored moorings, the automatic traps are used both anchored and free-drifting. Until recently, either chloroform was added to the traps or there were no additives.

The use of chloroform in sediment traps is not advisable because of a) the poor recovery of soluble particulate matter, e.g. pigments and lipids, b) the apparent attraction of swimmers and c) problems with the splitting of samples [Gundersen and Wassmann, 1990]. Recent observations on substantial respiratory losses of carbon in non-preserved sediment traps indicate that care must be taken to correct for these losses even during short-term deployments (days), especially at in situ temperatures greater than 10° C (Figure 3-12). Despite all the caveats introduced by the use of preservatives/poisons during sedimentation measurements, the use of additives is necessary to limit leakage and stop microbial degradation as well as zooplankton grazing. Therefore, an additive favorable to the parameters of interest has to be selected. Because bulk bioelements like C, N, pigments and Si are of main interest for PEBENOCO, formaldehyde (2% final concentration) in 40 ‰ NaCl solutions is used in the traps. The dense solution is introduced into the bottom after filling the traps with seawater by using a 250 mL syringe and a tube. Care is taken that the addition of the preservative containing brine does not decrease the aspect ratio of the traps by more than about 10-20%.

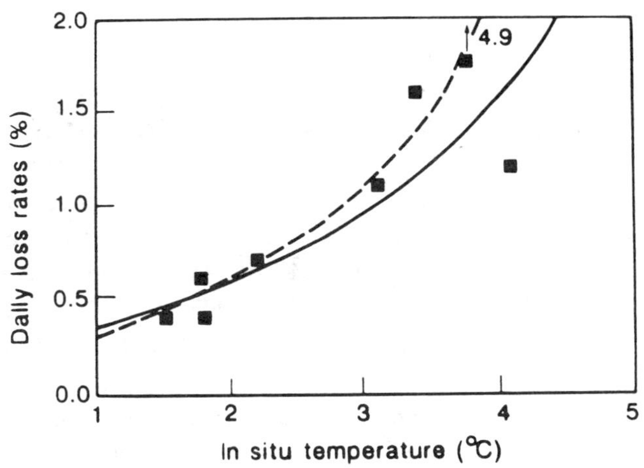

Fig. 3-12. The daily loss rates of sedimenting POC by respiration (carbon-specific ETS activity) in the Barents Sea as a function of in situ temperature. Best fit to the distribution out of three models is provided by exponential models. Data from Wassmann et al. (submitted, 1991).

3. Analysis

3.1. Dry Weight

Samples are taken on pre-weighed, 25 or 47 mm GF/F filters. Subsamples are taken from carefully hand stirred water bottles or sediment-water suspensions (bottle volume of 1 to 3 L) by using measuring cylinders or Bird pipettes. Subsample volumes range from 250 to 500 mL. The filters are washed after filtration with 20 mL of distilled water to remove salt, and stored in a freezer. The samples are dried at 60° C. Ash-free dry

Marine Particles: Analysis and Characterization
Geophysical Monograph 63

weight is determined at 450° C. This is lower than the Norwegian as well as other national standards which is normally 550° C. Magnesium rich carbonates evaporate at relatively low temperatures and combustion at too high temperatures gives rise to weight losses of inorganic matter which is interpreted as organic matter. For general discussions of dry weight determinations, see Dean [1974], Mook and Hoskin [1982] and Dankers and Laane [1983].

3.2. Carbon and Nitrogen

The filtration of water samples and sediment trap samples on GF/F glass fiber filters follows the general description (see above). Subsample volumes range from 25 to 250 mL. The filters are treated with fumes of concentrated HCl prior to analysis to remove carbonates. The analysis is carried out with a Carlo Erba analyzer or a Leeman Lab 440 analyzer.

3.3. Phosphorus

Filtration, as described above, on GF/F glass fiber filters. Subsample volume is 100 to 250 mL. A wet oxidation method in Sovirel bottles under pressure (1 atm) and at temperatures of 100° C is used following the descriptions of Grasshoff [1976]. This method is not able to differentiate between biologically available P and total P. The recovery of particulate P from sediment traps is poor, especially during lengthy periods of trap exposure due to rapid dissolution inside the traps. P analysis of sedimented matter is, therefore, not routinely executed by PEBENOCO.

3.4. Silicate

Water and sediment trap samples are filtered onto Nuclepore filters (0.4 to 1.0 μm pore size) and stored in Nunc tubes for analysis. Care is taken that all the equipment used is free of glass. Subsample volumes range from 25 to 100 mL. The digestion of particulate silicate (PSi) is described by Paasche [1980] and Krause et al. [1983]. Analysis of dissolved silicate follows Grasshoff [1976].

4. Comments

4.1. Screening and Mucilaginous Algae

Screening to remove certain particles from the suspensions has never been used. However, care is taken that larger "swimmers" are removed from sediment trap samples prior to analysis. In the areas studied by PEBENOCO contamination of sediment traps by "swimmers" is often of minor significance during major parts of the year. However, during summer, "swimmers" can be present in traps in larger amounts. The main reason for not using screens is the presence of large sized detritus and algae particles which clog screens.

In the coastal and oceanic waters of Norway large amounts of the gelatinous, colonial algae *Phaeocystis pouchetii* and their aggregates make filtration difficult due to clogging of the filter by mucilage [Lancelot and Mathot, 1985]. Presumably most of this mucilage will not be retained by the filters. C sedimentation rates during *P. pouchetii* blooms are, therefore, underestimates because it is not clear to what respect DOC associated with the particulate flux is missed from the filters [Wassmann et al., 1990]. Care was taken that the negative pressure during filtration does not increase to more than 0.4 kg cm^2. Filtration of each sample does usually take less than 20 minutes. While water samples are stored cool and in the dark, filtration is normally completed less than 6 hours after sampling.

The role of DOC associated with particulate matter is an important issue which has to be addressed in future sedimentation studies. What part of the DOC can be considered as "particulate" and how can we study it? Work with mucus-rich algae as e.g. *Phaeocystis pouchetii* reveals that the pools and fluxes of carbon are significantly underestimated by a yet unknown amount of DOC.

4.2. Variances Between Replicate Sediment Traps and Subsamples

One of the most significant problems when analyzing the particulate composition of sediment trap samples is the variance between replicate sediment traps and samples. The number of sample replicates is often limited. Studies with replicate sediment traps are rare. The variability between samples and traps is mostly unknown and, as a consequence, it is normally impossible to decide whether sedimentation rates between stations and depths are significantly different. As a result, comparative presentations of sedimentation rates often are over-interpreted.

In a study of the vertical flux at three depths and five stations in a land-locked fjord in western Norway it was found that the variance between subsamples was generally greater than that between traps, depths and stations. The only way to study these differences in sedimentation is to improve the precision of the analytical procedures in the laboratory including the precision of splitting and subsampling.

4.3. Deployment Time

The effect of deployment time on the samples has to be considered with great care. A comparison of the sedimentation rate of both POC and chlorophyll *a* during consecutive time intervals ranging from 7 to 21 days in a small, west-Norwegian fjord indicates that the rates of short term intervals (average of 7d + 7d, 7d + 11d and 7d + 7d +7d) were, in 20 out of 24 cases, higher than rates derived from the longer exposure times (14d, 18d and 21d) [Gundersen and Wassmann, 1990]. This indicates that an increasing percentage of particulate organic matter is lost with increasing deployment time, even when preservatives are added. Particulate losses seem to be increasingly significant for deployment times greater than 10 days where loss rates of up to 60% have been recorded (Figure 3-13).

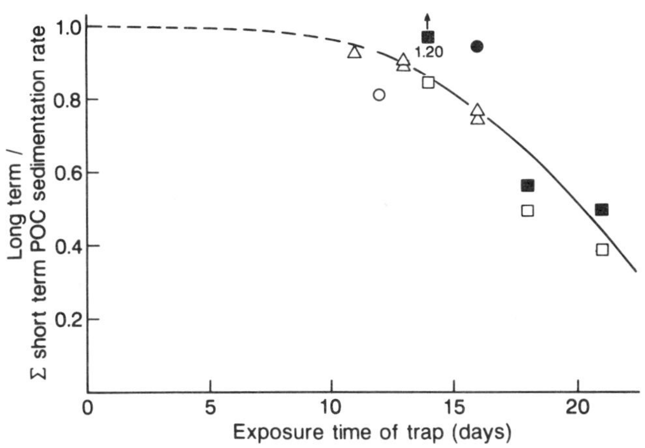

Fig. 3-13. Ratios of POC sedimentation rates derived from long term (11-21 days) versus Σ of short term measurements (2-11 days) as a function of time. Data from five studies, see Figure 7 in Gundersen and Wassmann [1990] for further information. The given line is fitted by eye.

4.4. Mineralization in Sediment Traps

The proportion of carbon which is daily respired in unpreserved trap material has to be carefully studied, even at low temperatures and short deployment periods. Obviously the turnover rates for the sedimenting carbon are significantly different in different investigations. In the Barents Sea the proportion of the daily respired carbon was small and ranged between 0.4% to about 5% with an average of 1.4% (Figure 3-12). An accurate elemental budget and the interpretation of sedimentation estimates during trap deployments in the Barents Sea during summer are, therefore, easily established. However, temperature and the biochemical composition play a significant role for the breakdown of organic matter in general and hence also in sediment traps. This is indicated by daily loss rates as high as 30% in tropical environments and the generally high loss rates during summer in boreal environments. The temperature dependence of carbon breakdown rates is evident even in the present study with temperature variations less than 2.6° C (Figure 3-12). Consequently, the in situ temperature experienced by sediment traps during deployment must be carefully considered. Measures should be taken to prevent decomposition even during short-term deployments, especially at higher temperatures, or decay rates must be calculated and taken into account when carbon sedimentation rates are determined.

References

Dankers, N., and R. Laane, A comparison of wet oxidation and loss on ignition of organic material in suspended matter, Environ. Technol. Lett., 4, 283-290, 1983.

Dean, W. E., Determination of carbonate and organic matter in calcareous sediments and sedimentary rocks by loss on ignition: comparison with other methods, J. Sediment. Petrol., 44, 242-248, 1974.

Grasshoff, K., Methods of Sea-water Analysis, 419 pp., Verlag Chemie, New York and Weinheim, 1976.

Gundersen, K., and P. Wassmann, Use of chloroform in sediment traps: caution advised, Mar. Ecol. Prog. Ser., 64, 187-195, 1990.

Krause, G. L., C. L. Schelske, and C. O. Davies, Comparison of three wet-alkaline methods of digestion of biogenic silica in water, Freshwater Biol., 13, 73-81, 1983.

Lancelot, C., and S. Mathot, Biochemical fractionation of primary production by phytoplankton in Belgian coastal waters during short- and long-term incubations with ^{14}C-bicarbonate. II. *Phaeocystis pouchetii* in colonial populations, Mar. Biol., 86, 227-232, 1985.

Mook, D. H., and C. M. Hoskin, Organic determinations by ignition: caution advised, Estuarine Coastal Shelf Sci., 15, 697-699, 1982.

Paasche, E., Silicon content of five marine plankton diatom species measured with a rapid filter method, Limnol. Oceanogr., 25, 474-480, 1980.

Wassmann, P., M. Vernet, B. G. Mitchell, and F. Rey, Mass sedimentation of *Phaeocystis pouchetii* in the Barents Sea, Mar. Ecol. Prog. Ser., 66, 183-195, 1990.

CHAPTER 4. ORGANIC INVENTORIES

WORKSHOP REPORT

Introduction by
Chair: Cindy Lee, Rapporteur: Susan Henrichs, Assistant Rapporteur: Ricardo Letelier

1. Introduction

Recent improvements in techniques for the analysis of organic compounds have greatly improved our knowledge of the composition of marine particulate matter on a molecular level. This, in turn, has led to improved understanding of the sources, transformations, and fate of particulate matter in the oceans. The complexity and diversity of organic compound structure requires a wide variety of techniques for the determination of sample composition. Detailed descriptions of analytical protocols for many organic compound classes appear as individual contributions in this chapter from several of the laboratories currently conducting these analyses. The Working Group, after reviewing these contributions, identified several broad areas in which new approaches and techniques are needed to obtain information essential to developing a fundamental understanding of processes affecting marine particles. Clearly, there are important gaps in our knowledge of both the nature of organic matter in marine particles and how organic matter affects interactions between particles and their surroundings. These questions can be addressed either by better application of existing analytical techniques or by development of new techniques. Much of the technical development required could be achieved within about 5 years.

Information on the quantity of total organic matter in a sample provides a quantitative context for the various preservation, transformation, and degradation processes affecting individual compounds and compound classes. The Working Group Report discusses the progress we have made toward a complete characterization of the organic material in marine particles, and many of the individual contributions discuss the progress made in the analysis of major compound classes. A complete, or nearly complete, molecular level inventory relative to organic carbon is not currently feasible, because so much of the macromolecular organic material cannot be characterized with present techniques. However, the Working Group noted that no single sample has been exhaustively analyzed by all of the best available techniques, so the fraction of particulate organic matter that can be currently characterized is unknown. Isotopic and elemental analysis of the total organic content is possible. Also, considerable information on functional groups and structural units present can be obtained. For many purposes, exhaustive molecular level characterization of the organic matter present is unnecessary. However, many studies would benefit from exploiting available techniques to obtain more complete information on the organic content of samples.

Quantitative comparisons of the results of analyses of organic materials in marine particulate material are greatly needed. In the interest of gaining as much information as possible about the total spectrum of compounds present, investigators from different laboratories have tended to work together to analyze different compounds in a sample rather than to intercompare similar techniques for the same compounds. An exception to this lack of interlaboratory comparisons is the analysis of chlorophyll pigments. We can currently measure the major chlorophyll and carotenoid pigments with sensitivity and precision if appropriate precautions are taken. These compounds are very labile and one of the major problems with their analysis is degradation, both during sampling and handling. Contributions by both Bidigare and Repeta (this chapter pages 121 and 127) discuss some of the analytical pitfalls with these methods.

A basic problem inhibiting the intercomparison of analytical methods is that, in many cases, authentic standards do not exist and, in most cases, no standard reference materials exist that are suitable for organic analysis. Appropriate standards and standard reference materials must be made available to assure data quality and comparability, especially in the case of large research programs in which several laboratories conduct analyses.

One of the key points brought out both in the Working Group sessions and the individual contributions was that our present need to hydrolyze or oxidize most polymers to their component monomers before analysis is both a strength and a weakness. Clearly, from the papers in this chapter by Hedges, Henrichs, and Mopper and Furton, (pages 131, 141 and 155), the amino acids and sugars are two of the most successfully analyzed compound classes, but only as the monomers. Analyses

Marine Particles: Analysis and Characterization
Geophysical Monograph 63

of sugars and amino acids in particulate material are usually conducted after acid hydrolysis of carbohydrates and proteins. Hydrolysis procedures are plagued by the competing biases of incomplete hydrolysis and degradation. The problem is more severe for the carbohydrates, which may partially degrade under even the mildest hydrolysis conditions. Some of the lipid compounds, such as triglycerides and wax esters, can be analyzed as macromolecules, but many require hydrolysis with its attendant problems (Wakeham and Volkman, this chapter, page 175). Oxidative techniques have been particularly useful for breaking down lignin and cutin polymers into their structural units, as discussed by Hedges (this chapter, page 131), but do not result in a complete characterization of these polymers at the molecular level. Monomer analysis provides useful information not always seen in the analysis of whole macromolecules. However, as with all hydrolytic and degradative techniques, information on the linkages between the structural units is inevitably lost. Advances in analysis of intact macromolecules have been made but it is clear that progress is still needed. Tanoue (this chapter, page 167) describes two methods, sodium dodecyl sulfate polyacrylamide gel electrophoresis and isoelectric focusing, for the analysis of intact proteins in particulate matter. Such approaches will be crucial to developing our understanding of particulate organic matter, particularly the influence of macromolecules on particle surface properties and aggregation and disaggregation processes.

Since organic compounds in particles are present in different matrices, sequential hydrolysis or extraction procedures can give valuable information on the linkages or binding mechanisms within the particle. Chemical fractionation techniques in use depend very much on the compound or compound class being investigated. The speed and efficiency of these extraction and separation techniques need to be improved and their efficiency and reproducibility documented. New and less degradative techniques, such as the supercritical fluid extraction discussed later by Mopper and Furton (page 155), should be further investigated. Other fractionation techniques being used by chemical oceanographers are based on size separation. Filters, and to a lesser extent screens, have long been used to differentiate between various sizes of particles. Recent developments in ultrafiltration and gel permeation techniques offer new opportunities in this direction. Benner and Brownawell (this chapter pages 185 and 191) describe some of these new techniques to separate bacteria, viruses, colloids, and high molecular weight material. Flow cytometry shows promise in separating particles by size as well as by type (Henrichs, this chapter, page 141).

Another area discussed in both individual contributions and by the Working Group was the development and calibration of new instrumental techniques. Popp (this chapter, page 203) discusses the powerful new "on-line" techniques being developed in the field of isotope-ratio mass spectrometry. These techniques will improve our ability to conduct isotope analysis of individual compounds, vastly improving the sensitivity and accuracy of isotope tracer techniques in geochemical studies. Benner and Repeta, pages 185 and 127, each discuss the NMR techniques becoming so important to functional group analysis of organic matter, particularly macromolecules. Also useful for functional group analysis are the pyrolysis GC-MS methods discussed by de Leeuw (this chapter, page 211). These methods have already given us new insights into the structures of biomacromolecules and have led to a better understanding of the large quantity of polymeric material present in bulk organic matter. The contribution by Carlson (this chapter, page 199) describes in situ optical techniques, using fiber optrodes, which show promise of giving a radically new view of the ocean by providing fine-scale data on particle and macromolecule distributions. Carlson also describes fluorescence depolarization methods using fluorescent dyes as molecular probes to study the rheological and chemical properties of seawater.

These and other new chemical fractionation and instrumental techniques will likely be necessary to investigate surface properties, one of the areas of particle analysis which has received less attention from chemical oceanographers. Surface properties of particles are strongly affected by organic matter. The processes of adsorption-desorption and aggregation-disaggregation are difficult to study but crucial to particle formation and persistence in the sea. Many of the techniques available to study these processes have been adapted from studies concerned with wastewater treatment and organic pollutant transport. While useful, these methods are best developed for very high particle and solute concentrations and for nonpolar molecules. There are large gaps in knowledge and fewer techniques available for polar, ionic, and biologically active molecules, particularly with regard to our understanding of macromolecule adsorption. In addition, the probable small-scale heterogeneity in composition of particles and particle surfaces has significant but unknown consequences for particle surface processes, particularly those involving microorganisms. Overall, the Working Group concluded that a mechanistic understanding of solute-particle and particle-particle interactions should be a goal for future research. The variety of particle surfaces and dissolved organic molecules in seawater precludes a purely empirical approach.

The Working Group discussed sampling and analytical strategies in terms of relevant time and space scales. The methods now available can sometimes produce an overwhelming quantity of data. Techniques for handling such large data sets must be adapted and used in organic analytical laboratories. However, this sort of expensive data should be collected only when it is necessary to answer the question being asked. For larger scale global questions, it may be necessary to sacrifice detail for a more cursory but extensive view of the world oceans. For example, high-performance TLC can quickly provide data on the composition of lipid compound classes in a sample (Wakeham and Volkman, this chapter, page 175); detailed gas chromatographic analyses of these fractions are not essential in some applications. New approaches promise to provide a substantial amount of structural information rapidly enough to permit analyses of large numbers of samples. Common features of such techniques include high sensitivity to permit analysis of small samples, minimal sample preparation, automation of sample and data processing, and on-line coupling of two or more instruments.

Because of the diversity and complexity of organic compound structure, the analysis of organic matter in marine particles is challenging. However, the diverse and complex structures contain a wealth of information; the organic composition of particles is sensitive to virtually every chemical, biological, and physical process affecting particles. We can better solve the puzzles of marine particle behavior by fully exploiting the existing approaches to the detailed analysis of compound classes, while developing new techniques for the characterization of macromolecules and for rapid, qualitative analysis of total organic content.

2. The Workshop Report by Members of the Working Group: Organic Inventories

2.1. Organic Inventories

An inventory of the organic components of marine particles is valuable in addressing many key questions related to particle dynamics and distribution. For example, knowledge of the preservation, transformation, and degradation of specific particulate organic compounds is more useful when placed in the context of the fate of other organic materials in the particle. In the following discussion of organic inventories in analyses of the composition of marine particles, we use the term "inventory" to refer to any broad assessment of the components of a material, without the restriction that this compilation be complete or quantitative for all compound classes. The materials in question include all varieties of particles and colloids that might be found in the ocean and recently deposited sediments.

One useful way to structure a discussion of organic inventories is based on the size of the structural unit in the sample that is analytically resolved. For chemical inventories of naturally occurring organic materials, the range of scale considered is extensive. Included are qualitative assessments of the particle types present, analyses of elemental and isotopic compositions, measurements of functional groups, assessments of the quantity and variety of small molecules and compound classes, and characterizations of bio- and geomacromolecules. This spectrum of characterizations involves a wide variety of sampling and analytical techniques that often can be applied to only a portion of the measurement range. Application of many of these analytical techniques at present involves a compromise between the fraction of the total material that can be characterized and the detail of the information that can be obtained for that component. In the following sections we will discuss the types of inventories that presently can be made at these various analytical scales, and some of their advantages and limitations.

2.1.1. Particle type inventories.

Marine particles, or their aggregates, can be investigated microscopically to define their apparent sources and histories. Microscopic examination is often neglected in chemical studies. Preparative-scale methods for separating different particle types before chemical analysis can help to better-define the characteristic organic compositions of individual components in complex environmental mixtures.

Individual particles within a bulk particulate sample often have very different organic chemical compositions; these distinct compositions can be used as sensitive tracers of particle sources or transformations. Often, knowledge of the particle types present (e.g., cell wall debris, whole plankton, mineral debris) and their individual compositions will aid interpretation of compositional data. Sensitive specific fluorescent dye and immunofluorescent staining techniques are potentially useful to differentiate between particle types based on constituent macromolecules [Porter and Feig, 1980; Hara and Tanoue, 1989; Ward, 1990]. A limitation of particle-type inventories, however, is that some kinds of particles are better-preserved and more easily identified (e.g., diatom frustules) than others (e.g., dinoflagellates). The identifiable components of a sample may sometimes represent only a small fraction of the total organic matter present. Also, the amount of organic C present in each individual particle type is not usually known well, so that numbers of particles cannot be easily related to total particulate carbon.

2.1.2. Organic elemental inventories.

Total amounts of the major elements comprising particulate organic matter (C, H, O, P, N and S) can be readily measured with high precision using different types of elemental analyzers, provided that they are present in simple matrices (see chapter 3, page 33, Major Bioelements). Assessments of elemental inventories of organic materials, however, pose challenges such as the analysis of organic hydrogen and oxygen and precise discriminations between organic and inorganic sources of C, H, N, and P.

The major advantage of making organic inventories at the elemental level is that this provides the only practical way to quantify the total amount of organic matter in the complex particle mixtures that are common in the marine water column and sediments. Due to the challenges of separating, identifying and directly quantifying the many different types of compounds comprising particulate marine organic material, a complete inventory at the molecular level is not yet possible.

Another important use of organic carbon content is to provide a normalizing term for expressing the abundance (or concentrations) of individual compound types from a mixture. This is often the only way that the fraction of a sample represented by measurements at the molecular level can be assessed and evaluated in terms of being characteristic of the history or properties of the total particulate material. Carbon-normalized concentrations also provide the advantage that they can be more meaningfully compared across a broader variety of particle types than simple mass-normalized concentrations that are sensitive to independent variations in the inorganic constituents. Nitrogen content can be used to estimate the total amount of nitrogenous organic matter in a complex mixture. In general, the fraction of total organic C and N that can be accounted for by direct measurements of the more abundant biomonomers (e.g., amino acids, carbohydrates, lignin phenols) varies inversely with the extent of diagenetic alteration of particulate material, thus providing a qualitative indicator of organic matter "freshness".

When it is possible to measure the C, H and O content of an organic mixture, crude estimates of the major biochemical components (e.g., polysaccharides, lipids, lignins, tannins and protein) can be made by use of van Krevelen-type plots of H/C vs O/C, or equivalent numerical treatments. N/C can be included as an additional parameter representative of nitrogenous components. The basic approach is to establish corresponding elemental ratios for the major categories of biopolymers in living organisms and to determine how these components must be mixed to give the bulk characteristics exhibited by a natural sample [Reuter and Perdue, 1984]. One advantage of this approach is that elemental analyses provide averaged characteristics of all organic matter in a sample, so that major components can be detected and crudely quantified even if they have suffered moderate structural alteration or are not evident at more molecular scales of characterization. The elemental content of organic mixtures also reflects the average oxidation state of the material and therefore its potential caloric content or oxygen demand for heterotrophic respiration.

Elemental and isotopic analyses also provide broad constraints on the types, relative amounts, sources, and residence times of the different organic materials in environmental mixtures. For example, in some environments C/N ratios can be used as indicators of the relative amounts of marine and terrigenous organic matter in an environmental mixture. Stable isotopic compositions of bulk particulate materials are also useful indicators of marine vs. terrestrial sources [e.g., Jasper and Gagosian, 1990] and of the availability of dissolved CO_2 for phytoplankton production [e.g., Rau et al., 1989].

The stable nitrogen isotope composition of marine plankton and their particulate remains can reflect both trophic structure [e.g., Fry, 1988; Altabet, 1989] and the extent of community nitrogen fixation. The ^{14}C content of marine particles has been a useful tracer for detecting rapid transport of organic matter into the deep ocean, and conversion of deep-water dissolved carbon into particulate form [Druffel and Williams, 1990]. The breadth of potential applications of ^{14}C as a tracer has been greatly increased over the last decade by accelerator mass spectrometry (AMS) techniques that allow analysis of sub-milligram carbon amounts.

The major limitation of organic inventories at the elemental level is their low information content. In particular, such data have little sensitivity to changes in the composition of minor components that are indicative of specific sources or processes. Thus, even though it is possible to determine the absolute amounts of organic matter with elemental analysis, often little is learned about the sources or structural characteristics of the particulate material. More practically, many of the elements are subject to error when measured in anything but a purely organic matrix. For example, CHN analyses of hydrogen are subject to large errors from water released by thermal breakdown of hydrous minerals or incomplete drying of the sample. Organic oxygen cannot be measured accurately by difference in mineral-rich samples and must be analyzed separately by pyrolytic release of CO_2. An example of the complexity of even simple determinations is the problem of quantifying dissolved organic carbon (DOC) in seawater, a problem that has challenged oceanographers for over 50 years. Because the concentration of DOC predominates over that of particulate organic carbon (POC) in seawater, and both biological and purely physical interconversions occur, the present uncertainty in the measurement of seawater DOC translates into large uncertainties in meaningful budgets for POC in seawater.

2.1.3. Functional group inventories.

Analyses of the functional groups in organic compounds represent the next higher level of characterization than elemental analysis. Characterizations can be made by a wide variety of means, including titration (for organic acids and bases) functional-group-specific reagents, and determinations of spectral properties (IR and NMR). The term functional group is applied here to all carbon forms including C-C and C-H bonds and is not limited only to heteroatomic (X, O, N, P, S, etc.) assemblages.

Functional group analyses have the advantage that they provide structural information and give a reasonably comprehensive characterization of an entire sample. These methods also are often rapid, nondestructive and insensitive to minor amounts of inorganic material. Some biochemicals are characterized by the presence of a particular type or pattern of functional groups that can be used to establish limits on their abundances in complex environmental mixtures. Functional group measurements are often sensitive to components that might not be detectable at the molecular level [e.g., tannins; Benner et al., 1990b]. Specific functional groups often impart properties (e.g., the ability to complex certain ligands) that can be assessed without more detailed structural information. The main limitation of functional group analyses is lack of sufficient resolution to assign characteristics to specific compound types. Most functional groups are common to a wide variety of biochemicals and may have very different environmental reactivities depending on the chemical and physical properties of the host molecule.

2.1.4. Molecular-level inventories.

Qualitative analyses of many compounds may help define key substances that indicate their biosynthetic origins and enable us to trace transformation processes. Low molecular weight, extractable organic compounds present in particles can often be analyzed using capillary gas chromatography-mass spectrometry (GC-MS). Routine, compound-specific isotopic analysis is now possible using isotope-ratio-monitoring gas chromatography-mass spectrometry (irmGCMS). A large variety of lipid classes can be analyzed, including hydrocarbons, acyclic alcohols, sterols, wax esters, sterol esters, acyclic ketones, steroidal and triterpenoidal ketones, acids, hydroxy acids, mono-, di-, and triglycerides, tocopherols, some carotenoids, etc. More polar extractable lipids such as phospho- and sulfolipids are also identifiable by GC-MS after relatively mild hydrolysis and appropriate derivatizations. Pigments such as carotenoids and chlorophylls are normally separated by liquid chromatography and identified by comparison of retention time and visible spectra with that of standard compounds. Isotopic analysis of pigments isolated and purified by liquid chromatography, while more time-consuming than irmGCMS analysis, is a powerful tool for defining the isotopic composition of the source organism [Bidigare et al., 1991].

In addition to qualitative information on the types of organic molecules present in complex environmental mixtures, it is desirable to know the absolute concentrations of all biomolecular components so that a mass balance approach can be used to investigate their sources, transformations and fates. The array of molecular level analyses required to characterize all organic components in environmental samples is formidable, and it is currently beyond the analytical capabilities of any individual laboratory. The total fraction of organic carbon and nitrogen in water column particulate matter characterized at the molecular level is typically less than 50% and 75%, respectively, and is even lower for sedimentary organic matter [Cowie and Hedges, in press].

It is not possible with present analytical methods to routinely characterize all organic components in environmental samples or to achieve a complete (100%) molecular level characterization of the total organic carbon and nitrogen in a single sample. The questions arise, however, as to what fraction of the total organic carbon and nitrogen could be analyzed at the molecular level and what is missed by currently used analytical methods. There are a variety of important biomolecules, such as tannins and aliphatic polymers, for which there are currently no molecular level analyses, and these components probably comprise an important fraction of the uncharacterized material. Most of the hydrolytic and degradative techniques used to depolymerize samples for molecular level analysis are not 100% efficient or specific in the cleavage of targeted linkages. Moreover, these treatments often result in the modification of target molecules to the extent that the biomolecular sources of these components can no longer be defined (see section 2.2.2.).

Fresh and decomposed mangrove leaves have been analyzed using several state-of-the-art techniques [Benner et al., 1990a, b]. About 30% of the total tissue carbon was characterized at the molecular level. Elemental composition (CHNO) and solid-state ^{13}C NMR were used to provide internally consistent bulk chemical information which indicated that a major group of biomolecules, the tannins, comprised about 20% of the total carbon. Also, about half of the total carbohydrate content, or about 25% of the tissue carbon, was not identified at the molecular level using a standard hydrolysis method to determine neutral sugar content. This information signals the need for additional molecular level techniques and for defining the limitations of currently used methods.

2.1.5. Recommendations.

1. Qualitative/semi-quantitative microscopic assessments of particle types are underutilized and should be conducted whenever possible.
2. The organic carbon and nitrogen contents of particulate organic material in marine samples should be routinely measured when more detailed molecular-level analyses are to be made.
3. Procedures for minimizing errors in the analyses of the elemental content of marine particulate material should be determined and followed. These would include methods for better separating inorganic and organic carbon, nitrogen and hydrogen (possibly by particle separation prior to analysis).
4. Determining an accurate standard method for the quantification of seawater DOC is imperative.
5. Techniques for functional group analysis such as in-source desorption/pyrolysis mass spectrometry, NMR, and FTIR spectrometry are capable of rapid and comprehensive characterization of small amounts of marine particulate materials and should be further developed.
6. A select group of environmental samples should be analyzed to determine what fraction of the organic matter can be identified using present techniques and to identify the types of organic components that are not now characterized at the molecular level. All biomolecules that are currently analyzed, including carbohydrates, amino acids, lipids, nucleic acids, lignin-derived phenols and cutin-derived acids, should be included. Bulk chemical properties, including elemental and functional group composition, should also be determined to place constraints on the chemical composition of the fraction of material that is not identified at the molecular level.

2.2. Comparability of Techniques

Comparability between techniques is important for compiling accurate inventories of organic matter at all levels of characterization, for interpreting large data sets compiled over periods of years to decades, and for assessing what fraction of organic carbon in environmental samples is truly uncharacterized. Comparability between techniques requires an understanding of the limitations and biases inherent in any chemical pretreatment and instrumental analysis of the sample. Many different techniques are required to analyze the 30 to 70% of the total organic carbon that has been characterized in some environmental samples. Each technique extracts a somewhat different portion of organic matter from the sample, and measures a different subset of the extracted material. For example, amino acid analyses by GC and HPLC may measure slightly different analyte windows and may not be strictly comparable.

In practice, intercomparison also requires readily available calibration standards and standard reference materials. Important aspects of analytical intercomparability, as it pertains to organic geochemical analysis of particulate matter samples, are discussed in the following sections.

2.2.1. Standards and standard reference materials.

There is a critical need for authentic internal and external standards for quantifying organic compounds in marine particles and dissolved organic matter. This need will grow in the near future as the resolving powers of analytical techniques improve. Quantification of organic compounds associated with marine particles is often complicated by the number of analytes of interest, as well as the number of extraction, clean-up, concentration, and analysis steps that are involved. The use of "internal" or procedural recovery standards that have physical and chemical properties similar to the target analytes is recommended to monitor recoveries of target compounds through various procedures, and also as a possible tool for quantification through use of relative responses determined by chromatography. While standards are currently available for many of the common fatty acids, sterols, amino acids, fatty alcohols, nucleic acids and carbohydrates, there is an urgent need for certain pigment, polar lipid, carbohydrate, polysaccharide, and (high) molecular weight standards. The availability of authentic standards represents an important first step in achieving comparability between marine organic laboratories. The increased emphasis on compound-specific stable isotope determinations in biogeochemical studies also dictates a need for well characterized individual organic compounds that can be used as stable carbon, nitrogen and sulfur standards. Once a wide range of pure standards becomes available, then inter-laboratory comparisons can be initiated using standard reference materials (SRMs). To date, intercalibration exercises have been coordinated for the analysis of phytoplankton pigments [Kennicutt et al., 1990] and dissolved organic matter (in progress); similar exercises should be performed for lipids, amino acids, carbohydrates and stable isotopes.

Intercomparison of results among different investigators can be complicated by the lack of, or inappropriate use of, standards. Furthermore, results are of greater use to other investigators if more precise and accurate concentrations or relative abundances were reported, even if the primary goals were qualitative assessments. Examples include identification of compounds in particles, or characterization of spectroscopic properties of particles or particle extracts. Results are often reported in terms of the response of a particular instrument, e.g., integration units, relative fluorescence, etc. Results reported in this fashion hinder attempts at replication and intercomparison. With only modest effort, instrument responses, even for unknown compounds, can be converted to meaningful units. The simplest approach is to calibrate the instrument response relative to readily available standard substances. For example, in the study of fluorescence spectra of unknown compounds in particle extracts, fluorescence intensity can be expressed relative to a suitable standard, such as quinine sulfate. It is important that the same instrumental conditions are used for both the unknown sample and reference solutions, and that corrected spectra are used when this type of information is presented. Of course, a full description of the standard preparation procedure should be given.

In addition to authentic standards, there is a requirement for standard reference materials to assess analytical performance (e.g., extractions, separations, quantification, etc.), create inventories of organic components and facilitate inter-laboratory comparisons. SRMs can be divided into two groups: standard reference cultures/tissue and standard reference samples. The first group of SRMs should include, at a minimum, representative phytoplankton such as diatoms, cyanobacteria, dinoflagellates, prochlorophytes, prasinophytes and prymnesiophytes; bacteria, i.e., Gram-positive and Gram-negative eubacteria, archaebacteria, and photosynthetic strains; zooplankton; and vascular plants, i.e., woody and non-woody tissues, angiosperms, gymnosperms, etc. Organisms maintained in reference culture collections represent a small sub-set of the biota found in natural assemblages. Thus, there is a pressing need to extend such collections to include the most abundant species in nature and to systematically characterize the organic monomers as well as the low and high molecular weight macromolecules in these organisms. This information is critical for the meaningful interpretation of complex organic matter distributions observed in nature. Potential sources for some of these SRMs include: organizations that routinely grow and harvest phytoplankton and bacteria for natural products (e.g., Cyanotech Inc., Microbio Resources, Inc., Center for Culture of Marine Phytoplankton); and the National Institute of Standards and Technology for plant tissues.

The second group of SRMs should include two to four archived samples of each type to be used for a quantitative inventory of organic components. These materials could include: marine sediments from oxic and anoxic basins; marine particulate matter; and sediment trap collections. These samples should contain enough material (tens of g to kg) so that organic analyses could be performed by a relatively large group of laboratories for intercomparison. We also recommend large sample sizes so that these samples could be reanalyzed when new or modified methods are introduced. Ideally, the maintenance, distribution and certification of SRMs would be coordinated by a governmental agency such as the Environmental Protection Agency or the National Institute of Standards and Technology.

2.2.2. Chemical fractionation techniques.

Particulate matter samples are usually extracted before analysis to selectively transfer a fraction of the total organic matter into solution where it can be more easily concentrated and analyzed. However, extraction introduces a bias into the analysis that must be considered in any attempt to make quantitative assessments of organic matter. Details of extraction technique depend on the type of material to be analyzed (matrix), the compounds to be analyzed, and the type of analyses that ultimately will be performed. The ideal extraction technique should be quantitative, rapid, result in no measurable changes in molecular properties (such as isotope fractionation, structural and stereochemical changes) and be compatible with a broad suite of compound classes and analyses. In practice, no method fulfills all these requirements. In order to increase extraction efficiency, "harsher" extraction conditions are often used, e. g., higher temperature, longer extraction time, stronger or more polar solvents. These conditions may result in increased degradation of some compound classes. Sequential extraction using a series of increasingly powerful solvents and conditions can remove labile compounds that might otherwise be degraded. Sequential extraction may also provide some insight into the nature of the matrix binding the organic matter.

As illustrated in the individual contributions in this chapter on the analyses of specific classes of organic compounds, a wide variety of methods differing in solvent, temperature, apparatus (Soxhlet, sonic extractor, bomb, sealed evacuated ampule), and extraction time are used. These methods can have variable and poorly defined extraction efficiencies, and are often tailored to one compound class. Nearly all methods used to date have been developed for low molecular weight compounds, including the monomers resulting from hydrolysis during extraction. Direct analyses of biopolymers will require the development of new, milder extraction techniques. Mopper and Furton (this chapter, page 155) describe one such approach using supercritical CO_2 as the extraction solvent. The technique is rapid and gentle, with mild temperature and non-reactive solvent, fast, may be optimized by the addition of modifiers, can be made selective by the addition of functional group specific reagents, and shows promise of higher recoveries than conventional organic solvent based techniques. The technique can be used in an "on-line" mode by directly coupling the extractor effluent to chromatographic or spectrometric instrumentation, thereby reducing analysis time. Another potentially useful technique that has not been widely employed is aqueous extraction in the presence of detergents. This technique can solubilize some high molecular weight biomacromolecules under relatively mild conditions [Helenius and Simons, 1975; Tanoue, this chapter, page 167].

Since most extractions are not quantitative, some of the organic matter of interest is invariably left in the solid residue. Techniques to analyze this residue directly, including pyrolysis GC and GC-MS, solid state NMR, reflectance spectroscopy, and direct chemical ionization mass spectrometry, have not been widely used and need to be more fully developed. Pyrolysis GC and GC-MS have provided the most comprehensive data on residue composition to date, but there are still relatively few analyses of both the extract and residue of the same sample, and the nature of the residual organic matter remains largely unknown.

2.2.3. Chemical modification and instrumental analysis.

Extracts of particulate matter samples are often chemically modified or derivatized to enhance detection by instrumental analyses. Experimental differences in chemolytic, pyrolytic and derivatization techniques can result in both qualitative and quantitative differences between different organic analyses. A number of techniques exist for the chemolysis of specific functional groups including the depolymerization of organic sulfides, ether cleavage by BCl_3, and enzyme-catalyzed ester hydrolysis, but it is generally accepted that no one technique quantitatively reacts with all constituents of each compound class. The problem is most severe for the chemolytic degradation of high molecular weight biomolecules including proteins, complex carbohydrates, heteropolysaccharides and nucleic acids. While reproducible, recovery of monomers from some simple and well-characterized biomacromolecules such as cellulose and lignin is not quantitative [Fengel and Wegener, 1979; Garves, 1979], even though both the structure of the biopolymer and mechanism of hydrolysis are well-understood. Chemolysis of natural samples is assumed to be even less quantitative than for well-characterized reference materials. Techniques employing acid hydrolysis are commonly used for natural particulate matter samples. Acid hydrolysis is tailored to the monomers of interest, and must be considered as a compromise between yield and specificity of the components released. Organic esters are solubilized by acid catalyzed trans-esterification in BF_3/MeOH, or by saponification with KOH/MeOH. Simple organic acids and alcohols released can be subsequently identified by LC or GC-MS. For example, cutins, suberins, and polyhydroxy-alkanoates can be characterized from their fatty acid and hydroxy fatty acid composition. More specific chemical reagents are used to depolymerize lignin. Oxidation of lignin by CuO has been used to degrade the polymer to simple aromatic products that can be analyzed by GC or GC-MS.

Enzymatic hydrolysis has been less widely used, but may offer some advantages in certain analyses. In specific applications, enzyme hydrolysis may be more selective and milder than purely chemolytic techniques. Partial degradation of biomacromolecules by enzymes, in combination with separation/detection devices (MS, LC[FAB]MS) to identify relatively high molecular weight analytes should be explored for better characterization of macromolecules. It should be noted however, that enzymatic reactions in-vitro are typically inefficient, and introduce their own problems with reaction conditions and contamination. Advances in immobilized enzyme technology may serve to ameliorate some of these limitations.

Pyrolysis is another method used to depolymerize high molecular weight organic substances. It is complementary to chemolysis because the technique identifies macromolecules for which presently there exist no chemolytic reagents. For example, aliphatic biomacromolecules present in plant cuticles, barks and seeds and in some algal cell walls have been identified by pyrolysis, whereas they went undetected using standard chemolytic techniques.

It should be emphasized that the results of these analytical techniques must be compared and evaluated qualitatively and quantitatively. For example, polyesters such as cutin can be characterized from the monomeric substituents released by saponification, pyrolysis, or CuO-oxidation. Although one method may be superior over the others when applied to isolated, purified standards, another method may be preferred for natural samples due to reactions with other substances present, or unforeseen catalytic effects.

A wide variety of highly selective derivatization procedures are used to enhance detection, extraction, and chromatographic separation. While many derivatization techniques are quantitative (greater than 98% yield), it is important to recognize that derivatization reactions are accompanied by some loss of sample, and that this loss must be assessed by the incorporation of internal standards into the analytical scheme. Most reactions have limitations when broadly applied to all compounds having a specific functional group. For example, acylation of alcohols by acid halides or anhydrides is nearly quantitative for 1° and 2° alcohols, but gives low and variable yields for 3° alcohols. The problem of nonquantitative reaction is most acute for highly functionalized compounds such as biomacromolecules where yields are sensitive to the number and spatial relationship of the functional groups. High yields obtained for the derivatization of simple sugars may not, for example, be directly applicable to oligo- and polysaccharides. Care must be exercised when selecting the derivatization reagent and conditions for any analysis, and when comparing results when different derivatization procedures are used.

A number of different instrumental techniques are available to quantitatively measure a single compound or compound class, and it is important to consider differences in instrumental bias when comparing data derived from two different methods. Most analytical schemes employ either a highly efficient chromatographic pre-separation to purify an analyte before passage to a universal, non-selective, detector (e.g., flame ionization detector), or employ little to moderate pre-separation before passage of the analyte to a highly specific detector (e.g., spectrofluorometric detector). Each approach requires different preparation of the samples and quantifies a different property of the compound. It is therefore very important to consider what molecular property is characterized in an analysis, and when preparing results thoroughly report all pertinent analytical conditions, particularly detector calibration procedures and the source and integrity of the calibration standards. Intercomparison of two different methods for the same analyte is desirable, but is not practiced often enough (see section 2.2.1.). Analysis by multiple methods may identify specific biases and problems in methodology and needs to be more widely employed.

2.2.4. Recommendations.

1. Quality assurance procedures (i.e., internal standards, blanks, spiked blanks, etc.) should be incorporated into routine analytical protocols to ensure that results are accurate, precise and free of contamination.
2. A standard reference culture and tissue collection, representing marine and terrestrial organisms, needs to be established for source characterization, assessing analytical performance, and inter-laboratory comparisons.
3. A set of standard reference samples, representing natural mixtures of organic matter, should be

established for inter-laboratory comparisons and creating an inventory of organic components using different analytical approaches.

4. There is a critical need for many pigment, carbohydrate (including polysaccharide), polar lipid, and high molecular weight standards. Synthetic and natural products that can serve as standards need to be identified and, when possible, the spectral characteristics and source of these compounds should be reported.
5. The distribution of biomarker compounds in representative marine phytoplankton, bacteria, and zooplankton needs to be more exhaustively surveyed.
6. Selective chemolytic and enzymatic techniques for the controlled degradation of macromolecular organic matter need to be more fully developed. Given the wide use of acid and base hydrolysis in many analytical schemes, the limitations of these techniques need to be more fully studied.
7. Detailed accounts of all pertinent analytical procedures, including detector calibration and the source and integrity of calibration standards, need to be reported along with data.
8. Intercomparison of results obtained by more than one analytical technique can provide more reliable quantitation or identify important problems and biases in a method. Such comparisons are now possible for a wide variety of organic compounds and need to be made on a wider scale.
9. The application of chemolytic agents (tests of new agents, comparisons of similar agents) should be optimized using mixtures of standards and standard reference materials.
10. Extraction techniques are known to be non-quantitative and can sometimes produce artifacts. New techniques (e.g., supercritical fluid, aqueous detergents) need to be more fully investigated.
11. Better techniques for analysis of the solid-phase extracted residue need to be developed. In particular, methods for the selective analysis of specific subfractions of a particulate matter sample need to be emphasized.

2.3. Surface Chemical Properties and Interactions of Particle Surfaces With Their Environment

A knowledge of surface properties is essential to understanding several key biogeochemical processes, including adsorption and desorption of solutes, aggregation and disaggregation of particles, and biological production and decomposition of particulate material. Surface properties will also influence light scattering and absorption by particles, photochemical alteration of particles, catalysis of reactions, sound scattering by particles, and effects of particles on water structure. A broad definition of a particle surface is: the region which interacts with the environment surrounding the particle. By this definition, the thickness of the surface region would vary depending on the particular interaction being considered. Key properties of particle surfaces include the physical properties of surface area and charge density. Important chemical properties include the types and numbers, spatial distribution and orientation, and reactivity of surface functional groups. Reactions of the functional groups can be described in terms of mechanisms, rates, products, and reversibility.

Organisms have a central role in the production, modification, and consumption of POM. Most marine particles are complex mixtures of clay minerals, biogenic minerals, detrital organic matter, living organisms, and exopolymers; exopolymers appear to make up a large volume fraction of marine snow [Riley, 1963; Alldredge and Silver, 1988]. It is often experimentally difficult to distinguish between the surface properties and behavior of living and dead components of natural particulate matter, although progress has been made in the area of metal adsorption by algae [Crist et al., 1988, 1990].

Neihof and Loeb [1972] found that mineral particles suspended in seawater, regardless of their initial composition or surface charge in pure water, developed a net negative surface charge. This was attributed to the adsorption of dissolved organic matter from seawater. This observation, confirmed by subsequent work [e.g., Hunter and Liss, 1982], has led to the hypothesis that many surface properties of particles in the ocean are largely controlled by the properties of the adsorbed organic matter. However, very little is known about the organic matter adsorbed at particle surfaces beyond indirect evidence that it contains mucopolysaccharides or glycoproteins [Baier et al., 1974].

In the following discussion, we will consider adsorption-desorption and aggregation-disaggregation in terms of present knowledge of the composition and reactivity of particle surfaces, major unknowns, and new approaches and techniques that have the potential to improve understanding of these processes. We will also consider particle surface heterogeneity, and the potential importance of microenvironments in determining particle surface properties.

2.3.1. Adsorption/desorption.

Sorption reactions of metal and organic solutes with particulate phases control particle charge, colloid stability, reaction rate, and the speciation of surface-active compounds between dissolved and sorbed phases. Solute-particle interactions (adsorption-desorption) differ from particle-particle interactions (aggregation-disaggregation) in that solutes ultimately approach particle surfaces via molecular diffusion. Other processes, e.g., Brownian motion, turbulent shear, and differential settling, affect particle aggregation.

There is a large body of research on the aqueous sorption, and to a lesser extent desorption, of metal and organic solutes by particulate and colloidal materials. Organic matter, which exists as biogenic phases, detritus, or coatings on mineral particles, is known to be important for sorption of several classes of organic compounds [Karickhoff, 1984] and certain metals (e.g., Cu). The surface properties and functional groups of organic matter determine the extent of sorption reactions. It is unknown how the composition of DOM adsorbed by particles and coating particle surfaces differs from DOM in solution and how the selectivity of DOM adsorption might depend on the composition of the particle. Aluminosilicate particles, for example, might develop an organic coating which differs in composition from that of biogenic silica or calcite. Only very limited, qualitative information exists on the types of functional groups present at particle surfaces.

Sorption of metal and small (less than 350 Daltons) organic solutes by organic matter can often be conceptualized as a phase distribution process, i.e., binding into a three dimensional phase, while surface adsorption of larger oligomers and polymers is likely dominated by surface interactions. The discrimination is significant, because bulk properties of particles and organic matter may determine small solute speciation, and accessible surface moieties may determine polymer adsorption. Mechanistic studies of solute/surface interactions have been conducted to understand metal speciation in seawater systems [Mantoura et al., 1978; Hering and Morel, 1989]. Various electrostatic and surface complexation models to describe metal-organic complexation have been proposed and examined. The use of model ligands and associated stability constants has also been explored.

The study of organic compound sorption has largely been driven by environmental concerns [Farrington and Westall, 1986; Westall, 1987], and mechanistic understanding is limited to smaller organic solutes that include but are not limited to nonpolar compounds, chlorinated phenols, aromatic bases, and ionic detergents. The approach of linear free energy or structure-activity relationships has been used to predict and understand distributions. The best example is for nonpolar organic solutes whose speciation can be described from solubility and organic carbon content of particles or organic colloids. Marine studies have tested models of organic solute speciation in the case of PCBs [Brownawell and Farrington, 1986]. Work has also been done on amino acids [Henrichs and Farrington, 1987], methylamines [Wang and Lee, 1990], and short-chain carboxylic acids [Sansone et al., 1987]. These studies are inherently difficult to conduct due to the problems of separating physical adsorption from biological uptake and transformation.

Given the large variety of organic solutes and particles that occur in marine environments, it is essential to develop an understanding of sorption and desorption reactions from a mechanistic view that allows for prediction of distributions of solutes between solution and surface phases. The ultimate goal is to describe, within a conceptual framework, the contributions of electrostatic, hydrophobic, and specific chemical interactions to the adsorption energy in terms of measurable properties of surfaces, solutes, and solution.

Potentially useful properties of surfaces include: organic carbon content; hydrophobicity, which can be estimated by several methods such as C:O, flotation, contact angle, and interfacial surface tension for macromolecular organic matter; and acidity, measured by titration. These provide broad measures of surface properties useful in making generalizations concerning particle behavior. Another approach to the study of particle surfaces is the measurement of electrophoretic mobility [Neihof and Loeb, 1972; Hunter, 1980; Hunter and Liss, 1982]. Electrophoresis provides information on surface properties that control the adsorption of charged solutes.

Spectroscopic techniques can be used to elucidate surface functional groups or, potentially, to probe various properties of surface environments or binding mechanisms. Examples include surface Raman spectroscopy to determine surface functional groups; time-resolved fluorescence spectroscopy with carefully designed probes sensitive to particular surface properties, e.g., sorbate orientation or polarity near the chromophore, pH, or redox state; and internal reflectance infrared spectroscopy to elucidate the nature of binding sites upon adsorption of sorbate. Unfortunately, the application of surface spectroscopic techniques to natural particles is difficult because of the compositional heterogeneity of the particles.

A variety of histological stains (light absorbing and fluorescent) are available commercially. These have been shown to react with varying degrees of specificity with functional groups on organic matter and consequently have been used to quantify the gross biochemical composition (i.e., carbohydrate, protein and lipid) of tissues, soils and related materials. In some instances, stains can give information about specific subgroups within these major biochemical divisions, e.g., uronic acids as a subgroup of carbohydrates. Stains of varying molecular size can be used to mimic the surface binding behavior of a range of organic molecules. Stains can thus potentially yield qualitative and semiquantitative information about the organic composition of particle coatings and the material responsible for the adhesion of particles, i.e., algal and bacterial exopolymers and other glue substances within particle matrices. Two studies [Gordon, 1970; Johnson, 1974] have used stains specific for carbohydrates, proteins, and lipids to find that the matrix material binding marine particle assemblages was dominated by carbohydrates. Use of specific fluorescent stains in conjunction with flow cytometry has good potential for providing information on particle composition as a function of size. Stains and staining techniques designed to affect only surfaces of particles could be a useful new approach to surface characterization.

A direct approach for the investigation of chemical properties of surfaces and of surface interactions with solutes is to use tracer molecules having a radioactive or fluorescent tag to measure the adsorption of solutes. Such experiments can provide specific information on a particular solute-particle interaction, or can be designed to provide broader information on the kinds of functional groups present and their reactivity, by using a suite of tracers with varying properties. Reversibility of adsorption is relatively easy to determine using this approach, for example, by adding a large excess of adsorbed solute or another substance expected to compete for adsorption sites. Tracer studies appear particularly useful because they can be conducted with a minimum of alteration of the particle or its environment.

2.3.2. Aggregation and disaggregation of particles.

One of the most important properties of particles in the aggregation process has been termed "stickiness", the probability that particles will remain associated if they collide [Alldredge and McGillivary, in press, 1991]. "Stickiness" is likely related to surface contact area and the chemical properties of the particle surface (e.g., types, reactivities, and density of functional groups) because particle association represents the formation of chemical bonds between the particle surfaces. The nature of the chemical bonds formed is probably an important factor in determining the stability or strength of the particle association. The importance of carbohydrates and mucopolysaccharides in the adhesion and flocculation of marine particles has been indicated in a number of laboratory studies [Harris and Mitchell, 1973; Corpe, 1980; Calleja, 1984]. Hydrogen bonding between hydroxyl groups is most likely one of the primary factors in causing particles to adhere. However, the strength of the particle, i.e., the ability of the aggregate to resist disaggregation by shear forces, will most likely be related to the formation of stronger bonds, such as covalent linkages between functional groups as well as metal ion bridges. The formation of these stronger bonds will in turn be related to the surface density of sterically unhindered, reactive functional groups, especially amine, carbonyl, and carboxy groups.

Beyond our knowledge of the general chemical and physical properties of particles in the ocean, we have little knowledge of the specific chemical bonds formed when particles associate. The literature on waste treatment, although typically describing systems with much higher particle density and lower salinity than in the marine environment, offers valuable information on aggregation processes [Harris and Mitchell, 1973; Thompson and Forster, 1983]. Studies have been conducted in the laboratory with flocculation of hematite colloids that have been modified by humic substances and model ligands [Liang, 1988] and of natural lake particles [Ali et al., 1985]. There has been less study of particle or colloid aggregation/disaggregation in marine systems. Increased ionic strength can promote aggregation of river-borne particles in estuaries [e.g., Edzwald et al., 1974] although in some other cases ionic strength variations did not have a major influence on particle aggregation.

There have been no studies directly addressing the effects of bacteria on the aggregation and disaggregation of marine particles. However, studies from the waste treatment field [Harris and Mitchell, 1973; Calleja, 1984] strongly suggest that their impact is significant. For example, small (less than 100 μm), chemically unreactive particles may have little tendency to aggregate. However, if they become colonized by bacteria, exopolymers produced may enhance adhesion (coagulation) of these particles. On the other hand, bacterial colonization of aggregates may result in disaggregation. For example, diatom flocs may become colonized by bacteria. As a result, the diatom-produced matrix will be degraded by the bacteria, resulting in disaggregation. Alternatively, the original diatom exopolymer could be replaced with bacterial polymer [Kiørboe et al., 1990].

The methods described above for the characterization of surfaces (Section 2.3) are also potentially useful for identifying the surface properties and functional groups that affect interactions between particles. There are also techniques specific to particle aggregation. One approach is to use well-characterized particles, with known surface properties, as tracers of certain particle aggregation processes [Newman et al., 1990a, b]. Also, the effect of certain conditions or reagents, such as shear, electrolyte concentrations, pH, cosolvents or detergents, on disaggregation of particles could provide information on the nature of particle association, both for well-characterized, artificial and for field-collected particles. Time-course studies of particle aggregates can provide useful information, especially on biological modification of particle constituents such as binding exopolymers and fecal pellet coatings. Flow cytometry can provide information on particle size distribution and changes under experimental conditions in particle aggregation studies [Newman et al., 1990a, b].

2.3.3. Microenvironments.

The existence of microscale environments surrounding and within marine particles may also affect the properties of their surfaces. Oxygen and pH gradients have been observed within particles [Alldredge and Cohen, 1987; Paerl and Prufert, 1987]. Increased nutrient concentrations within particles due to biological activity [Shanks and Trent, 1979] may affect physical and chemical characteristics of a given surface. Oxygen, pH, and nutrient gradients also have an indirect effect on the characteristics of surfaces by controlling biological processes. Anaerobic microenvironments of particles have been proposed as the site of processes such as nitrogen fixation, denitrification, and methanogenesis in generally oxic water columns [Kirchman and Ducklow, 1987]. Other surface characteristics besides oxygen and nutrient concentrations are also important to bacteria. Samuelsson and Kirchman [1990] have shown that the surface energy of a particle, defined as the work of adhesion of water, controls the growth of attached bacteria through the availability of adsorbed protein. The organic coating on natural surfaces seems to be largely made up of adsorbed protein and glycoprotein [Little, 1985].

Several concerns arise with respect to the probable existence of small-scale compositional variation on particle surfaces and within particles. The distance scales and concentration gradients most relevant to microorganisms are missed when bulk samples containing many particles (or even single particles) are analyzed. Particles probably contain environments that differ radically from those of the bulk solution or the bulk particle. This could result in consumption or production of chemical species that could not be predicted or understood without characterization of the microenvironments.

There are several existing techniques that can improve knowledge of compositional variations within particles. Microelectrodes have been used to measure oxygen and pH gradients [Alldredge and Cohen, 1987], but the particle must be manipulated for study. The tetrazolium dye technique [Paerl and Prufert, 1987] for the detection of reduced microzones does not have that disadvantage, but does not provide quantitative data on oxygen concentrations. A wide variety of probes (e.g., microelectrodes; fiber-optic probes), designed by biochemists for measurement of intracellular concentrations, are under development, and it is expected that these will soon be applicable to the study of marine particles.

2.3.4. Recommendations.

1. There should be an increased emphasis on studies of the surface chemistry of marine particles, since it is essential to the understanding of many oceanic processes.
2. More data on and a better theoretical understanding of sorption reactions of classes of important organic chemicals are necessary. These classes include organic bases and other ionic compounds; very weakly soluble lipids, such as di- and triterpanes; and organic macromolecules in general.
3. More field studies of organic compound and metal speciation are needed to test models derived from theory and from laboratory experiments.
4. The kinetics and reversibility of sorption reactions need to be further studied; this is especially true for macromolecules.
5. Aggregation and disaggregation of particles is a key process controlling the flux of materials in the ocean. Both chemical and biological processes of aggregation and disaggregation merit intensive study.
6. Exopolymers may play a key role in particle surface properties and particularly particle aggregation and persistence of aggregates. Information on the occurrence of exopolymers in particles, their composition, their surface properties, and their role in particle aggregates is urgently needed.

7. A major goal in studies of both adsorption and aggregation must be a mechanistic understanding of these processes, to allow modelling and prediction of the behavior of a wide variety of solutes and particles. The enormous number of solute and particle types present in seawater make purely empirical, case-by-case studies impractical.
8. Microenvironments on particle surfaces and within particles deserve additional study due to their effect on biological transformation and degradation and chemical exchanges between particles and the environment.

2.4. Optimization of Sampling and Analytical Strategies

Time and space scales of biogeochemical processes in the ocean span many orders of magnitude [Lee and Wakeham, 1988]. Therefore sampling strategies represent a major challenge. Experiments and analytical procedures must be carefully designed so that the resulting data have scales appropriate to the processes being studied. For example, investigations of the kinetics of adsorption and desorption of short-lived, labile organic species may require sampling and analyses on the order of seconds to minutes and at scales of microns to centimeters [Wakeham et al., 1986; Farrington and Westall, 1986]. On the other hand, studies addressing questions of global scale may require measurements on a basin-wide scale designed to cover decades to millennia [Brassell et al., 1986; Prahl and Muehlhausen, 1989]. The character of suspended particles, sinking particles and sediments all reflect processes occurring over different temporal and spatial scales.

Ship-time, sampling methods, analytical technology and the complexity of marine organic matter often limit the rate at which a data base can be generated. To maximize the information on distributions of organic compounds in marine environments that can be obtained using current methods, there are two complementary approaches. One is to measure many compounds of diverse structural features at a few sites, in experiments designed to identify local processes. An alternative approach is to focus on a few key compounds in a larger suite of samples, using more rapid, compound-specific techniques, in experiments designed to understand basin-wide or global processes. An example that illustrates the information that can be obtained from both approaches is the analysis of photosynthetic pigments in seawater. Shipboard spectrophotometric and fluorometric analyses of chlorophyll, and more recently, remote sensing of ocean color, have provided information on global distributions of phytoplankton biomass in the upper ocean. Such data are invaluable for studies of productivity, carbon flux and other biogeochemical processes [NAS, 1984; Berger, 1989]. More detailed investigations using recently introduced HPLC-based techniques have revealed a great diversity of pigments [Bidigare et al., 1990]. These studies have identified new photosynthetic pigments that have highlighted the importance of previously unrecognized phytoplankton such as picoplanktonic eukaryotes and prokaryotes.

However, most organic constituents of POM present greater, and as yet unresolved, analytical challenges. Chromatographic techniques, the major methods for specific compound analysis, are time-consuming. In general, it is difficult to foresee advances in chromatographic techniques that will lead to substantial decreases in analysis times. Because current techniques are slow, studies have often focused on specific compound classes. However, specific compound classes such as sterols or fatty acids rarely provide sufficient information to fully characterize biogeochemical processes. Thus, it is becoming a more common practice to analyze a wider range of compounds in each sample.

To achieve an adequate database to answer questions related to global processes, techniques that both provide detailed information on molecular structure and are rapid enough to permit analysis of many samples are essential. In the foreseeable future it should be feasible to perform a rapid semi-quantitative characterization of POM that will allow broad intercomparisons of POM composition from many times and locations. Any study of the sources, transport, and reactivity of organic matter would benefit from better spatial or temporal coverage. Such a comparative analysis of organic matter in particles might also lead to the recognition and monitoring of fast processes in particles due to mineralization, biotransformation, and changes in the nature and degree of primary production; the construction of regional data bases of organic matter compositions in POM to observe global-scale processes; and a better understanding of the relation between organic matter distributions and other parameters.

To realize these goals, an analytical technique must have the following characteristics: no or minimal sample pretreatment; good reproducibility; short analysis time; high sensitivity, to enable analysis of small samples; identification of a large number of characteristics per analysis; and automated analyses of the data sets obtained. One of the methods that may meet these requirements is "in-source desorption/pyrolysis mass spectrometry" which has been recently used to characterize biomacromolecular standards such as cellulose and lignin and compounds present in natural samples such as algal polysaccharides and plant roots and cuticles (de Leeuw, this chapter, page 211). This method provides qualitative information on molecular structure. Spectroscopic methods such as ^{13}C-NMR and FTIR can provide complementary information on molecular structure. A combination of these methods should have considerable potential for comparative studies of POM samples.

2.4.1. Recommendations.

1. Experiments and analytical procedures must be designed so that the data collected are appropriate to the time and space scales of the processes under study.
2. For large distance scales and for many rapid processes, new analytical techniques and approaches are needed, to increase the number of samples for which detailed compositional data can be obtained in a reasonable period of time.

References

Ali, W., C. R. O'Melia, and J. K. Edzwald, Colloidal stability of particles in lakes: measurement and significance, Water Sci. Technol., 17, 701-712, 1985.

Alldredge, A. L., and Y. Cohen, Can microscale chemical patches persist in the sea? Microelectrode study of marine snow, fecal pellets, Science, 235, 689-691, 1987.

Alldredge, A. L., and P. McGillivary, The attachment probabilities of marine snow and their implications for particle coagulation processes in the ocean, Deep-Sea Res., in press, 1991.

Alldredge, A. L., and M. W. Silver, Characteristics, dynamics, and significance of marine snow, Prog. Oceanogr., 20, 41-82, 1988.

Altabet, M. A., A time-series study of the vertical structure of nitrogen and particle dynamics in the Sargasso Sea, Limnol. Oceanogr., 34, 1185-1201, 1989.

Baier, R. E., D. W. Goupil, S. Perlmutter, and R. King, Dominant chemical composition of sea-surface films, natural slicks, and foams, J. Recher. Atmospher., 8, 571-600, 1974.

Benner, R., P. G. Hatcher, and J. I. Hedges, Early diagenesis of mangrove leaves in a tropical estuary: bulk chemical characterization using solid-state ^{13}C NMR and elemental analyses, Geochim. Cosmochim. Acta, 54, 2003-2013, 1990a.

Benner, R., K. Weliky, J. I. Hedges, Early diagenesis of mangrove leaves in a tropical estuary: molecular-level analyses of neutral sugars and lignin-derived phenols, Geochim. Cosmochim. Acta, 54, 1991-2001, 1990b.

Berger, W. H., Global maps of ocean productivity, in Productivity of the Ocean: Present and Past, edited by W. H. Berger, V. S. Smetacek, and G. Wefer, pp. 429-455, Wiley Interscience, New York, 1989.

Bidigare, R. R., M. C. Kennicutt II, M. E. Ondrusek, M. D. Keller, and R. R. L. Guillard, Novel chlorophyll-related compounds in marine phytoplankton: distributions and geochemical implications, Energy Fuels, 4, 653-657, 1990.

Bidigare, R. R., M. C. Kennicutt II, W. L. Keeney-Kennicutt, and S. A. Macko, Isolation and purification of chlorophylls *a* and *b* for the determination of stable carbon and nitrogen isotope compositions, Anal. Chem., 63, 130-133, 1991.

Brassell, S. C., G. Eglinton, I. T. Marlowe, U. Pflaumann, and M. Sarntheim, Molecular stratigraphy: a new tool for climatic assessment, Nature, 320, 129-133, 1986.

Brownawell, B. J., and J. W. Farrington, Biogeochemistry of PCBs in interstitial waters of a coastal marine sediment, Geochim. Cosmochim. Acta, 50, 157-169, 1986.

Calleja, G. B., Microbial Aggregation, 276 pp., CRC Press, Inc., Boca Raton, Florida, 1984.

Corpe, W. A., Microbial surface components involved in adsorption of microorganisms onto surfaces, in Adsorption of Microorganisms to Surfaces, edited by G. Bitton, and K. C. Marshall, pp. 105-144, John Wiley & Sons, New York, 1980.

Cowie, G. L., and J. I. Hedges, Sources and reactivities of amino acids in a coastal marine environment, Limnol. Oceanogr., in press, 1991.

Crist, R. H., J. R. Martin, P. W. Guptill, J. M. Eslinger, and D. R. Crist, Interactions of metals and protons with algae. 2. Ion exchange in adsorption and metal displacement by protons, Environ. Sci. Technol., 24, 337-342, 1990.

Crist, R. H., K. Oberholser, D. Schwartz, J. Marzoff, D. Ryder, and D. R. Crist, Interactions of metals and protons with algae, Environ. Sci. Technol., 22, 755-760, 1988.

Druffel, E. R. M., and P. M. Williams, Identification of a deep source of particulate organic carbon using bomb ^{14}C, Nature, 347, 172-174, 1990.

Edzwald, J. K., J. B. Upchurch, and C. R. O'Melia, Coagulation in estuaries, Environ. Sci. Technol., 8, 58-63, 1974.

Farrington, J. W., and J. Westall, Organic chemical pollutants in the oceans and groundwater: a review of fundamental chemical properties and biogeochemistry, in The Role of the Oceans as a Waste Disposal Option, edited by G. Kullenberg, pp. 361-425, D. Reidel Pub., Dordrecht and Boston, 1986.

Fengel, D., and G. Wegener, Hydrolysis of polysaccharides with trifluoroacetic acid and its application to rapid wood and pulp analysis, in Hydrolysis of Cellulose: Mechanisms of Enzymatic and Acid Catalysis, edited by R. D. Brown Jr., and L. Jurasek, pp. 145-158, American Chemical Society, Washington, D. C., 1979.

Fry, B., Food web structure on Georges Bank from stable C, N, and S isotopic compositions, Limnol. Oceanogr., 33, 1182-1190, 1988.

Garves, K., Rapid hydrolysis of cellulose in homogeneous solution, in Hydrolysis of Cellulose: Mechanisms of Enzymatic and Acid Catalysis, edited by R. D. Brown Jr., and L. Jurasek, pp. 159-165, American Chemical Society, Washington, D.C., 1979.

Gordon, D. C., A microscopic study of non-living organic particles in the North Atlantic Ocean, Deep-Sea Res., 17, 175-185, 1970.

Hara, S., and E. Tanoue, Simultaneous staining with three fluorescent dyes of minute plankters on an agarose gel filter, Deep-Sea Res., 36, 1777-1784, 1989.

Harris, R. H., and R. Mitchell, The role of polymers in microbial aggregation, Ann. Rev. Microbiol., 27, 27-50, 1973.

Helenius, A., and K. Simons, Solubilization of membranes by detergents, Biochim. Biophys. Acta, 415, 29-79, 1975.

Henrichs, S. M., and J. W. Farrington, Early diagenesis of amino acids and organic matter in two coastal marine sediments, Geochim. Cosmochim. Acta, 51, 1-15, 1987.

Hering, J. G., and F. M. M. Morel, Slow coordination reactions in seawater, Geochim. Cosmochim. Acta, 53, 611-618, 1989.

Hunter, K. A., Microelectrophoretic properties of natural surface-active organic matter in coastal seawater, Limnol. Oceanogr., 25, 807-822, 1980.

Hunter, K. A., and P. S. Liss, Organic matter and the surface charge of suspended particles in estuarine waters, Limnol. Oceanogr., 27, 322-335, 1982.

Jasper, J. P., and R. B. Gagosian, The sources and deposition of organic matter in the Late Quaternary Pigmy Basin, Gulf of Mexico, Geochim. Cosmochim. Acta, 54, 1117-1132, 1990.

Johnson, R. G., Particulate matter at the sediment-water interface in coastal environments, J. Marine Res., 32, 313-330, 1974.

Karickhoff, S. W., Pollutant sorption in aquatic systems, J. Hydraul. Eng. Div. ASCE, 10, 707-735, 1984.

Kennicutt, M. C., II, R. R. Bidigare, and M. E. Ondrusek, Quality assurance procedures for the analysis of photosynthetic pigments: JGOFS intercalibration, EOS Trans. Am. Geophys. Union, 71, 103, 1990.

Kiørboe, T., K. P. Andersen, and H. G. Dam, Coagulation efficiency and aggregate formation in marine phytoplankton, Mar. Biol., 107, 235-245, 1990.

Kirchman, D. L., and H. W. Ducklow, Trophic dynamics of particle-bound bacteria in pelagic ecosystems: a review, in Detritus and Microbial Ecology in Aquaculture, edited by D. J. Moriarty, and R. S. V. Pullin, pp. 54-82, ICLARM, Manilla, Phillipines, 1987.

Lee, C., and S. G. Wakeham, Organic matter in seawater: biogeochemical processes, in Chemical Oceanography, second ed., edited by J. P. Riley, pp. 1-51, Vol. 9, Academic Press, London, 1988.

Liang, L., Role of surface chemistry in the coagulation of hematite, Ph.D. dissertation, 159 pp., Cal. Inst. Technol., Pasadena, CA, 1988.

Little, B., Factors influencing the adsorption of dissolved organic matter from natural waters, J. Colloid Interface Sci., 108, 331-340, 1985.

Mantoura, R. F. C., A. Dickson, and J. P. Riley, The complexation of metals with humic materials in natural waters, Estuarine Coastal Mar. Sci., 6, 387-408, 1978.

National Academy of Sciences, Global Ocean Flux Study. Proceedings of a Workshop, National Academy Press, Washington D.C., 1984.

Neihof, R. A., and G. I. Loeb, The surface charge of particulate matter in seawater, Limnol. Oceanogr., 17, 7-16, 1972.

Newman, K. A., S. L. Frankel, and K. D. Stolzenbach, Flow cytometric detection and sizing of fluorescent particles deposited at a sewage outfall site, Environ. Sci. Technol., 24, 513-519, 1990a.

Newman, K. A., F. M. M. Morel, and K. D. Stolzenbach, Settling and coagulation characteristics of fluorescent particles determined by flow cytometry and fluorometry, Environ. Sci. Technol., 24, 506-513, 1990b.

Paerl, H. W., and L. E. Prufert, Oxygen-poor microzones as potential sites of microbial N_2 fixation in nitrogen-depleted aerobic marine waters, Appl. Environ. Microbiol., 53, 1078-1087, 1987.

Paerl, H. W., Microbial attachment to particles in marine and freshwater ecosystems, Microbial Ecol., 2, 73-83, 1975.

Porter, K. G., and Y. S. Feig, The use of DAPI for identifying and counting aquatic microflora, Limnol. Oceanogr., 25, 943-953, 1980.

Prahl, F. G., and L. A. Muehlhausen, Lipid biomarkers as geochemical tools for paleoceanographic study, in Productivity of the Ocean: Present and Past, edited by W. H. Berger, V. S. Smetacek, and G. Wefer, pp. 271-289, Wiley Interscience, New York, 1989.

Rau, G. H., T. Takahashi, and D. J. Des Marais, Latitudinal variations in plankton $\delta^{13}C$: implications for CO_2 and productivity in past oceans, Nature, 341, 516-518, 1989.

Reuter, J. H., and E. M. Perdue, A chemical structural model of early diagenesis of sedimentary humus/proto-kerogen, Mitt. Geol.-Paleont. Inst., 56, 249-262, 1984.

Riley, G. A., Organic aggregates in seawater and the dynamics of their formation and utilization, Limnol. Oceanogr., 8, 372-381, 1963.

Samuelsson, M., and D. L. Kirchman, Degradation of adsorbed protein by attached bacteria in relationship to surface hydrophobicity, Appl. Environ. Microbiol., 56, 3643-3648, 1990.

Sansone, F., C. C. Andrews, and M. Y. Okamoto, Adsorption of short-chain organic acids in marine sediments, Geochim. Cosmochim. Acta, 50, 99-105, 1987.

Shanks, A. L., and J. D. Trent, Marine snow: microscale nutrient patches, Limnol. Oceanogr., 24, 850-854, 1979.

Thompson, I. M., and C. F. Forster, Nutrient conditions in relation to bacterial flocculation, extracellular polymer production and sludge settlement, Biotechnol. Lett., 5, 761-766, 1983.

Wakeham, S. G., E. A. Canuel, and P. H. Doering, Geochemistry of volatile organic compounds in seawater: mesocosm experiments with ^{14}C-model compounds, Geochim. Cosmochim. Acta, 50, 1163-1172, 1986.

Wang, X.-C., and C. Lee, The distribution and adsorption behavior of aliphatic amines in marine and lacustrine sediments, Geochim. Cosmochim. Acta, 45, 2759-2774, 1990.

Ward, B. B., Immunology in biological oceanography and marine ecology, Oceanogr. Mag., 3, 30-35, 1990.

Westall, J. C., Adsorption mechanisms in aquatic surface chemistry, in Aquatic Surface Chemistry, edited by W. Stumm, pp. 3-32, John Wiley & Sons, New York, 1987.

Williams, P. M., E. Druffel, and K. L. Smith, Dietary carbon sources of deep-sea organisms as inferred from their organic radiocarbon activities, Deep-Sea Res., 34, 253-266, 1987.

ANALYSIS OF ALGAL CHLOROPHYLLS AND CAROTENOIDS

Robert R. Bidigare

Department of Oceanography, University of Hawaii, Honolulu, Hawaii 96822

1. Sample Collection

Water samples should be collected with Go-Flo bottles to avoid the contamination problems associated with the latex rubber closure mechanism of Niskin bottles [Williams and Robertson, 1989]. Price et al. [1986] have documented that latex rubber is extremely toxic to marine phytoplankton. It is recommended that only silicone tubing be used to dispense seawater into sample bottles. Seawater is dispensed into opaque Nalgene sample bottles to avoid light shock to deep-living phytoplankton assemblages. If pigment samples are collected to examine the dynamics of the diatoxanthin-diadinoxanthin (DT-DN) xanthophyll cycle, then casts should be performed quickly, i.e. less than 6 minutes, to minimize DT-DN interconversion prior to filtration [Welschmeyer and Hoepffner, in press; Welschmeyer, personal communication, 1990].

2. Filtration

Seawater samples (500 to 2000 mL for chlorophyll *a* concentrations ranging from 0.05 to 2.00 mg m^{-3}) are immediately filtered through either 0.4 μm Nuclepore polyester filters (47 mm) or GF/F glass fiber filters (47 mm, nominal pore size = 0.7 μm). Nuclepore polycarbonate filters should not be used as they contain a yellow pigment which elutes in the vicinity of chlorophyll *c* and peridinin, depending on column choice. Samples should be filtered in duplicate under mild vacuum (0.2 bar) to minimize cell breakage. If desired, samples can be prefiltered through 200 μm mesh to remove zooplankton. However, if large colonies of phytoplankton (e.g., *Trichodesmium* sp.) are present they may be retained by the mesh.

3. Sample Storage

Filters are stored under liquid nitrogen prior to analysis to minimize pigment alteration and increase extraction efficiency. For storage in liquid nitrogen, filters can be folded in half and placed in labeled (volume filtered, station and depth) aluminum foil packets.

Marine Particles: Analysis and Characterization
Geophysical Monograph 63

4. Extraction

Filters are removed from the liquid nitrogen, briefly thawed (~ 1 min), and placed in 12 mL graduated glass centrifuge tubes for extraction. Nuclepore filters retain very little seawater and can be directly extracted in 2 mL 90% acetone:water (v:v, HPLC grade solvents). GF/F filters on the other hand retain ~ 0.7 mL seawater per 47 mm filter due to their porous nature. To achieve a ~ 90% final acetone concentration with 47 mm GF/F filters, the centrifuge tube is brought to a final total volume (solvent plus filter) of 6.1 mL with 100% acetone. This is equivalent to a final extraction volume of 6.0 mL since the displacement volume of a 47 mm GF/F filter is 0.1 mL. It is recommended that 100 μL of an internal standard (canthaxanthin in acetone) be added to each sample to correct for experimental losses and solvent evaporation. Samples are then sonicated (0° C, subdued light) to break cell walls, sealed with Parafilm to minimize solvent evaporation, and allowed to extract for at least 4 hrs in the dark at -20° C. Alternatively, filters can be ground using an ice-cooled tissue homogenizer under subdued light [JGOFS, 1990]. Following extraction, samples are centrifuged for 5 min to remove cellular debris.

5. HPLC Pigment Separations

Pigment extracts should be routinely analyzed by reverse-phase C_{18} high-performance liquid chromatography (HPLC) since "acidic" normal-phase silica columns can produce artifacts. The reverse-phase HPLC methods described by Mantoura and Llewellyn [1983], Wright and Shearer [1984], Gieskes and Kraay [1986], and Bidigare et al. [1989] all provide good separations of most pigments of interest. The use of ion-pairing agents, e.g. ammonium acetate/tetrabutylammonium acetate [Mantoura and Llewellyn, 1983], greatly enhances the separation of the dephytolated chlorophylls and their Mg-free derivatives. Solvents (HPLC grade) are degassed with helium prior to analysis, and the HPLC system should be flushed and equilibrated with 100% MeOH when not in use. It is also recommended that a C_{18} guard column be used in order to extend column life. If an auto-sampler is used for the continuous analysis of samples, it should be refrigerated (4° C) to minimize pigment alteration.

The algal pigments of interest for most oceanographic studies include: chlorophylls *a*, *b*, c_1, c_2, c_3; Mg 2,4-divinylphaeoporphyrin a_5 monomethyl ester; fucoxanthin; 19′-hexanoyloxyfucoxanthin; 19′-butanoyloxyfucoxanthin; zeaxanthin; prasinoxanthin; alloxanthin; peridinin; diadinoxanthin; diatoxanthin; lutein; α- and β-carotene; chlorophyllide *a*; phaeophytin *a*; phaeophorbide *a* and its derivatives; and divinyl chlorophyll *a*. Unfortunately, potential co-elution problems exist for several of these pigments including, chlorophyll c_1, chlorophyll c_2 and Mg 2,4-divinylphaeoporphyrin a_5 monomethyl ester, separable by chromatography on polyethylene [Jeffrey and Wright, 1987]; lutein and zeaxanthin, separable by the reverse-phase HPLC method of Wright and Shearer [1984]; and (monovinyl) chlorophyll *a* and divinyl chlorophyll *a,* separable by normal-phase HPLC [Gieskes and Kraay, 1983].

6. Peak Detection and Identification

Eluting pigment peaks are detected by absorbance spectroscopy, at 436 nm for carotenoids and chlorophylls, and fluorescence spectroscopy, Ex: 400 to 460 nm, Em: >600 nm for phaeophytins and phaeophorbides. It should be further noted that the Soret band of divinyl chlorophyll *a* is shifted to the red by 8 nm relative to that of (monovinyl) chlorophyll *a* [Gieskes and Kraay, 1983], and that serious over-estimations of "chlorophyll *a*" are obtained if peaks are monitored at 440 nm. For this reason it is recommended that peaks be monitored at 436 nm, i.e., halfway between the Soret bands of chlorophyll *a* and divinyl chlorophyll *a,* to minimize this problem [Ondrusek et al., 1991]. Peak identities are assigned by comparing their retention times with those of pure standards (co-injection analysis) and algal extracts of known pigment composition which are available from the Center for Culture of Marine Phytoplankton, West Boothbay Harbor, ME 04575. Peak identities of representative samples should be routinely confirmed by "on-line" diode array or stopped-flow absorbance spectroscopy (400 to 700 nm).

The detection limits for chlorophylls/carotenoids, using absorbance spectroscopy, and phaeopigments, using fluorescence spectroscopy, as analyzed by HPLC are on the order of 1 ng and 3 ng, respectively. Exact values will depend upon flow rate and detector configuration.

7. Quantification

The commercial availability of pigment standards at the present time is poor. Standards can be obtained from Sigma Chemical Co. (chlorophylls *a* and *b*; α- and β-carotene; lutein) and Hoffmann-LaRoche (zeaxanthin, lutein and possibly fucoxanthin). Standards which are not commercially available can be purified by thin-layer chromatography [TLC; Jeffrey, 1981] or preparative-scale HPLC. Standards should be calibrated spectrophotometrically in the appropriate solvents using published extinction coefficients (Table 4-1). Absorption is measured at the appropriate wavelength (usually at λ_{max}) and 750 nm (to correct for light scattering). Concentrations of the standards are calculated as follows:

$$\text{Pigment Conc. (mg L}^{-1}) = \frac{(\text{Cx1000 (mg)})}{\text{E(L g}^{-1}\text{cm}^{-1})\text{x(cm) x(g)}} \text{x} \frac{1000 \text{ mg}}{1 \text{ g}}$$

After determining pigment concentrations, standards are injected onto an equilibrated HPLC system to calculate standard response factors. It is important that the linearity of the HPLC system be established before samples are analyzed. Response factors should be calculated as the weight of standard injected (determined spectrophotometrically) divided by the area of the parent pigment plus areas of structurally-related isomers if present since they contribute to the absorption measurement described above. This assumes that the derivatives/isomers have the same weight-specific spectral absorption coefficients. In fact, they do not and only limited data are available in the literature to estimate the errors associated with this assumption. If for example, the absorption coefficient for cis-fucoxanthin differs from that of fucoxanthin by 10% and cis-fucoxanthin was present at 10% of the total fucoxanthin-like pigment concentration then a ca. 1% error would be introduced. If the area of cis-fucoxanthin was not included in the response factor calculation then a 10% error would be introduced. Once response factors and system linearity are established, then pigment concentrations of samples can be calculated with knowledge of the sample volume filtered, the injection, or sample loop, volume and the extraction volume, determined with an internal standard.

8. Quality Assurance Procedures

If high-quality data are desired, quality assurance (QA) procedures should be employed to insure accurate, precise and representative results. All samples should be

TABLE 4-1.
Recommended List of Extinction Coefficients to be Used For the Calibration of Pigment Standards

Pigment	Wavelength (solvent)	$E_{1cm}(L\ g^{-1}\ cm^{-1})$	Reference
Chlorophyll *a*	663 nm (acetone)	88.15	(1)
Chlorophyll *a*	664 nm (90% acetone)	87.67	(1)
Chlorophyll *b*	647 nm (90% acetone)	51.36	(1)
Chlorophyll c_1	631 nm (90% acetone)[1]	44.8	(2)
Chlorophyll c_2	631 nm (90% acetone)[1]	40.4	(2)
Chlorophyll c_{1+2}	631 nm (90% acetone)[1]	42.6	(2)
Chlorophyllide *a*	664 nm (90% acetone)	128.0	(3)
Phaeophorbide *a*	665 nm (90% acetone)	69.8	(3)
Phaeophytin *a*	665 nm (90% acetone)	49.5	(3)
Fucoxanthin	449 nm (EtOH)	160.0	(4)
Peridinin	466 nm (acetone)	134.0	(5)
Peridinin	469 nm (MeOH)	136.0	(5)
Peridinin	472 nm (EtOH)	132.5	(5)
Diadinoxanthin	446 nm (EtOH)	262.0	(6)
Diatoxanthin	449 nm (EtOH)	262.0	(6)
Lutein	445 nm (EtOH)	255.0	(7)
Zeaxanthin	450 nm (EtOH)	254.0	(7)
Neoxanthin	439 nm (EtOH)	224.3	(7)
Violaxanthin	443 nm (EtOH)	255.0	(7)
β-carotene	453 nm (EtOH)	262.0	(7)
19′-butanoyloxyfucoxanthin	446 nm (EtOH)	(160.0)[2]	(NA)
19′-hexanoyloxyfucoxanthin	447 nm (EtOH)	(160.0)[2]	(NA)
Alloxanthin	453 nm (EtOH)	(262.0)[3]	(NA)
Prasinoxanthin	454 nm (EtOH)	(160.0)[4]	(NA)

References: (1) Jeffrey and Humphrey [1975]; (2) Jeffrey [1972]; (3) Lorenzen and Downs [1986]; (4) Jensen [1966]; (5) Jeffrey and Haxo [1986]; (6) Goodwin [1955]; (7) Davies [1976]; and (NA) not available.

Footnotes:

[1] plus 1% pyridine

[2] value for fucoxanthin [cf. Vesk and Jeffrey, 1987]

[3] value for diatoxanthin [cf. Gieskes and Kraay, 1984]

[4] value for fucoxanthin

analyzed in duplicate, and triplicate analyses should be performed at a rate of 10% (every tenth sample) to assess representativeness and precision. Fortified samples should be analyzed at a rate of 5% (every twentieth sample) as part of the QA effort. Fortified samples are prepared in duplicate by spiking a sample with known quantities of the analytes of interest at concentrations within the range of concentrations expected in the samples. Fortified samples are used to assess the method's precision and accuracy in the presence of a typical sample matrix. In addition, system and spiked blanks should be routinely analyzed. A system blank consists of a filter, reagents, and the glassware and hardware utilized in the analytical scheme. The system blank is quantified under identical instrumental conditions as the samples and is analyzed by appropriate quantitative methods. The system blank may not contain any of the analytes of interest above the MDL (see below) or corrective action is taken. A spiked blank is defined as a system blank plus an authentic external standard containing the analytes of interest. Each set of samples should be accompanied by a spiked blank and is quantified under the same instrumental conditions as the samples.

In addition to daily calibration of the HPLC system with external standards, an internal standard should be used to correct for experimental losses and volume changes. The internal standard should have properties similar to the analytes of interest, however, it should be completely resolved from all analytes of interest and any method interferences. The internal standard should be added to the sample prior to extraction and used to correct for sample processing losses and sample volume changes. When new external and internal standards are prepared they should be verified against previous standards and a standard reference solution if available.

To insure comparability with other laboratories, intercalibrations and analyses of standard reference materials (available from the U.S. Environmental Protection Agency) should be routinely performed. In multi-ship/investigator studies, replicate samples should be collected and archived for future intercalibration checks. If desired, the method detection limit (MDL) for the analytes of interest can be determined by seven replicate standard injections [Glaser et al., 1981]. The standard deviation of the seven replicate measurements is calculated and the MDL is computed as:

$$MDL = t_{(N-1,\ 1-\alpha\ =\ 0.99)} \times S_c$$

where, $t_{(N-1,\ 1-\alpha\ =\ 0.99)}$ is the student's t value for a one-tailed test at the 99% confidence level with N-1 degrees of freedom. S_c is the standard deviation of the seven replicate analyses.

References

Bidigare, R. R., O. Schofield, and B. B. Prezelin, Influence of zeaxanthin on quantum yield of photosynthesis of Synechococcus clone WH7803 (DC2), Mar. Ecol. Prog. Ser., 56, 177-188, 1989.

Davies, B. H., Carotenoids, in Chemistry and Biochemistry of Plant Pigments, edited by T. W. Goodwin, pp. 38-165, Academic Press, New York, 1976.

Gieskes, W. W. C., and G. W. Kraay, Unknown chlorophyll *a* derivatives in the North Sea and the tropical Atlantic Ocean revealed by HPLC analysis, Limnol. Oceanogr., 28, 757-766, 1983.

Gieskes, W. W. C., and G. W. Kraay, Phytoplankton, its pigments, and primary productivity at a central North Sea station in May, July and September 1981, Neth. J. Sea Res., 18, 51-70, 1984.

Gieskes, W. W. C., and G. W. Kraay, Analysis of phytoplankton pigments by HPLC before, during and after mass occurrence of the microflagellate Corymbellus aureus Green during the spring bloom in the northern North Sea in 1983, Mar. Biol., 92, 45-52, 1986.

Glaser, J. A., D. L. Foerst, G. D. McKee, S. A. Quave, and W. L. Budde, Trace analyses for wastewaters, Environ. Sci. Technol., 15, 1426-1435, 1981.

Goodwin, T. W., Carotenoids, in Modern Methods in Plant Analysis, edited by K. Paech and M. Tracey, Vol. 3, pp. 272-311, Springer, Berlin, 1955.

Jeffrey, S. W., Preparation and some properties of crystalline chlorophyll c_1 and chlorophyll c_2 from marine algae, Biochim. Biophys. Acta, 279, 15-33, 1972.

Jeffrey, S. W., An improved thin-layer chromatographic technique for marine phytoplankton pigments, Limnol. Oceanogr., 16, 191-197, 1981.

Jeffrey, S. W., and F. T. Haxo, Photosynthetic pigments of symbiotic dinoflagellates (zooxanthallae) from corals and clams, Biol. Bull., 135, 149-165, 1986.

Jeffrey, S. W., and G. F. Humphrey, New spectrophotometric equations for determining chlorophylls *a, b, c* in higher plants, algae, and natural phytoplankton, Biochem. Physiol. Acta, 167, 191-194, 1975.

Jeffrey, S. W., and S. W. Wright, A new spectrally distinct component in preparations of chlorophyll *c* from the micro-alga *Emiliania huxleyi* (Prymnesiophyceae), Biochim. Biophys. Acta, 894, 180-188, 1987.

Jensen, A., Algal carotenoids. V. Iso-fucoxanthin-rearrangement product of fucoxanthin, Acta Chem. Scans., 20, 1728-1730, 1966.

JGOFS, Pigments and chlorophylls, in Core Measurement Protocols, Report No. 6, Chapter 9, pp. 17-19, 1990.

Lorenzen, C. J., and J. N. Downs, The specific absorption coefficients of chlorophyllide *a* and phaeophorbide *a* in 90% acetone, and comments on the fluorometric determination of chlorophyll and phaeopigments, Limnol. Oceanogr., 31, 449-452, 1986.

Mantoura, R. F. C., and C. A. Llewellyn, The rapid determination of algal chlorophyll and carotenoid pigments and their breakdown products in natural waters by reverse-phase high-performance liquid chromatography, Anal. Chim. Acta, 151, 297-314, 1983.

Ondrusek, M. E., R. R. Bidigare, S. T. Sweet, D. A. Defreitas, and J. M. Brooks, Distribution of algal pigments in the North Pacific Ocean in relation to physical and optical variability, Deep-Sea Res., 38, 243-266, 1991.

Price, N. M., P. J. Harrison, M. R. Landry, F. Azam, and K. J. F. Hall, Toxic effect of latex and Tygon tubing on phytoplankton, zooplankton and bacteria, Mar. Ecol. Prog. Ser., 34, 41-49, 1986.

Vesk, M., and S. W. Jeffrey, Ultrastructure and pigments of two strains of the picoplanktonic alga *Pelagococcus subviridis* (Chrysophyceae), J. Phycol., 23, 322-336, 1987.

Welschmeyer, N. A., and N. Hoepffner, Xanthophyll cycling I: Dynamics of rapid, light-induced pigment changes in marine phytoplankton, Limnol. Oceanogr., in press, 1991.

Williams, P. J. Leb., and J. I. Robertson, A serious inhibition problem from a Niskin sampler during plankton productivity studies, Limnol. Oceanogr., 34, 1300-1305, 1989.

Wright, S. W., and J. D. Shearer, Rapid extraction and high performance liquid chromatography of chlorophylls and carotenoids from marine phytoplankton, J. Chromatogr., 294, 281-295, 1984.

ANALYSIS OF CAROTENOIDS AND CHLOROPHYLLS IN SUSPENDED PARTICULATE MATTER, SEDIMENT TRAP SAMPLES, AND SURFACE SEDIMENTS

Daniel J. Repeta

Woods Hole Oceanographic Institution, Department of Chemistry, Woods Hole, Massachusetts 02543

1. Sampling and Extraction

For shipboard analysis of suspended particulate matter, 0.5-1 L of seawater is filtered through a pre-extracted or pre-combusted (450° C) 47 mm Whatman GF/F glass fiber filter under low vacuum. The filter is immediately placed in a glass grinding tube at 0° C, 2 mL of acetone and 0.1 mL of the internal standard (Zn pyrophaeobutyn-a or canthaxanthin in acetone) added, and the sample ground at high speed for 2 min. The sample can be stored at 0° C for a short period of time (<4 hr) at this stage without measurable degradation of the pigments. After centrifugation, the sample is ready for analysis by high pressure liquid chromatography (HPLC). Sediment trap samples can likewise be filtered and handled in the same manner. If the sample is to be stored for future analysis, the filter is folded and placed inside a heat sealed plastic bag and rapidly frozen to -20° C. Storage above -20° C, or slow freezing of filtered algal samples results in considerable pigment degradation.

Pigments in sediments occur as very complex mixtures of both nonpolar and polar components, and are more difficult to extract. Frozen samples of 1-5 g sediment are thawed, placed in 35 mL centrifuge tubes and sonic extracted 3x with methanol (MeOH) (25 mL, 10-20 min each). Samples are centrifuged, the supernate decanted, and fresh solvent added following each sonic extraction. Unless the sediments are particularly rich in organic matter, the third MeOH extract is somewhat lighter than the second extract. A final three extractions are then made with methylene chloride ($MeCl_2$) (25 mL, 10-20 min each). Extracts are combined, spiked with 1 mL of an internal standard, and evaporated under reduced pressure to approximately 10-20 mL. Pigments can then be partitioned with diethyl ether or hexane/ethyl acetate and 10% NaCl (aq). The organic extracts are washed with 1% $NaHCO_3$ followed by distilled water, and evaporated to dryness.

Marine Particles: Analysis and Characterization
Geophysical Monograph 63

2. Preparative Chromatography

Suspended particulate matter samples and material collected by sediment traps can usually be analyzed directly by HPLC without further treatment. Pigment extracts from sediments are complex, often containing several hundred pigments, and need to be fractionated before HPLC analysis if a very detailed knowledge of the total pigment mixture is desired. If only specific pigments are to be measured, the samples can be analyzed directly by HPLC provided the appropriate separation conditions are known. We generally make an initial fractionation by polarity using silica or reverse phase (ODS) flash chromatography to separate carotenes, phaeophytins, monohydroxy- and keto-carotenoids, carotenoid diols, and more polar xanthophylls and phorbins [Repeta, 1989]. Subfractions are then further separated into carotenoids, metallophorbins, and phaeophorbides by gel permeation chromatography (GPC) on cross-linked polystyrene [Repeta, 1989]. Subfractions from GPC are then analyzed by HPLC.

3. High Pressure Liquid Chromatography

Simple mixtures of pigments from suspended particulate matter and sediment trap samples are analyzed after the method of Zapata et al. [1987]. For rapid analysis of seawater samples we use a 10 cm ODS column (adsorbosphere HS, 3 μm) eluted with a linear gradient of 0-100% B (A = 20/80 0.5 N NH_4Ac (aq)/MeOH; B = 20/80 acetone/MeOH) over 10 min at 1.5 mL/min. For more detailed analysis we use a 15 cm column and a 30 min gradient. Pigments extracted from sediments are run on 15 cm columns, but using a slower program through a much smaller gradient to achieve the desired separation. If possible, all samples are analyzed by diode array spectroscopy (350-700 nm) and fluorescence detection.

Individual components can be quantified providing the correct identification of the pigment has been made, and the appropriate detector calibration factor is known. Normalization of the data to the internal standard

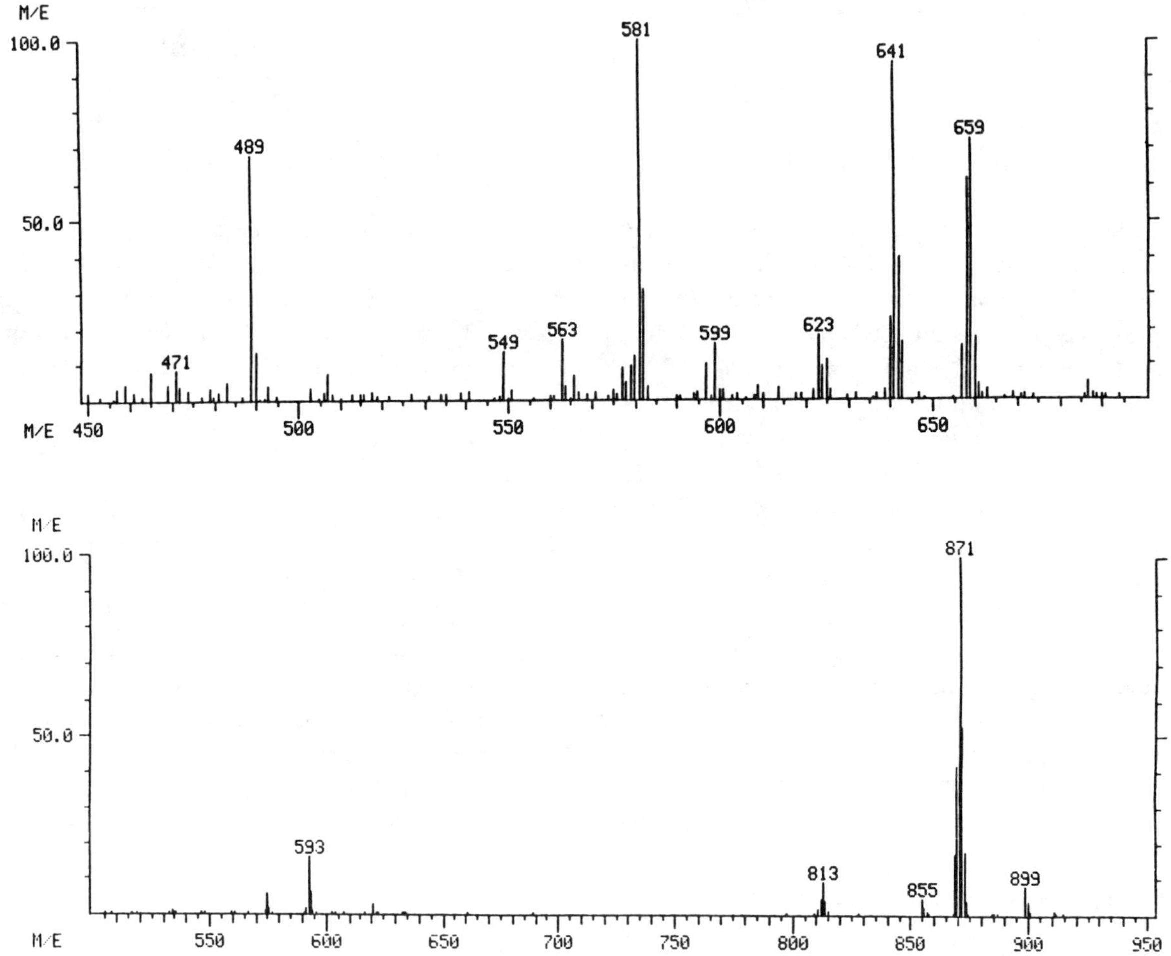

Fig. 4-1. Chemical ionization mass spectra of fucoxanthin (top) and phaeophytin-a (bottom). Spectra were collected on a gold-tipped direct insertion probe as described in the text. Assignments are as follows: fucoxanthin (M+1) 659, (M+1-18) 641, (M+1-18-18) 623, (M+1-60) 599, (M+1-60-18) 581, (M+1-60-18-18) 623, (M+1-170) 489, (M+1-170-18) 471; phaeophytin-a (M+1) 871, (M+1-CH2CO2) 813, (M+1-phytene) 593.

permits correction for evaporative losses, differences in water retention by filters, and nonquantitative transfers during handling. We calibrate our system by isolating pure pigments, making quantitative standards using a solvent in which the extinction coefficient for the compound has been reported, and carrying out a series of analyses over the calibration range of interest. Depending on the pigment's spectral properties, and the wavelength of detection, calibration factors for individual pigments may differ by up to a factor of 3. For calibration in the field, concentrated solutions of standards are prepared in benzene, and 1 mL aliquots dispensed to 10 mL volumetric flasks. The flasks are flushed with N_2, capped, and frozen at -20° C. Randomly chosen standards are stored in the laboratory for spectroscopic analysis at a later date. The remaining flasks are stored shipboard until use. For calibration, the standards are returned to room temperature and acetone of methanol added to 10 mL. Aliquots are then analyzed by HPLC. In the U.S. JGOFS program, over 30 replicate standards prepared and analyzed in this way, showed a SD of <10% in replicate HPLC analyses after three months of storage and an additional three months of shipboard analysis.

4. Mass Spectral Analysis

Mass spectrometry is used for the tentative identification of unknown pigments or to confirm identification when ambiguity exists for "known" compounds. It is important that strict precautions be taken to avoid contamination from colorless lipids during sample handling and purification. All solvents should be of the highest purity and only the minimum amount of solvent needed to affect the manipulations should be used. All

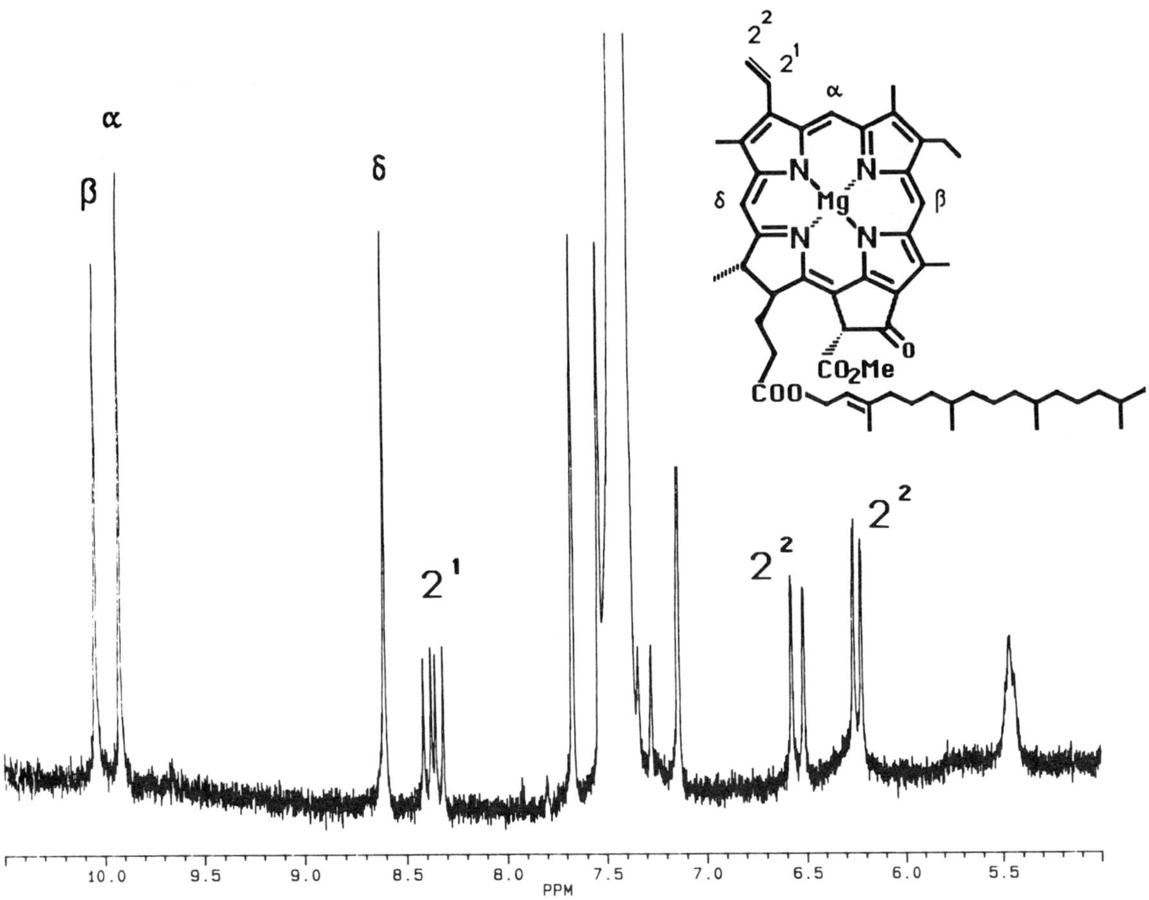

Fig. 4-2. PNMR of 52 μg chlorophyll-a in benzene-D6; 24877 scans collected overnight. The solvent peak is at 7.4 ppm, with peaks for water at 0.65 and 4.5 ppm.

glassware and thin layer chromatography (TLC) plates should be thoroughly cleaned and solvent rinsed before use. Pigments from suspended particulate matter samples are routinely analyzed by collecting the HPLC column eluant, evaporating it to dryness, and redissolving the pigment in $MeCl_2$ for placement on the probe. Pigments collected from HPLC analysis of sediments must be further purified by TLC before mass spectral analysis. Typically components collected from the HPLC are evaporated to dryness under a stream of N_2, redissolved in acetone and spotted on a prewashed (50/50 $MeCl_2$/acetone) ODS (Merck) TLC plate. The plate is developed in 100% acetonitrile, the major band collected and the pigment eluted with acetone. Pigments are analyzed by chemical ionization mass spectrometry using a gold-tipped, direct insertion probe, heated ballistically from ambient to 350° C in 10-15 sec. It is our experience that the faster the sample is heated, the better the resulting spectrum. Using this technique, good spectra can be routinely collected for demetallated phorbins (100 ng) and carotenoids (20-100 ng) (Figure 4-1). We are unable to collect mass spectra of metallochlorins; however, spectra have been reported using a similar sample introduction technique with soft ionization methods [Constantin et al., 1981].

5. Nuclear Magnetic Resonance Spectrometry (NMR)

Nuclear magnetic resonance spectrometry provides detailed structural information necessary for full structural characterization of unknown pigments. Good PNMR spectra can be collected on <50 μg of pigment with an overnight run on a 300 MHz instrument (Figure 4-2). Longer run times or high magnetic field strengths can reduce the sample requirement to < 10 μg. The same strict precautions against contamination that are necessary for mass spectral analysis apply to analysis by NMR. In addition, NMR tubes must be of very high quality and thoroughly cleaned, particularly for small samples. Deuterated solvents must also be of the highest quality.

References

Constantin, E., Y. Nakatami, G. Teller, R. Hernbert, and G. Ourisson, Electron impact and chemical ionization mass spectrometry of chlorophyll, phaeophytin, and phaeophorbides by field desorption on a gold support, Bull. Soc. Chim. Fr. (pt 2), 303-305, 1981.

Repeta, D. J., Carotenoid diagenesis in recent marine sediments: II. Degradation of fucoxanthin to loliolide, Geochim. Cosmochim. Acta, 53, 699-707, 1989.

Zapata, M., A. M. Ayala, J. M. Franco, and J. L. Garrido, Separation of chlorophylls and their degradation products in marine phytoplankton by reversed-phase high performance liquid chromatography, Chromatographia, 23, 26-30, 1987.

LIGNIN, CUTIN, AMINO ACID AND CARBOHYDRATE ANALYSES OF MARINE PARTICULATE ORGANIC MATTER

John I. Hedges

School of Oceanography, WB-10, University of Washington, Seattle, Washington 98195

1. Introduction

Our group at the University of Washington has specifically designed methods for the analysis of lignin compounds [Hedges and Ertel, 1982], cutin acids [Goñi and Hedges, 1990a], amino acids [Cowie and Hedges, in press, 1991a, b] and various carbohydrates, including aldoses [Cowie and Hedges, 1984], cyclitols [Hedges and Weliky, 1989] and uronic acids [Walters and Hedges, 1988; Bergamaschi and Hedges, in preparation], in particulate samples from aquatic environments. All of these procedures are derivatives of previous methods that we have adapted for application to complex natural mixtures and tested on a variety of sample types, such as plankton, woods, soils and sediments, for precision, accuracy and yield efficiencies. All the methods are written up in detail and only will be summarized in the following sections. The remaining discussion, covering the various compound types in the order given above, will focus on unpublished procedural developments for each technique, special problems that are unique to each method and related tricks of the trade. Lignin analysis will be treated in most detail because it is the method with which we have had the longest and most detailed experience.

2. Lignin Analysis

Lignins are large, structurally complex phenolic polymers unique to vascular land plants where they comprise up to 30 wt% of woody tissue mass [Sarkanen and Ludwig, 1971]. The polymers are comprised of p-hydroxyl, vanillyl (guaiacyl), syringyl and cinnamyl structural units which occur in differing abundance patterns based on plant and tissue type. These intrinsically stable structural units are cross-linked by relatively strong carbon-carbon and ether bonds, making lignin polymers resistant to both biological and chemical degradation. Lignin polymers are among the most stable components of vascular plant tissues where they typically are selectively preserved in both terrestrial [Bates and Hatcher, 1989] and aquatic [Hedges et al., 1985; Hedges and Weliky, 1989] environments. The characteristics of high concentration, chemical stability, and unique origin among vascular plant tissues of different fundamental types make lignin compounds useful tracers of land-derived organic matter in aquatic environments [e.g., Hedges et al., 1986; 1988].

2.1. *Published Analytical Method*

Our lignin analytical procedure was adapted from wood chemistry [Sarkanen and Ludwig, 1971]. It involves basic CuO oxidation that degrades lignin polymers to produce up to eight vanillyl, syringyl and cinnamyl phenols which are unique to lignins and closely related compounds in vascular plant sources [Hedges and Ertel, 1982].

The sample, typically 20 mg of plant material or 250 to 500 g of sediment, is heated along with 1 g of powdered CuO, 100 mg of $Fe(NH_4)_2(SO_4)_2{\cdot}6H_2O$ and 7 mL of 8 wt% NaOH (O_2-free) within a small (≈ 10 mL) stainless steel minibomb. Four such minibombs are heated together in a larger (200 mL) high pressure bomb mounted on a shaker table. The big bomb slips into a brass cylinder that is wrapped with heating tape and insulation and maintained at 170° C throughout a 3 hour reaction period.

The reaction mixture is spiked before heating with ^{14}C- labelled p-hydroxyacetophenone which has been shown to be a representative tracer for the recovery of lignin-derived phenols [Hedges, 1975]. Upon termination of the reaction, the liquid phase from the individual minibombs is acidified, extracted with ethyl ether, dried over Na_2SO_4, evaporated to dryness and stored in a freezer. At the time of analysis, the residue is dissolved in pyridine and a split of this solution is removed, dried and archived in a freezer. The remaining solution is spiked with ethylvanillin (GC internal standard) in pyridine, mixed with an equal volume of Regisil, plus 1% trimethylchlorosilane, and heated at 60° C for 10 minutes to form trimethylsilyl (TMS) derivatives.

The silylated phenol mixture is analyzed on a 30 m by 0.25 mm i.d. fused silica capillary column coated with methyl silicone (DB-1) liquid phase. The sample is injected into a helium gas stream at a 1:50 split and the

Marine Particles: Analysis and Characterization
Geophysical Monograph 63

GC oven is temperature programmed from 100° C, with no initial delay, to 270° C at 4° min^{-1} and held at the upper temperature for at least 16 minutes. Standards containing all eight lignin-derived phenols, ethylvanillin and selected other compounds, e.g., p-hydroxyl phenols, are analyzed about every fifth sample to establish relative response factors. In our hands this method has a sensitivity of 0.1 mg, a precision of about 10% and excellent long term repeatability. Four samples are processed at a time, with a maximal throughput rate of about two samples per day.

2.2. *Procedural Changes*

A number of significant changes in the previously outlined analytical procedure have been made since the 1982 publication. These alterations are highlighted sequentially in the following paragraphs in their approximate sequence within the procedure.

Bomb temperature is a key variable in the lignin CuO degradation procedure. Based on recent detailed studies (De Villiers and Hedges, in preparation) we know that the "ideal" heating conditions for maximal return of geochemical information represent a compromise between temperatures which are sufficiently severe to give high yields of vanillyl phenols yet mild enough to allow efficient recovery of the more thermally sensitive cinnamyl phenol products. We have found from experience with other groups who have set up their own lignin procedures, that our analytical results can be consistently repeated only if they closely copied our bomb set up.

Based on the previous observations, we recently converted to an improved heating apparatus which is more robust, better insulated and uses dual probes to monitor and record the temperatures of both the heating sleeve (as before) and the water directly jacketing the minibombs (new procedure). Based on an extensive series of experiments with both a sediment and NBS standard citrus leaves, we have now established that the usual oxidation procedure (170° C heating sleeve temperature) is nearly ideal in terms of geochemical information return, and corresponds to a final internal bomb temperature of 155° C (De Villiers and Hedges, in preparation). It's this latter effective temperature and the corresponding heating profile that must be reproduced by various labs using the CuO technique to make it possible for all of us to draw on a common lignin oxidation product data base. De Villiers' study will also facilitate the interpretation and intercomparison of CuO analytical results by providing additional information on the effects of heating rate, O_2, $Fe(NH_4)_2(SO_4)_2 \cdot 6H_2O$, and CuO texture on the yields of lignin phenols and other reaction products.

Omitting the radiolabelled recovery standard might be desirable to some workers who wish to avoid the trouble and expense involved with radioactive material. Although we have not done this, others (e.g., Benner, UTMSI) have spiked ethylvanillin into the reaction mixtures directly following the CuO oxidation so that this compound can essentially serve as a recovery standard. Under these circumstances, cinnamic acid can be used as a GC internal standard to monitor the absolute recovery of the ethylvanillin and to check for preferential loss of carbonyl products, e.g., ethylvanillin and vanillin, that can occur in nitrogen-rich samples [Hedges and Ertel, 1982]. We routinely use cinnamic acid as a second GC internal standard to check for possible selective losses of ethylvanillin, which is more volatile and more reactive toward a broader range of chemicals.

Alternate derivatization procedures can be quite useful for the preparation of phenolic TMS derivatives. For example, deuterated TMS (DTMS) reagents are helpful for determining the number of acidic functional groups in an unknown reaction product (indicated by an additional 9 amu/acidic hydrogen vs. conventional TMS derivatives). Because deuterated reagent is about 1000 times more expensive, we have gone to a "within-syringe" derivatization method (M. Goñi invention). In this procedure equal volumes ($\approx 0.5\ \mu L$) of the sample in pyridine and of DTMS reagent are drawn into a GC syringe that is then placed in a 60° C oven for 10 min before injecting the sample. This approach cuts reagent and sample expenditures by a factor of 100 and gives equivalent results to the conventional within-vial derivatization. Theoretically, this trick also could be used to derivatize and inject extremely small samples.

With the advent of relatively inexpensive, less than $5000, GC automatic injectors, such as the Shimadzu AOC 14, this capability has come within the reach of many labs. We have recently adapted our Hewlett Packard model 5890 GCs for compatible use with the AOC 14 and have confirmed that the TMS derivative solutions of most CuO reaction product mixtures are stable at room temperature for sufficiently long periods, up to at least 24 hrs, to allow uncompromised use of a small, less than 24 sample, automatic injector.

Simultaneous dual-column gas chromatography can be a useful tool for the analysis of compositionally complex CuO reaction product mixtures. Although methyl silicone, e.g., DB-1, is one of the best and most commonly used liquid phases for the analysis of lignin

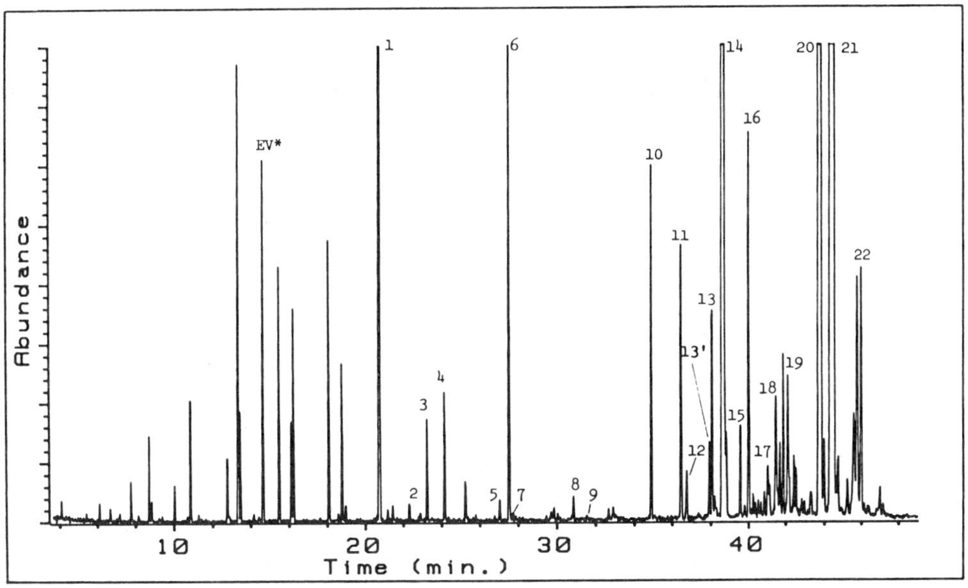

Fig.4-3. GC-MS trace (total ion current) of CuO oxidation products obtained from purified apple cuticle [Goñi and Hedges, 1990a]. Peak numbers are according to Table 4-2. Abbreviations: EV*, ethylvanillin (internal standard).

oxidation products by capillary gas chromatography, the TMS derivatives of some phenols, particularly p-hydroxybenzaldehyde and ferulic acid, may coelute with other reaction products. This is especially true of lignin-poor materials such as open ocean and bloom-period sediment trap samples. In the past we have compensated for this problem by separately analyzing lignin-poor samples using a second, more polar liquid phase, typically DB-1701, 14% cyanopropylphenyl-86% methylsilicone. To facilitate this procedure, we now have fitted the splitting injector on one HP 5890 GC with dual column connectors to accommodate DB-1 and DB-1701 capillary columns on which lignin phenol mixtures can be simultaneously analyzed. The outputs from the two flame ionization detectors (FID) are processed by separate integrators (HP 3390). The data are then entered into a computer program which allows the results to be visually compared to obtain the best overall match, from which any needed corrections for coelution can be readily made. We presently are in the process of more fully automating the integration and signal processing steps using a Lab Calc chromatography data system.

3. Cutin Analysis

Gas chromatograms of CuO reaction product mixtures from sediments typically contain approximately 100 easily discernible peaks. One of our current lines of research is to identify the additional components of these complex mixtures and to relate them to biochemical precursors which carry geochemical information. The advantage of this approach is that any information which can be obtained from the non-lignin products comes with essentially no additional effort or sample loss and is available for all previously analyzed samples.

One example of this effort has been the identification of a suite of up to 17 cutin-derived hydroxy acids (Table 4-2) among the CuO reaction products of many different plant and sediment samples [Goñi and Hedges, 1990a, b, c]. These late-eluting acids (Figure 4-3) are unique to cutin, a polyester that occurs as coverings only on nonwoody vascular plant tissues such as leaves, stems and fruits [Holloway, 1982; Goñi and Hedges, 1990b]. We have found that the cutin acids obtained with the CuO analytical procedure are essentially the same as those produced by simple base hydrolysis [Goñi and Hedges, 1990a], thereby allowing direct comparisons to existing literature.

A particular advantage of cutin acids (Table 4-2) is their ability to complement the simultaneously produced lignin-derived phenols as indicators of the types and amounts of vascular plant tissues in sedimentary mixtures. Cutin acids are particularly useful in such applications because they are produced only from nonwoody vascular plant tissues, not from woods, and allow distinctions among potential sources, e.g.,

TABLE 4-2.
Composition of Bulk Leaves and Purified Cuticles From English Holly (*Ilex aquifolium*)

RRT	Acid Name	Leaves (%)	Cuticles (%)	Peak#
	C_{16} ACIDS			
1959	16-Hydroxyhexadecanoic[3]	3.3	1.3	10
2037	Hexadecane-1,16-dioic[3]	0.8	0.5	11
2154	8,16-Dihydroxyhexadecanoic[3a]	2.9	2.4	14
2154	9,16-Dihydroxyhexadecanoic[3b]	3.0	4.0	14
2154	10,16-Dihydroxyhexadecanoic[3c]	35.1	29.8	14
2221	7-Hydroxyhexadecane-1,16-dioic[2a]	3.3	2.1	16
2221	8-Hydroxyhexadecane-1,16-dioic[2b]	4.3	6.1	16
	C_{18} ACIDS			
2119	18-Hydroxyoctadec-9-enoic[2]	6.5	2.8	13
2197	9-Octadecene-1,18-dioic[1]	0.3	0.2	15
2277	Dihydroxyoctadecedienoic[1]	0.4	0.2	17
2300	Dihydroxyoctadecenoic[1]	1.1	0.6	18
2333	8,18-Dihydroxyoctadecanoic[1a]	tr	0.2	19
2333	9,18-Dihydroxyoctadecanoic[1b]	tr	0.2	19
2333	10,18-Dihydroxyoctadecanoic[1c]	0.4	0.7	19
2436	9,10,18-Trihydroxyoctadec-12-enoic[2]	2.3	1.4	20
2478	9,10,18-Trihydroxyoctadecanoic[3]	35.8	46.3	21
	OTHER ACIDS			
2050	10, 15-Dihydroxypentadecanoic[2]	0.5	0.5	12
2110	17-Hydroxyheptadecanoic[2]	tr	0.6	13′
	Σ CA	5.7	66.8	
	%OC	48.1	63.4	
	C:N	43.5	117.0	

[1]Determined from mass spectral interpretation of TMS, DTMS or TMS/Me derivatives.
[2]Determined from matching with published mass spectra.
[3]Determined from direct retention time and mass spectral matching versus a commercial standard.
a,b,c, abundances of coeluting structural isomers determined from a mass spectrum at apex of peak
Abbreviations: RRT, relative retention time index of TMS ester/ether derivatives with respect to silylated normal fatty acid standards on a 30 m by 0.25 mm i.d. SE-30 column, temperature program from 100° C to 270° C at 4° C/min and isothermal at 270° C for 16 min; Peak#, chromatographic peak number in GC trace (Figure 4-3); tr, trace (less than 0.1% of total acids); ΣCA, total yield of all acids in units of mg/100 mg OC; %OC, weight percent organic carbon; C:N, atomic ratio.

monocots vs. dicots and cedar needles vs. fir and pine needles, that are not possible with lignin phenols alone [Goñi and Hedges, 1990b]. Our preliminary studies of the early diagenetic behavior of cutins from fir, hemlock and cedar needles indicate that the component acids have comparable overall stabilities similar to that of the bulk tissue [Goñi and Hedges, 1990c]. Thus, cutin acids in these tissue types are not susceptible to selective diagenetic alteration to the extent that lignin-derived phenols can be [Hedges and Weliky, 1989]. Work now underway to identify other non-lignin CuO reaction products is quite promising for several biochemical categories. Based on these results we fully expect that the CuO method will soon yield geochemical information on as broad a suite of biochemicals as is now possible with pyrolysis and afford the added advantages that these results are quantitatively interpretable and largely independent of sample matrix effects.

3.1. Procedural Considerations

Analyses of cutin acids, and other non-lignin CuO reaction products, involve some special considerations, or alterations, versus the standard phenol technique. An important note is that all GC runs should be allowed to go to full time (58.5 min) in order to elute cutin acids and other high molecular weight reaction products. In addition, integration programs should be set up to allow accurate measurements of as many peak areas as possible throughout the GC run, because many of the data may be utilized in the future. As new compound types are identified, it will also be necessary to add them to standard solutions and to incorporate proxies into recovery standard mixtures. Another general aid in identifying and cross comparing the many chromatography peaks in these reaction product mixtures is the use of standard fatty acid mixtures (C_4-C_{24}, Sigma) as retention time reference points so that individual reaction products can be assigned Kovats-type indices for different GC liquid phases. We are now working toward a system that will incorporate a flexible chromatography data system and data base manager to handle the substantial flow of information generated by simultaneous dual-column analyses of automatically injected samples containing 100+ major components each.

4. Amino Acid Analysis

As a part of his dissertation research on the comparative geochemistries of the major biochemicals, Greg Cowie developed a modified method for amino acid analysis which involves an abbreviated hydrolysis, automatic injection of precolumn derivatives, and charge-matched recovery standards (Cowie and Hedges, in press-1991a). These alterations were made to streamline conventional procedures and allow for varying recoveries of acidic, neutral and basic amino acids which undergo differential losses due to a variety of physical and chemical processes. This method has now been applied to a wide variety of samples including sediment trap and core samples [Cowie and Hedges, in press-1991b], plankton and fecal pellets [Cowie, 1990]. The following discussion will highlight the more novel aspects of this procedure.

The abbreviated hydrolysis procedure involves sealing O_2-free samples, ≈ 0.1 mg of protein, and a recovery standard solution (see below) in glass vials along with 1 to 2 mL of 6 N HCl. The samples are then maintained at 150° C for 70 min in a temperature-controlled heating block. Following hydrolysis the sample solution is dried and redissolved in water containing o-methylthreonine as a chromatography internal standard. Use of a Savant centrifugal evaporator, with a heating lamp warmer directed through a plexiglass top, greatly decreases sample processing time versus that possible with conventional rotoevaporators. Recovery comparisons with both pure protein and a marine sediment show that the high-temperature procedure gives essentially identical results, except for higher methionine yields, to those obtained with a more conventional hydrolysis of 6 N HCl for 22 hrs at 110° C.

Automated sample injection is accomplished with a Waters Model 510 HPLC fitted with a precolumn mixing chamber, a reverse-phase column (C_8 or C_{18}), a fluorescence detector and a computer-based data system and gradient controller. The "WISP" autosampler simultaneously injects known volumes of the hydrolysate mixture along with an o-phthaldialdehyde (OPA) reagent solution [Smith and Panico, 1985], thereby rigidly controlling the timing of formation of the various amino acid-OPA derivatives, which have short, hours to days, and variable lifetimes. HPLC analyses of the entire suite of naturally occurring protein and nonprotein amino acids and corresponding charge-matched recovery standards, see below, are carried out in less than 30 minutes (Figure 4-4). The entire analytical procedure (including hydrolysis) has a practical detection limit of about 1 picomole and a reproducibility for major amino acid components of 5 to 10% [Cowie and Hedges, in press-a]. Twelve samples are processed at a time, with a maximal average throughput rate of four samples per day.

Charge-matched recovery standards that are chromatographically separable from common amino acid hydrolysis products of natural samples (Figure 4-4) have now been identified for acidic, neutral, intermediate (histidine) and basic protein amino acid counterparts (Table 4-3). Tests with amino acid/recovery tracer mixtures that were spiked before hydrolysis into sediments and wood, or simply partitioned versus different clay mineral mixtures, demonstrate substantial and characteristically different recoveries among the above four charge categories. Within each of the individual charge classes, however, the recovery standards in Table 4-3 behave similarly to the corresponding natural amino acids and thus are good proxies for their recoveries [Cowie and Hedges, in press-a]. Based on these results it is evident that amino acid yields can be measured more accurately employing a suite of charge-matched recovery standards than with only one neutral recovery standard, as is presently used in almost all other procedures [see review by Mopper and Dawson, 1986].

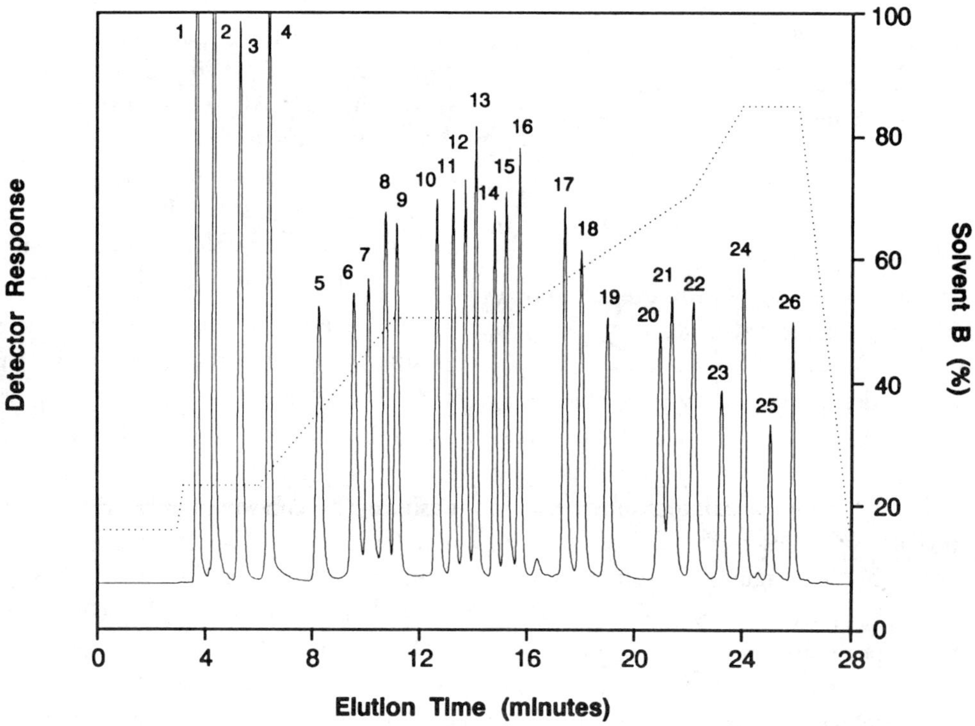

Fig. 4-4. Chromatographic trace (Cowie and Hedges, submitted, 1991) of a standard amino acid mixture, including a typical solvent gradient (dotted line). Solvent A is 30-55 mM sodium acetate plus 1.5-2.8% tetrahydrofuran, adjusted to pH 5.9-6.5. Solvent B is methanol. Peak labels: 1, aspartic acid; 2, glutamic acid; 3, α-aminoadipic acid; 4, α-aminopimelic acid; 5, serine; 6, histidine; 7, 1-methylhistidine; 8, glycine; 9, threonine; 10, arginine; 11, alanine; 12, β-alanine; 13, tyrosine; 14, o-methylthreonine; 15, γ-aminobutyric acid; 16, α-aminobutyric acid; 17, methionine; 18, valine; 19, phenylalanine; 20, p-fluorophenylalanine; 21, isoleucine; 22, leucine; 23, δ-hydroxylysine; 24, γ-methylleucine; 25, ornithine; 26, lysine.

5. Carbohydrate Analysis

We presently employ two different analytical procedures for carbohydrates in natural samples. The first of these is a method designed specifically for quantifying aldoses in natural samples by gas chromatography of equilibrated isomeric mixtures [Cowie and Hedges, 1984]. This method will be referred to as the "aldose" technique. The second procedure was more recently developed for the simultaneous determination of uronic acids and aldoses in natural samples by gas chromatography of N-hexylaldonamide and alditol acetates [Walters and Hedges, 1988] and will be called the "uronic acid" technique. Both methods also allow the simultaneous measurement of cyclitols as well as O-methyl ethers of aldoses, uronic acids and cyclitols [Hedges and Weliky, 1989; Bergamaschi and Hedges, in preparation]. Each carbohydrate method will be outlined below and discussed with respect to analytical topics of particular interest for marine chemists.

5.1. The Aldose Method

This procedure [Cowie and Hedges, 1984] involves pretreatment of the sample, $\approx$ 25 mg of plant material or 250 to 500 mg sediment, with 72 wt% H_2SO_4 for 2 hrs at room temperature to "soften" crystalline cellulose. The acid is then diluted to 1.2 M and the sample hydrolyzed for 3 hrs at 100° C in a stirred heating block. The hydrolysate mixture is spiked with a pyridine solution of adonitol and sorbitol (1 to 5 $\mu g\ \mu L^{-1}$), which function as recovery standards. The solution is neutralized with solid $Ba(OH)_2$ removed as $BaSO_4$, deionized on a mixed ion-exchange resin bed, and evaporated gently to dryness. The dried sample is dissolved in pyridine and an aliquot is split out for drying and frozen storage. The aliquot for analysis is mixed with an equal volume of 0.4% w/v $LiClO_4$ in pyridine and heated to 60° C for 48 hours to insure isomeric equilibration. TMS derivatives are then formed at 60° C with Regisil (plus 1% trimethylchlorosilane) and analyzed by gas chromatography using split

TABLE 4-3.
Amino Acid Recovery Standards

Class	Recovery Standard	Symbol	Relative RT	Potential Coeluting Peaks
ACIDIC	α-aminoadipic acid	AAAA*	0.33	
	α-aminopimelic acid	AAPA*	0.39	
NEUTRAL	o-methylthreonine	MTHRN	1.00	
	γ-methylleucine	MLEU*	1.51	
	ρ-fluorophenylalanine	FPHE*	1.34	
	homoserine	HSER	0.60	MHIS
	β-chloroalanine	CALA	0.98	MTHR
	norvaline	NVAL	1.25	PHE
	norleucine	NLEU	1.42	LEU
	b-hydroxynorvaline	HNOR	0.96	MTHR
INTERMEDIATE	1-methylhistidine	MHIS*	0.63	
BASIC	δ-hydroxylysine	HLYS*	1.49	
	homoarginine	HARG	0.86	ALA
	a-aminoguanidinopropionic acid	AGPA	0.53	HIS

*routinely used as relative recovery standards
† absolute recovery standard, retention time = 15-17 min

Key: Relative RT, retention time relative to o-methylthreonine; PHE, phenylalanine; LEU, leucine; ALA, alanine; HIS, histidine.

(1:50) injection and a 30 m by 0.25 mm i.d. fused silica capillary column coated with DB-1 liquid phase. Typically, the GC oven is temperature programmed from 140° C (4 min initial delay) to 270° C at 4° min^{-1} and held at the upper temperature for a few minutes at most (≈ 30 min total run time). Quantification is versus similarly equilibrated standards for which the percentage of each anomeric peak is known [Table 4-4, Cowie and Hedges, 1984]. Thus only one of the multiple isomeric peaks of each sugar must be resolved to quantify the total sample amount. The practical sensitivity of this method is about 0.1 mg of monosaccharide, with a precision of 5 to 10%, or better for major sugars. Six samples are analyzed at a time, with an average throughput of about three per day.

Procedural changes for this method have been fairly minimal. We have come, however, to recognize the critical importance of maintaining absolutely dry pyridine solutions during the equilibration and derivatization procedures and now use vacuum desiccation to completely dry the deionized hydrolysis products of carbohydrate-rich samples such as woods. A second problem, mentioned in the original [1984] paper, is the resolution of any ribose isomer peak in hydrolysate mixtures poor in this component. This task usually can be accomplished by going to a lower initial oven temperature, e.g., 125° C, and programming from there with no delay at 6 or 8° min^{-1}.

This analytical method has an advantage over those involving aldose reduction to alditols in that formation of ambiguous derivatives (e.g., arabitol from both arabinose and lyxose) does not occur. There are the disadvantages, however, that a complex chromatographic trace, with many peak overlaps, is created and that about 5% of the samples do not come to isomeric equilibrium. These problems are exacerbated by the fact that a variety of cyclitols, O-methyl aldoses, and heptoses may occur in some samples [e.g., Hedges and Weliky, 1989] and must also be resolved.

Our approach to these challenges has been to analyze the complete suite of peaks for coelutions and effective equilibration. For those aldoses which typically give a pair of resolved isomers (e.g., rhamnose, mannose, galactose and glucose), we check to confirm that the peak area ratio matches the equilibrium value within 5%. Typically, failure to equilibrate is a general phenomenon and this test is definitive. If the sample is well equilibrated, the next step is to calculate the areas of the isomers of all the common aldoses and individually subtract them from the total measured areas of every peak in the trace. The remaining peak areas of any magnitude correspond to other mixture components (e.g., cyclitols, other aldoses, O-methyl sugars, or heptoses) which can then be detected and often identified and quantified. In practice, both of these manipulations can be made by computer simulations where the results can be visually represented.

5.2. The Uronic Acid Method

A uronic acid is an aldehydic acid that differs from an aldose by having a terminal carboxyl carbon opposite the C-1 carbonyl. We initially worked out an analytical method for this compound class because of its wide distribution [e.g., Mopper and Larsson, 1978; Bergamaschi and Hedges, in preparation] and potential as biomarkers for microbial sources [Keene and Lindberg, 1983; Moers et al., 1990].

Our analytical method [Walters and Hedges, 1988], an adaptation of the method of Lehrfeld [1985], is the most complex procedure that we routinely practice. Briefly, the method involves hydrolysis of the sample, roughly 20 mg of tissue or 300 mg of sediment, in 0.5 N trifluoroacetic acid (TFA) for 2 hrs at 140° C. The hydrolysate supernatant is spiked with recovery standards, allose and ribonolactone, evaporated to dryness, and redissolved and heated (1 hr at 30° C) in 0.1 M Na_2CO_3 to split uronic acid lactones. The carbonyls of both aldoses and uronic acids are then reduced with sodium borohydride and the solutions are deionized and evaporated to dryness to completely convert all aldonic acids to aldonolactones. The aldonolactones are then converted to N-hexylaldonamides which are subsequently made into peracetate derivatives with acetic anhydride. After adding a perseitol peracetate GC internal standard, the peracylated alditols and aldonolactones are quantified by split (1:50) injection onto a 30 m by 0.25 mm i.d. GC capillary column coated with DB-1701 liquid phase. Conventionally, the GC oven is programmed from 220 to 270° C at 1° C min^{-1}. The method has a practical detection limit of 0.2 mg for both aldoses and uronic acids and an average precision of 10 to 15% for major components. Ten samples are processed per batch, with a maximal average throughput of five samples per day.

Procedural changes for this method include streamlining the sample handling procedure by carrying out all the post-hydrolysis reactions in one vial. This has been made possible by use of a 10-port vortex evaporator, with heating lamp warmer, which directly accepts test tubes and dries much faster than conventional rotoevaporators. In addition, we presently use a capillary column coated with DB-23 (50% cyanopropyl-50% methylsilicone) liquid phase to obtain optimal separation of the different aldose, cyclitol and uronic acid derivatives. This column is used with a multiramped program in which oven temperature is increased from 140° C at 1.5° C min^{-1} until the alditols all elute (280° C) and then at 4° C min^{-1} to 280° C, where the various N-hexylaldonamides elute isothermally (personal communication, B. Bergamaschi, 1990).

O-methyl ether aldoses and uronic acids can also both be quantified by this method and hold promise as indicators of both microbial and specific higher plant sources [Moers et al., 1990; Bergamaschi and Hedges, in preparation]. Using the approach of Klok et al. [1982] we have now synthesized and identified a wide variety of O-methyl carbohydrate standards, a number of which have now been detected in hydrolysate mixtures from microorganisms, higher plants and modern marine sediments (Bergamaschi and Hedges, in preparation). Although the uronic acid method measures only a fraction ($\approx$ 5%) of cellulose-bound glucose and is intrinsically less precise than the simpler aldose counterpart [Walters and Hedges, 1988], it holds great promise for comprehensive characterizations of the complex carbohydrate mixtures in natural samples.

Acknowledgements. I especially acknowledge and thank those graduate students, John Ertel, Greg Cowie, Jeff Walters, Miguel Goñi, Brian Bergamaschi and Stephanie de Villiers, who have developed and refined the analytical procedures described above and the National Science Foundation (Grants OCE 7818260, OCE 8023970, OCE 8219294; OCE 8421023, OCE 8613868, and OCE 8716481) which has funded essentially all the contributing research.

References

Bates, A. L., and P. G. Hatcher, Solid-state ^{13}C NMR studies of a large fossil gymnosperm from the Yallourn Open Cut, Latrobe Valley, Australia, Org. Geochem., 14, 609-617, 1989.

Cowie, G. L., Marine organic diagenesis: a comparative study of amino acids, neutral sugars and lignin, Ph.D. dissertation, 182 pp., Univ. of Washington, Seattle, 1990.

Cowie, G. L., and J. I. Hedges, Determination of neutral sugars in plankton, sediments, and wood by capillary gas chromatography of equilibrated isomeric mixtures, Anal. Chem., 56, 497-504, 1984.

Cowie, G. L., and J. I. Hedges, Improved quantification of amino acids in natural samples: Charged-matched recovery standards and reduced analysis time, Mar. Chem., in press, 1991a.

Cowie, G. L., and J. I. Hedges, Sources and reactivities of amino acids in a coastal marine environment, Limnol. Oceanogr, in press, 1991b.

Goñi, M. A., and J. I. Hedges, Cutin-derived CuO reaction products from purified cuticles and tree leaves, Geochim. Cosmochim. Acta, 54, 3065-3072, 1990a.

Goñi, M. A., and J. I. Hedges, Potential applications of cutin-derived CuO reaction products for discriminating vascular plant sources in natural environments, Geochim. Cosmochim. Acta, 54, 3073-3081, 1990b.

Goñi, M. A., and J. I. Hedges, The diagenetic behavior of cutin acids in buried conifer needles and sediments from a marine coastal environment, Geochim. Cosmochim. Acta, 54, 3083-3093, 1990c.

Hedges, J. I., Lignin compounds as indicators of terrestrial organic matter in marine sediments, Ph.D. dissertation, 137 pp., Univ. Texas at Austin, 1975.

Hedges, J. I., and J. R. Ertel, Characterization of lignin by gas capillary chromatography of cupric oxide oxidation products, Anal. Chem., 54, 174-178, 1982.

Hedges, J. I., and K. Weliky, The diagenesis of conifer needles in a coastal marine environment, Geochim. Cosmochim. Acta, 53, 2659-2673, 1989.

Hedges, J. I., G. L. Cowie, J. R. Ertel, R. J. Barbour, and P. G. Hatcher, Degradation of carbohydrates and lignins in buried woods, Geochim. Cosmochim. Acta, 49, 701-711, 1985.

Hedges, J. I., W. A. Clark, P. D. Quay, J. E. Richey, A. H. Devol, and U. de M. Santos, Compositions and fluxes of particulate organic material in the Amazon River, Limnol. Oceanogr., 31, 717-738, 1986.

Hedges, J. I., W. A. Clark, and G. L. Cowie, Organic matter sources to the water column and surficial sediments of a marine bay, Limnol. Oceanogr., 33, 1116-1136, 1988.

Holloway, P. J., The chemical constitution of plant cutins, in The Plant Cuticle, Linnean Society Symposium Series 10, edited by D. F. Cutler, K. L. Alvin, and C. E. Price, pp. 45-85, Academic Press, London, 1982.

Keene, L., and B. Lindberg, Bacterial polysaccharides, in The Polysaccharides 2, edited by G. O. Aspinal, pp. 98-108, Academic Press, New York, 1983.

Klok, J., H. V. Cox, J. W. de Leeuw, and P. A. Schenck, Analysis of synthetic mixtures of partially methylated alditol acetates by capillary gas chromatography, gas chromatography-electron impact mass spectrometry and gas chromatography-chemical ionization mass spectrometry, J. Chromatogr., 253, 55-64, 1982.

Lehrfeld, J., Simultaneous gas-liquid chromatographic determination of aldonic acids and aldoses, J. Anal. Chem., 57, 346-348, 1985.

Moers, M. E. C., M. Baas, J. W. de Leeuw, J. J. Boon, and P. A. Schenck, Occurrence and origin of carbohydrates in peat samples from a red mangrove environment as reflected by abundances of neutral monosaccharides, Geochim. Cosmochim. Acta, 54, 2463-2472, 1990.

Mopper, K., and R. Dawson, Determination of amino acids in seawater - recent chromatographic developments and future directions, Sci. Total Environ., 49, 115-131, 1986.

Mopper, K., and K. Larsson, Uronic and other organic acids in Baltic Sea and Black Sea sediments, Geochim. Cosmochim. Acta, 42, 153-163, 1978.

Sarkanen, K. V., and C. H. Ludwig, Lignins, 916 pp., Wiley-Interscience, New York, 1971.

Smith, R. J., and K. A. Panico, Automated analysis of *o*-phthaldialdehyde derivatives of amino acids in physiological fluids by reverse-phase high-performance liquid chromatography, J. Liq. Chromatogr., 8, 1783-1795, 1985.

Walters, J. S., and J. I. Hedges, Simultaneous determination of uronic acids and aldoses in plankton, plant tissues, and sediment by capillary gas chromatography of N-hexylaldonamide and alditol acetates, Anal. Chem, 60, 988-994, 1988.

METHODS OF SAMPLE HANDLING AND ANALYSIS FOR DISSOLVED AND PARTICULATE AMINO ACIDS AND CARBOHYDRATES IN SEAWATER

Susan M. Henrichs

Institute of Marine Science, University of Alaska, Fairbanks, Alaska 99701

1. Blank Control

Achieving low and consistent levels of contamination in sampling and analysis is crucial because of the low concentrations of amino acids and carbohydrates in marine samples. In seawater, concentrations range from less than 5 nM to 1 μM for dissolved free amino acids (DFAA), 0.1 to 10 μM for dissolved combined amino acids (DCAA), and 0.05 to 1 μM for suspended particulate combined amino acids (PCAA) [e.g., Lee and Bada, 1977; Mopper and Lindroth, 1982; Lee and Cronin, 1984; Henrichs and Williams, 1985; Coffin, 1989]. Concentrations in sediment-trap collected particles generally range from 10 to 30% of carbon [Lee and Cronin 1982, 1984]. Since often less than 10 mg of carbon is available for analysis, only 1 mg, or 10 μM, of combined amino acids are analyzed. Dissolved and particulate carbohydrate concentrations have been studied less, but the reported concentration ranges are similar to those for amino acids [Burney et al., 1979; Mopper et al., 1980; Cowie and Hedges, 1984a; Henrichs and Williams, 1985].

For water and suspended particle sampling, standard Niskin or Go-Flo bottles are satisfactory, but they should be cleaned with detergent and water before use and protected from contamination between hydrocasts. Bottle nozzles and sampling tubes, which should be clean and reserved for organic sampling only, should not be handled with bare hands; PVC or polyethylene gloves should be worn during all sampling procedures, but should never come in direct contact with the samples.

Glass sample containers, filtration apparatus, and other items are easily cleaned by washing with soap and water, rinsing with distilled water, and heating at 450° C for 8 hours. Alternatively, glass and Teflon can be cleaned by soaking for 8 hours in cold chromic acid or rinsing with 80° C chromic acid, then rinsing thoroughly with low-organic water. Containers can be closed with glass stoppers or with Teflon-lined screw caps. Polyethylene and polypropylene containers are also satisfactory for most purposes. Blanks should be checked before use for open ocean samples. Such containers should be new or previously used for storing seawater and should have polypropylene, not Bakelite, caps. They should be soap-and-water washed, rinsed with 1 N HCl, rinsed five times with low-organic water, and leached with three or more changes of filtered seawater or low-organic water, each for at least 1 day. Gloves should be worn at all times during cleaning and rinsing procedures.

Water used for rinsing, reagent preparation, etc. should contain low levels of organic compounds. Standard deionized or single-distilled water is not usually acceptable. Redistillation of deionized or single-distilled water from basic potassium permanganate or from persulfuric acid solution in an all-glass still usually gives sufficiently pure water. HPLC- grade water and water from special purification systems (e.g., Milli-Q) can often be satisfactory, but blanks should be checked before use.

The techniques described are sufficient for virtually all types of samples, but procedural blanks (including sample containers) should always be measured. Less intensive cleaning is often adequate in specific applications, particularly with sediments and larger (>0.1 g) sediment-trap samples.

2. Filtration

Filtration produces an operational distinction between particles retained by the filter and particles and dissolved substances which pass through the filter. Ideally, the filter and filtration technique used for a sample should be chosen based on the questions being addressed. Examples of common goals are (a) to separate large particles which sink through the water column from very small particles likely to exhibit colloidal behavior, 1 μm diameter or smaller, and dissolved substances, (b) to remove living organisms; 0.2 μm is the common cut-off, although some marine bacteria may be smaller [Torella and Morita, 1981], or (c) to divide particles into size-classes, which can be chosen based on particle hydrodynamic properties or size distributions of organisms or their food.

Glass fiber filters, including Gelman AE and Whatman GF/C and GF/F, are used almost exclusively for removing particles from seawater when the goal is to measure hydrolyzable amino acids or other organic molecules in the retained particulate matter. These

Marine Particles: Analysis and Characterization
Geophysical Monograph 63

filters can be pre-combusted at high temperature, 450° C for 8 hr, to remove organic contaminants, resulting in negligible blank yields to acid hydrolysis or solvent extraction. However, the glass fiber filters have the disadvantage that they do not have a defined pore size. Rather, particle removal results from trapping in the tortuous paths through the filter created by the random array of overlapping fibers [Brock, 1983]. Retention can be predicted statistically, but not absolutely. The median retention diameter of Whatman GF/C is about 0.7 μm [Sheldon, 1972]. GF/F retain finer particles. Experiments with samples of Bering Sea surface water found that >90% of the radiolabeled amino acids assimilated by bacteria and retained on 0.2 μm Nuclepore filters were also retained on GF/F (Henrichs, unpublished data, 1987).

If the goal is to separate particles into two or more size ranges, Nuclepore-type polycarbonate membrane filters are preferred. Of the common filters, Nuclepore-type filters most closely approximate sieves, in that the pores are vertical, cylindrical, and fairly uniform in shape and size [Brock, 1983]. In addition, the depth/width ratio of the pores is relatively low, making particle interactions with the walls less likely, although not unlikely; see below. Nuclepore filters have the significant disadvantage of very low particle capacity, due to their small pore density (3×10^8 cm^{-2}; manufacturer's literature). As little as 10 mL of some surface water samples will clog a 25 mm, 0.2 μm Nuclepore filter. These filters have very low dissolved free and combined amino acid and carbohydrate blanks in the filtrate if rinsed thoroughly. A 100 mL rinse with triply-distilled water was employed by Henrichs and Williams [1985] for 25 mm filters used for DCAA and dissolved carbohydrate analysis. However, a smaller rinse volume probably would be adequate in many applications. Three, 5 mL rinses with prefiltered seawater are sufficient for DFAA analysis (Henrichs, unpublished data, 1987). Unfortunately, Nuclepore filters disintegrate under acid hydrolysis, and thus are not suitable for direct analysis of retained particulate material. An alternative approach [Henrichs and Williams, 1985] is to determine particulate matter concentration as the difference in concentration between filtered and unfiltered samples. This approach will not work when particulate concentrations are low, i.e., less than 10% of the total.

Sheldon [1972], using seawater samples of moderate particulate content collected in the Bedford Basin, showed that the retention characteristics of filters could vary substantially from manufacturer-stated pore sizes. Millipore filters with nominal pore sizes from 0.5 to 8 μm all had an effective pore size near 0.5 μm. Although the differences in structure of glass fiber and Nuclepore filters suggest that the latter should give more accurate and precise size separations, this may not be true in practice. The effective pore size of Nuclepore filters, the size at which 50% of particles were retained, correlated quite closely with the manufacturer-stated pore size [Sheldon, 1972]. Surprisingly, however, the standard deviation of the particle-size distribution collected by glass fiber and Nuclepore filters did not differ significantly. This could be due to several factors. First, there is some variability in pore sizes of Nuclepore filters. Second, partial clogging of Nuclepore filters could cause them to retain particles smaller than the filter pore diameters, although Sheldon [1972] found that the effective pore size of Nuclepore filters remained essentially constant until they became blocked. Third, Nuclepore filters and virtually all other microporous materials, retain many particles by adsorption to the filter surface or the walls of pores; these particles can be much smaller than the pore diameter. For example, Zierdt [1979] found that 47 mm-diameter Nuclepore filters up to 8 μm pore-size retained most bacteria ranging in diameter from about 1 to 3 μm. The retention of small particles declined with increasing loading over a range which would correspond to the bacteria content of less than 1 to 500 mL of seawater. Adsorption is probably electrostatic, as it can be reversed with surfactants [Zierdt, 1979] and decreases with increasing ionic strength of the solution [Brock, 1983].

Given the potential problems mentioned above, it is perhaps surprising that size fractionation of particles in seawater samples has given apparently consistent results. In one of the most extensive studies of the particle-size distribution of PCAA, Henrichs and Williams [1985] used 8, 1, 0.2, 0.1, and 0.05 μm Nuclepore filters to find that most particulate combined amino acids in surface water and sea surface film samples collected off Baja California were in the size ranges 0.2 to 1 and 1 to 8 μm. This was the size range of major contributors to biomass in the samples, bacteria and small flagellates [Williams et al., 1986], and a significant fraction of the PCAA were probably in living organisms (range 15 to 95%, mean 50%). Centrifuging samples at speeds sufficient to sediment marine bacteria, which are mostly between 0.2 and 1 μm in diameter, gave DCAA concentrations in the supernatant between those in 0.2 and 1 μm filtrates, indicating that the effective and nominal pore sizes of those filters were similar [Henrichs and Williams, 1985].

An additional serious concern is that dissolved substances can be released from cells during the filtration process. The most extensive study [Fuhrman and Bell, 1985] on the release of DFAA found that filtration of greater than 50 mL of water, and the use of Whatman GF/F and 1.0 μm Nuclepore filters caused DFAA release, while 0.2 μm Nuclepore, 0.22 μm Millipore, and 0.45 μm Millipore did not. Moderate variation in the vacuum applied, 3 to 64 cm Hg, had little effect. Other work (Henrichs, unpublished data, 1988) found no difference between DFAA concentrations in GF/F and 0.2 μm Nuclepore-filtered samples. However, DFAA concentrations were found to increase substantially and rapidly when samples were exposed to fluorescent lighting in the laboratory.

A new type of inorganic membrane filter has recently become available, Anopore aluminum oxide filters. These have much greater pore density than Nuclepore filters and consequently higher flow rates [Jones et al., 1989]. Altabet [1990] compared the amounts of particulate carbon and nitrogen removed from seawater by Anotec Anopore and glass and quartz fiber filters. For seawater collected near Bermuda at depths less than or equal to 200 m, Whatman GF/D, QM-A, and GF/F were stacked over 0.2 μm Anopore filters. The glass and quartz filters passed an average of 40% of the particulate carbon retained by the entire stack. The large fraction of "small" particulate carbon was attributed to the high proportion of bacterial biomass in oligotrophic waters [Fuhrman et al., 1987].

The following filtration procedures are recommended:

a) Glass fiber filters are preferred when the goal is to approximately separate sinking from suspended particles, because they are the most generally non-contaminating filters for organic analysis and because their particle capacity and flow rate are high. Nuclepore filters are best for obtaining information on particle-size distributions. Neither 0.2 μm Nuclepore nor GF/F can be relied on to remove all bacteria, but both probably remove most metabolically-active bacteria.

b) Sample volumes should be less than the volume which results in a significant drop in flow rate through the filter, which indicates clogging. However sample volumes less than 5 mL could increase the effects of adsorption of dissolved molecules and small particles by the filters.

c) If dissolved intracellular pools are a significant part of the total concentration, then the sample volume should be less than 20 mL, for a 25 mm filter; vacuum should be kept to a minimum, ca. 100 mm Hg; the filter should not be "sucked dry"; and exposure of the sample to light, natural or artificial, should be avoided.

3. Sample Storage

Samples for DFAA analysis should be filtered as soon as possible, within 15 minutes, and unfiltered samples should be kept near in situ temperature in the dark. Filtered samples should be frozen quickly, within 30 minutes, and usually can be stored at -20 to -30° C for up to about 2 months without major changes in composition or concentration. However, for storage longer than a week, several control samples analyzed at the time of sampling should be stored and reanalyzed to check for storage artifacts. Samples should be thawed quickly, in a cold water bath, and analyzed immediately after thawing.

Similar precautions are appropriate for DCAA and suspended particulate amino acid samples on glass fiber filters, although these types of samples can probably be stored longer without significant degradation. For example, Henrichs and Williams [1985] obtained very similar results for DCAA concentrations in samples stored 6 to 10 months and 1 to 4 months. They also found that acidifying seawater samples before frozen storage had no effect on DCAA concentration, but probably was of no particular benefit.

4. Amino Acid Analysis

4.1. Hydrolysis

One hydrolysis procedure, with minor variations, has been employed for most analyses of combined amino acids in sediments, sediment-trap collected particles, or small, suspended particles [e.g., Henrichs et al., 1984; Burdige and Martens, 1988; Lee and Cronin, 1982; Henrichs and Williams, 1985; Chapman et al., 1988]. Combined amino acids are generally thought to be predominantly proteins or peptides, but could include free amino acids adsorbed on particles or amino acids bound in humic substances. The usual hydrolysis procedure is to heat the sample with a large excess of 6 N HCl at 110° C for 16 to 24 hours in a reflux apparatus or in sealed ampules under nitrogen or vacuum. Samples are sometimes oven-dried or freeze-dried before addition of 6 N acid; other investigators add concentrated (10 or 12 N) acid and avoid drying. Glass-distilled (6 N) or ultrapure HCl is probably essential when hydrolyzing 0.1

to 10 nmoles of CAA, the quantity dissolved in 1 mL of seawater or contained in the suspended particles of 100 to 500 mL seawater samples. Reagent grade acid is adequate for most sediment samples, containing 10 to greater than 100 μM amino acids g^{-1}. After hydrolysis, the excess acid can be evaporated under vacuum or neutralized with sodium hydroxide. The samples can then be analyzed by HPLC without further treatment. If they are to be analyzed by gas chromatography, they must be desalted by ion exchange chromatography (see below).

The two major concerns about the hydrolysis procedure are that hydrolysis will be incomplete or that it will result in degradative or other losses of amino acids. Acid hydrolysis of proteins has been evaluated in detail by biochemists [Blackburn, 1978]. After hydrolysis for 24 hours, sulfur-containing amino acids, cystine and methionine, are found to be largely decomposed, although a product of methionine decomposition, methionine sulfoxide, can be used to estimate methionine. Tryptophan is also destroyed. Losses of serine, threonine, and tyrosine are generally 5 to 15% but can be much larger, depending on protein composition, sample size, and presence of certain metal ions. The conclusion is that, for these sensitive amino acids, recoveries should be estimated for each type of sample by back-extrapolation from the concentration decrease observed with increasing hydrolysis times.

Chapman et al. [1988] evaluated hydrolysis conditions for seawater samples. They recommend hydrolysis in 6 N HCl in evacuated, sealed tubes for 16 hr at 110° C. They found complete destruction of free amino acids hydrolyzed under these conditions in nitrogen-flushed tubes. Such destruction of amino acids in nitrogen-flushed, sealed ampules was not found by Henrichs and Williams [1985] or Robertson et al. [1987]. However, in samples containing low concentrations of amino acids and relatively high concentrations of nitrate, almost complete destruction of amino acids can occur during hydrolysis, presumably due to oxidative degradation by HNO_3 [Robertson et al., 1987]. Low DCAA and high nitrate are characteristic of most sub-euphotic zone seawater samples. For analysis of 0.5 μM DCAA in seawater containing up to 40 μM nitrate, losses of amino acids can be prevented by addition of ascorbic acid (110 μM) [Robertson et al., 1987].

Few systematic studies of hydrolysis times for sediments or marine particles have been done. Times of greater than 48 hrs would be required, since sediment combined amino acid hydrolysis is not complete before that time. Some indirect evidence suggests that large losses of most amino acids do not occur. First, large proportions of the total sediment nitrogen in some organic-rich sediments are hydrolyzable amino acids [e.g., 60 to 70% in some Peru Upwelling Region sediments; Henrichs et al., 1984]. Second, mole fractions of serine, threonine, and tyrosine in hydrolyzed sediment amino acids are generally greater than or equal to those in phytoplankton, marine bacteria, or other organisms [Mopper and Degens, 1972; Henrichs, 1980].

On the other hand, some peptide bonds are not broken by a 24 hr hydrolysis, mostly because of steric hindrance by bulky side-chains of valine, leucine, isoleucine, and phenylalanine. For example, a second 24 hr hydrolysis of an organic-rich marine sediment from the Pettaquamscutt River Estuary yielded 18%, 11%, 14%, and 8% additional isoleucine, leucine, valine, and phenylalanine, but only 3 to 5% additional for other amino acids [Henrichs, 1980]. A lower organic-content sediment from Buzzards Bay yielded only 8% additional isoleucine and proportionally less of the other amino acids.

Another process potentially leading to incomplete recoveries of amino acids from sediment particles is adsorption. Basic amino acids are adsorbed to a greater extent than either neutral or acidic amino acids [Henrichs, 1980; Doyle, 1988]. However, comparison of the adsorption capacity of sediments to the concentrations of amino acids in hydrolyzates indicates that losses should be minimal, especially since experiments with radiolabeled amino acids show that adsorption is largely reversed by addition of 1 N acid [Doyle, 1988].

Overall, there is no strong evidence that the "standard" hydrolysis method is unsatisfactory, except when nitrate concentrations are high relative to amino acid concentrations. However, evaluation of amino acid recoveries by addition of free amino acid standards and by testing a range of hydrolysis times is advisable, especially when the weight fraction of amino acids in the sample is low.

4.2. Gas Chromatography

Analysis of amino acids by gas chromatography is much more time-consuming than analysis by liquid chromatography, but has the advantage of relatively straightforward compound identification by GC/MS. Analytical precision can be comparable to HPLC techniques [e.g., Henrichs, 1980; Mague et al., 1980] although the additional desalting and derivatization steps are a potential source of greater errors. Asparagine and glutamine are esterified during the necessary derivatization and are not distinguishable from aspartic and glutamic acids. Also, arginine and histidine, which form unstable derivatives, are more difficult to measure using GC, although proline and hydroxyproline, which are difficult using the OPA method, are determined easily.

A gas chromatographic procedure for measuring amino acids extracted from sediment and dissolved in interstitial waters was described in detail by Henrichs [1980] and briefly by Henrichs et al. [1984]. Samples were desalted by cation-exchange chromatography on a 15 cm^3 BioRad AG 50W-X8 (50 to 100 mesh) resin column. The resin was brought to the H^+ form by eluting with 20 mL of 6 N HCl and then rinsed with water. The sample, up to 25 mL volume, with norleucine added as an internal standard, was applied to the column. The column was washed with 20 mL of water. Ammonium hydroxide solution (2 N) was added to the top of the column, and the eluate was discarded until the base front, identified by a warm zone, just reached the bottom of the column. The first 70 mL of basic eluate was collected, evaporated to dryness under vacuum at 40° C, and the residue redissolved in 0.1 N HCl until analyzed by GC. Recoveries of amino acid standards from distilled water were greater than 90% except for the basic amino acids and tyrosine, with recoveries near 80%, and tryptophan and cystine, with low and variable recoveries. Tyrosine recovery was poor from seawater and methionine recovery variable. To minimize blanks, the resin was exhaustively rinsed with 6 N ammonium hydroxide, 6 N HCl, and water before use. Resin can be reused indefinitely, rinsing with acid, water, ammonium hydroxide, and water between samples. Glass-distilled 6 N HCl and ammonium hydroxide prepared by bubbling ammonia gas through glass-distilled water were used. This desalting method is not recommended for samples with amino acid concentrations less than 1 μM, or containing less than 0.025 μM total amino acids; amino acids can be extracted from low-level samples using Cu-saturated Chelex 100 resin [Siegel and Degens, 1966; Lee and Bada, 1977].

Amino acids are derivatized for gas chromatography using a method adapted from Roach and Gehrke [1969]. An aliquot of the 0.1 N HCl solution, containing 0.01 to 1 μM amino acids, is placed in a reaction vial and the water evaporated at 100° C under a stream of N_2. Methylene chloride (0.2 mL) is then added and evaporated to ensure complete removal of water. After cooling, 0.2 mL of 3 N HCl in n-butanol and 0.05 mL methylene chloride are added. The vial is sealed, sonicated for 15 minutes, and then heated at 100° C for 30 minutes. The excess reagent is evaporated at 60 to 70° C under N_2 until ca. 10 μL remains, then 0.2 mL methylene chloride is added and evaporation continued to dryness at room temperature. Heptafluorobutyric anhydride in acetonitrile (0.1 mL of a 20% solution) is added and the solution heated at 110° C for 15 minutes. After cooling, this reagent is evaporated under an N_2 stream. The derivatives, (N,O)-heptafluorobutyryl-n-butyl esters (HFBBE), are dissolved in methylene chloride for injection into the gas chromatograph. A wide variety of other amino acid derivatives can be used for GC analysis, but the HFBBE are less volatile, thus minimizing evaporative losses, and less polar than most, e.g., (N,O)- trifluoroacetyl-n-propyl esters.

Henrichs [1980] used a 32-m x 0.3 mm-i.d. SE-54 glass capillary column. Fused silica capillary columns with similar properties would be used now. If separation of enantiomers is desired, Chirasil-Val capillary columns can be used. In Henrichs [1980], the Hewlett-Packard 5840 gas chromatograph was equipped with a splitless injector and a flame ionization detector. Typical analysis conditions were: He flow rate, 1 to 3 mL min^{-1}; injector temperature 250° C; initial column temperature, 40° C; temperature increased at 30° C min^{-1} to 70° C and then at between 2 and 4° C min^{-1} to 250° C; FID temperature 250° C.

4.3. High Performance Liquid Chromatography

High performance liquid chromatography after precolumn derivatization with *o*-phthaldialdehyde [Lindroth and Mopper, 1979; Jones et al., 1981] has become the preferred method for amino acid analysis because of its speed, relative simplicity, and excellent sensitivity. The application of this technique to seawater has been reviewed by Mopper and Dawson [1986]. There are nearly as many variations of this technique as there are analysts. The method described below is similar to that used by Henrichs and Williams [1985] and is the one presently used in Henrichs' laboratory.

4.3.1. Seawater, DFAA.

The filtered samples (2 to 5 mL) and standards analyzed should be a fairly constant temperature before reagent addition. Significant temperature differences affect the rate of derivative formation and decomposition, and sometimes happen inadvertently when samples from a depth profile or recently thawed samples are being analyzed. For each 1 mL of sample, 50 μL of saturated sodium tetraborate buffer solution, α-amino adipic acid as an internal standard, and 25 μL of mixed OPA reagent are added. The mixed OPA reagent consists of 10 mg *o*-phthaldialdehyde, 10 μL of mercaptoethanol, and 20 mg of sodium lauryl sulfate, all dissolved in 1 mL methanol. This reagent is prepared fresh daily. The lauryl sulfate greatly improves sensitivity and reproducibility for lysine and ornithine. The pH of the mixture is around 9.2; care should be taken that it is not too high because this rapidly damages the column packing of ODS columns. The sample with added buffer and reagent is mixed and allowed to react for 1.5 minutes. During this time, the sample is drawn into a glass syringe, and a 13 mm stainless steel "Swinnex"-type filter holder with a 0.6 μm nylon filter is attached. The syringe and filter apparatus have been rinsed with methanol and ultrapure water. At precisely 1.25 minutes, injection of the sample into the HPLC sample loop is begun. Sample loop volumes used range from 0.5 to 2.0 mL. At 1.5 minutes, the sample is injected.

It is desirable in some applications to inject a very large excess of sample, about three times the nominal volume, through the sample loop. Because of the small diameter of the tubing in the loops, they apparently do not flush efficiently. This is not crucial when peaks are quantified relative to an internal standard, but is very important when the injected volume must be known accurately. An example is when the specific activity of amino acids is being measured in conjunction with isotope dilution studies [Fuhrman, 1987].

The blanks of "clean" or "certified" racked plastic pipet tips are very low, and these can be used to dispense reagents. The methanol used is glass-distilled or HPLC grade. The other reagents are obtained from Aldrich Chemical or Sigma and are usually used without further purification. Mercaptoethanol sometimes requires purification by vacuum distillation. When blank problems are encountered, try a new sample or batch of purified water, the borate buffer, the mercaptoethanol, and other solvents and reagents, in that order.

The following LC system is currently in use: Spectra-Physics SP8700 solvent delivery system, Kratos FS 950 Fluoromat fluorescence detector, Spectra-Physics SP4270 integrator, Gilson Model 201 fraction collector. However, any binary or ternary gradient system can be used for amino acid analysis. Newer systems where control and data processing are accomplished with software written for standard PC's are preferred.

Conditions of analysis are varied depending on the degree of resolution desired, the column installed, and its condition, so the following are only examples. Solvent A is 0.025 M phosphate buffer, containing equimolar amounts of monobasic and dibasic sodium phosphate and prepared in glass-distilled water; solvent B is glass-distilled or HPLC grade methanol; and the solvent flow rate is 1 mL min^{-1}. The detector is set at about 1/5 of its maximum sensitivity (for seawater DFAA analysis), with excitation wavelengths at 330 to 375 nm and the emission cut-off at 418 nm. The column is 25 cm x 0.4 cm i.d., containing 5 μm Spherisorb ODS or 10 μm PRP-1 (Hamilton) packing. The PRP-1 column does not resolve glycine and threonine or leucine and isoleucine, but is extremely durable. Usually, a 3 cm cartridge-type precolumn with the same packing is installed before the column. The "short" gradient, used to minimize processing time with little sacrifice of resolution, is: initial solvent, 80% A/20% B; program linearly in 20 minutes to 20% A/80% B; hold 5 minutes; return, instantly, to initial solvent composition; hold 5 minutes before next injection. Initial solvent composition is varied in the range of about 17 to 22% A to improve resolution and peak shape of early peaks. The gradient time is increased to 30 to 40 minutes in some cases to improve resolution of taurine/β-alanine/alanine/α-aminobutyric acid. Although we have tried additions of THF and acetonitrile (1 to 2%) to the methanol [Jones et al., 1981] and a variety of stepped gradients, we have not obtained consistently good resolution of glycine and threonine, especially as columns age. Proline and hydroxyproline are not measured.

Concentrations of about 0.5 nM can be measured reproducibly ($\pm$ 25%) under these conditions, with very careful attention to blanks. At 10 nM, the analytical precision for most amino acids is about $\pm$5%. Reproducibility for glycine and lysine is about 10%. Quantification and identification are with reference to standard amino acid solutions run under the same conditions as samples. It is preferable to prepare these standards in a low amino acid seawater, e.g., open ocean deep water, or use a standard addition approach if no amino acid-free seawater is available. The retention times of early-eluting amino acids, e.g., aspartic and glutamic acids, differ for distilled water and seawater. Also, Mopper and Dawson [1986] report that the molar response for certain amino acids is much greater in seawater, although we have not observed any large differences in our work. Useful variations on the method include: (a) 50 mM acetate buffer, pH 5.8 to 6.8, replacing the phosphate buffer [e.g., Fuhrman, 1987; Burdige and Martens, 1988]; (b) 3 μm ODS columns and very rapid solvent gradients, as little as 16 minutes analysis time [Manahan, 1989]; (c) "quenching" of the derivatization reaction by addition of acetic acid to lower the pH to less than 7.5, improving the stability of some derivatives and reducing the blank [Fuhrman and Bell, 1985].

If samples, e.g., porewater samples, contain much higher concentrations of ammonium than amino acids, ammonia must be stripped from the samples before analysis. For unknown reasons, ammonia produces multiple peaks [Haberstroh and Karl, 1989] which can be misidentified as late-eluting amino acids such as lysine and ornithine or as unknown amines. Ammonia can be removed by adding the borate buffer to the sample, heating it to about 40° C, and stripping with nitrogen for 30 minutes, although longer times or slightly higher temperatures may be necessary if mM ammonium is present. This treatment has no effect on the concentration of amino acids in porewater samples (Henrichs, unpublished data, 1985; see also Burdige and Martens [1990]).

4.3.2. Seawater, DCAA.

After hydrolysis, the analysis is the same as for DFAA except that the required detector sensitivity is less, or a smaller sample can be injected, and the amount or pH of the added buffer may need to be changed. It is worth the effort to check the pH of each sample with a minielectrode or micro-range pH paper, and adjust it to fall within the range 9.3 to 9.5.

4.3.3. Seawater, particles.

Again, after hydrolysis and careful adjusting of the sample pH, the analysis is the same as for DFAA. Samples often should be extensively diluted, if the system is being used for low-level analyses concurrently.

5. Carbohydrate Analysis

5.1. Hydrolysis

Burney and Sieburth [1977] hydrolyzed dissolved polysaccharides in seawater by adding 1 mL of 1 N HCl to a 10 mL aliquot of the sample and heating in sealed ampoules at 100° C for 20 hours. Although not mentioned in the original paper, the solution and ampoule headspace should be sparged with nitrogen before the ampoule is sealed. These mild hydrolysis conditions caused no destruction of a glucose standard and gave good recoveries, 80 to 100% for several tested oligosaccharides and polysaccharides; lower recoveries (ca. 50%) were found for sodium alginate and agar, however. These hydrolysis conditions also gave the highest total carbohydrate concentrations of those tested for Narragansett Bay water [Burney and Sieburth, 1977]. Most other studies of dissolved, combined carbohydrates in seawater have used hydrolysis conditions similar to those recommended by Mopper [1977] based on yields from North Sea surface waters, 1.0 N HCl for 3 hours at 100° C. Similar hydrolysis conditions have been found to be optimal for sediment samples: 2 hour pretreatment with 72% sulfuric acid, followed by dilution to 1.86 N and hydrolysis at 100° C for 4 hours [Mopper, 1977]; the same pretreatment with dilution to 1.2 M and hydrolysis at 100° C for 3 hours [Cowie and Hedges, 1984b]; and hydrolysis in 0.5 N trifluoroacetic acid at 135° C for 2 hours [Walters and Hedges, 1988]. All hydrolyses were carried out under nitrogen. No hydrolysis conditions can be expected to give 100% yields of sediment carbohydrates; losses are due both to destruction of sugar residues and incomplete hydrolysis, and no conditions have been found which avoid both for all polysaccharides. For example, Cowie and Hedges [1984b] found ca. 80% yield of glucose from α-cellulose; Walters and Hedges [1988] found 36 and 47% recoveries of alginic acid and polygalacturonic acid, added to sediments.

The evidence indicates that hydrolysis conditions must be chosen for the particular type of sample and polysaccharide to be analyzed. The conditions recommended by Mopper [1977] or by Cowie and Hedges [1984b] are appropriate for determination of combined aldoses in sediments and large-particle (e.g., sediment trap, plankton tow) samples. The method of Walters and Hedges [1988] is a useful alternative if uronic acids are of special interest. Hydrolyses of dissolved and small-particle carbohydrates should probably be carried out under milder conditions [Burney and Sieburth, 1977], although the best conditions for such samples have not been as thoroughly investigated as for sediments.

5.2. MBTH Method

Johnson and Sieburth [1977] and Burney and Sieburth [1977] report a method for measuring total dissolved carbohydrates in seawater. The method has the advantage that the response per mole is nearly equal for a wide variety of monosaccharides and related compounds, i.e., hexoses, pentoses, alditols, and uronic acids. Deoxysugars give about half the signal per mole as aldoses. Most other compounds tested, with the important exception of the amino acid serine, do not interfere with the MBTH assay. Another advantage is that it requires no specialized equipment.

Monosaccharides can be measured without any pretreatment of the water sample except that filtration is recommended. Hydrolysis (see above) is necessary before applying the MBTH method if total, monosaccharide plus polysaccharide, concentrations are desired. After hydrolysis, the acid should be neutralized by addition of 1 mL of 1 N NaOH. The following procedure is that given by Johnson and Sieburth [1977]. Six 1 mL aliquots of sample are dispensed into glass, screw-cap tubes or vials. Polyethylene screw caps are much better than Teflon-lined Bakelite, since the chips of this phenol-formaldehyde polymer are serious contaminants [Johnson et al., 1981]. To reduce aldoses to alditols, 50 μL of ice-cold 100 mg/5 mL potassium borohydride in distilled water is added to each of the 1 mL subsamples. The borohydride solution must be prepared within about 15 minutes of use. The samples are incubated in the dark at room temperature for 4 hours, then the excess borohydride is destroyed by adding 50 μL of 0.36 N HCl. After at least 10 min, the alditols in three of the subsamples are oxidized by adding 0.1 mL of 0.025 M periodic acid and reacting for 10 minutes in the dark. The oxidation is terminated by addition of 0.1 mL of 0.25 M sodium arsenite, and the tubes are allowed to stand for at least 10 minutes. In the other three subsamples, the controls, the alditols are not oxidized; instead, 0.2 mL of premixed (1:1) periodic acid and sodium arsenite solutions are added. The analysis can be interrupted at this point and the samples and controls stored at room temperature in the dark for 1 to 2 days [Johnson et al., 1981].

A colored product is formed from the formaldehyde produced by the alditol oxidation by adding 0.2 mL of MBTH reagent (276 mg/ 10 mL of 0.1 N HCl) to all six subsamples, heating in a boiling water bath for 3 min, and cooling immediately in a room-temperature water bath. Ferric chloride (0.2 mL of a 5% aqueous solution) is added and the reaction is allowed to proceed at room temperature for 30 min in the dark. Acetone (1 mL) is added and the absorbance of the colored solutions is measured at 635 nm. The average absorbance of the three unoxidized controls is subtracted from the absorbances of the three oxidized samples. Calibration curves can be generated by using glucose or other monosaccharide solutions of known concentration in distilled water or by standard addition to seawater; results are similar [Johnson and Sieburth, 1977]. As for amino acid analysis, scrupulous cleaning of glassware is essential, and low-organic water must be used in the preparation of all solutions. The sensitivity of the method is about 0.5 μM of total monosaccharides, and the precision ± 0.04 μM [Johnson et al., 1981].

No intercomparisons of the MBTH method and methods which measure specific monosaccharides in seawater or seawater hydrolyzates have been done. The MBTH method does give monosaccharide concentrations in the same range as specific techniques [e.g., Mopper et al., 1980], but as the concentration range is at least an order of magnitude, comparing different samples is not very informative. Comparison of MBTH total dissolved carbohydrate carbon to DOC gives ratios of 0.08 to 0.25 [Burney et al., 1979] and 0.1 to 0.22 [Henrichs and Williams, 1985]. MBTH particulate carbohydrate carbon to POC ratios were 0.11 to 0.48 in one study of surface water and sea surface microlayer samples [Henrichs and Williams, 1985]. These ratios appear reasonable and are comparable to those obtained by specific techniques.

The MBTH method appears to give an accurate measure of total dissolved or suspended particulate carbohydrates in seawater. However, intercomparison with one or more of the specific methods would be desirable. Also, since the MBTH method is sensitive to a wide variety of compounds, it is suitable only for applications where this broad sensitivity is acceptable. The MBTH method has not been evaluated for use in sediments or sediment trap materials.

6. Flow Cytometry

Flow cytometry is a technique for the rapid measurement of the optical properties of particles in a moving fluid. Modern flow cytometers can also sort particles based on their optical properties; this is accomplished by giving an electrical charge to the water droplet containing a particle and moving the droplet as it passes between high-voltage deflector plates [Mendelsohn, 1980]. Examples of particle variables which can be measured and used as a basis for sorting include size, which is related to forward light scatter, natural fluorescence at several wavelengths, and fluorescence due to reactivity with specific stains. Useful stains include DAPI (4,6-diamidino-2-phenylindole-2-HCl), which reacts with DNA, and FITC (florescein isothiocyanate), which reacts with protein. See Yentsch et al. [1983] and Cucci et al. [1985] for early marine applications.

One instrument currently being used for oceanographic work is a Cytofluorograf 2S, operated by Dr. Donald Button and Ms. Betsy Robinson of the University of Alaska Fairbanks. This flow cytometer is capable of sorting particles based on both size and fluorescence and can characterize particles in the size range of typical marine bacteria, 0.01 to 1 μm^3 volume and with recalibration, larger particles from 1 to ca. 30 μm diameter. This instrument would appear to have potential for use in the chemical characterization of suspended particles. One approach would be simply to obtain greater information on the nature of the particulate material in aliquots of samples to be filtered and analyzed by conventional methods. Many specific biochemical stains are available [Shapiro, 1983], and using these, considerable information on the chemical composition of particles could be obtained directly. Also, it should be possible to use the sorting capability of the instrument to isolate specific subpopulations of particles for analysis. For example, one could select particles in a given size range containing a particular fluorescent pigment. There are some significant limitations for marine organic chemical applications. Only about 1 mL sample volume per 90 minutes can be sorted in the size range of bacteria, and ideally the number of particles per mL should be on the order of 1,000,000. Preconcentration by centrifugation can be used if the particle concentration is too low. Thus, only 0.01 to 0.1 μg of particulate material might be available for analysis. This is an adequate amount for measurement of amino acids, but analysis of most other compound classes in such a small sample would be difficult. Also, analytical interferences could severely limit the use of fluorescent stains to tag cells which will be collected for further characterization.

Despite the limitations, flow cytometry could be a useful technique for addressing significant questions in marine organic chemistry, for example: Is the chemical composition of non-living particulate matter significantly different from that of viable suspended particles? Are certain biochemicals specific markers for particular groups of organisms? What are the fates of different biological and chemical categories of particles under experimental conditions intended to promote aggregation or perhaps disaggregation? Are there interactions between particles and fluorescently tagged solutes?

Acknowledgements. I thank NSF for support of my work in marine organic chemistry, most recently under grant OCE-8900362.

References

Altabet, M. A., Organic C, N, and stable isotopic composition of particulate matter collected on glass-fiber and aluminum oxide filters, Limnol. Oceanogr., 35, 902-909, 1990.

Blackburn, S., Sample preparation and hydrolytic methods, in Amino Acid Determination, edited by S. Blackburn, pp. 7-37, Marcel Dekker, New York, 1978.

Brock, T. H., Membrane Filtration: A User's Guide and Reference Manual, 381 pp., Science Tech, Madison, Wisconsin, 1983.

Burdige, D. J., and C. S. Martens, Biogeochemical cycling in an organic-rich coastal marine basin: 10. The role of amino acids in sedimentary carbon and nitrogen cycling, Geochim. Cosmochim. Acta, 52, 1571-1584, 1988.

Burdige, D. J., and C. S. Martens, Biogeochemical cycling in an organic-rich coastal marine basin: 11. The sedimentary cycling of dissolved, free amino acids, Geochim. Cosmochim. Acta, 54, 3033-3052, 1990.

Burney, C., and J. McN. Sieburth, Dissolved carbohydrates in seawater. II. A spectrophotometric procedure for total carbohydrate analysis and polysaccharide estimation, Mar. Chem., 5, 15-28, 1977.

Burney, C. M., K. M. Johnson, D. M. Lavoie, and J. McN. Sieburth, Dissolved carbohydrate and microbial ATP in the North Atlantic: concentrations and interactions, Deep-Sea Res, 26, 1267-1290, 1979.

Chapman, J., J. Braven, M. Liddicoat, and E. Butler, Optimization of the reaction conditions for the direct determination of the dissolved combined amino acid levels in seawater, Sci. Total Environ., 71, 215-224, 1988.

Coffin, R. B., Bacterial uptake of dissolved free and combined amino acids in estuarine waters, Limnol. Oceanogr., 34, 531-542, 1989.

Cowie, G. L., and J. I. Hedges, Carbohydrate sources in a coastal marine environment, Geochim. Cosmochim. Acta, 48, 2075-2087, 1984a.

Cowie, G. L., and J. I. Hedges, Determination of neutral sugars in plankton, sediments, and wood by capillary gas chromatography of equilibrated isomeric mixtures, Anal. Chem., 56, 497-504, 1984b.

Cucci, T. L., S. E. Shumway, R. C. Newell, R. Selvin, R. R. L. Guillard, and C. Yentsch, Flow cytometry: a new method for characterization of differential ingestion, digestion, and egestion by suspension feeders, Mar. Ecol. Prog. Ser., 24, 201-204, 1985.

Doyle, A., Respiration and adsorption of alanine, glutamic acid, lysine, and glucose in sediments from Resurrection Bay, Alaska, M.S. thesis, 123 pp., University of Alaska, Fairbanks, 1988.

Fuhrman, J., Close coupling between release and uptake of dissolved free amino acids in seawater studied by an isotope dilution approach, Mar. Ecol. Prog. Ser., 37, 45-52, 1987.

Fuhrman, J., and T. M. Bell, Biological considerations in the measurement of dissolved free amino acids in seawater and implications for chemical and microbiological studies, Mar. Ecol. Prog. Ser., 25, 13-21, 1985.

Fuhrman, J., T. Sleeter, and C. Carlson, Oligotrophic ocean biomass is dominated by nonphotosynthetic bacteria, even in the euphotic zone, EOS Trans. Am. Geophys. Union, 68, 1729, 1987.

Haberstroh, P. R., and D. M. Karl, Dissolved free amino acids in hydrothermal vent habitats of the Guaymas Basin, Geochim. Cosmochim. Acta, 53, 2937-2945, 1989.

Henrichs, S. M., Biogeochemistry of dissolved free amino acids in marine sediments, Ph.D. thesis, 253 pp., Woods Hole Oceanographic Inst.-M.I.T. Joint Program, 1980.

Henrichs, S. M., and P. Williams, Dissolved and particulate amino acids and carbohydrates in the sea surface microlayer, Mar. Chem., 17, 141-163, 1985.

Henrichs, S. M., J. W. Farrington, and C. Lee, Peru upwelling region sediments near 15° S. 2. Dissolved free and total hydrolyzable amino acids, Limnol. Oceanogr., 29, 20-34, 1984.

Johnson, K. M., and J. McN. Sieburth, Dissolved carbohydrates in seawater. I. A precise spectrophotometric analysis for monosaccharides, Mar. Chem., 5, 1-13, 1977.

Johnson, K. M., C. M. Burney, and J. McN. Sieburth, Doubling the production and precision of the MBTH spectrophotometric assay for dissolved carbohydrates in seawater, Mar. Chem., 10, 467-473, 1981.

Jones, B. N., S. Pääbo, and S. Stein, Amino acid analysis and enzymatic sequence determinations of peptides by an improved o-phthaldialdehyde precolumn labeling procedure, J. Liquid Chromatogr, 4, 565-586, 1981.

Jones, S. E., S. A. Ditner, C. Feeman, C. J. Whitaker, and M. A. Lock, Comparison of a new inorganic membrane filter (Anopore) with a track-etched polycarbonate membrane filter (Nuclepore) for direct counting of bacteria, Appl. Environ. Microbiol., 55, 529-530, 1989.

Lee, C., and J. L. Bada, Dissolved amino acids in the equatorial Pacific, the Sargasso Sea, and Biscayne Bay, Limnol. Oceanogr., 22, 502-510, 1977.

Lee, C., and C. Cronin, The vertical flux of particulate organic nitrogen in the sea: decomposition of amino acids in the Peru upwelling area and the equatorial Pacific, J. Mar. Res, 40, 227-251, 1982.

Lee, C., and C. Cronin, Particulate amino acids in the sea: Effects of primary productivity and biological decomposition, J. Mar. Res., 42, 1075-1097, 1984.

Lindroth, P., and K. Mopper, High performance liquid chromatographic determination of subpicomole amounts of amino acids by precolumn fluorescence derivatization with o-phthaldialdehyde, Anal. Chem., 51, 1667-1674, 1979.

Mague, T. H., E. Friberg, D. J. Hughes, and I. Morris, Extracellular release of carbon by marine phytoplankton; a physiological approach, Limnol. Oceanogr., 25, 262-279, 1980.

Manahan, D. T., Amino acid fluxes to and from seawater in axenic veliger larvae of a bivalve (*Crassostrea gigas*), Mar. Ecol. Prog. Ser., 53, 247-255, 1989.

Mendelsohn, M. L., The attributes and applications of flow cytometry, in Proceedings, IVth International Symposium on Flow Cytometry (Pulse Cytophotometry), edited by O. D. Laerum, T. Lindmo, and E. Thorud, pp. 15-27, Universitetforlaget, Oslo, Norway, 1980.

Mopper, K., Sugars and uronic acids in sediment and water from the Black Sea and North Sea with emphasis on analytical techniques, Mar. Chem., 5, 585-603, 1977.

Mopper, K., and R. Dawson, Determination of amino acids in seawater - recent chromatographic developments and future directions, Sci. Total. Environ., 49, 115-131, 1986.

Mopper, K., and E. T. Degens, Aspects of the biogeochemistry of carbohydrates and proteins in aquatic environments, Technical Report 72-68, 118 pp., Woods Hole Oceanographic Institution, 1972.

Mopper, K., and P. Lindroth, Diel and depth variations in dissolved free amino acids and ammonium in the Baltic Sea determined by shipboard HPLC analysis, Limnol. Oceanogr., 27, 336-347, 1982.

Mopper, K., R. Dawson, G. Liebezeit, and V. Ittekkot, The monosaccharide spectra of natural waters, Mar. Chem., 10, 55-66, 1980.

Roach, D., and C. W. Gehrke, Direct esterification of the protein amino acids: gas-liquid chromatography of N-TFA n-butyl esters, J. Chromatogr., 44, 269-278, 1969.

Robertson, K. J., P. M. Williams, and J. L. Bada, Acid hydrolysis of dissolved combined amino acids in seawater: a precautionary note, Limnol. Oceanogr., 32, 996-997, 1987.

Shapiro, H. M., Multistation multiparameter flow cytometry: a critical review and rationale, Cytometry, 3, 227-243, 1983.

Sheldon, R. W., Size separation of marine seston by membrane and glass fiber filters, Limnol. Oceanogr, 17, 494-498, 1972.

Siegel, A., and E. T. Degens, Concentration of dissolved amino acids from saline waters by ligand-exchange chromatography, Science, 151, 1098-1101, 1966.

Torella, F., and R. Y. Morita, Microcultural study of bacterial size changes and microcolony and ultramicrocolony formation by heterotrophic bacteria in seawater, Appl. Environ. Microbiol., 41, 518-527, 1981.

Walters, J. S., and J. I. Hedges, Simultaneous determination of uronic acids and aldoses in plankton, plant tissues, and sediment by capillary gas chromatography of N-hexylaldonamide and alditol acetates, Anal. Chem., 60, 988-994, 1988.

Williams, P. M., A. F. Carlucci, S. M. Henrichs, E. S. Van Vleet, S. G. Horrigan, F. M. H. Reid, and K. J. Robertson, Chemical and microbiological studies of sea-surface films in the southern Gulf of California and off the west coast of Baja California, Mar. Chem., 19, 17-98, 1986.

Yentsch, C. M., P. K. Horan, K. Muirhead, Q. Dortch, E. Haugen, L. Legendre, L. S. Murphy, M. J. Perry, and D. A. Phinney, Flow cytometry and cell sorting: a technique for analysis and sorting of aquatic particles, Limnol. Oceanogr., 28, 1275-1280, 1983.

Zierdt, C. H., Adherence of bacteria, yeast, blood cells, and latex spheres to large-porosity membrane filters, Appl. Environ. Microbiol., 38, 1166-1172, 1979.

EXTRACTION AND ANALYSIS OF POLYSACCHARIDES, CHIRAL AMINO ACIDS, AND SFE-EXTRACTABLE LIPIDS FROM MARINE POM

Kenneth Mopper[1] and Kenneth G. Furton[2]

[1]Chemistry Department, Washington State University, Pullman, Washington 99164-4630 (carbohydrates and amino acids)
[2]Department of Chemistry, Florida International University, University Park, Miami, Florida 33199 (supercritical fluid extraction of lipids)

1. Polysaccharide Composition

As major constituents of oceanic particulate organic matter (POM), carbohydrates function in several key roles: food storage, structural, osmoregulation, environmental conditioning, and "glue" to hold particles together. Structures of typical carbohydrates are given in Figure 4-5. Sources of polysaccharides in the sea include algae [Percival, 1970; Percival and McDowell, 1981] and diatoms in particular [Allan et al., 1972; Haug and Myklestad, 1976; Myklestad et al., 1972], bacteria [Fazio et al., 1982; Christensen et al., 1985; Kennedy and Sutherland, 1987], and marine fungus [Szaniszlo et al., 1968; Wessels and Sietsma, 1981]. Past studies have found important compositional and structural differences in polysaccharides between these groups (Table 4-4). These differences are expected to be reflected in oceanic particles from different sources as well as different stages of particle degradation in the water column and in particle traps.

1.1. Filtration and Fractionation Procedures

Polysaccharides and mucopolysaccharides particle samples can be divided into three broad categories: exocellular polymers, cell storage products soluble in hot water, and structural components, including water-insoluble mucilages, as shown in Figure 4-6. Non-covalently bonded exocellular polysaccharides and mucopolysaccharides are separated from cells and aggregates by gentle filtration [Allan et al., 1972; Myklestad et al., 1972; Sandford, 1979; Kennedy and Sutherland, 1987; Christensen et al., 1985]. Vacuum filtration through 0.45 μm membrane filters, with a pressure differential of < 200 mm Hg, has been shown effective in removing fragile cells from seawater without causing disruption or stress-induced release of internal pools of organic matter [Fuhrman and Bell, 1985]. An alternative mild filtration technique is tangential flow filtration, which has been used for separating microorganisms from seawater [Delong et al., 1989; Giovannoni et al., 1990]. These filtration procedures, should result in only minor release of intracellular polymers from organisms residing in the particles [Myklestad et al., 1972; Allan et al., 1972].

Exopolymers and water soluble POM extracts can be fractionated further according to molecular weight (size) using either ultrafiltration membranes [Wheeler, 1976] or size exclusion chromatography [Gershey, 1983; Sakugawa and Handa, 1985]. Low, non-specific protein binding, i.e., low protein adsorbing, ultrafiltration membranes (Millipore or Amicon) are available for tangential flow systems. The latter is preferred over stirred cell systems because of the much shorter contact time and faster filtration rate, which in turn minimizes adsorption effects. Calibration should be performed with polysaccharides and proteins of known molecular weight added to seawater [Wheeler, 1976].

1.2. Polysaccharide Hydrolysis and Depolymerization Techniques

One of the major deficiencies of past studies of polysaccharides in POM and seawater has been the use of nonselective degradation methods which largely obliterate information about the identity of polymers originally present. This problem can be avoided by use of partial acid hydrolyses and selective depolymerizing enzymes, i.e. enzymes specific to individual polysaccharides. Currently, we are testing the latter approach on algal exudates and POM. Emphasis is being placed on polysaccharides for which depolymerizing enzymes are commercially available. The cleavage products are analyzed by HPLC and capillary GC, discussed in the following section. It should be pointed out that enzyme efficiencies may be significantly lower for natural POM samples than for purified substrates as a result of matrix effects; therefore, polysaccharide concentrations may be underestimated. Despite this, the enzymatic approach is still valuable because it provides direct evidence for the presence of specific polysaccharides and bond linkages. Polysaccharides which can be studied by this method include amyloses, 1,4 linked glucans, which are characteristically formed by the *Rhodophyceae*, *Cyanophyceae* and *Chlorophyceae*; and laminarins (1,3 or 1,6 linked glucans), which are characteristic of the *Phaeophyceae*

Marine Particles: Analysis and Characterization
Geophysical Monograph 63

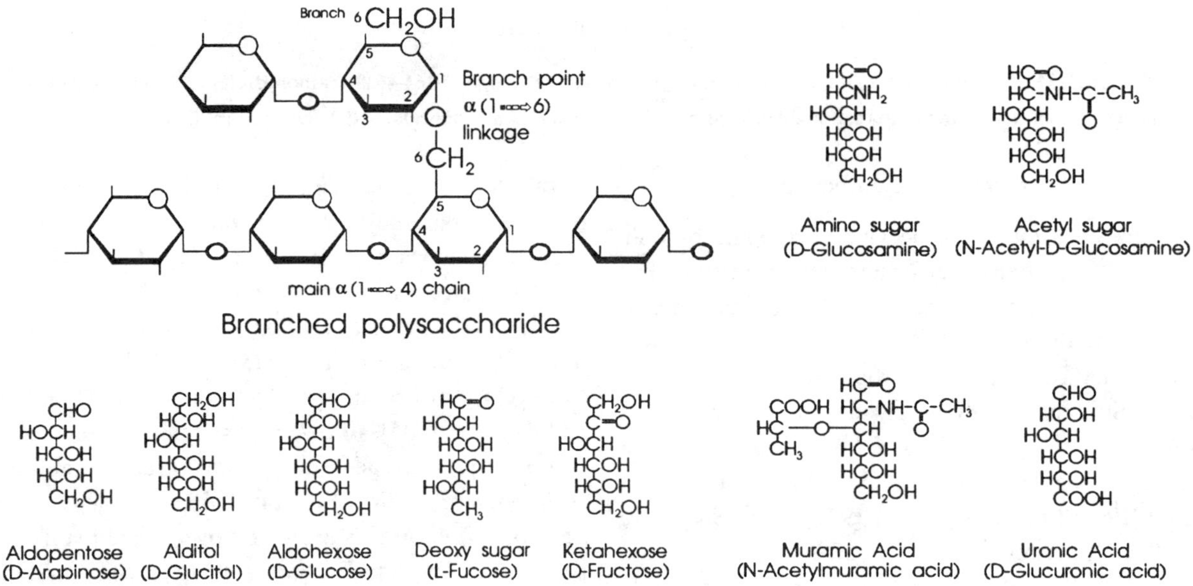

Fig. 4-5. Structures of some saccharides discussed in this section.

Table 4-4.
Carbohydrates as Source Indicators of Marine POM

Plants	
Structural: linear polysaccharides	
Land plants	cellulose (1,4 b-Glu)
diatoms	glucuromannans
crustaceans	chitin (N-acetyl GluNH2)
green algae	mannans
Structural: mucilaginous polysaccharides	
Red algae	agarose, carrageenan (Gal; 3,6-anhydro Gal; 6-O-me-Gal)
Brown algae	alganic acid (GalA-ManA), fucoidans
Food reserves: starches and LMW sugars	
Reds, blue-greens, green	a-1,4-glucans (amyloses)
Brown, Golden	a-1,3- and 1,6-glucans (laminarins)
Green, some diatoms	fructans
Brown	mannitol
Bacteria	
Structural:	muramic acid, o-methylated sugars, -acetylated sugars, branched sugars, deoxy sugars, 7-carbon (heptuloses), 8-carbon (octuloses), teichoic acid
Food reserve:	trehalose, melibiose

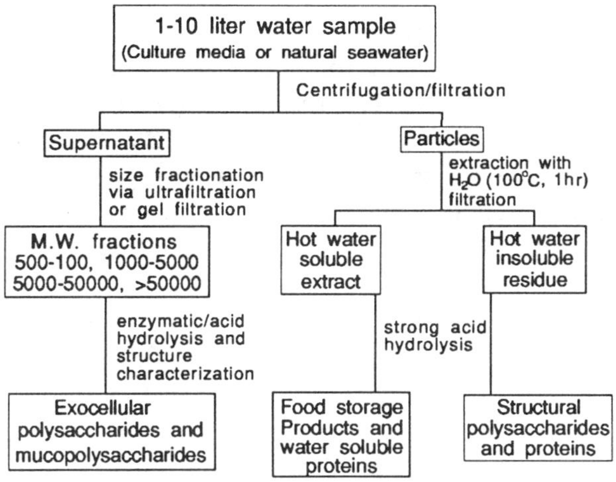

Fig. 4-6. Extraction of polysaccharides from marine POM.

and *Chysophyceae*. Commercially available laminarinases, amylases, amyloglucosidases, and pullulanase can be used in digestion procedures similar to those of Englyst and Cummings [1984].

Mucilaginous polysaccharides, such as agarose, carrageenan, alginic acids and fucoidans, are produced copiously by Red and Brown algae groups [Mackie and Preston, 1974]; and rhamnoidans, fucoidans, and polyglucuronates are characteristic products of diatoms [Lewin, 1955; Allan et al., 1972; Myklestad et al., 1972]. Fructans also may be important, particularly in Green algae and some diatoms [Craigie, 1974; Ittekkot et al., 1982]. However, the enzymatic approach can not be used readily to study these saccharides because of lack of commercially available enzymes. However, these polysaccharides can be analyzed satisfactorily by acid hydrolysis techniques because the resulting monosaccharides are characteristic of the original polysaccharide, e.g., agarose and carrageenans release 3,6-anhydrogalactose and various O-methylated galactoses, while alginic acids release almost exclusively mannuronic and guluronic acids [Mackie and Preston, 1974; Percival, 1970].

Partial acid hydrolysis techniques yield important information about the polysaccharide composition of POM; e.g., nature of the linkages [Mopper and Larsson, 1978; Sakugawa and Handa, 1985]. In particular, these techniques have proven useful for studying heteropolysaccharides, e.g., hemicelluloses, which are widespread in marine algae [Handa and Tominaga, 1969; Mopper and Larsson, 1978; Percival, 1970; Mackie and Preston, 1974]. Published permethylation and reducing terminus tagging procedures [Hase et al., 1978; Sakugawa and Handa, 1985] can be used in conjunction with partial acid hydrolysis techniques in order to obtain detailed structural information.

Strong acid hydrolysis techniques are recommended for determining the overall carbohydrate composition, both major and minor sugars, of the three main sample fractions (Figure 4-6): exocellular polymers (size fractionated), cell storage products, and structural components. Optimization and application of strong acid hydrolysis techniques to marine POM and DOM have been reported by several investigators [Mopper, 1977; Cowie and Hedges, 1984; Walters and Hedges, 1988].

1.3. Chromatographic Methods

Most past studies of carbohydrates in marine POM and seawater samples focussed on major sugar components, i.e. monosaccharides released after strong acid hydrolysis. Thus, valuable information which could have been obtained from minor constituents was lost. The full suite of major and minor saccharide components can be determined using high resolution chromatographic techniques.

1.4. Gas Chromatography

Saccharides in hydrolyzed extracts can be determined as their alditol acetate derivatives by capillary gas chromatography and flame ionization detection (FID) [Klok et al., 1981; Englyst and Cummings, 1984]. The resolution and selectivity with this GC technique are excellent, as shown by Klok et al. [1984] who identified about 70 sugars in sediment hydrolysates. Furthermore, with slight modifications of the derivatization procedure, several classes of carbohydrates, e.g. aldoses, aldonic acids, alditols, uronic acids, amino sugars, and muramic acid (Figure 4-6), can be run either simultaneously or alone using aliquots from the same sample extract [Lehrfeld, 1981, 1985; Walters and Hedges, 1988]. These modifications greatly simplify sample work-up procedures since time consuming fractionation steps [Mopper, 1977] and isomer equilibration steps [Cowie and Hedges, 1984] are avoided. Unfortunately, sugar pairs lyxose-arabinose and gulose-glucose yield the same alcohols upon reduction. Also, ketoses cannot be determined by this technique.

1.5. Liquid Chromatography

Other disadvantages of GC techniques for sugars are that the derivatization steps are time-consuming and require removal of water. Also, oligosaccharides, molecules containing 3-10 sugar units, which are the main products released by enzymatic and partial acid hydrolyses, must be hydrolyzed further to constituent monosaccharides before they can be analyzed by GC. Therefore, in order to complement GC procedures, we have modified a recently developed liquid chromatographic technique for use in determination of mono- and oligosaccharides in aqueous samples [Mopper et al., submitted, 1991]. The method is based on anion exchange chromatography at pH greater than 12 followed by electrochemical detection in the pulsed amperometric mode [Rocklin and Pohl, 1983]. Reducing and non-reducing mono-, di-, and oligosaccharides are separated readily and detected by this technique. Chromatographic selectivity can be controlled by gradient elution and by use of eluant modifiers. Aqueous samples, including seawater and aqueous POM extracts, can be injected directly after a simple desalting step. Acid hydrolysates require only a simple neutralization step prior to injection. The method is highly sensitive; we detect simple sugars in seawater down to a concentration of about 5 nanomoles per liter, or about one picomole per injection (Mopper et al., submitted, 1991). The main drawback of this HPLC method is that the resolution is significantly lower than that of capillary GC techniques, therefore, it is not as well suited for determination of minor sugars.

1.5.1. LC system description.

The LC system we are using currently was assembled for less than half the cost of equivalent commercial carbohydrate systems. A detailed description follows. A schematic of the sugar analyzer system is shown in Figure 4-7. The chromatograph consists of an E-Lab Model 2010 (OMS Tech, Miami, FL) gradient controller and data acquisition system. Both gradient control and data acquisition/integration are performed on an IBM-compatible PC containing the E-Lab circuit board. Mobile phase proportioning is done by a low pressure solenoid valve module (OMS Tech) situated before an Eldex Model 100A single piston pump (Eldex Laboratories, Menlo Park, CA). A 1.5 mL polypropylene column containing Teflon chips is placed between the solenoid valve and the pump. This column serves as a static mixer and as a debubbler. Pump pulsations are dampened with a 1.5 m coil of PEEK tubing (0.125 in. o.d., 0.0625 in. i.d.) teed in immediately after the pump. Sample injection is performed with an electronically activated valve (Valco Instrument Co. Houston, TX), with either a 100 or 250 μL PEEK sample loop. Timing of sample injection with the start of the run is programmed using the E-Lab system. A carbonate retarder column (Model ATC-1, Dionex Corp., Sunnyvale, CA) is placed between the pump and injector to remove traces of carbonate from the eluents, which, if present, would alter retention times. The analytical column, Model AS6 (Dionex), is 250 mm by 4.6 mm i.d. and contains 10 μm anion exchange packing material in the hydroxide form. A precolumn, Model AG6 (Dionex), was placed between the injector and the analytical column. To minimize baseline drift during gradient elution, post column addition of strong base, 0.5 N NaOH, is performed using gas pressure. About 25 psi He is applied to a 2 L pressure-tight polyethylene bottle (Dionex) and the effluent is forced through a 10 psi flow restrictor and an electronically activated shut-off valve. Sugars are detected by an EG&G (Princeton, NJ) model 200 amperometric detector in the pulsed amperometric detection (PAD) mode. The detector is equipped with a standard flow cell containing a single gold electrode and a Ag/AgCl reference electrode.

1.5.2. HPLC conditions.

The eluent for isocratic separation of mono- and disaccharides is 20-25 mM NaOH, made fresh every 1-2 days. In order to minimize column deactivation, only "carbonate-free", freshly deionized water and commercial high purity NaOH solution, not pellets, are used. The eluent is stored under carbonate-free conditions, e.g., blanket of helium or nitrogen. The normal flow rate is 1 mL min^{-1} and the column temperature is ambient. Sodium hydroxide (1 M) should be pumped through the column for 20 to 30 min once a day, or as needed between samples, to remove any carbonate that has accumulated. The detector must be disconnected and not reconnected until the system has been flushed with normal eluent for 10 min.

1.5.3. PAD conditions.

Typical voltage (E) and pulse time (T) settings for the EG&G detector are:

E1 = 50 mV T1 = 100 (1.666 sec)
E2 = 700 mV T2 = 32 (0.533 sec)
E3 = -600 mV T3 = 19 (0.316 sec)

The full scale current (I) range (0.1 - 100 μA) will depend on the concentrations to be measured. Examples of the ranges we use are: less than 0.1 μA is generally too noisy for this application; 0.2 μA for nanomolar monosac charides; and 5.0 μA for micromolar monosaccharides. The offset, background current, is between -0.200 and 0.200 μA under normal operating conditions. Outside this range the baseline tends to drift noticeably.

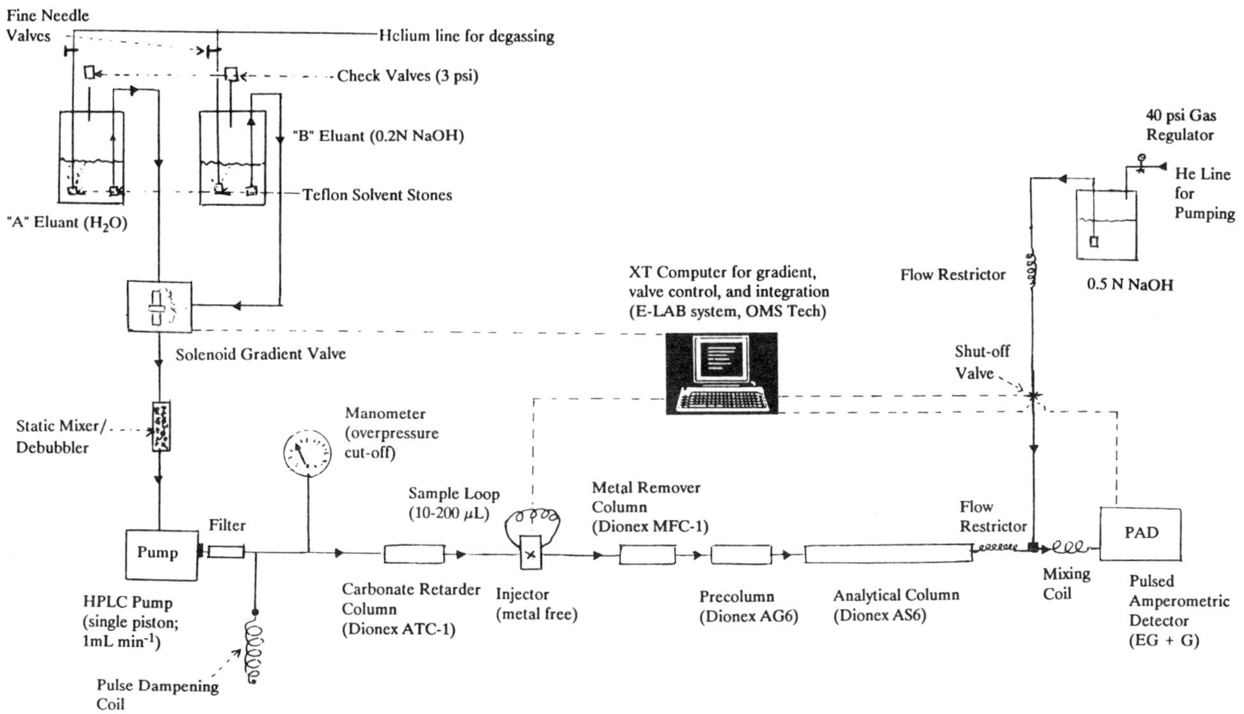

Fig. 4-7. Schematic of HPLC sugar system.

1.5.4. Sample preparation for LC analysis.

Free sugars and oligosaccharides in seawater and particle extracts. Two procedures have been developed for desalting seawater samples and aqueous particle extracts [Mopper et al., submitted, 1991]. The first method, based on ion exchange, is preferred because it is milder and less expensive.

1) *Desalting with ion exchange resins:* A 5 mL mixed bed of anion (carbonate form, AG2-X8, 20-50 mesh) and cation (hydrogen form, AG50-X8, 50-100 mesh) exchange resins is prepared in a column equipped with a bottom frit and stopcock. Resins of different mesh sizes facilitate their separation after use for regeneration. Excess water is blown out of the column with compressed high purity nitrogen. The resin bed is rinsed quickly, less than 5 sec each rinse, with three 0.5 mL aliquots of seawater sample containing an appropriate internal standard, i.e., a sugar not found free in the sample such as lyxose or gulose. The rinses are blown out with compressed nitrogen, prolonged rinses will exhaust the bed's exchange capacity. A 1 to 1.5 mL aliquot of seawater sample containing an internal standard is added to the column and allowed to sit in the resin bed about 5 min or until effervescence stops. The sample is eluted into a vial using compressed nitrogen, followed by injection into the HPLC system. The desalted sample can be stored frozen at least 2 weeks.

2) *Desalting with silver cartridges:* A silver cartridge (On-Guard Ag; Dionex) is rinsed with deionized water at a flow rate of about 1 mL min^{-1} for approximately 1 hr. Excess water is blown out with compressed nitrogen. About 3 mL of seawater sample containing an internal standard is passed through the Ag cartridge under gravity, discarding the first mL and collecting the second mL in a centrifuge vial. A Ag cartridge can remove chloride from approximately 3 mL of seawater. Calcium and sulfate are then removed by addition of 600 μL of 3 M $Ba(OH)_2$. After about 1 min at room temperature, the precipitate is spun down at about 3000 rpm (700 g) for about 5 minutes in a microcentrifuge and the supernatant is injected into the HPLC system through a 0.2 μm syringe filter.

Sediment hydrolysates. Water or particulate samples are hydrolyzed in 0.85M H_2SO_4 for 4 to 6 hours at 100° C under nitrogen [Mopper, 1977]. Approximately 1.5 equivalents of $CaCO_3$ powder, combusted at 300-400° C for about 12 hours to remove organic impurities, are added to neutralize the acid and precipitate sulfate. The sample is reacted for about 30 min at room temperature under constant stirring. Alternatively, 1 equivalent of barium hydroxide powder, combusted as above, can be used with only a five minute reaction. The precipitate is spun down at about 3000 rpm for five minutes and the supernatant is injected into the HPLC system through a 0.2 μm syringe filter. If the sample contains humic substances, amino acids or other potential interferences, the supernatant should be desalted as in procedure 1 above prior to injection.

2. Determination Of Chiral Amino Acids

2.1. Gas Phase Acid Hydrolysis

Only gas phase hydrolysis techniques will be discussed. Standard acid hydrolysis procedures as applied to marine particles will not be covered here since they are discussed in detail in contributions by Hedges, page 131 and Henrichs, page 141. However, it is recommended that protein hydrolysis procedures currently being used by biochemists be systematically examined for their applicability to marine particle samples.

For the analysis of dissolved combined amino acids (DCAAs) in seawater and soluble particle extracts, high temperature, gas phase acid hydrolysis techniques are preferred over conventional liquid hydrolysis techniques for several reasons. They are more rapid, less contaminating and less destructive toward released amino acid monomers [Tsugita et al., 1987; Keil and Kirchman, in press, 1991]. In general, small sample volumes, e.g., 50-200 μL, are used in the procedure. The samples are pipetted into muffled 1 mL micro test tubes, precombusted 4 hours at 450° C, and evaporated to dryness under a stream of high purity nitrogen. Up to 10 sample tubes are placed in the precleaned hydrolysis vessel (Waters Associates, Milford, MA) and 50 μL per tube of constant boiling sequanal grade HCl (Pierce) is carefully pipetted into the bottom of the vessel. The vessel is sealed under vacuum and placed in an oven at 150° C for 40 minutes. The vessel is allowed to cool to room temperature and then opened slowly to air. HCl in the tubes is evaporated under a stream of nitrogen. Amino acids in the tubes are redissolved in 500 μL of deionized water and derivatized for HPLC, as described below.

2.2. HPLC Conditions

The LC method for determining chiral amino acids described here is a modification of procedures of Nimura and Kinoshita [1986] and Buck and Krummen [1987]. The method is based on precolumn fluorescence derivatization with *o*-phthalaldehyde and a chiral thiol, N-acetyl-L-cysteine. A radial compression module with a Nova-Pak C18, 5 mm cartridge (Waters Associates, Milford, MA) is used to achieve the HPLC separations. The injector is fitted with a 250 μL loop. A fluorescence detector model 121 (Gilson Medical Electronics, Middleton, WI) is equipped with a 305-395 nm excitation filter and 450 ± 10 nm emission filter. Eluent A: 925 mL of 0.045 M sodium acetate, 0.005 M sodium dihydrogen orthophosphate mixed with 75 mL of methanol, pH adjusted to 7.4 with 5 M sodium hydroxide. Eluent B: Methanol. The following binary gradient is used at a total flow rate of 1.2 mL min^{-1}.

Time (m)	%B	Time (m)	%B	Time (m)	%B
0	3	25	32	33.5	100
4	3	27.5	41	35	3
10	27	29.5	41		
24	32	29.5	100		

2.3. Standards

A stock standard solution (20% methanol in deionized water) is prepared containing 300 μM of each amino acid. The stock solution is diluted in seawater to an appropriate concentration prior to injection.

2.4. OPA/NAC Reagent

Twenty five mg *o*-phthalaldehyde (OPA) is dissolved in 2.5 mL methanol; then 61.5 mg N-acetyl-L-cysteine (NAC) is added and made up to 20 mL with sodium borate buffer (0.5 M, pH 9.5). Allow to age about one day before use. The reagent is stable for about one week if stored under argon.

2.5. Derivatization and Calibration

In a clean glass or Teflon vial, an aliquot of the stock standard solution is added to seawater to give a total volume of 2 mL, then 200 μL of the OPA/NAC reagent is added. The mixture is shaken and allowed to react for 4 min before injected into the HPLC system. Calibration is performed using at least three levels of standard additions, usually ranging from 20 to 200 nM. The same procedure is used for samples, except no standard solution is added.

2.6. Blank

The sensitivity of the method is limited mainly by amino acid contamination in the reagent and in the deionized water used to determine the blank. The reagent background actually improves with age as fluorescent amino acid derivatives originally present in the fresh reagent gradually rearrange to nonfluorescent compounds [Simons and Johnson, 1978]. In order to evaluate the reagent background, at least three levels of standard additions of reagent to deionized water are

TABLE 4-5.
Critical Points and Typical Properties of Common Fluids

Properties	He	CO_2 (low r)	CO_2 (high r)	H_2O
Critical temperature (°C)	-268	313	31	374
Critical pressure (psi)	33	1071	1071	3199
Operating temperature (°C)	200	100	35	20
Operating pressure (psi)	20	1200	3000	1000
Aggregation	gas	supercritical fluids		liquid
Density (g/mL)	10^{-4}	0.1	0.8	1.0
Diffusivity (cm^2/S)	10^{-1}	10^{-3}	10^{-4}	10^{-5}
Viscosity (cP)	0.02	0.02	0.1	1.0

made. A reaction time of 20 sec is used instead of 4 min. The only significant background interferences are serine and glycine, usually corresponding to 1-5 nM, and ammonium, 10-30 nM.

3. Supercritical Fluid Extraction

The extraction and recovery of organic species from environmental and biological matrices are critical and often the limiting steps in analysis schemes for a wide variety of natural systems. Extraction of hydrophobic compounds, e.g., fatty acids, steroids, hydrocarbons, etc., is performed usually with liquid solvents using liquid-liquid extraction techniques such as Soxhlet extraction and sonication or solid-liquid extraction techniques such as solid-phase extraction (SPE). These traditional extraction methods generally take several hours to perform, require relatively large volumes of ultrapure and often hazardous solvents, making subsequent concentration steps necessary, and may yield incomplete recovery and/or degradation of sample species. Analytical supercritical fluid extraction (SFE) can overcome many of the problems associated with current liquid-liquid and solid-liquid extraction procedures for recovering compounds from complex matrices.

Supercritical fluids have densities comparable to liquids with higher diffusivities and lower viscosities approaching that seen for gases. Critical points and typical properties of a common gas, liquid, and supercritical fluid are given in Table 4-5.

The inherent supercritical fluid extractability of analytes from solid sorbents is controlled by a variety of factors including the affinity of the analytes for the sorbent, vapor pressure of the analytes, and solubility and diffusion coefficient of the analytes in the supercritical fluid. In addition, SFE efficiencies are controlled by

TABLE 4-6.
Variables Affecting Supercritical Fluid Extraction Efficiencies

analyte type (polarity, molecular weight, etc.)
extraction cell agitation
extraction cell dead volume
extraction cell geometry
extraction cell size
extraction fluid
extraction time
fluid flow rate
fluid modifiers (solvents, derivatizing agents, etc.)
mode of analyte accumulation
modifier concentrations
pressure (density)
restrictor type (simple, feedback regulator, etc.)
sample condition (humidity, pH, etc.)
sample matrix
sample particle size
sample size
temperature

a complex relationship between many experimental variables (Table 4-6).

One of the advantages of supercritical fluids is the ease with which the density, and effectively the solvating power, of the fluid can be changed by varying temperature and pressure. Carbon dioxide is the most widely used supercritical fluid because it is inexpensive, non-toxic, non-flammable, and has a relatively low critical temperature and pressure. Supercritical carbon dioxide can be operated at a low density of around 0.1 $g\,mL^{-1}$ up to a density of nearly 1 $g\,mL^{-1}$. Compared to liquids, supercritical fluids have improved mass transfer properties and more efficient penetration of sample matrix resulting in more rapid extraction of analytes. Additionally, most useful supercritical fluids are gases at room temperature which simplifies analyte concentration steps and minimizes potential losses of volatile sample constituents. Alternatively, SFE can be used in a continuous on-line fashion by combination with high-performance liquid chromatography, supercritical fluid chromatography, gas chromatography, or mass spectrometry, thereby eliminating losses due to sample handling. Analytical-scale SFE using pure supercritical CO_2 has been applied to the recovery of a variety of nonpolar analytes from solids, adsorbents, and flavor and fragrance compounds with up to an order of magnitude increase in the rate of extraction reported compared to conventional methods [Hawthorne and Miller, 1987; Furton and Rein, 1990].

The recovery of polar compounds requires use of fluids more polar than pure carbon dioxide. This can be achieved by use of fluid modifiers, i.e., entrainers, such as water, methanol, and reverse micelles, to increase solvent strength and selectivity of low polarity fluids. Nonpolar compounds, such as lipids and saturated fatty acids, can be extracted rapidly with pure supercritical carbon dioxide, whereas polar constituents, e.g., functionalized fatty acids and steroids, can be extracted with modified supercritical carbon dioxide or nitrous oxide. Solid samples can be placed easily within microextractor cells and extracted off-line into a solvent or on-line directly onto the head of a GC column for subsequent GC/MS analysis [Furton and Rein, 1990]. Aqueous samples can be extracted directly by modifying microextractor cells presently used, or alternatively, they can be passed through a microextractor cell containing a sorbent, analogous to a high pressure Solid Phase Extraction cartridge, and the collected analytes extracted with supercritical fluids as described above. Examples of the application of CO_2 SFE to the determination of lipids in sediment trap samples are shown in Figures 4-8 and 4-9. The former shows that SFE can be performed simultaneously with functional group derivatization to enhance extraction efficiency and selectivity.

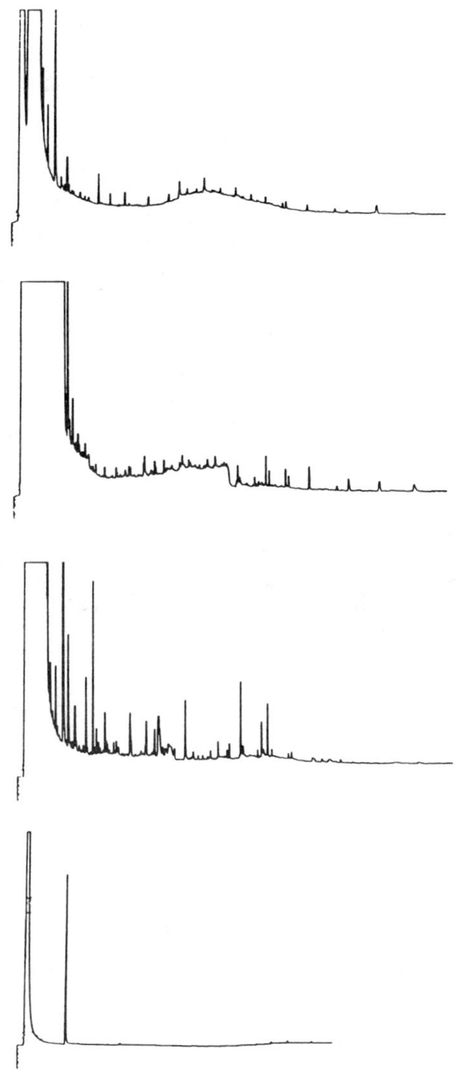

Fig. 4-8. (Courtesy of J. W. Hills and H. H. Hill, Washington State University). Extraction of sediment trap material from Daboub Bay (WA). Sample donated by J. I. Hedges. (A) Sample extracted with CO_2 alone for 60 min at 300 atm, 60° C; (B) Same as A but silylated with trimethylsilane (TMS, in pyridine) prior to injection into GC; (C) Re-extraction of sample with SCF CO_2 containing silylating reagent (TMS in pyridine); (D) Procedural blank. GC conditions: DB-225, 30 m, 60° C hold 5 min, ramp 5° /min, 235° C hold 30 min; FID.

References

Allan, G. G., J. Lewin, and P. G. Johnson, Marine polymers. IV. Diatom polysaccharides, Bot. Mar., 15, 102-108, 1972.

Buck, R. H., and K. Krummen, High-performance liquid chromatographic determination of enantiomeric amino acids and amino alcohols after derivatization with *o*-phthaldialdehyde and various chiral mercaptans, J. Chromatogr., 387, 255-265, 1987.

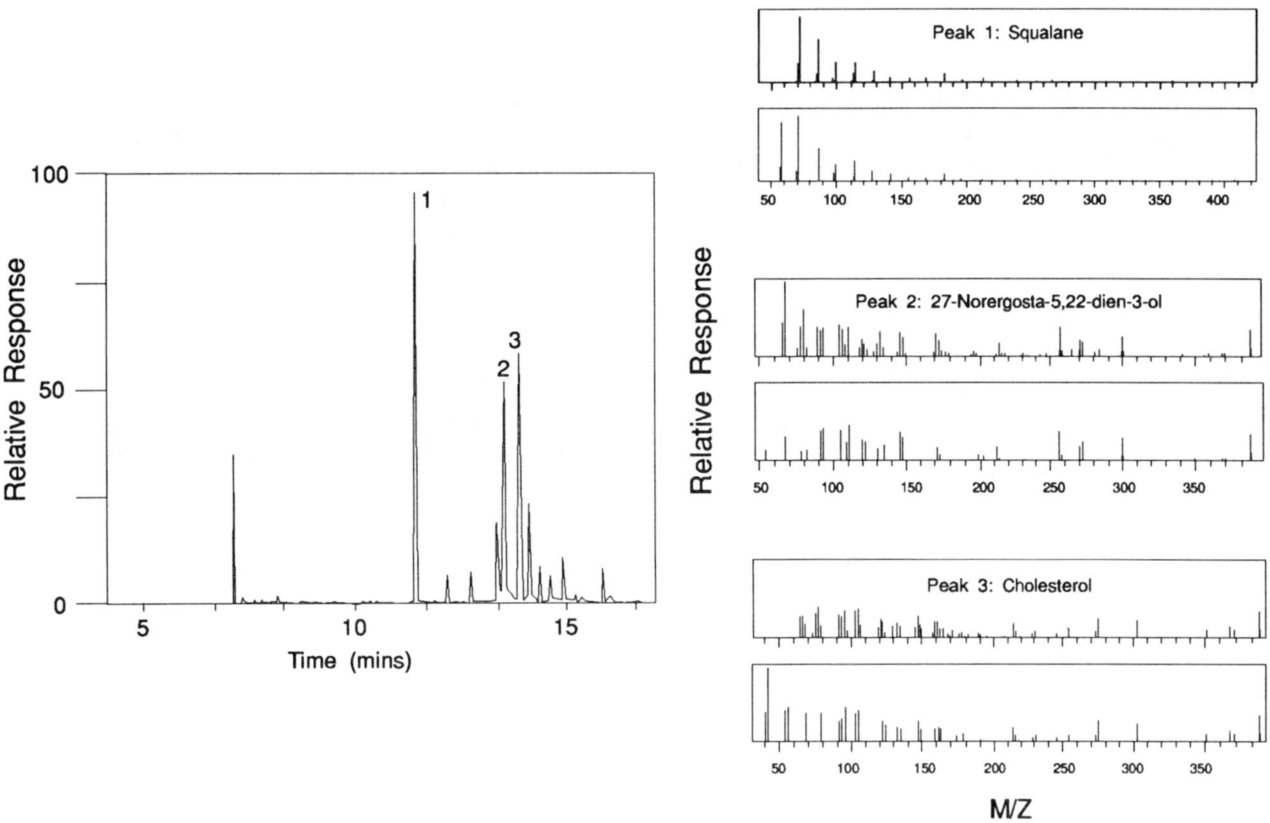

Fig. 4-9. SCF extraction of sediment trap material from the Mediterranean Sea (donated by G. Cauwet). Sample was extracted with CO_2 for 10 min at 4500 psi, 40° C. The extract was collected off-line into methylene chloride and analyzed by GC-MS. Left: Total ion chromatogram of the extract. Right: Mass spectra of the three main peaks. Lower spectra are library reference compounds.

Christensen, B. E., J. Kjosbakken, and O. Smidsrod, Partial chemical and physical characterization of two extracellular polysaccharides produced by marine, periphytic *Pseudomonas sp.* strain NCMB 2021, Appl. Environ. Microbiol., 50, 837-845, 1985.

Cowie, G. L., and J. I. Hedges, Determination of neutral sugars in plankton, sediments, and wood by capillary gas chromatography of equilibrated isomeric mixtures, Anal. Chem., 56, 497-504, 1984.

Craigie, J. S., Storage products, in Algal Physiology and Biochemistry: Botanical Monographs, edited by W. D. P. Stewart, Vol. 10, pp. 206-235, Blackwell Scientific Publications, Oxford, 1974.

Delong, E. F., T. M. Schmidt, and N. R. Pace, Analysis of single cells and oligotrophic picoplankton populations using 16S rRNA sequences, in Recent Advances in Microbial Ecology, edited by T. Hattori, Y. Ishida, Y. Maruyama, R. Morita, and A. Uchida, pp. 697-701, 1989.

Englyst, H. N., and J. H. Cummings, Simplified method for the measurement of total non-starch polysaccharides by gas-liquid chromatography of constituent sugars as alditol acetates, Analyst, 109, 937-942, 1984.

Fazio, S. A., D. J. Uhlinger, J. H. Parker, and D. C. White, Estimations of uronic acids as quantitative measures of extracellular and cell wall polysaccharide polymers from environmental samples, Appl. Environ. Microbiol., 43, 1151-1159, 1982.

Fuhrman, J. A., and T. M. Bell, Biological considerations in the measurement of dissolved free amino acids in seawater and implications for chemical and microbiological studies, Mar. Ecol. Prog. Ser., 25, 13-21, 1985.

Furton, K. G., and J. Rein, Trends in techniques for the extraction of drugs and pesticides from biological specimens prior to chromatographic separation and detection, Anal. Chim. Acta, 236, 99-114, 1990.

Gershey, R. M., Characterization of seawater organic matter carried by bubble-generated aerosols, Limnol. Oceanogr., 28, 309-319, 1983.

Giovannoni, S. J., E. F. Delong, T. M. Schmidt, and N. R. Pace, Tangential flow filtration and preliminary phylogenetic analysis of marine picoplankton, Appl. Environ. Microbiol., 56, 2572-2575, 1990.

Handa, N., and H. Tominaga, A detailed analysis of carbohydrates in marine particulate matter, Mar. Biol., 2, 228-235, 1969.

Hase, S., T. Ikenaka, and Y. Matsushima, Structure analyses of oligosaccharides by tagging of the reducing end sugars with a fluorescent compound, Biochem. Biophys. Res. Commun., 85, 257-263, 1978.

Haug, A., and S. Myklestad, Polysaccharides of marine diatoms with special reference to *Chaetoceros* species, Mar. Biol., 34, 217-222, 1976.

Hawthorne, S. B., and D. J. Miller, Directly coupled supercritical fluid extraction gas chromatographic analysis of polycyclic aromatic hydrocarbons and polychlorinated biphenyls from environmental solids, J. Chromatogr., 403, 63-76, 1987.

Ittekkot, V., E. T. Degens, and U. Brockmann, Monosaccharide composition of acid-hydrolyzable carbohydrates in particulate matter during a plankton bloom, Limnol. Oceanogr., 27, 770-776, 1982.

Keil, R. G., and D. L. Kirchman, Dissolved combined amino acids in marine waters as determined by a vapor-phase hydrolysis method, Mar. Chem., in press, 1991.

Kennedy, A. F. D., and I. W. Sutherland, Analysis of bacterial exopolysaccharides, Biotechnol. Appl. Biochem., 9, 12-19, 1987.

Klok, J., E. H. Nieberg-Van Velzen, J. W. de Leeuw, and P. A. Schenck, Capillary gas chromatographic separation of monosaccharides as their alditol acetates, J. Chromatogr., 207, 273-275, 1981.

Klok, J., H. C. Cox, M. Baas, P. J. W. Schuyl, J. W. de Leeuw, and P. A. Schenck, Carbohydrates in recent marine sediments--I. Origin and significance of deoxy- and O-methyl-monosaccharides, Org. Geochem., 7, 73-84, 1984.

Lehrfeld, J., Differential gas-liquid chromatography method for determination of uronic acids in carbohydrate mixtures, Anal. Biochem., 115, 410-418, 1981.

Lehrfeld, J., Simultaneous gas-liquid chromatographic determination of aldonic acids and aldoses, Anal. Chem., 57, 346-348, 1985.

Lewin, J. C., The capsule of the diatom *Navicula pelliculosa*, J. Gen. Microbiol., 13, 162-169, 1955.

Mackie, W., and R. D. Preston, Cell wall and intercellular region polysaccharides, in Algal Physiology and Biochemistry: Botanical Monographs, Vol. 10, edited by W. D. P. Stewart, pp. 40-85, Blackwell Scientific Publications, Oxford, 1974.

Mopper, K., Sugars and uronic acids in sediment and water from the Black Sea and North Sea with emphasis on analytical techniques, Mar. Chem., 5, 585-603, 1977.

Mopper, K., and K. Larsson, Uronic and other organic acids in Baltic Sea and Black Sea sediments, Geochim. Cosmochim. Acta, 42, 153-163, 1978.

Myklestad, S., A. Haug, and B. Larsen, Production of carbohydrates by the marine diatom *Chaetoceros affinis* var. Willei (Gran) Hustedt. II. Preliminary investigation of the extracellular polysaccharide, J. Exp. Mar. Biol. Ecol., 9, 137-144, 1972.

Nimura, N., and T. Kinoshita, *o*-Phthalaldehyde-N-acetyl-L-cysteine as a chiral derivatization reagent for liquid chromatographic optical resolution of amino acid enantiomers and its application to conventional amino acid analysis, J. Chromatogr., 352, 169-177, 1986.

Percival, E., Algal polysaccharides, in The Carbohydrates: Chem. and Biochem., edited by W. Pigman and D. Horton, second ed., Vol. IIB, pp 537-568, Academic Press, New York, 1970.

Percival, E., and R. H. McDowell, 2 algal walls - composition and biosynthesis, in Plant Carbohydrates II: Extracellular Carbohydrates, edited by W. Tanner and F. A. Loewus, pp. 277-316, Springer-Verlag, New York, 1981.

Rocklin, R. D., and C. A. Pohl, Determination of carbohydrates by anion exchange chromatography with pulsed amperometric detection, Liq. Chromatogr., 6, 1577-1590, 1983.

Sakugawa, H., and N. Handa, Isolation and chemical characterization of dissolved and particulate polysaccharides in Mikawa Bay, Geochim. Cosmochim., 49, 1185-1193, 1985.

Sandford, P. A., Exocellular, microbial polysaccharides, Adv. Carbohydr. Chem. Biochem., 36, 265-313, 1979.

Simons, S. S., Jr., and D. F. Johnson, Reaction of *o*-phthalaldehyde and thiols with primary amines: fluorescence properties of 1-alkyl (and aryl) thio-2-alkylisoindoles, Anal. Biochem., 90, 705-725, 1978.

Szaniszlo, P. J., C. Wirsen Jr., and R. Mitchell, Production of a capsular polysaccharide by a marine filamentous fungus, J. Bacteriol., 96, 1474-1483, 1968.

Tsugita, A., T. Uchida, H. W. Mewes, and T. Atake, A rapid vapor-phase acid (hydrochloric acid and triluoroacetic acid) hydrolysis of peptide and protein, J. Biochem., 102, 1593-1597, 1987.

Walters, J. S., and J. I. Hedges, Simultaneous determination of uronic acids and aldoses in plankton, plant tissues, and sediment by capillary gas chromatography of N-hexylaldonamide and alditol acetates, Anal. Chem., 60, 988-994, 1988.

Wessels, J. G. H., and J. H. Sietsma, Fungal cell walls: a survey in Plant Carbohydrates II: Extracellular Carbohydrates, edited by W. Tanner, and F. A. Loewus, pp. 352-394, Springer-Verlag, New York, 1981.

Wheeler, J. R., Fractionation by molecular weight of organic substances in Georgia coastal water, Limnol. Oceanogr., 21, 846-852, 1976.

ELECTROPHORETIC SEPARATION OF PARTICULATE PROTEINS IN SEAWATER

Eiichiro Tanoue

Geochemical Laboratory, Meteorological Research Institute, Nagamine 1-1, Tsukuba, 305, Japan

1. Introduction

Proteins are one of the most important organic constituents of all organisms. A major part of particulate organic matter (POM) in seawater is considered to be proteinaceous [Riley, 1970], and a large proportion of the particulate nitrogen (PN) is in the form of proteins or peptides [e.g., Degens, 1970; Packard and Dortch, 1975]. Carbon and nitrogen in amino acids from acid-hydrolyzed proteins account for approximately 30% and 50% or more of the particulate organic carbon (POC) and particulate nitrogen, respectively [Handa, 1970; Siezen and Mague, 1978; Tanoue and Handa, 1979; Tanoue, 1985]. Thus, analysis of amino acids in proteins has been cited in terms of "combined amino acids" or "proteinaceous compounds", and measurements of total amino acid and of the amino acid composition of POM after hydrolysis have received considerable study [e.g., Siezen and Mague, 1978; Lee and Cronin, 1984; Liebezeit and Bölter, 1986; Müller et al., 1986]. However, most proteins in marine organisms, the major source of POM, have similar amino acid compositions [Strickland, 1965; Degens, 1970] even though they differ greatly in biochemical properties. Twenty amino acids make up the basic building blocks of most proteins [Haschemeyer and Haschemeyer, 1973]. Therefore, a comparison of the amino acid composition of samples may not provide much information as to the fate of particulate proteins, i.e., their production, transportation and decomposition, in marine environments.

Attempts to make a more direct estimation of particulate protein have been undertaken by fluorometric or colorimetric methods, using the Lowry method [Parsons and Strickland, 1962], the biuret reaction [Danilenko, 1977], the fluorescamine assay [Packard and Dortch, 1975; Garfield et al., 1979] and the Coomassie blue assay [Setchell, 1981]. Concentrations of particulate protein have been shown to be significantly correlated with levels of chlorophyll, electron transport system (ETS) activity and PN in the euphotic layer [Garfield et al., 1979]. These findings indicate that amounts of particulate protein are closely related to biological activity throughout the water column. However, the amounts of particulate protein in these measurements have been expressed in terms of the total protein that is equivalent to an amount of an appropriate standard protein. Apart from sensitivity, there are methodological uncertainties associated with such analyses due to the selection of a standard protein, the reactivity of different types of protein and interference by other organic compounds in POM [Roth, 1971; Udenfriend et al., 1972; Bradford, 1976; Pollard et al., 1978; Chu and Casey, 1986]. Furthermore, such assays provide only limited information concerning the nature and fate of protein in the sea.

No information is available on the occurrence of protein molecules, especially in deep water, and virtually nothing is known about the nature of the protein in POM. One of the reasons why the molecular structure of particulate proteins has received so little attention is that the chemical nature of POM, especially in deep water, has been assumed to resemble that of humic-like materials. POM is considered to be derived from dissolved organic matter [DOM; e.g., Riley, 1970; Cauwet, 1978; Mopper and Degens, 1979]. POM may be a significantly older and chemically amorphous geopolymer which consists of degraded, biologically transformed, relatively stable macromolecules and refractory materials [Cauwet, 1978; Rashid, 1985]. The lack of a method for detecting specific proteins in POM has hindered our efforts to assess the occurrence of various proteins and to clarify the molecular structure of the proteinaceous material in POM.

2. Methods

2.1. Sodium Dodecyl Sulfate - Polyacrylamide Gel Electrophoresis (SDS-PAGE)

Some proteins consist of single subunits but many consist of multiple dissociable subunits, each one composed of a single chain; two or more chains may be linked by covalent bonds (e.g., disulfide bonds) or associated as a result of noncovalent interactions. The anionic detergent sodium dodecyl sulfate (SDS) dissociates polymeric proteins into their constituent polypeptide chains, it does not cleave disulfide bonds. Individual charge patterns of each polypeptide are influenced by the

Marine Particles: Analysis and Characterization
Geophysical Monograph 63

binding of the SDS anion, rendering all molecules negatively charged [Pitt-Rivers and Impiombato, 1968; Weber and Osborn, 1969; Reynolds and Tanford, 1970a]. The SDS-polypeptide complexes form extended rod-like particles, the length of which is proportional to the molecular weight of the protein [Fish et al., 1970; Reynolds and Tanford, 1970b; Helenius and Simons, 1975]. Polyacrylamide gel electrophoresis in the presence of SDS separates polypeptide chains according to their molecular weights [Shapiro et al., 1967; Weber and Osborn, 1969; Neville, 1971]. Thus, the size of the polypeptide chains of a given protein can be determined by a comparison of their electrophoretic mobilities on SDS gel to the mobilities of marker proteins with subunits of well-characterized molecular weights.

SDS-PAGE was applied to particulate proteins as follows. POM on a glass-fiber filter was lyophilized and ground with a mortar and pestle with Tris-HCl buffer, SDS, and 2-mercaptoethanol. The slurry was transferred to an Eppendorf tube and heated in a boiling-water bath for 3 min. After cooling, urea was added. In these steps, solubilization, complete denaturation and reduction of any disulfide bonds in particulate proteins were accomplished. Solubilization of the proteins was further facilitated by the presence of urea [Miyazaki et al., 1978]. During these procedures, the native "configuration" of the proteins in POM was destroyed, since covalently linked polypeptide subunits were effectively dissociated. Any disulfide bonds or other noncovalent bonds, such as ionic and hydrogen bonds and apolar (hydrophobic) interactions, were eliminated. Under these conditions, particulate proteins formed a negative charge and specific shapes which were dependent on their molecular weights but independent of their native shapes or ionic charge.

There are basically two types of gel, disc-, and slab-gel. In each gel electrophoresis, discontinuous [Laemmli, 1970] and continuous [Weber and Osborn, 1969] gel systems are applicable. In this paper, results obtained by the discontinuous slab gel system are shown. Samples and molecular marker proteins were loaded in the gel and bromophenol blue was added as tracking dye. Electrophoretic separation was performed until the tracking dye reached the bottom of the separation gel, at which point electrophoresis was terminated.

2.2. *SDS-Isoelectric Focusing (SDS-IEF)*

SDS-PAGE is a powerful tool for separation of insoluble proteins on the basis of molecular weight classes. However, with this method of separation, heterogeneity due to charge differences within the same molecular weight class cannot be detected. Isoelectric focusing on polyacrylamide gels, on the other hand, separates proteins based on charge differences but is difficult to apply without modification because of the insoluble nature of particulate protein. Samples prepared for SDS-PAGE are not readily amenable to analysis by isoelectric focusing. O'Farrell [1975] found that a sample applied to isoelectric focusing gel can contain SDS without destroying the separation. SDS bound to protein comes off the protein in the presence of excess non-ionic detergent (e.g., Nonidet P-40) and forms mixed micelles with the detergent which migrate to the acidic end of the gel during electrophoresis. In the presence of excess Nonidet P-40, particulate protein prepared for SDS-PAGE regains its original charge and IEF techniques can be successfully applied (data not shown).

2.3. *Detection of Proteins on the Gel*

After electrophoresis, the proteins on the gel are first fixed and then visualized by staining the protein using appropriate dyes. Many fixing and staining procedures have been reported using, e.g., Amido Black in acetic acid, Coomassie blue R in trichloroacetic acid, Coomassie blue R in alcohol, Coomassie blue G in perchloric acid [see review by Wilson, 1983] and silver staining [e.g., Switzer et al., 1979]. These procedures are satisfactory if the chief aim is to locate proteins on a gel, but there is no stain that is fully satisfactory for all PAGE and IEF systems or for all proteins [Wilson, 1983]. These procedures are also applicable to particulate protein. An electrophoretogram stained by Coomassie blue R250 is shown here because of its good reproducibility and relatively high sensitivity (Figure 4-10).

2.4. *Quantification of Protein on the Gel*

The PAGE technique is a very promising approach for separation and identification of proteins in POM but has the disadvantage that quantification is difficult. There are essentially two different methods used for quantifying protein on gels. In the first, stained proteins remaining on the gel are quantified by densitometric scanning or by solvent-extraction of dye from the dye-protein complex [Ball, 1986; references cited therein]. In the second method, the protein is extracted from the gel using solvent [e.g., Tsugita, 1987], electroelution, or electroblotting [e.g., Hunkapiller et al., 1983]. Each of the above-mentioned methods are useful for various particular applications. The second method will likely lead to more detailed characterization of the proteins in POM, but recovery of protein from the gel is usually low

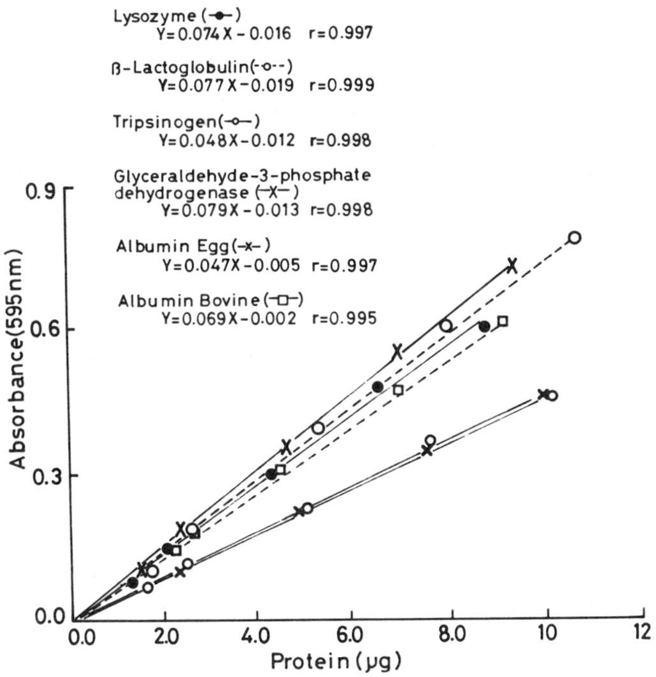

Fig. 4-10. Plot of absorbance (595 nm) of dye eluted from dye-protein complexes on the gel versus amounts of six standard proteins on the gel. The following reference proteins, with their molecular weights given in Daltons were used: lysozyme, 14,300; β-lactoglobulin, 18,400; trypsinogen, 24,000; glyceraldehyde-3-phosphate dehydrogenase, 36,000; egg albumin, 45,000; and bovine serum albumin, 66,000.

and the extraction behavior is dependent on the nature of the protein.

A simple method for the extraction of Coomassie blue R from stained protein as described by Ball [1986] can be applied to quantifying particulate protein. This method requires no unusual equipment and is suitable for measurement of multiple samples. After staining, the gel is cut into small pieces and soaked in extraction solvent. Absorbance of eluted dye is measured at 595 nm by spectrometry. Blank values obtained from a piece of gel of similar size are subtracted from experimental values. The assay was applied to six marker proteins for construction of standard curves in the range of 0 to 11 μg protein (Figure 4-10). All six proteins gave straight lines when absorbance at 595 nm was plotted against the amount of protein although the slopes of the lines were different.

3. Application of PAGE Techniques

Figure 4-11 shows an example of an SDS-PAGE electrophoretogram showing Coomassie blue R stained

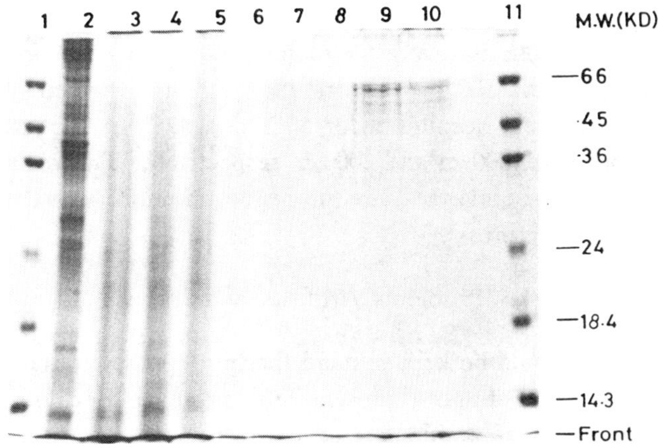

Fig. 4-11. Electrophoretograms of proteins in POM (lanes 3-10) and plankton (lane 2) samples from a Bering Sea station (52° 58.0′N; 177° 30.2′E). Standard marker proteins are shown for comparison (lanes 1 and 11). POM samples were collected from depths of 0 m (lane 3), 25 m (lane 4), 50 m (lane 5), 100 m (lane 6), 200 m (lane 7), 500 m (lane 8), 1000 m (lane 9), and 2000 m (lane 10).

proteins in POM and plankton from a station in the Bering Sea. The electrophoretograms of POM in the surface waters and the plankton were essentially the same. Many proteins were observed over a wide range of molecular weights from more than 66,000 to less than 14,300 Daltons. A large number of proteins, each at a low level, overlapped with each other in the gel to give rise to uniform staining over a wide range of molecular weights. Significant amounts of low-molecular-weight proteins were concentrated at the migration front. Clear bands of protein were found over a wide range of molecular weights. These clear bands were superimposed on the "background proteins" and rapidly decreased with depth being barely evident in POM collected from waters deeper than 100 m. However, some intense specific bands of proteins were seen in the molecular-weight range from 45,000 and 66,000 Daltons in the intermediate and deep POM.

Each sample lane on the gel was cut at 5 mm intervals and the dye from dye-protein complexes in the gel was eluted and quantified. Figure 4-12 shows the relative abundance of the proteins in each 5 mm fraction. With the exception of the fraction at the migration front, each protein fraction was uniformly stained and did not exceed 10% of the total protein in the plankton or POM shallower than 50 m. Specific proteins in the molecular-weight range of 45,000 to 66,000 Daltons (shaded fraction) were not apparent in the plankton and surface

POM, whereas these proteins were visible in POM below 100 m depth and were the major proteins in POM from deeper than 500 m. Contributions of these proteins to total protein accounted for 50% and 42% of the total protein at 1000 m and 2000 m respectively. The results from these analyses are discussed in more detail in Tanoue [in press].

4. Problems, Artifacts, and Promise

It should be kept in mind that proteins separated by SDS-PAGE do not represent intact proteins. Because of the result of manipulations performed on POM, it is impossible to distinguish between the native proteins and their subunits. This drawback does not allow a simple interpretation of the electrophoretograms. For example, three possibilities could explain the occurrence of the specific proteins in intermediate and deep POM:(1) a single kind of protein made up of the subunits which are dissociated and separated by SDS-PAGE occurs in POM, (2) the measured subunits are derived from many kinds of protein encountered in POM, or (3) both.

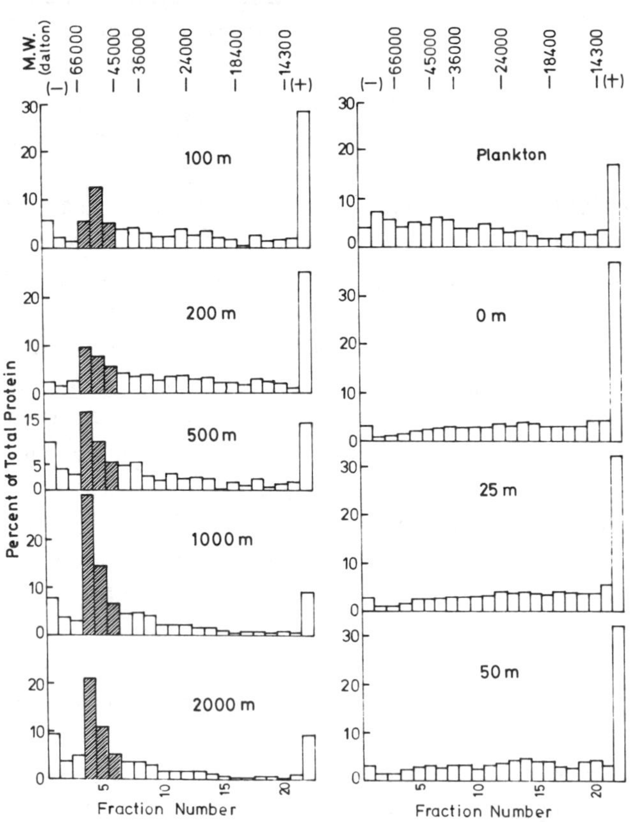

Fig. 4-12. Relative abundance of the proteins in individual 5 mm pieces of gel determined by the dye elution method. The shading indicates fractions where specific proteins were clearly visible.

There are significant protein-to-protein variations in the efficiency of the formation of the dye-protein complexes, as found in the present study (Figure 4-11) and in the literature [Pollard et al., 1978; Ball, 1986]. Of the six standard proteins examined here, glyceraldehyde dehydrogenase was the most active and egg albumin and tripsinogen were the least active, with bovine serum albumin (BSA) being intermediate with respect to the efficiency of formation of the dye-protein complex. It should be emphasized that the quantity of protein determined by the dye elution method is a relative value. To determine an absolute value, a standard curve using the protein of interest must be prepared. There is no single procedure which will satisfactorily quantify all proteins in POM.

Small peptides and amino acids are eliminated during electrophoresis, staining and destaining processes. Coomassie blue R specifically binds proteins but not small peptides and amino acids [Bradford, 1976; Setchell, 1981]. The amount of carbon and nitrogen in particulate protein, quantified using a BSA standard, is 17-24% of the POC and 29-55% of the PN upper 50 m at the station reported here. These values are lower than values determined based on amino acid contents in POM samples from the same oceanic area [Tanoue and Handa, 1979]. The SDS-PAGE/dye-elution method specifically measures protein only but should be considered an approximation and may be an underestimate.

Approximately 100 L of seawater are required to detect proteins in POM from deep water by SDS-PAGE techniques when Coomassie blue R is used as the staining agent. A more sensitive detection method would be preferable. Blank glass-fiber filters (GF/F) treated in the same manner as particulate samples show no visible protein when Coomassie blue R is used, so blanks are not a problem. Silver staining techniques can dramatically increase sensitivity 50-100 times that of Coomassie blue R [e.g., Switzer et al., 1979; Merril, 1990], but can result in artifactual bands with molecular weights ranging from 50-68 kDa [Ochs, 1983; Marshall and Williams, 1984]. In the present study, faint but sharp bands can sometimes be visualized with silver staining of the blank as well as sample gels. Locations of the artifactual bands on the gel are close to those for the specific protein bands found in intermediate and deep water POM. Evidence suggests that the contaminating bands are due to keratin-type proteins [Ochs, 1983]. These bands may be enhanced by the presence of a reducing agent, such as mercaptoethanol [Merril, 1990]. Although these artifactual bands are distinguishable from those of the specific proteins in POM and are not visualized with Coomassie blue R staining, special attention should be made to avoid contamination by skin protein.

Presently, the simplest possible procedures have been employed because no information exists on the nature of POM protein. No preconcentration or desalting was used and SDS was used to prevent protein from sticking to the extraction vessels or the electrophoresis cell. Various pretreatments before electrophoresis are necessary to better understand the nature of protein in POM. For example, with SDS, most if not all of the hydrophobic as well as hydrophilic proteins are solubilized but their native structure and charge are completely destroyed. The non-ionic detergents such as Nonidet P-40, or the essentially identical Triton X-100, have a relatively low solubilization efficiency, but in general do not interfere with the surface charge of the solubilized specimen. The zwitter-ionic surfactants can solubilize the protein least efficiently, but are non-denaturing, i.e., they do not perturb the secondary or tertiary structure of the protein [Satta et al., 1984; references therein]. The selection of a detergent or sequential extraction technique will give important information on the character of the protein in POM. In particular, separation of the native protein in POM without denaturation will be one of the most important approaches to identifying its source.

The resolution of one-dimensional electrophoresis is inadequate for the total separation of proteins in surface POM. In two-dimensional separations, proteins spread in two directions and according to different principles. Thus, while one dimension of electrophoresis can be in the presence of SDS which separates proteins on the basis of their molecular weight, the separation in the other dimension could be on the basis of protein charge [O'Farrell, 1975]. With modification, this technique should be applicable to high-resolution separation of protein in POM.

After separation of a particular protein is satisfactorily accomplished, specific staining methods are available, such as periodic acid-Schiff (PAS) staining for glycoprotein or immunological detection after blotting the proteins from the gel to an appropriate blotting membrane. A method for extraction or blotting of a particular protein from the gel is essential for further study of amino acid composition or for sequencing the protein. Sequencing is the most versatile and promising approach to identifying protein in POM and is now feasible due to recent developments. Current sequencing techniques are able to analyze sub-nanomole amounts of protein which is sensitive enough for Coomassie blue R stained protein bands [Tsugita, 1987]. Identifying some of the major proteins in POM can lead to a more realistic picture of the fate of biogenic organic compounds in the water column.

References

Ball, E. H., Quantitation of proteins by elution of Coomassie brilliant blue R from stained bands after sodium dodecyl sulfate-polyacrylamide gel electrophoresis, Anal. Biochem., 155, 23-27, 1986.

Bradford, M. M., A rapid and sensitive method for the quantitation of microgram quantities of protein utilizing the principle of protein-dye binding, Anal. Biochem., 72, 248-254, 1976.

Cauwet, G., Organic chemistry of seawater particulates. Concepts and developments, Oceanol. Acta, 1, 99-105, 1978.

Chu, F. F., and B. B. Casey, A comparison of protein assays for oyster larval proteins using two different standards, Mar. Chem., 19, 1-7, 1986.

Danilenko, A. F., Some results of protein analysis in Black Sea suspended matter, Oceanol., 17, 529-531, 1977.

Degens, E. T., Molecular nature of nitrogenous compounds in seawater and recent marine sediments, in Organic Matter in Natural Waters, edited by D. W. Hood, pp. 77-106, Univ. of Alaska, Institute of Marine Science Occasional Publication 1, 1970.

Fish, W. W., J. A. Reynolds, and C. Tanford, Gel chromatography of proteins in denaturing solvents. Comparison between sodium dodecyl sulfate and guanidine hydrochloride denaturants, J. Biol. Chem., 245, 5166-5168, 1970.

Garfield, P. C., T. T. Packard, and L. A. Codispoti, Particulate protein in the Peru upwelling system, Deep-Sea Res., 26, 623-639, 1979.

Handa, N., Dissolved and particulate carbohydrates, in Organic Matter in Natural Waters, edited by D.W. Hood, pp. 128-152, Univ. of Alaska, Institute of Marine Science Occasional Publication 1, 1970.

Haschemeyer, R. H., and E. V. Haschemeyer, Proteins: A Guide to Study by Physical and Chemical Methods, 445 pp., John Wiley and Sons Inc., New York, 1973.

Helenius, A., and K. Simons, Solubilization of membranes by detergents, Biochim. Biophys. Acta, 415, 29-79, 1975.

Hunkapiller, M. W., E. Lujan, F. Ostrander, and L. E. Hood, Isolation of microgram quantities of proteins from polyacrylamide gels for amino acid sequence analysis, in Methods in Enzymology, edited by C. H. W. Hirs, and S. N. Timasheff, vol. 91, pp. 227-236, Academic Press, New York, 1983.

Laemmli, U. K., Cleavage of structural proteins during the assembly of the head of bacteriophage T4, Nature, 227, 680-685, 1970.

Lee, C., and C. Cronin, Particulate amino acids in the sea: effects of primary productivity and biological decomposition, J. Mar. Res., 42, 1075-1097, 1984.

Liebezeit, G., and M. Bölter, Distribution of particulate amino acids in the Bransfield Strait, Polar Biol., 5, 199-206, 1986.

Marshall, T., and K. M. Williams, Artifacts associated with 2-mercaptoethanol upon high resolution two-dimensional electrophoresis, Anal. Biochem., 139, 502-505, 1984.

Merril, C. R., Silver staining of proteins and DNA, Nature, 343, 779-780, 1990.

Miyazaki, K., H. Hagiwara, Y. Nagao, Y. Matuo and T. Horio, Tissue-specific distribution of non-histone proteins in nuclei of various tissues of rats and its change with growth of rhodamine sarcoma, J. Biochem., 84, 135-143, 1978.

Mopper, K., and E. T. Degens, Organic carbon in the ocean: nature and cycling, in The Global Carbon Cycle, edited by B. Bolin, E. T. Degens, S. Kempe, and P. Ketner, Scope Report 13, pp. 293-316, John Wiley & Sons, Chichester, 1979.

Müller, P. J., E. Suess, and C. A. Ungerer, Amino acids and amino sugars of surface particulate and sediment trap material from waters of the Scotia Sea, Deep-Sea Res., 33, 819-838, 1986.

Neville, D. W., Molecular weight determination of protein-dodecyl sulfate complexes by gel electrophoresis in a discontinuous buffer system, J. Biol. Chem., 246, 6328-6334, 1971.

Ochs, D., Protein contaminants of sodium dodecyl sulfate-polyacrylamide gels, Anal. Biochem., 135, 470-474, 1983.

O'Farrell, P. H., High-resolution two-dimensional electrophoresis of proteins, J. Biol. Chem., 250, 4007-4021, 1975.

Packard, T. T., and Q. Dortch, Particulate protein-nitrogen in North Atlantic surface water, Mar. Biol., 33, 347-354, 1975.

Parsons, T. R., and J. D. H. Strickland, Oceanic detritus, Science, 136, 313-314, 1962.

Pitt-Rivers, R., and F. S. A. Impiombato, The binding of sodium dodecyl sulphate to various proteins, Biochem. J., 109, 825-830, 1968.

Pollard, H. B., M. R. Brandt, M. A. Pazoles, C. J. Creutz, and A. Romu, Application of Bradford's protein assay to adrenal gland subcellular fractions, Anal. Biochem., 86, 761-763, 1978.

Rashid, M. A., Geochemistry of Marine Humic Compounds, 300 pp., Springer-Verlag, New York, 1985.

Reynolds, J. A., and C. Tanford, Binding of dodecyl sulphate to proteins at high binding ratios. Possible implications for the state of proteins in biological membrane, Proc. Natl. Acad. Sci. U. S. A., 66, 1002-1007, 1970a.

Reynolds, J. A., and C. Tanford, The gross conformation of protein-sodium dodecyl sulfate complexes, J. Biol. Chem., 245, 5161-5165, 1970b.

Riley, G. A., Particulate organic matter in sea water, Adv. Mar. Biol., 8, 1-118, 1970.

Roth, M., Fluorescence reaction for amino acids, Anal. Chem., 43, 880-882, 1971.

Satta, D., G. Schapira, P. Chafey, P. G. Righetti, and J. P. Wahrmann, Solubilization of plasma membranes in anionic, non-ionic and zwitter-ionic surfactants for iso-dalt analysis: a critical evaluation, J. Chromatogr., 299, 57-72, 1984.

Setchell, F. W., Particulate protein measurement in oceanographic samples by dye binding, Mar. Chem., 10, 301-313, 1981.

Shapiro, A. L., E. Vinuela, and J. V. Maizel Jr., Molecular weight estimation of polypeptide chains by electrophoresis in sds-polyacrylamide gels, Biochem. Biophys. Res. Commun., 28, 815-820, 1967.

Siezen, R. J., and T. H. Mague, Amino acids in suspended particulate matter from oceanic and coastal waters of the Pacific, Mar. Chem., 6, 215-231, 1978.

Strickland, J. D. H., Production of organic matter at the primary stages of the marine food chain, in Chemical Oceanography, edited by J. P. Riley, and G. Skirrow, Academic Press, London, 1965.

Switzer, R. C., C. R. Merril, and S. Shifrin, A highly sensitive silver stain for detecting proteins and peptides in polyacrylamide gel, Anal. Biochem., 98, 231-237, 1979.

Tanoue, E., Distribution and chemical composition of particulate organic matter in the Pacific sector of the Antarctic Ocean, Trans. Tokyo Univ. Fish., 6, 43-57, 1985.

Tanoue, E., Occurrence and characterization of particulate proteins in the Pacific Ocean, Deep-Sea Res., in press, 1991.

Tanoue, E., and N. Handa, Distribution of particulate organic carbon and nitrogen in the Bering Sea and northern North Pacific Ocean, J. Oceanogr. Soc. Japan, 35, 47-69, 1979.

Tsugita, K., Developments in protein microsequencing, Adv. Biophys., 23, 81-113, 1987.

Udenfriend, S., S. Stein, P. Bohlem, W. Dairman, W. Leimgraber, and M. Weigle, Fluorescamine: a reagent for assay of amino acids, peptides, proteins and primary amines in the picomole range, Science, 178, 871-872, 1972.

Weber, K., and M. Osborn, The reliability of molecular weight determinations by dodecyl sulfate-polyacrylamide gel electrophoresis, J. Biol. Chem., 244, 4406-4412, 1969.

Wilson, C. M., Staining of proteins on gels: comparisons of dyes and procedures, in Methods in Enzymology, edited by C. H. W. Hirs, and S. N. Timasheff, vol. 91, pp. 236-247, Academic Press, New York, 1983.

SAMPLING AND ANALYSIS OF LIPIDS IN MARINE PARTICULATE MATTER

Stuart G. Wakeham[1] and John K. Volkman[2]

[1]Skidaway Institute of Oceanography, Box 13687, Savannah, Georgia 31416

[2]CSIRO Division of Oceanography, GPO Box 1538, Hobart, Tasmania 7001, Australia

1. Sampling

Particle samples for lipid analyses are typically collected via sediment traps, for "large, sinking particles" [e.g. de Baar et al., 1983; Wakeham and Canuel, 1988] or by in situ filtration, for "small, suspended particles" [e.g. Wakeham and Canuel, 1988]. In most cases, the low concentrations of particles in the water column, especially below the photic zone, mean that water bottle samples collected with Go-Flo or glass Bodman bottles provide too little material unless only major compounds are of interest. Sufficient material for studies of fatty acids, sterols and other major lipids may be obtained from about 50 L of seawater. Thus, for CTDs utilizing Niskin or Go-Flo bottles it is necessary to pool water from several bottles.

Particulate organic matter is collected by filtering seawater through glass fiber filters. For small volumes of water (up to about 10 L) 47 mm diameter glass fiber filters (e.g. Whatman GF/F, nominal pore size 0.7 μm) are suitable. For larger volumes, it is necessary to use larger diameter filters to speed the filtration. Filters should be heated in a muffle furnace before use to remove any organic contamination and then stored in solvent-washed aluminum foil. Blank filters should always be carried through all handling procedures.

Sampling of suspended particles has usually been done using some form of in situ filtration device. There is no consensus as to optimum design. A small in situ pump called Seastar is available from Seastar Instruments, British Columbia. These self-contained pumps operate on D cell batteries and can be deployed on a hydrowire or mooring [Nichols et al., 1989]. The water is sucked through a 142 mm diam. glass fiber filter contained in a Teflon filter holder and then it passes through a Teflon column packed with adsorbent, typically XAD resin, which collects a fraction of the DOC. This system provides very clean samples of particulate matter, and swimming organisms are rarely sampled due to the low flow rate used which is less than 10 L hr^{-1}.

Disadvantages include the high capital cost, and the fact that relatively small volumes of water are filtered. Earlier models could not be deployed below 300 m, but more recent versions are rated to 2000 m.

Wakeham and coworkers [Wakeham et al., 1983; Wakeham, 1987; Wakeham and Canuel, 1988] have collected particles for lipid analyses from up to 5000 L of seawater to depths of 2000 m with a stream-powered filtration system, and Conte and Bishop [1988] collected large volume samples to depths of 2000 m with a pump powered from the ship. Filtration times ranged from a few hours in surface waters to 12 hours in deep waters. Particles were collected on 293 mm quartz or glass fiber filters (e.g. type A/E; ashed before use) having nominal pore sizes in water of about 1 μm. Particles have been size fractionated in situ using a 53 μm stainless steel screen or a 53 μm Nitex screen positioned above a glass fiber filter. Some pre-screening during filtration may be desired to remove zooplankton.

Various sediment trap designs have been used to collect sinking particles for lipid studies. While hydrodynamic considerations have not always been strictly adhered to, the exact design is most often determined by the amount of material needed, the expected particle flux, and the deployment period. Thus in highly productive surface waters having high particle fluxes, e.g. Peru upwelling area, traps of relatively small collecting areas, e.g. 0.13 m^2 [Staresinic, 1980; Wakeham et al., 1983, 1984a] may be adequate even on diel time scales. In the deep-sea, larger collecting areas, e.g. 0.25-1.5 m^2 [Wakeham et al., 1980, 1984b; Wakeham and Canuel, 1988] with long deployment periods (weeks - months) are generally needed. In general, traps with small collection areas, for example the Moss Landing PITS with 0.005 m^2 openings [Knauer et al., 1979], do not collect sufficient material for most lipid analyses.

The question of how to preserve sediment trap material for subsequent lipid analysis during long deployment periods is only now receiving adequate attention. In the past, $HgCl_2$ and NaN_3 have been used as poisons (for killing bacteria) and formalin as a preservative to fix the sample and reduce bacterial activity. Only recently have the effectiveness of both types of treatments been systematically compared [Knauer et al., 1984; Lee et al., in press]. A potentially more difficult

Marine Particles: Analysis and Characterization
Geophysical Monograph 63

problem for lipid analyses resulting from the use of some toxic treatment is the inclusion of "swimmers" in trap samples [Lee et al., 1988]. Screening and hand picking of swimmers from formalin-treated traps have been used to "correct" POM fluxes [Karl and Knauer, 1989], but picking is extremely subjective and may still yield inaccurate fluxes, especially for biochemicals which are specifically derived from zooplankton. Picking of swimmers from $HgCl_2$ or azide treated traps is difficult because the bodies are soft and easily disintegrate. The best solution for this whole problem is to isolate the poison/preservative in a cod end which prevents swimmers from reaching the poison and being killed (Peterson et al., in preparation).

For all particle sampling, extreme caution is needed to minimize contamination, especially from petroleum hydrocarbons which are abundant in and around ships and in their atmosphere. Hydrocarbon fractions of particle samples often contain a signature typical of petroleum contamination, i.e., UCM and homologous series of alkanes with no even/odd predominance. In these cases, it is unclear whether the samples are contaminated from sampling procedures or whether the particles themselves contain a background of petroleum-derived material. Even in cases where there may be obvious petroleum contamination of the hydrocarbon fraction, other lipid fractions may not be adversely affected. Particle samples should be stored at or below -20° C. However, some autolysis may occur even at this temperature, so it may be preferred to extract a sample, thereby denaturing enzyme systems, as soon as possible after collection and store the lipid extract in solvent and in a freezer until subsequent analysis. Air freighting of solvent extracts, if in large volumes, may be a problem.

2. Lipid Extraction and Fractionation

Details of analytical workup need to be set up depending on the specific application (e.g. Figure 4-13). Filters are usually extracted with mixtures of organic solvents. A common procedure is extraction with chloroform-methanol-water (chloroform:methanol (2:1, v/v) or 1:2:0.8, v/v/v) using ultrasonication, in a modification of the Bligh and Dyer [1959] extraction procedure designed for fish lipids [e.g. Nichols et al., 1989]. For large filters, Soxhlet extraction may be required. The filters are extracted several times to ensure good recoveries of organic compounds. The extracts are combined and then chloroform and purified water from a Milli-Q system or 5% NaCl solution are added to bring the chloroform-methanol-water ratio to 1:1:0.9 (v/v/v). The aqueous and solvent layers are allowed to separate over a few hours. The aqueous phase containing salts and other water soluble material is usually discarded and the lipids are recovered in the lower chloroform phase. The solvents are removed to just dryness by rotary evaporation and the lipid extract can then be stored under nitrogen at -20° C until analysis. To minimize hydrolysis of polar lipids and losses of more labile lipids such as polyunsaturated fatty acids, it is usually advisable to analyze the extract within a few days.

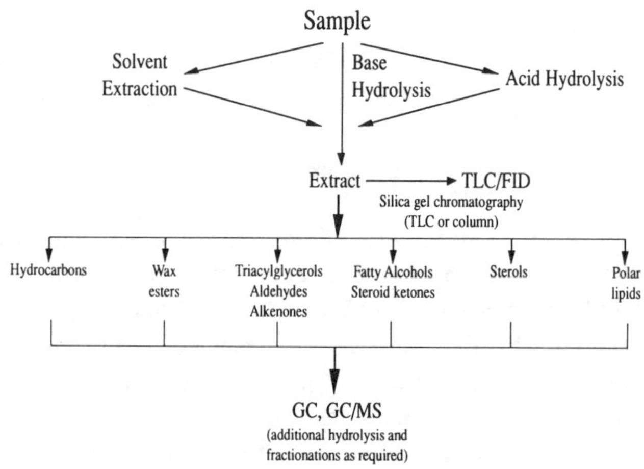

Fig. 4-13. Typical flowchart for lipid analyses. Details of the workup will depend on the specific application.

3. Compositional Analysis

3.1. *Class Composition*

"Bulk" lipid classes (Figure 4-14) may be estimated by gravimetry using an electrobalance following separation by adsorption chromatography or TLC. This step is not routinely used, however, partly because sample sizes (weights) often are prohibitively small. On the other hand, the combination of the high resolving power of thin-layer chromatography (TLC) with a relatively sensitive detection system has offered new techniques for determining lipid class composition in the form of TLC-flame ionization detection (TLC-FID) and high performance TLC-densitometry detection (HPTLC).

3.2. *TLC-FID*

The Iatroscan-Chromarod TLC-FID system is based on separations of lipid classes on reusable silica gel-coated quartz rods [Okumura et al., 1975; Ackman, 1981] using conventional TLC development followed by "scanning" the rods in the FID of the Iatroscan analyzer

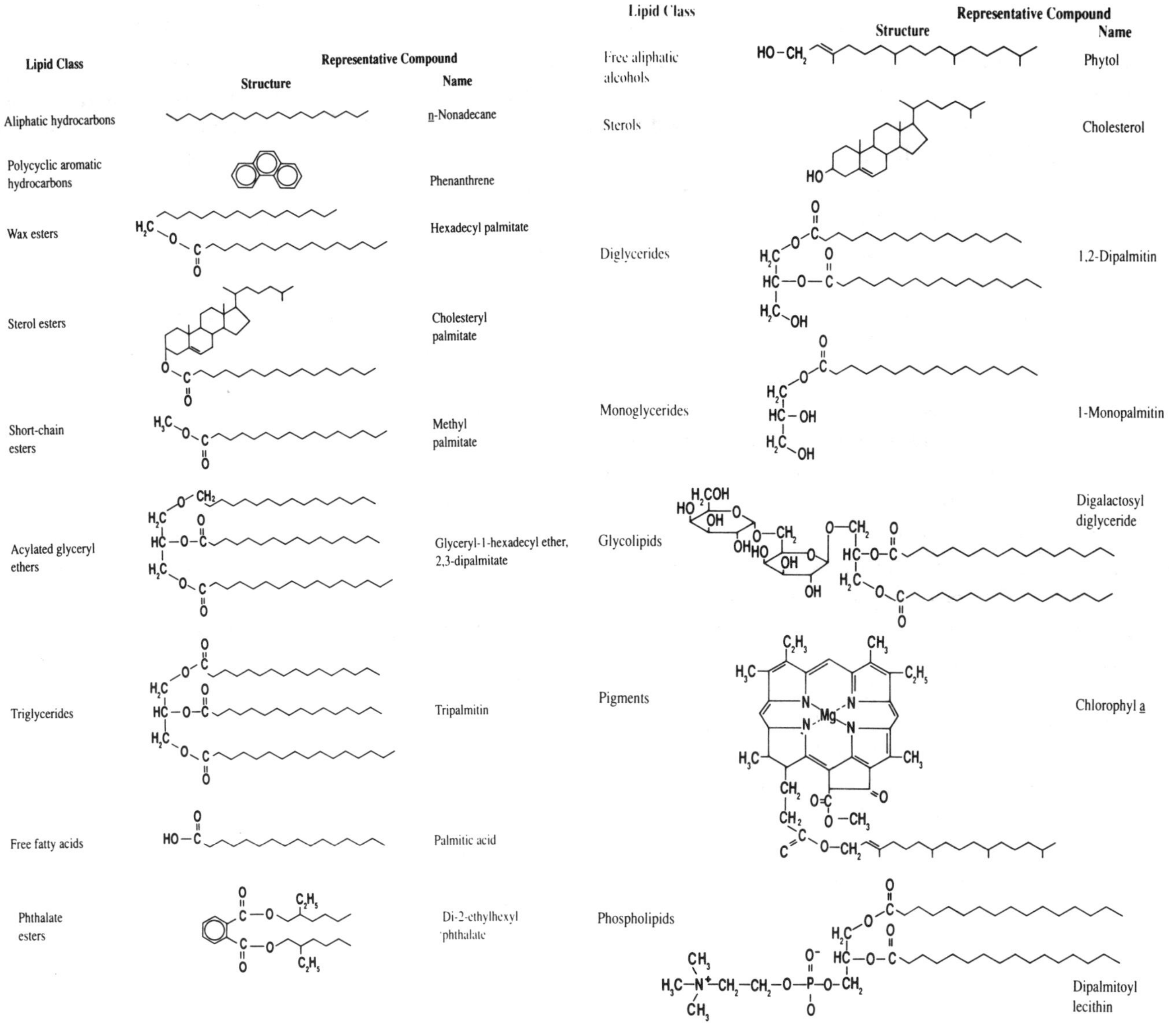

Fig. 4-14. Molecular structures of major lipid class compounds.

(Iatron Laboratories, Japan). This technique has gained wide acceptance as an invaluable tool for the analysis of marine lipids [e.g. Parrish and Ackman, 1983, 1985; Delmas et al., 1984; Volkman et al., 1986; Zhou and Wangersky, 1987; Parrish, 1987a, b; Parrish and Wangersky, 1988; and references cited therein].

A textbook on TLC-FID principles and applications is available [Ranný, 1987]. Samples are applied ("spotted") via microliter syringe or micropipette to the origin of the Chromarod in the minimum volume (0.5-1.0 μL being typical; up to 10 μL, but this can be tedious) so as to minimize band spreading. Excess solvent is evaporated and the rods are developed in glass tanks lined with pre-extracted filter paper (the TLC component). Various solvent systems may be employed depending on the desired resolution. After development, the Chromarods are oven-dried for 5 min at 80 to 100° C and immediately analyzed in the FID of the Iatroscan. Typically the FID is operated with a hydrogen flow rate of 160 mL min^{-1} and an air flow-rate of 2000 mL min^{-1}. Scan speeds may be varied, but a 30 second scan may be most useful. Output from the FID may be recorded using an electronic recording integrator or a personal computer; various integrating and data reporting packages are available if desired. A set of 10 Chromarods are held in a metal rod-holder so that 10 samples may be developed simultaneously and scanned sequentially (and automatically).

Much work has gone toward establishing solvent systems for desired separations. For example, Volkman et al. [1986] chose hexane-diethyl ether-acetic acid (60:17:0.15, v/v/v) for analysis of nonpolar lipids, while chloroform-methanol-water (80:15:2) was used for separating polar lipids. In cases where separations of wax esters and steryl esters from hydrocarbons were required, hexane-diethylether (96:4) was used. A much more complicated development scheme using multiple partial developments and partial scans has been described by Parrish [Parrish and Ackman, 1983; and Parrish, 1987a]. Three different solvent systems of increasing polarity were used to obtain three chromatograms per rod. Hydrocarbons, wax esters, methyl esters and ketones were separated using hexane-diethyl ether-formic acid (99:1:0.05); triacylglycerols, free fatty acids, fatty alcohols, sterols and diglycerides were separated using hexane-diethyl ether-formic acid (80:20:0.1); various polar lipids were separated using methylene chloride-methanol-water (5:4:1). The rods were scanned partially between each development. The solvent system of choice depends on the application and other solvent systems may be found in references cited by Parrish [1988] and Volkman et al. [1986] or available from Iatron Laboratories. In addition, the form of the silica may be altered to fit a specific separation requirement, for example by boric acid or silver nitrate impregnation to resolve unsaturated components [e.g. Sebedio and Ackman, 1981; Tatara et al., 1983].

TLC-FID is several orders of magnitude less sensitive than GC-FID. Typically, detection limits are about 0.25 μg. The FID may be linear up several tens of μg [Volkman et al., 1986; Parrish, 1987a], but at low loadings, less than 1-2 μg, a curved response may be observed, often with a non-zero abscissa. For complex lipid mixtures, internal standards are not possible. It is imperative that calibration curves using external standards which cover the range of analyte mass be constructed as samples are analyzed. Ideally the standards should have similar mobilities on the Chromarods as "unknowns". Different lipid classes often have different FID-responses and hence different calibration curves. Reproducibility of replicate analyses may be 20-30%; however reproducibilities as low as 8% are possible with extensive practice [Volkman et al., 1986]. The major source of error appears to be spotting samples onto the rods. Samples should therefore be run in duplicate, or better, in triplicate.

The quality of the rods greatly affects the resolution obtained. Immediately prior to use, rods are activated and cleaned of any contaminants adsorbed from the atmosphere by scanning them in the FID. Chromarod performance may be improved by equilibration in humidity chambers [e.g. Delmas et al., 1984], but this step has not gained wide acceptance. Chromarods are generally robust and may be used for up to perhaps 100 analyses, although they must be handled with care and may be scratched by the syringe needle. Residue remaining after scanning may be burned off by repeated scanning of the rods or they may be soaked in strong acid, e.g. 35% nitric acid or chromic acid, followed by water and distilled water rinses. Rods should be stored in a desiccator between use; even so, they tend to pick up contaminants from the atmosphere.

3.3. HPTLC

TLC followed by scanning densitometic detection of derivatized lipids may offer a more sensitive alternative to Iatroscan TLC-FID for marine lipids. Like TLC-FID, the technique of HPTLC has been more broadly used in biomedical analysis. Typically, lipids are separated on silica gel HPTLC plates (Merck) and then converted to fluorescence-inducing derivatives which then are detected by a scanning (fluorescence) densitometer, often available in medical laboratories. Conte and Bishop [1988] have used HPTLC for analysis of lipids extracted from seawater particulate matter. They report the method to be 1-2 orders of magnitude more sensitive, with a range of detection from 10 to 500 ng, and more reproducible, ±2-6%, than Chromarod-Iatroscan TLC-FID.

4. Molecular Composition

It is usually not practical to analyze a lipid extract directly for individual compounds due to the diversity of compounds present, so the extract is often hydrolyzed using acid or base to convert esterified lipids to free fatty acids and neutral constituents. Acid hydrolysis and transesterification can be combined into a single operation. This involves reducing the sample extract to dryness under N_2 and adding methanol-hydrochloric acid-chloroform (10:1:1, 3 mL). The solution is heated for 2 hours at 80° C, after which distilled water (3 mL) is added. The reaction products can be extracted with chloroform-hexane (1:4, 3 x 3 mL). This procedure esterifies free fatty acids, transesterifies triacylglycerols and more polar esters, and liberates any neutral lipids bound by ester, glycosidic or amide bonds.

With base hydrolysis, the extract is saponified at 80° C for 3 hours in a methanol-water (4:1) solution containing KOH at a pH of 12. After hydrolysis, neutral constituents are extracted with hexane:chloroform (4:1) and derivatized with BSTFA (as required) for GC and GC-MS analysis. The solution is then acidified to pH 1 with concentrated HCl and the free fatty acids that have been liberated by the base hydrolysis are extracted with hexane-chloroform (4:1) and converted to methyl esters as described above. An alternative to hydrolysis is to isolate the individual lipid classes by thin-layer or column chromatography.

4.1. Thin Layer Chromatography

The extracts are applied in chloroform to silica gel 0.25 mm G-25 UV254 pre-coated plates (Machery-Nagel or equivalent) which are then developed in hexane:diethyl ether:acetic acid (50:50:1). Typical R_f values using this system are: hydrocarbons-0.71; methyl ketones-0.51; C_{30} alcohols-0.41, hexadecanol-0.36; 4-methyl sterols and C_{30} alkyl keto-ols-0.34; 4-desmethyl sterols-0.27; C_{30} alkyl diols-0.22. Lipid classes are usually insufficiently abundant to be seen readily, so bands are visualized by spraying with 0.1% 2′,7′-dichlorofluorescein in methanol. They can then be extracted with chloroform, derivatized as appropriate and analyzed by GC and GC-MS.

4.2. Column Chromatography

Column chromatography is perhaps the simpler technique and so has been more widely employed, although with thin-layer chromatography it is possible to obtain much finer separations. Glass columns containing 3 or 7 g of silicic acid deactivated with 5% water provide good separations of most compound classes of interest. The columns are eluted with increasingly polar solvents starting with hexane, hexane-toluene, hexane-ethyl acetate mixtures, chloroform and finally methanol. Details are provided in Wakeham et al. [1980] and Volkman et al. [1983].

4.3. Capillary Gas Chromatography and Mass Spectrometry

Detailed information about the compounds present in seawater can be obtained using capillary gas chromatography combined with suitable detectors such as the flame ionization detector (FID) for general purpose work, electron capture detector (ECD) for halogenated compounds, flame photometric detector (FPD) for sulfur compounds or mass spectrometer (MS) for compound identification.

Samples are dissolved in a suitable solvent such as hexane or chloroform, and then a small portion (0.2-2 μL) is injected onto the capillary column. Commonly used columns include nonpolar methyl silicone and polar cyanopropylsilicone fused-silica capillary columns, 25 or 50 m in length with an internal diameter of 0.2 to 0.32 mm i.d. For best resolution, the 50 m columns are preferred, but analysis times are longer and the columns are quite expensive. Oven temperature programs are variable depending on the desired application. It is often necessary to maintain the final temperature (e.g., 320° C) for about 20 minutes or to program the oven to higher temperatures, up to 360° C for fused silica or even higher with aluminum-clad columns, to elute high boiling lipids such as wax esters and C_{37}-C_{39} alkenones. Glass capillary columns with cross-linked silicone phases have been heated up to 370° C for analyses of intact wax esters and triacylglycerols [Wakeham and Frew, 1982]. Cooled on-column injection is preferred to minimize loss of higher boiling components. Hydrogen is a good carrier gas as it yields shorter retention times relative to an equivalent head pressure of helium, and has the advantage that it can be produced from laboratory generators. Helium is considerably more expensive.

Analysis by gas chromatography-mass spectrometry (GC-MS) is essential for identifying organic compounds isolated from seawater. Small benchtop quadrupole instruments such as the Hewlett Packard Mass Selective Detector or Finnigan Incos 50 provide adequate sensitivity and mass range for most applications. More rigorous identifications can be obtained from magnetic sector mass spectrometers using metastable reaction monitoring, or from triple quadrupole instruments.

5. Compound Classes

5.1. Hydrocarbons

Hydrocarbons are generally relatively minor lipids in marine particles, except for cases where the particles are contaminated by petroleum residues. It is often difficult to know whether any contamination results from "natural" incorporation of these residues into the particles or whether they are artifacts picked up during sampling; molecular distributions in both cases may be similar. Thus data showing homologous series of n-alkanes over the carbon number range C_{17}-C_{40} and without an odd predominance are usually attributed to petrogenic sources [Saliot et al., 1985].

Biogenic hydrocarbons associated with particles are of marine origins, e.g. C_{15}, C_{17}, C_{19}, pristane, branched

C_{20} and C_{25} isoprenoids, or of terrestrial origins, e.g. C_{23}, C_{25}, C_{27}, C_{29}, and C_{31}. Even in the presence of small amounts of petrogenic hydrocarbons, these biogenic compounds are often detectable. In some nearshore areas characterized by sufficient biomass in surface waters and/or lateral advection of continental material, the biogenic compounds may actually dominate, as was observed in the Cariaco Trench [Wakeham, 1990] and Black Sea [Wakeham et al., in press]. Hydrocarbons which can be attributed to bacterial sources, e.g., branched C_{25}, C_{30}, and C_{40} isoprenoids, sterenes, and hopenes, can be detected if present in relatively high concentrations, especially in particles collected from suboxic or anoxic waters [Wakeham 1987; Wakeham, 1990; Wakeham et al., in press, 1991]. In some cases these bacterial compounds are the major hydrocarbons present in particles.

5.2. Fatty Acids and Esters

Fatty acids are often one of the major lipid classes in seawater particles [Tanoue and Handa, 1980; de Baar et al., 1983; Wakeham and Canuel, 1988; Nichols et al., 1989]. These compounds usually give poor peak shapes on non-polar capillary columns so it is desirable to convert them to methyl esters before GC analysis. Acid-catalyzed transesterification with chloroform-hydrochloric acid-methanol or with 14% BF_3/methanol is commonly employed.

Fatty acids are denoted by X:Y(n-Z) where X is the number of carbon atoms, Y the number of double bonds and Z the position of the ultimate double bond from the methyl end of the molecule. The presence or absence of specific unsaturated fatty acids can be used as an indicator of major sources. For example, the distribution of fatty acids in most diatoms is dominated by 14:0, 16:(n-7), 16:0 and 20:5(n-3). Dinoflagellates, however, contain 16:0, 20:5(n-3), 22:6(n-3) and significant amounts of C_{18}-polyunsaturated fatty acids [Volkman et al., 1980, 1989].

The presence of intact esterified lipids suggests that a proportion of the lipids are associated with living or recently-dead cells. Esters break down rapidly after cell death due to enzymatic reactions and chemical hydrolysis releasing free fatty acids. Triacylglycerols are common energy storage lipids in microalgae and marine animals, but they are not common in bacteria. Wax esters can be the major lipid in some crustaceans, particularly calanoid copepods, but are very rare in microalgae.

5.3. Alcohols and Sterols

Sterols in particulate matter have been the subject of many investigations because of their potential as biomarkers for assigning the major sources of the organic carbon [Gagosian, 1976; Gagosian and Nigrelli, 1979; Gagosian et al., 1983; Saliot et al., 1983; Volkman, 1986; Wakeham and Canuel, 1988]. Sterols and alcohols contain a free hydroxy group and so must be derivatized to acetates using acetic anhydride-pyridine or to trimethylsilyl ethers using bis(trimethylsilyl)-trifluoroacetamide (BSTFA) or another silylating reagent before analysis. Acetates have the advantage of being more stable than TMS-ethers which are readily hydrolyzed by traces of water. However, their mass spectra provide fewer diagnostic ions of side-chain cleavage which can make precise identifications of some components more difficult.

Cholesterol is often the major sterol in seawater particles, presumably reflecting the importance of marine animals and their feces as sources of particles in seawater. Microalgae can have distinctive sterol patterns which in favorable cases can be used to assign the relative contributions for the different algal classes [Volkman, 1986]. Such data can compliment inferences drawn from pigment analyses [Everitt et al., 1990; Bidigare, this volume, page 121].

Alcohols are usually much less abundant than sterols, although significant amounts of C_{14}-C_{22} saturated and monounsaturated alcohols can occur when there are substantial amounts of copepod-derived feces and particles [Wakeham et al., 1983]. An exception is the unsaturated isoprenoid alcohol phytol which is often abundant either free or esterified in chlorophyll *a*. Free phytol rapidly breaks down in seawater to phytenes, phytadienes, a C_{18}-isoprenoid ketone and other products [Volkman and Maxwell, 1986].

5.4. Wax Esters and Triacylglycerols

Marine wax esters (long chain alkyl esters) are esters of long chain fatty acids (C_{14}-C_{24}) and long chain fatty alcohols (C_{14}-C_{26}), yielding intact esters of carbon numbers ranging from C_{26}-C_{50}. Terrestrial wax esters typically contain 40-60 carbon atoms. Triacylglycerols (triglycerides) are generally composed of C_{14}-C_{26} fatty acids esterified to glycerol, yielding intact triacylglycerols having 42-60 carbon atoms. Both the acyl and alkyl moieties vary in their degree of unsaturation. In wax esters, mono- and di-unsaturated moieties are usually most abundant while in triacylglycerols polyunsaturated fatty acids (e.g. 20:5 and 22:6) are common. Steryl esters and alkyldiacylglycerols have also been observed. Wax esters and triacylglycerols are major energy storage lipids in phytoplankton and zooplankton (wax esters are most abundant only in calanoid copepods). They are thus often the major neutral lipids in particles collected in surface waters, but generally less abundant in deeper waters as the esters decompose by autolysis after cell death [Wakeham, 1982, 1985].

Intact wax esters and triacylglyerols are high boiling and not amenable to capillary gas chromatography under usual conditions. Most data on marine wax esters and triacylglycerols consist of fatty acid and fatty alcohol distributions determined after hydrolysis. Wakeham and Frew [1982] developed a (non-routine) high temperature capillary gas chromatography-mass spectrometry technique which was then applied to determining intact wax esters and triacylglycerols, along with steryl esters and alkyldiacylglycerols, in particulate matter [Wakeham, 1982, 1985].

5.5. Polar Lipids

The generic class of polar lipids usually comprises most (70-90%) of the extractable lipids. As a group, polar lipids include, amongst others, phospholipids, glycolipids, lipopolysaccharides, glycosides, and sphingolipids. All are fatty acid esters of various structures and many are involved in membrane structure. Polar lipids are not amenable to analytical techniques common in marine organic chemistry, except hydrolysis and analysis of the fatty acids thus released and TLC-FID [Parrish, 1987a, b] and have received little attention. Nevertheless, the fatty acid composition of polar lipids, in this case predominantly phospholipids, has been used to evaluate bacterial community structure in sediments [Bobbie and White, 1980] and freshwater particulate matter [Fredrickson et al., 1986].

References

Ackman, R. G., Flame ionization detection applied to thin-layer chromatography on coated quartz rods, in Methods in Enzymology, vol. 72, edited by M. J. Lowenstein, pp. 205-252, Academic Press, London, 1981.

Bligh, E. G., and W. J. Dyer, A rapid method of total lipid extraction and purification, Can. J. Biochem. Physiol., 37, 912-917, 1959.

Bobbie, R. J., and D. C. White, Characterization of benthic microbial community structure by high resolution gas chromatography of fatty acid methyl esters, Appl. Environ. Microbiol., 39, 1212-1232, 1980.

Conte, M. H., and J. K. B. Bishop, Nanogram quantification of non-polar lipid classes in environmental samples by high performance thin layer chromatography, Lipids, 23, 493-500, 1988.

de Baar, H. W., J. W. Farrington, and S. G. Wakeham, Vertical flux of fatty acids in the North Atlantic Ocean, J. Mar. Res., 41, 19-41, 1983.

Delmas, R. P., C. C. Parrish, and R. G. Ackman, Determination of lipid class concentrations in seawater by thin-layer chromatography flame ionization detection, Anal. Chem., 56, 1272-1277, 1984.

Everitt, D. A., S. W. Wright, J. K. Volkman, D. P. Thomas, and E. J. Lindstrom, Phytoplankton community compositions in the western equatorial Pacific determined from chlorophyll and carotenoid pigment distributions, Deep-Sea Res., 37, 975-997, 1990.

Fredrickson, H. L., T. E. Cappenburg, and J. W. de Leeuw, Polar lipid ester-linked fatty acid composition of Lake Vechten seston, FEMS Microbiol. Ecol., 38, 381-396, 1986.

Gagosian, R. B., A detailed vertical profile of sterols in the Sargasso Sea, Limnol. Oceanogr., 21, 702-710, 1976.

Gagosian, R. B., and G. E. Nigrelli, The transport and budget of sterols in the western North Atlantic Ocean, Limnol. Oceanogr., 24, 838-849, 1979.

Gagosian, R. B., G. E. Nigrelli, and J. K. Volkman, Vertical transport and transportation of organic compounds from a sediment trap experiment off the coast of Peru, in Coastal Upwelling, Its Sediment Record, Part A, edited by E. Suess, and J. Thiede, pp. 241-272, Plenum Press, New York, 1983.

Karl, D. M., and G. A. Knauer, Swimmers: a recapitulation of the problem and a potential solution, Oceanogr. Mag., 2, 32-35, 1989.

Knauer, G. A., J. M. Martin, and K. W. Bruland, Fluxes of particulate carbon, nitrogen and phosphorus in the upper water column of the northeast Pacific, Deep-Sea Res., 26, 97-108, 1979.

Knauer, G. A., D. M. Karl, J. M. Martin, and C. N. Hunter, In situ effects of selected preservatives on total carbon, nitrogen and metals collected in sediment traps, J. Mar. Res., 42, 445-462, 1984.

Lee, C., S. G. Wakeham, and J. I. Hedges, The measurement of oceanic particle flux - Are swimmers a problem?, Oceanogr. Mag., 1, 34-36, 1988.

Lee, C., J. I. Hedges, S. G. Wakeham, and N. Zhu, Effectiveness of poisons and preservatives in retarding bacterial activity in sediment trap material and their effect on the collection of swimmers, Limnol. Oceanogr., in press, 1991.

Nichols, P. D., J. K. Volkman, and D. A. Everitt, Occurrence of *cis*-6-hexadecenoic acid and other unusual monounsaturated fatty acids in the lipids of oceanic particulate matter, Oceanol. Acta, 12, 393-403, 1989.

Okumura, T., T. Kadono, and A. Iso'o, Sintered thin-layer chromatography with flame ionization detector scanning, J. Chromatogr., 108, 329-336, 1975.

Parrish, C. C., Separation of aquatic lipid classes by Chromarod thin-layer chromatography with measurement by Iatroscan flame ionization detection, Can. J. Aq. Fish., 44, 722-731, 1987a.

Parrish, C. C., Time series of particulate and dissolved lipid classes during spring phytoplankton blooms in Bedford Basin, a marine inlet, Mar. Ecol.Prog. Ser., 35, 129-139, 1987b.

Parrish, C. C., Dissolved and particulate marine lipid classes: a review, Mar. Chem., 23, 17-40, 1988.

Parrish, C. C., and R. G. Ackman, Chromarod separations for the analysis of marine lipid classes by Iatroscan thin-layer chromatography-flame ionization detection, J. Chromatogr., 262, 103-112, 1983.

Parrish, C. C., and R. G. Ackman, Calibration of the Iatroscan-Chromarod system for marine lipid class analyses, Lipids, 20, 521-535, 1985.

Parrish, C. C., and P. J. Wangersky, Iatroscan-measured profiles of dissolved and particulate marine lipid classes over the Scotian slope and in Bedford basin, Mar. Chem., 23, 1-15, 1988.

Ranný, M. D., Thin-Layer Chromatography with Flame Ionization Detection, 198 pp., Reidel Publishing Co., Dordrecht, 1987.

Saliot, A., C. Andrie, M. Goutx, A. Fevrier, D. Tusseau, and J. C. Marty, Large size suspended matter: analysis of some organic compounds, associated concentrations and evolution in the water column, Mar. Chem., 12, 236, 1983.

Saliot, A., C. Andrie, R. Ho, and J. C. Marty, Hydrocarbons in the Mediterranean Sea: their occurrence and fate in the sediment and in the water column, as dissolved and associated with small and large size particles, Int. J. Environ. Anal. Chem., 22, 25-46, 1985.

Sebedio, J. L., and R. G. Ackman, Chromarod-S modified with silver nitrate for the quantitation of isomeric unsaturated fatty acids, J. Chromatogr. Sci., 19, 552-557, 1981.

Staresinic, N., The vertical flux of particulate organic matter in the Peru upwelling as measured by a free-drifting sediment trap, Ph.D. thesis, 255 pp., WHOI/MIT Joint Program in Biological Oceanography, 1980.

Tanoue, E., and N. Handa, Vertical transport of organic materials in the northern North Pacific as determined by sediment trap experiment, J. Oceanogr. Soc. Jpn., 36, 231-245, 1980.

Tatara, T., T. Fujii, T. Kawase, and M. Minagawa, Quantitative determination of tri-, di-, monooleins and free oleic acid by the thin layer chromatography-flame ionization detector system using internal standards and boric acid impregnated chromarod, Lipids, 18, 732-736, 1983.

Volkman, J. K., A review of sterol markers for marine and terrigenous organic matter, Org. Geochem., 9, 83-99, 1986.

Volkman, J. K., and J. R. Maxwell, Acyclic isoprenoids as biological markers, in Biological Markers in the Sedimentary Record, edited by R. B. Johns, pp. 1-42, Elsevier, Amsterdam, 1986.

Volkman, J. K., R. B. Johns, F. T. Gillan, G. J. Perry, and H. J. Bavor, Jr., Microbial lipids of an intertidal sediment. I. Fatty acids and hydrocarbons, Geochim. Cosmochim. Acta, 44, 1133-1143, 1980.

Volkman, J. K., J. W. Farrington, R. B. Gagosian, and S. G. Wakeham, Lipid composition of coastal marine sediments from the Peru upwelling region, in Advances in Organic Geochemistry, 1981, edited by M. Bjorøy et al., pp. 228-240, Wiley, Chichester, 1983.

Volkman, J. K., D. A. Everitt, and D. I. Allen, Some analyses of lipid classes in marine organisms, sediments and seawater using thin-layer chromatography-flame ionization detection, J. Chromatogr., 356, 147-162, 1986.

Volkman, J. K., S. W. Jeffrey, P. D. Nichols, G. I. Rogers, and C. D. Garland, Fatty acid and lipid composition of ten species of microalgae used in mariculture, J. Exp. Mar. Biol. Ecol., 128, 219-240, 1989.

Wakeham, S. G., Organic matter from a sediment trap experiment in the equatorial north Atlantic: wax esters, steryl esters, triacylglycerols and alkyldiacylglycerols, Geochim. Cosmochim. Acta, 46, 2239-2257, 1982.

Wakeham, S. G., Wax esters and triacylglycerols in sinking particulate matter in the Peru upwelling area (15° S, 75° W), Mar. Chem., 17, 213-235, 1985.

Wakeham, S. G., Steroid geochemistry in the oxygen minimum zone of the eastern tropical North Pacific Ocean, Geochim. Cosmochim. Acta, 51, 3051-3069, 1987.

Wakeham, S. G., Algal and bacterial hydrocarbons in particulate matter and interfacial sediment of the Cariaco Trench, Geochim. Cosmochim. Acta, 54, 1325-1336, 1990.

Wakeham, S. G., and E. A. Canuel, Organic geochemistry of particulate matter in the eastern tropical North Pacific Ocean: implications for particle dynamics, J. Mar. Res., 46, 183-213, 1988.

Wakeham, S. G., and N. M. Frew, High temperature glass capillary gas chromatography/mass spectrometry of wax esters, steryl esters, and triacylglycerols, Lipids, 17, 831-843, 1982.

Wakeham, S. G., J. W. Farrington, R. B. Gagosian, C. Lee, H. de Baar, G. Nigrelli, B. W. Tripp, and S. O. Smith, Organic matter fluxes from sediment traps in the equatorial Atlantic Ocean, Nature, 286, 798-800, 1980.

Wakeham, S. G., J. W. Farrington, and J. K. Volkman, Fatty acids, wax esters, triacylglycerols and alkyldiacylglycerols associated with particles collected in sediment traps in the Peru upwelling, in Advances in Organic Geochemistry, 1981, edited by M. Bjorøy et al., pp. 185-197, Wiley, Chichester, 1983.

Wakeham, S. G., J. W. Farrington, and R. B. Gagosian, Variability in lipid flux and composition of particulate matter in the Peru upwelling region, in Advances in Organic Geochemistry, 1983, edited by P. A. Schenck, J. W. de Leeuw, and G. W. M. Lijmbach, Pergamon Press, New York, Org. Geochem., 6, 203-215, 1984a.

Wakeham, S. G., C. Lee, J. W. Farrington, and R. B. Gagosian, Biogeochemistry of particulate organic matter in the oceans: results from sediment trap experiments, Deep-Sea Res., 31, 509-528, 1984b.

Wakeham, S. G., J. A. Beier, and C. H. Clifford, Organic matter sources in the Black Sea as inferred from hydrocarbon distributions, edited by J. W. Murray, and E. Izdar, NATO Advanced Studies Institute volume on Black Sea Oceanography, in press, 1991.

Zhou, X., and P. J. Wangersky, Characterization of marine organic matter by thin-layer chromatography with flame ionization detection, Mar. Chem., 20, 211-218, 1987.

ULTRA-FILTRATION FOR THE CONCENTRATION OF BACTERIA, VIRUSES, AND DISSOLVED ORGANIC MATTER

Ronald Benner

Marine Science Institute, University of Texas at Austin, Port Aransas, Texas 78373-1267

1. Concentration of Bacteria by Ultrafiltration

During the past decade heterotrophic bacteria have become increasingly recognized as important components of the oceanic community. Bacterial abundance typically ranges from 2 to 30 x 10^8 cells L^{-1}, and marine bacteria (avg. 0.06 μm^3) contain about 20 fg C and 5 fg N $cell^{-1}$ [Lee and Fuhrman, 1987 and references therein]. Recent estimates of bacterial biomass in surface waters of the central Pacific Ocean indicate that bacteria account for 40% of the total particulate organic carbon (POC) [Cho and Azam, 1988]. Similar estimates of the quantitative significance of bacterial biomass have been made in the Sargasso Sea [Fuhrman et al., 1989], the Southern California Bight and the Santa Monica Basin [Cho and Azam, 1990]. Bacteria contain >70 and >80% of the euphotic zone microbial C and N in the Sargasso Sea and > 90% of the biological surface area [Fuhrman et al., 1989]. The significance of these bacterial parameters is even greater when the entire water column is considered. About 40% of the primary production is required to maintain the estimated growth and respiration rates of bacteria in oceanic waters [see review by Cole et al., 1988].

It should be noted that inclusion of bacteria in the POC pool is logical, but many bacteria are smaller in dimensions than the arbitrary 0.45 μm boundary chosen for the operational distinction between POC and DOC. Moreover, the glass fiber filters commonly chosen to separate the POC and DOC pools have even larger nominal pore sizes (nominally 0.7 μm for the removal of particles from water) and pass most of the bacteria found in oceanic environments. It would be much more appropriate to use filters that retain bacteria (0.2 μm pore-size) considering that bacteria can make up >70% of the total biomass in seawater. Inert aluminum oxide filters (Anopore) with pore sizes in the 0.02 to 0.2 μm range recently became available. The retention characteristics of these filters presented by the manufacturer are excellent. Our experience with these filters and natural waters indicates that they have very slow filtration rates.

Marine Particles: Analysis and Characterization
Geophysical Monograph 63

1.1. Methods

Over the past three years several investigators have attempted to concentrate bacteria from natural water samples using ultrafiltration. The objective of most of these studies was to obtain sufficient biomass for molecular biological studies [Fuhrman et al., 1988; Barkay et al., 1989; Giovannoni et al., 1990]. One study, however, concentrated bacterial biomass for determination of stable carbon isotope compositions in an attempt to distinguish the sources of carbon fueling bacterial growth in estuaries [Coffin et al., 1990]. Two different ultrafiltration systems, available from Millipore and Amicon Corporations, have been used. I will discuss the available data on filtration rates and recovery efficiencies for these systems, and present the methods used in my laboratory.

The Pellicon tangential flow ultrafiltration unit (Millipore Corp.) has been used by several investigators to concentrate bacteria from seawater. Filter cassettes with 5 ft^2 of surface area are used with this unit and are available in a variety of pore sizes. Up to five cassettes can be used in series, and a peristaltic pump is used to maintain flow through the system. Coffin et al. [1990] used a Pellicon system with 1 to 3 filter cassettes (100,000 nMW) to concentrate bacteria from a variety of estuarine waters. Water samples (50-100 L) were prefiltered through Millipore cartridge filters (1 μm pore-size) to remove most eucaryotes. Prefiltration did not significantly reduce the abundance of bacteria. Bacterial abundance was determined using the epifluorescence direct count method [Hobbie et al., 1977]. Recoveries of bacteria in the concentrates ranged from 16 to 88% and averaged 47%. The authors noted that recoveries of bacteria varied with filtration rates. Higher recoveries were obtained when the ratio of recycled water to filtrate was above 15 (about 30 L hr^{-1} filtration rate). A Pellicon system was also used by Giovannoni et al. [1990] to concentrate bacteria from oceanic waters. Two filter cassettes (10 ft^2) with 0.1 μm pore-sizes were used to process large water samples (1,700 and 8,000 L). Processing rates of 150 L hr^{-1} were reported. Water samples were pre-sieved through Nytex screening (10 μm pore size) and cells were enumerated using epifluorescence microscopy. Recoveries of cells were about 37%. The authors noted that the recoveries were low because of occasional line ruptures during which 10 to 20% of the concentrate volumes were lost. Scanning electron microscopy of microbial cells in unfiltered seawater and in ultrafiltered concentrates revealed the same types of morphotypes in similar proportions, suggesting that the ultrafiltration did not appear to bias the collection of particular types of cells.

Fewer data are available in the literature on the use of Amicon ultrafiltration systems for the concentration of bacteria from natural waters. A small (0.3 ft^2) hollow-fiber filter (0.1 μm pore size) was used to concentrate cells from estuarine water samples (4 L). A 96% recovery of cells was found in the concentrate (30 mL). A larger (5 ft^2) hollow-fiber filter (0.1 μm pore size) was used by Fuhrman et al. [1988] to concentrate bacteria from a seawater sample (150 L). A 38% recovery of cells was found in the concentrate (3 L). We have used similar Amicon hollow fiber ultrafilters to concentrate bacteria from seawater. The following observations on methodology are offered as a guideline for other potential users. Filtration rates for seawater tend to be two- to tenfold lower than rates for deionized water published in Amicon literature. Most of the losses of bacteria (and other particles) during filtration occur due to retention on the filter. It is important to maintain high recirculation rates to minimize retention (polarization) on the filter. As polarization increases, the filtration rate will decrease. We have observed, as might be expected, that polarization is more of a problem with high surface area filters. At the end of filtration, the recovery of bacteria can be significantly improved by clamping off the filtrate line, removing all back pressure, and running the inlet pressure to the maximal operating value. The effect of this procedure is to cause a large pressure differential along the length of the fibers. Using this procedure we typically recover 50 to 80% of the total bacterial population, whereas without cleaning, recoveries range from 30 to 60%.

The very recent reports of high viral concentrations in seawater (see below) indicate that it may be wise to choose ultrafilters with a pore size of 0.1 μm for the concentration of bacteria. This pore size will retain most (greater than 99%) bacteria and minimize the concentration of viral particles. This is particularly important when bacteria are concentrated for molecular biological studies or other studies utilizing the nucleic acid fraction. It is also important to note that Koike et al. [1990] recently reported that >95% of the sub-micron particles (0.38-1.0 μm size range) in the upper ocean are non-living. Ultrafiltration may prove to be a valuable tool for concentrating these particles for chemical characterization, but methods for separating living from non-living particles need to be devised.

2. Concentration of Viruses by Ultrafiltration

Viruses were, until recently, thought to be relatively insignificant components of the oceanic community. Past methods used for the enumeration of viruses in natural waters were based on bioassay techniques, and it is now apparent that these techniques were highly selective and thereby greatly underestimated the number of viruses. Direct counts of viruses using transmission electron microscopy (TEM) indicate, however, that concentrations of viruses in natural waters are several orders of magnitude higher than previously estimated [Bergh et al., 1989; Proctor and Fuhrman, 1990].

Estimates of viral abundance in seawater range from 0.003 to 460 x 10^9 particles L^{-1} [Bergh et al., 1989; Proctor and Fuhrman, 1990; Suttle et al., 1990a]. Most of the viruses in seawater have head sizes in the 30 to 60 nm size range [Bergh et al., 1989; Børsheim et al., 1990].

Reliable estimates of the elemental composition of natural marine viruses are not yet available. Viruses are composed of nucleic acids encapsulated in a protein coat and therefore are nitrogen-rich particles. The average biovolume of viruses was found to be about 100-fold lower than the average biovolume of bacteria [Børsheim et al., 1990].

2.1. Methods

Hollow fiber ultrafilters (Amicon Corp.) have been used for the concentration of viruses from natural waters [Belfort et al., 1974, 1976]. Filters with nMW cutoff values of 10,000 to 100,000 have been used, but it is possible that the 100,000 nMW cutoff filters pass some of the smaller viruses. Backwashing, reversing flow through the filter, of filters at the end of concentration was recommended by Belfort et al. [1976] to improve recoveries. Total recoveries of polio virus ranged from 50-75% with backwash recoveries contributing 20-45% of the total recovery. No viruses were observed in the ultrafiltrates [Belfort et al., 1976].

Proctor and Fuhrman [1990] reported an 81% recovery of bacteriophage T5 using an Amicon hollow fiber filter (100,000 nMW). Similar recoveries (80-90%) of viral particles are reported by Suttle et al. [1990b] using Amicon hollow fiber (100,000 nMW) and spiral cartridge (30,000 nMW) filters.

3. Concentration of DOM by Ultrafiltration

The chemical composition of the bulk of dissolved organic matter (DOM) in the oceans remains largely unknown because it has not been possible to isolate a significant fraction of the total DOM for characterization. A variety of methods for isolating DOM have been used with varying success [Thurman, 1985]. The adsorption of dissolved humic substances onto nonionic XAD resins is currently the favored method, but this isolation procedure typically recovers only 5-15% of the total DOM from seawater, is chemically selective, and subjects the organic matter to severe changes in pH. Ultrafiltration has been used to provide molecular weight ranges of DOM [Carlson et al., 1985], but, to my knowledge, has not been used to isolate DOM from seawater because of very slow filtration rates. Recent advances in ultrafiltration technology have made this method more attractive for DOM isolation, and ultrafiltration has the advantages of being primarily a physical method of DOM concentration and no severe pH changes are required in the isolation procedure.

To date, ultrafiltration of seawater samples has primarily been used to estimate the molecular weight distribution of DOM. It is very important that the DOC in the initial water, the concentrate, and the filtrate be measured for analysis of the percentage of DOM retained by a particular ultrafilter. The need for this mass balance approach was demonstrated by Carlson et al. [1985] who found that many ultrafilters "bleed" varying amounts of organic carbon that can contaminate both the filtrate and the concentrate. Loss of DOC due to adsorption within the system can also be monitored using the mass balance approach. Many of the earlier ultrafiltration studies of the molecular weight distribution of DOM relied on the analysis of DOC in the filtrates alone, and this data should therefore be interpreted with caution.

Carlson et al. [1985] report that most (60-70%) of the total DOM in seawater have molecular weights less than 1000 d. Only 6% of the DOM had molecular weights greater than 30,000 d and only 1% had molecular weights greater than 100,000 d. Sugimura and Suzuki [1988] recently reported a different molecular weight distribution of DOM based on gel-filtration for MW separation and a high- temperature (680° C) DOC oxidation. These authors found that only 16% of the total DOM was less than 1800 d and that about 40% was greater than 20,000 d. In part, these differences are due to the higher overall DOC recoveries reported by Sugimura and Suzuki [1988]. They also analyzed the various MW size fractions using a standard persulfate oxidation and found that 40% of the total DOM was less than 1800 d and that about 46% was greater than 20,000 d. If the higher DOC values presented by Sugimura and Suzuki [1988] are correct, about 80% of the DOM is greater than 1800 d.

3.1. Methods

The stirred cell and thin-channel ultrafiltration systems that are commercially available do not provide sufficient flow-rates to be practical for the concentration of sufficient quantities of DOM for chemical and isotopic analyses. Larger scale systems utilizing high surface area (10-50 ft^2) cartridge filters are required to process the volumes (50-200 L) of seawater needed for recoveries of tens to hundreds of milligrams of DOM. We have experimented with a high capacity system from Amicon (DC-10L) using a 1000 d nMW cutoff filter. This filter has the lowest nMW cutoff of any commercially available cartridge filter and was chosen to give maximal recoveries of DOM from seawater. Our methods are undergoing continuing refinement, and the results presented herein should be considered as a demonstration of the utility of ultrafiltration for DOM concentration rather than a refined technique.

All water samples are prefiltered through Nuclepore cartridge filters with 3.0 μm and 0.2 μm pore-sizes prior to ultrafiltration. A mass balance approach is used to determine the percentage of total DOC that is concentrated by ultrafiltration. The entire volume of filtrate is saved for determination of DOC concentration because we observed that the concentration of DOC passing the ultrafilter increases during processing. Filtration rates with a 10 ft^2 filter (1000 d cutoff) are about 9 L hr^{-1} for seawater. Desalting of samples was achieved by diafiltration with 6 to 9 volumes (equivalent to the volume of concentrate) of distilled or Milli-Q treated water. Diafiltration of these volumes of water should allow for greater than a 99% exchange of salts. The filtrate from diafiltration is also saved for DOC analysis because we have found that 1 to 10% of the DOC in the concentrate is lost during diafiltration. Samples for measurement of DOC were analyzed using a high temperature, (680° C) Pt catalyzed oxidation (Shimadzu TOC 500 or TOC 5000).

A total of 17 samples (50-100 L each) from the Amazon River and its major tributaries were processed for DOM concentration. The average recovery of DOC was 76 ± 7% and an average of 104 ± 6% of the total DOC in the initial water was recovered in the concentrate and filtrate. For comparison, Ertel et al. [1986] recovered 41 ± 6% of the initial DOC from the same locations using XAD resins. A total of 7 samples (40-50 L each) have been processed from the hypersaline Laguna Madre in Texas. The average recovery of DOC was 45 ± 6% and an average of 103 ± 4% of the total DOC in the initial water was recovered in the concentrate and filtrate. This past July and August the ultrafiltration system was used to concentrate DOM samples from the open Gulf of Mexico across the salinity gradient of the Mississippi River plume and from the River itself. Recoveries of DOM were highest from Mississippi River water (45%) and decreased offshore to values of about 25% for open Gulf of Mexico water. These results indicate that riverine DOM, which is largely soil derived, is of considerably higher molecular weight than marine DOM. These results are consistent with those of Carlson et al. [1985] who found that approximately 70% of marine DOC had molecular weights less than 1000 d.

4. Characterization of Complex Mixtures by Solid-State ^{13}C NMR

Detailed molecular-level analyses of dissolved and particulate organic matter can provide valuable information on the contributions of specific sources to pools of organic matter as well as the pathways and mechanisms of transformation. Molecular-level analyses, however, often fail to provide a comprehensive view of the chemical composition of complex substrates. This problem is particularly frustrating for analysts who wish to follow nutrient flow in ecosystems rather than the diagenetic fate of specific compounds.

Recent analyses of vascular plant tissues in marine waters and sediments [Hedges and Weliky, 1989; Benner et al., 1990a, b] demonstrate the difficulty in achieving a comprehensive view of chemical composition based on molecular-level analyses. Often times samples contain components for which no molecular-level analysis exists (e.g. tannins), or the analyst is unaware of the classes of compounds that comprise the uncharacterized fraction. Solid-state ^{13}C NMR is one of the few characterization methods available with sufficient breadth and detail to provide comprehensive information on the complex organic mixtures that comprise most naturally occurring organic matter.

The purpose of this section is to provide the reader information on the potential benefits of solid-state ^{13}C NMR rather than details of NMR spectrometry [see Wilson, 1987 for details].

One major advantage of solid-state ^{13}C NMR spectroscopy over standard degradative methods is the ability of NMR spectroscopy to provide a measure of the relative abundances of carbon atoms associated with major classes of compounds in a single, nondestructive analysis. The applicability of solid-state ^{13}C NMR is limited, however, by the concentration of organic carbon in the sample. The lower limit for organic carbon concentration is about 3%, and the lower limit for total organic carbon in the sample is about 10 mg. The limit of detection of carbon can be affected by the presence of paramagnetic ions (e.g. Fe^{+3}).

In our studies of the early diagenesis of mangrove leaves in marine sediments [Benner et al., 1990a, b], solid-state ^{13}C NMR spectroscopy has provided the following information that was not available from a suite of molecular-level analyses of carbohydrates, lignin, cutin, and lipids. A large aromatic component (≈ 20 wt%) of leaf tissues was identified as tannins by use of the dipolar dephasing technique which highlights nonprotonated and methyl carbons. NMR analyses also disclosed the presence of a major carbohydrate component (≈ 25 wt%) of leaf tissues that was not identified by neutral sugar analysis. This information has directed our research towards the molecular-level analysis of other potential carbohydrate components such as uronic acids. The presence of nonsaponifiable, polymethylene-like polymers in the cuticle of leaves was also identified by NMR spectroscopy. These results using a defined substrate demonstrate the potential of solid-state ^{13}C NMR spectroscopy to identify unknown components of complex organic mixtures as well as to direct the application of molecular-level analyses to obtain a more comprehensive view of the chemical composition of particles.

Acknowledgements. I thank the National Science Foundation (grant BSR 8910766) and the National Oceanographic and Atmospheric Administration Office of Sea Grant (grants NA 89AA-D-SG139, and NA 90AA-D-SG689) for funding this research.

References

Barkay, T., C. Liebert, and M. Gillman, Hybridization of DNA probes with whole-community genome for detection of genes that encode microbial responses to pollutants: mer genes and Hg^{2+} resistance, Appl. Environ. Microbiol., 55, 1574-1577, 1989.

Belfort, G., Y. Rotem, and E. Katzenelson, Virus concentration using hollow fiber membranes, Water Res., 9, 79-85, 1974.

Belfort, G., Y. Rotem, and E. Katzenelson, Virus concentration using hollow fiber membranes-II, Water Res., 10, 279-284, 1976.

Benner, R., P. G. Hatcher, and J. I. Hedges, Early diagenesis of mangrove leaves in a tropical estuary: bulk chemical characterization using solid-state ^{13}C NMR and elemental analyses, Geochim. Cosmochim. Acta, 54, 2003-2013, 1990a.

Benner, R., K. Weliky, and J. I. Hedges, Early diagenesis of mangrove leaves in a tropical estuary: molecular-level analyses of neutral sugars and lignin-derived phenols, Geochim. Cosmochim. Acta, 54, 1991-2001, 1990b.

Bergh, Ø., K. Y. Børsheim, G. Bratbak, and M. Heldal, High abundance of viruses found in aquatic environments, Nature, 340, 467-468, 1989.

Børsheim, K. Y., G. Bratbak, and M. Heldal, Enumeration and biomass estimation of planktonic bacteria and viruses by transmission electron microscopy, Appl. Environ. Microbiol., 56, 352-356, 1990.

Carlson, D. J., M. L. Brann, T. H. Mague, and L. M. Mayer, Molecular weight distribution of dissolved organic materials in seawater determined by ultrafiltration: a re-examination, Mar. Chem., 16, 155-171, 1985.

Cho, B. C., and F. Azam, Major role of bacteria in biogeochemical fluxes in the ocean's interior, Nature, 332, 441-443, 1988.

Cho, B. C., and F. Azam, Biogeochemical significance of bacterial biomass in the ocean's euphotic zone, Mar. Ecol. Prog. Ser., 63, 253-259, 1990.

Coffin, R. B., D. J. Velinsky, R. Devereux, W. A. Price, and L. Cifuentes, Stable carbon isotope analysis of nucleic acids to trace sources of dissolved substrates used by estuarine bacteria, Appl. Environ. Microbiol., 56, 2012-2020, 1990.

Cole, J. J., S. Findlay, and M. S. Pace, Bacterial production in fresh and saltwater ecosystems: a cross-system overview, Mar. Ecol. Prog. Ser., 43, 1-10, 1988.

Ertel, J. R., J. I. Hedges, A. H. Devol, J. E. Richey, and M. den G. Ribeiro, Dissolved humic substances of the Amazon River system, Limnol. Oceanogr., 31, 739-754, 1986.

Fuhrman, J. A., D. E. Comeau, A. Hagstrom, and A. M. Chan, Extraction from natural planktonic microorganisms of DNA suitable for molecular biological studies, Appl. Environ. Microbiol., 54, 1426-1429, 1988.

Fuhrman, J. A., T. D. Sleeter, C. A. Carlson, and L. M. Proctor, Dominance of bacterial biomass in the Sargasso Sea and its ecological implications, Mar. Ecol. Prog. Ser., 57, 207-217, 1989.

Giovannoni, S. J., E. F. Delong, T. M. Schmidt, and N. R. Pace, Tangential flow filtration and preliminary phylogenetic analysis of marine picoplankton, Appl. Environ. Microbiol., 56, 2572-2575, 1990.

Hedges, J. I., and K. Weliky, The diagenesis of conifer needles in a coastal marine environment, Geochim. Cosmochim. Acta, 53, 2659-2673, 1989.

Hobbie, J. E., R. J. Daley, and S. Jasper, Use of nuclepore filters for counting bacteria by fluorescence microscopy, Appl. Environ. Microbiol., 33, 1225-1228, 1977.

Koike, I., S. Hara, K. Terauchi, and K. Kogure, Role of sub-micrometer particles in the ocean, Nature, 345, 242-244, 1990.

Lee, S., and J. A. Fuhrman, Relationships between biovolume and biomass of naturally derived marine bacterioplankton, Appl. Environ. Microbiol., 53, 1298-1303, 1987.

Proctor, L. M., and J. A. Fuhrman, Viral mortality of marine bacteria and cyanobacteria, Nature, 343, 60-62, 1990.

Sugimura, Y., and Y. Suzuki, A high-temperature catalytic oxidation method for the determination of non-volatile dissolved organic carbon in seawater by direct injection of a liquid sample, Mar. Chem., 24, 105-131, 1988.

Suttle, C. A., A. M. Chan, and M. T. Cottrell, Infection of phytoplankton by viruses and reduction of primary productivity, Nature, 347, 467-469, 1990a.

Suttle, C. A., A. M. Chan, and M. T. Cottrell, The use of ultrafiltration to isolate viruses from seawater which are pathogens to marine phytoplankton, Appl. Environ. Microbiol., in review, 1990b.

Thurman, E. M., Organic Geochemistry of Natural Waters, 497 pp., Kluwer Academic, Hingham, Massachusetts, 1985.

Wilson, M. A., NMR Techniques and Applications in Geochemistry and Soil Chemistry, 353 pp., Permagon Press, Oxford, 1987.

METHODS FOR ISOLATING COLLOIDAL ORGANIC MATTER FROM SEAWATER: GENERAL CONSIDERATIONS AND RECOMMENDATIONS

Bruce J. Brownawell

Marine Sciences Research Center, SUNY, Stony Brook, New York 11794-5000

1. Introduction

Characterization of the distribution and chemical nature of macromolecular (or colloidal) organic matter in seawater has long remained an elusive goal of marine chemists despite its recognized importance in affecting a number of important biogeochemical processes. The recent measurements of marine DOM by the platinum-catalyzed high temperature combustion method developed by Suzuki and coworkers [Suzuki et al., 1985; Sugimura and Suzuki, 1988] have profound implications with respect to our understanding of the cycling and role of colloidal organic matter. The results of these workers suggest that commonly utilized methods for DOC determination have underestimated the concentration of DOC by two or three times and that the missing DOC is comprised of thermally stable fractions that are comprised primarily of substances having molecular sizes greater than 4000 MW. These authors have effectively argued that the high molecular weight fractions are biologically labile and are important in controlling heterotrophic activity at depths below the euphotic zone. Previous studies had indicated that macromolecular organic matter accounted for a much smaller fraction (5 to 45%, typically 15 to 20%) of DOC in marine environments, see reviews by Carlson et al. [1985a], and Sugimura and Suzuki [1988]. Some aspects of those and other studies are discussed below. The importance of the new DOC story which is emerging calls now more than ever for better understanding of the cycling of the macromolecular organic matter in seawater.

There is no precise or necessarily useful definition for colloidal organic matter in seawater. Instead colloidal or high molecular weight fractions of organic matter are operationally defined by the isolation methods employed. High-molecular weight DOM has been typically isolated by filtration of water through 0.4 to 1.0 μm pore-sized filters followed by ultrafiltration [Carlson et al., 1985a], by gel-permeation chromatography [Sugimura and Suzuki, 1988], or as marine "humic substances" by solid phase extraction on hydrophobic surfaces [Stuermer and Harvey, 1974; Kerr and Quinn, 1975, 1980; Harvey et al., 1983]. The DOM isolated by all these procedures are of size between 0.001 and 1.0 μm, which are the dimensions which have been defined as colloidal by Vold and Vold [1966]. The use of the term colloidal to define substances that are in such operational size fractions is useful only in the most general sense. It cannot be implied that such definitions account for any demarcation between "truly dissolved" compounds and "particulate matter". It may be more useful to consider marine organic matter to be represented by a continuum of sizes and behaviors. Organic matter isolated as colloids is distinguished by its apparently high concentrations relative to low molecular weight material and POC, its tendency to be surface active, and its negligible Stokes settling velocities.

Given the newfound significance of organic colloids in the organic carbon cycle, and their importance in the many geochemical processes listed above, reliable methods for sampling and isolating high molecular weight are critically important. The requirements for isolation of colloid samples depend on the nature of study. At least three general purposes for isolating organic colloids are apparent: (1) characterization of concentrations and oceanographic distributions of various size classes of DOC; (2) structural characterization, including molecular size/weight determination; and (3) studies of biogeochemical function, e.g., biological utilization, photochemical reactions, metal complexation, organic solute sorption, etc. The requirements for isolation conditions and volumes of water needed for these purposes can vary tremendously and dictate the most appropriate method to use. For example, only small volume samples are necessary to determine the concentrations of DOC in various size fractions determined by size exclusion chromatography [Sugimura and Suzuki, 1988] and ultrafiltration with Amicon Centricon microseparation devices (Chin and Gschwend, submitted, 1991). Some methods that are available for characterizing structural features of organic colloids, e.g., the direct pyrolysis mass spectrometry method of Zsolnay [1982], require very small volumes(less than 1 mL) for analysis. Larger amounts of colloids are necessary for detailed chemical analysis or for some experimental studies. Harvey and

Marine Particles: Analysis and Characterization
Geophysical Monograph 63

coworkers [1983] isolated 0.10 to 2.2 g of humic substances from seawater in the Gulf of Mexico, such that they could determine proton NMR spectra and additional chemical studies, e.g., analysis of ozonolysis products. Large volumes of seawater may need to be concentrated to obtain colloid concentrations that are sufficient for some physical studies of colloid behavior, e.g., binding studies with solutes that are only weakly associated at ambient concentrations of organic colloids. It is suggested here that one goal of marine chemists should be to obtain a reference standard of marine colloidal organic matter, analogous to the Suwanee River humic substances that are distributed by the Humic Society. Such a reference material could be used for intercalibration of experimental procedures, structural analysis, and instrumental conditions. To acquire 100 g of marine organic colloids would ultimately require isolation and enrichment from tens of thousands of liters of seawater. There are definite needs for enriching or extracting colloids from large volumes of water.

Not only are there different volume requirements associated with various purposes for colloid sampling and isolation, but with different purposes in mind there can be different requirements for extraction/enrichment procedures because such procedures may alter the physical form of colloids from that of their native state. The conformation or aggregation state of colloidal material may be important in affecting a number of geochemical processes. For example, the structure of large organic polyelectrolytes would affect adsorption to particle or air-water interfaces, and could affect binding affinity for metals or organic solutes. Accurate estimates of molecular size/weight distributions require that sampling and isolation procedures do not disturb the native structures. On the other hand some chemical analyses of colloids may be relatively insensitive to changes of conformation or aggregation state. Chemical structure analysis would, however, be sensitive to chemical hydrolysis, condensation, or dehydration reactions that could potentially be caused by concentration, desiccation, or extreme changes in pH or ionic strength that are used in some methods of colloid isolation.

A wide variety of methods has been used to determine molecular weight distributions, quantify, or isolate high molecular weight organic matter in seawater. It must be cautioned that generalizations or comparisons of results from most studies are complicated by the different isolation methods used. The most commonly used techniques for concentrating and isolating colloidal organic matter in aquatic environments have been reviewed by Aiken [1985]. Most often, marine macromolecular organic matter has been studied as operationally defined humic substances [Stuermer and Harvey, 1974; Kerr and Quinn, 1975; Harvey et al., 1983; Meyers-Schulte and Hedges, 1986] or by retention by various size ultrafilters [Sharp, 1973; Ogura, 1977; and Carlson et al., 1985a]. More recently colloids have been separated into various size fractions in parallel with gel-filtration columns containing Biogel-P beads of variable pore size [Sugimura and Suzuki, 1988]. The conclusions that have been drawn in most organic colloid studies should be questioned if the results of Sugimura and Suzuki [1988] prove to be accurate and representative of conditions in other regions of the world's oceans. In the remaining discussion I will: (1) critically review some of the available methods for colloid separation and isolation for their ability to meet desired criteria; (2) summarize methods for molecular weight determinations of colloids; (3) discuss one method for sampling and enriching organic colloids from coastal seawater that the author has experience with, and make suggestions on how this method could be improved or verified; and (4) make some recommendations concerning sampling and concentration of organic colloids from seawater, as well as a subset of experiments that need to be done to confirm the ability of such methods to provide representative colloid samples.

2. Isolation Methods for Organic Colloids in Seawater

2.1. Humic Substance Extraction

One approach to the study of macromolecular organic matter in seawater has been to follow the lead of freshwater chemists and isolate operationally-defined humic substances from acidified seawater on hydrophobic support materials such as hydrophobic microporous resins (XAD-2 and XAD-8) [Stuermer and Harvey, 1974, 1977; Mantoura and Riley, 1975; see also review in Thurman, 1985], nylon [Sieburth and Jenson, 1968], or charcoal [Kerr and Quinn, 1975, 1980]. Marine humic substances isolated in this manner are hydrophobic acids, which are base soluble, and yellow colored. Marine humic substances do not include organic colloids which are too hydrophilic to adsorb to the support at a pH of 2.0 [Thurman and Malcolm, 1983; Ram and Morris, 1982] or colloids which are not reversibly eluted with base (sometimes in methanol) from the support. It is suggested that a significant amount of organic colloids in estuarine and marine waters is hydrophilic [Sigleo et al., 1982; Barger and Means, 1985; Sugimura and Suzuki,

1988]. Irreversible sorption of humic substances can be a problem when charcoal is used [Kerr and Quinn, 1975]. Polyacrylate XAD-8 resins adsorb humic substances more reversibly than do polystyrene-based XAD-2 resins, but the latter are used extensively for marine applications due to the high bleed of organic carbon from the former resin (J. Ertel, Univ. Georgia, personal communication, 1990).

Humic substances are comprised of fulvic (acid and base soluble) and humic acids (base soluble) and together typically make up 5-30% [reviewed by Thurman, 1985] of DOC, determined by persulfate oxidation, in seawater. However, in the study by Harvey et al. [1983] a higher fraction (12-100%) was observed. Marine humic substances are primarily made up of fulvic acids, a large fraction of which have apparent molecular weights less than 700 amu [Stuermer and Harvey, 1974].

Isolation of organic colloids by solid phase adsorption appears to have two advantages: (1) colloids from large volumes of seawater can be easily concentrated on adsorbent cartridges and (2) solid phase adsorption is directly comparable to the method which is used in freshwater environments, where humic substances make up the bulk of the DOM [Thurman, 1985].

Disadvantages of humic substance extraction as a method to isolate marine colloids are manifold. (1) The method concentrates on the basis of nonspecific interactions with the support phase, not on the basis of molecular size, which is part of any definition of colloidal materials. (2) Large volumes are necessary to accumulate amounts of DOM that are significantly above system blanks. (3) There is a risk of acid or base catalyzed hydrolysis or elimination reactions during the adsorption and desorption steps. For example, there is evidence for some ester hydrolysis of humic substances during base extraction, and irreversible acid-base titrations of humic substances are commonly observed (Patrick MacCarthy, Colorado School of Mines, personal communication, 1990). (4) Changes in the conformation or physical state of organic molecules caused by changes in pH, adsorption, or methanolic extraction are to be expected; it is not known to what extent such changes are reversible if the humic substance is to be used in a subsequent study.

Thus the isolation of organic colloids as humic substances is not recommended here, except for studies which are aimed at tracing the fate of terrigenous organic matter in the ocean. For this purpose, the effect of salinity on the isolation procedure should be well documented.

2.2. Ultrafiltration

Filtration through ultrafiltration membranes has been widely used to isolate colloidal fractions from seawater. These techniques separate organic materials on the basis of molecular size and can also separate on the basis of interactions with the membranes used. Separation with ultrafiltration membranes does not result in narrow molecular weight cutoffs. For instance soil fulvic acids of about 1,000 amu are highly retained by an ultrafiltration membrane of reported 50,000 molecular weight cutoff and a hollow fiber filter of 10,000 MW cutoff [Kwak and Nelson, 1977). The use and limitations of stirred cell ultrafiltration to fractionate DOM are discussed by Buffle et al. [1978]. Retention by ultrafiltration can be affected by: (1) aggregation of concentrated colloids in the retentate; (2) adsorption to the filter; and (3) "solute polarization" which is the buildup of organic solutes near the membrane surface and can affect filtration characteristics. Techniques using cross-flow ultrafiltration, such as thin-channel cells, hollow fiber and spiral membrane cartridges, attempt to minimize solute polarization with a high shear at the membrane interface, and are recommended for separation of marine colloids from large volumes. Thin-channel systems have an advantage in that the membranes can be physically cleaned of adsorbed colloids. Hollow fiber and spiral membrane cartridges are useful for concentrating colloids from samples greater than a couple of liters.

It is difficult to compare the results from molecular size fractionation studies that have been conducted with ultrafiltration for two reasons: (1) a wide range of ultrafiltration systems and membranes have been used; the membranes have different size cutoffs and are made of materials of variable hydrophobicities and have different tendencies to adsorb organic macromolecules; and (2) many studies up till now have determined the carbon in the colloid fractions with methods that may not oxidize much of the organic colloid carbon.

There were several early reports of organic colloids above 100,000 amu size in seawater [Sharp, 1973; Ogura, 1974, 1977; Wheeler, 1976]. However, Carlson and coworkers [1985a] found that an average of 34% of the DOM in the North Atlantic was of molecular weight above 1,000, but that only 6% had molecular weights greater than 30,000. They reviewed most of the ultrafiltration studies that had been conducted up to that time and concluded that many reports of larger colloids could have been due to procedural artifacts. In light of the recent work on the high temperature catalytic oxidation method (HTCO) for DOC measurements in seawater, it is time to go back and calibrate and intercompare the many methods of ultrafiltration that are available for size fractionating colloids in seawater.

Characterization of marine colloids isolated by ultrafiltration shows them to be distinct from those isolated as humic substances [Maurer, 1976; Zsolnay, 1979, 1982; Sigleo et al., 1982]. These studies indicated that marine colloids are rich in amino acids and carbohydrates. In contrast, marine humic substances contain very small amounts of those components and are lower in total nitrogen content [Gagosian and Stuermer, 1977; Harvey and Boran, 1985].

2.3. Gel Permeation

Gel permeation or size exclusion chromatography has been utilized increasingly as a method to determine molecular size distributions of marine organic colloids [Carlson et al., 1985b; Sugimura and Suzuki, 1988; Gschwend, Chin, and coworkers, presented at 1990 Boston ACS Meeting]. Sephadex gels, Bio-Gel beads, and modified silica columns have been used in low and high pressure systems. Size exclusion chromatography can be complicated by interactions of colloids with the stationary phase, namely adsorption and charge exclusion. Gschwend and coworkers have noted the importance of matching the ionic strengths of the sample and mobile phase to minimize artifacts arising from the latter effect. Reid et al. [1990] compared size exclusion chromatography in 0.1 M NaCl, ultracentrifugation, and other methods reported in the literature for reference humic substances and concluded that size exclusion chromatography using Sephadex G75 and Bio-Gel P60 supports provided less than satisfactory results. Further work is required to calibrate size exclusion and other methods of molecular size determinations in seawater.

An additional problem for all methods of determining mass average molecular weight distributions that should be further addressed is intercalibration of various macromolecules that are used as model solutes in molecular weight calibrations. Globular proteins and polystyrenesulfonates that are used do not have identical molecular size characteristics, and have different tendencies to change conformation as a function of pH and ionic strength.

3. Molecular Size Determinations

There are several methods available for determining molecular size and weight distributions of organic colloids. The most commonly used methods for marine colloids are ultrafiltration and size exclusion chromatography, which were discussed above. Other methods that have been used to estimate molecular weights of humic substances in freshwaters include small-angle X-ray scattering and vapor-phase osmometry which cannot be used in seawater. These four methods are critically reviewed by Aiken [1985].

Two additional methods have been applied recently to molecular weight estimation of aquatic organic matter and show some promise for characterizing marine colloids. These methods are flow field-flow fractionation [Beckett et al., 1987] and ultracentrifugation [Reid et al., 1990]. Flow field-flow fractionation (FFFF) separates organic solutes along thin channels with a parabolic axial flow subject to a crossflow, such that separation can be achieved based on the diffusion coefficients of the solutes; higher molecular weight materials preferentially accumulate closer to the semipermeable accumulation wall where the flow is slower, such that they are retained longer. The method of Beckett has been applied to describe the retention of model polystyrenesulfonate polymers and humic substances of known molecular weights.

Another method for determining molecular weight has recently been proposed by Reid and coworkers [1990]. By coupling ultracentrifugation with a UV scanner, it is possible to calculate molecular weight distributions from sedimentation equilibria. The results for IHSS humic substances agreed well with estimates for the same materials obtained by FFFF [Beckett et al., 1987]. Potential limitations of this approach are that it requires a knowledge of the molar absorptivities of colloids as a function of molecular size, and that treatment of polydisperse molecular weight assemblages would be difficult in practice. Both ultracentrifugation and FFFF have advantages over ultrafiltration and gel permeation with respect to minimizing interactions with interfaces that can bias fractionation.

4. Concentration of Organic Colloids by Hollow Fiber Ultrafiltration

I have employed a hollow fiber ultrafiltration method similar to that described by Means and Wijayaratne [1982] to isolate and concentrate organic colloids from coastal seawater sampled at Chemotaxis Dock, Woods Hole Oceanographic Institution on Nantucket Sound, MA, USA. Colloidal fractions were enriched for subsequent headspace partitioning experiments that were conducted to examine the association of PCB congeners with organic colloids [Brownawell, 1986].

Twenty-liter water samples were collected at about 1 meter depth by direct lowering of an acid, then seawater, rinsed 20 L stainless steel filtration bomb from the end of the pier, or by repeated lowerings of a Teflon plug-stoppered 5.0 L glass round bottom flask to the same depth to avoid surface microlayer contamination. Water was then immediately pressure-filtered at 5-10 psi through 142 mm diameter Nuclepore 0.4 μm membrane filters with Gelman Type AE glass fiber filters serving as prefilters in stainless steel filter holders. Glass fiber filters had been combusted at 500° C overnight, 5 L of permanganate distilled water were filtered through each of the filters to remove DOC contamination derived from the Nuclepore filters. Filtrations required 3-6 hr to complete and necessitated changing the Nuclepore filter three times to process the 20 L samples. Filtrates were collected in 5 gallon glass carboys which had been rinsed with 1.2 M HCl and six times with permanganate distilled water. At the beginning of filtration, 2.0 mL of 1% NaN_3 were added to the filtrate to minimize microbial transformations of DOM during subsequent ultrafiltration or partitioning experiments. In hindsight, this poisoning dose was probably too low to inhibit microbial activity. A NaN_3 level of 10 mM is recommended for inhibiting degradation of labile organic compounds, such as linear alkylbenzenesulfonates (Zhang et al., in preparation).

Ultrafiltrations of filtered seawater were performed within 24 hr of filtrations using recirculating Amicon H1P5 hollow fiber ultrafiltration cartridges having a nominal molecular weight cutoff of 5000 MW. The ultrafiltration apparatus was also rinsed with two changes of distilled water and 500 mL of filtered seawater prior to each use. The cartridges were stored in 15 mM NaN_3 at 4° C between uses to prevent membrane fouling. Ultrafiltration rates for seawater were about 60 mL min^{-1} such that ultrafiltration of 20 L took 5 to 7 hr. It is important to note that there was a reproducible and irreversible threefold decrease in filtration rate when a new cartridge was changed from 0.5 M NaCl to filtered seawater. It is hypothesized that Ca and Mg interacted specifically with the membrane and, perhaps by swelling, the polymer changed its filtration characteristics. Final retentate volumes were 205 to 500 mL, corresponding to enrichments in the retentate of 31.7 to 75.2 times.

DOC was determined on samples of filtered seawater, ultrafiltrate, and retentate using a wet persulfate oxidation method adapted from Stainton [1973]. Colloidal organic carbon concentrations were calculated from the difference between retentate and ultrafiltrate and the enrichment factor. The precision of the DOC measurements was 5%; the values of DOC reported are in question given the conclusion that persulfate oxidation can severely underestimate DOC in seawater. The concentrations for COC (colloidal organic carbon) and total DOC are provided in Table 4-7. For the samples analyzed before July 1985, the sum of ultrafiltrate and COC was 0.26 ± 0.15 mg L^{-1} greater than the original filtered water samples. This may have been due to contamination coming from the silicone tubing that was used in the peristaltic pump used to recirculate water through the hollow fibers. Pyrolysis GC/MS on enriched colloid samples obtained from two of these samples indicated that there was organic silicone contamination in dialyzed retentate samples. The peristaltic pump was replaced on June 13, 1985 by a stainless steel/Teflon gear pump with stainless steel and glass tubing to minimize DOC contamination. The two samples isolated after this date exhibited a small deficit (0.17 ± 0.12 mg L^{-1}) of DOC in the retentate + ultrafiltrate when compared to the original filtered water. It is recommended here to use stainless steel or Teflon gear pumps or diaphragm pumps in place of peristaltic pumps to avoid this potential contamination problem. Mass balances of DOC should be routinely checked on any isolation method that is used for colloids to check for contamination or potential adsorption/degradation losses of DOC.

The same procedures, except under inert atmosphere conditions, were used to isolate organic colloids from interstitial water DOC on a composite (2-15 cm) sediment grab sample obtained from Station M in Buzzards Bay (Brownawell et al., in preparation). Sampling and centrifugation of sediments were performed within 8 hr of collection using procedures described elsewhere [Brownawell and Farrington, 1986]. Filtration and ultrafiltration of interstitial waters were performed in a nitrogen-purged glove bag.

DOC, COC, and other properties of six Chemotaxis water samples are reported in Table 4-7. The DOC concentrations range for 0.95 to 2.38 mg L^{-1} and are similar to DOC determinations of 1.98 and 2.77 mg L^{-1} on unfiltered samples reported previously at this site [Stuermer and Harvey, 1974]. POC was 0.32 and 0.28 mg L^{-1} and are slightly higher than the concentrations of COC which were 0.12 to 0.29 mg L^{-1} and averaged 18% of the total DOC. It is not known whether some of these results are biased by silicone contamination, although the sample obtained after removing the tubing on July 23, 1985 was not significantly different. Stuermer and Harvey [1974] reported that 3.7 and 4.7% of the DOC in

TABLE 4-7.

Results From Filtration and Ultrafiltration of Seawater Sampled at Chemotaxis Dock on Nantucket Sound, Massachusetts

Date (mon-day-yr)	Temp (° C)	pH	DOC (mg L^{-1})	COC (mg L^{-1})	f_{coc}	POC (mg L^{-1})
3-21-1984	3.3	7.91	1.15	0.29	0.25	
8-29-1984	22.8	8.11	2.38	0.27	0.11	
4-11-1985	6.3	8.05	1.18	0.14	0.12	
6-13-1985	13.5		0.95	0.27	0.28	0.318
7-09-1985	21.8	8.21	1.45	0.12	0.084	
7-23-1985	23.2	7.98	1.20	0.27	0.23	0.28
Average Chemotaxis Dock DOC			1.38	0.23	0.18	
pore water composite (2-15 cm) from Stn. M	21.6	7.88	10.3	8.03	0.78	18,000*

*estimated from sediment properties at Station M

two samples at this site were isolated as dissolved humic substances.

The DOC and organic colloid concentrations measured in the interstitial water sample from Station M are higher than levels found in the water column. Seventy eight percent of the DOC was enriched as a colloidal fraction by ultrafiltration. This result is consistent with other studies where ultrafiltration was performed on interstitial water from reducing marine sediments [Krom and Sholkovitz, 1977; Orem and Gaudette, 1984].

This is an example of an application of a cross-flow ultrafiltration method which could be used to concentrate colloidal organic carbon from seawater, such that the COC could then be either characterized chemically or be used with minimal disturbance in subsequent "process-oriented" experiments. However, several steps are required before this or similar methods can be used with confidence: (1) DOC should be determined by HTCO or the most efficient oxidation method available; (2) effective doses of poison should be determined by measuring any changes in known labile components of COC, such as protein amino acids; (3) molecular size/weight cutoffs of the hollow fiber cartridge in seawater should be calibrated with globular proteins and other monodisperse polymers with known MW and geometry; and (4) possible effects of filtration and ultrafiltration on the molecular size distributions due to concentration and/or aggregation should be examined by monitoring size distributions with an independent method, such as size exclusion chromatography or Centricon ultrafiltration.

5. Recommendations

1. The gel permeation and HTCO methods of Suzuki and coworkers should be more fully tested. Examples of useful experiments include: dialyzing the colloid fractions isolated by gel permeation chromatography with ultrafiltration and combusting the dried organic matter; and comparison of molecular weight distributions determined by gel permeation chromatography with those measured by ultrafiltration, followed by DOC analysis by HTCO.
2. Sampling and handling of water for colloids should be as gentle as possible to minimize aggregation/disaggregation effects; water for colloid analysis should be pressure filtered instead of vacuum filtered to avoid nucleation on bubbles; possible aggregation/disaggregation effects should be examined by varying surface to volume of the sampling bottle and subjecting filtered water to variable agitation.
3. Various methods for molecular size separation and isolation should be rigorously calibrated with different types of model compounds and intercalibrated with each other; FFFF and ultracentrifugation should be pursued as methods that separate strictly on the basis of size and largely eliminate fractionation caused by charge exclusion or interactions with surfaces.
4. Effects of isolation methods and storage on size distribution or changes in native form of organic colloids should be evaluated with the best means available. Measures of size distribution, for instance, can be determined at any step in isolation or storage procedure. Changes in the function of DOM should

be examined before and after separation procedures where possible.

5. Thought should be given to establishing a reference colloid material for open ocean seawater. Such a dialyzed, dried material would need to be tested for changes in its properties before and after concentration.

References

Aiken, G. R., Isolation and concentration techniques for aquatic humic substances, in Humic Substances in Soil, Sediment, and Water, edited by G. R. Aiken, D. M. McNight, R. L. Wershaw, and P. MacCarthy, pp. 363-385, Wiley-Interscience, New York, 1985.

Barger, W. R., and J. C. Means, Clues to the structure of marine organic material from the study of physical properties of surface films, in Marine and Estuarine Geochemistry, edited by A. C. Sigleo, and A. Hattori, pp. 47-68, Lewis Publishers, Chelsea, Michigan, 1985.

Beckett, R., Z. Jue, and J. C. Giddings, Determination of molecular weight distributions of fulvic and humic acids using flow field flow fractionation, Environ. Sci. Technol., 21, 289-295, 1987.

Brownawell, B. J., The role of colloidal organic matter in the marine geochemistry of PCBs, Ph.D. thesis, 318 pp., WHOI/MIT Joint Program in Oceanography, 1986.

Brownawell, B. J., and J. W. Farrington, Biogeochemistry of PCBs in interstitial waters of a coastal marine sediment, Geochim. Cosmochim. Acta, 50, 157-169, 1986.

Buffle, J., P. Delodoey, and W. Haerdi, The use of ultrafiltration for the separation and fractionation of organic ligands in fresh waters, Anal. Chim. Acta, 101, 339-357, 1978.

Carlson, D. J., M. L. Brann, T. H. Mague, and L. M. Mayer, Molecular weight distribution of dissolved organic materials in seawater determined by ultrafiltration: a re-examination, Mar. Chem., 16, 155-171, 1985a.

Carlson, D. J., L. M. Mayer, M. L. Brann, and T. H. Mague, Binding of monomeric organic compounds to macromolecular dissolved organic matter in seawater, Mar. Chem., 16, 141-153, 1985b.

Gagosian, R. B., and D. H. Stuermer, The cycling of biogenic compounds and their diagenetically transformed products in seawater, Mar. Chem., 5, 605-632, 1977.

Harvey, G. R., and D. A. Boran, Geochemistry of humic substances in seawater, in Humic Substances in Soil, Sediment, and Water, edited by G. R. Aiken, D. M. McNight, R. L. Wershaw, and P. MacCarthy, pp. 233-247, Wiley-Interscience, New York, 1985.

Harvey, G. R., D. A. Boran, L. A. Chesal, and J. M. Tokar, The structure of marine fulvic and humic acids, Mar. Chem., 12, 119-132, 1983.

Kerr, R. A., and J. G. Quinn, Chemical studies on the dissolved organic matter in seawater. Isolation and fractionation, Deep-Sea Res., 22, 107-116, 1975.

Kerr, R. A., and J. G. Quinn, Partial chemical characterization of estuarine dissolved organic matter, Org. Geochem., 2, 129-138, 1980.

Krom, M. D., and E. R. Sholkovitz, Nature and reactions of dissolved organic matter in the interstitial waters of marine sediments, Geochim. Cosmochim. Acta, 41, 1565-1573, 1977.

Kwak, J. C. T., and R. W. P. Nelson, Ultrafiltration of fulvic and humic acids, a comparison of stirred cell and hollow fiber techniques, Geochim. Cosmochim. Acta, 41, 993-996, 1977.

Mantoura, R. F. C., and J. P. Riley, The analytical concentration of humic substances from natural waters, Anal. Chim. Acta, 76, 97-106, 1975.

Maurer, L. G., Organic polymers in seawater: changes with depth in the Gulf of Mexico, Deep-Sea Res., 23, 1059-1064, 1976.

Means, J. C., and R. D. Wijayaratne, Role of natural colloids in the transport of hydrophobic pollutants, Science, 215, 968-970, 1982.

Meyers-Schulte, K. J., and J. I. Hedges, Molecular evidence for a terrestrial component of organic matter dissolved in ocean water, Nature, 321, 61-63, 1986.

Ogura, N., Molecular weight fractionation of dissolved organic matter in coastal seawater by ultrafiltration, Mar. Biol., 24, 305-312, 1974.

Ogura, N., High molecular weight organic matter in seawater, Mar. Chem., 5, 535-549, 1977.

Orem, W. H., and H. E. Gaudette, Organic matter in anoxic marine porewater: oxidation effects, Org. Geochem., 5, 175-181, 1984.

Ram, N. M., and J. C. Morris, Selective passage of hydrophilic nitrogenous organic materials through macroreticular resins, Environ. Sci. Technol., 16, 170-174, 1982.

Reid, R. M., A. E. Wilkinson, E. Tipping, and M. N. Jones, Determination of molecular weights of humic substances by analytical (UV scanning) ultracentrifugation, Geochim. Cosmochim. Acta, 54, 131-138, 1990.

Sharp, J. H., Size classes of organic carbon in seawater, Limnol. Oceanogr., 18, 441-447, 1973.

Sieburth, J. McN., and A. Jenson, Studies on algal substance in the sea, I. Gelbstoffe (humic material) in terrestrial and marine waters, J. Exp. Mar. Biol. Ecol., 2, 174-189, 1968.

Sigleo, A. C., T. C. Hoering, and G. R. Helz, Composition of estuarine colloidal material: organic components, Geochim. Cosmochim. Acta, 46, 1619-1626, 1982.

Stainton, M. P., A syringe-gas stripping procedure for gas chromatographic determination of dissolved inorganic and organic carbon in fresh water and carbonates in sediments, J. Fish. Res. Board Can., 30, 1441-1445, 1973.

Stuermer, D. H., and G. Harvey, Humic substances from seawater, Nature, 250, 480-481, 1974.

Stuermer, D. H., and G. Harvey, The isolation of humic substances and alcohol-soluble organic matter from seawater, Deep-Sea Res., 24, 303-309, 1977.

Sugimura, Y., and Y. Suzuki, A high-temperature catalytic oxidation method for the determination of non-volatile dissolved organic carbon in seawater by direct injection of a liquid sample, Mar. Chem., 24, 105-131, 1988.

Suzuki,Y., Y. Sugimura, and T. Itoh, A catalytic oxidation method for the determination of total nitrogen dissolved in seawater, Mar. Chem., 16, 83-97, 1985.

Thurman, E. M., Organic Geochemistry of Natural Waters, 497 pp., Kluwer Academic Pub., Hingham, MA, 1985.

Thurman, E. M., and R. L. Malcolm, Structural study of humic substance: new approaches and methods in Aquatic and Terrestrial Humic Materials, edited by R. F. Christman, and E. T. Gsessing, pp. 1-23, Ann Arbor Science, Ann Arbor, 1983.

Vold, R. D., and M. J. Vold, Colloid chemistry, in Encyclopedia of Chemistry, pp. 263-265, Rheinhold, New York, 1966.

Wheeler, J. R., Fractionation by molecular weight of organic substances in Georgia coastal water, Limnol. Oceanogr., 21, 846-852, 1976.

Zsolnay, A., Coastal colloidal carbon. A study of its seasonal variation and the possibility of river input, Estuarine Coastal Mar. Sci., 9, 559-567, 1979.

Zsolnay, A., A new pyrolysis-mass spectrometry approach to organic marine chemistry using chemical ionization, J. Anal. Appl. Pyrol., 4, 47-58, 1982.

IN SITU EXPLORATION OF MACROMOLECULAR AND PARTICULATE MATERIALS IN SEAWATER

Dave Carlson

College of Oceanography, Oregon State University, Corvallis, Oregon 97331-5503

As a lad I filtered for DOC.
Now, I leave whole for rheology.
Am I older and wiser?
Perhaps also a miser?
Or just hoping to avoid contro'versy?

1. Introduction

Suppose that, despite our best efforts, macromolecular materials in seawater cannot be extracted or otherwise isolated without irreversibly altering their tertiary structure, so that any sample of these materials in a trap, on a filter, or at the bottom of a centrifuge tube is unavoidably of a different effective size, reactivity, and perhaps even composition than the original. There are some hints of these effects in the literature and in our cumulative experience, worrisome but not yet sufficient to cause us to suspend work. If this supposition is correct, then we need to develop in situ or non-extractive techniques to analyze these materials, in whole samples and perhaps directly in the ocean. Even if the supposition is only partially or slightly correct, it still behooves us to take alternate approaches, to crack the analytical and intellectual constraints imposed by our present modes of operation, to throw off our filtration and sediment trap blinders. A direct, non-extractive approach to macromolecular materials will necessarily examine the ocean on very small scales and it has always been true that as attention moves from larger to smaller scales, from coarse to fine resolution, unexpected patterns and dynamics appear.

2. Possible In Situ Method

I suggest two alternate methods for considering macromolecular materials, one optical and one fluorometric. Neither of these in their present form are fully ready or suitable for the dilute oceanic mixture, but they show promise. Even if they fail directly, they may allow us to consider less disruptive means of isolation or concentration.

The first of these techniques involves fiber optrodes, long light-transmitting fibers with tiny tips and small illumination volumes. Because of their extent and flexibility, these fibers can be inserted directly into oceanic environments, taking the optics to the mixture as an alternate to the conventional practice of bringing samples of the mixture to the optics. Because of their small effective illumination volumes, these optrodes can monitor very small regions of the ocean with very high resolution. Direct or chemically-induced fluorescence techniques are well suited to the transmittance properties and sensitivity requirements of optrode measurements and several have been adapted to oceanographic measurements. Figure 4-15 shows an example of high vertical resolution of phycoerythrin-containing particles [Cowles et al., 1990]. In contrast to state-of-the-art CTD-based fluorometers, the fiber optic techniques detect a great deal of cm-scale structure of the particles associated with cm-scale structure of temperature and turbulence. These measurements approach the scales on which the particles accumulate and interact, abiologically or biologically, with each other and with the ocean environment. These scales can not be sampled or monitored by conventional techniques.

The second technique for discussion also involves fluorometry, specifically the depolarization of fluorescence from molecular probes in response to the rheological and chemical properties of their immediate environment. As presently practiced, these fluorescence depolarization techniques involve adding probes to a sample removed from seawater but not otherwise filtered or extracted. A description of the technique used to measure seawater viscosity has been published [Carlson et al., 1987]. I describe here the implications of that rheological work for determining macromolecule abundance in seawater and the compositional information that apparently becomes available when one uses multiple probes of various molecular sizes.

We have extended our original application of fluorescence depolarization to seawater to include more probes and more fluorescent dyes. For probes, we are using fluorescent dyes covalently bound to linear dextrans - these are available in considerable variety from several commercial sources but are also relatively easy to synthesize.

The fluorochromes need to have high fluorescent yields and short (but not too short) room-temperature

Marine Particles: Analysis and Characterization
Geophysical Monograph 63

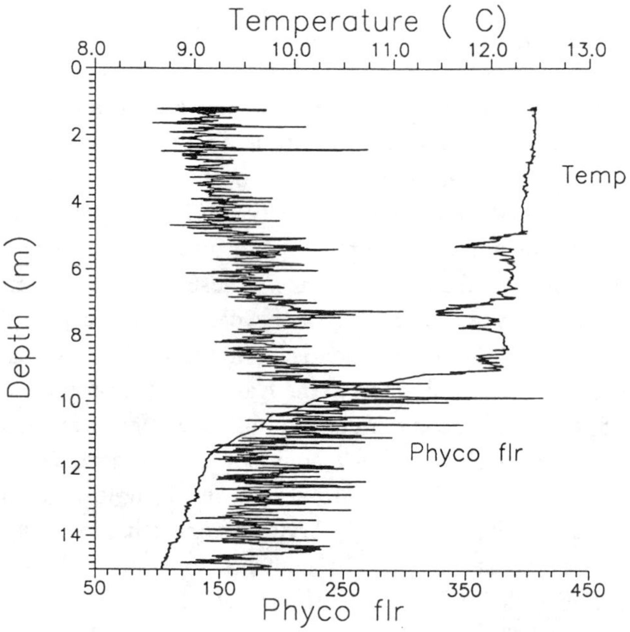

Fig. 4-15. Vertical profile of temperature and phycoerythrin fluorescence (585 nm) as measured with a laser/fiber optic microstructure profiler using 514 nm Argon excitation. The fluorescence signal was extracted from fluorescence emission spectra (540-700 nm) collected 30 times sec^{-1} during a free-fall drop of the profiler, yielding approximately 2 cm resolution of the fluorescence signal coincident with the physical data.

fluorescence lifetimes; several commercially available fluorochromes, e.g., fluorescein isothiocyanate, dansyl chloride, rhodamine B, satisfy both requirements. Some of these fluorochromes are also pH sensitive so one might simultaneously measure local pH by direct fluorescence and local viscosity by fluorescence depolarization. Dextrans can be prepared in a wide range of molecular weights, covering approximately 5k to 5000k Daltons; commercial preparations typically have molecular weights of 10k, 40k, 70k, and 150k Daltons. A high ratio of fluorochrome to dextran is desirable. We find that probes supplied by Molecular Probes (Eugene, OR) have better ratios than most of those we prepare or those available from Sigma. We have used dextrans because of their high solubility and their stability in storage and in seawater, but there are other macromolecular carriers (e.g. polylysines) waiting to be explored.

A typical fluorescence depolarization experiment involves adding small, nanomolar, amounts of a probe to a whole, unfiltered, unaltered, sample of seawater. One could apply several probes simultaneously if each has a different non-interfering fluorochrome. The degree of depolarization is then determined for each probe. Depending on the fluorochrome combinations, two or more probes can be in the sample. In our lab, all depolarization measurements are done with an automated polarization accessory on a research grade spectrofluorometer. We also use a high precision thermocirculator, working against a constant temperature cooling bath, to maintain accurate and precise temperatures in the cuvettes and to precondition samples awaiting analysis.

Examples of the response of three probes, e.g., dansyl on a 70k Dalton dextran, fluorescein on a 10k Dalton dextran and rhodamine B alone, to large globular and small linear proteins and to a disaccharide are shown in Figure 4-16. In order to normalize the separate linear depolarization-to-viscosity relationships of the individual probes the data are presented as fluorescence depolarization-determined viscosity in mPas, all at 30° C, versus concentration, in artificial seawater, of each organic material.

The fluorescence depolarization responses of the three probes to the three organic components are very different: two of the probes show large changes in the presence of bovine serum albumin, i.e. globular, molecular weight about 70k Daltons, in the upper panel, but no response to insulin with linear, molecular weight about 6k Daltons, in the middle panel. These probes did respond to the presence of sucrose (MW = 342 Daltons; Figure 4-16, lower panel), but in a different pattern than their response to BSA.

The probes were also less sensitive to rheological effects induced by the presence of sucrose than to those induced by BSA, even on a per carbon basis (sucrose concentrations were 0.1 to 0.4 M while BSA concentrations were 0.003 to 0.1 mM, a factor of 1000 or greater, while the sucrose to BSA molecular weight and carbon ratios are only about 200). In oceanographic terms, these particular probes were sensitive, perhaps with a diagnostic pattern, to rheological changes induced by as little as 10 mg protein C per liter.

These measurements represent an ability to detect the presence of and perhaps identify the composition of materials sufficiently large and/or abundant to modify the rheological properties of a mixture, i.e. macromolecules, colloids, aggregates, and particles, in other words, on hitherto unmeasurable scales. For oceanography, the measurements represent the potential to make the same determinations directly in a whole unaltered sample and the possibility to eventually make

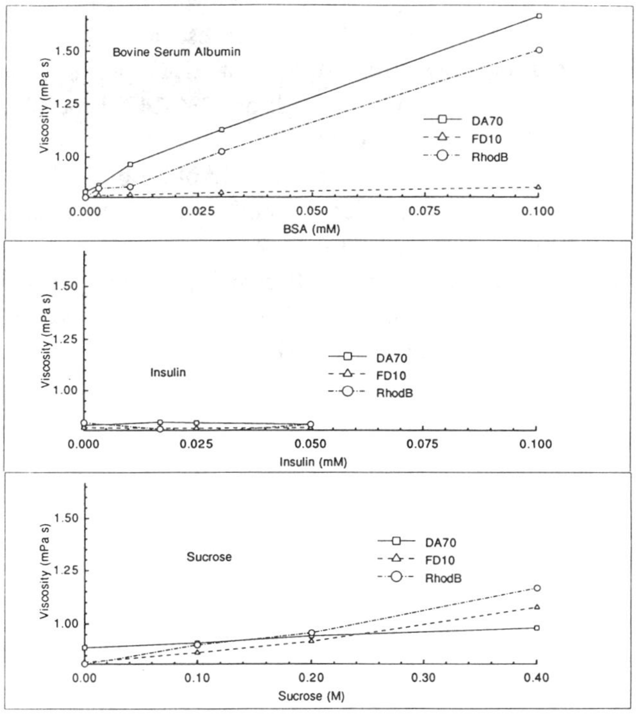

Fig. 4-16. Examples of responses of three molecular probes.

the same measurements directly in the ocean by coupling fluorescence depolarization with the in situ optical technology described earlier.

Are these potentials and possibilities realistic, and are they useful? Using single fluorescence depolarization probes, we have observed apparent viscosity variations of 5 to 20% in samples of bulk seawater and variations approaching a factor of 2 in samples from particle-rich regions of the ocean such as surface microlayers. In those single-probe measurements, major portions of the materials affecting the fluorescence depolarization signals were removed by Whatman GF/F glass fiber filtration, indicating that the natural filterable materials in seawater, macromolecules, aggregates, colloids, particles, cells, or whatever, do influence small-scale rheological properties and are detectable because of that influence. Ours are not the only reported measurements of altered rheological properties in the ocean: Jenkinson [1986] and Jenkinson and Biddanda [in press, 1991] have observed very large modifications of rheological properties using conventional techniques applied to cell cultures and samples from coastal waters. It seems likely that we have underestimated or ignored the incidence and information content of these modified rheological conditions, partly for lack of attention and partly for lack of techniques to make the measurements on sufficiently small scales. Now, however, with the ability to monitor small volumes in situ and with fluorescence depolarization and other techniques to determine the properties of such small volumes, we have an opportunity to consider macromolecular components on scales and in ways that we will never achieve with our water bottles, sediment traps, transmissometers or conventional in situ fluorometers.

If we had sensitive rheological probes and a fiber optic system to use them directly in the ocean, what could we learn and would it be useful or relevant to understanding particle-related processes in the ocean? Some argue that the techniques are neither sensitive or informative. Those are honest doubts about real uncertainties. It may be unwise, however, to reject the possibilities of new information on new smaller scales without having explored them. Because we have no experience of measuring viscosity on molecular scales, even though it is fundamentally a molecular property, our exploration will have to be empirical rather than theoretical. Such an exploration is not without some analytical precedent: polymer chemists make intense and extensive use of conventional viscosity measurements to determine important properties of their polymer mixtures. An exploration of viscosity in seawater will necessarily be on very fine scales, much finer than our present CTD-related profiles of particle and particle-related properties or our sediment trap profiles of particle fluxes. The initial hints from the in situ fiber optic fluorometry and from viscosity measurements suggest that the ocean has a great deal of texture on those scales. Texture, that is, on the scales of important particle-related processes: flocculation, aggregation, scavenging, etc. High resolution small scale explorations may help bridge the gaps between long integrative measurements and large-scale samples and the molecular-scale processes that are the fundamental determinants of the ocean particle field and its effect on light and sound propagation. Fine scale explorations may even help us improve our definitions of complex components of the seawater mixture, allowing us to replace length-based, e.g., colloids, or operationally-based, e.g., particulate organic carbon, descriptors with more accurate descriptions based on the degree and extent to which the materials constitute an alternate phase.

All of these consequences and supposed significances are too much burden to impose on nascent measurements of as yet uncertain significance. In the absence of funding constraints, I hope we would be bold enough and clever enough to undertake these explorations. In the real world of funding constraints and commitment to known and tested (even if not true) techniques, I hope we allow our imaginations freedom to explore.

References

Carlson, D., L. E. Morrill, and J. E. Brophy, Techniques of fluorescence depolarization for measuring seawater viscosities, Limnol. Oceanogr., 32, 1377-1381, 1987.

Cowles, T. J., R. A. Desiderio, J. M. Moum, M. Myrick, D. Garvis, and M. S. Angel, Fluorescence microstructure using a laser/fiber optic profiler, in Ocean Optics X, edited by R. W. Spinrad, pp. 336-345, Proc SPIE 1302, Orlando, Fla., 1990.

Jenkinson, I. R., Oceanographic implications of non-Newtonian properties found in phytoplankton cultures, Nature, 323, 435, 1986.

Jenkinson, I. R., and B. A. Biddanda, Viscosity of seawater during a bloom of *Noctiluca* and *Phaeocystis*, Oceanol. Acta, in press, 1991.

ISOTOPIC ANALYSES OF INDIVIDUAL COMPOUNDS

Brian N. Popp

University of Hawaii, School of Ocean and Earth Science and Technology,
Department of Geology and Geophysics, 2525 Correa Road, Honolulu, Hawaii 96822

1. Introduction

The complexity of molecular components which make up suspended and sinking organic particles in the ocean has recently been recognized [e.g., Wakeham et al., 1980, 1984; Lee et al., 1983; Wakeham and Canuel, 1988; Lee and Wakeham, 1988]. The variety and complexity of these molecules indicate not only multiple origins of these compounds but also multiple pathways of degradation. Such complex mixtures of molecules in sedimentary organic matter have recently been studied using compound-specific isotopic analyses [e.g., Hayes et al., 1987, 1989, in press, 1991; Popp et al., 1989; Boreham et al., 1989; Ocampo et al., 1989; Freeman et al., 1990; Jasper and Hayes, 1990]. This contribution reviews methods used for isotopic analyses of individual compounds of which many are new and have yet to be used to study marine organic particles.

Several workers have determined carbon- and nitrogen-isotopic compositions of individual compounds [e.g., Abelson and Hoering, 1960; Parker, 1962; Macko et al., 1987] and even carbon-isotopic compositions of individual carbon atoms within molecules [e.g., Monson and Hayes, 1980, 1982a, b]. Most compound-specific isotopic analyses have been used to elucidate biosynthetic pathways or to determine isotope effects associated with biosynthetic pathways in organisms grown in culture. Only a few studies have used compound-specific isotopic analyses to determine origins of molecules in complex mixtures such as sedimentary organic matter. In all studies, compounds must be separated and purified prior to combustion to a gas suitable for mass spectrometric analyses. These methods can be separated into two categories: 1) off-line separation followed by combustion, and 2) on-line separation and combustion. These methods have specific advantages and disadvantages and each will be treated separately.

2. Off-Line Separation of Compounds for Isotopic Analysis

Early studies utilized off-line, column chromatographic techniques for the separation and purification of individual molecules for isotopic analyses. Most notable are the carbon-isotopic studies of Abelson and Hoering [1960, 1961] and Parker [1962]. Abelson and Hoering [1960, 1961] used resin chromatography to separate individual amino acids, whereas Parker [1962] used high-temperature, packed-column gas chromatography with a thermal conductivity detector (i.e., non-destructive) to isolate individual esterified fatty acids. The purified compounds were then combusted to produce CO_2. Each method yielded compound-specific isotopic analyses, but had unacceptably large errors of $\geq 1^{o}/_{oo}$. In these studies, incomplete recovery of compounds during chromatographic separations was documented as the source of largest error in the procedure. Abelson and Hoering [1960, 1961] also documented isotopic fractionation during decarboxylation of amino acids by ninhydrin.

To determine the isotope effect associated with biosynthesis of fatty acids and to elucidate mechanisms controlling the isotopic composition of acetyl-coenzyme A, Monson and Hayes [1980, 1982a, b] determined the abundance of ^{13}C at the methyl and carboxyl positions of fatty acids as well as molecular average isotopic compositions of fatty acids. Although techniques for determining intramolecular isotopic ordering are fascinating, they are beyond the scope of this report and the interested reader is referred to the series of papers by Monson and Hayes.

Monson and Hayes [1980, 1982a, b] presented a reliable method for the isolation and isotopic analyses of individual fatty acids. Briefly, lipids extracted using the method of Bligh and Dyer [1959] were saponified under nitrogen using 0.5 M KOH in methanol. To remove the nonsaponified lipid fraction, the mixture was extracted with petroleum ether. The mixture containing fatty acids was then acidified with 12 N HCl and the fatty acids recovered by extraction in chloroform. No significant isotopic fractionation was observed associated with the Bligh-Dryer extraction, saponification, neutral lipid extraction or acidic lipid extraction procedures [see Table 4, p. 143, Monson and Hayes, 1982a].

Monson and Hayes [1980, 1982a, b] separated individual fatty acids by reverse-phase HPLC. The method, using methanol-tetrahydrofuran-water (85/10/5, v/v/v), resolved all but a few unsaturated short-chain

Marine Particles: Analysis and Characterization
Geophysical Monograph 63

fatty acids. These short-chain fatty acids were collected and further isolated using HPLC after hydrogenation to produce fully saturated fatty acids [Monson and Hayes, 1980, 1982a, b]. Reproducibility for HPLC separation on an oleic acid standard was within 0.5 ‰ of the isotopic composition of the starting material. These authors stressed the importance of verifying that separation procedures did not cause isotopic fractionation.

Macko et al. [1986, 1987] recently devised a preparative scale, liquid chromatographic technique to isolate milligram quantities of pure amino acid for carbon- and nitrogen-isotopic analyses. Amino acids were added to a column containing cation-exchange resin that had been carefully washed, sized and fractionated, and were eluted with hydrochloric or acetic acid [Macko et al., 1986, 1987]. Utilizing acids eliminated possible contamination that may be present in organic solvents. Individual amino acids were detected using ion exchange chromatography on aliquots of the eluant [Hare, 1972] or using the level of fluorescence in aliqouts of eluant reacted with *o*-phtaldialdehyde reagent [Macko et al., 1987]. The chromatographic procedure was complete in less than 48 hours depending on the number of compounds separated. The separated compounds as hydrochlorides of amino acids were converted to CO_2 and N_2 using a Dumas, closed-tube combustion technique [Macko et al., 1982]. Briefly, the compounds were transferred to a quartz tube containing high-purity copper and precombusted copper oxide wire, the tube evacuated, sealed and combusted at 850° C to 900° C for 1 to 8 hours, followed by 1 hour at 550° C. This latter cooling assured reduction of all nitrogen oxides to N_2.

Macko et al. [1987] documented carbon- and nitrogen-isotopic fractionation during ion-exchange chromatography. They noted that isotopic contrasts existed between the first and second half of peaks; the difference being largest between the tails of peaks. Macko et al. [1986, 1987] stressed that complete base-line separation of components and quantitative recovery were required in chromatographic procedures to obtain uncertainties of ± 0.5 for $\delta^{15}N$ and of ± 0.3 for $\delta^{13}C$. They also cautioned that uncertainty in isotopic analyses increased with increasing quantities of contamination from analytical blanks (e.g., from column bleed) or co-eluting compounds. To avoid problems associated with analytical blanks, large samples were prepared so that, even if a blank was present, it constituted only a minor proportion of the sample.

Engel and Macko [1984] presented a method for the separation of amino acid enantiomers by HPLC for carbon- and nitrogen-isotopic analyses. D- and L-amino acids were separated by reverse-phase HPLC using an aqueous chiral mobile phase at a column temperature of 0° C. The mobile phase, consisting of copper acetate and *N*, *N*-di-*n*-propyl-L-alanine, was prepared using the method of Weinstein et al. [1982]. Again, base-line separation was needed to insure that no fractionation was associated with the chromatographic procedure [Engel and Macko, 1984]. Prior to combustion and isotopic analyses, the isolated D- and L-amino acids were separated from the mobile phase using essentially the same technique of Macko et al. [1987].

Boreham et al. [1989] and Ocampo et al. [1989] isolated and analyzed carbon-isotopic compositions of microgram quantities of individual porphyrins from the Cretaceous marine Julia Creek Oil Shale from Australia. A comparison of the results of these studies is presented in Hayes et al. [in press, 1991]. The individual samples came from three different, stratigraphically equivalent locations and were prepared in three different laboratories using two slightly different techniques (see below). In spite of the fact that different techniques were used, isotopic compositions of individual porphyrins were similar, strongly suggesting that imposition of isotopic fraction during preparation of these samples was low [Hayes et al., in press, 1991].

Boreham et al. [1989] used the chromatographic procedures of Boreham and Fookes [1989] to isolate and purify individual porphyrins. The bulk solvent extract from the crushed shale was separated into crude nickel and vanadyl porphyrin fractions by column chromatography on silica. The vanadyl porphyrins were either demetallated with methanesulfonic acid and nickel inserted into the free-base porphyrins with Ni $(acetylacetonate)_2$ or analyzed as a vanadyl complex. No difference in the isotopic composition of vanadyl porphyrins was observed when samples were prepared using both techniques. The crude Ni-porphyrin fractions were further purified by TLC prior to separation of individual compounds using semi-preparative reverse-phase HPLC [Boreham and Fookes, 1989]. Ocampo et al. [1989] used the reverse-phase HPLC chromatographic procedures of Fookes [1983]. In this study, all porphyrins were analyzed as Ni-porphyrins, the vanadyl complexes being similarly transmetallated prior to HPLC separation. In both studies, the purity of individual porphyrins was determined by NMR to be $\geq 95\%$ [Boreham et al., 1989; Ocampo et al., 1989].

Although samples of porphyrins were determined to be "NMR-pure", contamination by non-pigment carbon inevitably occurred during handling and transfers subsequent to NMR analyses. This analytical blank was removed and the purity of pigments determined by further purifying each sample by TLC prior to acquisition of a UV/Vis spectrum. To determine the isotopic composition of contamination associated with solvent impurities and handling, three to four samples and a standard containing 0.3 to 2.0 μM C were spotted on silica-gel TLC. After elution, individual spots were transferred to a vial and taken to dryness under a stream of N_2 at 25° C. The mass of the analytical blank associated with TLC and subsequent transfers was determined by scraping a clean portion of each TLC plate with an area and an R_f approximately equal to that of the samples and was taken through the following procedures as a sample. Spectral analysis was determined on each sample in 0.5 mL or 1.0 mL of solvent using a cuvette with a path length of 5 mm. Maximum absorbance of peaks in which reliable extinction coefficients are available were monitored. Uncertainty for the determination of the absorbance and the volume of the solvent were typically less than 5×10^{-3} AU and less than 1×10^{-5} L (one standard deviation), respectively. Individual samples were transferred to a quartz tube, ~ 200 mm x 9 mm, using a minimal amount of solvent, the solvent evaporated at 25° C, less than 20 mg of cupric oxide added and the tube evacuated and sealed prior to combustion at 850° C for greater than 6 h. The samples were opened under vacuum and the CO_2 from combustion cryogenically purified. The yield of CO_2 (sample plus analytical blank) was determined manometrically. The uncertainty for such measurements using electronic manometers is typically less than 20 nmol CO_2 (one standard deviation). The purity [≡ 100(porphyrin carbon)/(total carbon)] of each sample was determined by comparison of the results of spectrometric and manometric analyses [see Hayes et al., in press,1991; Popp et al., in press, 1991].

Although contamination associated with post-NMR transfers and handling was removed by TLC, further contamination by non-porphyrin carbon likely occurred during transfers associated with spectrophotometric analyses and during combustion. Because isotopic compositions can be sensitive to the presence of impurities, the abundance and isotopic composition of the analytical blank were determined by carrying a suitable standard of known isotopic composition through the TLC, spectrophotometry and combustion steps in parallel with the samples [see Boreham et al., 1989; Ocampo et al., 1989; Hayes et al., in press, 1991]. Hayes [1983] has shown that blank-corrected isotopic compositions can be determined using the expression

$$\delta_s = (n_T\delta_T - n_b\delta_b)/(n_T - n_b) \qquad (1)$$

where δ_s is the true isotopic composition of some unknown sample, n_T is the total molar quantity of CO_2 analyzed isotopically, and δ_T is the isotopic composition of the sample plus blank.

The result derived by solution of equation 1 is uncertain because of uncertainties resulting from the inherent variability in n_b and δ_b. Gelwicks and Hayes [1990] showed that the uncertainty associated with a blank-corrected isotopic composition can be determined quantitatively using a propagation of error calculation derived from partial differentiation of equation 1.

$$\sigma^2_{\delta_s} = (\partial\delta_s/\partial n_T)^2\sigma^2_{nT} + (\partial\delta_s/\partial\delta_T)^2\sigma^2_{\delta T} + (\partial\delta_s/\partial n_b)^2\sigma^2_{nb} + (\partial\delta_s/\partial\delta_b)^2\sigma^2_{\delta b} \qquad (2)$$

Analytical solution of the partial derivatives is as follows:

$$\partial\delta_s/\partial n_T = n_b(\delta_b - \delta_T)/(n_t - n_b)^2 \qquad (3)$$
$$\partial\delta_s/\partial\delta_T = n_T/(n_t - n_b) \qquad (4)$$
$$\partial\delta_s/\partial n_b = n_T(\delta_T - \delta_b)/(n_t - n_b)^2 \qquad (5)$$
$$\partial\delta_s/\partial\delta_b = -n_b/(n_t - n_b) \qquad (6)$$

Using this procedure, Boreham et al. [1989] and Ocampo et al. [1989] determined the isotopic composition of the analytical blank (δ_b) by solving equation 1 for δ_b using the results of analyses of the internal porphyrin standard. The analytical blank represented contamination from solvents and handling and, in these studies, ranged in size from 38-85 nmol CO_2 (average 55 nmol CO_2) whereas sample size ranged from 0.5 to 3.5 micromoles of carbon. Once the isotopic composition of the blank was determined, the quantity of CO_2 from the analytical blank was subtracted from that of the sample to yield the true sample size and the isotopic composition of the sample calculated using equation 1. Replicate analyses of a porphyrin standard processed in parallel with the natural isolates over a time-span of approximately 18 months yielded a blank-corrected result of -27.46 ± 0.10 (mean, standard deviation of population, n = 22). In 19 of these 22 cases these authors found that the assigned 95% confidence interval, determined using equation 2, i.e., 2 s.d. = $2(\sigma^2_{\delta s})^{1/2}$, included the known isotopic composition of the standard. In the other three cases the confidence intervals missed the correct result by 0.02, 0.02 and 0.05. Boreham et al. [1989] and Ocampo et al. [1989] similarly used equation 2 to assign 95% confidence intervals to each isotopic analysis reported.

Recently, Bidigare et al. [1991] described a method to isolate and purify milligram quantities of chlorophylls *a* and *b* from plant tissues for carbon and nitrogen isotopic analyses. The procedure used required that pigments first be precipitated from MeOH:dioxane:water (10:1:1) at -20° C. The suspension was filtered and the residue dissolved in 2 mL of acetone. Complete baseline separation of chlorophylls *a* and *b* was obtained by HPLC when pigments were eluted through a Dynamax preparative-scale C_{18} column using MeOH at a flow rate of 10 mL min^{-1}.

The procedure of Bidigare et al. [1991] is fast and produces only a minimum isotopic fractionation. Although the method described required a large quantity of starting material (approximately 15 gram dry weight of bulk material), these authors suggested that the method can be scaled down for use with smaller sizes of bulk starting materials.

Off-line techniques for separation and analyses of individual compounds have several advantages. First, there is virtually no limit to the type of compound that can be analyzed; if a compound can be isolated and burned, it can be analyzed isotopically. This is especially useful because typically larger compounds have greater specificity of origin. Second, techniques not currently used to isolate compounds for isotopic analyses may be modified for this purpose. And lastly, relatively small quantities of compounds can be analyzed.

There are disadvantages. Off-line chromatographic separation techniques tend to be complicated and time-consuming and can impose isotopic fractionation if baseline separation of compounds is not achieved. Contamination from extraneous carbon and nitrogen sources can strongly affect isotopic compositions of samples, but, in many instances, the mass and isotopic compositions of analytical blanks can be measured and samples corrected for the contribution of the blank. Finally, when developing complicated separation procedures for compound-specific isotopic analyses, it is imperative that a test standard of known isotopic composition be carried through the entire sample-processing procedure in order to be sure that no isotopic fractionation is imposed by the method or that if any contamination of samples occurs, that it is constant and the effects can be corrected.

3. On-Line Compound-Specific Isotopic Analyses

Over the last thirty years, improved electronics in mass spectrometers have increased accuracy and precision of isotopic analyses. This improvement has led to increased sensitivity of instruments and, thus, popularity in the use of stable isotopic analyses mainly because researchers have been able to investigate isotopic variations in various materials at progressively smaller scales. Recently, the interfacing of a gas chromatograph to an isotope ratio mass spectrometer via a combustion oven has allowed on-line separation and isotopic analyses of individual organic compounds. This technique, called isotope-ratio-monitoring gas chromatography-mass spectrometry (irmGCMS), currently permits determination of $^{13}C/^{12}C$ ratios of individual volatile compounds in complex mixtures such as sedimentary organic matter [e.g., Matthews and Hayes, 1978; Freeman et al., 1990; Hayes et al., in press, 1991; Jasper and Hayes, 1990].

The irmGCMS system facilitates the measure of the natural abundance of ^{13}C of individual organic compounds eluting from a gas chromatograph. Complex mixtures of compounds are injected into a gas chromatograph (GC) where chemical separation based on size and structure in a capillary column occurs [see Freeman et al., 1990; Hayes et al., in press, 1991]. Normally, compounds eluting from the capillary GC (as individual compounds or mixtures of compounds, depending on the efficiency of separation) are directed to a detection system which determines their abundance. In irmGCMS systems, the effluent is commonly split between a conventional flame ionization detector (FID) and the mass spectrometer. It is conceivable that, in many laboratories, an ion trap detector or quadrupole mass spectrometer may substitute for the FID. In this configuration, structural information could be collected simultaneously with isotopic data. Compounds directed towards the mass spectrometer are first combusted in a micro combustion furnace at approximately 800° C in the presence of cupric oxide and the water of combustion is continuously removed by diffusion through a permeable membrane or by cryogenic trapping. Isotopic abundances are determined from continuous collection of masses 44, 45 and 46 in three Faraday cups connected to high-speed amplifiers on eluting CO_2 "peaks". To determine δ-values, the measured ratios are compared to internal standards whose isotopic composition is well characterized using conventional techniques. The irmGCMS is capable of determining carbon-isotopic compositions on compounds containing less than 5 nM of carbon. Current development of irmGCMS systems is directed at on-line, compound-specific nitrogen-isotopic determinations (J. M. Hayes, personal communication, 1990; S. A. Macko, personal communication, 1990), and on-line, compound specific analyses of derivatized compounds [e.g., see Engel et al., 1990].

The performance of irmGCMS analyses is limited by chromatographic resolution, accuracy of calibration of isotopic abundance relative to international standards, and adequacy of corrections for background CO_2 [Freeman et al., 1990]. Contamination of samples occurs as a result of combustion of column bleed. This contribution of CO_2 to an eluting peak is easily corrected by subtracting background ion currents measured directly before and after an eluting peak. This procedure has been written into most software packages and is done automatically. Unfortunately, automatic routines can sometimes measure background signals on nearby peaks. Therefore, it is critical to review all "combustograms" at high resolution. High resolution editing should be available in the software of all commercially available irmGCMS systems.

Separations of compounds in a capillary column are subject to vapor-pressure isotope effects which cause molecules containing carbon-13 to elute slightly before those containing only carbon-12 [see examples, Freeman et al., 1990; Hayes et al., in press, 1991]. This fractionation is avoided by integrating ion currents at masses 44, 45 and 46 across the width of each peak. However, when peaks overlap, significant errors in δ-values can occur. Freeman et al. [1990] suggested that software techniques for deconvolving contributions of unresolved peaks may soon be commercially available. But, at present, as with off-line techniques, base-line separation of compounds is necessary. Typically, coelution of peaks can be recognized by monitoring the instantaneous mass 45/mass 44 ratio [Freeman et al., 1990; Hayes et al., in press, 1991]. Unfortunately, two peaks which nearly completely co-elute are often times difficult to resolve. Therefore, it is imperative to obtain structural information from conventional GCMS on all samples analyzed using irmGCMS. Freeman et al. [1990] achieved greater chromatographic resolution by analyzing separately branched and cyclic hydrocarbon fractions.

The calibration scale of isotopic abundances is typically determined by comparison with co-injected standards. Freeman et al. [1990] utilized deuterated normal alkanes for standards in their analyses of complex mixtures of cyclic and branched hydrocarbons from the Eocene Messel Shale. Whereas, Jasper and Hayes (personal communication, 1990) used nC_{37}, nC_{38} and nC_{41} alkanes as standards for the determination of a C_{37} diunsaturated alkenone which was prepared for irmGCMS analyses using the technique outlined in Jasper [1988]. It was particularly important for these workers to achieve base-line separation of standards during irmGCMS analyses as these molecules were used to determine the δ-value of other compounds in the sample. It is common to use multiple standards with retention times bracketing retention times of the sample compounds. It is anticipated by the author of this report that the National Bureau of Standards will become a clearinghouse for isotopic standards for irmGCMS analyses in the same manner in which they distribute standards for conventional isotopic analyses.

Procedures for compound-specific isotopic analyses of derivatized volatile compounds by irmGCMS are currently being developed [Engel et al., 1990; S. G. Wakeham, personal communication, 1991]. These analyses require that isotopic compositions of derivatizing agents and stoichiometry of derivatization reactions be known so that isotopic compositions of the compounds of interest can be determined by mass balance (e.g., eq. 1). In principal, any compound for which it is possible to convert a chemically active functional group, e.g., hydroxyl or amine group, to an inactive species can be analyzed isotopically by irmGCMS. As with other procedures, a compound of known isotopic composition should always be carried through the derivatization reaction to verify that the procedure did not cause isotopic fractionation. In fact, it is good practice to spike each sample with such a compound for use as an internal check on the procedure. Once a procedure has been verified, it is important to exercise quality control to assure that isotopic compositions of derivitizing agents do not change over time.

Isotope-ratio-monitoring gas chromatography-mass spectrometry has distinct advantages over off-line techniques. Samples separated using off-line techniques such as HPLC often contain contamination which must be measured and corrected for. The use of the on-line irmGCMS technique minimizes sample contamination and maximizes the variety of natural and derivatized volatile compounds that can be analyzed. Further, the size of samples which can be analyzed by irmGCMS is at least an order of magnitude smaller than that which can be measured using off-line techniques. It is anticipated that on-line, nitrogen-isotopic analyses of individual compounds will soon be possible.

Isotopic analysis of organic matter has been a powerful tool in the study of bulk organic particles. Compound-specific isotopic analysis, because of its greater specificity has the potential to provide new perspectives on the origin and pathways of degradation of compounds constituting marine organic particles. Biochemical mechanisms underlying the observed relationships between the carbon-isotopic abundances of marine plankton and the dissolved inorganic carbon pools and paleoclimatological reconstructions based on the sedimentary record have recently been elucidated through compound-specific isotopic analyses [e.g., Hayes et al., 1989; Popp et al., 1989; Jasper and Hayes, 1990]. Furthermore, the origin of sedimentary biomarkers and the extent of recycling of these compounds by the sedimentary microbial community have recently been investigated using compound-specific isotopic analyses [e.g., Hayes et al., 1987, in press, 1991; Boreham et al., 1989; Ocampo et al., 1989; Freeman et al., 1990]. These are only a few of the investigations that have become possible given the sensitivity and specificity of compound-specific isotopic analyses.

Acknowledgements. - (SOEST Contribution Number 2440).

References

Abelson, P. H., and T. C. Hoering, The biogeochemistry of the stable isotopes of carbon, Carnegie Inst. Washington Yearb., 59, 158-165, 1960.

Abelson, P. H., and T. C. Hoering, Carbon isotope fractionation in formation of amino acids by photosynthetic organisms, Proc. Nat. Acad. Sci., 47, 623-632, 1961.

Bidigare, R. R., M. C. Kennicutt II, W. L. Keeney-Kennicutt, and S. A. Macko, Isolation and purification of chlorophylls *a* and *b* for the determination of stable carbon and nitrogen isotope compositions, Anal. Chem., 63, 130-133, 1991.

Boreham, C. J., and C. J. R. Fookes, Separation of Ni (II) alkyl porphyrins by reverse-phase high-performance liquid chromatography-methodology and application, J. Chromatogr., 467, 195-208, 1989.

Boreham, C. J., C. J. R. Fookes, B. N. Popp, and J. M. Hayes, Origins of etioporphyrins in sediments: evidence from stable carbon isotopes, Geochim. Cosmochim. Acta, 53, 2451-2455, 1989.

Bligh, E. G., and W. J. Dyer, A rapid method of total lipid extraction and purification, Can. J. Biochem. Physiol., 37, 911-917, 1959.

Engel, M. H., and S. A. Macko, Separation of amino acid enantiomers by high-performance liquid chromatography for stable nitrogen and carbon isotopic analyses, Anal. Chem., 56, 2600-2602, 1984.

Engel, M. H., S. A. Macko, and J. A. Silfer, Carbon isotope composition of individual amino acids in the Murchison meteorite, Nature, 348, 47-49, 1990.

Fookes, C. J. R., Identification of a homologous series of nickel (II) 15, 17-Butanoporphyrins from an oil shale, J. Chem. Soc., Chem. Commun., 1983, 1474-1476, 1983.

Freeman, K. F., J. M. Hayes, J.-M. Trendel, and P. Albrecht, Evidence from GC-MS carbon-isotopic measurements for diverse origins of sedimentary hydrocarbons, Nature, 353, 254-256, 1990.

Gelwicks, J. T., and J. M. Hayes, Carbon-isotopic analyses of dissolved acetate, Anal. Chem, 62, 535-539, 1990.

Hare, P. E., Ion exchange chromatography in lunar organic analyses, Space Life Sci., 3, 354-359, 1972.

Hayes, J. M., Practice and principals of isotopic measurements in organic geochemistry, in Organic Geochemistry of Contemporaneous and Ancient Sediments, edited by W. G. Meinschein, pp. 5-1 - 5-31, Society of Economic Paleontologists and Mineralogists, Bloomington, Indiana, 1983.

Hayes, J. M., K. H. Freeman, B. N. Popp, and C. H. Hoham, Carbon-specific isotopic analyses: a novel tool for the reconstruction of ancient biogeochemical processes, Organic Geochemistry 1989, Advances in Organic Geochemistry, in press, 1991.

Hayes, J. M., B. N. Popp, R. Takigiku, and M. W. Johnson, An isotopic study of biogeochemical relationships between carbonates and organic carbon in the Greenhorn Formation, Geochim. Cosmochim. Acta, 53, 2961-2972, 1989.

Hayes, J. M., R. Takigiku, R. Ocampo, H. Callot, and P. Albrecht, Isotopic compositions and probable origins of organic molecules in the eocene messel shale, Nature, 329, 48-51, 1987.

Jasper, J. P., An organic geochemical approach to problems of glacial-interglacial climatic variability, Ph.D. dissertation, 312 pp., Woods Hole Oceanographic Institution, 1988.

Jasper, J. P., and J. M. Hayes, A carbon-isotope record of CO_2 levels during the late Quaternary, Nature, 347, 462-467, 1990.

Lee, C., and S. G. Wakeham, Organic matter in seawater: biogeochemical processes, in Chemical Oceanography, edited by J. P. Riley, pp. 1-51, Academic, London, 1988.

Lee, C., S. G. Wakeham, and J. W. Farrington, Variations in the composition of particulate organic matter in a time-series sediment trap, Mar. Chem., 13, 181-194, 1983.

Macko, S. A., W. Y. Lee, and P. L. Parker, Nitrogen and carbon isotope fractionation by two species of marine amphidods: laboratory and field studies, J. Exp. Mar. Biol. Ecol., 63, 45-49, 1982.

Macko, S. A., M. L. Fogel Estep, M. H. Engel, and P. E. Hare, Kinetic fractionation of stable nitrogen isotopes during amino acid transamination, Geochim. Cosmochim. Acta, 50, 2143-2146, 1986.

Macko, S. A., M. L. Fogel Estep, P. E. Hare, and T. C. Hoering, Isotopic fractionation of nitrogen and carbon in the synthesis of amino acids by microorganisms, Chem Geol. (Isot. Geosci. Sect.), 65, 79-92, 1987.

Matthews, D. E., and J. M. Hayes, Isotope-ratio-monitoring gas chromatography-mass spectrometry, Anal. Chem., 50, 1465-1473, 1978.

Monson, K. D., and J. M. Hayes, Biosynthetic control of the natural abundance of carbon 13 at specific positions within fatty acids in *escherichia coli.* Evidence regarding the coupling of fatty acid and phospholipid synthesis, J. Biol. Chem., 255, 11,435-11,441, 1980.

Monson, K. D., and J. M. Hayes, Carbon isotopic fractionation in the biosynthesis of bacterial fatty acids. Ozonolysis of unsaturated fatty acids as a means of determining the intramolecular distribution of carbon isotopes, Geochim. Cosmochim. Acta, 46, 139-149, 1982a.

Monson, K. D., and J. M. Hayes, Biosynthetic control of the natural abundance of carbon 13 at specific positions within fatty acids in *saccharomyces cerevisae.* Isotopic fractionation in lipid synthesis as evidence for peroxisomal regulation, J. Biol. Chem., 257, 5568-5575, 1982b.

Ocampo, R., H. J. Callot, P. Albrecht, B. N. Popp, M. R. Horowitz, and J. M. Hayes, Different isotope compositions of C_{32} dpep and C_{32} etioporphyrin iii in oil shale. Origin of etioporphyrin iii from heme?, Naturwiss, 76, 419-421, 1989.

Parker, P. L., The isotopic composition of the carbon of fatty acids, Carnegie Inst. Washington Yearb., 61, 187-190, 1962.

Popp, B. N., R. Takigiku, J. M. Hayes, J. W. Louda, and E. W. Baker, The post-paleozoic chronology and mechanism of ^{13}C depletion in primary marine organic matter, Am. J. Sci., 289, 436-454, 1989.

Popp, B. N., C. J. Boreham, and J. M. Hayes, Microscale determination of the spectral characteristics and carbon-isotopic compositions of porphyrins, Energy Fuels, in press, 1991.

Wakeham, S. G., and E. A. Canuel, Organic geochemistry of particulate organic matter in the eastern tropical north Pacific Ocean: implications for particle dynamics, J. Mar. Res., 46, 183-213, 1988.

Wakeham, S. G., J. W. Farrington, R. B. Gagosian, C. Lee, H. de Baar, G. E. Nigrelli, B. W. Tripp, S. O. Smith, and N. M. Frew, Organic matter fluxes from sediment traps in the equatorial Atlantic Ocean, Nature, 286, 798-800, 1980.

Wakeham, S. G., C. Lee, J. W. Farrington, and R. B. Gagosian, Biogeochemistry of particulate organic matter in the oceans: results from sediment trap experiments, Deep-Sea Res., 31, 509-528, 1984.

Weinstein, S., M. H. Engel, and P. E. Hare, The enantiomeric analysis of a mixture of all common protein amino acids by high-performance liquid chromatography using a new chiral mobile phase, Anal. Biochem., 121, 370-377, 1982.

RECENT DEVELOPMENTS IN CHROMATOGRAPHIC, PYROLYTIC AND SPECTROMETRIC TECHNIQUES

Jan W. de Leeuw

Technical University of Delft, De Vries Van Heysplantsoen 2, 2628 RZ Delft, The Netherlands

1. Liquid Chromatography-Gas Chromatography (-Mass Spectrometry)

Very recently on-line LC-GC instrumentation has become commercially available. According to the specifications and some application notes this type of instrumentation may become a very powerful tool in the qualitative and quantitative analysis of DOM and POM, because it combines several significant features: a) the separation power for complex samples is high; b) the sensitivity goes up by three orders of magnitude due to the procedure of interfacing followed [Grob et al., 1984]; c) optimal conditions of separation for specific groups of compounds can be selected; d) direct analysis of (sea)water samples with only a limited concentration is possible [Noy et al., 1988; Pouwelse et al., 1988]. This latter aspect is important because the tricky sample preparation steps of (sea)water samples can be simplified and less artificial effects are thus expected. Preliminary results using these techniques for the trace analysis of organotin compounds in water are promising (J. Leenderts, U. A. Th. Brinkman, personal communication, 1990).

An extension of this on-line LC-GC method can be made in several ways: a) the system can be coupled in the "classic" way to a high- or low-resolution mass spectrometer or to any other specific detector (ECD, FPD, NPD); b) by using known on-line methods in HPLC to derivatize certain families of compounds, the LC-GC approach can also be operated in the on-line version with intermediate derivatization. Especially for the analysis of polar compounds, it is of interest to develop on-line esterifications, silylations, etc. [Brinkman et al., 1989].

High-molecular weight compounds can also be separated by LC- methods. Before GC-analysis, however, they "have to be cut into specific pieces". This can be performed by means of pyrolysis or "chemolysis" (selective chemical break-down). Hence, the advantages of on-line LC-GC (-MS) are lost. However, the same equipment can still be used, although in a non-hyphenated version, for analysis.

Marine Particles: Analysis and Characterization
Geophysical Monograph 63

2. Flash-Pyrolysis

Flash pyrolysis-gas chromatography (Py-GC), flash pyrolysis-mass spectrometry (Py-MS), and flash pyrolysis-gas chromatography-mass spectrometry (Py-GC-MS) are now well-accepted methods to analyze and characterize in detail a large variety of soluble and insoluble high-molecular-weight substances, either as the pure compound or as present in the incredibly complex mixtures encountered in natural samples like humic substances, sediments, soils, DOM and POM. Over the last few years efforts have been made to better understand the depolymerization mechanisms during flash pyrolysis using well-selected standards. As a result of those studies the pyrolytic behavior of virtually all common biomacromolecules is presently known, enabling us to reconstruct properly (mixtures of) biomacromolecules present in natural samples. A very short overview is given here.

2.1. Natural Polyesters

Flash pyrolysis of the naturally occurring plant polyesters cutin and suberin yield almost exclusively the monomeric building blocks. Mono- and di-unsaturated C_{16} and C_{18} fatty acids are generated via a six-membered ring rearrangement; these compounds are easily separated and identified by GC-MS [Tegelaar et al., 1989]. If the mid-chain hydroxyl group is not esterified (a non-crosslinking unit), mono-unsaturated mid-chain-hydroxy C_{16}- and C_{18}- fatty acids are generated. The relative amounts of these latter pyrolysis products are an indication of the degree of crosslinking. This pyrolytic behavior of natural polyesters seems to be common because similar observations were made for bacterial wax esters [Wieten et al., 1984].

2.2. Lignins

Py-GC-MS and Py-MS analyses of lignins indicate that depolymerization yielding monomeric and dimeric building blocks is the dominant mechanism [e.g. Saiz-Jimenez and de Leeuw 1984, 1986; Scheijen and Boon, 1989, 1991]. Lignin characteristics such as their nature (p-hydroxyl, guiacyl- or syringyl type) and degree of biodegradation are recognized via analysis of the pyrolysates. The results are in full agreement with and complimentary to chemical degradation and spectroscopic characterization of lignins [Saiz-Jimenez et al., 1987; Hedges, this volume, page 131].

2.3. Sporopollenins

Only a few pyrolysis studies have been performed with sporopollenins [Schenck et al., 1981; P. van Bergen, unpublished results, 1991]. The results indicate that at least two types of sporopollenins exist: a highly aliphatic one comparable with algaenan, suberan and cutan, see hereafter, generating homologous series of n- α-, ω-alkadienes, n-alk-1-enes and n-alkanes, and a phenylpropanoid type which releases alkylphenols and benzaldehydes partly comparable to lignin pyrolysis products. It should be noted that these findings are in agreement with recent spectroscopic and chemical degradation studies of sporopollenins [Osthoff and Wierman, 1987; Guilford et al., 1988] and that carotenoid derivatives, previously thought to make up sporopollenins, are not at all involved.

2.4. Highly Aliphatic Non-saponifiable Biomacromolecules

Recently a new class of aliphatic biomacromolecules has been discovered in algal cell walls (algaenan), in plant cuticles (cutan) and in barks (suberan) [see Tegelaar et al., 1989 for a review]. Upon flash pyrolysis this type of biopolymer yields extended homologous series of n- α-, ω-alkadienes, n-alk-1-enes and n-alkanes ranging from C_6 to C_{32}. The highly aliphatic polymethylenic character of these biomacromolecules is also concluded from ^{13}C-NMR and FT-IR spectra. From geochemical investigations it has become clear that these biomacromolecules are extremely resistant so that they are selectively preserved, even in aquatic environments [Proceedings NSF-Marine Organic Geochemistry Workshop, January 1990, Honolulu, in press, 1991].

2.5. Resins

Polymeric structures present in angiosperms, e.g., polycadinenes in damars, and gymnosperms, e.g., polysesquiterpenoids, release highly characteristic monomeric, dimeric and oligomeric units upon flash pyrolysis [B. G. K. van Aarssen et al., 1990; and unpublished results, 1991]. Cadinenes, muurolenes, bicadinenes, tricadinenes are identified by GC-MS analysis of flash-pyrolysates of damar resins and specific tricyclic sesquiterpenoids are identified by GC-MS as major products in pyrolysates of gymnosperm resins.

2.6. Polyhydroxyalkanoates

These linear polyesters occur as storage materials in Eubacteria and are composed of β-hydroxybutanoate and -pentanote units [Capon et al., 1983]. Their pyrolysis products consist of a limited series of butenoic acids and methylbutanals in the case of poly-β-hydroxybutanoate [Boon et al., 1986]. This is in perfect agreement with chemical degradation reactions [Capon et al., 1983].

2.7. Polysaccharides

Much work has been performed to understand the general mechanisms of flash pyrolysis of polysaccharides of many different origins. The excellent work by Pouwels et al. [1989] concerning the flash pyrolysis behavior of cellulose by several pyrolysis methods such as Py-GC-MS and pyrolysis desorption chemical ionization mass spectrometry (Py-DCI-MS) has shown that the formation of anhydro-oligosaccharides (nonamers are detected) is the most important pathway. Helleur and Guevremont [1989] used Py-DCI-MS in combination with Py-DCI-MS-MS to characterize polysaccharides such as starch, dextran, laminarin and agar. They also identified anhydrosugars as the major pyrolysis compounds. Their study indicated that linkage information and differentiation of monosaccharide stereoisomers are possible by this approach. Tas et al. [1989] using Py-DCI-MS-MS came to the same conclusions based on investigations of amylose, dextran and the fungi *Penicillium italicum*. The above mentioned studies show that flash-pyrolysis methods yield a wealth of information about the nature of polysaccharides that is not easily obtained in other ways.

2.8. Proteins

Proteins yield highly characteristic flash pyrolysis products, reflecting the amino acid moieties to some extent and giving some information concerning the sequence [Boon and de Leeuw, 1987; Munson and Fetterolf, 1987]. Because other methods are preferred to sequence amino acid moieties in proteins it is understandable that virtually no effort is being made at present to improve the flash pyrolysis approach.

2.9. Future Applications of Pyrolysis

From the above it is clear that most known biomacromolecules can be characterized by flash pyrolysis in combination with MS-techniques or with GC-MS. It is therefore attractive to "screen" natural samples like DOM and POM with flash pyrolysis methods in order to characterize the consortium of biomacromolecules present, before more time-consuming and more specific analyses are undertaken. Actually, the flash pyrolysis results can help to optimize the analytical chemical strategy to further analyze the samples of interest. Because of the relatively high reproducibility of Py-(GC)-MS analyses one can "fingerprint" large series of samples and deconvolute chemical changes in the samples by application of multivariate analyses [e.g., Eglinton et al., submitted 1991].

To improve the Py-GC-MS analysis for the separation and identification of polar pyrolysis products, the on-line derivatization procedure recently described by Challinor [1989] is very promising. Interestingly, the described on-line esterification procedure is performed in an aqueous environment and works surprisingly well. It can therefore be expected that this on-line derivatization procedure is powerful in the analysis of (sea)water organics. Moreover, a total removal from (sea)water is not necessary using this method so that sample work-up procedures may be simplified somewhat.

To analyze and compare large series of POM or DOM samples at a molecular level the Pyrolysis/Desorption-Cl-MS approach (Py/D-Cl-MS) has great potential. Therefore a more detailed description of this recently developed technique is provided.

Ca. 10 μg of the sample to be analyzed is applied to a Pt/Rh wire which is introduced into a DCI or EI source of a high-resolution mass spectrometer. The metal wire is resistively heated to ca. 350° C at a rate of 20-50° C/sec. During the first stage, low-molecular components (predominantly lipids) are volatilized and ionized resulting in a low-molecular CI mass spectrum depending on the mode of ionization used. By using these types of ionization, fragmentation of the molecular ions is suppressed, which aids recognition and tentative identification of distribution patterns of different compound classes. The wire is further heated to ca. 1000° C to generate characteristic pyrolysis products of the high-molecular-weight constituents present in the sample. As a consequence, mass spectra mainly consisting of molecular ions of these products are produced. By using special wires it has been shown that it is also possible to release inorganic materials such as metals and metaloxides and to visualize them as their mass spectra. Mass spectra can be recorded in both a low- and high resolution fashion and the mass range can be selected as required. (In the case of cellulose, mass spectra up to mass 2000 were recorded and nonamers were detected). The whole analysis is carried out in ca. 5 minutes and the nature of the (mass) data allows for automated storage and data manipulations including multivariate analysis.

Examples of this procedure concern the analysis of an artist's oil painting [Eijkel et al., in preparation, 1991] and of plant rootlets [Van der Heijden et al., in press, 1991].

3. High-Temperature GC(-MS)

The quality of the so-called high-temperature GC-columns is continuously improving and it is now possible to separate and analyze relatively apolar, thermostable compounds with ca. 100 C-atoms such as complex lipids (without or with derivatization). Because relatively high molecular-weight complex lipids probably make up a part of (sea)water organic matter, this new version of GC in combination with MS-analysis and HPLC preparation is certainly worth exploring.

4. "Chemolysis"

Specific chemical degradation techniques are potentially worthwhile, especially when they can be carried out in (sea)water media. Recently we have found that MeLi/MeI is highly selective in desulfurization of di- and polysulfide-compounds which can be abundantly present in very recent sediments or anaerobic (sea)water columns. Relatively high-molecular weight organic matter cross-linked by S_x (x>2) bridges is released and the building blocks are easily analyzed as methylthioethers by GC-MS [Kohnen et al., in press, 1991]. Unfortunately, this reaction cannot be used in an aqueous medium. Emphasis should be laid on the chemolysis approach in aqueous media; the reaction can then be performed in a concentrated amount of (sea)water and via an appropriate extraction procedure, these products or their derivatized counterparts can be isolated from the aqueous medium and subsequently analyzed by GC- or LC-MS. Along the same lines we may have to explore the possibility of using enzymes or mixtures of enzymes to very specifically depolymerize DOM or POM in a (sea)water medium. After such treatments, released compounds can be extracted as such or after modification.

5. Spectroscopic Methods

Although NMR (^{1}H, ^{13}C, ^{31}P) and FTIR methods do not give sufficient information concerning molecular structures, they are very helpful for quantitative detection of functional groups if applied carefully. These methods should be used in a complementary fashion with another non-quantitative approach.

References

Boon, J. J., and J. J. de Leeuw, Amino acid sequence information in proteins and complex proteinaceous material revealed by pyrolysis-capillary gas chromatography - low and high resolution mass spectrometry, J. Anal. Appl. Pyrol., 11, 313-327, 1987.

Boon, J. J., B. Brandt-de-Boer, W. Genuit, J. Dallinga, and E. Turkstra, Characterisation of particulate organic matter, in Organic Marine Geochemistry, ACS Symposium Series 305, edited by M. L. Sohn, pp. 76-90, ACS, Washington, 1986.

Brinkman, U. A. T., R. W. Frei, and H. Lingeman, Post-column reactors for sensitive and selective detection in high- performance liquid chromatography: categorization and applications, J. Chromatogr., 492, 251-298, 1989.

Capon, R. J., R. W. Dunlop, E. L. Ghisalberti, and P. R. Jeffries, Poly-3-hydroxyalkanoates from marine and fresh-water cyanobacteria, Phytochemistry, 22, 1181-1184, 1983.

Challinor, J. M., A pyrolysis-derivatization-gas chromatography technique for the structural elucidation of some synthetic polymers, J. Anal. Appl. Pyrol., 16, 323-333, 1989.

Grob, K., D. Fröhlich, B. Schilling, H. P. Neukom, and P. Nägeli, Coupling of high-performance liquid chromatography with capillary gas chromatography, J. Chromatogr., 295, 55, 1984.

Guilford, W. J., D. M. Schneider, J. Labovitz, and S. J. Opella, High resolution solid state ^{13}C NMR investigation of sporopollenin from different plant taxa, Plant Physiol., 86, 134-136, 1988.

Helleur, R. J., and R. Guevremont, Initial application of desorption chemical ionization-pyrolysis tandem mass spectrometry to structural analysis of carbohydrates, J. Anal. Appl. Pyrol., 15, 85-95, 1989.

Kohnen, M. E. L., J. S. Sinninghe-Damsté, A. C. Kock-Van Dalen, and J. W. de Leeuw, Identification of acyclic di(poly)sulphide linkages in sulphur-rich geomacromolecules by using a novel selective chemical degradation method, Geochim. Cosmochim. Acta, in press, 1991.

Munson, T. O., and D. D. Fetterolf, Evidence for the formation of 2,4-imidazolidinediones and pyrrolidino [1,2a]-3,6-piperazinediones in human hair pyrolysate by pyrolysis-gas chromatography-mass spectrometry, J. Anal. Appl. Pyrol., 11, 15-24, 1987.

Noy, T., E. Weiss, T. Herps, H. Van Cruchten, and J. Rijks, On-line combination of liquid chromatography and capillary gas chromatography. Preconcentration and analysis of organic compounds in aqueous samples, J. High. Resolut. Chromatogr. Chromatogr. Commun., 11, 181-186, 1988.

Osthoff, K. S., and R. Wierman, Phenols as integrated compounds of sporopollenin from Pinus pollen, J. Plant Physiol., 131, 5-15, 1987.

Pouwels, A. D., G. B. Eijkel, P. W. Arisz, and J. J. Boon, Evidence for oligomers in pyrolyzates of microcrystalline cellulose, J. Anal. Appl. Pyrol., 15, 71-84, 1989.

Pouwelse, A., D. De Jong, and J. H. M. Van Den Berg, Use of a polar coated retention gap for the introduction of large volumes of polar solvents in on-column capillary gas chromatography, J. High. Resolut. Chromatogr. Chromatogr. Commun., 11, 607-609, 1988.

Proceedings NSF-Marine Organic Geochemistry Workshop, January 1990, Honolulu, in press, 1991.

Saiz-Jimenez, C., and J. W. de Leeuw, Pyrolysis-gas chromatography-mass spectrometry of isolated, synthetic and degraded lignins, Org. Geochem., 6, 417-422, 1984.

Saiz-Jimenez, C., and J. W. de Leeuw, Lignin pyrolysis products: their structures and their significance as biomarkers, in Advances in Organic Geochemistry 1985, edited by D. Leythaeuser, and J. Rullkötter, Org. Geochem., 10, 869-876, 1986.

Saiz-Jimenez, C., J. J. Boon, J. I. Hedges, J. K. C. Hessels, and J. W. de Leeuw, Chemical characterization of modern and buried woods by analytical pyrolysis; comparison of pyrolysis data with NMR and chemical degradation data, J. Anal. Appl. Pyrol., 11, 437-450, 1987.

Schenck, P. A., J. W. de Leeuw, G. Van Graas, J. Haverkamp, and M. Bouman, Analysis of recent spores and pollen and of thermally altered sporopollenin by flash pyrolysis-mass spectrometry and flash pyrolysis-gas chromatography-mass spectrometry, in Organic Maturation Studies and Fossil Fuel Exploration, edited by J. Brooks, pp. 225-237, Academic Press, London and New York, 1981.

Scheijen, M. A., and J. J. Boon, Characterization of tobacco lignin preparations by Curie-point pyrolysis-mass spectrometry and Curie-point pyrolysis-high-resolution gas chromatography/mass spectrometry, J. Anal. Appl. Pyrol., 15, 97-120, 1989.

Scheijen, M. A., and J. J. Boon, Micro-analytical investigations on lignin in enzyme-digested tobacco lamina and midrib using pyrolysis-mass spectrometry and Curie-point pyrolysis-gas chromatography-mass spectrometry, J. Anal. Appl. Pyrol., in press, 1991.

Tas, A. C., A. Kerkenaar, G. F. La Vos and J. Van der Greef, Pyrolysis-direct chemical ionization mass spectrometry of some biopolymers in the positive and negative ionization mode, J. Anal. Appl. Pyrol., 15, 55-70, 1989.

Tegelaar, E. W., J. W. de Leeuw, and P. J. Holloway, Some mechanisms of flash pyrolysis of naturally occurring higher plant polyesters, J. Anal. Appl. Pyrol., 15, 289-295, 1989.

Tegelaar, E. W., S. Derenne, C. Largeau, and J. W. de Leeuw, A reappraisal of kerogen formation, Geochim. Cosmochim. Acta, 53, 3103-3107, 1989.

van Aarssen, B. G. K., C. H. Cox, P. Hoogendoorn, and J. W. de Leeuw, A cadinene biopolymer in fossil and extant dammar resins as a source for cadinanes and bicadinanes in crude oils from South East Asia, Geochim. Cosmochim. Acta, 54, 3021-3031, 1990.

Van der Heijden, E., J. J. Boon, and M. A. Scheijen, Pyrolysis-mass spectrometric mapping of selected peats and peatified plant tissues, in Proceedings of Peat 90, edited by R. Sopo, in press, 1991.

Wieten, G., J. J. Boon, D. G. Groothuis, F. Portaels, and D. E. Minnikin, Rapid detection of mycobacterial contamination in batches of purified *Mycobacterium leprae* by pyrolysis gas chromatography, FEMS Microbiol. Lett., 25, 289-293, 1984.

CHAPTER 5. INORGANIC AND RADIOACTIVE PROPERTIES

WORKSHOP REPORT

by

Co-Chairs: Robert Anderson and George W. Luther, III, Rapporteur: Robert W. Collier,
Assistant Rapporteur: Peter Sedwick and Members of the Working Group

1. Introduction

Classic papers by Goldberg [1954] and Krauskopf [1956] discussed the importance of particulate matter in regulating the concentration of many inorganic chemical species in the ocean. These authors noted that the most important mechanism removing certain elements from the deep ocean involves adsorption onto particulate matter. Dissolved species are at the mercy of advective and diffusive transport processes that largely follow along isopycnals. Consequently, it may take hundreds of years for dissolved species to be transported between the top and the bottom of the ocean. Particulate matter on the other hand, most notably large aggregates formed in surface waters, can short-circuit thermohaline transport pathways and transit the entire water column in a matter of days to weeks. Particles, therefore, not only regulate the concentrations of many inorganic chemicals through their adsorptive properties, but they are also the principal mechanism for rapid vertical transport in the ocean. This is even more true if living "particles" are included in this generalization.

Many investigators have endeavored to develop systematic methods of describing particle dynamics and the scavenging behavior of particles. About 20 years after the concept of marine scavenging was introduced [Goldberg, 1954], Schindler [1975] described a quantitative model for the removal of metallic trace elements from seawater in which he pointed out that the stability constants for adsorption of metals to silica gel are proportional to the hydrolysis constants of the metals. From this it could be inferred that if scavenging takes place largely by adsorption to oxide surfaces, and if other oxides behave like silica, then the residence time of an element in the ocean should be proportional to one of its fundamental thermodynamic properties, namely its hydrolysis constant. It should, therefore, be possible to predict the residence time of an element in the ocean based on its thermodynamic properties and furthermore, given a knowledge of the supply of the element to the ocean, it should be possible to predict its concentration and distribution in the ocean as well.

Surface complexation models and relationships between adsorption behavior and thermodynamic properties were later extended to include additional oxides as well as selected organic phases [e.g., Balistrieri et al., 1981]. Adsorption-desorption equilibrium, a prerequisite for application of complexation models, was shown to take place in the deep sea [Bacon and Anderson, 1982]. Complexation models have been further developed and extended, and additional models of particle behavior have been constructed to describe coagulation (of colloids), aggregation (of fine particles), and disaggregation of all sizes of aggregates or composite particles [e.g., Nozaki et al., 1987; Honeyman and Santschi, 1989; Clegg and Sarmiento, 1989; Clegg and Whitfield, 1990; Murnane et al., 1990].

This type of modelling approach is appealing because one of our fundamental goals as oceanographers is to achieve a sufficient understanding of marine processes that models can be developed to describe and predict distributions of chemical species in the ocean. However, despite the advances noted above, we do not yet see anyone publishing maps of model-generated distributions of trace elements in the oceans. Why? Is it because we lack the thermodynamic stability constants for the dissolved inorganic species that exist in the sea? Is it because marine particulate matter is so chemically heterogeneous that its surface chemical properties may never be described by a limited set of fundamental constants? Is it because as yet uncharacterized organic matter and other coatings of natural particles give them surface chemical properties totally unlike those of the model oxides and organic phases used to demonstrate simple relationships between adsorption behavior and thermodynamic stability constants? Are these the foremost questions that need to be addressed at a workshop on the analysis and characterization of marine particulate matter, or are there still more fundamental questions that need to be addressed first?

Each of the above questions warrants examination, and the answers to these questions are not beyond our experimental capabilities. However, in reading over the papers appended to this chapter, it seems that some

Marine Particles: Analysis and Characterization
Geophysical Monograph 63

more basic issues of particle collection need to be dealt with in advance. There may not be a "right" way to collect a marine particle, but there are certainly methods that are wrong, or inappropriate, for the achievement of desired goals. This can be illustrated by reviewing some basic aspects of particle collection about which most investigators in this group agree, and then considering those aspects about which opinion differs.

Points of agreement were:

1) All investigators noted special precautions taken to clean sampling equipment, from collection bottles to the filters themselves, to prevent contamination.
2) Nearly all investigators noted the importance of carrying unused filters through the sample collection and processing steps, not only to evaluate analytical blanks, but also simply to obtain proper weights of particulate matter collected; e.g., to correct for effects of variable humidity, temperature, etc. during weighing.
3) Use of membrane filters (e.g., Nuclepore) was recommended to get good (stable) weights of particulate matter collected when small amounts of material (100's of μg) are involved.
4) Use of alpha-emitting antistatic devices was also recommended to obtain good weights.
5) Finally, there seems to be universal agreement that precautions need to be taken to collect the dregs, i.e., large particles that settle to the bottom of sampling bottles during the course of collection and filtration.

Despite the widespread recognition of the need to collect the dregs from sample bottles, no one indicated that any consideration had been given to the potential consequences of the methods used to recover the dregs. Some investigators immediately transfer the entire contents of the sample to a more appropriate filtration vessel, and then filter the entire sample. Other investigators employ some kind of device to agitate (stir or shake) the collection bottles during the filtration process to keep the large particles homogeneously distributed throughout the sample.

Most of us learned early in our career what happens when filtered seawater is shaken and then refiltered. New particles appear. Could similar coagulation/aggregation processes be taking place during the simple transfer of a sample from one container to another, or when a sample bottle is agitated to maintain large particles in suspension? A recent comparison of particulate nitrogen concentrations obtained by filtration in situ with corresponding results obtained by processing water collected by conventional bottle techniques suggests that such problems may, in fact, occur (see contribution by Bishop in this volume, page 229).

Another fundamental potential source of error lies in how filters are processed after collection. Is it necessary, or even appropriate, to wash residual seawater, and hence sea salt, from filters before drying them? This is the one point which finds the greatest disagreement among investigators. On the one hand, residual sea salt could easily overwhelm the mass of particulatc matter collected, and thereby preclude obtaining accurate dry weights of particles. On the other hand, washing filters to remove sea salt dissolves significant amounts of labile particulate phases (see e.g. contribution by Collier in this volume, page 235). Is it better to allow sea salt to remain on the filters and measure the major ions when processing the filters to make corrections for sea salt? Or, is it preferable to rinse the filters and collect the rinse solution which can then be analyzed for the species of interest which may have dissolved during the rinsing procedure?

Virtually all issues of concern raised by participants in the Inorganic and Radioactive Properties group in their written contributions as well as during group discussions focussed on sample collection and handling procedures rather than with procedures for instrumental analysis. The nature of their concerns results from a couple of general factors:

1) Most of the instrumental techniques used for the analysis of inorganic elements and compounds have been tested and used extensively outside of the field of oceanography. Precision of instrumental techniques is generally better than 5 to 10%, and isotope dilution procedures can improve precision to better than 1%. Certainly, as new methods come on line they require evaluation and testing. Most analyses, though, suffer more from problems or uncertainties associated with the collection and handling of samples than from precision of instrumental analyses.
2) It is well known that the nature and amount of particulate matter collected by filtration is operationally defined by the sampling method used. Factors such as manipulation of water samples prior to filtration, filter type, pore size, flow rate, degree of filter loading, and filter washing influence the nature of particulate matter collected when filtering seawater. Sampling methods can destroy existing particles, create new ones, and seriously contaminate those that are recovered.

Because of the situation mentioned above, the participants in the Inorganic and Radioactive Properties group devoted most of their attention to methods of collection and handling of marine particles. This was not, however, to the exclusion of other topics of concern. Results of group discussions of appropriate standards and intercalibrations, of blanks and the methods by which they are evaluated, of methods for achieving total dissolution of samples for inorganic analysis, and of the relative merits of several instrumental methods for elemental analysis follow this introduction. Results of group discussions of the properties of various filters used in marine applications and of the relative merits of in situ filtration versus sampling from a bottle have been incorporated into Chapter 1 (Particle Sampling and Preservation).

2. Standard Reference Materials and Intercalibration for Total Element Analysis

Precision and accuracy of analytical procedures are usually evaluated by analysis of known samples. Standard reference materials (SRM's) have been developed for this purpose for a number of marine environmental matrices such as seawater, marine sediments and biological tissues [see Cantillo, 1989 for a list of materials and sources]. Although none have been developed specifically for marine suspended particles, marine sediment SRM's are useful for the purpose of evaluating procedures for analysis of particles. For example, the NBS estuarine sediment standard (SRM 1643) has been used by Windom and Smith (see Table 5-14 in contribution by Windom and Smith, this chapter, page 319) to evaluate precision and accuracy of trace metal analysis of small (~ 20 mg) samples treated as suspended particle samples. Even though NBS only certifies the homogeneity of this SRM for aliquots of greater than 100 mg, it and other sediment SRM's appear to be suitable for the calibration of particulate analyses.

Another way to evaluate analytical procedures is through interlaboratory comparisons. The International Council for the Exploration of the Sea (ICES) Working Group on Marine Sediments initiated the first phase of a SPM intercomparison exercise in 1989-90. The objective of this exercise was to test the ability of laboratories to analyze small size samples precisely and accurately. Certified sediment reference materials from the National Research Council of Canada were distributed and 19 laboratories participated. The elements analyzed were Cu, Pb, Zn, Cd, Al, Li, Mn, Ni, and Cr using total dissolution techniques. The results indicate that the performance of the laboratories depends on element, sample size and element concentration [Hovind and Skei, 1990].

ICES has also conducted two intercomparison exercises that investigated the ability of laboratories to collect and analyze actual coastal water suspended particle samples. The first of these [Bewers et al., 1985] was a shipboard intercomparison of seawater filtration procedures of six different laboratories. Homogenized water samples were filtered by each participant using his standard filtration procedures and filtrates analyzed for trace metals by all participants. In the second experiment [Yeats and Dalziel, 1987], filters, each containing approximately 1 mg of coastal particles, were distributed to eight laboratories and the samples were analyzed for 12 major and minor elements. The results showed that reasonable agreement could be achieved on trace metal concentrations in coastal particles.

3. Procedural and Analytical Blanks

3.1. Procedural

3.1.1. Sediment traps.

No simple procedural blanks have been developed for checking contamination during deployment. Corrosion of trap hardware may contribute to blanks, although this has not been evaluated. Sediment traps with multi-cup sample-collection carousels have been deployed at sea where selected cups did not rotate into collection position. These cups return without detectable amounts of particulate matter (i.e., negligible particle blanks), and solutions within the sample cups contain negligible amounts of those species which dissolve from particulate matter normally collected by traps (e.g., organic carbon, phosphorus, manganese; Collier and Dymond, unpublished, 1990).

3.1.2. In situ filtration.

One approach has been to load acid-washed, unused filters into the pump system and deploy it without pumping at a selected depth for a time comparable to sampling times. These filters are analyzed in the same fashion as sample filters. Another approach has been to pair filters during sampling with the downline filter serving as a blank. However, small particles and organics may pass through the upper filter and bias this type of blank analysis. See chapter 6, Mineralogy and Surface Properties, for additional discussion.

Also, concern was expressed for adsorption of selected chemical species on filters during deployment. It has been suggested that downstream filters may be used to identify adsorption problems [Bishop and

Edmond, 1976]. Potential contamination by adsorption can arise from surface slicks as demonstrated for Pb (Bishop, unpublished, 1991) and lipids such as 22-1 omega-9 fatty acids [Conte, 1989]. A mitigating factor with respect to the blank problems described here is the relatively high particle loading (10's to 100's of mg) achieved with in situ pumps.

3.1.3. Bottle filtration.

There are two commonly used bottle types, Go-Flo and Niskin. The Go-Flo bottles can be lowered below the sea surface closed so that surface slick contamination is avoided.

Wong et al. [1983] have discussed some general considerations for contamination by bottle samplers. True measurement of procedural blanks is difficult. One approach is to fill the sample bottle with distilled, deionized water and pass it through an acid-washed filter which is subsequently analyzed. A second approach is to filter increasing volumes of the same sample through different filters. For example, a 30 L sample may be shaken and 1 L, 5 L, and 15 L volumes processed independently. Then, concentrations of elements of concern (e.g. Zn, Pb or other trace elements) can be ratioed to major elements, such as Fe or Al, to identify anomalous ratios when small loadings of particles are processed. This may not work when sedimentation bias is important or in the case of organic matter.

3.1.4. Sources of contamination.

Potential sources of contamination include: hydrowire (Kevlar is sometimes used to avoid potential contamination by metallic hydrowires), sacrificial Zn anodes on the rosette or the ship, over the side discharges of waste water, bilge water or other debris. Memory effects have been noted in bottle samplers when sampling shifts from an area of high to low suspended matter concentrations (see also chapter 1, Particles Sampling and Preservation).

a) Bottle cleaning and inspection. To avoid contamination in bottle samplers, they should be acid-washed (dilute HCl) followed by a distilled water rinse and seawater conditioning. Go-Flo bottles seem to be preferred when analyses of dissolved constituents will be carried out. The seawater conditioning step is best achieved by lowering the bottles to a depth of 1 to 2 km so that the bottle is in contact with seawater for about 1 hour. During the acid wash and distilled water rinse, manipulation of ball-valves, spigots and other moving parts is recommended to remove any entrapped particles.

Before deploying the bottle, check the state of the Teflon coating on the interior, seals and O-rings, handles, any internal caulking or other components (surfaces inside the bottle may lead to adsorption of radionuclides and surface reactive species), and interior springs. Teflon and epoxy coatings sometimes show small voids where the coating did not cover the surface, or cracks due to wear. For new bottles, complete disassembly is recommended and all parts should be washed with a high-quality laboratory detergent (e.g. Micro). After washing with detergent, the parts and the bottle should be rinsed with distilled, deionized water and then cleaned with dilute HCl and rinsed again. After reassembly, the bottle is conditioned with seawater as described above.

With concern for contamination and detection limits, the following elements are currently being analyzed on samples collected with bottles, but not on a single sample: Al, Ti, Ca, Sr, Ba, Si, Fe, Mn, I, Cu, V, radionuclides (Pb-210, Po-210, Th-234), As, Se, and bioelements (C, N, P and S).

b) Pump system cleaning. Plastic parts of pump systems, including tubing and filter holders, are cleaned with a high quality detergent, rinsed with distilled, deionized water, then soaked in warm 1 N HCl and again rinsed with distilled, deionized water before deployment. Between deployments, the filter holder parts are rinsed with distilled, deionized water. It is recommended that the entire pump be covered when not in use.

3.2. Analytical Blanks

Determination of filter and acid blanks is described in detail in the individual papers. Basically, acid-washed filters are treated in the same manner as a sample.

4. Total Dissolution of Suspended Sediments

Many instrumental techniques for elemental analysis (e.g., AA, ICP-ES, ICP-MS, colorimetric and spectrofluorimetric) require that samples be dissolved prior to analysis. Other methods require sample dissolution prior to chemical purification of the element(s) of interest. Procedures that work on the principle of isotope dilution (e.g., TIMS, ICP-MS, radioactive counting) likewise require sample dissolution in order to achieve isotopic equilibrium between added tracers and naturally-occurring isotopes. Consequently, considerable effort has gone into developing methods for dissolution of marine particles (see individual papers in this volume by Fleer and Bacon, page 223, Fleisher and Anderson, page 221, Collier, page 235, Landing and Lewis, page 263, Sherrell, page 285, Trefry and Trocine, page 311, Windom and Smith, page 317, and Yeats, page 321).

Two general methods are employed to achieve total dissolution: fusion and mixed-acid dissolution.

4.1. Fusion

Total dissolution of dried powdered rock by fusion with lithium metaborate flux is commonly used by rock geochemists. More recently sediment trap material is being dissolved by fusion - Fleer and Bacon (this chapter, page 223), Manganini and Honjo (unpublished, 1990) and Falkner (unpublished, 1990). The fusion process converts sediments into a homogenous glass bead which then easily dissolves in dilute acid. Fusion can simplify analysis as it avoids the need for HF and $HClO_4$ treatments for the solubilization of the sample. It is also the method of choice for rare earth elements (REE), Zr and Ti as refractory minerals are reliably dissolved. It has also been used for Ba, U and Th. Once the glass bead is dissolved in acid, major and minor elements can be determined by a variety of analytical techniques (i.e., AAS, ICP-ES and ICP-MS).

Fusion is normally used on samples weighing 2 mg or more. Flux to sample ratios are typically 4:1 to 7:1. The material is then shaken in the crucible and heated over a bunsen burner or in a furnace until the crucible is red hot. The use of smaller samples will require scaling down the fusion while maintaining total recovery. The use of fusion for trace elements depends on the blank of the flux relative to the concentration of the element in the sample. Ultraclean fluxes are commercially available from AESAR (Johnson Matthey). Obviously the solid solution will contain high levels of Li and B which might result in matrix effects having to be assessed for each element and each analytical technique. The large linear working ranges (up to 10^6) of ICP-MS and ICP-ES offer distinct advantages for multielement analysis of fusion solutions. For ICP-MS, low detection limits permit dilution of fusion solutions helping to further minimize matrix problems (Falkner, this chapter, page 243). Other methods such as AAS can be performed on these solutions as well.

An inherent problem of fusion techniques is the loss of volatile elements (e.g. Pb). Significant losses of Na and Ca can occur when fusing $CaCO_3$ rich samples. On the other hand, an important advantage of fusions in contrast to HF dissolutions is that Si and Al are not lost. Fusion will also decompose filters used to support the suspended matter samples. In summary, fusion coupled with ICP methods provides a relatively rapid and simple means of total dissolution and analysis of many elements. Micro-fusion techniques need further development for sub-milligram sized samples.

4.2. Acid Dissolution Techniques

There are two general types of non-fusion acid dissolution techniques: those done in open vessels and those done in closed vessels. The former type can be done quickly and with simpler apparatus but invariably loses some elements such as Si to volatilization. Thus it cannot produce a total analysis and is of limited usefulness as in determining transition metals in various types of samples.

Closed-vessel dissolutions typically use some combination of HCl, HNO_3, HF and perchloric acid. Vessels are typically Teflon. Older methods housed a Teflon vessel in a steel jacket to withstand the considerable pressures that are generated on heating to about 200° C for an hour or more. Newer methods use microwave ovens with Teflon pressure released vessels which allow for venting of volatiles if excessive pressures are generated. The greater speed (on the order of 1 min) of microwave ovens precludes the need for a steel jacket. Screw-cap or stoppered Teflon vessels, simply heated in a water bath or on a hot plate, have also been used successfully (see Sherrell, page 285, and Trefry and Trocine, page 311).

Incomplete dissolution is a potential problem with both methods. A mixture of acids is used in order to dissolve the wide variety of materials in marine particles, but there are some samples that produce an insoluble residue. Sometimes residual solids and precipitates are visible to the naked eye but frequently are not visible. In addition, $CaCO_3$ rich samples dissolved in HF commonly cause CaF_2 to precipitate, which removes variable amounts of Al, Pb, Fe, Cr, REE, radionuclides and alkaline earth elements. Boric acid is commonly added to redissolve the fluoride precipitates. For these reasons, as little HF as possible should be used. HNO_3/HF ratios in published methods range from 25 to 1 depending on Si content of the samples.

Organic matter can be dissolved first with base such as ammonia (which also dissolves polycarbonate Nuclepore filters) or with HNO_3 followed by perchloric acid. Perchloric acid also helps dissolve residual fluorides.

A new paper by Loring and Rantala (submitted, 1991) provides a detailed manual for the dissolution of sediments and suspended matter by acids.

5. Instrumental Methods of Analysis

Chemical analysis of inorganic components in particulate matter may be made using a variety of instrumental methods. The choice of analytical method depends on a number of factors, including the range of elements to be quantified, concentration(s) in the sample and the amount of sample available. The main instrumental methods are outlined below. Please see the sections in this chapter by Falkner (page 243) and Collier for other comparisons(page 235).

5.1. Atomic Absorption Spectroscopy (AAS)

This method, which has been widely used for analysis of marine particulate matter, is mainly suited to group I and II elements, most transition elements and some metalloids. In some cases, simple flame atomization may be possible, especially for major and minor elements where sample size is not severely limited. However for most elements, graphite furnace atomization is used to attain sufficient analytical sensitivity. Hydride generation methods can be applied to several metalloids (As, Te, Bi, Se, Sb, Sn) with high sensitivity. Hg can be readily determined in the gas phase as Hg^0 after reduction, e.g. with $SnCl_2$, followed by concentration on Au-coated quartz. GFAAS has the advantage of high sensitivity, but has low sample throughput (20 to 40 single element determinations per day).

5.2. Inductively-Coupled Plasma Emission Spectroscopy (ICP-ES)

This method combines classical atomic emission spectroscopy with the argon plasma torch as an ion source. In general terms, ICP-ES is intermediate in sensitivity to flame and graphite furnace AAS. It is thus applicable to the analysis of most elements in marine particulate matter provided there is a sufficient amount of the element present in the sample digest. ICP-ES has the advantages of multi-element capability, high sample throughput, and large working range.

5.3. Inductively-Coupled Plasma Mass Spectrometry (ICP-MS)

This method combines mass spectrometric element detection with the argon plasma ion source. It has the advantages of high sensitivity, high sample throughput and multi-element capability. ICP-MS is highly suitable for elements of atomic mass greater than 80, including the rare earths and platinum group metals. However because of spectral interferences in the argon plasma, it cannot be easily applied to analysis of Ca, P, Fe and Si. Chloride rich matrices can pose isobaric interferences. Isotope dilution analysis may also be performed by ICP-MS with routine precisions of 1-3%, not as precisely as is possible by TIMS. The application of ICP-MS to marine particle analysis is discussed in this chapter by Falkner, page 243.

5.4. Thermal Ionization Mass Spectrometry (TIMS)

In this technique, a purified solution of the analyte is applied directly to the filament of a mass spectrometer and ions generated for mass analysis by resistively heating the filament. The technique is most commonly applied to the measurement of isotopic ratios, e.g. for B, Th, Sr, Nd, U and Pb. For elemental analysis, quantification is carried out by isotope dilution. Detection limits are often limited by blanks, with values in the low picogram range having been reported for REE, U, Th and Pb.

5.5. Instrumental Neutron Activation Analysis (INAA)

This multi-element method involves neutron activation of the sample in a pelletized form followed by high resolution gamma-counting. The procedure is discussed in this volume by Fleer and Bacon, page 223. INAA is not suitable for the analysis of Pb, Cd, Hg, Si and Zr. Collier (this chapter, page 235) presents a comparison of INAA with AAS for marine particle analysis.

5.6. Radiochemical Analysis and Nuclear Counting

Counting techniques are used for analysis of most natural radionuclides. For isotopes with half-lives longer than that of Ra-226 (1620 years), it is considered preferable to use TIMS rather than nuclear counting. Techniques used for marine particles are summarized in this chapter by Fleer and Bacon, page 223.

5.7. X-Ray Fluorescence (XRF)

Thin-film X-ray fluorescence is ideally suited to the non-destructive analysis of marine particles collected by filtration methods. The technique is capable of detecting most elements of atomic number greater than 11. XRF is available in two different operating modes, wavelength-dispersive XRF in which elements are analyzed sequentially and energy-dispersive XRF in which x-ray emission from all elements is determined simultaneously using a multi-channel counter. Detection limits for most elements are 1 ppm in the sample by wavelength-dispersive XRF and 10 ppm by energy-dispersive. In this chapter, XRF techniques applied to marine particles are discussed by Feely et al., page 251,

Sholkovitz, page 295, and Skei, page 303. The main elements determined are Al, Si, P, K, Ca, Ti, V, Mn, Fe, Cu, Sr, Br, Ba and I.

5.8. Miscellaneous Techniques

Colorimetric and spectrofluorimetric techniques are applicable to many of the more abundant elements in marine particulate matter, e.g. Al, Si, Fe and Mn. For cosmogenic and anthropogenic isotopes present in extremely low concentrations, accelerator-mass spectrometry can be applied. AMS is presently used for the analysis of ^{14}C, ^{7}Be and ^{10}Be, and extension of the technique to other isotopes is anticipated.

References

Bacon, M. P., and R. F. Anderson, Distribution of thorium isotopes between dissolved and particulate forms in the deep sea, J. Geophys. Res., 87, 2045-2056, 1982.

Balistrieri, L., P. G. Brewer, and J. W. Murray, Scavenging residence times of trace metals and surface chemistry of sinking particles in the deep ocean, Deep-Sea Res., 28, 101-121, 1981.

Bewers, J. M., P. A. Yeats, S. Westerlund, B. Magnusson, D. Schmidt, H. Zehle, S. S. Berman, A. Mykytiuk, J. C. Duinker, R. F. Nolting, R. G. Smith, and H. L. Windom, An intercomparison of seawater filtration procedures, Mar. Poll. Bull., 16, 277-281, 1985.

Bishop, J. K. B., and J. M. Edmond, A new large volume filtration system for the sampling of oceanic particulate matter, J. Mar. Res., 34, 181-198, 1976.

Cantillo, A. Y., Standard reference materials for marine science, NOAA Technical Memorandum NOS OMA 51, NOAA NOS, Rockville, MD, 1989.

Clegg, S. L., and J. L. Sarmiento, The hydrolytic scavenging of metal ions by marine particulate matter, Prog. Oceanogr., 23, 1-21, 1989.

Clegg, S. L., and M. Whitfield, A generalized model for the scavenging of trace metals in the open ocean-I. Particle cycling, Deep-Sea Res., 37, 809-832, 1990.

Conte, M. H., The biogeochemistry of particulate lipids in warm core Gulf Stream ring systems, Ph.D. thesis, 584 pp., Columbia University, 1989.

Goldberg, E. D., Marine geochemistry. I- Chemical scavengers in the sea, J. Geol., 62, 249-265, 1954.

Honeyman, B. D., and P. H. Santschi, A Brownian-pumping model for oceanic trace metal scavenging: Evidence from Th isotopes, J. Mar. Res., 47, 951-992, 1989.

Hovind, H., and J. Skei, Intercomparison exercise on the trace metals in suspended particulate matter (SPM). Phase 1 The capabilities of the laboratories to analyze small samples accurately and reproducibly, Preliminary report presented at WGMS meeting in Lisbon Portugal, April 2 to 5 1990, WGMS 3/1/1990, 1990.

Krauskopf, K. B., Factors controlling the concentrations of thirteen rare metals in seawater, Geochim. Cosmochim. Acta, 9, 1-32, 1956.

Murnane, R. J., J. L. Sarmiento, and M. P. Bacon, Thorium isotopes, particle cycling models, and inverse calculations of model rate constants, J. Geophys. Res., 95, 16,195-16,206, 1990.

Nozaki, Y., H. S. Yang, and M. Yamada, Scavenging of thorium in the ocean, J. Geophys. Res., 92, 772-778, 1987.

Schindler, P. W., Removal of trace metals from the oceans: a zero order model, Thalassia Jugosl., 11, 101-111, 1975.

Wong, C. S., E. Boyle, K. W. Bruland, J. D. Burton, and E. D. Goldberg, Trace Metals in Sea Water, 920 pp., Plenum Press, New York, 1983.

Yeats, P. A., and J. A. Dalziel, ICES intercalibration for metals in suspended particulate matter, J. Cons., Cons. Int. Explor. Mer., 43, 272-278, 1987.

PARTICULATE MATTER DIGESTION (FROM MG TO 10'S OF G) AND RADIONUCLIDE BLANKS

M. Q. Fleisher and R. Anderson

Lamont-Doherty Geological Observatory (LDGO), Columbia University, Palisades, New York 10964

1. Introduction

Our laboratory is involved in trace metal and radionuclide research that requires the total digestion of particulate matter samples ranging from a few mg's of sediment trap material to 10 or more grams of surface sediments necessary for Pu measurements using alpha spectrometry. This note illustrates a sample digestion technique that can be modified to process samples where the sample size ranges over three orders of magnitude. The analytical techniques we currently use include atomic absorption spectrometry (flame and graphite furnace), alpha spectrometry, and DC plasma spectrometry. We are just beginning to exploit thermal ionization mass spectrometry for U-Th series and Pu measurements.

A general procedure is outlined below first for dissolving a 1-2 gram sample, then for a 5-20 mg sample. Our procedure at Lamont follows the general procedure outlined in Anderson and Fleer [1982]. Please see this document for additional details. Using this procedure, we have never encountered significant blanks in measuring U, Th, Pa, ^{210}Pb, or Pu by alpha spectrometry. In this note we also give a brief description of some future modifications which will reduce our blanks to levels that will allow us to more fully utilize the mass spectrometric methods for trace analyses.

2. Basic Digestion Technique Using 1 to 2 g of Material

The sample is weighed in an acid cleaned 400 mL Teflon TFE beaker. Abrasive cleansers or brushes should not be used to clean the inside surfaces of these beakers. When using isotope dilution techniques, our spikes are calibrated gravimetrically, rather than volumetrically. It is important to add the spikes, which are in acid solutions, to the beaker so as not to contact the sample directly, especially when processing carbonate samples. The following digestion procedures use strong acids and require a perchloric acid hood. Extreme care should be exercised throughout this procedure to avoid acid burns. This precaution is especially true when using HF; it should always be added slowly to digestion solutions.

The sample is brought into solution as follows:

1) Cover the sample with distilled water and add about 1 mL of concentrated nitric acid (HNO_3). This solution attacks carbonates, metal oxide coatings, and labile organic matter. Samples can sit overnight in this state. The sample is then heated uncovered, until the sample is dry or nearly so.
2) HNO_3 and perchloric acid ($HClO_4$) are added to the sample in a mixture of 3 mL HNO_3:12 mL $HClO_4$. The sample is then heated until dense, white $HClO_4$ fumes are visible. Cover the sample with a Teflon watch glass to reflux for thirty minutes. This step digests everything except silicates and the more refractory mineral phases.
3) At this point, the sample is taken from the hot plate, cooled for a minute or so, and then the walls of the beaker are washed down with hydrofluoric acid (HF). HF attacks the silicate phases and drives off Si as a volatile component in the reaction. Take note: if Si is to be measured in the samples in addition to radionuclides, this method cannot be used. After adding the HF, the sample is heated to $HClO_4$ fumes a second time. The HF addition is repeated and the sample is heated to $HClO_4$ fumes a third time.
4) The sample is covered with a Teflon watch glass and refluxed for an additional fifteen minutes.
5) The watch glass is removed and the walls of the beaker are washed down with HNO_3. The sample is now heated to $HClO_4$ fumes a fourth time to drive off the last of the HF (care must be taken not to precipitate fluorides, especially with larger sample sizes, since fluorides scavenge metals quite efficiently). The walls of the beaker are washed down with HNO_3 a second time.
6) The sample is now heated to a blob of $HClO_4$ residue. When the volume of the blob is about 2 mL, a few mL of HNO_3 is added, quickly followed by 20 to 40 mL of distilled water.
7) This solution is then heated for a few minutes to help dissolve the blob of $HClO_4$ residue.

Marine Particles: Analysis and Characterization
Geophysical Monograph 63

At this point, the next step is determined by the kind of analyses to be carried out. For trace element analyses, the sample is centrifuged to remove any very refractory mineral phases that remain such as Ti- and Zr-rich materials. The sample is then diluted to a known volume prior to analysis by flame AAS. For radionuclide analyses, an Fe coprecipitation step takes place followed by radiochemical purification.

3. Modification For Smaller Samples

In order to study the major and trace element composition of sediment trap samples using DC plasma spectroscopy, it was necessary to modify the above method to accommodate the smaller amounts of material available, aiming for a sample size of 5 mg. We modified the above method by changing the volumes of acids used, and controlling the total volumes more accurately using pipettes to add the acids. To test the validity of the method, 100 mg aliquots of both standard materials and surface sediments were digested using our acid dissolution procedure and the lithium metaborate fusion method used by Dr. Langmuir's group. The agreement between these batches was within the few per cent precision for each element of interest. This result encouraged us to then compare the results from digesting 5 mg samples by our modified method with the fusion procedure. Since the results of this comparison agreed within 10% or better, we have used this method to prepare sediment trap samples and other samples where there is only a limited amount of material for DC plasma spectroscopy analysis.

All of the acids used for the radiochemical analyses by alpha spectrometry described above have been reagent grade acids. To analyze small sediment trap samples by DC plasma spectroscopy, and to more fully realize the capabilities of the growing number of solid source mass spectrometers at LDGO, we are beginning to test the degree to which blank levels can be reduced by working in a class 100 clean lab or in a laminar flow bench, by using redistilled acids and water to lower reagent blanks, and by adopting procedures that generally work on a smaller scale (e.g., less reagents, smaller ion exchange columns, etc.).

The need for improved blanks is best illustrated using ^{238}U and ^{232}Th, where detection limits are reduced from $\approx 10^{-7}$ grams by alpha spectrometry to less than 10^{-12} g by mass spectrometry. Some of the early mass spectrometric analyses completed at LDGO, which included seawater samples processed for ^{232}Th and small sediment trap samples processed for ^{238}U by "old" methods, each had blank levels of $\sim 10^{-9}$ g. Adoption of "clean" techniques reduced blanks to 0.5 to 2×10^{-11} g for ^{238}U and $\approx 5.0 \times 10^{-12}$ g for ^{232}Th.

Additional improvements in the analysis of refractory elements such as Th by mass spectrometry are obtained by using secondary ionization techniques (SIMS). Whereas detection limits are greatly improved through increased ionization efficiency, new interferences are introduced as well because SIMS produces ions of various oxides of many metals that may remain with Th in the sample, metals that presented no problem for analysis by alpha spectrometry or by thermal ionization mass spectrometry. Consequently, analysis by SIMS requires not only that low blanks be obtained for the elements of interest, but careful purification as well from other metals whose oxides introduce isobaric interferences with the mass spectra of the Th isotopes or other elements of interest.

References

Anderson, R. F., and A. P. Fleer, Determination of natural actinides and plutonium in marine particulate material, Anal. Chem., 54, 1142-1147, 1982.

NOTES ON SOME TECHNIQUES OF MARINE PARTICLE ANALYSIS USED AT WHOI

Alan P. Fleer and Michael P. Bacon

Department of Chemistry, Woods Hole Oceanographic Institution, Woods Hole, Massachusetts 02543

1. Introduction

Methods of particle analysis that have been used in our laboratory can be placed into three categories: (1) gravimetric determination of total mass (dry weight), (2) determination of elemental composition by instrumental neutron activation analysis (INAA), and (3) determination of radionuclide content by radiochemical analysis and alpha- or beta-counting or by non-destructive gamma-counting. The methods have been applied both to suspended particles collected by filtration of seawater and to settling particles collected with sediment traps. Much of our methodology is documented in published literature. The material that follows updates the published information, discusses some special problems, and notes certain precautions that need to be observed.

2. Gravimetric Analysis

Early work used a stacked pair of 47 mm Millipore HA filters (0.45 μm pore size) so that the washout loss of soluble material from the bottom filter could be used to correct for losses from the top filter [Spencer and Sachs, 1970]. Shortly later the superiority of Nuclepore filters for gravimetric work was realized. Brewer et al. [1976, 1987] describe the techniques used during the Atlantic GEOSECS expedition, which involved pressure filtration through 37 mm Nuclepore filters (0.6 μm pore size). Some special problems with this technique should be noted. The effect of static electricity on the weighing of plastic filters stored in polystyrene boxes must be appreciated. Anti-static strips are effective at reducing this problem. Our anti-static strips contain 500 μCi ^{210}Po and are supplied by NRD, Inc., Grand Island, NY (Model 2U500). Two strips are used to speed the process when a batch of filters is being weighed as it takes time for the static charge to be neutralized. Each filter is transferred by forceps from the box to the first strip, then to the second strip, and finally to the balance pan of a Mettler M5 microbalance, and read to the nearest μg. It is advisable that a consistent "filter shiny side up" convention be adopted to prevent possible confusion. It is also extremely important that unused filters be set aside as weighing blanks. Even in a nominally humidity-controlled room, blanks can be $\pm$ 20 μg for a 15 mg filter. Blanks are fairly consistent, $\pm$0.2 μg, within any one filter lot, but can differ significantly between lots treated exactly the same. Thus careful recording of lot numbers is essential. More recently, to reduce the amount of sample handling, the pressure-filtration technique has been abandoned in favor of vacuum filtration directly from 30 L Niskin bottles through 47 mm Nuclepore (0.4 μm pore size) aerosol filters held in Millipore in-line filter holders into calibrated and evacuated glass carboys. Volumes can be determined to better than $\pm$ 0.2 liters, resulting in a less than 2% error. Alternatively, weights can be taken with a digital scale/computer system we have built for weighing water samples at sea. Sample volumes are generally 15 to 16 L, but sometimes less for surface or inshore waters. Every fifth filter is set aside as a weighing blank.

3. Instrumental Neutron Activation Analysis

The use of INAA for the determination of La, Au, Mn, Hg, Cr, Sb, Ag, Sc, Fe, Zn, Co, Ba, Ti, Sr, Mn, Mg, Cu, V, Al, Ca and I is detailed in Spencer et al. [1972, 1978] and in the manuscript by Brewer et al. [1976], with changes in the procedure as noted here. Short-irradiation standards are now made up chlorine-free to reduce the effects of the ^{38}Cl ($t_{1/2}$ = 37.2 min) gamma peaks at 2.17 and 1.64 MeV. Although these chlorine peaks do not specifically interfere with other peaks, they do increase the background continuum and the magnitude of the dead-time corrections. To the aliquoted stock solutions, redistilled nitric acid is added. The solution is heated in a proper chemical hood to a small volume to drive off the HCl, and subsequent dilution and pipetting of aliquots onto filters are as before.

Another problem with the short-irradiation standard was that the stock Mn standard solution was originally prepared from potassium permanganate, the Mn reduced with SO_2, and the solution then boiled to expel the excess SO_2. It was found that precipitation of barium sulfate occurred and led to erratic results. Preparing the Mn standard separately and aliquoting onto the filter separately eliminated this problem.

Marine Particles: Analysis and Characterization
Geophysical Monograph 63

In the early 1980's a change in the manufacturing process for Nuclepore filters resulted in significant contamination with bromine. The activation products ^{80}Br ($t_{1/2}$ = 17.6 min) and ^{82}Br ($t_{1/2}$ = 6.1 min) with gamma peaks at 616 and 777 keV, respectively, made activation analysis by short irradiation very difficult for low-energy gamma emitters Ba, Ti, Sr and I. The old stock number for the 47 mm (0.6 μm pore size) clear plain regular filters was N060CPR04700, and lots 73F1I 15 and 16, among others, had very low bromine. The new stock No. 111108, lot 7302A10, was contaminated. A switch was made to Nuclepore 0.4 μm trace-metal-free aerosol filters, Stock No. 111130. These were found to have acceptable levels of bromine, but still contained much more bromine than the original filters.

In earlier work hardened steel pellet presses were used to pelletize the filters for activation analyses. These presses contributed to blank levels, particularly for iron, manganese, and vanadium. Later work was done with presses made of delrin, with 1/8 and 5/32 inch bores to accommodate 37 and 47 mm filters, respectively (Figure 5-1). To make the pellet, the filter is carefully packed into the hole in the cylinder and the top is pressed down with about 100 lbs. of force. The pellet is removed by withdrawing the top and the slotted spacer and then by pushing the cylinder onto the base. There is no measurable contribution of elements of interest to the blank. Delrin will tolerate brief immersion in 1 M HCl for cleaning.

In preparing ground sediment-trap samples for activation analysis, material is weighed onto a creased Nuclepore filter, very carefully folded with stainless steel forceps and gloved hands, and pelletized as above. Triplicate weighings and analyses of sediment-trap samples, greater than 100 μg, are ordinarily performed.

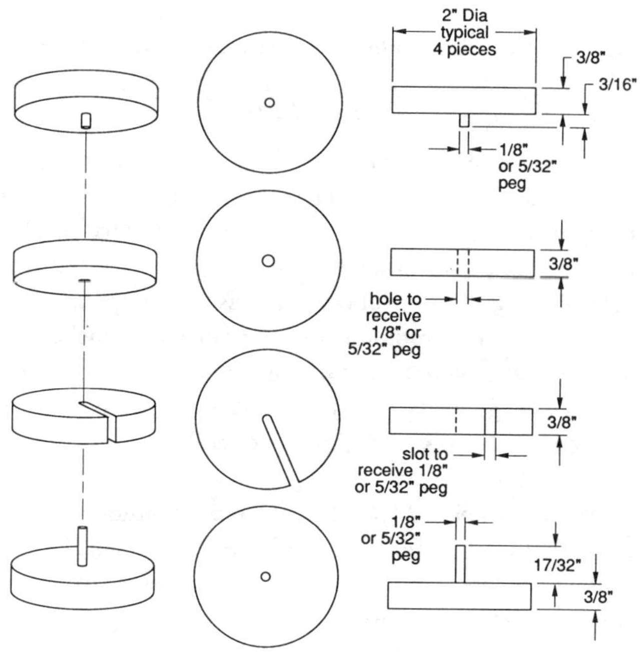

Fig. 5-1. Drawing of delrin pellet press used to prepare particle sample on Nuclepore filters for INAA (after a design by B. Ray, University of Rhode Island).

4. Radiochemical Analysis and Nuclear Counting

Procedures for measuring Th isotopes, U isotopes, ^{231}Pa, ^{210}Pb, ^{210}Po, and $^{239+240}Pu$ are described by Anderson and Fleer [1982], Fleer and Bacon [1984] and Fleer (this chapter, page 227). In the analysis of ^{210}Pb and ^{210}Po, tracer ^{208}Po has been replaced with ^{209}Po, resulting in better resolution of the two peaks in the alpha-particle spectrum and insignificant tailing corrections. In the ion exchange procedure for thorium, the most critical step is the evaporation of the chloride column effluent and its preparation for loading on the nitrate column. Both Si and Al pass through the chloride column with the thorium. If this fraction is allowed to precipitate or go to dryness in the presence of nitric acid, it may be impossible to re-dissolve in 8 N HNO_3. In the analysis of protactinium, glass econo-columns have been replaced by 8 mm x 10-cm polyethylene columns with polyethylene frits. The combined Pa fractions are taken to dryness, the sides of the beaker washed down 3 times with concentrated nitric acid, and the resulting solution diluted with an equal volume of water. The solution is then loaded on a nitrate column, the column rinsed with 9 N HCl to remove traces of thorium, and the Pa eluted with 9 N HCl plus 0.13 N HF. The eluate is evaporated to near dryness and electroplated in the standard way with the addition of an equal activity of ^{241}Am to serve as an internal standard. These changes are designed to: 1) reduce the significance of uneven distribution of Pa on the planchet surface, 2) reduce variations in alpha detector efficiency, and 3) eliminate solvent extraction and evaporation steps from the original procedure.

Another method for ^{234}Th determination is by gamma-counting of the 63 and 93 keV doublets having abundances of 3.9 and 3.0%, respectively. We use a Canberra GCW2022 high purity germanium closed end coaxial well detector of 98 cc active volume and a 20 x 38-mm well in a high purity magnesium end cap. Although the efficiency for counting the thorium gammas is high, ~ 55%, and geometry effects on less than 2 mL volume are small, the low gamma abundance, combined with relatively high background due to the large crystal size, limits its use to fresh large volume particles or sediment trap materials.

The report by Sachs et al. [1989] describes a system for in situ collection of large-volume particles on Nuclepore filters or Cuno polypropylene cartridges of 0.5 or 1 μm nominal rating (Model DPPPZ or DPPPY, respectively]. The Nuclepore filters are analyzed by the procedure of Anderson and Fleer [1982], with modifications described above. Cuno cartridges are dried at 100° C, melted in a 400 mL glass beaker at 200° C, and ashed at 500° C. In ashing, the temperature is slowly raised to 400° C, maintained there for at least an hour, slowly raised to 500° C, and again maintained there for an hour. To prevent fire, the muffle furnace is not opened during the high temperature steps. The pause at 400° C is to minimize the chances of the polypropylene bubbling over. The ash is digested in a perchloric acid hood in the glass beaker with 40 mL perchloric acid and 20 mL nitric acid after the addition of the appropriate spikes. The sample is transferred to a 250 mL Teflon beaker, the glass being rinsed with 6 N HCl plus 0.26 N HF to remove the 10% or greater amount of protactinium that sticks to the glass. Twenty-five mg iron as $FeCl_3$ solution, previously cleaned by extraction into isopropyl ether and back extraction into 0.1 M HCl, and 10 mL 49% HF are added, and the sample is heated to fuming perchloric acid. The sides of the beaker are washed down with nitric acid followed by heating to perchloric fumes three times. Swirling the sample helps prevent it from bumping. Extreme caution is required during this procedure.

In addition to making practical the sampling of large volumes of water, the in situ filtration method avoids the possibility of errors due to enclosure of the water sample in a bottle or to the delay and additional sample handling that often accompany on-deck filtration. For example, delay in filtration of seawater samples collected with bottles may result in the transfer of significant amounts of ^{210}Pb and ^{210}Po from dissolved to particulate forms [Bacon, 1975].

Livingston [1985] has reported a comparison between Nuclepore and Cuno filters, both of 1 μm nominal pore size, in their efficiency for collecting particulate Th, Pu, and Am by large volume, in situ filtration. The results show that the Cuno filters consistently collected an average of 60% more particulate activity than did the Nuclepore filters. Cuno filters may act as depth filters and consequently collect sub-micron particles more efficiently.

5. Other Notes on Sediment-Trap Samples

Silicon is measured in sediment-trap samples by atomic absorption spectroscopy after a fusion and dissolution procedure. About 5 mg ground sample is mixed with 200 mg lithium tetra- and metaborate [Johnson Matthey Spectraflux 100A Lot 54899/C) in a 1 mL platinum thimble, heated in a Meeker burner oxidizing flame for a few minutes. Care must be taken not to volatilize any lithium, which would be indicated by a red emission from the flame. The resulting glassy material is then dissolved in 3% HNO_3 added slowly to the thimbles with stirring by micro stir bars, the solution transferred into tared 30 mL plastic bottles, the stir bar being held in the thimble with a magnet, and taken to a final volume of 15 mL. Aluminum and potassium are also measured to compare with the activation analysis results. This solution is also suitable for analysis of other elements by ICP emission spectroscopy.

Most of the sediment-trap samples analyzed in this laboratory have been obtained from S. Honjo's program. A detailed account of their sample handling procedures will be available (S. Manganini, in preparation). Trap samples over the years have been collected in various sealed containers: 1 liter acrylic, 1 liter PVC, 1 liter Belart polypropylene bottle, 250 mL Nalgene wide mouth HDPE, and 250 mL IEC polypropylene centrifuge tubes, roughly in chronological order. A sample of the supernate is filtered and saved for nutrient, inorganic and dissolved organic carbon analyses. The remaining supernate is used in 1 mm sieving of the sample. The two size fractions are then split in quarters, the less than 1 mm fraction with a rotating delrin vane splitter with Plexiglas trays and polypropylene spigots into disposable polypropylene cups and the greater than 1 mm fraction with a plankton splitter.

The seawater used in splitting the samples has been filtered Sargasso Sea or Vineyard Sound (ESL) surface water that is filtered again just before use, after treatment with one of the following: (no treatment), 2% sodium azide, saturated mercuric chloride, or 3% formalin with 0.1% sodium borate buffer, again roughly in chronological order. The less than 1 mm fraction can be further divided by quarters or tenths, the latter splitter using glass or plastic scintillation vials as collection containers. Large (~ 100 mg) duplicate fractions (three pairs) weighed in this laboratory have agreed to better than 3%.

We filter our split onto a 25 mm Nuclepore aerosol filter (0.4 μm pore size) Stock No. 110630, held in a

Millipore glass microanalysis frit support, Catalog No. XX1002500, with 15 mL funnel capacity. Samples are rinsed with 1 or 2 mL portions of distilled water three times to remove excess sea salt. It is important not to let the sample cake go dry as it will begin to contract and the wash will be less effective, and care must be taken to insure that the lighter pteropods and forams do not stick to the funnel wall or otherwise get omitted from the sample. The sample is dried at 60° C, cooled in a desiccator and weighed. The filter is carefully peeled away and the sample ground in an agate mortar. Even with very small (less than 5 mg) samples, the 25 mm filter can always be peeled away with minimal losses. When small samples are loaded on larger filters, the filter will not peel away, risking large losses and fractionation, or necessitating the inclusion of the filter in the sample analysis.

References

Anderson, R. F., and A. P. Fleer, Determination of natural actinides and plutonium in marine particulate material, Anal. Chem., 54, 1142-1147, 1982.

Bacon, M. P., Applications of Pb-210/Ra-226 and Po-210/Pb-210 disequilibria in the study of marine geochemical processes, Ph.D. thesis, 165 pp., Massachusetts Institute of Technology/Woods Hole Oceanographic Institution, 1975.

Brewer, P. G., D. W. Spencer, P. E. Biscaye, A. Hanley, P. L. Sachs, C. L. Smith, S. Kadar, and J. J. Fredericks, The distribution of particulate matter in the Atlantic Ocean, Earth Planet. Sci. Lett., 32, 393-402, 1976.

Brewer, P. G., D. W. Spencer, and P. E. Biscaye, Measurements of concentration of suspended particulate matter, in Chapter 1, Description of Sampling and Measurement Methods, GEOSECS, Atlantic, Pacific and Indian Ocean Expeditions, edited by H. G. Ostlund et al., National Science Foundation, Washington, D.C., 1987.

Fleer, A. P., and M. P. Bacon, Determination of lead-210 and polonium-210 in seawater and marine particulate matter, Nucl. Instr. Methods Phys. Res., 223, 243-249, 1984.

Livingston, H. D., 1984 Annual Progress Report, Subseabed Disposal Program, Contract 47-5722 with Sandia National Laboratories, 24 pp., 1985.

Sachs, P. L., T. R. Hammar, and M. P. Bacon, A large-volume, deep-sea submersible pumping system, WHOI Tech. Rept.-89-55, 53 pp., Woods Hole, Massachusetts, 1989.

Spencer, D. W., and P. L. Sachs, Some aspects of the distribution, chemistry, and mineralogy of suspended matter in the Gulf of Maine, Mar. Geol., 9, 117-136, 1970.

Spencer, D. W., P. G. Brewer, and P. L. Sachs, Aspects of the distribution of and trace element composition of suspended matter in the Black Sea, Geochim. Cosmochim. Acta, 36, 76-86, 1972.

Spencer, D. W., P. G. Brewer, A. P. Fleer, S. Honjo, S. Krishnaswami, and Y. Nozaki, Chemical fluxes from a sediment trap experiment in the deep Sargasso Sea, J. Mar. Res., 36(3), 493-523, 1978.

UPDATED DETERMINATION OF PARTICULATE AND DISSOLVED THORIUM-234

Alan P. Fleer

Woods Hole Oceanographic Institution, Woods Hole, Massachusetts 02543

The determination of particulate and dissolved ^{234}Th is similar to the procedure of Anderson and Fleer [1982]. Samples are collected using 30 L Niskin bottles with Teflon- or epoxy-coated internal springs. On deck, the sample is pumped with a delrin impeller pump through a 0.45 μm pore size 147-mm diameter Millipore filter and into a pre-rinsed 6 gallon plastic cubitainer held in a plastic milk crate. An in-line plastic water meter records volumes in gallons. The particulate sample filter is folded twice and stored in a polyethylene sample bag. To the ~20 L filtered sample is added: 30 mL reagent grade 16 N HNO_3; 500 mL ^{230}Th tracer of ~30 dpm mL^{-1}; and 5 mL 50 mg mL^{-1} iron carrier previously cleaned by extraction into isopropyl ether from an 8 M HCl solution and back-extracted into 0.1 M HCl. The acidified sample is allowed to equilibrate for from one day to a maximum of several days. The sample is weighed on a Heathkit digital scale and the pH is adjusted to approximately 8 with about 40 mL 10 M NH_4OH to precipitate iron hydroxide, which carries the thorium and uranium from the solution. The precipitate is allowed to settle for 12 to 24 hours. The supernate is drawn off, and the precipitate is spun down in a centrifuge tube to about an 8 mL volume. The precipitate is resuspended in distilled water and spun down again, then dissolved in three times its volume with 12 N HCl to make a 9 N HCl solution. A 1.5 cm x 12 cm ion exchange column is filled with AG1 X 8 100-200 mesh resin and conditioned with 9 N HCl. The sample solution is run slowly through.

The iron and uranium are retained on the column; the thorium passes through in the effluent and is collected in a Teflon beaker. An additional 30 mL 9 N HCl is used to rinse the centrifuge tube and resin. The collected thorium-containing effluent is evaporated to near dryness and cooled slightly and 3 to 4 times that volume of concentrated HNO_3 is added. After an initial effervescence subsides, the sample is heated, the sides of the beaker being rinsed down twice with concentrated HNO_3, until all the HCl has evaporated but before precipitates begin to form. While still hot, the sample, now in concentrated HNO_3, has an equal volume of distilled water added, resulting in an 8 N HNO_3 solution. If the solution is not perfectly clear, more 8 N HNO_3 can be added and/or the solution heated slightly.

A 0.7 cm x 12 cm ion exchange column is filled with AG1 X 8 100-200 mesh resin and conditioned with 8 N HNO_3. The Th sample solution is run through, plus an additional 20 mL 8 N HNO_3 to rinse the beaker and resin. The effluent is discarded and the thorium is eluted with three 10 mL portions of 9 N HCl. The eluate is collected in a Teflon beaker. The solution is heated to near dryness with the addition of a few mL concentrated HNO_3. If the final volume of sample is less than drop-size before becoming solid, then as it starts to solidify, only 1 mL 0.01 M HNO_3 is added. If the sample begins to solidify as a drop sized or larger mass, a repeat of the second nitrate column is necessary to purify the sample sufficiently. To the sample in 0.01 M HNO_3 is added 1 mL 2 M NH_4Cl. An additional 2 mL 2 M NH_4Cl is used to transfer the sample to an electroplating cell with the following characteristics: 1 inch diameter polished silver planchet as the cathode with 3/4 inch diameter center portion exposed, Teflon walls, a coiled platinum anode 3/16 inch from silver disk, and a Teflon stir bar above the anode to prevent bubble formation on the electrodes. A 0.8 amp current at 6 volts is applied while stirring. The potential may initially drop but will begin to rise to 6 volts. As the electroplating nears completion, the current will drop. The procedure is complete when the current reaches 0.2 or 0.3 amp, usually after one hour. However, plating should be terminated if the current drops and then begins to rise, for the platinum anode is dissolving and plating on the sample. To end the plating, a few drops concentrated NH_4OH is added to the plating cell and the leads to the electrodes are disconnected. The plating cell is rinsed out and the silver disk removed, rinsed, and gently flamed, care being taken not to melt the silver.

To the particulate sample in a Teflon beaker, 1 mL 50 mg mL^{-1} iron carrier, 0.5 mL ^{230}Th tracer, 10 mL HNO_3, 3 mL concentrated $HClO_4$, and 2 mL HF are added. The sample is heated and evaporated to dryness, redissolved in 2 N HCl, and the iron precipitated with

Marine Particles: Analysis and Characterization
Geophysical Monograph 63

concentrated NH_4OH. The precipitate is centrifuged, dissolved with 9 N HCl, reprecipitated with concentrated NH_4OH, centrifuged and decanted to remove any traces of $HClO_4$ in the sample. The procedure from this point on is the same as for the dissolved thorium samples except that a smaller 0.7 cm x 12 cm first column can be used due to the smaller amount of iron added to the sample.

The recovery of the ^{230}Th tracer is measured by alpha counting in a Nuclear Measurements Corporation 2-π gas-flow proportional counter at 1100 volts bias. A ^{230}Th standard is counted between samples to correct for short-term variations in counting efficiency.

^{234}Th is measured as the beta activity of ^{234}Th and its short-lived beta emitting daughter ^{234}Pa ($t_{1/2}$ = 1.18 min). Beta activities are measured in low-background gas flow beta detectors with lead shielding and anti-coincidence [Lal and Schink, 1960]. The higher energy beta particles of ^{234}Pa are counted with about 50% efficiency, and the lower energy betas of ^{234}Th with less than 5% efficiency. The sample planchets are held in a delrin ring with a plastic window of about 10 mg mL^{-2}. Backgrounds range from 0.2 to 0.8 cpm and are measured weekly. An electroplated ^{238}U standard is counted weekly to correct for short-term variations in counting efficiency. For every sample set, two or more ^{234}Th standards are run to determine overall beta counter efficiencies. ^{234}Th standards are prepared by separation from a U standard solution (prepared from NBS 950A) by the same separation procedures discussed above and by using the same ^{230}Th tracer as in the samples.

Each sample is counted several times, and a plot of $e^{-\lambda t}$ vs. net cpm is prepared. Net cpm is background corrected, and t = 0 on the plating date. Three fits to the data are made, and in all the points are weighted according to their statistical errors. The first fit allows a y intercept other than zero. The second forces the fit through the origin. The third is a weighted average of the counts decay-corrected to the plating date. If there are no significant interferences from other beta emitters and no significant residual U in the sample, the three fits should be essentially the same. The y intercept in cpm at $e^{-\lambda t}$ = 1 is the net cpm at the time of plating.

For dissolved thorium samples, a correction is made for the ingrowth of ^{234}Th from ^{238}U between sampling and separation. Uranium concentrations are calculated from salinity [Chen et al., 1986].

The activities are calculated in the following manner for dissolved ^{234}Th samples:

$$^{234}\text{Th dpm kg}^{-1} = \frac{[(\text{net cpm at } t_3)/e^{-\lambda(t_3-t_2)}(\text{effic})(\text{recov})] - (^{238}\text{U})(\text{weight})(1-e^{-\lambda(t_2-t_1)})}{e^{-\lambda(t_2-t_1)}(\text{weight})}$$

where:

t_1 = time of sampling

t_2 = time of separation from U = time of first column

t_3 = time of plating

^{238}U = ^{238}U concentration in dpm/kg

λ = 0.028755 d^{-1}

weight = weight of sample in kg of seawater

effic = efficiency of beta detector

recov = chemical recovery of Th through the procedure

For particulate ^{234}Th samples:

$$^{234}\text{Th dpm/kg} = \frac{(\text{net cpm at } t_3)}{e^{-\lambda(t_3-t_1)}(\text{effic})(\text{recov})(\text{weight})}$$

References

Anderson, R. F., and A. P. Fleer, Determination of natural actinides and plutonium in marine particulate material, Anal. Chem., 54, 1142-1147, 1982.

Lal, D., and D. R. Schink, Low background thin-wall flow counters for measuring beta activities of solids, Rev. Sci. Instrum., 31, 395-398, 1960.

Chen, J. H., R. L. Edwards, and G. J. Wasserburg, ^{238}U, ^{234}U and ^{232}Th in seawater, Earth Planet. Sci. Lett., 80, 241-251, 1986.

GETTING GOOD WEIGHT

James K. B. Bishop

Lamont-Doherty Geological Observatory, of Columbia University, Palisades, New York 10964

1. Introduction

In spite of twenty years of progress there are significant, i.e. order of magnitude, problems remaining in the accurate determination of many fundamental aspects of suspended matter. In this paper, I suggest that many of these problems occur during the steps leading up to the isolation of particles on a filter and not necessarily during sample handling or analysis.

2. Bottles and Pumps

There are currently two dominant approaches to quantifying suspended particulate mass and elemental concentrations in the ocean. Sediment traps collect particulate matter but do not allow estimation of particle concentration. The first and most widely used method is the use of Niskin or Go-Flo bottles to isolate water at depth and return the water samples for filtration aboard ship [e.g. Brewer et al., 1976; Nelson et al., 1985; Murray et al., 1989]. Typical volumes filtered range from 1 to 30 L. A major factor facilitating the use of bottle techniques is the wide-spread availability of good water sampling equipment aboard oceanographic ships. The second approach utilizes electrically powered (battery or ship's electricity) in situ filtration devices which filter 1000 L volumes of water (some filter 20,000 to 30,000 L). There are many kinds of in situ pumps operated by individual investigators (Table 5-1). Due to the fact that filtration takes place at depth, there are no handling steps involved prior to the entrapment of particles on filter(s). Beyond this fundamental advantage over bottles, a major advantage of the in situ filtration approach is that the 30-1000 fold greater amount of material collected enables a vastly expanded suite of biological, morphological, inorganic and organic determinations to be made on each sample.

The issue of filter pore size and type versus particle catch will not be discussed here since there is an extensive literature in the early 1970's dealing with this topic. Also, current methods used employ filters in the 0.4 to 1 μm size range. Compared with most of the differences described below, this variation due to filter pore size is a relatively minor contributor to the story.

Marine Particles: Analysis and Characterization
Geophysical Monograph 63

3. Suspended Particulate Matter Mass

GEOSECS invested considerable effort in obtaining accurate representation of the suspended particulate matter (SPM) distributions in the ocean [Brewer et al., 1976]. Prior to this time there was considerable disagreement in estimates of even the concentration of suspended matter filterable in a liter of seawater. For example, estimates based on Russian data [e.g. Lisitzin, 1972] suggested that SPM levels were rarely lower than 1 mg L^{-1}. On the other hand, GEOSECS Atlantic data showed that surface SPM concentrations ranged from 50 to 100 μg L^{-1} in oligotrophic waters to nearly 1 mg L^{-1} in biologically productive areas. In the majority of the oceanic water column deeper than 100 m, SPM values were in the 10 μg L^{-1} range in oligotrophic regions and several tens of μg L^{-1} in more productive areas. As the sea floor is approached, concentrations rise to several hundred μg L^{-1} (and occasionally to as high as several μg L^{-1} in regions of strong bottom turbulence).

It turns out that the 10-100 fold differences between US and Russian SPM estimates may be attributable, in large part, to a sedimentation bias. This is illustrated by a comparison of different bottle methods carried out during a 1972 cruise of the R/V Dmitri Mendeleev in the eastern tropical Pacific.

The GEOSECS method involved collecting samples using 30 L Niskin bottles fitted with Teflon-coated internal springs. Ten to 20 L samples were drained from the Niskin into acid-cleaned, distilled water rinsed polyethylene containers and subsequently pressure filtered in a closed system under nitrogen through preweighed (5 μg precision) 0.4 μm Nuclepore (polycarbonate) membrane filters. The samples were extensively washed under suction (7-10 times, total = 10-50 mL) with distilled water to remove seasalt. GEOSECS filter handling was always performed in either laminar flow benches or in plastic glove boxes. The washing procedure had to be performed as soon as possible after collection to prevent precipitation of relatively insoluble sea salt components such as $CaSO_4$. Samples were air dried individually in the precleaned plastic boxes used to individually store the filters. A second method used during GEOSECS for near bottom samples differed only in the fact that 10-30 L volumes of

TABLE 5.1: CAPABILITIES OF LARGE VOLUME IN SITU FILTRATION SYSTEMS

System	Number in use	Depth Capability (m)	Largest Volume (Liters)	power source	weight in air (kg)	area projected to flow (cm^2)	time per sample (hrs)	Filter diam (mm)	Filter pore size (μm)	Filter type
1. BENTHOS	1	>5000	200	battery	~100	800	3	142	0.45	MP
2. GEOSECS	2-3	>5000	5,000	battery	~200	~2,000	8-12	293	1.2	T
								293	1.0	MP
3. LVFS	1	400*	25,000	ship's	430	4,600	3-4	293	1.0	2xGF
								293	53	N
4. LVFS-4	1	1500*	25,000	ship's	500	5,200	3-4	293	1	2xGF
								293	53.	N
5 . WHOI	2-3	>5000	5,000	battery	~200	n.a.	8-12	293	1.0	NP
								75	62.	N
								Cartridge	<1.	MnO_2/N
6. MULVFS	12	1000**	25,000	ship's	36	1,000	3-4	293	1.0	2xQF
								293	53.	N/SS
7. DWPS	1	>5000	1,000	battery	450	5,200	1	293	1.0	NP
							(3 max)	142	30.	P
8. WHISP	3-4	1500	4,000	propeller	~25	11,500	8-14	142	1.0	GF
								142	53	N
9. WHOI	8	6000	5,000	battery	~200	4,500	8-12	Cartridge.	0.5	PP
								Cartridge.	<0.5	2xMnO_2/PP
10. RAPPID	3-4	>5000	1,500	battery	~100	~1,000	3-4	142	1.0	NP
								142	53.	P
11. ESAUP	1	>5000	500	battery	~200	~4,000	1-2	293	1.0	MP
								Cartridge	<1.0	

Notes: * Depth capability is limited only by Electromechanical Cable length. **MULVFS being upgraded to 2000 m.
Filter type: MP - Millipore; T - Tamil POLY; GF - Glass Fiber Filters; NP- Nuclepore; MnO_2/N - manganese treated nitex; QF - Whatman quartz fiber filters; N/SS Nitex or Stainless steel mesh (organics); P -Polyester mesh; PP - Polypropylene.

Filtration Systems Listed in Table 1:

(1) Spencer and Sachs [1970] major and minor elements;
(2) Krishnaswami et al. [1976] deployed on trawl wire. radionuclides;
(3) Bishop and Edmond [1976], Bishop et al. [1977] end of special electromechanical (EM) cable. Major, minor and trace elements.;
(4) Bishop et al. [1980] special EM cable / 4 filters per deployment.;
(5) Bacon and Anderson [1982] on trawl wire, radionuclides.;
(6) Bishop et al. [1985] multiple pumps on EM cable. Winch and level wind required. Major, minor and trace inorganic and organic chemical species.;
(7) Simpson et al. [1987] conducting trawl wire/ 4 filters. rare earth elements.;
(8) Wakeham and Canuel [1988] trawl wire. towed. organic geochemistry.;
(9) Sachs et al. [1989] trawl wire. radionuclides.;
(10) Sherrell and Boyle [1988] on hydrowire. trace metals.;
(11) Fernendez, J. M. [1989] 2 Parallel Independently Controlled Pumps.

water were suction filtered directly from the Niskin bottles through an in-line filter holder into calibrated glass jugs used for radon extraction. These two schemes yielded indistinguishable SPM data (± 10-20%) with replicate samples at the same depth indicating a precision of ~ 5 μg L^{-1}. Quality control on the data was maintained through replicate sampling, unused filter blanks, and procedural blanks in which unused filters were taken through all of the steps used for samples. The first GEOSECS method was used during the Russian cruise with the exception that filter handling was not performed in a laminar flow bench environment but rather in a clean area of one of the ship's laboratories.

The Russian approach was to use a single 200 or 500 L stainless steel "bathometer" for water sampling. Once on board, 1-3 L samples were drawn into glass conical flasks from a spigot mounted in the bottom lid of the sampler. From there, they were carried into the lab and poured into an open filter funnel and drawn under suction through preweighed membrane filters, similar to Millipore, which had been preweighed (0.01 mg precision). Following this step, the samples were rinsed with distilled water and air dried.

The results using the US method gave SPM ranging from 5 to 10 μg L^{-1} in the intermediate water column and up to 100 to 200 μg L^{-1} in surface waters, more or less in agreement with GEOSECS Pacific SPM data and with the L-DGO nephelometer profiles obtained concurrently during the cruise. These SPM results were obtained in spite of the fact that filter handling was not performed in a laminar flow bench environment.

Later work showed that the GEOSECS sampling scheme was biased low due to sedimentation of particles within the water sampler prior to filtration. This fraction not exiting the side-mounted valves on the Niskin (and therefore not sampled) approaches 50% of the particles originally present in near surface waters and 10 to 20% deeper [Bishop and Edmond, 1976; Gardner, 1977]. The former study compared SPM profiles by in situ filtration with SPM by the second GEOSECS method. The latter study quantified the fraction of material found on the bottom of the water sampler.

We never did find out what the Russian SPM numbers were; however, based on these studies, I suspect that the major cause of difference between US and Russian SPM estimates originated in the fact that the 1 to 3 L sample probably also contained the sedimented material from the whole 200 to 500 L bathometer sample. This also may explain why the Russian systematics of major phase composition of SPM (e.g. percentages of opal, carbonate, organic carbon, etc.) appeared reasonable while concentrations did not.

SPM in the 1990's

Murray et al. [1989] in their paper on dissolved and particulate thorium isotope distributions in the eastern tropical Pacific show SPM profiles ranging from 600 to 700 μg L^{-1} in surface waters to 100 μg L^{-1} in the water column below 100 m. Their dissolved and particulate thorium data are in excellent agreement with other workers. However, it is not clear why SPM concentrations are 5 to 20 times higher than those obtained in the same ocean region using either GEOSECS or in situ filtration methodology. A contamination with residual sea salt is suspected.

In situ filtration methods [e.g. Bishop et al., 1977] are capable of quantifying SPM concentrations to a precision of several percent, independent of depth in the water column. This is because the volume of seawater which is filterable rises as particle concentration decreases. Within the constraints posed by sedimentation bias, the GEOSECS Atlantic SPM data and in situ filtration SPM data agree within a factor of 2.

4. Particulate Organic Matter

GEOSECS did not focus on estimating particulate organic matter (POM), particulate organic carbon (POC) or particulate organic nitrogen (PON). Estimates of POC published for the Hudson '70 section down the western Atlantic [Wangersky, 1976] nearly everywhere exceeded total SPM concentrations obtained during GEOSECS one year later [Brewer et al., 1976]. Methodology used during the Hudson '70 expedition for POC determination involved bottle collection, (gravity filtration of 4 L volumes of seawater through 0.8 μm silver filters) followed by drying and analysis by carbon analyzer. While the Hudson '70 POC results suggested a much higher SPM level than GEOSECS, Gordon [1977] filtered 10-30 L volumes of seawater directly from bottles and obtained POC results consistent with GEOSECS SPM levels (POC $\leq$ SPM/2.5).

Organic Matter in the 1990's

Is POC or PON reliably determined today? Recent comparisons among pump (in situ filtration) and bottle techniques during the Warm Core Rings Experiment (WCRE) suggests that it is difficult to answer this question with complete confidence. PON levels were up to 10 times higher from bottles than pumps; ^{15}N data showed up to 30 per mill differences; C/N ratios were up to a factor of 2 to 3 different [Altabet et al., in press,

1991]. Best agreement between methods was found in near surface waters. We don't know the exact reason for such large differences, but of potential problems examined (sample handling prior to filtration, filter pore size, pressure differential during filtration, particle loading on the filter, filter blanks, distilled water rising and the analysis method used), sample handling prior to filtration appeared to be the major factor.

5. Other Elements

Of the elements determined in GEOSECS samples by instrumental neutron activation analysis (Al, Au, Ba, Ca, Co, Cr, Cu, Fe, I, Mn, Ni, Sc, Sb, Zn) only Al, Ba, Ca, I, Sc and possibly Mn survived scrutiny of the past 2 decades of research [see, e.g. Landing and Bruland, 1987; Bishop and Fleisher, 1987; Bishop, 1989; Sherrell, 1989; Lewis and Landing, 1990]. Of this list, the most reliable GEOSECS data were obtained in the Atlantic, early in GEOSECS, since the Pacific data suffered from $CaSO_4$ precipitation artifacts. The major factor eliminating most of the other elements on the GEOSECS list was contamination from hydrowire, bottles, and ship. Methods of sample analysis do not appear to have been a major source of error since care was exercised in standardization and blanks.

Biogenic opal estimates from bottles have shown little evidence of either sample handling or contamination artifacts. Data from the 1970's [Copin-Montegut and Copin-Montegut, 1972; Nelson and Goering, 1977; Bishop et al., 1977] remain consistent with results from current methodology [Nelson et al., 1985; Bishop et al., 1986]. This is also independent of whether samples are bottle or pump derived. The GEOSECS samples, if they still exist, could be analyzed for Si with little or no problem apart from the sedimentation bias already described. Little problem also seems to exist so far with any of the particulate radiotracer data, e.g. ^{210}Pb, ^{210}Po, ^{234}Th, determined using either bottles or pumps since contamination sources are minor.

6. Summary

The GEOSECS data show that good results can be obtained for SPM and for a limited suite of other elements using bottles. However, the good results are obtainable only when extreme attention is paid to filter preparation, bottle preparation, water sample handling prior to filtration, sedimentation biases, procedural blanks, filter blanks and post filtration analysis.

I have used both bottles and pumps extensively for particle sampling. Large volume in situ filtration methodology is much more effective at returning consistent estimates of SPM properties and concentration than any method utilizing bottles. It does this mainly by virtue of the elimination of all handling of the water prior to the isolation of particles on filters and by returning so much sample that contamination artifacts are minimized. In situ filtration systems exist in a variety of configurations (Table 5-1) ranging in weight from 25 to 500 kg. The lighter pump units are as easy to handle as bottles and they have the distinct advantage of having a much higher SPM sampling rate/hour of ship time ratio than bottles. A wide deployment of pumps would contribute much towards an accurate representation of SPM in the ocean.

7. Bibliography

The following papers give details of the particle sampling and analysis techniques that have been employed by my laboratory.

7.1. LVFS Specific Papers

Bishop, J. K. B., The effect of filtration speed in large particle catch during large volume in situ filtration, EOS Trans. Am. Geophys. Union, 63, 46, 1982.

Bishop, J. K., and J. M. Edmond, A new large volume filtration system for the sampling of oceanic particulate matter, J. Mar. Res., 34, 181-198, 1976.

Bishop, J. K. B., J. M. Edmond, D. R. Ketten, M. P. Bacon, and W. B. Silker, The chemistry, biology, and vertical flux of particulate matter from the upper 400 m of the equatorial Atlantic Ocean, Deep-Sea Res., 24, 511-548, 1977.

Bishop, J. K. B., D. R. Ketten, and J. M. Edmond, The chemistry, biology and vertical flux of particulate matter from the upper 400 m of the Cape Basin in the S.E. Atlantic Ocean, Deep-Sea Res., 25, 1121-1161, 1978.

Bishop, J. K. B., R. W. Collier, D. R. Ketten, and J. M. Edmond, The chemistry, biology and vertical flux of particulate matter from the upper 1500 m of the Panama Basin, Deep-Sea Res., 27A, 615-640, 1980.

Bishop, J. K. B., J. C. Stepien, and P. H. Wiebe, Particulate matter distributions, chemistry and flux in the Panama Basin: response to -environmental forcing, Prog. Oceanogr., 17, 1-59, 1987.

7.2. MULVFS Papers

Bishop, J. K. B, The barite-opal-organic carbon association in oceanic particulate matter, Nature, 332, 341-343, 1988.

Bishop, J. K. B., and M. Q. Fleisher, Particulate manganese dynamics in Gulf Stream warm-core rings and surrounding waters of the N.W. Atlantic, Geochim. Cosmochim. Acta, 51(10), 2807-2826, 1987.

Bishop, J. K. B., D. Schupack, R. M. Sherrell, and M. Conte, A Multiple Unit Large Volume in situ Filtration System (MULVFS) for sampling oceanic particulate matter in mesoscale environments, in Mapping Strategies in Chemical Oceanography, edited by A. Zirino, pp. 155-175, Advances in Chemistry Series, Vol. 209, American Chemical Society, Washington, D. C., 1985.

Conte, M. H., The biogeochemistry of particulate lipids in warm-core Gulf Stream ring systems, Ph.D. thesis, 584 pp., Columbia University, 1989.

Conte, M. H., and J. K. B. Bishop, Nanogram quantification of lipid classes in oceanic particles by high-performance thin layer chromatography, Lipids, 23, 493-500, 1988.

7.3. Other Papers Related to SPM Analysis

Bishop, J. K. B., Determination of Ba in seawater using Vanadium/Silicon modifier and direct injection graphite furnace atomic absorption spectrometry, Anal. Chem., 62, 553-557, 1990.

Bishop, J. K. B., and P. E. Biscaye, Chemical characterization of individual particles from the nepheloid layer in the Atlantic Ocean, Earth Planet. Sci. Lett., 58, 265-275, 1982.

Acknowledgements. This work was supported by grant OCE 89-17784 from the National Science Foundation. LDGO contribution number 4786.

References

Altabet, M. A., J. K. B. Bishop, and J. J. McCarthy, Differences in particulate nitrogen concentration and isotopic composition for samples collected by bottles and large volume pumps in Gulf Stream warm core rings and the Sargasso Sea, Deep-Sea Res., in press, 1991.

Bacon, M. P., and R. F. Anderson, Distribution of thorium isotopes between dissolved and particulate forms in the deep sea, J. Geophys. Res., 87, 2045-2056, 1982.

Bishop, J. K. B., Regional extremes in particulate matter composition and flux: effects on the chemistry of the ocean interior, in Productivity of the Ocean Present and Past, edited by W. H. Berger, V. S. Smetacek, and G. Wefer, pp. 117-137, Dahlem Konferenzen, John Wiley and Sons Ltd., Chichester, 1989.

Bishop, J. K. B., and J. M. Edmond, A new large volume filtration system for the sampling of oceanic particulate matter, J. Mar. Res., 34, 181-198, 1976.

Bishop, J. K. B., and M. Q. Fleisher, Particulate manganese dynamics in Gulf Stream warm-core rings and surrounding waters of the northwest Atlantic, Geochim. Cosmochim. Acta, 51, 2807-2826, 1987.

Bishop, J. K. B., J. M. Edmond, D. R. Ketten, M. P. Bacon, and W. B. Silker, The chemistry, biology and vertical flux of particulate matter from the upper 400 m of the equatorial Atlantic Ocean, Deep-Sea Res., 24, 511-548, 1977.

Bishop, J. K. B., D. Schupack, R. M. Sherrell, and M. Conte, A multiple unit large volume in situ filtration system (MULVFS) for sampling oceanic particulate matter in mesoscale environments, in Mapping Strategies in Chemical Oceanography, edited by A. Zirino, Advances in Chemistry Series, Vol. 209, pp. 155-175, American Chemical Society, Wash. D.C., 1985.

Bishop, J. K. B., M. H. Conte, P. Wiebe, M. Roman, and C. Langdon, Particulate matter production and consumption in deep mixed layers; observations in a warm core ring, Deep-Sea Res., 33, 1813-1841, 1986.

Brewer, P. G., D. W. Spencer, P. E. Biscaye, A. Hanley, P. L. Sachs, C. L. Smith, S. Kadar, and J. Fredericks, The distribution of particulate matter in the Atlantic Ocean, Earth Planet Sci. Lett., 42, 399-411, 1976.

Copin-Montegut, C., and G. Copin-Montegut, Chemical analysis of suspended particulate matter collected in the northeast Atlantic, Deep-Sea Res., 19, 445-452, 1972.

Fernendez, J. M., ESAUP-6000, rapport provisoire, CEA-SMT/IFREMER CT, La Seyne Sur Mer, FR., 21 pp., 1989.

Gardner, W. D., Incomplete extraction of rapidly settling particles from water samples, Limnol. Oceanogr., 22, 764-768, 1977.

Gordon, D. C., Variability of particulate organic carbon and nitrogen on the Halifax-Bermuda section, Deep-Sea Res., 24, 257-270, 1977.

Krishnaswami, S., D. Lal, B. L. K. Somayajulu, R. F. Weiss, and H. Craig, Large volume in situ filtration of deep Pacific waters: mineralogical and radioisotope studies, Earth Planet. Sci. Lett., 32, 420-429, 1976.

Landing, W. M., and K. W. Bruland, The contrasting biogeochemistry of iron and manganese in the Pacific Ocean, Geochim. Cosmochim. Acta, 51, 29-43, 1987.

Lewis, B. L., and W. M. Landing, Dissolved and particulate trace metal speciation in the Sargasso Sea, EOS Trans. Am. Geophys. Union, 71, 153, 1990.

Lisitzin, A. P., Sedimentation in the world oceans, S.E.P.M., spec. publ. 17, 218 pp., 1972.

Murray, J. W., J. N. Downs, S. Strom, C.-L. Wei, and H. W. Jannasch, Nutrient assimilation, export production and Th-234 scavenging in the eastern equatorial Pacific, Deep-Sea Res., 36, 1471-1489, 1989.

Nelson, D. M., and J. J. Goering, Near-surface silica dissolution in the upwelling region off south west Africa, Deep-Sea Res., 24, 65-74, 1977.

Nelson, D. M., H. Ducklow, G. L. Hitchcock, M. A. Brzezinski, T. J. Cowles, C. Garside, R. W. Gould, Jr., T. M. Joyce, C. Langdon, J. J. McCarthy, and C. H. Yentsch, Distribution and composition of biogenic particulate matter in a Gulf Stream warm-core ring, Deep-Sea Res., 32, 1347-1371, 1985.

Sachs, P. L., T. R. Hammar, and M. P. Bacon, A large-volume deep-sea submersible pumping system, WHOI Technical Report 89-55, 53 pp., Woods Hole, Massachusetts, 1989.

Sherrell, R. M., The trace metal geochemistry of suspended oceanic particulate matter, Ph.D. thesis, 211 pp., MIT/WHOI Joint Program in Oceanography, Woods Hole, Massachusetts, 1989.

Sherrell, R. M., and E. A. Boyle, Suspended particulate trace metals distributions and implications for adsorptive equilibrium in the deep Atlantic and Pacific, EOS Trans. Am. Geophys. Union, 69, 1256, 1988.

Simpson, W. R., T. J. P. Gwilliam, V. A. Lawford, R. A. Lewis, and M. J. R. Fasham, An in situ deep water particle sampler and real-time sensor package with data from the Madiera Abyssal Plain, Deep-Sea Res., 34, 1477-1497, 1987.

Spencer, D. W., and P. L. Sachs, Some aspects of the distribution, chemistry, and mineralogy of suspended matter in the Gulf of Maine, Mar. Geol., 9, 117-136, 1970.

Wakeham, S. G., and E. A. Canuel, Organic geochemistry of particulate matter in the eastern tropical North Pacific Ocean: implications for particle dynamics, J. Mar. Res., 46, 183-213, 1988.

Wangersky, P. J., Particulate organic carbon in the Atlantic and Pacific oceans, Deep-Sea Res., 23, 457-466, 1976.

ANALYSIS OF PARTICULATE MATTER COLLECTED BY SEDIMENT TRAPS AND FROM SEDIMENT CORES

Bob Collier (with contributions from Jack Dymond, Roberta Conard, and Jim Robbins)
College of Oceanography, Oregon State University, Corvallis, Oregon 97331

1. Introduction

These contributions are primarily directed at elemental analyses for major and trace cations and silicon on particles collected by sediment traps and sediment cores; samples typically having at least several hundred milligrams available. The techniques outlined below and other wet chemical methods are reviewed in more detail in an OSU technical report by Robbins et al. [1984].

2. Filtration Methods and Materials

Typically, our group filters water samples for trace metal analyses immediately upon their arrival at the surface. The primary emphasis of these techniques focuses on the dissolved material. Since we have not had a full clean-lab available, we subsample the Niskins immediately using a semi-closed connection to large mouth bottles (LPE) which minimizes atmospheric exposure to contamination. The samples are then vacuum filtered within a portable laminar-flow hood (HEPA filtered) using plastic filtration "chimneys" (Millipore or Nuclepore) placed over a plastic vacuum chamber which contains the filtrate sample bottle. Filtration is carried out with 1 N HCl acid-leached Nuclepore or Poretics filters (0.4 μm) and the filtrate is collected directly into the final sample bottle. Both filtered and unfiltered subsamples are acidified to a pH$<$2 with 2 mL 6 N HCl L^{-1} sample using subboiling-distilled HCl, and the difference in concentration between the filtered and unfiltered sample is taken to represent an acid-labile particulate fraction.

These are the same filters that we have previously used for total suspended mass analyses from 30 L Niskin bottles and for INAA analyses of particles. Blanks are generally low for most elements although chromium has been reported to be a serious contaminant on Nuclepore filters (C. Measures, Univ. Hawaii, personal communication, 1990) which must be treated by a stronger, hot-acid leach such as 6 N HCl at 60° C. Note that such a leach considerably weakens the filters.

3. Loss of "Labile Substances" During Storage and Handling

Marine particulate matter tends to change its chemical composition after collection from the water column. This is due to a variety of factors which mostly fall into two categories: 1) on-going changes that were already operative in the water column such as excretion, enzymatic decomposition, etc.; 2) changes due to alterations of the temperature, pressure, re-equilibration of the particulate matter with its new "sampled" environment resulting in osmotic cell lysis/disruption, ion exchange, and contamination. These changes are most significant for the organic fractions associated with "fresh" biogenic samples such as plankton tows and may also be significant for minor elements associated with these organic fractions. These chemical changes translate into a loss or gain of some component in the sample when the sample is stored for some period of time in a liquid suspension followed by some final separation to a form where the particulate matter sample is "closed" analytically. This discussion focuses on this issue of loss/gain from the "total particulate matter" matrix and not on the equally important but separable question of the redistribution of elements between components or phases within the bulk matrix such as changes in oxidation state or phase recrystallization.

Sediment samples collected from cores are probably more "stable" with respect to these changes as long as they are stored wet, cold and closed from the atmosphere. However, I am aware of little quantitative evidence of this assumption. Since subsequent chemical analyses usually proceed on the total matrix including the pore waters, redistribution between these phases may not be significant to analyses focused on the total concentration of an element. Most of the particles sensitive to these losses are biogenic in origin. The lithogenous phases may also be labile to some degree, especially for minor elements such as P, Fe, Mn and exchangeable cations such as Ca. This lability may be of

Marine Particles: Analysis and Characterization
Geophysical Monograph 63

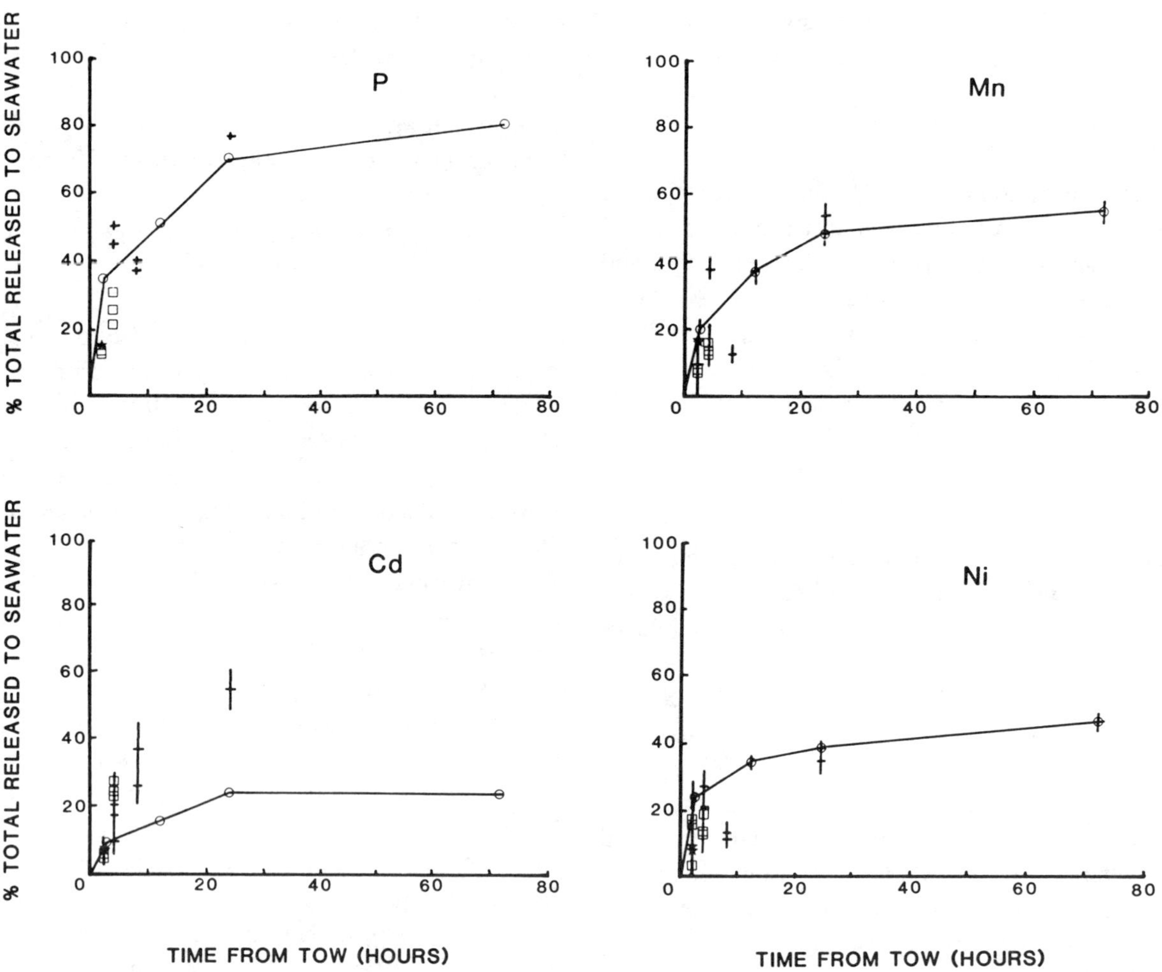

Fig. 5-2. Percentages of selected elements (P, Mn, Cd, Ni) in plankton released to the seawater suspension around the plankton as a function of time from collection of the tow [from Collier and Edmond, 1984]. Data points connected by the lines are replicate subsamples from a single plankton tow. Vertical lines represent the 95% confidence limits on the percentages. Plankton samples collected from various locations in the tropical Pacific.

particular importance for samples collected "near" their atmospheric or estuarine injection points.

3.1. Losses as a Function of Time

The release of components from particulate matter often proceeds according to some complex exponential approach towards an "equilibrium" partitioning between particle and solid. One outcome of this is that there is usually a set of changes which occurs very quickly (minutes to hours) and others that may manifest over long storage times (days to months). Figure 5-2 shows an example of the loss of P, Mn, Cd and Ni from particles to seawater as a function of time (hours) from collection by a net tow for a seawater/plankton suspension [Collier and Edmond, 1984]. Although organic carbon was not partitioned in this analysis, we can assume that it was also released (roughly in parallel with P). Figure 5-3 shows the fraction of total phosphorus found in the "solution fraction" of sediment trap samples collected and stored within the trap cup for several months before final recovery and isolation of the particles.

Obviously, the best approach to this problem is to complete the processing of the sample quickly with respect to the characteristic time scale for solubilization of the component of interest. Otherwise the analyst must determine the total concentration of material found in both the solution and solid fraction of the sample mixture which obviously requires that you keep both. This is now "routinely" done for elemental analyses of plankton [Collier and Edmond, 1984] and sediment trap materials [i.e. U.S. GOFS, 1989]. Still, this alternative may be difficult in cases when the particles are collected in situ or when the sample is "washed" or contained in a large volume of liquid (sensitivity problem). There is a progressive decrease in the labile fraction, starting from the plankton samples (Figure 5-2) followed by the trap samples of increasing depth (Figure 5-3). Independent of the mechanism behind this, which could be interpreted as some index of progressively refractory organic matter, it clearly demonstrates that the specific loss rate for an element is also dependent on the nature of the particles. Other losses for minor elements can be demonstrated for both the plankton and sediment trap samples (Figures 5-2, 5-4).

In the case of the sediment traps this labile fraction is easily captured and considered as part of the original particle-associated flux. Figure 5-5 shows the fraction of organic carbon lost to the cup solution and the relationship between the soluble carbon and the soluble phosphorus in one set of traps from the NE Pacific (Collier and Dymond, in preparation). The C/P ratio in this fraction is not dramatically different from Redfield ratios. With the advent of the new generation DOC analyzers, we should be able to do many more analyses on the labile as well as particulate fraction of suspended matter. However, many trap poisons (e.g. formaldehyde) and handling procedures are not compatible with characterizing the solubilized fraction. In the long run, the obvious requirement is to sample it properly (contain it) and then analyze all fractions. Furthermore, a more detailed chemical speciation would be helpful in

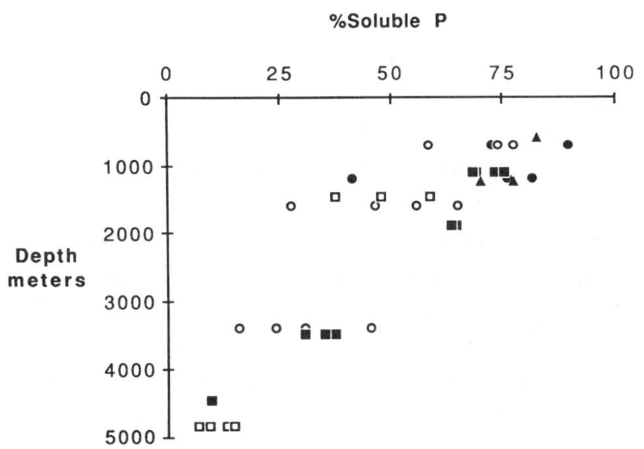

Fig. 5-3. Percentage of the total phosphorus (dissolved + particulate) released to the dissolved fraction contained by sediment trap sample cups as a function of depth of the sample. Samples are from various locations in the north Pacific.

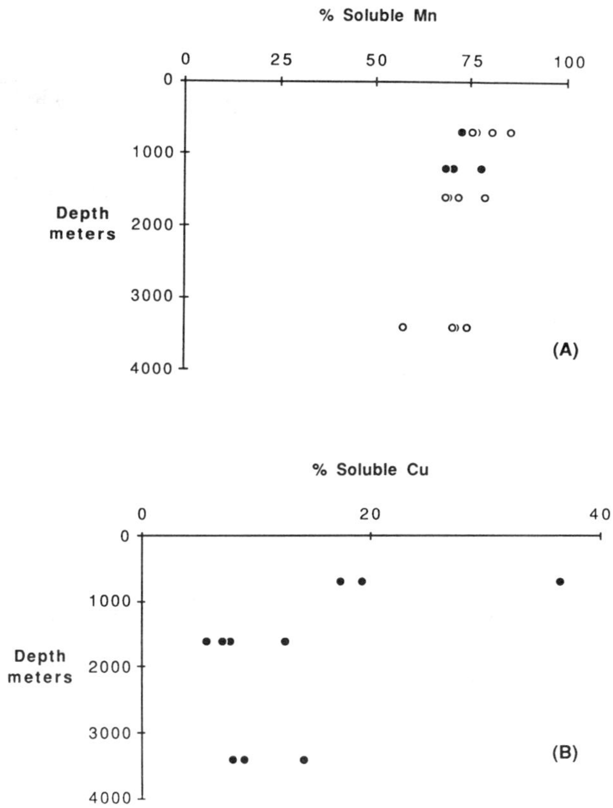

Fig. 5-4. Percentage of the total Mn (A) and Cu (B) found in the trap cup solutions as a function of depth in samples collected from the Pacific.

understanding the specific cellular origins and the likely fate of this fraction in the water column. Hopefully, this labile fraction contains useful information [Peng and Broecker, 1987] and is not just an analytical annoyance!

3.2. Losses as a Function of "Washing"

It has been demonstrated that washing samples with distilled water, which is a common practice to remove the analytical interferences of the major sea salts, causes a significant loss in materials from some marine particulate matter, especially plankton samples [Collier and Edmond, 1984]. This loss, which can also be seen to a lesser degree by washing with seawater and isotonic exchange solutions, is due to a combination of cell lysis and ion exchange. Table 5-2 shows loss to washing solutions (seawater, distilled water, NH_4Cl) for P, Ca, and several minor elements. Although this type of processing is often necessary for the standard chemical techniques, the wash solutions must be collected and

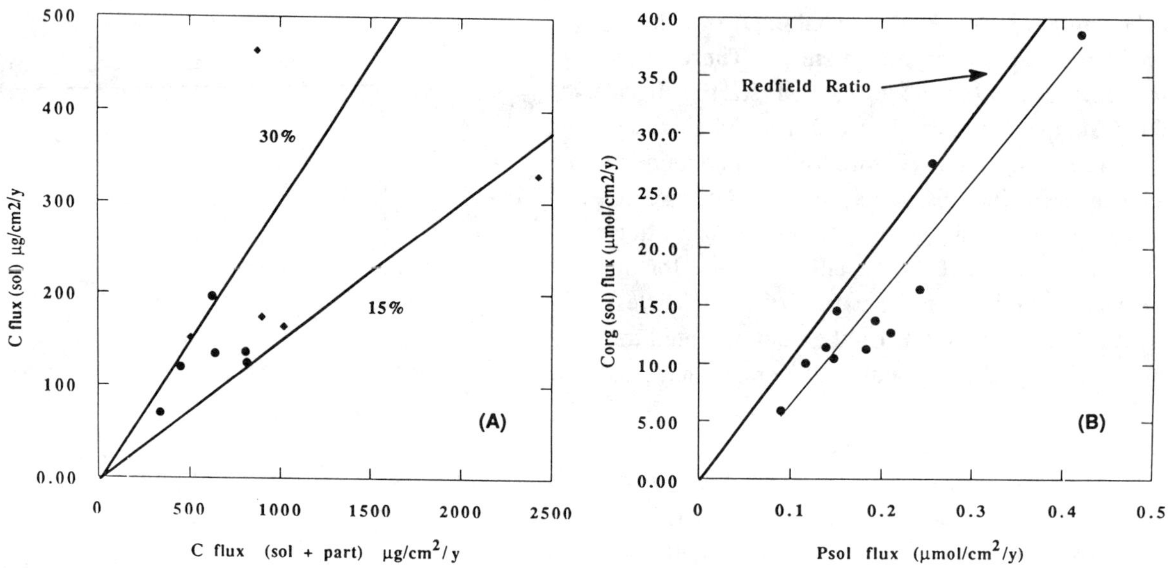

Fig. 5-5. (A) Comparison of the soluble organic carbon (sol) to the total organic carbon (soluble + Particulate) collected in sediment trap cups from the NE Pacific (Collier and Dymond, in preparation); (B) Comparison of the soluble organic carbon to soluble phosphorus in the sediment trap cups. The dark line represents the "Redfield ratio" and the light line is a regression through the data.

evaluated for any significant concentrations of the elements being analyzed in the particles.

4. Atomic Absorption (AA) Analysis

4.1 Sample Preparation

Total dissolution of solids using HF-HNO_3-H_3BO_3, for inorganic analysis by atomic absorption using flame or flameless atomization (see Table 5-3):

1) Transfer the dry, weighed sample (~ 200 mg) to a 125 mL, narrow-mouthed polypropylene bottle. Choice of quality bottles with good seals is important to the safety and quality of results; we use only Nalgene.
2) Moisten with 1 mL of distilled water to moderate rapid reaction of highly siliceous or carbonate samples.
3) Add 8 mL of a concentrated HF-HNO_3 (1:1) mixture and squeeze out ~ 1/3 of the air from the bottle to allow for gas generation during the oxidation/dissolution. Screw the cap on tightly.
4) Place the closed bottles in a safety tray and heat in a 60° C oven (± 5°) for 12-14 hours.
5) Remove and cool in hood for 15 minutes. Add 50 mL of 4% boric acid (H_3BO_3) and tighten cap. Return to the 60° C oven for 1 hour to dissolve any precipitated fluorides and neutralize excess HF.
6) Remove and cool in hood for 15 minutes. (If the samples contain >5% Mn (i.e. nodules) add 1 mL of a 1 M hydroxylamine hydrochloride ($NH_2OH \cdot HCl$) to solubilize Mn(IV) oxides).
7) Dilute to final volume/mass of 100 mL with or without the addition of atomic absorption matrix modifiers as necessary (e.g. Cs, La, etc.).
8) Proceed with AA analyses.

4.2 Sample Analysis

Analyses by this method must be standardized using appropriate blanks and matrix-adjusted standards. Standard additions should be performed on a few samples to verify slopes. NBS or USGS standards are usually run in parallel with each sample run to verify recovery and accuracy. The accuracy and precision of the technique are usually in the range of 1-5% depending on the composition and mass of the sample. Problems with the oxidation-solubilization step have been experienced with high organic matter samples such as plankton tows or shallow traps. In these cases a residual carbon residue remains and a more oxidizing attack using aqua regia or perchloric acid has been necessary as a first step. Also, in samples with high concentrations of barite or sulfides, the above solution may not solubilize all the primary or secondary barite.

TABLE 5-2.
Elemental Loss During Washing of Plankton Tow Sample
Percentage of Total Elements Released to Wash Solution and Metal/P Ratios in the Soluble Fraction

Element	Sea water		Distilled water		Isotonic NH_4Cl	
	%	Me/P	%	Me/P	%	Me/P
Ca	0		1.2		1.6	
P	27.5		35.9		19.2	
Cu	15.5	0.26	16.1	0.14	26.9	0.75
Ni	15.2	0.51	32.2	0.50	40	2.3
Cd	24.9	0.53	17.1	0.19	28	1.0
Fe	6.3	1.1	7.1	0.68	6.3	1.7
Mn	14.0	0.19	29.8	0.23	34.2	0.63

5. Instrumental Neutron Activation Analyses (INAA)

We frequently use instrumental neutron activation analysis to augment the single-element analyses provided by the atomic absorption technique. This approach is carried out on solid samples at the OSU Radiation Center (1 MW TRIGA). Elements of interest are often Cl, Br, Na, Al, Ba, V, Mn, Ca, Fe and a variety of rare earth elements. Again, we apply standard techniques for this method [De Soete et al., 1972] and we usually work with ~ 100 mg of dry sample. Depending on the specific elements needed, we combine a short and a long irradiation of the same sample with several counting intervals to optimize sensitivity and minimize background. In general, the precision and accuracy of the INAA technique is 5-10% for most of the applicable elements.

6. Intercalibration of AA and INAA Techniques

Because several elements can be detected by both methods, we have had the opportunity to make comparisons between the two techniques. Primarily, we view these as evaluations of the efficiency of the wet chemical dissolution steps which precede the AA analyses because the two detection techniques are comparable to within their individual precisions. Below we compare these analyses on a variety of sediment trap samples which have a wide range in compositions: near-surface organic rich samples, deep traps dominated by lithogenous resuspension, traps dominated by hydrothermal inputs of metal sulfides and oxides.

Figure 5-6 shows that analyses for Al, Fe and Mn are equivalent for both techniques. The AA analyses are generally more precise but the multielement INAA analysis can be more cost-effective and offers other elements not detectable by AA. Analyses for Ca are comparable to within the precision of the two techniques (Figure 5-7a). The AA analysis is considerably more precise and should be chosen, if possible. Figure 5-7b shows a comparison of $CaCO_3$ concentrations estimated by Ca analysis (AA) and by acid-soluble carbon analysis; in the latter approach, phosphoric acid dissolves the sample and liberates CO_2 which is detected by IR absorption. In general, the two estimates agree very well. The divergence at low $CaCO_3$ is probably due to Ca present in other phases such as organic matter [Bishop et al., 1978] and alumino-silicates.

We have had difficulties dissolving high concentrations of barium (as barite) from samples using the wet chemical technique outlined above. Although the precision of the INAA Ba analysis is poor (15-25%) at concentrations below 5000 ppm, a systematic trend of lower wet chemical recovery by AA detection compared

TABLE 5-3.
Compositions of Extractant Solutions.
Dilute quantities in column B with water to one L of total volume

Solutions used for sequential extractions and total dissolution of solids	B
1. *Na-acetate--HOAc, (pH 5.0):*	
Na-acetate [1.0 M]	136.1 g
HOAc [0.516 M] (mLs glacial HOAc)	516 meq (29.5 mL)
2. *Sodium Dodecyl Sulfate, -- $NaHCO_3$, (pH 8.8):*	
Sodium Dodecyl Sulfate [1% W/V]	10 g
$NaHCO_3$ [0.2 M]	16.8 g
EDTA	1 g
3. *$NH_2OH \cdot HCl$-- Na-Citrate, (pH 5.0):*	
$NH_2OH \cdot HCl$ [1.0 M]	69.5 g
Na-Citrate [0.175 M]	51.57 g
4. *HF/HNO_3 Dissolution Mixture:*	
(see section 4)	
HF [40% W/V]	40 mL
HNO_3 [concentrated]	40 mL
H_3BO_3 [4% W/V]	500 mL
NH2OH.HCl [1.0 M, as above]	5 mL

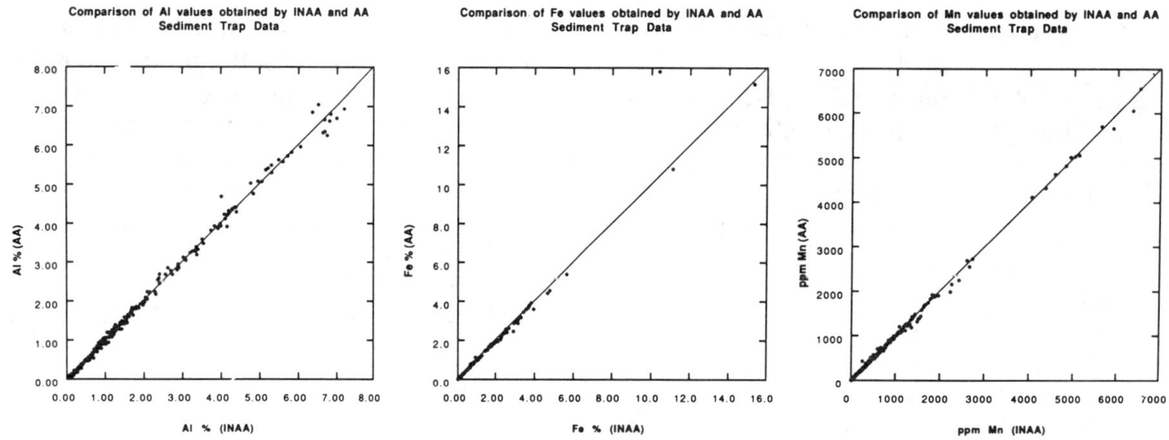

Fig. 5-6. Comparisons of selected total elemental analyses (Al, Fe, and Mn) of sediment trap particulate matter carried out by wet chemical dissolution / atomic absorption spectrometry (AA) and by instrumental neutron activation analysis of the solid sample (INAA).

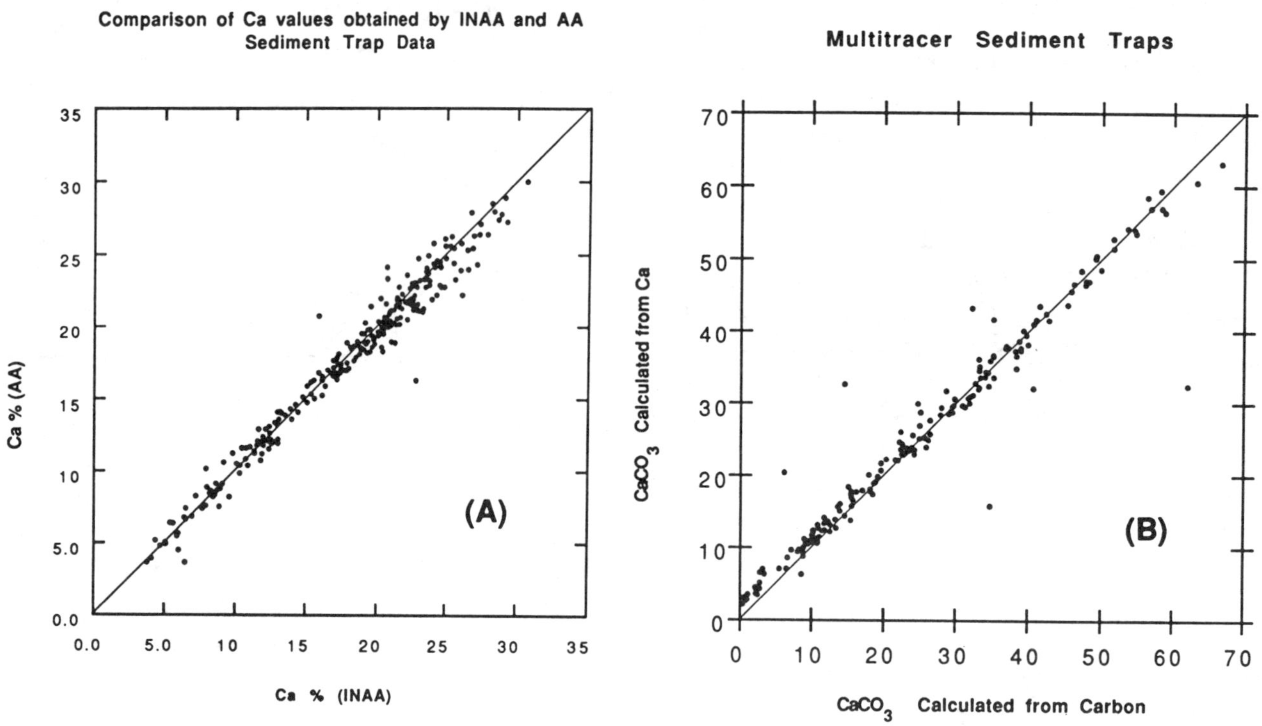

Fig. 5-7. (A) Comparison of Ca analysis by atomic absorption (AA) and neutron activation analysis (INAA) as in Figure 5-6. (B) Comparison of calculated $CaCO_3$ derived from AA analysis of Ca vs. acid dissolution of carbonate followed by CO_2 detection by IR absorption.

to INAA analysis is observed (Figure 5-8). Most of these "high Ba" samples are from hydrothermal systems precipitated at the vents. It is possible that the oxidation of sulfides to sulfate helped to supersaturate the dissolution matrix with respect to barite. We do not have a good comparison at low Ba concentrations, but interpretation of the data always suggests the wet chemical technique may be low by several tens of percent at all concentrations above 500 ppm. Bishop (LDGO personal communication, 1988) reports good recovery of

TABLE 5-4.
Optimized Sequential Extraction Procedure Sediment

EXTRACTION TREATMENTS to be followed for each extraction step (using 0.5 gram sample in 50 mL centrifuge tubes):

1. Add 20 mL of extractant.
2. Vortex mix (10 min.).
3. Ultrasonic bath (10 min.).
4. Centrifuge (10 min. at 15,000 g).
5. Pipet off supernate.
6. Transfer supernate to 100 mL plastic bottle, combining supernates from all repeats for a total of 100 mL.

Sequential Extractions Extractants (see Table 5-3)	Repeat	Resultant Fraction
1. NaOAc - HOAc, pH 5.0	×5	Carbonate, sorbed (and salt)
2. H_2O (DDW) wash (discard)	×1	
3. Sodium dodecyl sulfate - $NaHCO_3$, pH 8.8, at 100° C	×2	Organic
4. Sodium dodecyl sulfate - $NaHCO_3$, pH 8.8, at 25° C	×3	
5. $NaHCO_3$ (0.2 M) wash (discard)	×1	
6. $NH_2OH{\cdot}HCl$ - Na-Citrate, pH 5.0	×5	Oxyhydroxide
7. $HF\text{-}HNO_3$ digestion of residue		Residual (see section 4)
8. $HF\text{-}HNO_3$ digestion of total sample		Total (see section 4)

barite in suspended particulate matter samples by dissolution with dilute HCl.

7. Sequential Extraction Procedures

A method for partitioning elements between

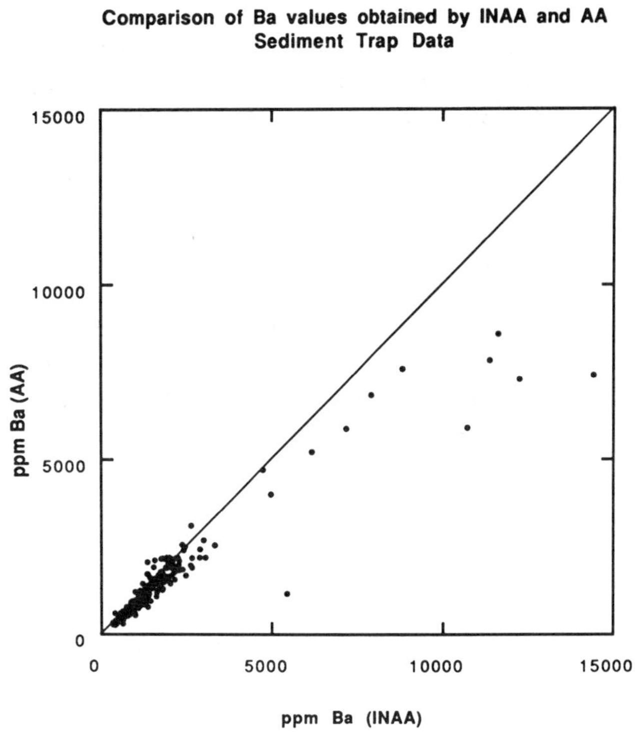

Fig. 5-8. Analyses of barium, primarily in the form of barite, by wet chemical dissolution (AA) and by activation analysis (INAA). The wet chemical technique outlined in Section 4 does not efficiently dissolve high concentrations of barite.

coexisting phases in marine sediments was developed by Robbins et al. [1984] and is outlined below. The primary reference should be consulted for a detailed discussion of these and related sequential extraction procedures [such as Chester and Hughes, 1967]. These techniques were evaluated using a wide range of sediments: calcareous sediments; siliceous clay; organic-rich hemipelagic; red clay; metalliferous sediments; ferromanganese nodules; nontronite; phillipsite; several "standard" clays, namely Ca-montmorillonite, hectorite, kaolin, and chlorite.

The fractions isolated are: 1) carbonate + adsorbed ions; 2) labile organic; 3) oxyhydroxide; 4) residual. Considerable effort was made to separate the two fractions of the first isolate, but this was not possible in all cases and the carbonate and adsorbed fractions were "combined".

In Tables 5-3 and 5-4 we present a summary of the extractions [Robbins et al., 1984]. Elemental analyses on the leach solutions were carried out by atomic absorption.

References

Bishop, J. K. B., D. R. Ketten, and J. M. Edmond, The chemistry, biology and vertical flux of particulate matter from the upper 400 m of the Cape Basin in the southeast Atlantic Ocean, Deep-Sea Res., 25, 1121-1161, 1978.

Chester, R., and M. J. Hughes, A chemical technique for the separation of ferro-manganese minerals, carbonate minerals and adsorbed trace elements from pelagic sediments, Chem. Geol., 2, 249-262, 1967.

Collier, R. W., and J. Edmond, The trace element geochemistry of marine biogenic particulate matter, Prog. Oceanogr., 13, 113-199, 1984.

De Soete, D., R. Gijbels, and J. Hoste, Neutron Activation Analysis, Vol. 34, 836 pp., Wiley-Interscience, London, 1972.

Peng, T.-H., and W. S. Broecker, C/P ratios in marine detritus, Global Biogeochem. Cycles, 1, 155-161, 1987.

Robbins, J. M., M. Lyle, and G. R. Heath, A sequential extraction procedure for partitioning elements among co-existing phases in marine sediments, College of Oceanography, Oregon State University, Reference #84-3, 64 pp., 1984.

U.S. GOFS Working Group, G. A. Knauer and V. Asper, Co-chairs, Sediment trap technology and sampling, Report No. 10, November 1988, 94 pp., Woods Hole, Massachusetts, 1989.

CHARACTERIZATION OF MARINE PARTICULATE MATERIAL BY ICPMS

Kelly Kenison Falkner

GRGS/CNES, 18 ave. Edouard Belin, Toulouse 31055 Cedex, France

1. Introduction

The viability of inductively coupled plasma quadrupole mass spectrometry (ICPMS) as an analytical tool was first demonstrated in the late 1970's [Douglas and Houk, 1985; Horlick et al., 1987]. Realization of its potential awaited the development of a practical interface for sampling ions from an ICP, which operates under atmospheric pressure at extremely high temperatures ($\approx$ 10,000° K), and a quadrupole mass analyzer requiring high vacuum conditions [Douglas and Houk, 1985; Horlick et al., 1987]. This was achieved by 1984 at which time the first commercial instruments became available. Since its introduction, ICPMS has been successfully applied to a wide variety of fields of study including mineral exploration, medicine, metallurgy, marine biology and geochemistry [for examples of the latter, see Beauchemin et al., 1988; Falkner, 1990; Garbarino and Taylor, 1987; Klinkhammer and Chan, 1990; Lichte et al., 1987; McLaren et al., 1985]. Its enthusiastic acceptance by the analytical community stems from its very high sensitivity, multi-element and high throughput capabilities. As specialized sample introduction methods, such as direct insertion, electrothermal ionization and laser ablation, are being implemented, the capabilities of ICPMS are being broadened considerably [i.e. Gray, 1985; Hall et al., 1988; Park et al., 1987a,b; Sing and Salin, 1989]. The following discussion focuses on the use of ICPMS in the marine sciences, beginning with a general introduction to the technique. Its capabilities and limitations are presented based on the experience of the author with VG instrumentation. The discussion is concluded with a brief survey of alternative sample introduction methods and how they might be expected to contribute to the study of marine particulate matter.

2. General Principles

In the conventional mode of operation (see Figure 5-9), aqueous samples are introduced into the ICPMS by peristaltic pumping through a concentric glass nebulizer. The aerosol created is subsampled by use of a spray chamber so that the droplet size presented to the ICP is small enough ($\leq$ 10 μm) to allow for complete desolvation, atomization and ionization within the sampling zone of the plasma. As a result, only about 1% of the sample enters the plasma with the rest going to waste. Although not particularly efficient, nebulization is the standard mode of sample introduction because it is convenient, reliable and relatively inexpensive.

The subsampled aerosol is borne by the nebulizer gas stream ($\approx$ 1 L Ar min^{-1}) into the flowing Ar gas ICP ($\approx$ 14 L Ar min^{-1}). While the dynamic plasma environment is most certainly not at thermodynamic equilibrium, useful estimates of the extent of ionization at a given point in the plasma have been made using the Saha expression, which presumes local thermodynamic equilibrium and Maxwell-Boltzman distributions:

$$\frac{(n_i\, n_e)}{(n_a)} = \frac{(2\pi m_e kT)^{3/2}}{(h^3)} \bullet \frac{(2z_i)}{(z_a)} \bullet \exp\left[\frac{(-E_i)}{(kT)}\right]$$

where n_i is the ion concentration, n_e the free electron concentration, n_a the atom concentration, m_e the mass of an electron, k the Boltzmann constant, T the temperature in Kelvin, h Planck's constant, E_i the ionization potential, z_i the partition function of the ion and z_a the partition function of the atom. To a first approximation, the electron population is "buffered" by the extent of Ar ionization and so is governed by the Ar ionization potential (15.8 eV). This results in the efficient (greater than 70%) single ionization of most elements of the periodic table under most plasma conditions. Notable exceptions include those elements whose ionization potentials exceed that of Ar and so are not efficiently ionized, such as the lighter noble gases, fluorine and chlorine. Elements with very low second ionization potentials, such as Ba, can be doubly ionized. In practice, doubly ionized species do not exceed a few percent of their ion populations in the analytical zone or central part of the plasma, for even the most susceptible elements. It is noted that the ICP used for ICPMS employs the same technology as is used for emission spectroscopy (ICPES), except that the plasma is mounted horizontally rather than vertically and is specially grounded for this application. For more

Marine Particles: Analysis and Characterization
Geophysical Monograph 63

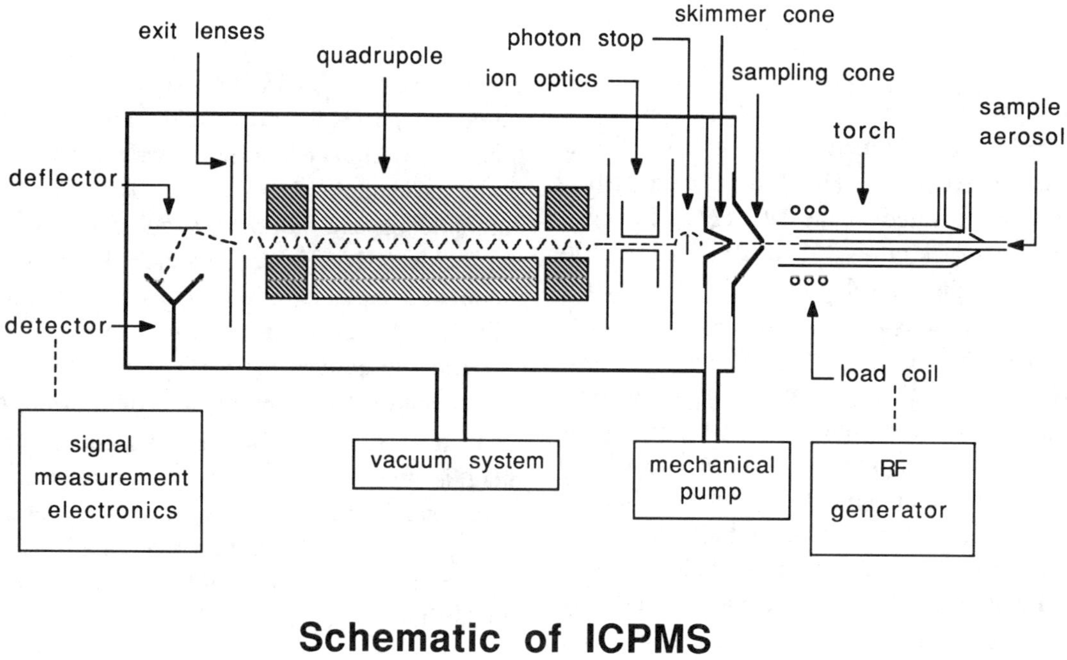

Fig. 5-9. Schematic of conventional inductively coupled plasma quadrupole mass spectrometry (not to scale).

detailed information concerning properties and operation of an ICP, see Montaser and Golightly [1987] and references therein.

The ICP is sampled through an orifice, approximately 1 mm in diameter, of a conductively cooled cone, generally machined from high purity Ni. Cones constructed of Al or having Pt tips have also been used successfully. The sampled plasma expands supersonically into an initial vacuum stage (1-2 mbar) where it is subsampled by a second cone with a somewhat smaller orifice (0.75 mm) into a second vacuum stage. Ion optics then direct ions to the high vacuum stage ($\approx 2 \times 10^{-6}$ mbar) housing a quadrupole mass analyzer. Here, mass separation is accomplished by applying AC and DC voltages to the rods of the quadrupole such that ions with the desired charge/mass ratio oscillate in the plane parallel to the rods and are transmitted to a detector. Ions with other charge to mass ratios and neutral species collide with the rods or are pumped to vacuum. By varying the amplitude of the applied voltages, it is possible to rapidly scan, up to 3000 amu sec^{-1} over the entire mass range. Alternatively, for greater sensitivity, the peak transmission energies of desired masses may be monitored. Mass resolution for the quadrupole is determined by the AC/DC voltage ratio and is typically set for unit mass resolution. Optimal achievable abundance sensitivities are a function of quadrupole design and are typically 1 in 10^6 on the high mass side of a peak and 1 in 10^5 on the low mass side.

Ions are normally detected with a continuous dynode electron multiplier which is usually utilized in a highly sensitive pulse mode having a dynamic range of 10^5. At a lower bias potential, it can also be used in a less sensitive proportional mode, further extending the dynamic range. Signals from the electron multiplier are routed to a preamplifier and are then either sequestered to a multichannel analyzer and passed to a computer or are passed directly to a computer for data manipulation.

In general, the spectra obtained by ICPMS are fairly simple, with far fewer interferences than encountered in ICPES. However, significant isobaric, background and matrix-induced molecular interferences do arise and must be considered. All elements of the periodic table (excepting In) have at least one isotope free from isobaric overlap, though it may be of lower abundance and involve a compromise in sensitivity. It is frequently possible to make a correction for an isobaric interferent based on a non-overlapping isotope of the interfering element. Argon from the plasma, the major constituents of the atmosphere and molecular combinations of these contribute significant background peaks to ICPMS spectra. The dominant peak, which is generally skipped to preserve the life of the multiplier, is that of the most

TABLE 5-5.
Comparative Absolute Detection Limits (ng)*

Element	ICPMS	ICPES	GFAA	INAA
Sc	0.08	1	nd	1
Ti	0.06	2	0.04	100
Y	0.02	1	10	400
Zr	0.03	3	nd	800
Nb	0.02	20	nd	3000
Ru	0.05	20	30	40
Rh	0.02	10	12	5
Pd	0.06	30	0.03	30
In	0.02	30	0.01	0.06
Ba	0.02(0.005)	0.5	0.04(0.005)	20
REE	0.01-0.04	30-100	0.01-0.05	0.03-250**
Hf	0.03	10	nd	0.6
Ta	0.02	20	nd	100
W	0.06	30	nd	4
Re	0.06	10	nd	0.8
Os	0.02	50	1	1000
Ir	0.06	20	0.2	0.3
Pt	0.08	30	0.1	100
Au	0.1(0.002)	4	0.01	0.5
Pb	0.05	30	0.01	nd
Th	0.02	100	nd	200
U	0.02	50	12	3

nd = not done
* = see text for details
** = not all rare earth elements can be analyzed by INAA.

abundant isotope of Ar at mass 40. Its presence precludes the determination of Ca at its most abundant mass. Other prominent background peaks include the dimer $ArAr^+$ at mass 80, making the routine analysis of Se difficult, and the oxide ArO^+ at mass 56, posing a problem for Fe determinations. Silicon, sulfur and phosphorus have interferences arising from molecular ions of nitrogen, oxygen and hydrogen. In contrast to the lower mass region, the region above mass 80 is generally free from background interferences. Matrix effects are discussed in more detail in the following section.

Detection limits based on nebulization (1 mL min^{-1}), single ion monitoring and 10 second integrations are generally on the order of 0.01-0.4 ppb. Approximately 2 mL of sample are required for the analyses, thus absolute detection limits are approximately 0.02-0.08 ng. Comparative detection limits for selected elements are given in Table 5-5 including ICPMS, ICPES, graphite furnace atomic absorption spectrophotometry (GFAA) and instrumental neutron activation analysis (INAA). The values are expressed as the absolute amount (in ng) of element required to produce a signal 3 times the standard deviation of the noise. This presumes 2 mL of solution for ICP work, 20 μL injections for GFAA and a 24 hour irradiation with a 1 x 10^{12} n cm^{-2} s^{-1} neutron flux for INAA and otherwise standard operating conditions. The ICPMS, ICPES and GFAA results were obtained by the author and the INAA results were derived from a table provided by a commercial analytical service. ICPMS values in parentheses were obtained by flow injection techniques described below. Improvements for the GFAA determination of the carbide forming Ba have been obtained using tantalum liners in graphite tubes (value in parentheses; D. Lea, personal communication, 1990) or through matrix modification [Bishop, 1990]. While ICPMS and GFAA detection limits are comparable, it is not possible to determine the more refractory elements by GFAA. ICPMS detection limits are generally two to three orders of magnitude better than obtained by ICPES and with few exceptions are far superior to those obtained by instrumental neutron activation analysis (INAA).

3. Application to the Marine Sciences

3.1. Matrix Effects

While ICPMS is relatively tolerant of the presence of matrix elements, three types of matrix-induced problems can arise. The first type can be classified as physical. Samples must be free of particles since they can cause nebulizer blockage. High dissolved inorganic solids can result in the build-up of material on the sampling cone orifice. The worst culprits are elements such as Al and Si which deposit refractory oxides that can eventually cause complete blockage. It is recommended that the total solid contents of such elements be kept below 0.1-0.2% w/v to avoid oxide build-up. The total solid content of less refractory elements should not exceed 1-2%. Thus, seawater, with a salinity of 35 ppt, or 3.5%, must be diluted at least a factor of two for routine analyses. Organic material can deposit on the nebulizer, spray chamber and torch walls and lead to memory effects and can also affect sample viscosity which affects sample flow rates. While this is generally not a problem for diluted seawater samples, it may present difficulties for particle analysis when organic matter has not been completely digested.

The second category of matrix problems is spectral interference. Large amounts of a matrix element with respect to an analyte can give rise to significant molecular and isobaric interferences. For example, hydrochloric acid solutions form molecular chlorides that interfere with the first row transition elements; and phosphoric and sulphuric acids do likewise. Tables of such interferences can be obtained from the manufacturers. The matrix of choice for ICPMS solutions is dilute nitric acid which does not introduce any species not already contained by the plasma. The dominant matrix derived interferences in seawater spectra are molecular chlorine and sulfur species.

The third type of problem, matrix induced signal suppression, is not yet completely understood. Signal suppression is observed to be mass dependant such that for a given matrix, lighter masses are suppressed more than heavier ones and for a given element, a heavier matrix element causes more suppression than a lighter one. Space charge effects at the sampling orifice account for some but not all of the observations [Gilson et al., 1988]. Until a better understanding is achieved, it must be assured empirically that standard and sample matrices are sufficiently matched or presuming no interferences are present, the technique of standard additions can be employed. An example of signal suppression is given in Figure 5-10. It can be seen that suppression of the Au signal intensity is not significant until Br exceeds 0.1 molar in concentration. Though seawater is approximately 0.5 molar NaCl, sodium and chlorine are lighter in mass and exert less suppression than Br.

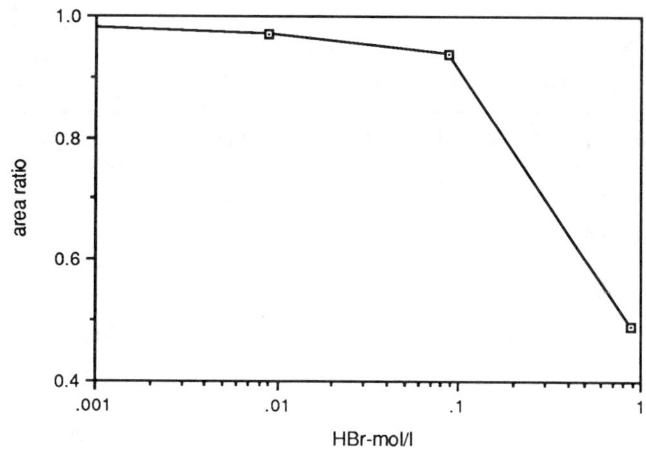

Fig. 5-10. Signal suppression of Au, expressed as the ratio of the peak areas with respect to a Au standard (0.25 nmol/L) in 10% AR, as a function of the bromine concentration. The same effect is observed for equal molar concentrations of HBr and Br_2.

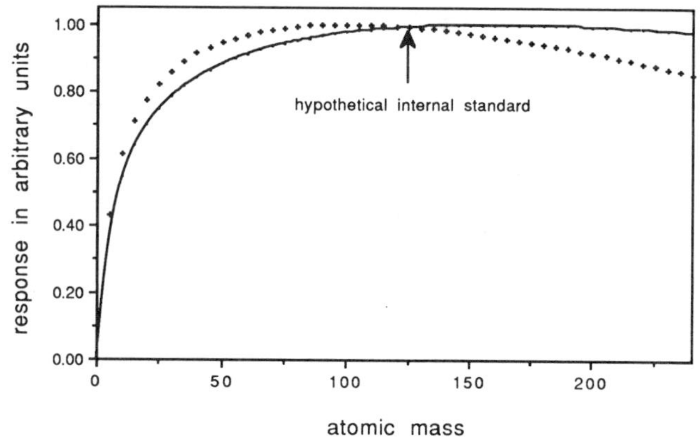

Fig. 5-11. Schematic showing typical abundance normalized response curves for ICPMS. The exact shape of the curve depends upon a variety of factors including quadrupole specifications and control parameters and lens settings. The arrow illustrates the problem of using a single internal standard over the entire mass range as referred to in the text. While the response for the internal standard remains unchanged despite a change in response curve shapes, the upper and lower mass regions require corrections to account for instrumental drift.

3.2. Standardization

As for all instrumentation, the accuracy of ICPMS analyses ultimately depends upon careful calibration of standards. Analyses of standard reference materials have shown ICPMS to compare favorably with other methods such as GFAA, INAA and ICPES, when all possible interferences, matrix effects and instrumental drift are accounted for [Horlick et al., 1987]. Drift is used to refer to gradual changes in sensitivity with time. Stepwise or very rapid changes constitute instability. Instrumental drift can arise from a number of causes and nearly always occurs to a measurable degree. It can manifest itself as a mass independent intensity change, a unidirectional mass dependent intensity change or a change in the shape of the response curve (Figure 5-11). The latter two possibilities mean that a single element as an internal standard cannot adequately account for drift over the whole mass range. One must use an element close in mass to the measured constituents as an internal standard. For example, using In as an internal standard for the analyses of Br and I in diluted hydrothermal fluids generated results which matched those obtained independently by chemical titration methods [Campbell and Edmond, 1989]. It would not be advisable to use In as an internal standard for Li or U.

3.3. Isotope Dilution

The availability of information at the isotopic level makes it possible to carry out isotope dilution analysis (ID) by ICPMS. The precision for isotope ratios ranging from 0.05-50, determined by ICPMS, is 0.1% at best and 1-2% routinely. Outside of this ratio range, instrumental precision worsens markedly. As for thermal ionization mass spectrometry (TIMS), accurate results require normalization to a known standard. Unlike TIMS, the mass bias observed in ICPMS is not a readily predictable phenomenon. Isotope ratios obtained by ICPMS can be higher or lower (generally $\leq 4\%$) than their true values depending upon a number of variables including plasma parameters, the tuning of the ion lenses, matrix effects and the method of signal monitoring [i.e. Longerich et al., 1987; Russ and Bazan, 1987]. Each time any of these parameters is changed, the system must be recalibrated by running a known standard in a matrix equivalent to that of the samples or correcting data to expected ratios of isotopes with fixed abundances. Figure 5-12 shows that matrix equivalency is readily achieved for the purposes of isotope dilution analyses (2% precision) in the upper mass range where the response curve tends to be flat. In the lower mass region, where its slope is more pronounced, matrix matching for isotope dilution becomes more critical. While ratios tend to be less prone to instrumental drift than absolute signals, drift can measurably affect ratios and should be monitored throughout a run.

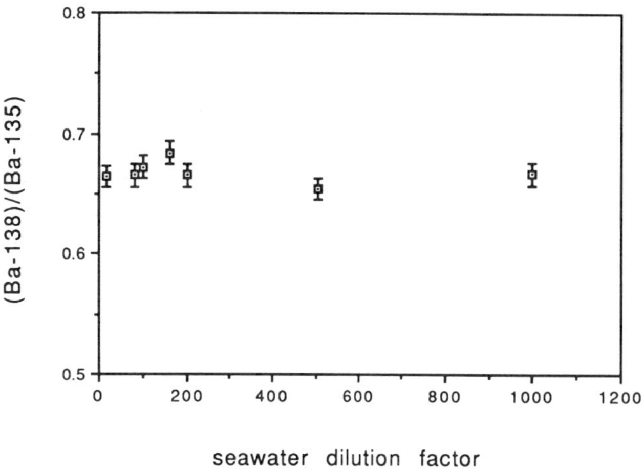

Fig. 5-12. The measured isotope ratio for a spiked Ba seawater solution as a function of the dilution factor (diluted with 0.2 N HNO_3). Absolute count rates did not exceed 12,000 cps. Error bars on the plotted ratios are 1 standard deviation for triplicate determinations. The ratio was effectively constant within 2 standard deviations over the range of 2 to 1000 fold seawater dilutions.

3.4. Particle Analyses

ICPMS offers several advantages for inorganic analyses of particles. With its comparatively low detection limits, analyses can be accomplished on smaller sample sizes. Semi-quantitative, multi-element compositions can rapidly be obtained for determining optimal conditions for quantitative analyses. It can provide information for a host of elements, including refractory elements, the platinum group elements and the entire suite of rare earth elements, which are otherwise difficult to determine.

In the conventional mode of sample introduction, particles must be dissolved for analyses. A variety of methods for solubilizing particulate matter, including acid dissolutions, conventional and microwave digestions and fusion, are directly applicable to ICPMS work. It is important to keep in mind the following caveats. Ordinarily, the ICPMS is intolerant of HF in a matrix since it will etch and destroy the glass nebulizer, spray chamber and torch. Dilute solutions of HF can be accommodated by use of a Teflon based sample introduction system, although the quartz torch will eventually be degraded with time. High dissolved solid loads will lead to a variety of problems as discussed above. In many cases, these problems can be overcome by sample dilution.

4. Alternative Modes of Sample Introduction

4.1. Flow Injection

Peristaltic pumping is employed for the conventional mode of sample introduction in order to control sample flow rates. The flexible tubing required for peristaltic pumping by necessity contains plasticizers which tend to act as adsorption sites for some elements (e.g. Ba and Au but not U). This leads to a slow rise to steady state signals and persistent memory effects. It can also lead to false blanks when the desorption rate of an element is matrix dependent. One inexpensive way to avoid these problems, yet still control sample flow rates, is to insert a Teflon flow injection valve downstream of th^ peristaltic pump and upstream of the nebulizer. The sample need only contact inert Teflon tubing before arriving at the nebulizer. This is particularly advantageous when sample size is limited since the size of the sample loop ($\geq 50\ \mu L$) can be readily customized for the application. Significantly improved detection limits were afforded by this technique for the analyses of Au in preconcentrated seawater samples [Falkner, 1990].

4.2. Liquid Chromatography

Coupling a liquid chromatograph to ICPMS methods can provide a convenient way to accomplish elemental separations when interferences pose a problem [Thompson and Houk, 1986]. A variety of column/eluent possibilities have been developed for this purpose and systems fully integrated with ICPMS are available.

4.3. Electrothermal Vaporization, Direct Insertion, Hydrides, Volatile Compounds, Laser Ablation

The conventional mode of sample introduction is only 1-2% efficient. Recirculating spray-chambers and ultrasonic nebulizers can be more efficient but tend to present memory and/or stability problems. There is no fundamental reason why a sample needs to be introduced into the plasma in an aqueous form, however, and a variety of ways of introducing the sample as an atomic or molecular vapor have been successfully applied [Gray, 1985; Hall et al., 1988; Park et al., 1987a, b; Powell et al., 1986; Russ et al., 1987; Sing and Salin, 1989]. Once optimized, these methods have afforded more efficient sample introduction and diminished molecular interferences, such as oxides and hydroxides associated with excess water vapor loading of the plasma [Long and Browner, 1988]. The result is one to three orders of magnitude improvement in detection limits depending upon the element and the particular mode of introduction. Although all of these methods require further development, in addition to improved sample size requirements, they present future possibilities for examining the more rare and refractory elements which may contribute unique information to our understanding of particle cycling in the oceans.

References

Beauchemin, D., J. W. McLaren, S. N. Willie, and S. S. Berman, Determination of trace metals in marine biological reference materials by inductively coupled plasma mass spectrometry, Anal. Chem., 60, 687-691, 1988.

Bishop, J. K. B., Determination of barium in seawater using vanadium/silicon modifier and direct injection graphite furnace atomic absorption spectrometry, Anal. Chem., 62, 553-557, 1990.

Campbell, A. C., and J. M. Edmond, Halide systematics of submarine hydrothermal vents, Nature, 342, 168-170, 1989.

Douglas, D. F., and R. S. Houk, Inductively coupled plasma mass spectrometry, Prog. Anal. At. Spectros., 8, 1-18, 1985.

Falkner, K. K., Determination of gold at femtomolar levels in natural waters by flow injection inductively coupled plasma quadrupole mass spectrometry, Anal. Chem., 62, 1477-1481, 1990.

Garbarino, J. R., and H. E. Taylor, Stable isotope dilution analysis of hydrologic samples by inductively coupled plasma mass spectrometry, Anal. Chem., 59, 1568-1575, 1987.

Gilson, G. R., D. J. Douglas, J. E. Fulford, K. W. Halligan, and S. D. Tanner, Nonspectroscopic interelement interferences in inductively coupled plasma mass spectrometry, Anal. Chem., 60, 1472-1474, 1988.

Gray, A., Solid sample introduction by laser ablation for inductively coupled plasma source mass spectrometry, Analyst, 110, 551-556, 1985.

Hall, G. E. M., J.-C. Pelchat, D. W. Boomer, and M. Powell, Relative merits of two methods of sample introduction in inductively coupled plasma mass spectrometry; Electrothermal vaporization and direct sample insertion, J. Anal. At. Spectrom., 3, 791-797, 1988.

Horlick, G., S. H. Tan, M. A. Vaughn, and Y. Shao, Inductively coupled plasma-mass spectrometry, in Inductively Coupled Plasmas In Analytical Atomic Spectrometry, edited by A. Montaser, and D. W. Golightly, pp. 361-398, VCH (Publishers), New York, 1987.

Klinkhammer, G. P., and L. H. Chan, Determination of barium in marine waters by isotope dilution inductively coupled plasma mass spectrometry, Anal. Chim. Acta, 232, 323-329, 1990.

Lichte, F. E., A. L. Meier, and J. G. Crock, Determination of the rare-earth elements in geological materials by inductively coupled plasma mass spectrometry, Anal. Chem., 59, 1150-1157, 1987.

Long, S. E., and R. F. Browner, Influence of water on conditions in the inductively coupled argon plasma, Spectrochim. Acta, 43B, 1461-1471, 1988.

Longerich, H. P., B. J. Fryer, and D. F. Strong, Determination of lead isotope ratios by inductively coupled plasma-mass spectrometry (ICP-MS), Spectrochim. Acta, 42B, 39-48, 1987.

McLaren, J. W., A. P. Mykytiuk, S. N. Willie, and S. S. Berman, Determination of trace metals in seawater by inductively coupled plasma mass spectrometry with preconcentration on silica-immobilized 8-hydroxyquinoline, Anal. Chem., 57, 2907-2911, 1985.

Montaser, A., and D. W. Golightly (Eds.), Inductively coupled plasmas, in Analytical Atomic Spectrometry, 660 pp., VCH, New York, 1987.

Park, C. J., J. C. Van Loon, P. Arrowsmith, and J. B. French, Design and optimization of an electrothermal vaporizer for use in plasma source mass spectrometry, Can. J. Spectros., 32, 29-41, 1987a.

Park, C. J., J. C. Van Loon, P. Arrowsmith, and J. B. French, Sample analysis using plasma source mass spectrometry with electrothermal sample introduction, Anal. Chem., 59, 2191-2196, 1987b.

Powell, M. J., D. W. Boomer, and R. J. McVicars, Introduction of gaseous hydrides into an inductively coupled plasma mass spectrometer, Anal. Chem., 58, 2864-2867, 1986.

Russ, G. P., III, and J. M. Bazan, Isotopic ratio measurements with an inductively coupled plasma source mass spectrometer, Spectrochim. Acta, 42B, 49-62, 1987.

Russ, G. P., III, J. M. Bazan, and A. R. Date, Osmium isotopic ratio measurements by inductively coupled plasma mass spectrometry, Anal. Chem., 59, 984-989, 1987.

Sing, R. L. A., and E. D. Salin, Introduction of liquid samples into the inductively coupled plasma by direct insertion on a wire loop, Anal. Chem., 61, 163-169, 1989.

Thompson, J. J., and R. S. Houk, Inductively coupled plasma mass spectrometric detection for multielement flow injection analysis and elemental speciation by reversed-phase liquid chromatography, Anal. Chem., 58, 2541-2548, 1986.

SAMPLING OF MARINE PARTICULATE MATTER AND ANALYSIS BY X-RAY FLUORESCENCE SPECTROMETRY

Richard A. Feely, Gary J. Massoth and Geoffrey T. Lebon

Pacific Marine Environmental Laboratory, National Oceanic and Atmospheric Administration, 7600 Sand Point Way N.E., Seattle, Washington 98115

1. Introduction

Over the past fifteen years our laboratory has been involved with a variety of research studies that required large numbers of analyses of samples of marine particulate matter from estuarine, coastal and open-ocean environments. The common theme of all of these studies has been to provide a better understanding of how chemical reactions involving marine particles affect the composition of seawater and the underlying sediments. During the course of these studies, we have endeavored to develop a variety of procedures for collection, preservation and analysis of several different types of particulate samples. These range from nearly-pure biogenic matter from plankton blooms to mostly inorganic fine-grained resuspended sediments. In some instances, the inorganic matter might be highly reactive, as is the situation when we study newly-formed Fe- and Mn oxyhydroxide and sulfide particles emitted from deep-sea hot springs on mid-ocean ridges. In other cases, the particulate samples may be very delicate and subject to rapid dissolution during recovery and processing. In all of these situations, we had to develop specialized techniques to maintain the physical and chemical integrity of the samples from the time of sampling to the completion of the analytical workup. What follows below is a general description of our basic methods for collection, preservation and analysis of particulate samples from the marine environment. Since most of our work has been involved with studies of the changes in their elemental concentration, we will confine this discussion to methods for analysis of the elemental composition of suspended matter.

2. Samples Collected From a Hydrowire

At present, the largest body of information about chemical variations of marine particulate matter has been derived with samples collected from a hydrowire. The samples are collected either by attaching Teflon-coated Niskin bottles at various levels on a standard hydrographic cable or a Kevlar line suspended from a ship, by attaching Niskin bottles to a rosette which is lowered to specified depths and triggered, or else by in situ filtration. The Niskin bottles are Teflon coated to minimize sample contamination.

Our laboratory uses both standard Niskin bottles equipped with silicon-based Silastic tubing and o-rings for closure of the end caps, or Go-Flo Niskin bottles when certain types of trace metal analyses are required. We generally attach the bottles to a twelve position rosette equipped with CTD-transmissometer, altimeter, and pinger. The rosette is lowered to predetermined depths and triggered via a signal from the deck unit on the ship. The continuous readout of temperature, salinity, and transmissometer readings aid in the determination of the final sample depths [Feely et al., 1990b]. Upon retrieval, the Niskin bottles are placed in a non-contaminating pressure filtration rack and the samples are pressure filtered through 37 mm diameter 0.4 μm pore size Nuclepore aerosol polycarbonate membrane filters. The filtration racks are designed to tilt the bottles at a 45° angle to insure that all of the sample is filtered with no "dregs" remaining in the bottle [Gardner, 1977; Calvert and McCartney, 1979]. We also periodically gently agitate the Niskin bottles to keep the particles in suspension during the filtration process. In most cases, we filter about 18-22 L of seawater through a 37 mm filter. The filter holders used during the filtration process are the Teflon Savillex filter holders equipped with a Teflon insert designed to concentrate the particles onto an acid-washed (4 N HNO_3 for 24 hours followed by 2% HCl for 24 hours with multiple rinses with distilled Milli-Q water in between treatments) filter. Tygon tubing with attached Teflon end pieces is used to transfer the water sample from the Niskin bottle to the filter holder. One Teflon end is attached to a Teflon stopcock on the Niskin bottle. The tube is flushed with 100 mL of sample and then attached to the filter holder in such a manner that any air in the head space is allowed to escape. When the air is out of the filter holder it is secured and the filtration begins. The standard air vent on the top of the Niskin bottle has been refitted with a nylon air vent equipped to receive a tygon hose from a nitrogen gas cylinder.

Marine Particles: Analysis and Characterization
Geophysical Monograph 63

Fig. 5-13. Photograph of the McLane Water Transfer System being deployed on a mooring at the North Cleft Segment vent field on the Juan de Fuca Ridge. During this deployment seawater was pumped through 37 mm 0.4 μm pore-size Nuclepore filters.

There is an air filter in the tygon line to prevent contamination of the sample from particles in the nitrogen gas. The samples are pressure filtered at 12 psi above ambient pressure. A 20 L sample takes about 2-8 hours for completion of the filtration process. The filtrate is directed into a precalibrated 20 L plastic jerrican bottle and the volumes are measured to the nearest 100 mL.

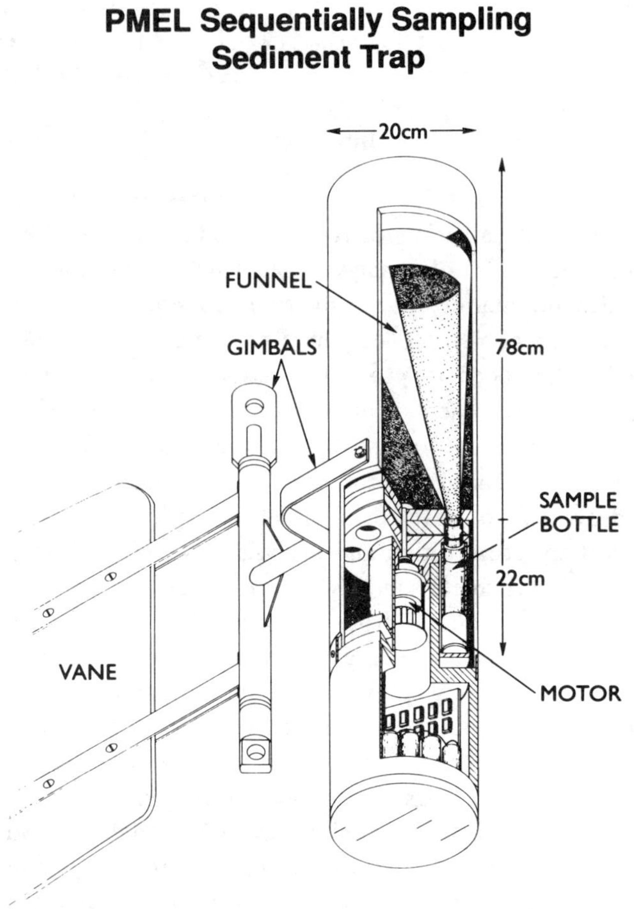

Fig. 5-14. Schematic drawing of the PMEL sequentially-sampling sediment trap. The electronic timer can be programmed to rotate the sample bottles at various time intervals from hours to months.

In our laboratory, in situ filtration is accomplished by deploying a Water Transfer System (McLane Research Laboratories, Falmouth, MA) on a rosette or on a mooring (Figure 5-13). The system is designed to filter up to eighteen samples sequentially through the same kind of 37 mm Nuclepore filters that we use for the sampling with the Niskin bottles. The sample acquisition times and durations are controlled by an internal computer system which is preprogrammed prior to deployment.

After the filtrations are completed, the filter holders are brought into a laminar flow hood where the samples are carefully rinsed with distilled water adjusted to pH 8.0 ± 0.3 with Ultrex ammonium hydroxide. The rinses are conducted from the outside of the filter to the center to insure that all of the filter is covered with the distilled water to remove any trace of residual seawater. Near the

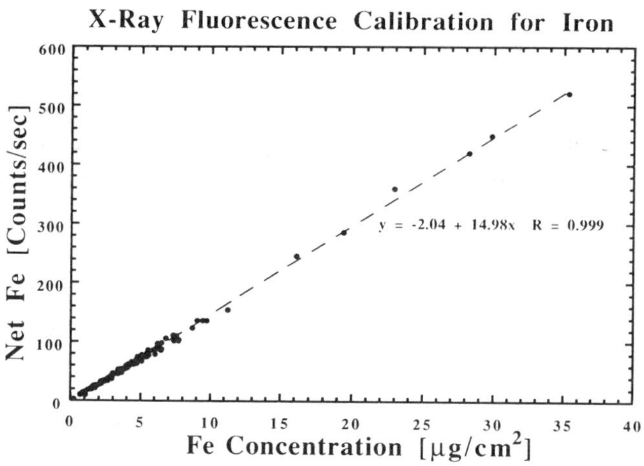

Fig. 5-15. X-ray fluorescence calibration for Fe using a selection of standard reference materials listed in Table 5-6.

edge of the filter where it makes contact with the Teflon insert requires special care, since this is the region where most of the residual buildup of sea salt occurs. It should be noted, however, that too many rinses can cause some of the delicate particles to dissolve in the rinse water. Consequently, we keep the number of rinses to the minimum required to remove all residual seawater, usually about 3-4 rinses.

After the filters are rinsed, they are put into prelabeled Millipore petri dishes and placed into a desiccator containing a silica gel drying agent which is covered to prevent dust contamination. The samples are dried under vacuum in the desiccator for at least 24 hours and packaged for shipment to the laboratory.

3. Sample Collection With Sediment Traps

Settling particulate matter is sampled with sequentially sampling sediment traps [Baker and Milburn, 1983] deployed at various heights above the bottom on moorings or in vertical stands on the sea floor [Baker et al., 1985; Feely et al., 1986, 1990b]. The sediment traps consist of a 1.0 by 0.2 m polyvinyl chloride (PVC) cylinder containing an asymmetric funnel which directs the particles into one of ten acrylic sample bottles (Figure 5-14).

The sample bottles are prewashed with 6 N HCl, rinsed with filtered deionized water (Milli-Q), and air dried. Each bottle contains a formalin solution mixed with filtered seawater (final solution 1% formalin in 80 parts per thousand salinity saltwater) to produce a continuous antibacteriological effect. The sediment traps are electronically programmed to collect ten samples during the duration of the deployment. During deployment and recovery the sample collection carousel is positioned in an open position to insure that no samples are lost.

TABLE 5-6.
X-ray Fluorescence Spectrometry: Standards and Values Used in the Calibration* Wherever Values are Given

Standard	Al Wt. %	Si Wt. %	K Wt. %	Ca Wt. %	Ti ppm	V ppm	Mn ppm	Fe Wt. %	Cu ppm	Sr ppm	Ba ppm
GXR-1							930	24.67	1300	280	560
GXR-4				0.90			140	2.97	6500	220	1350
GXR-6					4970	180		5.58	105		1100
MAG-1	9.23	25.33		0.98	4760		820	5.17		148	
BCSS-1				0.53	4490		234	3.35			
BHVO-1				8.10	16110		56580	8.55	140	420	
BCR-1								9.37		330	
GSD-1				3.32		120	930	5.13		520	950
GSD-4					5390	120	850	4.12			
GSD-5					5390	110	1160	4.08	135	200	
GSD-6			1.69		4670	140	1010	4.09	380	260	
GSD-7			2.02	1.19	4490	125		4.54		220	730
SGR-1			1.35	5.95	1440		35370	2.08		430	290

* Calibration of Ni and Zn, as well as of P and As, implemented by dissolving selected samples in a Teflon bomb and analyzing by atomic absorption and/or flow injection analysis.

TABLE 5-7.
X-ray Fluorescence Spectrometry: Minimum Detection Limits and Precision Using 37 mm Nuclepore Aerosol Polycarbonate Membrane Filters

Element	Minimum Detection Limit	Precision RSTD (%)
Al wt.%	0.019	13
Si wt.%	0.020	2.1
P wt.%	0.006	2.1
K wt.%	0.008	30
Ca wt. %	0.006	6.1
Ti ppm	28	28
V ppm	8	5
Mn ppm	8	24
Fe wt. %	0.006	1.1
Ni ppm	5	47
Cu ppm	6	4
Zn ppm	5	5
As ppm	9	4
Sr ppm	22	7
Ba ppm	26	27

4. Laboratory Analyses

In the laboratory, the samples are kept in a desiccator with the drying agent until they are weighed and analyzed. The samples are weighed using a Cahn Model 2600 Electrobalance in a controlled temperature room. The samples are weighed to the nearest microgram. Reference filters are weighed along with the samples and corrections are made to take into account local variations in the weight of the filters which are due to temporal changes in the temperature and humidity of the weighing room.

The elemental composition of the particulate matter is determined by thin-film x-ray primary- and secondary-emission spectrometry using a Kevex Model 8000-770 x-ray energy spectrometer. Samples are placed into a specially designed sample holder and excited with x-rays generated by a rhodium x-ray tube operated at a potential between 10 and 58 kV. We analyze the samples for Mg, Al, Si, P and Cl by direct excitation; K, Ca, Ti, V, Cr, Mn, Fe, Ni, Cu, Zn and Ba by germanium secondary excitation; As and Pb by zirconium secondary excitation; and Sr by silver secondary excitation, which maximizes the overall counting efficiency for the elements of interest. The procedure is non-destructive and has the capability of detecting most elements of atomic number greater than 11 and concentration levels greater than 10 ppm in the sample. This technique for analyzing particulate matter has been described in the literature [Price and Angell, 1968; Cann and Winter, 1971; Baker and Piper, 1976; Criss, 1976; Subramanian and d'Angeljan, 1976; Vanderstappen and Van Grieken, 1976; Sholkovitz and Price, 1980; Holmes, 1981; Feely et al., 1981, 1986, 1987, 1990a, b; Pietilainen et al., 1981].

Standardization is effected by analysis of "thin-film" deposits of geological reference standards of similar matrices on Nuclepore filters. In our laboratory, we use the U.S. Geological Survey Standard Sediment and Rocks (MAG-1, BCR-1, BHVO-1, SGR-1, GXR-1, -4 and -6), National Research Council of Canada Reference Materials (BCSS-1), and Chinese Institute of Geophysical and Geochemical Exploration Sediments (GSD-1, -4 through -7), suspended on Nuclepore filters identical to those used for sample acquisition (Figure 5-15). The reference values for the thin-film standards used in the calibrations are shown in Table 5-6. The reported values for trace elements in suspended particulate matter are calculated in the following manner:

$$E_i = \frac{C_i \cdot A}{W_t \cdot S}$$

where:

E_i = the concentration of element "i" in ppm,

C_i = the net counts/sec^{-1} cm^{-2}),

W_t = the mass of suspended matter on the filter in mg,

A = the effective area of suspended matter on the filter, and

S = slope of the calibration line for the standards.

For some elements, we double check the standard values using flow injection and atomic absorption procedures developed in our laboratory.

The minimum detection limits are determined from the counting statistics and are defined as:

$$MDL = 2 \cdot K \cdot \frac{1}{(T)^{1/2}} \cdot \frac{(I_b)^{1/2}}{I_p}$$

where:

K = standard concentration in the desired units (Wt. % or ppm),

T = analysis time in seconds,

Ib = Background intensity in counts per second,

Ip = net peak intensity in counts per second.

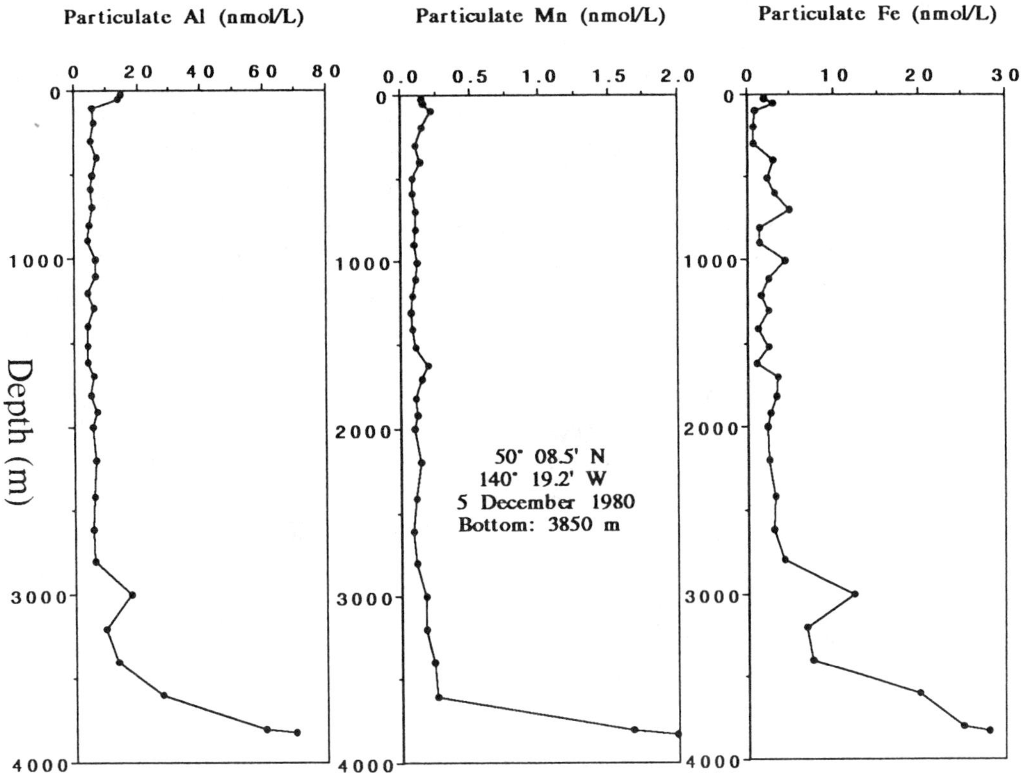

Fig. 5-16. Profiles of particulate Al, Mn, and Fe at a station located in the subarctic North Pacific (50° 68.5′N, 140° 19.2′W; 5 December 1980). The samples were analyzed by the PMEL x-ray fluorescence spectrometer.

The precision is given in terms of coefficient of variation:

$$C.V. = \frac{1\sigma_{error}}{\text{mean value}} \cdot 100$$

The minimum detection limits and precision for 26 replicate measurements of a typical deep-ocean sample are given in Table 5-7. Figure 5-16 shows representative profiles of particulate Al, Mn, and Fe from the subarctic region of the northeast Pacific based on our x-ray fluorescence technique. The general patterns of surface water enrichment due to atmospheric input and near-bottom enrichment due to resuspension of bottom sediments are clearly indicated in the data.

The concentration levels are similar to what has been observed in other profiles for the northeast Pacific [Orians and Bruland, 1986; Landing and Bruland, 1987]. The analytical results show a high degree of precision because the concentrations observed in these open-ocean samples are well above the minimum detection limits for most of the elements of interest. Since the x-ray fluorescence procedures are totally non-destructive, our group uses subsets of the analyzed samples for further studies of the chemical and physical parameters as described below.

Trace elements associated with poorly structured oxyhydroxide phases of the suspended matter are determined by the method described by Feely et al. [1986]. Desiccated samples are leached with 5 mL of a mixed reagent containing 0.04 N hydroxylamine hydrochloride in 25% Ultrex acetic acid at 50° C for 2 hr. The resulting supernate is filtered through an acid-cleaned Teflon filtration unit containing a 0.2 μm Nuclepore filter. The residue is rinsed and refiltered; the supernate is then combined with the original supernate, acidified with 0.5 mL of concentrated Ultrex HCl, diluted to 20 g total weight with quartz-distilled water, and analyzed for trace elements by flameless atomic absorption spectroscopy.

Morphological analysis of individual mineral phases in particulate samples is determined with an ISI DS130S Scanning Electron Microscope (SEM) equipped with a Kevex Extra Detector with a beryllium window or in the windowless mode. The SEM is operated at an energy of 20 kV and a magnification range of about 2000-27,000x. Sections of particulate samples on 0.4 μm pore size

Nuclepore aerosol polycarbonate membrane filters are fixed with carbon paint to the tops of carbon planchettes that are mounted on aluminum stubs. The samples are then sputter coated with a thin film of palladium for electrical conductivity. Palladium is chosen because its x-ray peaks do not interfere with the x-ray lines from samples of marine particulate matter we have analyzed. Size distribution analysis combined with qualitative chemical analysis is performed on the samples using a Kevex Model 8000 x-ray energy spectrometer, interfaced to the above mentioned SEM. Full field analysis is achieved using Kevex Automated Image Analysis software. The data are used to determine trace metal associations with specific particle phases as well as size changes of specific mineral phases in buoyant hydrothermal plumes [Feely et al., 1987, 1990b].

Acknowledgements. The authors wish to express their appreciation to Anthony J. Paulson and Dave Tennant of PMEL for their discussions about the work presented in this paper and to the officers and crew of the NOAA Ships Oceanographer and Discoverer for logistic support during sampling. This work was supported by the National Oceanic and Atmospheric Administration with funds from the NOAA VENTS Program. Contribution No. 1248 from Pacific Marine Environmental Laboratory/NOAA.

References

Baker, E. T., and H. B. Milburn, An instrument system for the investigation of particle fluxes, Cont. Shelf Res., 1, 425-435, 1983.

Baker, E. T., and D. Z. Piper, Suspended particulate matter; Collection by pressure filtration and elemental analysis by thin film x-ray fluorescence, Deep-Sea Res., 3, 181-186, 1976.

Baker, E. T., R. A. Feely, M. R. Landry, and M. F. Lamb, Temporal variations in the concentrations and settling flux of carbon and phytoplankton pigments in a deep fjord-like estuary, Estuarine Coastal Shelf Sci., 21, 859-877, 1985.

Calvert, S. E., and M. J. McCartney, A note on the effect of incomplete recovery of large particles from water samples on the chemical composition of oceanic particulate matter, Limnol. Oceanogr., 24, 532-536C, 1979.

Cann, J. R., and C. K. Winter, X-ray fluorescence analysis of suspended sediment in seawater, Mar. Geol., 11, M33-M37, 1971.

Criss, J. W., Particle size and composition effects in the X-ray fluorescence analysis of pollution samples, Anal. Chem., 48, 179-186, 1976.

Feely, R. A., G. J. Massoth, and W. M. Landing, Major and trace element composition of suspended matter in the northeast Gulf of Alaska, Mar. Chem., 10(15), 431-453, 1981.

Feely, R. A., G. J. Massoth, E. T. Baker, J. F. Gendron, and A. J. Paulson, Seasonal and vertical variations in the elemental composition of suspended and settling particulate matter in Puget Sound, Washington, Estuarine Coastal Shelf Sci., 22, 215-239, 1986.

Feely, R. A., M. A. Lewison, G. J. Massoth, G. Robert Baldo, J. W. Lavelle, R. H. Byrne, K. L. Von Damm, and H. C. Curl, Jr., Composition and dissolution of black smoker particulates from active vents on the Juan de Fuca Ridge, J. Geophys. Res., 92(11), 347-363, 1987.

Feely, R. A., G. J. Massoth, E. T. Baker, J. P. Cowen, M. F. Lamb, and K. A. Krogslund, The effect of hydrothermal processes on midwater phosphorus distributions in the Northeast Pacific, Earth Planet. Sci. Lett., 96, 305-318, 1990a.

Feely, R. A., T. L. Geiselman, E. T. Baker, G. J. Massoth, and S. R. Hammond, Distribution and composition of buoyant and non-buoyant hydrothermal plume particles from the ASHES vent field at Axial Volcano, Juan de Fuca Ridge, J. Geophys. Res., 95(B8), 12,855-12,874, 1990b.

Gardner, W. D., Incomplete extraction of rapidly settling particles from water samplers, Limnol. Oceanogr., 22 (4), 764-765, 1977.

Holmes, G. S., The limitations of accurate 'thin film' x-ray fluorescence analysis of natural particulate matter; problems and solutions, Chem. Geol., 33, 333-353, 1981.

Landing, W. M., and K. W. Bruland, The contrasting biogeochemistry of iron and manganese in the Pacific Ocean, Geochim. Cosmochim. Acta, 51, 29-43, 1987.

Orians, K. J., and K. W. Bruland, The biogeochemistry of aluminum in the Pacific Ocean, Earth Planet. Sci. Lett., 78, 397-410, 1986.

Pietilainen, K., F. Adams, H. Nullens, and P. Van Espen, Multi-element analysis of phytoplankton by tube-excited energy-dispersive x-ray fluorescence, X-ray Spectrom., 10(1), 31-40, 1981.

Price, N. B., and G. Angell, Determination of minor elements in rocks by thin-film x-ray fluorescence techniques, Anal. Chem., 40, 660-663, 1968.

Sholkovitz, E. R., and N. B. Price, The major-element chemistry of suspended matter in the Amazon Estuary, Geochim. Cosmochim. Acta, 44, 163-171, 1980.

Subramanian, V., and B. F. d'Angeljan, Methodology of x-ray fluorescence analysis of suspended matter in estuarine waters, Mar. Geol., 22, M1-M6, 1976.

Vanderstappen, V., and R. Van Grieken, Trace metal analysis of sediments and particulate matter in seawater by energy-dispersive x-ray fluorescence, Z. Anal. Chem., 282, 25-30, 1976.

SURFACE CHARGE AND SIZE SPECTRA OF MARINE PARTICLES

Keith A. Hunter
Department of Chemistry, University of Otago, Dunedin, New Zealand

1. Particle Electrophoresis

This technique is well established in colloid chemistry, but needs modification for application to marine particles. The technique measures the electrophoretic mobility of individual particles in suspension, i.e., their velocity per unit of applied field. It is closely related to other electrokinetic techniques: electro-osmosis, streaming potential and sedimentation potential.

The theory and practice of electrophoresis has been reviewed by Shaw [1969]. The electric field surrounding a particle in an electrolyte suspension is determined by several factors. These include charges developed at the particle surface through ionization processes, specific adsorption of ions in the inner and outer compact layers and the diffuse double layer. During electrophoretic motion, a viscous slipping plane develops within the double layer. This results in part of the double layer moving with the particle in a combined electrophoretic unit. Consequently, electrophoretic mobilities cannot be directly related to the charge on the surface of the particle.

Electrophoretic mobility is customarily converted to the zeta potential ζ which may be regarded as the potential at the slipping plane. The conversion factor depends on re-tardation, relaxation and surface conductance effects which must be estimated by numerical solution of the equations of motion [Wiersma et al., 1966]. The principal controlling factor is the ratio of double layer thickness κ^{-1} to particle radius of curvature a.

For very small particles having $\kappa a < 1$, the zeta potential may be calculated from the electrophoretic mobility u_E using the equation:

$$\zeta = 3\eta/2\varepsilon\varepsilon_o u_E$$

where η is the kinematic viscosity, ε the electric constant and ε_o the permittivity of free space.

For very large particles having $\kappa a > 300$, the Smoluchowskii equation is used

$$\zeta = \eta/\varepsilon\varepsilon_o u_E$$

At intermediate values, $1 < \kappa a < 300$, a more complex relationship must be used [Ottewill and Shaw, 1972]. Since the double layer thickness in normal seawater is only a few Angstroms, calculation of zeta potentials for marine particles is problematic.

Equipment required for particle electrophoresis is relatively simple. A suitable apparatus is manufactured by Rank Instrument Co., Cambridge, U.K. The main modifications needed for work with seawater electrolytes result from the high electric current that develops from the electric field. Hunter and Liss [1979, 1982] introduced a quartz cell of low cross-sectional area (4 x 0.5 mm) to reduce current and used Ag-AgCl electrodes to minimize electrode gassing and polarization. Hydrogen-charged palladium electrodes [Neihof and Schuldiner, 1960; Neihof, 1969] are also very useful, but less convenient to use and more limited in their current-carrying capacity.

Generally, measurements are made on at least 10 particles at each of the electro-osmotic stationary levels in the cell. Random errors affect the results, so that electrophoretic mobilities for electrokinetically homogeneous particles have experimental errors of about 0.08 x 10^{-8} m^2 s^{-1} V^{-1}. The most difficult measurements to make are those on particles of near-zero mobility, less than 0.2 x 10^{-8} m^2 s^{-1} V^{-1}, especially in seawater. A rotating prism attachment for measuring the mean mobility of a large number of particles is available for the Rank instrument. I have found that this is not very effective for natural particle populations because the variations in mobility from particle to particle make it very confusing to use.

The particle size range applicable to electrophoresis is approximately 0.5 to 5 μm radius. The upper size limit depends on density and is limited by the rate at which particles settle out of the field of view in the cell. The lower size limit is set by optical visibility and may be reduced by replacing the quartz-halogen lamp with a He-Ne laser. It is often useful to remove larger particles from the size range before electrophoretic measurements using differential sedimentation, e.g., in a tall glass measuring cylinder.

In seawater electrolyte, the aggregation of particles is relatively rapid. Thus, particle suspensions are normally dispersed by shaking or ultrasonication immediately before electrophoretic measurements are

Marine Particles: Analysis and Characterization
Geophysical Monograph 63

made. My experiments have indicated that this has no measurable effect on particle mobilities.

A minimum particle concentration of approximately 10^5 particles mL^{-1} is needed for reliable work. Generally, coastal and estuarine samples can be studied without preconcentration. Open-ocean waters, which have not been extensively studied, require concentration. High-speed centrifugation is the most obvious technique to use. I believe that particles could be redispersed into seawater from Nuclepore filters, e.g., by ultrasonication, but I have not investigated this approach yet. About 20 mL of final suspension is needed for a measurement. This includes water required for rinsing the electrophoresis cell.

Because of the importance of organic adsorption by particle surfaces, organic contamination is a major problem in making measurements on dilute suspensions such as offshore seawater. Plastic material must be avoided where possible. To date, I have not worked extensively with sub-surface samples and I do not have information as to whether conventional samplers, e.g., Niskin and Go-Flo bottles, introduce artifacts. Glassware and reagents must be extensively decontaminated of organics before use. I have found that heating glassware in an annealing oven is highly suitable. Aqueous solutions can be purified by extensive ultraviolet photo-oxidation [Hunter and Liss, 1982]. As a general rule, much of the plastic equipment necessary to avoid trace metal contamination is probably incompatible with electrophoresis measurements because of plasticizers that may leach from the equipment and coat the surface of particles. This may not be a problem with natural particle populations, whose surfaces are already coated with marine organics. However, it is a serious problem in the study of adsorbable marine organics by means of electrophoresis using synthetic surfaces added to the water.

My studies with coastal water and estuarine particles suggest that the electrophoretic mobilities of particle suspensions are relatively stable if the samples are stored for periods of weeks. I believe that this is principally a result of the blanketing effect of surface organic films. It is therefore possible that considerable remineralization of the particle matrix could occur without showing up in the electrophoretic studies.

Particle electrophoresis has been applied to marine particles in two different ways. Several studies have focussed on determining the mobilities of marine particles [e.g., Neihof and Loeb, 1972; Hunter and Liss, 1979, 1982; Hunter, 1983; Loder and Liss, 1984]. These studies, which have been largely confined to coastal and estuarine waters, have all emphasized the relatively uniform mobility distributions exhibited by marine particles. This has been attributed to the presence of adsorbed films of organic matter.

The nature of adsorbed films has also been investigated by electrophoresis [Loeb and Niehof, 1975; Neihof and Loeb, 1974; Hunter, 1980; Hunter and Liss, 1982]. A wide variety of surfaces become rapidly coated with an adsorbed organic film when exposed to particle-free seawater. Hunter [1980] showed that the mobility-pH and mobility-metal ion properties of the adsorbed films indicated the presence of carboxyl and phenolic functional groups. This study shows that well-defined particle suspensions, e.g., alumina, quartz or polystyrene, can be used as probes for examining the chemical nature of colloidal organic matter. In addition to being useful for looking at functional group chemistry, the microelectrophoresis of such suspensions can be useful in the development of specific adsorption schemes for the isolation of colloidal organics from larger volumes of seawater.

To date, relatively few studies have been made of the electrophoretic properties of oceanic particles. Specific aspects that need attention include the functional group composition of organic coatings on oceanic particles and whether these coatings are removed or modified as particles sink down the water column and are remineralized. Attention needs to be given to the suitability of different water samplers for electrophoresis measurements.

2. Organic Coatings on Particles

The studies just cited provide strong evidence for the existence of organic coatings on most marine particles. However, particle electrophoresis is unable to distinguish bulk properties of particles so that biogenic organic particles would appear similar to minerals with adsorbed coatings.

The reversibility of organic film adsorption is not yet well known, but has important implications for the effects of films on particle-reactive trace elements. Adsorbable organics isolated from natural waters, e.g. humic acids, exhibit well-characterized reversible adsorption on model particle surfaces. However it is likely that consolidated films may form on particles after repeated exposure to seawater. Films studied by Hunter [1980] were not removed by repeated washing in organic-free electrolytes. This suggests very strong adsorption, perhaps mediated by chemical bonding to the surface.

Specific questions that remain unanswered concerning organic films on particles are:

1) How common are films on deep ocean particles?
2) Are films irreversibly adsorbed, and how resistant are they to microbial or chemical degradation?
3) Do surface films mask reactive element adsorption to the particle matrix itself?
4) Is the film material a significant trace element scavenger?
5) Are trace elements important to the adhesion of films, e.g. through bridging mechanisms?

3. Sub-Micron Colloidal Particles

Considerable work has been carried out on the size spectra of particles larger than about 1 μm in seawater, particularly using the Coulter counter technique. These measurements have considerable relevance to an understanding of particle removal rates by settling processes. However, the size spectra of sub-micron colloidal particles in seawater are unknown. The colloidal fraction is important in mechanisms of particle growth into the settleable range, and in mechanisms for reactive element scavenging.

Recently, I have experimented with techniques based on that of Nomizu et al. [1988] for the determination of sub-micron particle size spectra of fresh water samples. This method involves the deposition of sub-micron particles onto a hydrophilic carbon film by ultracentrifugation. Subsequent analysis is carried out by transmission electron microscopy (TEM). The hydrophilicity of the carbon film is essential for promoting adhesion of the particles to the surface. If adhesion is weak, sub-micron particles are quickly detached after centrifugation is terminated because their Brownian displacement distance is relatively large.

Particles are deposited onto standard 3 mm Cu/Rh microscope grids of BS 135 mesh size (supplied by Graticules Ltd.). The grids were cleaned with detergent and water, then with Freon in an ultrasonic bath. After drying in a laminar-flow bench, the grids were coated with a Formvar plastic support film. This was found necessary to prevent the fracture of the carbon film. Carbon coating was carried out with a Hitachi HUS-5GB high vacuum evaporator to deposit a 20 nm layer of carbon. Grids were stored in covered petri dishes on top of clean filter paper. Just before deposition of particles (within 1 hour) the carbon surface was rendered hydrophilic by glow discharge for 1 minute at 3 kV under vacuum using a Hitachi HUS-3 instrument.

Water samples were collected in 250 mL Pyrex glass bottles which were precleaned with detergent and water and then baked at 450° C for 1 hour to destroy adhering organics immediately before collection. In experiments to date, some samples have been processed unfiltered, while some were prefiltered through 0.4 μm pore size Nuclepore filters to remove larger particles.

For particle deposition, grids were mounted in a custom-built Plexiglass holder with a Teflon collar supporting the grid. This combination was then placed in a 4.4 mL polyethylene centrifuge tube. Next, 3 mL of water sample were added and the tube centrifuged for 1 hour using a Beckmann L5-65 refrigerated ultracentrifuge and SW60 rotor head at 50,000 rpm. This generates a centrifugal force of 300,000 g. This geometry corresponds to a minimum particle diameter transferred over the length of the centrifuge tube to the grid surface of 9.4 nm diameter, assuming a particle density of 2 g mL^{-1}.

After centrifugation, the supernatant water was carefully removed using a Pasteur pipette inside a clean bench. The grid holder was then withdrawn using special tweezers. Finally, the grid was lifted out of the holder, again with tweezers, and placed on filter paper in a covered petri dish.

Sample handling after centrifugation has to be different with seawater samples to prevent the formation of salt crystals on the carbon surface. In this case, the water is removed by pipette to just above the surface of the grid, after which the centrifuge tube is refilled with deionized, organic-free water. The Plexiglass holder is designed to fill the bottom of the centrifuge tube as much as possible, so that the amount of seawater remaining is minimal. After the seawater and deionized water have been mixed, the water removal process is repeated once or twice more. The essential detail of this protocol is to avoid allowing the grids to become exposed to air while the water is saline.

After drying, the grids were examined by TEM using a Hitachi EM-002A microscope. Photographs of the grid sections (typically 15-20 per grid) were taken at different positions, mostly at 20,000 magnification. These were enlarged by 3.5 to give a typical magnification factor of 70,000. Particle size analysis was carried out manually. Photographs were distributed among eight people and the number and size of all individual particles determined manually. Particle size was reported as the maximum and minimum dimension of the particle image and converted to an equivalent particle radius by assuming ellipsoidal geometry. There is a certain amount of subjectivity involved in the measurement of particle size by this method, but hopefully any systematic bias introduced from this source is averaged out by using a number of different people to make the measurements.

The application of this method to seawater particles is still in its infancy. In my work, I have concentrated mainly on particle size determination and the presence of fractal aggregates in the sub-micron size range. The results obtained are consistent with theoretical expectations that Brownian motion plays a dominating role in determining the shape of the size distribution. Very recently, Ed Goldberg (Scripps Institution of Oceanography) has carried out TEM-EMP elemental analysis of individual particles by this method. Other grid materials, e.g. Be, would be desirable for this application. However, the technique has great promise for revealing details of particle dynamics in the important sub-micron colloidal size range.

References

Hunter, K. A., Microelectrophoretic properties of natural surface active organic matter in coastal seawater, Limnol. Oceanogr., 25, 807-822, 1980.

Hunter, K. A., On the estuarine mixing of dissolved substances in relation to colloid stability and surface properties, Geochim. Cosmochim. Acta, 47, 467-473, 1983.

Hunter, K. A., and P. S. Liss, The surface charge of suspended particles in estuarine and coastal waters, Nature (London), 282, 823-825, 1979.

Hunter, K. A., and P. S. Liss, Organic matter and the surface charge of suspended particles in estuarine waters, Limnol. Oceanogr., 27, 322-335, 1982.

Loder, T. C., and P. S. Liss, The role of organic matter in determining the surface charge of suspended particles in estuarine and oceanic waters, Thalass., 17, 433-447, 1984.

Loeb, G. I., and R. A. Neihof, Marine conditioning films, in Application Chemistry at Protein Interfaces, edited by R. E. Baier, Adv. Chem. Ser. 145, pp. 319-335, American Chemical Society, Washington D.C., 1975.

Neihof, R. A., Microelectrophoresis apparatus employing palladium electrodes, J. Colloid Interface Sci., 30, 128-133, 1969.

Neihof, R. A., and G. I. Loeb, The surface charge of particulate matter in seawater, Limnol. Oceanogr., 17, 7-16, 1972.

Neihof, R. A., and G. I. Loeb, Dissolved organic matter in seawater and the electric charge of immersed surfaces, J. Mar. Res., 32, 5-12, 1974.

Neihof, R. A., and S. Schuldiner, Simple non-degassing electrodes for use in electrophoresis, Nature (London), 185, 526-527, 1960.

Nomizu, T., K. Goto, and A. Mizuike, Electron microscopy of nanometer particles in freshwater, Anal. Chem., 60, 2653-2656, 1988.

Ottewill, R. H., and J. N. Shaw, Electrophoretic studies on polystyrene latices, J. Electroanal. Chem., 37, 133-142, 1972.

Shaw, D. J., Electrophoresis, 144 pp., Academic Press, London, 1969.

Wiersma, P. H., A. Loeb, and J. Th. G. Overbeek, Calculation of the electrophoretic mobility of a spherical colloid particle, J. Colloid Interface Sci., 22, 78-85, 1966.

COLLECTION, PROCESSING, AND ANALYSIS OF MARINE PARTICULATE AND COLLOIDAL MATERIAL FOR TRANSITION METALS

William M. Landing and Brent L. Lewis

Department of Oceanography, Florida State University, Tallahassee, Florida 32306

1. Introduction

Trace-metal analysis of marine particles can add tremendously to our understanding of the oceanic biogeochemical cycles of these elements. As examples, particulate Al can be used as a tracer of aeolian input, near-bottom resuspension, and offshore horizontal transport [Spencer and Sachs, 1970; Wallace et al., 1977; Feely et al., 1981; Landing and Bruland, 1987]. Redox cycling of Mn and Fe during suboxic diagenesis can be studied by comparing dissolved and particulate profiles [Landing and Bruland, 1987]. In fact, any cycling process which is suggested to transform dissolved metal to/from particulate form must be reflected in the chemistry of the particles. In addition, trace-metal analysis of rapidly settling material collected via sediment traps can be used to put limits on the rates of some biogeochemical trace-metal cycling processes. In the open ocean, the flux of metals out of the photic zone can be interpreted much like that of nitrogen, representing the effects of "new production" on trace-metal cycling. Changes in the flux of reactive metals such as Mn with depth can be used to calculate rates of in situ scavenging or regeneration [Landing and Bruland, 1987; Martin and Knauer, 1984]. Finally, the role played by colloidal-sized particles as intermediates between filterable (>0.4 μm) and truly dissolved metal species may be important for metals whose solution chemistry is dominated by hydrolysis reactions [Al, Fe, Th; Honeyman and Santschi, 1989].

The successful collection and processing of particulate samples for trace-metal analysis requires strict attention to contamination control, especially during sample collection and initial filtration or processing. Airborne dust, dirt and oils from handling, and exposure to ship-board sources of metals are all potential sources of contamination. Contamination problems are so severe for dissolved metals such as iron and lead that no reliable data for the marine concentrations of these elements existed in the literature prior to the mid- to late 1970's. While there is pervasive belief that such contamination problems are not so severe for particulate metals, we feel that the best approach is to take all reasonable precautions to avoid any contamination. In this way, we are able to collect samples for both dissolved and particulate trace-metal determinations with confidence that we will obtain accurate values. Thus, in our research, sample collection and processing are carried out under "trace-metal clean" conditions, following many of the procedures described by Bruland et al. [1979] and Bruland [1980].

2. Clean Lab Facilities

All sample collection, handling and processing procedures are carried out in "clean room" facilities. At Florida State University, a small existing room was converted to a clean lab. A 2" x 4" lumber framework was erected inside the room and covered with plastic sheeting. A small area just inside the room door was left to serve as a changing area. This clean lab is equipped with a Class-100 (HEPA) blower (NuAire) and a Class-100 laminar-flow, exhausting fume hood (Liberty Industries). Positive pressure in the clean lab helps to prevent dust from entering the clean area. All personnel working in the clean lab wear nylon hair nets and booties and powder-free polyethylene gloves; full-body coveralls are often used in the clean lab facilities of other trace-metal investigators. Any sample manipulations outside the clean lab are done in Class-100 table-top laminar flow hoods (Environmental Air Control).

3. Reagents

"Trace metal-free" water and reagents are prepared in the clean lab. Ultra-pure (18 megohm) water is produced from doubly-deionized water using a Barnstead Nano-Pure water system. Clean acids (concentrated Q-HNO_3 and Q-HAc; 6 M Q-HCl) are prepared by sub-boiling point redistillation of reagent grade (RG) acids in a quartz still. Concentrated RG-HCl is diluted 1:1 with Nano-Pure water prior to redistillation. Clean ammonia solutions (Q-NH_3, 5-7 M) are prepared by passive distillation of concentrated RG-NH_3 or by bubbling anhydrous ammonia through Nano-Pure water. Clean ammonium acetate (NH_4Ac) buffers are prepared by mixing appropriate amounts of dilute equimolar Q-HAc

Marine Particles: Analysis and Characterization
Geophysical Monograph 63

(acetic acid) and Q-NH_3 solutions. Final pH adjustment is done by addition of small aliquots of Q-HCl or Q-NH_3.

4. Cleaning Procedures

All plastic ware is subjected to an extensive cleaning procedure, depending on the chemical resistivity of particular plastics: an acetone rinse to remove organics, 10% Micro detergent (Cole-Parmer) wash, doubly-deionized water rinse, 3 M RG-HCl soak for more than one week, double distilled water rinse, 0.5 M Q-HCl soak for another 1-2 weeks, and a final rinse with Nano-Pure water. Bottles for sample storage are further cleansed by filling the bottles with fresh 0.5 M quartz-redistilled hydrochloric acid (Q-HCl) for at least one week. Sample bottles are then rinsed four times with Nano-Pure water, double-bagged in new ziplock storage bags, and stored in plastic garbage bags in the clean lab until packed for shipping. We do not use nitric acid in our bottle cleaning procedure due to potential oxidation artifacts associated with residual nitric acid which can leach out of the bottle walls. We have observed this problem with redox-sensitive reagent solutions, including colorimetric indicators and sulfide solutions.

Filters for ship-board sample filtration, 142 mm diameter, 0.4 μm pore size Nuclepore polycarbonate, are also pre-cleaned using the following recipe. The filters are batch washed for one week in 6 M Q-HCl, rinsed well with Nano-Pure water, batch washed for one week in 3 M Q-HNO_3, and finally rinsed well by holding each filter with plastic forceps and rinsing with a squirt bottle of Nano-Pure water. Filters are then laid flat to dry on polystyrene, 1 x 1 cm cell, "egg-crate" in plastic tubs in the clean lab. The tubs are covered but not tightly sealed to allow clean air to circulate freely.

Dried filters are pre-weighed on a Cahn C-30 microbalance and individually packaged unfolded in small, sandwich-size ziplock plastic bags. The balance is always calibrated with the same 200 mg calibration weight, and filters are weighed to the nearest 0.01 mg. Calibration filters are also weighed several times to account for any drift during the weighing period. To reduce the effects of static electricity, two ^{210}Po ionizing strips are used. One is used outside the balance chamber to reduce static on the filter before placing in the chamber. The other strip is placed under the balance pan in the weighing chamber.

5. Sample Collection

Dissolved and particulate trace-metal samples have been collected using approximately 28.5 L actual volume, Teflon-coated, Go-Flo bottles (General Oceanics) or similar Teflon-coated PVC ball-valve bottles constructed at the School of Oceanography at the University of Washington. These bottles are fitted with a Teflon bulkhead-mount spigot valve (1/2" M-NPT to 3/8" OD tube; Chemplast). A short length of Teflon tubing is run from the inside of this valve to the lower internal edge of the bottle near the lower ball-valve to ensure complete draining of the bottle contents. In the event that the filtration rate drops below 10 mL min^{-1}, a second filter holder will be substituted to complete filtration of the bottle contents. The results from the two filters are then added together.

Sampling has been conducted using non-metallic hydrowire such as Kevlar or polyethylene-coated cable, and on some occasions using a rosette/CTD system. Samples collected using corroded hydrowire and triggered by sliding messengers have been shown to be severely contaminated for metals such as Fe [Betzer and Pilson, 1975]. Sampling bottles with internal closing mechanisms may also be prone to contamination, however use of new silicone-rubber tubing or epoxy-coated stainless steel springs may be acceptable.

6. Ship-Board Sample Processing

Great care is taken throughout the sample processing procedures to prevent contamination. When bottles reach the deck the outer surface of the upper ball valve of each bottle is wiped with a clean Kimwipe towel and large plastic bags are placed over the top of each bottle. The top and bottom ball valves are never touched with bare hands, and the ship's rail where the bottles go over the side is cleaned prior to sampling, and usually covered with a thick layer of duct tape. The sampling bottle valves and gas fittings, the filter sandwiches, and plastic ware are carefully rinsed with Nano-Pure water at each stage of the sample processing. Polyethylene gloves are worn throughout the sample processing procedures and filters are handled with acid-washed plastic tweezers.

Upon recovery, the sampling bottles are taken into the clean area for sample processing. Portable "clean" vans provide a convenient area for processing samples for both dissolved and particulate trace metals. In the absence of such a facility, it is a simple matter to construct a "clean" area using 2" x 4" lumber and plastic sheeting, generally in one corner of the ship's lab space. The plastic sheeting is tacked down using a staple gun, and exposed staples are covered with duct tape. Positive-pressure, HEPA-filtered air supply is provided using inexpensive squirrel-cage blowers and Class-100 HEPA filters held in lightweight housings, available from many commercial laminar-flow bench suppliers. These can be positioned above or behind the work surface where filters are loaded into the Teflon holders, providing nearly-laminar, clean air flow across this surface. It is possible to simplify the construction of a clean area by taping the plastic sheeting directly to the walls and ceiling in the ship's lab. This will hold for a few days, however it is quite annoying to have such an enclosure collapse in the middle of sample processing.

Care must be taken to eliminate sources of dust and other air-borne contamination from the enclosure. Only those individuals actually involved in the subsampling and filtration should enter the enclosure during sample processing. Individuals involved in the sample processing should wear clean, non-shedding clothing; wool sweaters and blue jeans shed numerous, visible fibers. Smoking around this area should be strongly discouraged. A thick, damp mat such as a carpet remnant serves as an inexpensive doormat to reduce tracking in of dust. A curtain door can be made from long strips of the plastic sheeting.

Bottle samples are pressure-filtered using nitrogen gas at 10-12 psi over-pressure through the acid-washed, pre-weighed, 142 mm diameter, 0.4 μm Nuclepore filters held in Teflon/316 Stainless (Millipore) filter sandwiches. Filters are then allowed to blow dry with nitrogen for 5 to 15 minutes and are frozen in individual ziplock bags for return to the clean lab at FSU. To avoid possible artifacts associated with rinsing, we do not rinse these filters, however every effort is made to push all of the overlying excess seawater through the filter, leaving them reasonably dry. A salt correction is then made on the digestion solutions (see below).

7. Shore-Based Sample Processing

Upon return to the FSU clean lab, filters are thawed, placed in plastic tubs in the clean lab to dry, and reweighed to determine the mass collected. Blank filters are carried through the entire procedure including to/from the field to monitor trace-metal blanks as well as to correct for weighing artifacts such as variable humidity in the clean lab. A salt correction is made on the suspended matter weights via analysis of Na, K, Mg, and Ca, in the acetic acid leach solutions (see below).

7.1. Chemical Leaching Techniques

The three-step leaching/digestion procedure described here is an example of a sequential chemical digestion method, as opposed to "selective" methods which start with fresh material for each leach and attempt to dissolve metals from specific mineral phases. Whereas the latter methods require the ability to make very exact separations with very little overlap between different leaches, sequential methods produce useful information as long as they are applied in a consistent manner throughout and between data sets. In addition, the series of chemical leaches has been developed based on literature reports, personal experience, and ease of contamination avoidance.

The general approach has been to use varying concentrations of weak vs. strong acids to solubilize various fractions of the sample. Conclusions concerning metal geochemical cycling are based generally upon the suspended metal profile, the relative contributions of the leachable and refractory fractions to the total suspended particulate metal, and the metal/Al ratios of the total suspended material and in the various fractions. The basic premise is that refractory metals are associated with detrital mineral particles and not subject to low temperature alteration and cycling on short time scales, while metals in reactive, weak-acid soluble phases are largely formed by in situ processes such as biogenic production, metal-oxyhydroxide precipitation or scavenging/adsorption processes.

The most important factor to consider in the use of any chemical leaching treatment is whether metals released by a given treatment remain in solution. For example, use of dilute, acidified, pH 4 H_2O_2 has been suggested as a means of solubilizing metals associated with easily oxidizable organic matter in suspended matter, sediment trap, and sediment samples [Landing and Feely, 1981]. However, recent experiments where dissolved Al, Mn, and Fe were added to 10% H_2O_2 leaches of some Black Sea sediments, at pH values as low as 3.5, showed that most of the added metals were lost from solution (J. Wang and W. M. Landing, unpublished data, 1990). The metals were subsequently recovered in a dilute acetic acid leach of the sediments.

The leaching apparatus used for suspended matter filter samples consists of acid-washed polyethylene powder funnels fitted with Teflon stopcocks and suspended in a Plexiglas rack in the laminar flow fume hood.

Teflon tape is used to assure a tight funnel/stopcock seal and clean plastic lids loosely cover the funnel mouth. Funnels and stopcocks are rinsed three times at the beginning and between each leach with a 2 M Q-HCL/ 1 M Q-HNO_3 mixture, followed by three rinses with Nano-Pure water.

The leach sequence consists of 20 mL of 4.5 M Q-acetic acid for 4 hr at room temperature, followed by a dilute "aqua regia" mixture consisting of 20 mL 2 M Q-HCl/1 M Q-HNO_3, also for 4 hr at room temperature. The Q refers to high purity, reagent-grade material. The 25% acetic acid leach is reported to dissolve adsorbed cations, carbonate phases and reactive Mn-oxyhydroxides [Hodgson, 1960; Chester and Hughes, 1967; Bolger et al., 1978; Landing and Bruland, 1987], while the stronger "aqua regia" leach was developed to dissolve more resistant Fe-oxyhydroxides and metal sulfides [Lewis and Landing, in press, 1991]. Filters are placed in the funnels and immersed in the leach solution for 4 hours at room temperature, after which the leachate is drained into a Teflon beaker and taken to dryness over low heat on a hot plate in the laminar flow fume hood (12-14 hr). The residue is redissolved in 5 mL of 0.5 M Q-HCl/0.1 M Q-HNO_3 and transferred to acid-cleaned 30 mL CPE bottles for analysis.

The residual refractory material left on the filters following the strong-acid leach is then subjected to a total digestion at 100° C in sealed Teflon digestion bombs [Eggimann and Betzer, 1976]. The folded filter is placed in the bomb and covered with 3 mL of 6 M Q-HCl and 0.5 mL of concentrated Q-HNO_3. The bomb is sealed, heated in a boiling water bath for 90 minutes, then cooled in a freezer for 45 minutes. An additional 0.1 mL of concentrated Ultrex HF (J.T. Baker) is added and the procedure repeated. The cooled acid mixture is pipetted into tared 30 mL CPE bottles; the bomb is rinsed three times with 2 mL of Nano-water; the solution is made up to 10 grams total net weight; and stored until analysis. This procedure dissolves the refractory material not attacked by the weak- and strong-acid leaches, predominantly alumino-silicates. The efficiency of the digestion and the accuracy of the analyses is monitored by routine digestions of the NIST Estuarine Sediment standard reference material.

7.2. *Analytical Procedures*

Trace-metal concentrations in particulate matter leaches and digestions are measured by graphite furnace atomic absorption spectrometry (GFAAS), using a Perkin-Elmer 5000 Atomic Absorption Spectrophotometer (tungsten/deuterium background correction) equipped with an HGA-400 Graphite Furnace Atomizer and an AS-40 Autosampler. Matrix effects are accounted for using the method of standard additions using serial dilutions of commercially-available 1000 ppm AA standards, with Nano-Pure water or an appropriate acid mixture (matching the sample matrix) as the diluent. Standards are usually added to an equal volume of sample directly into the autosampler cups for convenience and cleanliness, and also to avoid the necessity of a correction for sample dilution. The autosampler cups are stored in 0.5 M Q-HCl, then rinsed three times with Nano-Pure water and allowed to air dry in a laminar-flow hood between use.

We use either regular non-pyrolytic graphite tubes or pyrocoated platform tubes with L'vov platforms. Uncoated tubes generally yield better precision, although with somewhat lowered sensitivity. The platforms have proven useful for analysis of refractory elements including Al.

In general, the furnace program conditions are close to the manufacturer's recommended parameters. Actual settings vary as a function of the salt and/or organic interferents in the sample and the type of graphite tube. The background correction instrumentation is inefficient above about 0.500 absorbance units, so the temperature and hold time of the char step are varied to maintain a background absorbance signal less than approximately 0.300.

Major cations (Na, K, Mg, Ca) are measured on the HAc leach solutions by ion chromatography using a Dionex 4500i-Fast Cat I/II system. This multi-element analytical technique offers generally inferior sensitivity to flame atomic absorption, however it requires less than 100 μL of the total 5 mL of digest solution for analysis of all four cations. Flame AAS generally requires much larger sample volumes.

7.3. *Blanks*

Successful trace-metal analysis requires that the blanks be very well constrained. Filter processing/analytical blanks are determined on filters which were cleaned, transported, stored and processed in the same manner as were the samples. While it is easy to determine the filter/processing blank, it is much more difficult to accurately determine the sample collection apparatus blank. Although the total sample collection plus processing blank is not routinely determined, it can be no greater than the lowest measured sample values.

Table 5-8 contains reagent and filter blanks from our three step sequential leaching procedure, and Table 5-9 chromatography column blanks from our colloid collection experiments (see section below). The error in the combined filter plus reagent blank can be used to estimate a detection limit by taking two times the standard deviation divided by a 28 L sample volume. In practice, a

TABLE 5-8.

Analytical Blanks for the Sequential Leaching Treatment of 142 mm, 0.4 μm Nuclepore Polycarbonate Filters; Data are Means and Precision is Expressed as ± 1 Standard Deviation (for Sample Sizes of 5 to 7)

	25% Q-Acetic Acid		2 M Q-HCl/1 M Q-HNO_3		HCl/HNO_3/HF (bomb)	
Metal	Reagent (ng)	Filter (ng)	Reagent (ng)	Filter (ng)	Reagent (ng)	Filter (ng)
Al	39.0 ± 4	2.0 ± 5	25.0 ± 7	35.0 ± 7	450. ± 120	200.0 ± 100
Mn	1.5 ± 0.3	1.8 ± 0.3	0.9 ± 0.3	2.0 ± 0.3	1.9 ± 1.1	2.9 ± 1.1
Fe	40 0 ± 5	47.0 ± 5	50.0 ± 5	55.0 ± 5	64.0 ± 4	72.0 ± 5
Co	1.5 ± 0.5	2.0 ± 0.5	0.7 ± 0.1	0.8 ± 0.1	1.8 ± 0.6	2.6 ± 0.6
Ni	<10.0	<10.0	<10.0	<10.0	<10.0	<10.0
Cu	21.0 ± 1	31 ± 2	13.0 ± 1	15.0 ± 1	43.0 ± 2	53.0 ± 3
Zn	28.0 ± 3	35.0 ± 5	9.0 ± 2	15.0 ± 3	37.0 ± 10	45.0 ± 10
Cd	25.0 ± 5	50.0 ± 8	3.0 ± 1	3.0 ± 1	3.0 ± 1	3.0 ± 1
Pb	3.0 ± 1	5.0 ± 2	0.7 ± 0.3	1.5 ± 0.4	4.0 ± 1	6.0 ± 2

Table 5-9

Blanks for 1.0 mL Columns of Toyopearl TSK Size-Exclusion Chromatography Resin in Polypropylene Columns (Biorad)

Metal	TSK Resin Column Blank (ng)
Mn	0.6 ± 0.2
Fe	9.0 ± 3
Co	1.6 ± 0.1
Ni	4.0 ± 2
Cu	6.0 ± 1
Zn	16.0 ± 6
Cd	0.3 ± 0.1
Pb	2.0 ± 0.6

more realistic "quantitation" limit is probably reflected by the blank value itself (converted to a molar concentration), since blank corrections of more than 50% of the measured value are generally undesirable from an analytical standpoint.

8. Particulate Metal Profiles in the Sargasso Sea

In April 1989, a vertical profile of dissolved and suspended particulate trace-metal samples was collected at the JGOFS-BATS (Bermuda Atlantic Time Series) station in the Sargasso Sea (31° 43.06′ N, 64° 01.27′ W). The suspended matter samples were treated using the three-step sequential leaching scheme described above.

The particulate Al profile (Figure 5-17A) shows generally low concentrations throughout the water column (less than 8 nM), with the exception of a near-bottom maximum (68 nM), presumably due to resuspension, and a small peak at 440 m. The sequential leaching treatment (Figure 5-17B) illustrates that the particulate Al profile is dominated by the refractory fraction throughout the water column except at 105 and 440 m where the HAc fraction reaches 60 and 40%, respectively. The HCl-HNO_3 fraction is also slightly elevated at these depths. The leachable/REF Al ratios in these samples (Figure 5-17C) represent evidence of Al geochemical cycling, either within the water column when in situ processes dominate or in the sediments assuming offshore transport dominates. The particulate Al around 100 m and at 440 m is significantly more reactive towards dilute acetic acid leaching than at surrounding depths.

The particulate Mn profile (Figure 5-18A) exhibited low concentrations in the upper 100-125 m (0.1-0.2 nM), increasing to higher values at depth (greater than 0.3 nM). Similar to reports from the Pacific [Landing and Bruland, 1987], the particulate Mn is dominated by the HAc fraction (Figure 5-18B, >80%), with that fraction decreasing somewhat towards the surface (70%, evidence for photochemical reduction) and towards the bottom (60%, due to resuspension of more resistant Mn-oxyhydroxide phases). This fractionation is accentuated by the ratios of leachable-to-refractory Mn (Figure 5-18C). In contrast to the particulate Al reactivity, the HAc/REF Mn ratios show a broad, mid-depth maximum from 100-600 m, a peak around 175 m, and much lower ratios in the surface and deep waters.

The particulate Fe profile (Figure 5-19A) looks superficially much like the Al profile, however features related to the geochemical cycling of Fe are evident from the sequential leaching results. In contrast to Mn, the HAc fraction is relatively low for Fe, generally less than 20% (Figure 5-19B). The HCl-HNO_3 fraction dominates from 75 to 150 m, while the refractory fraction approaches 90% at 4000 m. The HAc/REF and HCl-HNO_3/REF ratios both show a narrow peak from 90-125 m with a small maximum also at 440 m (Figure 5-19C).

The samples from 90 to 125 m are somewhat unusual, in that the profiles of total particulate Al and Fe show no unusual features while the reactivity of these two metals in this zone (as revealed by the leaching treatments) is distinctly different from the surrounding waters. The two metals also behave differently, with reactive Fe found in the harsher, HCl-HNO_3 treatment and Al in the weak HAc leach. While it isn't known exactly which processes are responsible for these features, the utility of "selective" leaching techniques for identifying zones of geochemical cycling is clearly demonstrated.

Another important feature of these profiles is the increased concentrations for all three metals in the deep waters, presumably due to resuspension. This clearly supports the notion that the deep water column in this region is strongly affected by horizontal mixing processes. In addition, the chemistry of the resuspended sediment found in the deepest sample is clearly different from that found above, in that the reactivities of Al, Mn, and Fe are all greatly reduced.

The samples from 440 m are especially notable, as other chemical species also exhibited unusual features at this depth in April, 1989. Dissolved Mn, dissolved Ni, hydrogen sulfide, and carbonyl sulfide, for example were all present at higher concentrations than anywhere else in the water column, however the sample does not appear to be contaminated with dissolved Fe, Zn, or Pb [Lewis and Landing, 1990]. One obvious explanation is that these observations represent an offshore transport signal, one that carries a chemical signature much like bottom water overlying organic-rich, nearshore sediments. The elevated total particulate Al and Fe concentrations would support such a conclusion, although the high degree of reactivity of both metals suggests that such resuspended sediments must have different chemical character than those collected at 4000 m. Evidence for offshore transport of material from the Bermuda platform to the JGOFS-BATS region at shallow depths has been found in sediment trap samples from 500 m (P. Betzer, University of South Florida, personal communication, 1991).

9. Colloid Collection Using Chromatography Resins

Landing et al. [1986] have proposed a sequential filtration-chromatography column scheme for the determination of dissolved trace-metal chemical fractionation in aqueous solutions. This technique has been further developed and tested in our lab at FSU and during field studies in the Black Sea, the Sargasso Sea and the Ochlockonee River estuary (Florida). The concept behind this method is to rapidly separate particulate and colloidal-sized species, anionic metal-organic and/or metal-sulfide dissolved complexes, and "free" hydrated metal cations and/or labile metal-major ion complexes.

The sample is filtered through a 0.4 μm Nuclepore filter to collect the "particulate" fraction, then pumped through a sequence of columns filled with various chromatography resins for fractionation of the "dissolved" metals. The resin sequence starts with a 1 mL column of unmodified Toyopearl TSK size-exclusion resin to trap rigid, colloid-sized species. This is followed by a 1-3 mL column of the weak-base, anion exchange diethyl-aminoethyl (DEAE) modified form of the TSK resin to collect anionic metal-ligand complexes. The final column in the sequence is a 1-3 mL column of the metal-chelating, cation exchange 8-hydroxyquinoline (TSK-8HQ) modified form of the TSK resin, synthesized in our lab, [Landing et al., 1986]. In an effort to avoid any re-equilibration artifacts, the samples are not altered in any way by pH adjustment, buffering, etc., other than being placed in 4 L cubitainers, prior to processing through the column sequence.

Attempting to first separate the free metals using cation exchange would result in some re-equilibration depending on the contact time of solution with the resin, the formation kinetics of ternary ligand-metal-resin complexes, and the dissociation kinetics of dissolved metal-ligand complexes. In addition, because the various resin columns all consist of the same resin matrix, each of the modified resins would also act to trap certain colloidal forms, thus the unmodified resin must be placed first in the sequence. By rapidly separating the species in the order described it was felt that the effects of re-equilibration would be minimized. It is important to stress that this, or any other, "operational" fractionation scheme is exactly that. We have attempted to characterize the column scheme by a combination of field and laboratory studies as described below. However, the

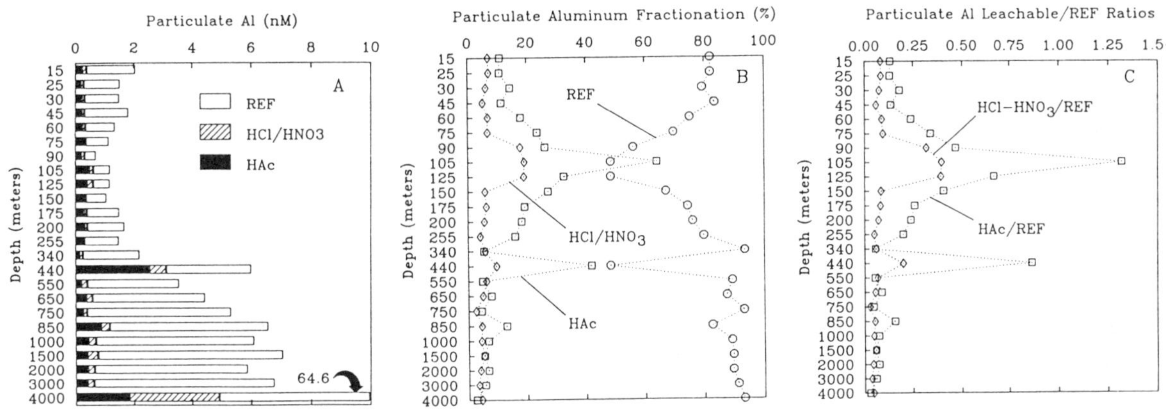

Fig. 5-17. A. Vertical profile of particulate Al from the JGOFS/BATS station in the Sargasso Sea, broken down into the three fractions measured using the selective leaching treatment described in the text. B. Profiles of the particulate Al fractionation (%) vs. depth illustrating the dominance of the refractory fraction. C. Profiles of the leachable-to-refractory Al ratios which illustrate zones of geochemical cycling behavior. Note the non-linear depth axes.

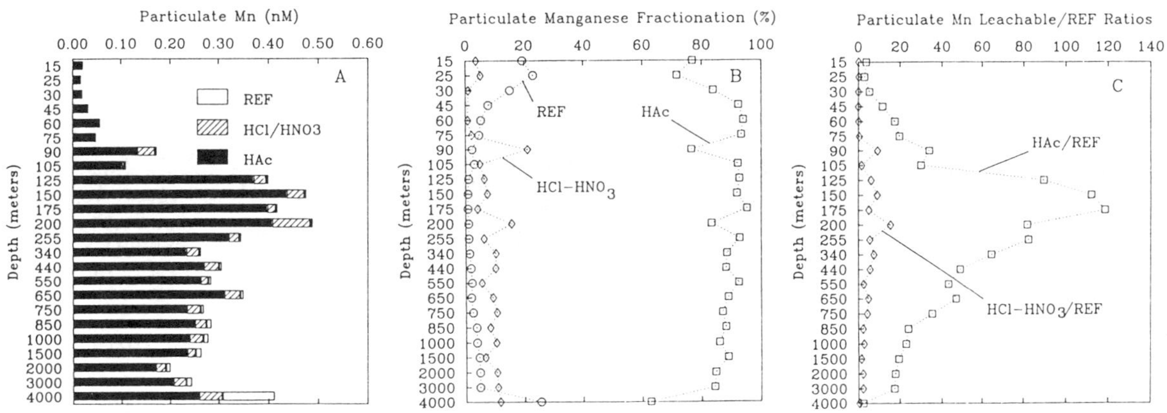

Fig. 5-18. A. Vertical profile of particulate Mn from the JGOFS/BATS station in the Sargasso Sea, broken down into the three fractions measured using the selective leaching treatment described in the text. B. Profiles of the particulate Mn fractionation (%) vs. depth illustrating the dominance of the acetic acid fraction. C. Profiles of the leachable-to-refractory Mn ratios which illustrate zones of geochemical cycling behavior. Note the non-linear depth axes.

interpretation of our field results must always be qualified by recognizing that we generally cannot "prove" which species are being collected or extracted. Thus we focus more on general trends or patterns in the data to make our conclusions. Use of such operational schemes in a controlled, consistent manner does, however, provide interesting variations within the data set which allow us to draw some conclusions regarding the important biogeochemical processes.

The cutoff between particulate and dissolved material has been traditionally defined by filtration through a 0.4-0.45 μm membrane filter. The nominal pore size of Nuclepore filters fairly accurately reflects the effective pore size and the filtration characteristics remain approximately constant nearly to the point of the filters becoming clogged [Laxen and Chandler, 1982]. Speciation studies have shown, however, that a substantial fraction of total "dissolved" metal is associated with colloidal material, which will pass through a 0.4 μm filter [Florence and Batley, 1980; Hart and Davies, 1981a, b; Laxen and Harrison, 1981; Laxen and Chandler, 1983]. In a study of the aqueous solubility of amorphous ferric hydroxide, Fox [1988] concluded that many of the studies in the literature had overestimated the dissolved iron concentration in rivers due to ineffective filtration methods. He further concluded that the dissolved iron in

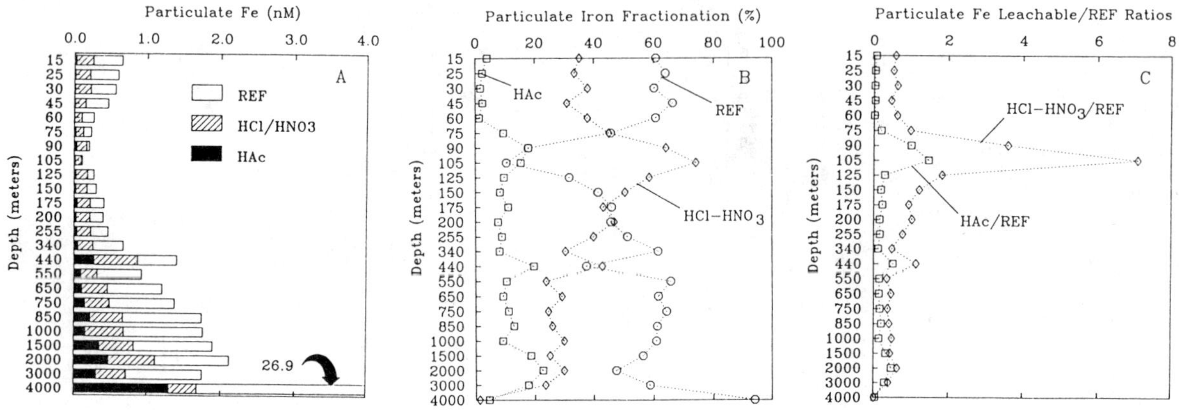

Fig. 5-19. A. Vertical profile of particulate Fe from the JGOFS/BATS station in the Sargasso Sea, broken down into the three fractions measured using the selective leaching treatment described in the text. B. Profiles of the particulate Fe fractionation (%) vs. depth illustrating the dominance of the refractory and HCl-HNO_3 fractions. C. Profiles of the leachable-to-refractory Fe ratios which illustrate zones of geochemical cycling behavior. Note the non-linear depth axes.

most rivers may be near equilibrium with colloidal amorphous Fe(III) hydroxide.

In our experiments, colloidal-sized material is defined to include those particles which pass through a 0.4 μm membrane filter but are retained on a 1 mL column of Toyopearl TSK-75F resin (Supelco; formerly "Fractogel" resin). TSK-75F is a vinyl polymer agglomerate size exclusion chromatography gel with a bead diameter of 32-63 μm and approximately 7.5 nm pores [Landing et al., 1986]. A hydrophilic resin, TSK has high mechanical and chemical stability, high porosity, and no significant cation or anion exchange capacity, nor is it believed to adsorb or release organic compounds. TSK is normally used in biochemical applications for the chromatographic separation of proteins and similar high molecular weight, easily "deformable" macromolecules. The resin should not physically trap flexible, colloidal-sized organic macromolecules (as would a dialysis or ultra-filtration membrane) but only slow them down relative to the eluent stream. Flexible organic macromolecules and metal-organic complexes would be expected to chromatograph their way through the column, eluting as fronts roughly according to their molecular size as the column becomes saturated for each species under the specific chromatographic conditions employed. To separate colloidal macromolecules making use of this principle, as is routinely done in standard biochemical research applications, it is necessary to first concentrate the macromolecules, then apply this concentrate as a layer to the top of a TSK column or any other size-exclusion resin.

However, information provided by the manufacturer suggested that the resin would trap more rigid colloidal-sized particles. Particles such as metal-oxyhydroxide and metal-sulfide colloids are expected to clog or "foul" the resin bed by clogging the pores and the inter-bead spaces near the top of the column. The column should thus act as a tortuous-path depth filter, efficiently trapping relatively rigid, natural metal colloids in a band at the top of the column.

Such banding was observed for Black Sea samples collected from the "dissolved" iron maximum in the anoxic zone [Lewis and Landing, in press, 1991]. The colloidal fraction accounted for from 30-60% of the total "dissolved" Fe(II). Laboratory tests have confirmed that this is not an artifact from the oxidation of Fe(II) to Fe(III) and subsequent precipitation of Fe-oxyhydroxides. The colloidal fraction appears to be a reactive Fe-sulfide with a nominal diameter of less than 50 nm based on some 50 nm Nuclepore filtration experiments performed on the Black Sea expedition.

In further investigations of the TSK resin properties, suspensions of monodisperse hematite (Fe_2O_3) particles were prepared according to the method of Penners and Koopal [1986] in size ranges of 30-50 and 90-110 nanometers diameter as determined by SEM. Dilute suspensions of 100-500 nM total Fe of these particles were passed through 1.0 mL TSK columns, yielding collection efficiencies greater than 90% for both size ranges and a visible band near the top of each column. Thus, the TSK resin acts as an efficient depth filter for rigid small particles, trapping them near the top of the column with little or no "bleed-through". Additions of the colloid suspension to natural surface seawater from the Sargasso did not reduce the collection efficiency of the TSK resin.

Additional experiments, in which the dilute hematite suspensions were mixed in varying proportions with organic-rich tea-colored Ochlockonee River (Florida) water, having 5 mgL^{-1} DOC as determined by high-temperature catalytic oxidation analysis, indicated that the collection efficiency was drastically reduced by organic surface coatings on the colloidal particles [Powell et al., 1990]. Thus, this colloid trapping scheme will probably not work effectively in organic-rich systems.

10. Sediment Trap Sampling, Processing, and Analysis

We have routinely used Soutar-type sediment traps to collect vertically settling material for trace-metal and radionuclide analysis [Bruland et al., 1981]. The trap consists of paired, 0.25 square meter, conical, Gel-coated, fiberglass cones equipped with a specially designed swimmer-avoidance cod-end [Coale, 1990]. The traps are baffled with polystyrene "egg crate" (1 x 1 cm cells). The traps are carefully pre-cleaned in the lab before shipment to the field using Micro detergent and Nano-Pure water. No acids are used, however great care is taken to eliminate any sources of particulate contamination. The traps are covered with plastic bags and/or sheeting and transported in large boxes equipped with internal plastic sheeting to act as a dust barrier.

The traps are deployed in free-floating arrays using sub-surface flotation with small surface floats, radio beacon, strobe, etc. The traps are allowed to fill naturally with ambient seawater as they descend to their deployment depth, avoiding the dissolution artifacts commonly observed when traps are filled with a density gradient solution. A small polyethylene bottle of buffered paraformaldehyde is placed in the cod-end immediately prior to deployment. Holes drilled in the bottle cap allow for slow dissolution of this mixture into the cod-end. Upon recovery, the cod-end is loosened to allow the overlying water in the trap to drain. The cod-end is then removed from the trap and taken immediately into the clean area where the supernatant solution is decanted and stored for later analysis to estimate the extent of solubilization of the analyte of interest. The trapped particles from the inner cod-end are decanted into acid-washed, pre-weighed, 50 mL polypropylene (PP) centrifuge tubes and spun at 500 x g for 45 minutes. The supernatant solution is again decanted and stored for analysis. The particles are then frozen in these tubes for return to the lab.

The trap samples are thawed and the sample wet-weight is determined. The tubes are then placed in a drying oven at 80° C for greater than 24 hours, or until constant weight is achieved, for determining the dry weight. A crude salt correction can be made at this time, knowing the salinity at the deployment depth of the trap. For trace-metal analyses, the dry trap samples are pulverized using a Teflon spatula, then ground with an agate mortar and pestle. The homogenized material is then subjected to sequential chemical leaching, as described earlier for the suspended matter samples.

Since the trap material is not deposited on a filter, the leaching is done in 50 mL PP centrifuge tubes using 30 to 50 mg of material. After the four hour leach, the material is centrifuged (500 x g, 20 minutes) and washed twice with Nano-Pure water. The leachate and rinses are filtered through acid-washed, pre-weighed, 25 mm, 0.4 μm Nuclepore filters held in PP Swinnex syringe filter holders (Millipore and Gelman), combined and taken to dryness in Teflon beakers in the clean lab as described above. After the sequential leaching, the residue in the centrifuge tube is collected on the same filter used above, re-weighed, and digested in the Teflon bombs as before. Trace metals are determined in these solutions using GFAAS, and major cations are measured using ion chromatography.

References

Betzer, P. R., and M. E. Q. Pilson, The effect of corroded hydrographic wire on particulate iron concentrations in seawater, Deep-Sea Res., 22, 117-120, 1975.

Bolger, G. W., P. R. Betzer, and V. V. Gordeev, Hydrothermally-derived manganese suspended over the Galapagos Spreading Center, Deep-Sea-Res., 25, 721-733, 1978.

Bruland, K. W., Oceanographic distributions of cadmium, zinc, nickel, and copper in the North Pacific, Earth Planet. Sci. Lett., 47, 176-198, 1980.

Bruland, K. W., R. P. Franks, G. A. Knauer, and J. H. Martin, Sampling and analytical methods for the determination of copper, cadmium, nickel, and zinc at the nanogram per liter level in seawater, Anal. Chim. Acta, 105, 233-245, 1979.

Bruland, K. W., R. P. Franks, W. M Landing, and A. Soutar, Southern California inner basin sediment trap calibration, Earth Planet. Sci. Lett., 53, 400-408, 1981.

Chester, R., and M. J. Hughes, A chemical technique for the separation of ferro-manganese minerals, carbonate minerals and adsorbed trace elements from pelagic sediments, Chem. Geol., 2, 249-262, 1967.

Coale, K. H., Labyrinth of doom: a device to minimize the "swimmer" component in sediment trap collections, Limnol. Oceanogr., 35, 1376-1381, 1990.

Eggimann, D. W., and P. R. Betzer, Decomposition and analysis of refractory oceanic suspended materials, Anal. Chem., 48, 886-890, 1976.

Feely, R. A., G. J. Massoth, and W. M. Landing, Major and trace element composition of suspended matter in the northeast gulf of Alaska; relationships with major sources, Mar. Chem., 10, 431-453, 1981.

Florence, T. M., and G. E. Batley, Chemical speciation in natural waters, CRC Crit. Rev., Anal. Chem., 9, 219-296, 1980.

Fox, L. E., The solubility of colloidal ferric hydroxide and its relevance to iron concentrations in river water, Geochim. Cosmochim. Acta, 52, 771-777, 1988.

Hart, B. T., and S. H. R. Davies, Trace metal speciation in the freshwater and estuarine regions of the Yarra River, Victoria, Estuarine Coastal Shelf Sci., 12, 353-374, 1981a.

Hart, B. T., and S. H. R. Davies, Trace metal speciation in three Victorian lakes, Aust. J. Mar. Freshwater Res., 32, 175-189, 1981b.

Hodgson, J. F., Cobalt reactions with montmorillonite, Soil Sci. Soc. Am. Proc., 24, 165-168, 1960.

Honeyman, B. D., and P. H. Santschi, A brownian-pumping model for oceanic trace metal scavenging: evidence from Th isotopes, J. Mar. Res., 47, 951-992, 1989.

Landing, W. M., and K. W. Bruland, The contrasting biogeochemistry of iron and manganese in the Pacific Ocean, Geochim. Cosmochim. Acta, 51, 29-43, 1987.

Landing, W. M., and R. A. Feely, The chemistry and vertical flux of particles in the northeastern Gulf of Alaska, Deep-Sea Res., 28, 19-37, 1981.

Landing, W. M., C. Haraldsson, and N. Paxeus, Vinyl polymer agglomerate-based transition metal cation-chelating ion-exchange resin containing the 8-hydroxyquinoline functional group, Anal. Chem., 58, 3031-3035, 1986.

Laxen, D. P. H., and I. M. Chandler, Comparison of filtration techniques for size distribution in freshwaters, Anal. Chem., 54, 1350-1355, 1982.

Laxen, D. P. H., and I. M. Chandler, Size distribution of iron and manganese species in freshwaters, Geochim. Cosmochim. Acta, 47, 731-741, 1983.

Laxen, D. P. H., and R. M. Harrison, A scheme for the physicochemical speciation of trace metals in freshwater samples, Sci. Total Environ., 19, 59-82, 1981.

Lewis, B. L., and W. M. Landing, Dissolved and particulate trace metal speciation in the Sargasso Sea, EOS Trans. Am. Geophys. Union, 71(2), 153, 1990.

Lewis, B. L., and W. M. Landing, The biogeochemistry of Mn and Fe in the Black Sea, Deep-Sea Res., in press, 1991.

Martin, J. H., and G. A. Knauer, VERTEX: manganese transport through oxygen minima, Earth Planet. Sci. Lett., 67, 35-47, 1984.

Penners, N. H. G., and L. K. Koopal, Preparation and optical properties of homodisperse haematite hydrosols, Colloids Surf., 19, 337-349, 1986.

Powell, R. T., B. L. Lewis, and W. M. Landing, Trace metal speciation in an organic rich estuary, EOS Trans. Am. Geophys. Union, 71, 99, 1990.

Spencer, D. W., and P. L. Sachs, Some aspects of the distribution, chemistry, and mineralogy of suspended matter in the Gulf of Maine, Mar. Geol., 9, 117-136, 1970.

Wallace, G. T., Jr., G. L. Hoffman, and R. A. Duce, The influence of organic matter and atmospheric deposition on the particulate trace metal concentration of northwest Atlantic surface seawater, Mar. Chem., 5, 143-170, 1977.

ELEMENTAL SULFUR ANALYSIS

George W. Luther III
College of Marine Studies, University of Delaware, Lewes, Delaware 19958

1. Introduction

Elemental sulfur as total zerovalent sulfur, S(0), is frequently measured in anoxic and/or sulfidic waters, and at the interface of oxic and anoxic waters. Examples of bodies of water with oxic-anoxic interfaces are the Black Sea, Kau Bay, the Cariaco Trench, the northern Chesapeake Bay, Saanich inlet, Framvaren fjord and the brines at the bottom of the Mediterranean Sea. The sampling and analytical procedures commonly used are given below.

2. Sampling Procedure

Samples are obtained from Go-Flo bottles off the CTD rosette system. The Go-Flo bottles are attached to a N_2 cylinder to maintain N_2 over the water during collection. A positive pressure is maintained so that water from the bottles would be able to be filtered hermetically through 0.2 μm Nuclepore filters directly into polypropylene or glass syringes. The syringes are equipped with three way stopcocks so that the lines from the Go-Flo bottles can be purged of air during sampling. This insures that no O_2 can enter the syringe or the filter device and cause sulfide oxidation resulting in elemental sulfur formation. The syringes are stored in N_2 filled glove bags or ziploc bags for subsequent analysis.

The Nuclepore filters are dried under N_2 and can be stored frozen in 10 mL polypropylene or glass vials for subsequent elemental sulfur analysis. The total volume of water sample filtered for elemental sulfur analysis is typically 100-500 mL. Larger volumes can be filtered to detect smaller quantities of elemental sulfur present in the water column.

3. Analytical Procedures

Elemental sulfur trapped on the Nuclepore filters is measured by one of two methods. These are the conversion of sulfur to thiosulfate by reaction with sulfite, and the conversion of sulfur to thiocyanate by reaction with cyanide.

Marine Particles: Analysis and Characterization
Geophysical Monograph 63

3.1. Thiosulfate Formation and Detection

In the first method, 1 mL of 0.01 M sodium sulfite and 9 mL of 0.1 M $NaClO_4$ are added to the filters in the polypropylene vials described above. The vials are capped and heated at 90° C for a minimum of two hours. Elemental sulfur quantitatively reacts with sulfite to form thiosulfate which is measured using square wave voltammetry (SWV) or other polarographic methods. Results are reported as total zerovalent sulfur, S(0), because the form of elemental sulfur, for example S6, S8, organic polysulfide etc., is unknown. The minimum detection limits (MDL) for S(0) is approximately 5 nM using SWV and depends on the volume of sample filtered because the filtration procedure is a preconcentration procedure. This method is similar to that used by Luther et al. [1985, 1991] and Boulègue et al. [1982]. Duplicate or triplicate filters should be analyzed at each depth. Standard curves are prepared by weighing elemental sulfur into vials and repeating the above procedure or by preparing standard calibration curves with thiosulfate in the $NaClO_4$ solution.

The determination of $S_2O_3^{2-}$ can be accomplished by the use of square wave voltammetry (SWV) or linear sweep voltammetry. SWV is more sensitive by about two orders of magnitude. An EG&G Princeton Applied Research Model 384B-4 in conjunction with a Model 303 static dropping mercury electrode or other suitable voltammetry system can be used for the analysis. A square wave pulse of 25 mV and a frequency of 100 Hz with a 2 mV scan increment is recommended. Voltage is scanned from -0.1 V to -0.6 V. Minimum detection limits in seawater or other aqueous solutions are estimated to be 200 nM for $S_2O_3^{2-}$. By stirring at a deposition potential set at -0.1 V for 60 sec for a cathodic stripping square wave voltammetry method, the MDL's can be lowered by a factor of about five . All potentials are reported versus the saturated calomel electrode as the reference electrode.

3.2. Thiocyanate Formation and Detection

The second commonly used procedure for the determination of zerovalent sulfur uses the reaction of sulfur with cyanide ion to form thiocyanate [Bartlett and Skoog, 1954; Troelsen and Jorgensen, 1982]. The thiocyanate ion is then reacted with Fe^{3+} to form the red $FeSCN^{2+}$ complex. This red complex is then measured by spectrophotometry at a wavelength of 460 nm.

Five mL of CS_2 are added to the filter in glass tubes for extraction of the elemental sulfur. The samples are stoppered tightly and shaken on a whirlmixer, then placed horizontally on a shaker overnight for extraction; whirlmixing is repeated once more after shaking. A 1 mL aliquot of the sample is quantitatively transferred to a glass vial where it is left to evaporate to dryness in a fume hood at room temperature. After evaporation, the S(0) is redissolved in 5 mL of cyanide reagent. The cyanide reagent is 1 g NaCN per liter of solvent which is 95% acetone and 5% water.

The cyanolysis reaction is allowed to proceed for 4 hr at room temperature. To 1.5 mL of the cyanolysis solution is added 1.5 mL of $FeCl_3$ (5 g $FeCl_3$ $6H_2O$ per liter of 95% acetone - 5% H_2O solvent) in a 1-cm cuvette. The absorption of the red complex is read at 460 nm. Standard curves are prepared by dissolving elemental sulfur in CS_2 and repeating the above procedure. Corrections should be made for blank absorption due to reagent color, pigments extracted from the filter and S(0) in the CS_2. Discard CS_2 if its blank value becomes too high. Avoid any contact of the solvents with rubber or plastic.

4. Conclusion

Both analytical methods have been used successfully in the water column of the Black Sea with similar results [Jorgensen et al., in press, 1991; Luther et al., in press, 1991]. The choice of methods depends on equipment availability.

References

Bartlett, J. K., and D. A. Skoog, Colorimetric determination of elemental sulfur in hydrocarbons, Anal. Chem., 26, 1008-1011, 1954.

Boulègue, J., C. J. Lord, and T. M. Church, Sulfur speciation and associated trace metals (Fe, Cu) in the porewaters of Great Marsh, Delaware, Geochim. Cosmochim. Acta, 46, 453-464, 1982.

Jorgensen, B. B., H. Fossing, C. O. Wirsen, and H. W. Jannasch, Sulfide oxidation in the anoxic Black Sea chemocline, Deep-Sea Res., in press, 1991.

Luther, G. W., III, R. Varsolona, and A. E. Giblin, Polarographic analysis of sulfur species in marine porewaters, Limnol. Oceanogr., 30, 727-736, 1985.

Luther, G. W., III, T. M. Church, and D. Powell, Sulfur speciation and sulfide oxidation in the water column of the Black Sea, Deep-Sea Res., in press, 1991.

Troelsen, H., and B. B. Jorgensen, Seasonal dynamics of elemental sulfur in two coastal sediments, Estuarine Coastal Shelf Sci., 15, 255-266, 1982.

THE APPLICATION OF CROSS-FLOW FILTRATION TO THE COLLECTION OF COLLOIDS AND THEIR ASSOCIATED METALS IN SEAWATER

S. B. Moran

Department of Chemistry, Woods Hole Oceanographic Institution, Woods Hole, Massachusetts 02543

1. Introduction

A variety of techniques have been used for the collection of colloidal material in natural waters; for example, centrifugation, filtration, ultrafiltration, dialysis, gel-permeation chromatography and field-flow fractionation [e.g. Karaiskakis et al., 1982; Laxen and Chandler, 1982; de Mora and Harrison, 1983; Salbu et al., 1985; Giddings, 1988]. There have been few studies of the size speciation of trace materials, organic or inorganic, in oceanic waters due primarily to problems of contamination and slow filtration rates (e.g. $mL\,hr^{-1}$) associated with conventional ultrafiltration equipment. A relatively new technique for marine applications, cross-flow filtration (CFF), has the potential to overcome both of these problems. Cross-flow, or tangential-flow, filtration has been used in such disciplines as medical research, sewage treatment and microbiology. However, the application of this technique to oceanographic studies has only recently been reported [e.g. Whitehouse et al., 1986, 1990; Moran and Moore, 1989].

The purpose of this paper is to provide a description of the Millipore Pellicon cross-flow system that has been modified and used specifically for sampling trace metals (e.g. Al, Th) in oceanic waters; general considerations regarding ultrafiltration are described in separate sections of this volume (cf. contributions by Brownawell, page 187, Benner, page 181 and Yeats, page 321). The method was developed to provide a means for sampling trace metals and radionuclides in the colloid size range in an attempt to investigate the role of colloids in oceanic trace metal scavenging.

2. Basic Principle and Experimental Set-Up

Cross-flow filtration differs from conventional filtration in that fluid flow is tangential as well as perpendicular to the surface of the filter membrane. Size separation occurs with fluid containing solutes smaller than the nominal filter pore size passing through the filter. Solutes larger than the filter pore size flow tangential to the filter surface [Millipore Corp.; Josephson, 1984]. The sweeping motion of fluid across the filter membrane reduces the retention of solutes on the filter, known as concentration polarization, thereby increasing filtration efficiency.

A schematic of the Millipore Pellicon cross-flow filtration system described in this report is shown in Figure 5-20. A water sample is first passed through a prefilter (e.g. 0.45 μm) and then into a 60 L sample reservoir. The prefiltered sample is then pumped through the cross-flow filter cell. Material smaller than the nominal pore size of the cross-flow filter (the filtrate, or permeate) passes through the filter and is collected in a separate carboy. Solutes larger than the filter pore size flow tangential to the filter surface (the retentate) and are recycled back into the sample reservoir. The system is run continuously in this concentrate mode until colloids in a sample are concentrated down to a volume limited only by the ca. 500 mL dead volume of the system.

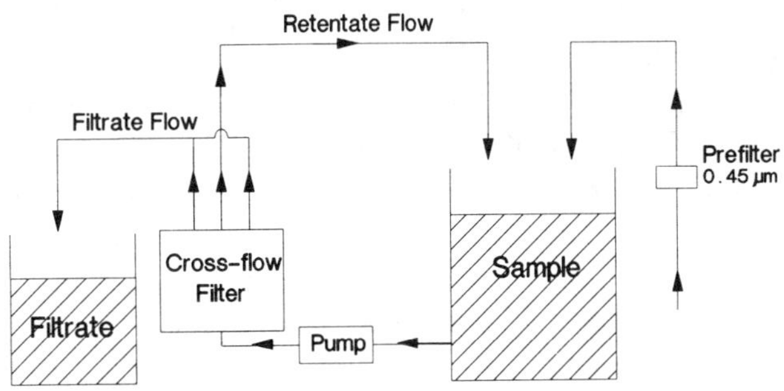

Fig. 5-20. Schematic of the Millipore cross-flow filtration system used by Moran and Moore [1989].

There are two main advantages in using the modified cross-flow filtration system to collect colloidal material in seawater. First, all wetted parts of the cross-flow system are constructed of materials that can be thoroughly cleaned, as required for trace inorganic and organic analyses. Second, the sustained high filtration rates of 25-50 $L\,hr^{-1}$ of this technique enable processing of 100's of L on board ship within a reasonable time. The latter is important because the concentration of trace metals associated with colloids in oceanic waters may be very low, possibly at or below the detection limit of present analytical methods, thereby requiring a means of

Marine Particles: Analysis and Characterization
Geophysical Monograph 63

concentrating the colloid fraction. In addition, the modular design of the Pellicon system allows one to easily adapt the filter cassette area for the required application.

The Millipore Pellicon cross-flow system was modified by installing Teflon valves and connecting tubing and an air driven pump, which are compatible for sampling trace metals. The filter membrane is installed and tested following Millipore guidelines and the entire system kept covered with plastic at all times to reduce potential sources of contamination. Although the modified cross-flow filtration system has been used mainly for processing seawater samples for determination of Al [Moran and Moore, 1989], it should be compatible for sampling many other oceanic trace metals and radionuclides.

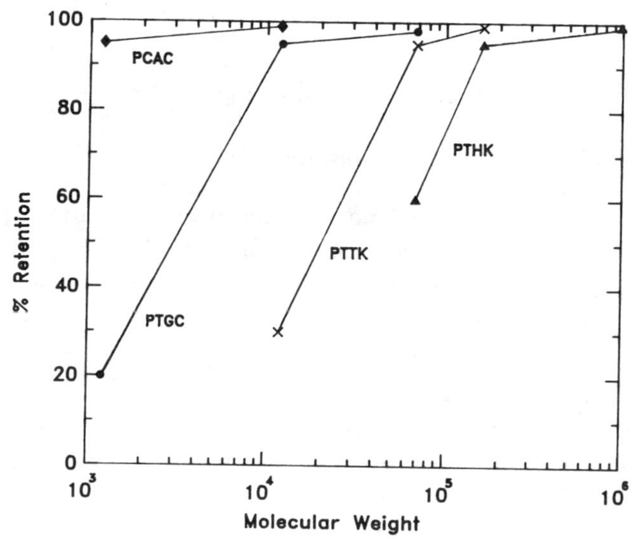

Fig. 5-21. Retention characteristics of Millipore polysulfone cross-flow filters determined using various macromolecules (vitamin B-12, human serum albumin, gamma globulin, and blue dextran). The operationally defined size cut-offs are 1,000 NMW (PCAC), 10,000 NMW (PTGC), 30,000 NMW (PTTK) and 100,000 NMW (PTHK). Data from Millipore Corp.

3. Cross-Flow Filters

Cross-flow filtration of seawater has been conducted [Moran and Moore, 1989] using a Millipore PTGC-000-05 filter cassette. This filter has a nominal molecular weight size (NMW) 10,000 cutoff, equivalent to a size range of about 1 to 10 nm, and an effective filtration area of 0.464 m^2. The cross-flow filter is made of polysulfone, has a polyethylene substrate, and is sealed around the edges with urethane. Millipore has recently introduced regenerated cellulose membranes (1000, 5000, 10,000, 300,000 NMW) that are reported to have considerably reduced protein binding characteristics compared to polysulfone membranes. Regenerated cellulose is only slightly less resistant to acids and bases than polysulfone, and the decreased sorption characteristics of the cellulosic material suggest it may be more suitable for marine applications.

The size cut-off of Millipore cross-flow filters is defined operationally as the molecular weight at which ca. 90% retention of spherical macromolecules occurs, reported in units of NMW. These size cutoffs are not precise; cross-flow filters retain material above and below their rated size cut-off (Figure 5-21). Retention of macromolecules will depend primarily on their size and structure. This fact should be recognized when these filters are used to separate and define natural colloidal matter because the molecular weight, structure, and hence the degree to which these natural materials may be retained by cross-flow filters, are largely unknown. For these reasons, the reported size cut-off should not be regarded as a precise indication of the molecular weight of natural colloids.

The procedure for installing and testing CFF filters is provided in detail by Millipore and summarized as follows. A new PTGC filter cassette is installed in the cross-flow cell between two acrylic plates held together by stainless steel connecting rods. The glycerin-formalin solution in which the filter is shipped is thoroughly rinsed (10-20 L) from the filter with Super-Q water. Satisfactory rinsing of the cross-flow filter is indicated when the concentration of dissolved organic carbon (DOC) in fresh Super-Q is equal to the DOC blank. Super-Q is then recirculated through the system for about 15-30 minutes to ensure that the filter is completely wetted.

The cross-flow filter is then tested to ensure that the filter is properly installed and that the filter does not pass air below a pressure of 5 psi (see the integrity test procedure described by Millipore Corp.). As the Dominator pump is not air-tight when pressurized to 5 psi, a peristaltic pump that does not leak air at this pressure can be used to conduct the integrity test. A 5 psi gauge is installed in one of the permeate lines between the pump and the cross-flow filter. Super-Q is recirculated through the system with the permeate outlet valves turned off. The feed-line to the cross-flow cell is removed from the water until air bubbles emerge from the submerged retentate tubing in the sample reservoir. The retentate outlet is then closed with the pump still running. The pressure in the filter cell is then increased to 5 psi at which point the pump is turned off. This test is passed when the filter can hold a pressure of 5 psi for 30 seconds (Millipore Corp.).

The clean water flux measurement (Millipore Corp.) is determined following the integrity test. This involves measuring the permeate flow rate of Super-Q water with the system configured in the concentrate mode. The retentate outlet valve is closed, thereby forcing all fluid to flow through the membrane. The permeate flow rate is measured using a graduated cylinder and stop watch. The clean water flux is measured after processing samples and serves as a check on the performance of the filter. A decrease in the clean water flux with volume filtered (or with time) would indicate fouling of the filter membrane, which should then be either cleaned or replaced.

4. Pump

The cross-flow filtration system is operated with a half-inch air driven double diaphragm pump (Dominator Maskin P15-PVT, Granzow Corp.). The Dominator pump was chosen for the cross-flow system based on the following criteria:

1) cleanable with acids and bases;
2) delivers total flow rates of 100-400 L hr^{-1} at discharge head pressures that may exceed ~ 30 psi;
3) low dead volume;
4) self-priming.

Wetted parts of the pump are polypropylene, with a Viton diaphragm and Teflon ball valves. The pump is not lubricated to reduce the possibility of contamination from lubricating oil venting into the air near the cross-flow sample reservoir. No problems have been encountered due to insufficient lubrication. It is recommended that for about every 2000 L of seawater filtered, the self-actuating air valve should be packed with white lithium grease (American Grease Stick Co.).

The pump has been operated in the laboratory with a compressed air supply and at sea using portable electric air compressors. The pump is connected to the air supply with quick-fit connectors crimped onto braided PVC air hoses. A single-stage regulator fitted with an air filter is used to control the pump. The pump operating pressure is usually set to 35-40 psi during cross-flow filtration. The pump operating pressure is chosen based on the following criteria:

1) operation of the pump at pressures greater than about 50 psi necessitates the use of a lubricant for the pump, which represents a potential source of contamination;
2) permeate flow rates of typically ca. 25-30 L hr^{-1} are sufficient at this pressure to cross-flow filter 100 L water samples at sea within several hours.

Peristaltic pumps equipped with silicone tubing may also be used with this system and are the pumps equipped with the CFF system sold by Millipore. These pumps have the advantage of being electric, so that a source of compressed air is not required and which may not be available at sea (there are electric diaphragm pumps, although I have not tested them). I have not used a peristaltic pump as they typically have lower flow rates and require considerably longer processing times. Also, the pump head may degrade the silicone tubing and small pieces enter the CFF system (B. Brownawell, personal communication, 1991). One attractive modification of CFF would be to interface a CFF system to an in situ pump; this should be attempted in the future as such a modification would considerably simplify the procedure.

5. Cleaning Procedure

The cleaning procedure used for sampling dissolved Al, which should be suitable for most other oceanic trace metals, is described as follows. The 60 L cross-linked polyethylene reservoir is filled with 2.5 M HCl and the acid recirculated through the system which includes the filter, pump and connecting lines. After 10 days, the acid is drained and the 60 L reservoir thoroughly rinsed with 0.22 μm-filtered Super-Q. Super-Q is filtered through an additional 0.22 μm filter to remove particulate matter that might pass through Super-Q filtration cartridges and lead to premature fouling of the cross-flow filter. The filter, pump and connecting tubing are then rinsed by pumping 6 to 8 L of filtered Super-Q through the system with the retentate and permeate lines going to waste.

6. Cross-Flow Filtration Procedure

Seawater samples are first prefiltered into the 60 L sample reservoir. The type of prefilter (e.g. material, diameter, pore-size) selected will depend on the particular sampling program. Prefiltration through clean 0.45 μm Gelman Mini Capsule filters and Teflon tubing has been conducted at sea by pressurizing Go-Flo bottles containing seawater with compressed N_2 [Moran and Moore, 1989]. Prefiltration of a 60-100 L sample in this manner is usually complete within 10 minutes. The cross-flow system is initially rinsed by pumping 2-3 L of prefiltered sample through the system with the permeate and retentate lines going to waste. The cross-flow filter is then preconditioned by soaking the filter in prefiltered seawater sample for about 30 minutes. This procedure is recommended by the Millipore Corporation to reduce retention of colloidal material by the filter during cross-flow filtration.

The cross-flow filtration system is then configured in the concentrate mode, with the permeate lines going to a separate carboy and the retentate line recirculated back to the sample reservoir (Figure 5-20). The pump is turned on by setting the operating pressure to 35 psi. Any air bubbles that accumulate in the lines during operation can be removed by tilting the cross-flow cell to allow the bubbles to exit through the permeate and retentate lines. The permeate is collected in a calibrated carboy that is also used to monitor the filtration rate. Subsamples of the permeate are taken directly from the permeate feed. The system is run continuously in the concentrate mode until the sample volume is concentrated down to the desired volume. The sample reservoir is kept covered with plastic during processing to reduce the possibility of contamination. Additional prefiltered water is added during operation by turning off the pump and recording the amount of water added to the sample reservoir. The pump is then turned on with the system set in the concentrate mode (Figure 5-20).

The pump is turned off when the total sample volume is concentrated down to the desired volume. Subsamples of the colloid concentrate are then transferred with a graduated cylinder to clean polyethylene bottles for analyses. An accurate account of prefiltered, cross-flow filtered and colloid concentrate volumes, or weights, is recorded to determine the concentration factor. The system is then rinsed with 2 to 3 L of Super-Q prior to processing the next sample. When not in use, the filter is stored in the cross-flow acrylic cell in Super-Q.

An alternative way to cross-flow filter samples is to operate the system with the retentate line going to a separate collection vessel rather than recirculating back to the feed reservoir. In this configuration, colloids are not concentrated; the CFF system only separates the water sample into two fractions (the permeate and retentate). This mode of operation should reduce any potential artifacts associated with particle aggregation occurring in the feed reservoir during processing in the concentrate mode. While the non-recirculating mode of operation would not increase the concentration of species in the colloid fraction, it should still be possible to determine for example, radionuclides (e.g. Th) in the colloid fraction by increasing the size of the sample processed.

7. Filtration Rate

The permeate filtration rate is monitored to determine the performance of the cross-flow filter. A steady decrease in permeate flow rate indicates fouling of the membrane with colloidal material according to the Millipore Corporation. The permeate filtration rate can easily be determined during cross-flow filtration by measuring the permeate flow rate with a graduated cylinder and stopwatch.

During cross-flow filtration, the filtration rate typically decreases initially by about 10 to 15% and subsequently remains nearly constant. An example of the permeate flow rate measured during cross-flow filtration is shown in Figure 5-22. These data suggest that

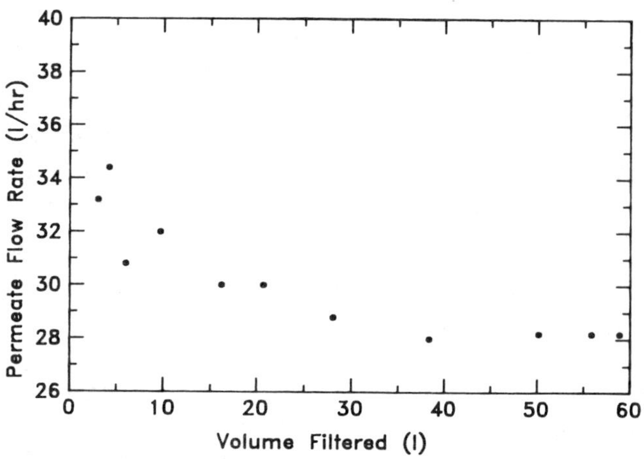

Fig. 5-22. Plot of the permeate flow rate against the cumulative volume of cross-flow filtered seawater.

colloidal material is initially removed by the filter, causing a decrease in the flow rate, and, subsequently, material is not removed during cross-flow filtration. These data may be explained in terms of physico-chemical attachment of amphiphilic surface-active material to the filter surface, resulting in a layer of hydrophilic groups oriented on the filter surface [Shaw, 1986]. Such material would have polar and non-polar functional groups. Further attachment to such material to the hydrophilic surface layer is impeded because surface-active material will preferentially remain in solution. This explanation may account for the decreased retention of colloidal material following preconditioning of filters observed during this study and also using Amicon ultrafiltration membranes [e.g. Niven, 1989].

8. Calculation of Cross-Flow Filtration Results

During cross-flow filtration, species smaller than the filter pore size pass through the filter and remain at their original concentration in the retentate and the permeate. To correct for the presence of truly dissolved species (less than 10,000 NMW) in the colloid fraction, the concentration of species in the colloid concentrate (10,000 NMW - 0.45 μm) is calculated by: 1) subtracting the concentration of species measured in the truly dissolved size fraction from the concentration of species measured in the colloid fraction; and 2) dividing the corrected colloid size fraction by the concentration factor.

The cross-flow filtration concentration factor (CF) is calculated by dividing the initial sample volume by the final sample volume after cross-flow filtration. The final volume is determined by carefully measuring all remaining sample in the sample reservoir, using either a graduated cylinder, or by weighing, and then adding to this the dead volume of the system.

The percentage of material in the colloid size range is calculated by dividing the concentration of species measured in the colloid concentrate (10,000 NMW - 0.45 μm) by the concentration of species measured in the prefiltered (PF) or "dissolved" (less than 0.45 μm) size [colloidal = (colloidal/PF) x 100]. As an example, Figure 5-23 shows colloidal organic C and Al data collected off Nova Scotia using the modified Millipore Pellicon CFF system.

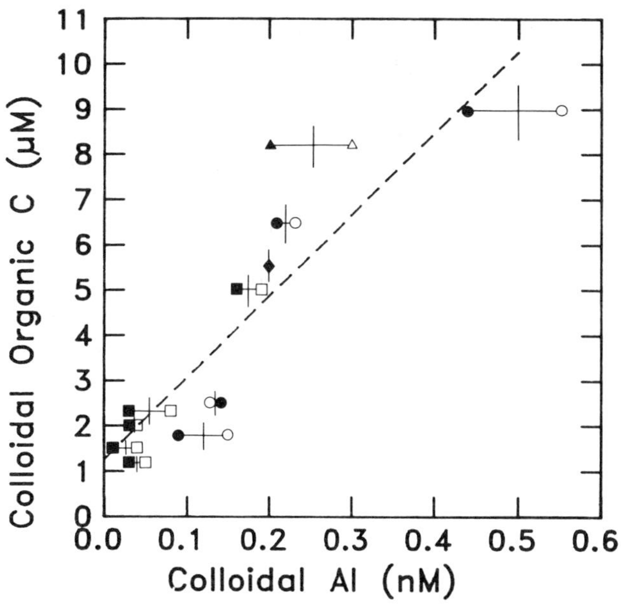

Fig. 5-23. Plot of organic carbon against aluminum samples measured in the colloid (10,000 NMW-0.45 μm) size range collected off Nova Scotia [Moran and Moore, 1989]. Solid symbols denote unacidified samples and open symbols denote samples acidified with 0.1 M Ultrex HCl for 2 hours (pH 2.6). Dashed line represents linear regression of data (r = 0.89, n = 22), (circles) Scotian Shelf, (squares) Gulf Stream, (triangles) Scotian Shelf break surface waters, (diamond) Northwest Atlantic surface waters.

9. Summary

This note describes the basic procedures used to modify and use the Millipore Pellicon cross-flow filtration system for sampling colloids and their associated trace metals in oceanic waters. Advantages of this system include:

1) compatibility with trace metal sampling;
2) high filtration rates;
3) shipboard capability.

The cost of Pellicon systems has decreased significantly over the past five years; a new Super Pellicon system including pump costs about $5,000. Membrane cassettes can be used for many samples and cost about $800 each. The cross-flow filtration system described here has been successfully used in a study of colloidal Al and organic C [Moran and Moore, 1989] in oceanic waters (e.g. Figure 5-23), and very recently to examine temporal variations of several radioisotopes (e.g. ^{234}Th, ^{210}Pb) in truly dissolved, colloidal, and "dissolved" phases during a phytoplankton bloom.

References

de Mora, S. J., and R. M. Harrison, The use of physical separation techniques in trace metal speciation studies, Water Res., 17, 723-733, 1983.

Giddings, J. C., Field-flow fractionation, Chem. Eng. News, 66, 34-45, 1988.

Josephson, J., Cross-flow filtration, Environ. Sci. Technol., 18, 375A-376A, 1984.

Karaiskakis, G. A., K. D. Caldwell, and J. C. Giddings, Sedimentation field-flow fractionation of colloidal particles in river water, Int. J. Environ. Anal. Chem., 12, 1-15, 1982.

Laxen, D. P. H., and I. M. Chandler, Comparison of filtration techniques for size distribution in freshwaters, Anal. Chem., 54, 1350-1355, 1982.

Moran, S. B., and R. M. Moore, The distribution of colloidal aluminum and organic carbon in coastal and open ocean waters off Nova Scotia, Geochim. Cosmochim. Acta, 53, 2,519-2,527, 1989.

Niven, S. E. H., Solid-solution partitioning of thorium in seawater, Ph.D. thesis, 175 pp., Dept. of Oceanog., Dalhousie Univ., Canada, 1989.

Salbu, B., H. E. Bjornstad, N. S. Lindstrom, E. Lydersen, E. M. Brevik, J. P. Rambaek, and P. E. Paus, Size fractionation techniques in the determination of elements associated with particulate colloidal material in natural fresh waters, Talanta, 9, 907-913, 1985.

Shaw, D. J., Introduction to Colloid and Surface Chemistry, third ed., 273 pp., Butterworths & Co., Ltd., England, 1986.

Whitehouse, B. G., G. Petrick, and M. Ehrhardt, Crossflow filtration of colloids from Baltic seawater, Water Res., 20, 1599-1601, 1986.

Whitehouse, B. G., P. A. Yeats, and P. M. Strain, Crossflow filtration of colloids from aquatic environments, Limnol. Oceanogr., 35(6), 1368-1375, 1990.

SAMPLING AND ANALYSIS OF PARTICLES FROM BUOYANT HYDROTHERMAL PLUMES

Michael J. Mottl

Department of Oceanography, University of Hawaii, 1000 Pope Road, Honolulu, Hawaii 96822

1. Objective

The objective of our studies has been to identify the chemical processes that occur in the buoyant part of hydrothermal plumes and to evaluate their role in determining the ultimate fate of the hydrothermal input to the oceans. Our first such effort is described by Mottl and McConachy [1990].

2. Sampling Methods

Because the buoyant plume is a small feature that contains very large physical and chemical gradients, we have sampled it from manned submersibles. We have used two different samplers, both manufactured by General Oceanics in Miami: the Go-Flo bottle and the Chopstick sampler. Four Go-Flo bottles of 1.7 L capacity can readily be mounted on most submersibles, vertically and in a forward position in sight of the pilot's viewport and video cameras, without interfering with other operations on a dive. On Alvin they have typically been mounted on the outside starboard edge of the basket. On Turtle they were mounted on the outside edge of the port manipulator. We chose Go-Flo rather than Niskin bottles because the latter are prone to spillage when the rods attached to the end caps are bumped against an object such as the seafloor, as often happens on a submersible dive. Go-Flo bottles are also more easily rigged for pressure filtration than are Niskins. The main disadvantage of Go-Flo bottles vs. Niskins for this application is the internal silicone rubber ring that holds the ball valves in place on each end of the Go-Flo. This ring tends to trap large particles that are then difficult to dislodge and collect. The rings are also difficult to clean between dives.

Both Go-Flo and Niskin bottles have the disadvantage of requiring flushing to obtain a representative sample. Because the submersible tends to rise buoyantly along with the particles once it enters the plume, this type of bottle will not flush adequately without intervention from the submersible pilot. We have addressed this problem by asking the pilot to trip a bottle only after holding the submersible stationary at a given depth for about ten seconds using the lift thrusters or, preferably, driving downward about 1 meter. The degree of success can best be assessed by recording the sampling on videotape, along with a data block displaying pressure-depth. The exact time that the sampler closes can better be seen if orange reflective tape has been attached to the top of the rubber tube that tensions the upper ball valve. The movement of this tape is then clearly visible on the videotape. Lacking this tape, it is still usually possible to see the upper ball valve close. Because the smoke in the buoyant plume is frequently thick enough to obscure visibility, it is advisable for the pilot to grasp the lanyard that has been rigged to close the bottle before entering the plume, using the submersible's manipulator.

Depending on visibility and the pilot's skill, there are two different ways that these samples can be collected. In the case of poor visibility, the submersible must be driven into the plume at the desired height once for each sample, and the bottle tripped as close as possible to the centerline of the plume. Where visibility is better and the pilot skilled with the manipulator, the submersible may enter the plume at its base and rise with the plume. The bottles are then triggered in succession at the desired heights and as close as possible to the centerline of the plume, pausing at each sampling height or driving downward slightly as described above.

In contrast to Go-Flo or Niskin bottles, Chopstick samplers require no flushing time. They collect water in a polyethylene bag. Other materials have also been available in the past, including a plastic-aluminum foil laminate that is impermeable to gases, including helium. Ideally the plastic bag is initially empty; on triggering the bag is pulled open, drawing water in, and then sealed by rapidly rolling the open top of the bag around a pair of aluminum rods, or "chopsticks". In practice, it is difficult to empty the bag completely prior to mounting it on the sampler. It is necessary to fill the bag with a few mL of water or seawater and then to flatten it, squeezing most of this water back out. This is an obvious disadvantage for many applications, but it becomes an advantage if one wishes to add some liquid reagent to a sample immediately on collection. The bag can be sent down already containing the reagent. The chopstick samplers we have used collect 1.5 L of water, but larger sizes are also available. We have mounted them on moorings attached to the submersible and on the submersible itself, with

Marine Particles: Analysis and Characterization
Geophysical Monograph 63

good results. On recovery the samples must be gravity, or vacuum-filtered, as pressure-filtration is not practical for these samplers.

In all cases it is critical to plan the dive so that no bottom sediment resuspended by the activities of the submersible is in the water column at the time that the plume is sampled. If sediment has been artificially resuspended, at least 20 to 30 minutes should elapse before sampling the plume, and preferably much longer. In practice, this means that the plume should be sampled and all the bottles tripped before sampling hot water or sulfides from the chimneys or mounds at the base of the plume. Unfortunately, this requirement tends to lengthen the time between plume sampling and processing of the samples on board the surface ship, increasing the time for reaction between plume solutions and particles within the sample bottles. An ideal solution is to sample a plume at which no other work has been done on a given dive, at the end of that dive, but optimal vent location and navigation are required to implement this solution.

3. Filtering the Plume Samples

Once on board the surface ship, the bottles should be filtered as soon as possible. We hang the bottles vertically on a wall rack and connect them to a manifold made of flexible PVC tygon tubing that allows us to filter all four bottles at once if necessary. The manifold is pressurized with nitrogen from a tank that is filtered on-line using a Gelman air filter. The air-inlet plugs at the top of the Go-Flo bottles are replaced with plastic nipple fittings that connect with the manifold tubing. The drain valve of each Go-Flo is connected by plastic tubing to an on-line Nuclepore filter holder containing a 47 mm diameter, 0.4 or 0.1 μm Nuclepore polycarbonate membrane filter. The samples are then filtered under nitrogen pressure into suitable containers, taking care to avoid contamination from atmospheric dust.

We have also tried Millipore cellulose-acetate-nitrate filters and have found them to be equally acceptable for bulk chemical analyses of plume "smoke" particles. The Millipore filters dissolve in acids used to dissolve the particles whereas the Nuclepore filters do not, but we have found the blank levels from both types to be very low for our work in the buoyant plume, where particle concentrations are relatively large. The chief advantage of the Nuclepore filters is for scanning electron microscopy (SEM), where they provide a cleaner substrate that is closer to an ideal membrane filter.

We filter nearly all aliquots taken of the plume water, including those for dissolved gases such as H_2S, oxygen, and methane. We do this because we wish to recover the particles quantitatively in order to calculate particle concentration and because smoke particles can interfere with some of the chemical analyses of the solutions. The Nuclepore filter holders, which are made of clear polycarbonate plastic, are definitely preferable to the opaque Millipore filter holders for this purpose, because their transparency allows one to see when all of the air has been removed from the filter holder after filtration has begun. The air can be removed by carefully manipulating the tubing and by rapping the filter holder sharply with your hand.

As the "smoke" is generally very fine-grained, it largely stays in suspension prior to filtration. As suggested by R. Feely, (personal communication, 1991) filtering a small volume of plume water yields a low particle concentration on a given filter that is suitable for SEM work. Too high a concentration causes the particles to aggregate on the filter and obscures the natural aggregates that may have formed in the plume prior to sampling. Other than this, it is preferable to collect as much of the total smoke as possible on a single filter, to facilitate measuring particle concentration by weighing, mineralogical analysis by x-ray diffraction (XRD), and bulk chemical analysis following dissolution. After the bottle has been drained to the level of the drain valve, we open the lower ball valve and dump the remaining solution and particles into a clean plastic tray. This material is then sucked into a clean plastic syringe and ejected through the same filter, discarding the solution. We then rinse the filter with 100 mL of pure water ejected through the syringe. The filter is then removed from the holder using Teflon-coated tweezers, its underside blotted on a Kimwipe, and placed in a plastic petri dish.

At this point we remove the lower ball valve from the Go-Flo bottle and rinse it into the same plastic tray. The Go-Flo bottle itself is then rinsed with about 250 mL of filtered surface seawater to remove any remaining particles, which are also collected in the tray. These particles, which have settled in the bottle and are therefore generally larger, are collected by syringe and filtered through a separate filter, using the same filter holder.

After being placed in petri dishes, all filters with their load of smoke particles are placed in a plastic desiccator which is immediately evacuated. Where practical, after a few days of sampling, the petri dishes are transferred to a second desiccator, which is kept evacuated and frozen. After their return to the shore-based laboratory, all samples are stored under vacuum and frozen, except when actually being worked on.

All filters should be preweighed ashore, under carefully controlled conditions of humidity, and loaded into filter holders, which are then sealed into labelled ziplock polyethylene bags. The drawback of this approach is that one requires a large number of filter holders, which are expensive. As the alternative is acid-cleaning filter holders on board ship, however, under less than clean conditions and with a limited supply of clean water, it is worthwhile to invest in a large number of filter holders.

4. Analyzing the Plume Solutions

We analyze the solutions immediately on board ship for oxygen by Winkler titration; H_2S, nitrite, and dissolved Fe by colorimetry; and pH and alkalinity by electrochemical titration with HCl. If time and manpower permit, we also analyze for dissolved Mn and Si. We collect an unfiltered aliquot that is acidified for "total dissolvable" Fe and Mn. All other aliquots are filtered. We collect an unacidified aliquot for chlorinity, sulfate, and other major and minor species. We collect an unacidified aliquot that is frozen for the nutrients nitrate, phosphate, and ammonia. We dilute an aliquot with 0.1 M HCl for Si analysis. We collect an acidified aliquot for dissolved metals. We acidify with sub-boiling distilled HCl (~ 6 M) at a rate of 4 mL acid per liter of sample, which results in a pH of about 1.5.

5. Analyzing the Particles

We determine the mineralogy of the particles with XRD, SEM, and optical microscopy, using both binocular and petrographic microscopes, the latter after mounting the particles in epoxy and polishing them. We determine mineral chemistry using the electron microprobe. We measure bulk chemistry by dissolving the particles in acids and analyzing the resulting solutions with a combination of inductively-coupled plasma atomic emission spectrometry (ICP-AES) and flame and flameless atomic absorption spectrophotometry.

References

Mottl, M. J., and T. F. McConachy, Chemical processes in buoyant hydrothermal plumes on the East Pacific Rise near 21° N, Geochim. Cosmochim. Acta, 54, 1911-1927, 1990.

COLLECTION OF OCEANIC SUSPENDED PARTICULATE MATTER FOR TRACE METAL ANALYSIS USING A NEW IN SITU PUMP

Robert M. Sherrell

Department of Earth Atmospheric and Planetary Sciences, Massachusetts Institute of Technology, Cambridge, Massachusetts 02139

1. Introduction

The interaction between dissolved chemical constituents and fine particles suspended in the ocean is an important control on the trace element composition of seawater. In order to understand this interaction, it is necessary to sample and analyze both seawater and suspended particles accurately. Techniques for collection and analysis of seawater samples for trace elements have advanced rapidly in the last 15 years, and we now have a basic knowledge of the concentration and distribution of many trace metals dissolved in the world's oceans [Bruland, 1983]. The distribution of particulate trace metals, which generally account for a small fraction of the dissolved concentration, is not as well known [Buat-Menard and Chesselet, 1979]. Accurate knowledge of the partitioning of trace metals between dissolved and particulate forms at several representative oceanic stations would lead to a better understanding of transformations between these two phases, and the processes which result in vertical flux of metals out of the oceanic system [Jickells et al., 1984; Dymond and Collier, 1987]. This paper describes a new device for collecting relatively large uncontaminated samples of suspended oceanic particulate matter, and a procedure for analysis of the particles for major constituent elements and trace metals.

One of the difficulties in collection of open ocean suspended matter for chemical analysis is the low mass concentration and consequent need to sample large volumes of seawater. The most commonly used technique is ship board filtration of up to a few tens of liters of water collected with conventional Niskin bottles [Brewer et al., 1976]. When clean room-type precautions and Go-Flo bottles are used in sample handling, this procedure has been shown to work very well for analysis of relatively abundant metals such as Al, Fe, and Mn [Landing and Bruland, 1987; Orians and Bruland, 1986] and for some more contamination-prone elements such as Zn (K. Bruland, personal communication, 1989). The disadvantage of this technique is that sample sizes are limited to a few hundred micrograms at typical deep ocean particle concentrations, restricting the number of different analyses or dissolution procedures that can be carried out on each sample. In addition, bottle artifacts, such as incomplete sampling due to rapid settling of particles in the samplers, [Gardner, 1977; Calvert and McCartney, 1979] can be minimized (bottle inversion, re-orientation of sampling orifice), but are difficult to evaluate. Hence we chose to avoid these limitations and increase sample size by 1-2 orders of magnitude over the bottle technique by filtering the particles in situ with a new submersible pump.

The use of pumps as oceanographic sampling devices is not new. Deck-mounted pumps have been used for some time for sampling particles in the upper water column by sucking water through various lengths of tubing [Beers et al., 1967; Lisitzin, 1972; Lenz, 1972; Jeffrey et al., 1973]. In situ pumps capable of collecting particles in the deep ocean have also been used to study the biological, mineralogical, chemical and radiochemical composition of suspended matter [Laird et al., 1967; Spencer and Sachs, 1970; Beer et al., 1974; Hess, 1977; Bacon and Anderson, 1982; Winget et al., 1982; Cochran et al., 1987]. Some such devices are powered from the surface by means of an electro-mechanical cable and are capable of filtering up to 25,000 L of seawater [Bishop and Edmond, 1976; Bishop et al., 1985]. Another instrument couples battery-powered filtration with sensors which relay temperature, depth, light transmission and particle size fractionation data to the ship as the instrument is deployed [Simpson et al., 1987]. The current design was conceived as a relatively simple sampling device, capable of being used on any vessel equipped with a standard hydrographic winch, and incorporating design elements intended to minimize trace metal contamination of the particulate sample.

Here I present a detailed description of the new pump, called the Rotating Automatic Pump for Particulate Inorganics Determination (RAPPID), and a procedure for analysis of oceanic particulate matter for major elements (Ca, Al, Fe), phosphorus, opaline silica, the trace metals Mn, Co, Cu, Zn, Ni, Cd, and Pb, and the radionuclides ^{210}Pb and ^{210}Po. Three of these devices have been constructed and used successfully on eight oceanographic cruises to collect over 65 samples of

Marine Particles: Analysis and Characterization
Geophysical Monograph 63

particulate matter from volumes of 200 to 1500 L to depths of 4000 m. Results of the chemical analyses indicate that concentrations of particulate metals in the open ocean are, for several elements, much lower than previously estimated [Sherrell, 1989; Buat-Menard and Chesselet, 1979]. Moreover, the observed variations with depth and oceanographic regime clarify the role of suspended matter in geochemical cycling and flux of metals in the open ocean [Sherrell, 1989].

2. Instrumentation

Two key features distinguish the new pump as a design especially suited for trace metal studies:

1) the entire unit mounts on a standard hydrowire so that it pivots like a weathervane, such that the filter holder and intake are oriented upstream of any particulate contamination from the wire or the metal components of the pump itself;
2) the filter holder and its mounting struts are constructed of plastics which have been shown to contribute minimal metal blanks in dissolved trace metal studies.

The main components of the pump are mounted within a double box-shaped aluminum frame (Figure 5-24).

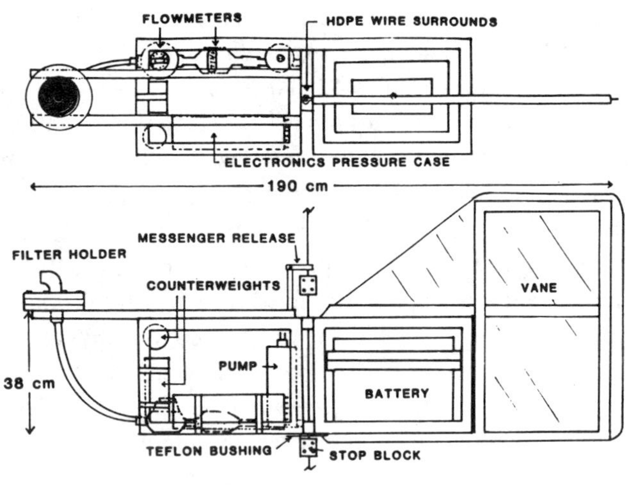

Fig. 5-24. Schematic of pump construction and principal components.

These consist of the battery unit, pump and motor unit, electronics pressure case, flow meters and, for certain configurations, stainless steel counterweights. The filter holder is mounted on Plexiglas struts at the front of this frame, so that the intake is located about one meter laterally distant from the hydrowire. A large plastic and aluminum vane mounted on the back end serves to orient the intake into the ambient current. When the pump is turned on, water is sucked through the filters, passes through two flow meters, into the pump impeller, and is exhausted at the rear.

This section describes each of these components and outlines the deployment procedure. A summary of the performance specifications is given in Table 5-10.

TABLE 5-10.
RAPPID Performance Specifications

Weight: 80 kg in air
Filter: 142 mm, 127 mm effective diameter
Maximum flow rate: ~ 7.5 L min^{-1}
Maximum face velocity: 1.0 cm sec^{-1}
Battery: 12V, 76AH or 24V, 38AH
~ 6 hrs pumping per charge cycle
3 year battery life span
Maximum deployment depth: 6500 m
(limited by pressure housing)
Winch speed: 50 m min^{-1}

2.1. *Structure and Mounting*

The frame consists of two symmetrical box-shaped units constructed of 1.5 x 1/8" structural aluminum (6061-T6 alloy) bolted together with 1/4" x 20 tpi stainless steel bolts (316 alloy). These are joined along one side and at the longitudinal axis, creating a lateral slot into which the hydrowire is positioned upon mounting (Figure 5-24). The wire is held at this central axis by four high-density polyethylene blocks, two of which are structural members in the frame, the other two being held in place with 1/4" toggle bolts which are inserted once the pump is in position. Stainless steel stop blocks are bolted firmly to the wire above and below the pump. The entire pump rests on the bottom stop block, with a Teflon sheet washer in between to decrease pivoting friction and minimize abrasion of the aluminum frame. The upper stop block prevents the pump from riding up the wire during lowering or heavy ship roll. In water, the pump is dynamically balanced about the central wire, minimizing resistance to rotation so that orientation can be achieved even in slow currents.

2.2. *Power*

The RAPPID pump is powered by a pressure-compensated Sea Battery (Deep-Sea Power and Light, San Diego, California). This unit consists of two 12V,

38AH lead-acid suspended electrolyte batteries mounted in a rugged polyethylene box with a flexible, transparent polyurethane diaphragm in the lid. The box is filled with high-viscosity mineral oil for pressure compensation. Power is brought out through doubled conductors in an underwater cable which passes through the diaphragm. Two versions of the pump have been used, based on 12V or 24V wiring configurations. The battery is charged with 12V, 10A automotive chargers, used singly for the 12V application or wired in series for charging the batteries in the 24V configuration. Line voltage to the chargers is controlled with a Variac transformer, allowing the charging amperage to be manually adjusted during the charge cycle. Charging is stopped when current falls below 1.0A, or when gas is observed emanating from the cells. A typical charging cycle took eight hours for near full discharge; the resulting charge was sufficient for 2-3 pump deployments.

2.3. Pump/Motor Unit

Two different pump/motor combinations have been used. The first is a 12V unit which runs in Bray oil and is fitted with a Jabsco flexible vane pump head (Pelagic Electronics, Falmouth, MA). A more recent version uses a superior 24V motor, pressure compensated with aviation-type hydraulic oil, mated to a rotating-disk positive displacement adjustable flow-rate pump head (Flotec) by means of an anodized aluminum housing with ceramic shaft seal (Oceanic Industries, Monument Beach, MA). These pumps have similar flow rate versus pressure differential characteristics, but the latter has proved somewhat more efficient (55 vs. 75 watts at full load) and more durable. Stainless steel counterweights are hose-clamped to the pump frame when the lighter 12V pump/motor is used, in order to maintain the fore/aft balance of the entire unit.

2.4. Flow Meters

Two flow meters mounted in series downstream of the filter holder record total volume sampled and flow rate versus time data. The first (Kent Meters, model C-700, Ocala, FL) was modified to send electrical pulses to the electronics unit by mounting a small magnet on the rotating central spindle of the index, and fixing a matching reed switch to the index face (T. Hammar, personal communication, 1984). Flow through the meter causes the switch to make and break a circuit to the control unit. These pulses are recorded by the computer at the rate of 200 state changes per gallon (52.8 L^{-1}). Flow rate decreases quasi-exponentially with time due to

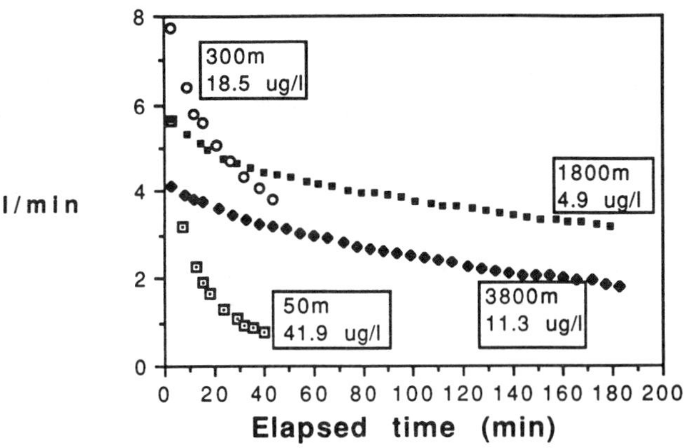

Fig. 5-25. Typical flow rate versus time curves for samples collected at various depths and less than 53 μm particle concentrations in the Sargasso Sea. Flow rate is a function of filter loading and the state of charge of the battery. The 3800 m sample is in the nepheloid layer, approximately 400 m above the bottom.

filter loading from maximum rates of 4.0 to 7.5 L min^{-1} to final rates of 0.5 to 4.0 L min^{-1} (Figure 5-25). Pumping rates also depend partially on battery charge. The record of flow rate evolution allows evaluation of the effect of filtration speed on sample integrity [Bishop, 1982], provides an indication of pump performance during the deployment, serves as a diagnostic tool for locating pump failures, and guides subsequent sampling so that the pump interval can be set to optimize sample volume within ship time constraints.

The second flow meter has a different design (Kent Meters, model PSM 190, Ocala, FL) which is not readily adapted for telemetering but has the advantage of a lower stall rate (1/8 gal min^{-1} vs. 1/4 gal min^{-1}). The two meters provide replicate measurements of sample volume, and give equivalent results within the manufacturer's stated precision: $\pm 1.5\%$, at flow rates exceeding stated stall rate. Discrepancies between the two were noted only when flow rate fell below 1 L min^{-1}. In these cases, volume indicated by the PSM 190 was assumed correct.

2.5. Electronic Control

The pumping interval is controlled by an independently powered Hewlett-Packard 41CX handheld computer mated to a small input/output/control unit (CMT-200 Data Acquisition and Control Unit, Corvallis Microtechnology, Corvallis, OR). Before pump deployment, the computer is pre-set for on/off times by programming control alarms actuated by the built-in time function in the HP 41CX. After some experience, the time required for deployment can be estimated to within 20 minutes, so that little ship time is lost. To save battery power, the computer is on only during the pump interval, using a "wake-up" function actuated by the control alarm. The control unit powers a relay, also contained within the pressure vessel, which carries primary current from the battery. An input channel on the control unit is connected to the flow meter circuit, and pulses are sub-totaled after a specified time interval, usually 5 min for a typical deployment, and recorded in data storage registers in the computer. All of these functions are controlled by an operation program stored in computer memory. Upon retrieval of the pump, the pressure case is opened, and a data reduction program transfers pertinent data which includes sub-totaling interval, flow counts for each interval, counts/minute during each interval, total counts, time on, time off, to a compatible thermal printer (Hewlett-Packard model 22143A).

2.6. Filter Holder

The RAPPID system is designed to be used with 142 mm diameter filters, which were judged to be large enough to collect sufficient sample in a reasonable length of time while avoiding the handling difficulties of the larger 294 mm size filter. The filter holder is custom designed to optimize effective filter area and to minimize use of potentially contaminating materials. The two-stage holder consists of three acrylic plastic sections (Figure 5-26). The base supports the primary filter on a 35 μm pore-size sintered polyethylene insert (Bel-Art). The middle section is simply an open ring which supports a prefilter, held only by tension around its perimeter. The top section incorporates a flow expansion chamber and a PVC intake elbow, with a 15.2 cm^2 orifice facing horizontally so that the pump does not act as a trap for rapidly sinking particles.

Silicone o-rings sit in glands machined in the upper faces of the bottom two pieces, such that they project 0.25 mm above the face of the holder. All three pieces are bolted together with six 1/2" Nylon bolts, and tightened with custom-made Lexan polycarbonate wrenches so that all faces contact firmly. This establishes a crisp edge to the sample area, so that subsamples can later be cut as accurate fractions of the total sample. Tightening also compresses the o-rings slightly, and water leakage is prevented by the o-ring's position on the low-pressure side of the primary filter. Effective filter

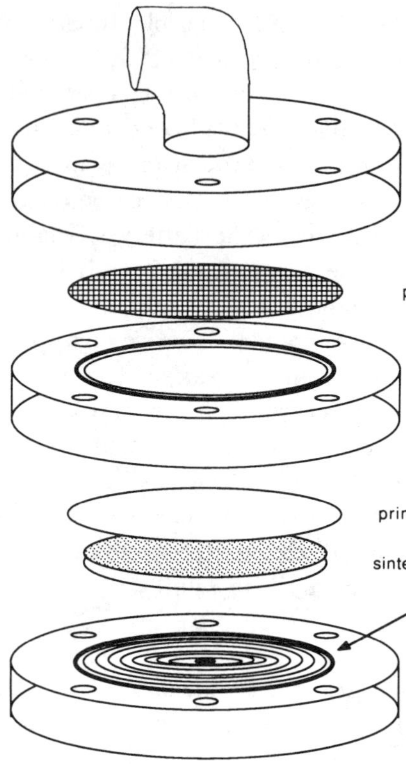

Fig. 5-26. Schematic of filter holder design.

area is 126.7 cm^2 (80% of total filter area). The assembled filter holder is connected to the flow meters by 1/2" polyethylene tubing and polypropylene compression fittings (Jaco).

In preparation for use, all parts of the filter holders are carefully washed, thoroughly rinsed, soaked in dilute HCl at 35° C for several days, and rinsed in distilled, deionized water. Between deployments at sea, filter holders are rinsed with distilled, deionized water from a polyethylene squeeze bottle, using polyethylene gloves for all stages of handling.

2.7. Deployment Procedure

The RAPPID is designed to be used in multiple-pump casts, so that several-depth profiles can be obtained in one cast. Up to three pumps have been deployed simultaneously in seas up to 6 m without undue difficulty. In calmer weather, it is expected that several more pumps could be deployed on a single cast, increasing sampling efficiency and allowing greater synoptic depth coverage.

In preparation for a cast, filter holders are loaded with clean filters in a laminar flow clean bench, bolted together, sealed with a clean polyethylene plug inserted in the intake, and stored temporarily in clean zip-lock polyethylene bags. Computers are set for pump interval and sealed inside the pressure cases, which are purged with dry Freon to prevent condensation of water vapor on the electronic components. Pressure cases are attached to the pump frames, initial flow meter readings are recorded, and the bottom stop block is affixed to the hydrowire. When the pump is ready for deployment, the filter holder is removed from its storage bag, attached to the struts with acid-cleaned Nylon wing-bolts, connected to the pump plumbing, and re-covered with a larger polyethylene bag. The weight of the pump requires it to be lifted into position on the wire using a block and tackle. The hydrowire is captured in position by fixing the removable plastic blocks in place, and the lifting tackle is removed. The top stop block is bolted to the wire, and a messenger is attached to a release mechanism on the pump frame (Figure 5-26).

To minimize the possibility of airborne contamination reaching the filters, the protective plastic bag and intake plug are removed only moments before the pump is submersed. Winch speed is limited to 50 m min^{-1} to avoid kiting, although faster speeds have occasionally been used without mishap.

The pumping interval varies from 40 minutes to 3 hours, depending on anticipated particle concentration and filter clogging rate. Sampling depth is estimated using reversing thermometers attached to Niskin bottles which are mounted 10 m above or below the pumps. These are tripped by dropping a messenger at the mid-point of the pumping interval. Additional bottles have been mounted at other depths to combine water and particle collection in a single cast. Upon retrieval, the plastic intake plugs are replaced immediately as the pumps emerge from the water, and the pumps are removed from the wire. Filter holders are immediately removed for disassembly in a laminar flow clean bench, the flow meter indexes are read, the pressure cases are removed and opened, and computers are withdrawn for data retrieval.

3. Sample Handling and Analysis

3.1. Filter Cleaning, Drying and Weighing

The filters used in this work are chosen for an optimum combination of low metal blank, low water absorption (for precise mass determination), and reasonably high flow rate. The standard filter pair includes a 1.0 μm pore size Nuclepore polycarbonate primary filter and a 53 μm mesh size square weave polyester prefilter cut from bulk material, Pecap, #7-53/2, Tetko, Briarcliffe Manor, NY, both 142 mm diameter. Control of trace metal blank levels requires that they be carefully cleaned and handled only with acid-cleaned Teflon forceps and polyethylene gloves in a laminar flow clean bench.

Filters are cleaned in batches of 25 by placing them in 1 L polyethylene bottles, filling with ~1 N HCl (reagent grade, in distilled deionized water), and heating in an oven at 60° C overnight. The acid is then decanted to waste, and the bottles filled with distilled deionized water and allowed to sit for ~1 hour. This solution is then decanted to waste, and the filters are rinsed five times in the bottles with distilled deionized water, swirling and inverting the bottles each time. The filters are then removed to a large, clean polyethylene beaker in a laminar flow clean bench and allowed to sit in additional distilled deionized water. They are then removed, drip-dried, and transferred one at a time to acid-cleaned 150 mm diameter polystyrene petri dishes. The lids are placed loosely ajar on the dishes and the filters are allowed to dry overnight at room temperature in the laminar flow clean bench.

Dried filters are weighed to constant weight (± 0.1 mg) on an analytical balance, controlling electrostatic interactions with two $\alpha-$source anti-static strips positioned in the weighing cabinet. Filters are prevented from contacting the stainless steel weighing pan by placing a 152 mm diameter acid-cleaned Teflon mesh disk on the pan or by using an easily constructed polystyrene "napkin-holder" to support the filter in a "u" shape without touching the pan. Typical masses for primary and prefilter are 180 and 640 mg, respectively. Filters are returned to the labeled petri dishes, which are then sealed individually in zip-lock polyethylene bags, and in batches in larger bags for transporting to sea.

4. Shipboard Procedures

At sea, working in a laminar flow clean bench, filters are mounted in the filter holders immediately before use. Upon recovery, any excess seawater retained in the filter holders is removed using a vacuum line attached to a clean length of polyethylene tubing, and can be recovered for determination of analytes lost from the particulate samples to solution. Filter holders are left assembled and sealed in zip-lock polyethylene bags to minimize precipitation of seasalt before rinsing. Within three hours of recovery, the holders are disassembled and the filters rinsed to remove residual seawater. For the primary filters, this is accomplished by sealing the filter holder base, with the filter left in position, to a vacuum chamber in which is placed an acid-cleaned polyethylene 250 mL bottle. The filter is rinsed by gently dripping ~ 125 mL of pH 8.3 distilled water (adjusted with pre-cleaned NaOH) through the filter in a repeating spiral fashion under moderate vacuum, so that the entire surface is uniformly rinsed. This method limits particulate redistribution to the scale of a rinsing drop. Rinse solutions are then acidified with 0.5 mL 3x Vycor-distilled 6 N HCl and treated as dissolved trace element samples. Loss of metal from the particulate matter during rinsing can be determined by analysis of these "rinsates".

The prefilters are then placed on the same holder base and rinsed in the same manner, still within the laminar flow bench. In some studies, clean 0.4 μm or 1.0 μm Nuclepore filters were placed under the prefilters before rinsing, to determine whether particulate material was rinsed through the prefilter, perhaps after disaggregation by the rinse water flow. In the oligotrophic eastern North Atlantic (Beta Triangle) we found that 13 to 350% but usually less than 50% of the mass of material left on the prefilter was rinsed off onto the Nuclepore "rinsing filter". While the total greater than 53 μm mass can be determined this way, meaningful analysis of the large particles requires analysis of both fractions, generally against unfavorable filter blanks. Not rinsing the prefilter may be a desirable option for preserving the physical and chemical integrity of the sample, although retained seasalt could result in an inaccurate mass determination unless much more material could be caught on the same filter area.

After rinsing, both filters are then returned to their original petri dishes and allowed to dry for several hours in the laminar flow bench before being sealed as before. During transport back to the laboratory, samples are kept level to prevent dislodging of particulate material from the filter surface.

5. Laboratory Filter Handling

Upon return to the laboratory, the filters are re-dried for several days in a desiccator. Filters are rapidly transferred from the desiccator, placed in a laminar flow bench, to the balance, and sample weights are determined by re-weighing the filters as before (this time in a desiccated weighing cabinet). Blank filters are carried through the entire operation, and average differences between pre- and post-sampling weights for these filters are used to correct final weights for changes in the mass of the filter material itself due to water absorption. This correction was less than +0.2 mg for Nuclepore filters, but was larger (+1.0 mg) for the more absorbent prefilters. This method was used instead of equilibration with ambient room air to prevent erroneously high sample dry weights due to water absorption by the particles or by residual seasalt. Within each weighing batch, several filters are re-weighed to determine reproducibility. Corrections for the mass of residual seasalt were made by analyzing filter sub-samples for Na (see below), and assuming seawater major ion composition. Salt corrections for the Nuclepore filters are generally less than 5% of particle mass and are often negligible (less than 1%). Typical sample masses were 4.0 to 18.0 ± 0.2 mg of less than 53 μm material, and 0.5 to 2.0 ± 0.5 mg of greater than 53 μm material.

We present here methods pertaining only to analysis of the less than 53 μm fraction, which contains most of the suspended mass in the open ocean water column [e.g. 75-94% of total particulate mass in the Sargasso Sea near Bermuda Sherrell, 1989]. Analysis of the greater than 53 μm fraction would enable a better understanding of exchange of metals among particle size classes, but blank levels are too high to allow accurate analysis of many elements of interest at the sample volumes collected in our work. Estimates of elemental abundances in the greater than 53 μm fraction relative to prefilter blank levels (Table 5-11) suggest that useful data could be acquired for Al, Fe, Mn, Pb, and perhaps Cd and perhaps other elements we have not examined, but only preliminary attempts to do this have been made to date. The problem might be avoided if a method could be devised to quantitatively or representatively remove the particles from the prefilter.

The primary filter was subsampled by using a small inverted, thin-walled polystyrene box as a template and a stainless steel scalpel to cut known-area squares for various treatments. Reproducibility of the analyses of replicate subsamples depends on:

1) homogeneity of particle distribution across the face of the filter and

2) analytical procedures.

Sub-sampling reproducibility was estimated at ± 10%, based on analyses of replicate sub-samples, subtracting the analytical error determined by replicate analyses of single sub-sample digest solutions. Occasionally, visible redistribution of the particles occurred because of rough handling during pump retrieval, and replicates on these samples gave ± 25% subsampling reproducibility. Use of a depth filter rather than the membrane filter used here might decrease redistribution of the sample for more precise chemical analyses using small subsamples. However, we have not

TABLE 5-11.
Filter Blanks, Reagent Blanks and Analytical Precision

Element	Filter Blank Primary[a]	(nmol/filter) Prefilter[a]	Reagent blank[b] (as nmol/filter)	Analytical[c] precision
Ca	ND	--	40	1%
Al	20	2.0	20	4%
	5	2.0	5	5%
Mn	0.1	0.8	0.1	2%
Co	0.04	--	0.04	4%
Zn	0.4	1.0	0.1	10%
Cu	0.3	0.25	0.3	4%
Ni	1.6	--	0.1	3%
Cd	0.001	0.002	ND	4%
Pb	0.03	0.015	0.02	3%

a) Primary is mean for unused Nuclepore 142 mm 1.0 μm pore size, acid cleaned. Prefilter is mean for unused 142 mm 53 μm mesh Pecap polyester.
b) Combined acid plus procedural blank, expressed as equivalent per filter.
c) Estimated percent uncertainty, based on replicate analyses of single digest solutions, primary filter only.
ND = not detectable, -- = not determined

TABLE 5-12.
Typical Suspended Particulate Trace Metal Concentrations in the Open Ocean[a]

Element	Al[b]	Fe	Mn	Cu	Zn	Co	Cd	Pb	Ni
Surface water[c]	1.2	0.3	34	6.3	5.2	0.9	0.3	1.2	5.8
Deep water[d]	3.6	1.2	150	15	7.4	1.5	0.06	2.1	6.0

a) All data from Sargasso Sea near Bermuda, five station occupations between 9/86 and 4/88.
b) Al and Fe in nmol L^{-1}, remainder in pmol L^{-1}.
c) 0-100 m mean, n=4.
d) 500-3000 m mean, n=17.

yet found a suitable depth filter with sufficiently low blank levels. In any case, analytical uncertainties generally limit the ultimate precision to a few percent, which is good enough for most geochemical problems of interest.

6. Analytical Procedures

Filter sub-samples were prepared for chemical analysis using several different dissolution procedures. For trace metals, Ca, Al, Fe, ^{210}Pb, and ^{210}Po analyses, two subsamples comprising 8.3% of total filter area, cut from radially opposed positions on the filter face, were placed in a 7 mL flat-bottom screw-cap Teflon PFA vial (Savillex). Acid-cleaned Teflon forceps were used to maneuver the filter squares within the vial such that they adhered by electrostatic attraction to opposite walls of the vial, particle side inward. With the filter pieces so positioned, 0.5 mL of 3x Vycor-distilled concentrated HNO_3 was added in 100 μL aliquots, dripping these down the faces of the filter to initiate dissolution and further adhere the filters to the wall. To assure complete dissolution of refractory mineral phases, 20 μL of sub-boiling distilled concentrated HF were then added. The vials were tightly capped, weighed, and placed on a hot-plate to fume/reflux for 4 hours at ~ 90° C. With proper temperature regulation, refluxing down the faces of the filter pieces did not lead to total destruction of the polycarbonate matrix, simplifying subsequent handling. After cooling, vials were re-weighed to correct for evaporative loss of acid (generally less than 0.5%), and were diluted with 2.0 mL of distilled, deionized water.

The dilution step was adopted after the discovery of a precipitate phase apparently present in some undiluted digest solutions. The precipitate was never visible, but its presence was suspected when poor reproducibility was noticed during analysis of replicate aliquots for Al. To test this suspicion, digest solutions were transferred to an acid-cleaned 1.5 mL microcentrifuge tube and spun for a few minutes. Small aliquots (10 μL) were then removed from the solution surface and from the conical bottom of the tube, where a pellet would be located if one were visible. These were analyzed for Al again. Al concentrations in the pellet aliquot were greater than 10 times higher than the supernatant aliquot, which had an Al concentration roughly equivalent to that of a dipped blank. Subsequent aliquots pipeted from the bottom of the same tube gave lower concentrations, approaching that of the supernatant. Analysis of calcium in the "pellet" aliquots indicated Ca/Al ratios of ~ 3.5 mol/mol, more than an order of magnitude higher than the ratio for shales [Turekian and Wedepohl, 1961]. This suggested that the solid phase was not the result of incomplete dissolution of refractory mineral particles. It was concluded that Al was probably adsorbing to a fine precipitate, most likely CaF_2. Dilution by a factor of five apparently caused a sufficient decrease in the solubility product to redissolve the solid. The centrifuge test on diluted solutions gave equal aluminum concentrations for "pellet" and "supernatant" aliquots.

Diluted digest solutions were analyzed for Na, Al, Fe, Mn, Co, Cu, Zn, Ni, Cd and Pb by graphite furnace atomic absorption spectroscopy, using additional dilutions as necessary and standardizing by the method of standard additions. Most analyses were carried out on a Perkin-Elmer 5000 equipped with continuum source background correction, HGA 400 furnace controller, and AS-40 autosampler. A Perkin-Elmer Zeeman/5000 with HGA 500 furnace controller was used for Co and some Zn determinations because furnace blanks were lower for these elements on this particular instrument (not because Zeeman effect background correction was necessary). Ca was determined by diluting 100 μL of digest solution to 5.0 mL in dilute HCl/HNO_3 (La added as an ionization suppressant), and analyzing against matrix-matched standard solutions using flame atomic absorption spectrophotometry (Perkin-Elmer 403). Long-term secondary standards were maintained throughout this work. Analytical precision was determined by replicate analyses of single solutions. Blanks were determined on unused filters and on "dipped" blanks, filters which were treated identically to a normal sample, including submersion on a non-operating pump. For the Sargasso Sea sample sequence, blanks were generally less than 10% of sample signals, often on the order of 1%, but occasionally exceeded 50% for Zn, Cu and Ni in near-surface samples. A summary of analytical precision and blank values is given in Table 5-11. Typical open ocean particulate trace metal concentrations are given in Table 5-12.

Losses of particulate metals during rinsing, a consequence of desorption, dissolution, or disaggregation, is important to evaluate. Rinsing losses for Sargasso Sea samples were 0-10% for Zn and 15-50% for Cd [very high values were rejected as contamination suspects, Sherrell, 1989]. Rinse solutions have not been analyzed for other elements, although Mn, at least, should be considered a likely candidate [Landing and Bruland, 1987].

For ^{210}Po and ^{210}Pb determination, portions of digest solutions from three separate subsample pairs, equivalent to 20% of the total sample, were pooled, spiked with ~ 10 dpm ^{209}Po yield monitor and 4.0 mL 70% $HClO_4$, and heated to near dryness in a Teflon beaker. After redissolution in 2.0 M HCl, ^{210}Po and ^{209}Po were autoplated on silver disks following the method of Fleer and Bacon [1984]. Alpha counting of the plated Po isotopes was performed on an alpha spectrometer fitted with low-background detectors. ^{210}Pb was determined after 6-12 months ingrowth by measuring the ingrown daughter ^{210}Po. Because ^{210}Po activity on oceanic particulate matter is generally several times greater than ^{210}Pb activity [Bacon et al., 1980], counting times of 1-2 days were sufficient to accumulate greater than 500 counts on the first plating. For the ingrown ^{210}Po determination, however, 3-5 days counting time was required to obtain adequate counting statistics. Uncertainty due to counting statistics was less than 5%, except for a few low-Pb surface samples (less than 10%). Stable Pb was determined on each digest solution before pooling, so that accurate $Pb/^{210}Pb$ ratios could be determined.

Particulate phosphorus and opaline silica were determined on separate, smaller pairs of subsamples (2.4% of total sample). Phosphorus was liberated as inorganic phosphate by wet oxidation in perchloric acid using a micro adaptation of the procedure of Strickland and Parsons [1968]. Amorphous silica was selectively dissolved by leaching in 0.7M Na_2CO_3 at 90° C for 4 hours [Eggimann et al., 1980]. This treatment should leave ordered silicates largely undissolved. Leachates were neutralized and analyzed colorimetrically in a narrow 1 cm cell, again using a micro adaptation of the standard seawater method [Strickland and Parsons, 1968]. Organic carbon content could not be determined directly, but was estimated by difference between total weight and summed inorganic components, assuming carbon is 50% of the total residual weight. Mass fractions for calcium carbonate, opaline silica, alumino-silicates, and organic matter were calculated from elemental concentrations using the following equations:

$$\%CaCO_3 = 100g/40g \times (Ca - (Ca/Al)crust \times Al) \times 100/(wt) \quad (1)$$

$$\%\ Opal = 70g/28g \times Si \times 100/(wt) \quad (2)$$

$$\%\ Alumino\text{-}silicate = 10 \times Al \times 100/(wt) \quad (3)$$

$$\%\ Organic\ matter = 100 - (\%CaCO_3 + \%opal + \%alumino\text{-}silicate) \quad (4)$$

where (wt) is total dry weight and all elemental quantities are expressed as μg per sample. The estimate of alumino-silicate-bound Ca is based on average Ca/Al for shales, rather than average crust, as the crustal value includes a carbonate rock contribution [Turekian and Wedepohl, 1961]. The conversion factor used in equation (2) assumes that opaline silica contains some hydration water not removed during drying [Moore, 1969]. Equation (3) is based on the assumption that alumino-silicates contain 10 wt% Al, somewhat higher than the crustal average [Taylor, 1964] but consistent with the relatively Al-enriched clays found in low-latitude nepheloid layers in the Atlantic [Bishop and Biscaye, 1982, and references therein].

7. Summary

A new in situ pump has been designed which collects uncontaminated particulate samples for trace metal analysis from volumes of the order of 1000 L. It is reliable and simple to use, requiring ordinary oceanographic deck hardware, and can be deployed to full ocean depth. Three such RAPPID pumps have been built, and have been used to collect greater than 65 particulate samples from the North Atlantic and North Pacific oceans.

These samples have been analyzed for a suite of trace metals, as well as major constituent elements, using wet digestion techniques combined with atomic absorption and UV/VIS spectrometry. This new technique has enabled the first accurate determination of the suspended particulate concentrations of several trace metals in the open ocean [Sherrell, 1989].

References

Bacon, M. P., and R. F. Anderson, Distribution of thorium isotopes between dissolved and particulate forms in the deep sea, J. Geophys. Res., 87, 2,045-2,056, 1982.

Bacon, M. P., D. W. Spencer, and P. G. Brewer, Lead-210 and polonium-210 as marine geochemical tracers; review and discussion of results from the Labrador Sea, in Natural Radiation Environment III, edited by T. F. Gesell, and W. F. Lowder, Vol. 1, pp. 473-501, U.S. Dept. Energy Rpt., CONF-780422, Springfield, Virginia, 1980.

Beer, R. M., J. P. Dauphin, and T. S. Sholes, A deep-sea in situ suspended sediment sampling system, Mar. Geol., 17, M35-M44, 1974.

Beers, J. R., G. L. Stewart, and J. D. H. Strickland, A pumping system for sampling small plankton, J. Fish. Res. Board Can., 24, 1811-1818, 1967.

Bishop, J. K. B., The effect of filtration speed on large particle catch during large volume in situ filtration, EOS Trans. Am. Geophys. Union, 63, 46, 1982.

Bishop, J. K. B., and P. E. Biscaye, Chemical characterization of individual particles from the nepheloid layer in the Atlantic ocean, Earth Planet. Sci. Lett., 58, 265-275, 1982.

Bishop, J. K. B., and J. M. Edmond, A new large volume filtration system for the sampling of oceanic particulate matter, J. Mar. Res., 34, 181-198, 1976.

Bishop, J. K. B., D. Schupack, R. M. Sherrell, and M. H. Conte, A multiple unit large volume in situ filtration system (MULVFS) for sampling oceanic particulate matter in mesoscale environments, in Mapping Strategies in Chemical Oceanography, edited by A. Zirino, Advances in Chemistry Series, Vol. 29, pp. 155-175, American Chemical Society, Washington, D.C., 1985.

Brewer, P. G., D. W. Spencer, P. E. Biscaye, A. Hanley, P. L. Sachs, C. L. Smith, S. Kadar, and J. Fredricks, The distribution of particulate matter in the Atlantic Ocean, Earth Planet. Sci. Lett., 32, 393-402, 1976.

Bruland, K. W., Trace elements in sea-water, in Chemical Oceanography, edited by J. P. Riley, and R. Chester, Vol. 8, pp. 158-220, Academic Press, London, 1983.

Buat-Menard, P., and R. Chesselet, Variable influence of the atmospheric flux on the trace metal chemistry of oceanic suspended matter, Earth Planet. Sci. Lett., 42, 399-411, 1979.

Calvert, S. E., and M. J. McCartney, The effect of incomplete recovery of large particles from water samplers on the chemical composition of oceanic particulate matter, Limnol. Oceanogr., 24, 532-536, 1979.

Cochran, J. K., H. D. Livingston, D. J. Hirschberg, and L. D. Surprenant, Natural and anthropogenic radionuclide distributions in the northwest Atlantic Ocean, Earth Planet. Sci. Lett., 84, 135-152, 1987.

Dymond, J., and R. Collier, Nares Abyssal Plain sediment flux studies, in Subseabed Disposal Project Annual Report, Physical Oceanography and Water Column Studies, edited by S. L. Kupferman, pp. B1-B24, Sandia National Laboratories, Livermore, California, 1987.

Eggimann, D. W., F. T. Mannheim, and P. R. Betzer, Dissolution and analysis of amorphous silica in marine sediments, J. Sediment. Petrol., 50, 215-225, 1980.

Fleer, A. P., and M. P. Bacon, Determination of Pb-210 and Po-210 in seawater and marine particulate matter, Nuc. Instrum. Methods Phys. Res., 223, 243-249, 1984.

Gardner, W. D., Incomplete extraction of rapidly settling particles from water samples, Limnol. Oceanogr., 2, 764-768, 1977.

Hess, F. R., Stream powered, large volume deep ocean sampler, Invention Disclosure 214-ERDA Docket S-47, 453, WHOI, Woods Hole, Massachusetts, 1977.

Jeffrey, L. M., A. D. Fredericks, and E. Hillier, An inexpensive deep-sea pumping system, Limnol. Oceanogr., 18, 336-340, 1973.

Jickells, T. D., W. G. Deuser, and A. H. Knap, The sedimentation rates of trace elements in the Sargasso Sea measured by sediment trap, Deep-Sea Res., 31, 1169-1178, 1984.

Laird, J. C., D. P. Jones, and C. S. Yentsch, A submersible batch filtering unit, Deep-Sea Res., 14, 251-252, 1967.

Landing, W. M., and K. W. Bruland, The contrasting biogeochemistry of iron and manganese in the Pacific Ocean, Geochim. Cosmochim. Acta, 51, 29-43, 1987.

Lenz, J., A new type of plankton pump on the vacuum principle, Deep-Sea Res., 19, 453-461, 1972.

Lisitzin, A. P., Sedimentation in the world ocean, S.E.P.M. Spec. Publ., 17, 218 pp., Tulsa, Oklahoma, 1972.

Moore, T. C., Radiolaria: change in skeletal weight and resistance to solution, Geol. Soc. Am. Bull., 80, 2103-2108, 1969.

Orians, K. J., and K. W. Bruland, The biogeochemistry of aluminum in the Pacific Ocean, Earth Planet. Sci. Lett., 78, 397-410, 1986.

Sherrell, R. M., The trace metal geochemistry of suspended oceanic particulate matter, Ph.D. thesis, 211 pp., MIT/WHOI, Woods Hole, Massachusetts, 1989.

Simpson, W. R., T. J. P. Gwilliam, V. A. Lawford, M. J. R. Fasham, and A. R. Lewis, In situ deep water particle sampler and real-time sensor package with data from the Madeira Abyssal Plain, Deep-Sea Res., 34, 1477-1497, 1987.

Spencer, D. W., and P. L. Sachs, Some aspects of the distribution, chemistry, and mineralogy of suspended matter in the Gulf of Maine, Mar. Geol., 9, 117-136, 1970.

Strickland, J. D. H., and T. R. Parsons, A Practical Handbook of Seawater Analysis, Bull. 167, 310 pp., Fisheries Research Board of Canada, Ottawa, 1968.

Taylor, S. R., Abundance of chemical elements in the continental crust; a new table, Geochim. Cosmochim. Acta, 28, 1273-1285, 1964.

Turekian, K. K., and K. H. Wedepohl, Distribution of the elements in some major units of the earth's crust, Geol. Soc. Am. Bull., 72, 175-192, 1961.

Winget, C. L., J. C. Burke, D. L. Schneider, and D. R. Mann, A self-powered pumping system for in situ extraction of particulate and dissolved materials from large volumes of seawater, Tech. Rpt. WHOI 82-8, 16 pp., WHOI, Woods Hole, Massachusetts, 1982.

FILTRATION TECHNIQUES FOR RIVER AND COASTAL WATERS WITH EMPHASIS ON TRACE ELEMENTS IN ANOXIC WATERS

Edward Sholkovitz

Chemistry Department, Woods Hole Oceanographic Institution, Woods Hole, Massachusetts 02543

1. Introduction

Since 1972 I have used a variety of filtration schemes to collect particles and filtrates. I have also used several analytical methods to determine the inorganic composition of particles. In this report I will present an overview of my practices. While many details do appear in my publications, I will add to these where appropriate. With my research primarily on lake, river and estuarine systems, filtration is mandatory. The importance of colloids in the geochemistry of these systems adds another complication vis-a-vis filtration techniques and potential artifacts. In addition, many of my projects have followed the geochemical response of trace elements to seasonal anoxia. Filtering anoxic water without introducing air and contaminants represents another filtration challenge. Lastly, organic-rich waters, whose estuarine geochemistry I researched in Edinburgh, present other problems. These include aging of the sample and the researcher as these waters quickly clog the filters and the filtration rate slows to an exasperating drop per minute through 47 mm filters of 0.45 μm pore size.

2. Sampling and Chemical Analysis

For my research into particle-rich systems of rivers, lakes and estuaries, the following factors come into play when designing a filtration technique:

1. Trace concentrations in the filtrate: for example, pmol kg^{-1} of rare earth elements (REE).
2. Large distribution constant (K_d): for example 10^6 for REE.

This means that any leakage of the particles around a filter will severely affect the concentration of the filtrate. For 1 L of estuarine water containing 10 mgL^{-1} of suspended particles of crustal material, there will be about 100 times more Nd (a typical REE) in the particles than in the filtrate. Hence if 1% of these particles leak into the filtrate then we have doubled the latter's concentration. A 0.1% leakage will raise the filtrate concentration by 10% while our analytical precision is 0.5%. The same problem applies to iron and other more common trace elements.

3. The need to filter 1-2 L quickly without introducing contamination at the pmol/kg level of concentration for the filtrate. Speed (10-30 min) is required because of ship time needs and in some cases because we want to sample at time scales much shorter than tidal mixing.
4. For anoxic waters it is mandatory to filter the water prior to its contact with air. This applies to elements both with (e.g., Fe, Mn) and without (e.g., REE, Pu, Pb, Th, P) redox couples, as the oxides of the former can scavenge many inorganic trace elements and organic compounds. These reactions would lead to constituent removal prior to filtration.
5. A system for estuaries is best which can be operated off small boats either by battery or AC power. For work off row boats, as we did in our study of the particle chemistry of Esthwaite Water [Sholkovitz and Copland, 1982] the system must be compact, light and waterproof.

Membrane filters have been my choice. They come as loose membranes, as sealed units and as large cartridges which are placed in a housing. My more recent studies of rare earth elements (REE) in rivers and estuaries have used sealed units, Millipore Millipak to be specific. My earlier studies of estuarine and lake chemistry have used the loose membranes in filter holders and my study of Pu in estuaries used cartridge filters. I will focus on a sealed unit system as I believe that this method has real advantages for lake and coastal systems. A brief discussion of the cartridge system will be presented first. In our study of Pu aquatic chemistry in Sholkovitz and Mann [1987], we have a detailed description of cartridge filtration.

2.1. Cartridge Filtration

To study the estuarine and lake geochemistry of Pu, particles and filtrates from 100 to 200 L of water were required from particle-rich, organic-rich and sometimes anoxic waters. Filtration through large (10") cartridge filters was the only alternative. Because plutonium is an artificial radionuclide, contamination from the tubing and the pump was not an important concern. All major

Marine Particles: Analysis and Characterization
Geophysical Monograph 63

manufacturers of filtration equipment have cartridge filters with different pore sizes, in different materials, in lengths from 4" to 12", and in different housing materials. Cartridges are fabricated out of a variety of materials, including paper, cotton, polypropylene, polycarbonate and even Teflon.

Using a small impeller pump, 100-150 L of turbid estuarine water can be filtered in 20 to 30 minutes, and from several depths at the same time. The filtrates are then processed by coprecipitation onto iron oxides which are then collected on a second cartridge filter. Measuring the composition of the particles and precipitates so collected is difficult as the cartridge filters are massive and have their filtering material sealed into a plastic structure. While it is possible to backflush, ultrasonicate and wash off some of the particles, the recovery will be poor and unknown. It is also impossible to measure the weight of filtered particles to determine a particle concentration. Hence, the downside of fast filtration rates, is that large surface areas lead to large tare weights and problems in recovering the particles for analysis. For radioisotope measurements, the whole cartridge is dried and then combusted in a large furnace. Prior to combusting, the plastic housing is cut away from the filtering material when possible. The filter material is then combusted and the ash analyzed. The cotton-wound type cartridges [CUNO Microwynd in Sholkovitz and Mann, 1987] combust rapidly; extreme care must be taken not to set your laboratory on fire. The polypropylene wound cartridges give off noxious fumes; the whole furnace must be placed in a fume hood. The furnace can't be used for anything else and heavy duty 4 L beakers are needed and replaced frequently. As it is not possible to quantitatively recover all the ash from the beaker, the total recovery has about a 5% uncertainty in the ash weight alone.

Returning to the cartridges themselves, we [Sholkovitz and Mann, 1987] reported that filtration of estuarine waters through a CUNO Microwynd cotton cartridge still left the filtrate with a milky appearance. We then used a CUNO Thin Sheet Media (TSM) cartridge and the filtrates of these same turbid waters were crystal clear. The use of the TSM cartridges reduced the Pu activity of the filtrate by 80% under certain river discharges. This incomplete reduction raises three fundamental issues. First, the pore size of cartridge filters appears to be less precisely characterized than many types of membrane filters. The nominal size for the Microwynd is 1 μm while the TSM is 0.5 to 0.7 μm. Millipore and Nuclepore also offer cartridge filters with a range of pore sizes. Second, the large mass of the particles being collected on the cartridges will probably alter the effective pore size as the filtration proceeds. This problem is inherent in all types of filtration but might be more severe for cartridge filters. And third, the adsorption of trace elements from the filtrate to the cartridge material itself is more likely to be a bigger problem because of the large surface area and mass of cartridges. This last problem can be quantified by refiltering a filtrate through a second cartridge and measuring the retention of a particular element on the filter; using this double refiltration procedure, Sholkovitz and Mann [1987] found that Pu had a small amount of adsorption.

Despite all the above negative remarks, when large volumes of filtrate and gram quantities of particles are required and must be collected quickly, the cartridge system does work well. With respect to contamination several points arise. For particles where the whole cartridge is ashed, the cartridge blank of a given element must be small relative to the concentration in the sample. For most trace elements (e.g., Fe, Mn, Pb, Zn, Cd, REE) on particles where only mg quantities of samples are required, cartridge filtration is not necessary and would most certainly lead to unacceptably high blanks. Where this system works well is for the artificial and natural radionuclides such as the isotopes of Pu, Am, Cs, Th, Pb and Ra. With respect to the filtrate, trace elements might be leached off the cartridge material to elevate their concentrations and/or be absorbed by the cartridge to reduce their concentration. The extent of this problem will be hard to quantify and will probably be variable. My practice is always to flush the tubing and cartridge extensively with the water sample prior to collecting an aliquot of the filtrate for analysis. This procedure should allow the filtrate to come into equilibrium with the cartridge material and reduce the leaching/adsorption problem.

For open ocean studies, cartridge filtration using in situ pumping systems works well and has been extensively used for the measurement of natural and artificial radioisotopes [Sachs et al., 1989]. Thousands of liters of water are pumped through cartridge prefilters and cartridge filters impregnated with precipitated Mn oxides. This tandem system collects both suspended particles and filtrates (Fleer and Bacon, this volume, page 223). With respect to the filtrate, the chemisorption efficiency must be known or measured for each radionuclide. To use chemisorption in estuaries is complicated, as the efficiency will depend on the salinity, pH and composition of the waters, all of which are variable. These sorption conditions would have to be well understood before use in the field.

2.2. Individual and Loose Membrane Filters

The collection of particles on individual and loose membrane filters is the most commonly used technique. There are obvious advantages to this method. These include the measurement of particle concentrations, ease of chemical dissolution prior to analysis and ability to make non-destructive visual and instrumental measurements such as the light microscope, electron microscope with EDAX, thin film x-ray fluorescence and diffraction. Membrane filters can also be precleaned for use with trace element sampling. For lakes, rivers and estuaries the main disadvantages are slow rates of filtration and the possibility of particle leakage into the filtrate. Again, if one only needs ten's of mL, then membrane filtration works well. For my recent research into REE, 2 liters of filtered estuarine water were needed. In Chesapeake Bay, filtration of this volume through a 47 mm, 0.22 μm membrane filter would take days. As noted earlier many trace elements have large K_d's. While not usually discussed in the literature, the leakage of particles into the filtrate worries me, especially when studying the REE, Fe and other trace elements. Given the arbitrary nature of the filtration process, it is extremely difficult to focus on and assess the magnitude of this one problem.

Because of their small tare weights, low blanks and robust properties, Nuclepore filters are favored for trace element studies. They are composed of polycarbonate material and filter through individual holes made by literally shooting them full of holes. In my own experience their thinness makes it more difficult to form a good seal in pressure or vacuum filtration units such as the Sartorius polycarbonate system. Sealing against an o-ring sometimes causes these filters to buckle if there is twisting action involved in tightening the unit. The release of the pressure or the vacuum also causes the filters to flex. I prefer and have extensively used Millipore Durapore filters (47 mm, 0.22 μm) for estuarine and pore water sampling. They are composed of a hydrophilic Teflon-like material, polyvinylidene difluoride, which means that they can be cleaned in strong acid and treated with strong acids to decompose or leach particles. They maintain very constant tare weights, and their added thickness makes them seal well against o-rings. While they weigh about 10 times that of a Nuclepore filter, in tests we have shown that their tare weights of 120 mg for 47 mm filter are reproducible to 0.2 mg after being used to filter prefiltered water. We have not tested their use for the open ocean where suspended particle concentrations are much lower; for this situation their larger tare weights are very disadvantageous. Like other frequently used membrane filters (e.g., Millipore cellulose type), they filter by collecting particles passing through a tortuous path.

To guard against particle leakage I use Millipore Millipak filters in which the filters are sealed into a plastic body. All major manufacturers of filtration equipment sell sealed units in a large variety of membrane and housing materials, pore sizes and filtration capacities.

2.3. Membrane Filtration as Applied to Rare Earth Element Studies

The following description covers the filtration technique and rationale which we have developed for the estuarine study of REE. This procedure worked well for a time-series study of Chesapeake Bay. A schematic diagram is shown in Figure 5-27.

The requirement is 2 L of filtrate from particle rich and anoxic waters. With the REE at pM kg^{-1} concentrations we needed a clean, yet rapid and simple method. We usually used the MP-60 Millipak filter units which have a surface area of 300 sq. cm and 0.22 μm pore size. In contrast, a single 47 mm filter has only 10 sq. cm of filter area. Millipaks range in surface area from 100 to 1000 sq. cm. The Millipaks are fabricated out of a polycarbonate housing and Durapore membrane filters arranged in a sandwich layering. For example, an MP-60 has three layers of back-to-back filters. With an MP-60 unit we could filter about 20 L of Bay water in an hour. The philosophy behind using a sealed filter unit is one of increasing the flow rate, eliminating the chance of particle leakage and reducing contamination.

Millipaks were chosen for several reasons. The Millipak housing is composed of polycarbonate and the filters are of Durapore. Both these materials are good ones for trace element work and can be cleaned with strong acids. The Millipaks also have a vent which is useful when pressure filtering. Different sized Millipaks should be chosen for different applications: very large filtration areas for extremely turbid waters or smaller areas for less turbid waters. At the turbid end of the scale, MP-60 and MP-100 (500 sq. cm) have been successfully used to filter river and delta water from the Amazon and Ganges/Bhramaputra systems.

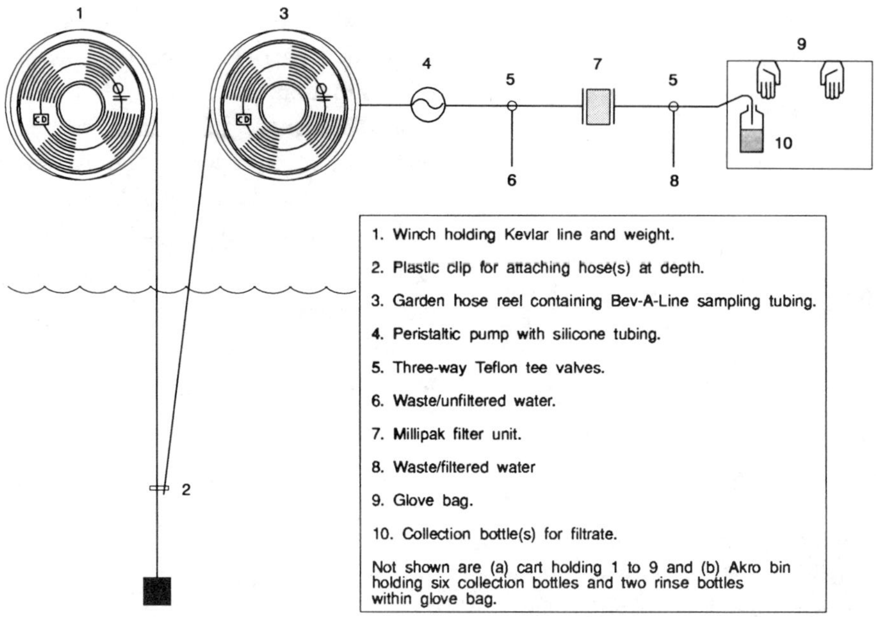

Fig. 5-27. Schematic diagram showing the filtration unit used by the author to collect suspended particles and filtrates from the water column of Chesapeake Bay.

The Millipaks were not precleaned prior to use. Instead they were extensively flushed with the sample in order to leach out any contaminants and to "equilibrate" the filter material with the sample. Typically 5 liters of water are pumped through the hose and passed to waste unfiltered. Then 5 liters of sample would be filtered through the Millipak and the next 2 L collected as the filtrate for REE analysis. The first 5 L of filtrate would either be pumped to waste or used for other studies. New Millipaks were used for each sample.

The heart of the system is a Cole-Parmer peristaltic pump. This type of pump is simple to use and keeps the water from touching parts of the pump. We used silicon tubing in the pump head and Bev-A-Line tubing elsewhere. Bev-A-Line (Cole Parmer) is tubing within tubing. A thin inner tube of polyethylene is molded to a thicker tube of ethyl vinyl acetate. This combination leads to a flexible and extremely strong tubing with a clean type of plastic in contact with the solution. We acid clean the Bev-A-Line. Flexibility with strength is important for our sampling technique where the hose is reeled in and out. Flexibility is needed in winter when other types of tubing become stiff.

To collect the filtrate cleanly on deck, we set up our own portable clean bag. Two L acid cleaned LPE narrow mouth sample bottles are preloaded into a Model X-27-37 glove bag manufactured by Instruments for Research and Industry. Six sample bottles are held in an AKRO bin which also contains two other 2 L bottles. These two bottles collect the filtrate used to flush out the tubing between samples; this waste filtrate comes from the next water sample. The whole system which includes pump, two heads, two Millipaks, Teflon valves and the glove bag containing the 8 bottles within an AKRO bin is placed on a Rubbermaid plastic cart. This cart is rolled on deck for sample collection, placed next to the winch holding the Kevlar line and the plastic garden hose reel holding the Bev-A-Line tubing. With two heads per pump we can filter samples from two depths. For the Bay study one person could run two systems thereby allowing the simultaneous pumping and filtration of four samples over the water column. At times we have also added a few mL of our purified HCl to the 2 L sample bottles prior to loading them in the glove bag. In this way the filtrates are acidified directly upon filtration, and another handling step on the ship is eliminated. In fact we usually load all the glove bags at WHOI so the whole system is ready to use.

This pumping system in modified form was used by Sholkovitz and Copland [1982] to study the composition of particles in lakes with large seasonal oxygen variations. We used a DC pump off a small row boat. More recently Dr. Greg Cutter of Old Dominion University has had great success using a portable system in his studies of lakes. Recently Dr. Lex van Geen of Stanford University ran a test to see if Millipaks would yield contamination free filtrates for trace metals. He chose Zn because it is easy to contaminate for Zn during handling of samples. The results in Figure 5-28 show flushing with trace metal free water releases Zn to the filtrate. The Zn concentration decreases sharply with the volume of rinse and reaches low values after 600 mL. For REE we flush with about 5 L of sample. It appears that pre-rinsing Millipaks with clean water or acid followed by flushing with the sample will allow them to be used for trace metal studies. This needs confirmation for other metals.

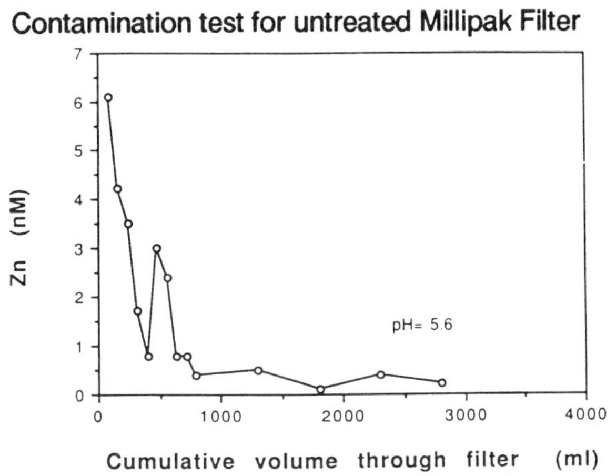

Fig. 5-28. Concentration of zinc as ultrapure water at pH of 5.6 is filtered through a new and untreated Millipore Millipak MP-60 filter unit. Data courtesy of Dr. A. van Geen.

3. Particle Recovery From Filters

Removing the particles from cartridge type filters presents a problem, as was discussed earlier. Similar problems exist when using Millipaks or other sealed units. We have some experience in this area. We have analyzed Chesapeake Bay particles collected on Millipaks for REE. Here we required about 0.5 g of dried material so we did need the large volume filtration of the Millipaks, as the Bay water contained about 10 to 50 mg L^{-1} suspended particles. We usually filtered 25-30 L through a MP-60 unit. Two techniques helped remove the particles, ultrasonication and freezing. We added clean water to a Millipak, sealed the ends and ultrasonicated in a water bath. We also tried using an ultrasonic probe to increase the energy. Repeated ultrasonication did remove the major part of the particle load. This is a visual observation only. If you make the large assumption that all the particles have the same composition, then you don't need 100% recovery to measure the composition of an element. Adding clean water and freezing the Millipak also released more particles upon thawing. Of course with both techniques there is a good probability that elements will be released to solution. The solution could be saved and analyzed with or without the particles. We combined the two phases for one analysis.

Because of their filter and housing material, Millipaks can withstand strong acids. Hence, leaching of suspended particles by acids or other types of solutions can be done directly within the Millipak units and the leachate or filtrate can be collected by applying pressure to the filter unit. We have done this in our laboratory under test conditions. In fact we have used aqua regia with no visual effect on the filter unit. Only when hot aqua regia is used does "melt down" start to occur. The unit will also withstand HF at room temperature. If the particle composition from solution leaches is needed and useful for a certain geochemical problem, then Millipak units offer a simple and direct experimental set-up assuming that blanks are acceptably low.

4. Analytical Methods

For particles collected by the Millipak system described in the last section, we used ICP-ES (Inductively Coupled Plasma Emission Spectroscopy) to measure both the REE and a suite of major elements (Al, Fe, Mn, Ca, Ti, and Zn). Methods are described in Sholkovitz [1990]. Three WHOI technical reports provide a detailed description of both the operating conditions for our ICP-ES [Bankston, 1988] and the methods specific to our determination on REE [Schneider, 1986; Schneider and Bankston, 1987].

It is important to analyze several Geochemical Reference materials along with samples. Also the type of dissolution can affect the results. We learned this for REE as HF-insoluble material such as zircon carries a large proportion of the heavy REE [Sholkovitz, 1990].

The author has extensively used thin-film x-ray fluorescence to measure the major element composition of particles and precipitates. The geochemistry was directed toward rivers, lakes and estuaries [Sholkovitz, 1976; Eckert and Sholkovitz, 1976; Sholkovitz et al., 1978; Sholkovitz, 1979; Sholkovitz and Price, 1980; Sholkovitz and Copland, 1982]. Thin-film XRF has been described in the following papers directed toward particles in natural waters: Baker and Piper [1976], Cann and Winter [1971], Holmes [1978, 1981], Price and Angell [1968], Price and Calvert [1973] and Pik et al. [1979]. Limitations of this method have been well documented in Criss [1976] and Holmes [1978, 1981]. XRF has the advantage of being a non-destructive technique which is fast and sensitive for many elements. For particles the idea is to deposit a thin film on a filter and directly analyze the sample by placing the whole filter in an XRF instrument. Hence, one can go from filtration in the field to the instrument and measure the composition of Si, Al, K, Ca, Mg, Ti, Fe and Mn. Depending on their concentrations other elements can also be measured. Self-adsorption and calibration are two problem areas. The above referenced papers go into details but essentially the particles must be "thin" so that the self-adsorption of fluorescencing x-rays is small or correctable. Precipitates fit the thinness criteria best. With respect to calibration, the NBS sells a set of thin glass film standards for XRF [Pella et al., 1986]. At Edinburgh University we prepared two types of standards. One type consisted of standard solutions spotted and dried on filters, the second of well analyzed finely ground rock powders which were suspended and filtered onto filters. Obsidian worked well because its glass texture gave it a uniform composition. Analysis by XRF is more commonly employed by European geochemists. With respect to particle analysis the thin-film method offers many advantages for determining the concentration of major elements and selected minor elements. The thin-film criteria falls apart when analyzing particles from turbid waters where large sized particles (greater than 5 μm) cause self-adsorption and calibration problems. The magnitude of these problems also depends on the element or more specifically the emitted energies of the analytical lines.

References

Baker, E. T., and D. Z. Piper, Suspended particulate matter: collection by pressure filtration and elemental analysis by thin film x-ray fluorescence, Deep-Sea Res., 23, 181-186, 1976.

Bankston, D. C., General guidelines for operating a rapid sequential inductively coupled plasma-atomic emission spectrometer, WHOI Tech. Memo. WHOI-4-88, 23 pp., Woods Hole, Massachusetts, 1988.

Cann, J. R., and C. K. Winter, X-ray fluorescence analysis of suspended sediment in seawater, Mar. Geol., 11, 33-37, 1971.

Criss, J. W., Particle size and composition effects in x-ray fluorescence analysis of pollution samples, Anal. Chem., 48, 179-186, 1976.

Eckert, J., and E. R. Sholkovitz, The flocculation of iron, aluminum and humates from river water by electrolytes, Geochim. Cosmochim. Acta, 40, 847-848, 1976.

Holmes, G. S., The chemistry of suspended particulate matter; Firth of Forth, Ph.D. thesis, 200 pp., Univ. of Edinburgh, Scotland, 1978.

Holmes, G. S., The limitations of accurate "thin film" x-ray fluorescence of natural particulate matter: problems and solutions, Chem. Geol., 33, 333-353, 1981.

Pella, P. A., D. E. Newbury, E. B. Steel, and D. H. Blackburn, Development of National Bureau of Standards thin glass films for X-ray fluorescence, Anal. Chem., 58, 1133-1137, 1986.

Pik, A. J., A. J. Cameron, J. M. Eckert, E. R. Sholkovitz, and K. Williams, The determination of metals at PPB levels by thin-film X-ray spectrometry after co-precipitation with a molybdenum carrier complex, Anal. Chim. Acta, 110, 61-66, 1979.

Price, N. B., and G. Angell, Determination of minor elements in rocks by thin-film x-ray fluorescence techniques, Anal. Chem., 40, 660-663, 1968.

Price, N. B., and S. E. Calvert, A study of the geochemistry of suspended particulate matter in coastal waters, Mar. Chem., 1, 169-189, 1973.

Sachs, P. L., T. R. Hammar, and M. P. Bacon, A large volume deep-sea submersible pumping system, WHOI Tech. Rpt. No. 89-55, 53 pp., Woods Hole, Massachusetts, 1989.

Schneider, D. L., Analytical protocol for separation of rare-earth elements from marine sediments, WHOI Tech. Memo. WHOI-1-86, 4 pp., Woods Hole, Massachusetts, 1986.

Schneider, D. L., and D. C. Bankston, Determination of rare-earth elements in marine sediments by ion-exchange separation and ICP emission spectrometry, WHOI Tech. Rpt. WHOI-87-56, 42 pp., Woods Hole, Massachusetts, 1987.

Sholkovitz, E. R., Fluctuation of dissolved organic and inorganic matter during the mixing of river water and seawater, Geochim. Cosmochim. Acta, 40, 831-845, 1976.

Sholkovitz, E. R., Chemical and physical processes controlling the chemical composition of suspended material in the River Tay and Tay Estuary, Estuarine Coastal Mar. Sci., 8, 523-545, 1979.

Sholkovitz, E. R., The rare earth elements in marine sediments and geochemical standards, Chem. Geol., 88, 333-347, 1990.

Sholkovitz, E. R., and D. Copland, The chemistry of suspended matter in Esthwaite Water, a biologically productive lake with seasonally anoxic hypolimnion, Geochim. Cosmochim. Acta, 46, 393-410, 1982.

Sholkovitz, E. R., and D. R. Mann, ^{239,240}Pu in estuarine and shelf waters of northeastern United States, Estuarine Coastal Shelf Sci., 25, 413-434, 1987.

Sholkovitz, E. R., and N. B. Price, The major-element chemistry of suspended matter in the Amazon Estuary, Geochim. Cosmochim. Acta, 44, 163-171, 1980.

Sholkovitz, E. R., R. van Grieken, and D. Eisma, The major-element composition of suspended matter in the Zaire River and estuary, Neth. J. Sea Res., 12, 407-413, 1978.

THE ANALYSIS AND CHARACTERIZATION OF MARINE PARTICLES IN ESTUARINE WATER

Jens Skei

Norwegian Institute for Water Research, P.O.Box 69, Korsvoll, N-0808 Oslo 8, Norway

1. Introduction

The Norwegian Institute for Water Research (NIVA) has been engaged in studies of fjords and coastal water for more than 20 years. Analyses of particulate matter in water were early included in these studies [Price and Skei, 1975], encompassing environmental work [Skei et al., 1973], geochemical problems [Skei and Melsom, 1982] and anoxic environments in particular [Anderson et al., 1987; Skei, 1988]. The chemistry of particulate matter in seawater elucidates chemical reactions, the origin of particles and in many cases the transport of water masses. Small particles in the size range 0.1 to 1.0 μm show an approximate conservative behavior and the chemistry of these particles may be used as a tracer for water masses.

A considerable change in methodology has taken place during the last 20 years. Faulty data were created during the first few years due to inadequate sampling methods, types of filters and analytical techniques. Experience has been gained through the years, but still unsolved problems remain.

2. Collection of Water

Water samples have been collected using HydroBios, Niskin and Go-Flo sample bottles. HydroBios is no longer used if chemical analyses are to be performed, due to risk of contamination. The bottles have been attached to a stainless steel hydrowire or a CTD rosette sampler. Analyses of dissolved trace metals in nearshore water have not indicated that water sampling is subject to contamination. Hence, it has been assumed that the sampling technique is also adequate for particulate matter studies.

3. Collection of Particulate Matter

Collection of particulate matter some 20 years ago was mainly done for gravimetric analyses. Glass fiber filters were used. Due to the thickness and the weight of the filter and the problem of non-specific pore size these filters were soon abandoned. Today these filters are only used for analyses of particulate carbon and nitrogen.

The next filter used was the Sartorius membrane filter with diameter 47 mm and pore size 0.45 μm. Due to problems of removing seasalt from these filters, they were also abandoned.

During the last 15 years Nuclepore membrane filters, 0.4 μm, have been used with satisfaction. They have been applied for gravimetric analysis, major and minor element analysis and microscopic work. Prior to filtration the filters are acid cleaned, rinsed with deionized water and kept in deionized water prior to use. Precleaning of filters is necessary. Tests have been performed at NIVA on contamination from filters for nutrients. Included in the test were Whatman GF/C and GF/F, Sartorius, Millipore and Nuclepore membrane filters. Deionized water was filtered through the various filters and the levels of phosphate and nitrate were recorded in the filtrate. Figure 5-29 indicates the relative leaching of nutrients from the filters. Tests on methods of cleaning the filters using HCl and/or water indicate that the filters should remain in acid or water at least 24 hrs before being used (Figure 5-30). The filtration of water is now always performed under nitrogen pressure unless particulate matter is collected for carbon and nitrogen analysis.

The filtration is normally done immediately after collection of water. The water is transferred from the sampling bottle to the filtration device. If small volumes of water are needed because of high particulate load, a 0.5 L Sartorius plastic pressure filtration unit is used. If larger volumes are to be filtered, a home built pressure filtration unit (Plexiglas) with a capacity of 3 L is used. If the sample contains high concentrations of particulate matter a magnetic stirrer is used to keep particles in suspension in the water reservoir. If only a part of the water in the bottle is filtered, the bottle is shaken vigorously before the water is poured into the filtration vessel to obtain a homogeneous sample. Sometimes the water contains sulfide and to avoid oxidation the transfer of water from the bottle to the vessel is done under a stream of nitrogen.

The pressure filtration is done at ~ 15 psi nitrogen pressure. At the end of the filtration the filter, while still wet in the filter holder, is rinsed with minimum 50 mL of

Marine Particles: Analysis and Characterization
Geophysical Monograph 63

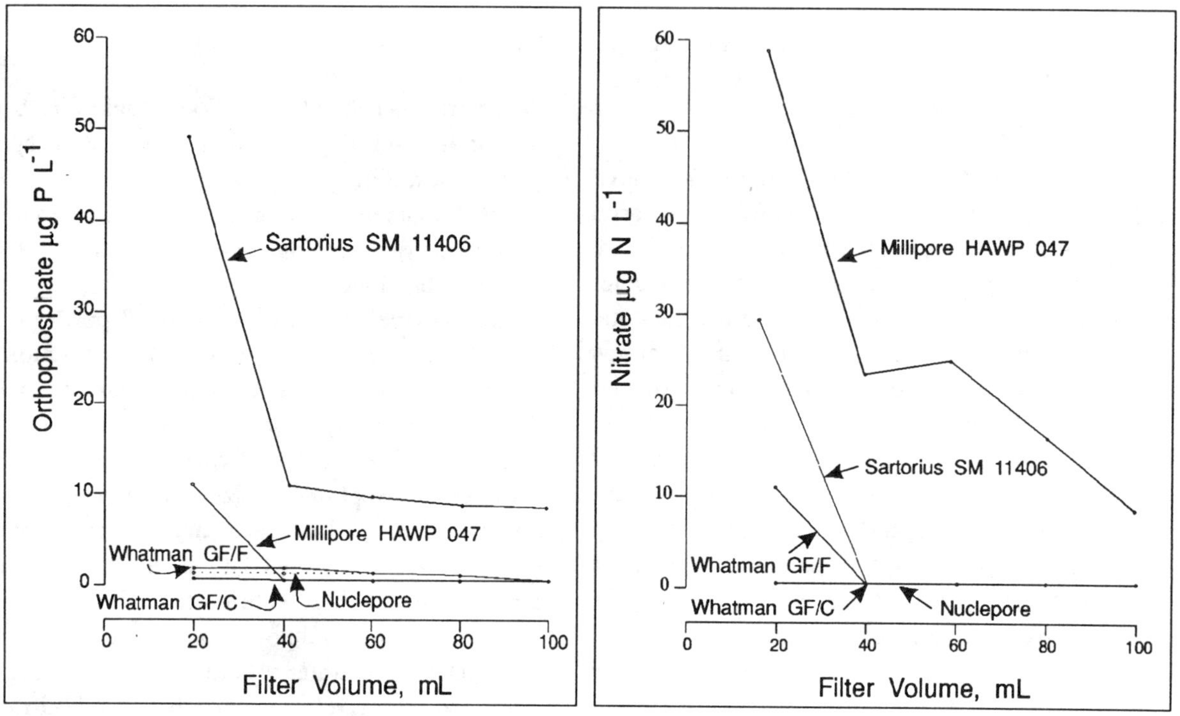

Fig. 5-29. Leaching of phosphate and nitrate from various types of filters.

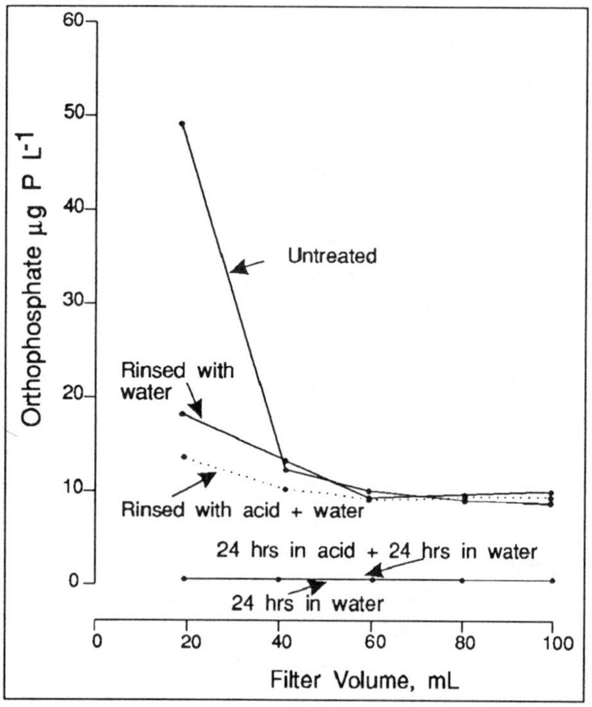

Fig. 5-30. Leaching of phosphate from Sartorius filters following various cleaning procedures.

deionized, prefiltered water. The removal of seasalt by rinsing with distilled water causes concern due to possible loss of lipids and desorption from particles. Experiments have shown that this is a problem [Price and Skei, 1975].

The loading of the filters is kept at a minimum as the filtration characteristics change during the buildup of particles on the filter. As soon as the filters start to clog, the filtration is terminated. A maximum of 1-2 mg on a 47 mm diameter filter is optimal. This is sufficient for weighing and for most analytical work.

Collection of particulate matter for carbon and nitrogen analyses is done on 10 mm GF/F glass fiber filters using suction filtration. Similarly, 20 mm GF/F filters are used for phosphorus analyses.

4. Sample Handling

When still in the filter holder the sample is flushed by nitrogen to remove moisture. It is removed using a plastic tweezer and placed in a clean plastic petri dish. With the lid off, the petri dish is left in a desiccator for 24 hrs to obtain a constant weight. Glass fiber filters for C/N analyses are immediately frozen.

5. Gravimetric Analyses

The amount of particulate matter in fjords and coastal water in Norway varies between 0.1 to 10 mg L^{-1}. Normally the concentrations are between 0.5 and 1.0 mg L^{-1}. The Nuclepore filter is preweighed after cleaning and drying in a desiccator and reweighed again after the same procedure using a Sartorius 4503 micro balance with a minimum detectable weight of 0.001 mg. A radioactive source removes the electrostatic forces. The reproducibility of the weighings at 1 mg is better than 10%.

6. Major and Trace Element Analyses

A thin film technique using X-ray fluorescence has been adapted for major elements [Price and Calvert, 1973]. The method was later slightly modified by Skei and Melsom [1982]. A 37 mm diameter 0.4 μm Nuclepore filter is used and preparation of standards is described in Price and Skei [1975] and Skei and Melsom [1982]. The elements which have been analyzed with this method include major elements (Al, Si, Fe, Mn, Ti, Mg, K, Ca, P and S) as well as some trace elements (Zn, Pb, Cu and Ba). A filter with a diameter of 37 mm fits into the sample holder of Phillips X-ray fluorescence instruments. The advantage of this technique is that it is non-destructive and the filters can be used for other purposes after the XRF analyses. Additionally, the method shows high reproducibility.

Analytical precision based on the percentage deviation of six replicate analyses gave the following results: Al = 2%, Si = 6%, Fe = 1.5%, Mn = 4%, Ti = 5%, Mg = 3%, K = 2%, Ca = 2.5%, P = 4% and S = 3% [Price and Skei, 1975]. The detection limit for the elements is 1 μg for Fe, Si and Al and 0.3 μg and 0.2 μg for P and Mn, respectively [Skei and Melsom, 1982].

The main problems with this technique are standards and accuracy. Few certified standards on filters exist and each laboratory makes up its own set of standards to produce a calibration curve. Price and Skei [1975] determined Zn and Pb by referring their respective peak intensities (i.e. Zn $K_{\alpha 1}$ and $PbL_{\alpha 1}$) to those of a series of synthetic standards of preweighed deposits of $Zn_2Fe(CN)_6$ and $PbCrO_4$ diluted into an argillaceous base of 1 μm grain size. For major elements standard reference rock samples have been used. Skei and Melsom [1982] processed these standards in an agate ball mill to obtain a particle size smaller than 20 μm. A sample of the powder of 0.5 to 3.0 mg was transferred to 50 mL of filtered, distilled water and the suspension was homogenized in an ultrasonic cleaning bath. After addition of 250 mL filtered and distilled water the suspension was filtered through Nuclepore filters. Calibration curves were prepared by plotting the intensity of the K_{α} – line against the known weight of the corresponding element on each filter.

To minimize the source or error connected to establishing calibration curves, a large number, 40-50, of calibration standards was prepared. After recording the X-ray intensity as a function of weight of the corresponding elements, 15 of the calibration filters were analyzed for Fe by instrumental neutron activation analysis (NAA). From the results of Fe, the weight of the standard on the filters was determined and compared with the weight added to the solution [Skei and Melsom, 1982]. Some samples showed large variation (greater than 20%) from a linear intensity vs. weight relationship. The NAA control showed that the solid in the suspension had not been quantitatively transferred in these cases. On the other hand, samples that fitted the linear relationship turned out to contain the expected amount of standard. This agreement shows that major systematic errors have been avoided.

A different method of producing standards was tried out by Skei and Melsom [1982]. Known amounts of the elements to be analyzed in aqueous solutions were transferred to Nuclepore filters using pipettes. Care was taken to obtain a uniform distribution of small drops on the entire area irradiated by X-rays. The filters were mounted in holders during the period of preparation and dried under a lamp. Standards prepared from solutions and the standards based on international rock standards showed good agreement.

The results of the measurements of the elemental composition of the particulate matter may be expressed as μg L^{-1} or as μg g^{-1} dry material, if preweighed filters are used. The latter allows comparisons with data from sediment traps and bottom sediments.

Recently wet chemical methods have been adapted to analyze trace metals in suspended matter. A total dissolution technique using a Teflon bomb has been developed by Loring and Rantala [1990] and has been proven to give reproducible results. The ICES Working Group on Marine Sediments in Relation to Pollution (WGMS) ran the first phase of a suspended particulate matter (SPM) intercomparison exercise from 1989 to 1990 [Hovind and Skei, 1990]. A total of 29 laboratories requested intercomparison samples, consisting of three certified sediment reference standards for trace metals purchased from the National Research Council of Canada (A = PACS-1, B = MESS-1, C = BCSS-1). Approximately 5 mg of sample A, 15 mg of sample B and 20 mg of sample C were distributed to the participants. They were asked to weigh out 1 mg of sample A, 3 mg of sample B and 5 mg of sample C, with 3 replicates of each sample. The participants were requested to analyze the samples for Cu, Pb, Zn, Cd, Al, Li, Mn, Ni and Cr using a total dissolution technique. Phase 1 of this intercomparison should indicate the laboratories ability to analyze quantities similar to those collected on filters. A total of 19 laboratories submitted results. Table 5-13 summarizes the results. N is the number of laboratories returning results. Mean is the grand mean of the mean value reported by each laboratory, and SD is the standard deviation of the grand mean. The performance of the laboratories depends both on element, sample size and concentration. Figures 5-31 and 5-32 illustrate the spread in the results of Cd and Cu, respectively.

TABLE 5-13.
Summarized Results From the ICES SPM-Intercomparison Exercise

Element		Sample A			Sample B			Sample C		
		N	Mean	SD	N	Mean	SD	N	Mean	SD
Al	%	10	5.80	0.72	10	5.49	0.65	10	5.66	0.84
Fe	%	15	4.40	0.91	16	2.62	0.59	18	3.01	0.53
Cd	$mg\ kg^{-1}$	12	2.51	0.65	12	0.78	0.28	14	0.36	0.14
Cr	$mg\ kg^{-1}$	11	94.4	28.8	12	56.5	20.5	15	102.6	39.9
Cu	$mg\ kg^{-1}$	15	377	118	15	24.7	5.5	17	20.4	9.3
Mn	$mg\ kg^{-1}$	15	427	114	14	445	112	17	218	51.3
Ni	$mg\ kg^{-1}$	9	50.6	15.7	11	25.5	9.1	13	52.1	14.2
Pb	$mg\ kg^{-1}$	15	381	119	16	32.3	9.4	17	25.8	14.3
Zn	$mg\ kg^{-1}$	13	796	196	13	178	62	15	107.4	31.6
Li	$mg\ kg^{-1}$	3	40.9	14.3	3	44.8	6.6	3	42.9	7.5
Co	$mg\ kg^{-1}$	1	19.5		2	12.8		2	12.6	

The true values for samples A, B, and C for the respective elements are as follows:

Element		Sample A		Sample B		Sample C	
		Mean	SD	Mean	SD	Mean	SD
Al	%	6.47	0.12	5.83	0.20	6.26	0.22
Fe	%	4.87	0.08	3.05	0.17	3.29	0.10
Cd	$mg\ kg^{-1}$	2.38	0.20	0.59	0.10	0.25	0.04
Cr	$mg\ kg^{-1}$	113	8	71	11	23	14
Cu	$mg\ kg^{-1}$	452	16	25.1	3.8	18.5	2.7
Mn	$mg\ kg^{-1}$	470	12	513	25	229	15
Ni	$mg\ kg^{-1}$	44.1	2.0	29.5	2.7	55.3	3.6
Pb	$mg\ kg^{-1}$	404	20	34.0	6.1	22.7	3.4
Zn	$mg\ kg^{-1}$	824	22	191	17	119	12
Co	$mg\ kg^{-1}$	17.5	1.1	10.8	1.9	11.4	2.1

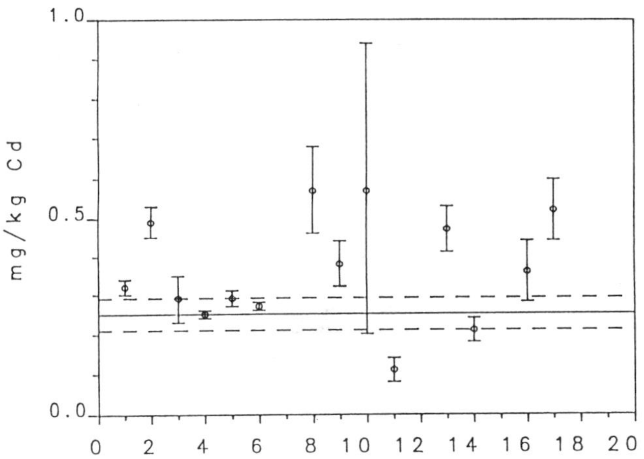

Fig. 5-31. Results for Cd, sample C. ICES SPM-intercomparison, 1990.

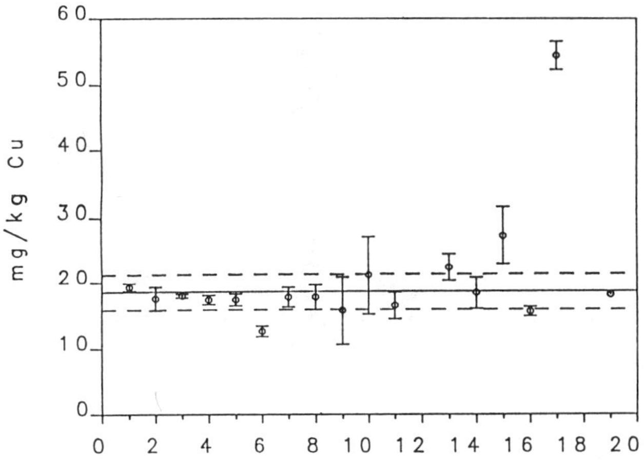

Fig 5-32. Results for Cu, sample C. ICES SPM-intercomparison, 1990.

7. Analyses of Single Particles

Beside the objective of measuring the elemental composition of bulk SPM samples it is also of interest to study the chemistry and the morphology of single particles. This type of work will help us to understand the origin of particles and the associations among elements.

Following the XRF-analyses, selected filters are cut in 1 cm circles, mounted on a brass pad with double-sided tape and coated with carbon or gold. By using a JEOL Scanning Electron Microscope (SEM) coupled to an Energy Dispersive X-ray (EDAX), qualitative analyses of single particles can be obtained [Skei and Melsom, 1982; Skei, 1988]. The SEM-analyses also indicate the homogeneity of the sample surface, which again is related to the success of filtration. Additionally, information on particle size and form; single particles or clusters; inorganic detritus or plankton fragments is valuable when interpreting the bulk chemistry. Figures 5-33 and 5-34 illustrate the SEM/EDAX facility.

If quantitative analyses of single particles are desirable, an electron microprobe instrument may be used. Filters which first have been analyzed by SEM/EDAX, and where particles of special interest have been singled out, are used for electron microprobe [Jacobs et al., 1985]. By analyzing a large number of particles of the same type an average composition of the particle may be established. If this concerns mineral phases the stoichiometry may be ruled out [Skei, 1988]. This technique is particularly useful when small sample sizes exclude X-ray diffraction analyses of minerals.

8. Analyses of Carbon, Nitrogen and Phosphorus

Particulate matter in coastal water is rich in organic material of both terrestrial and marine origin. It is essential to quantify the levels of carbon and nitrogen to make comparison with the bottom sediments. This will help to understand mineralization rates of organic matter.

Carbon and nitrogen collected on 10 mm GF/F glass fiber filters are analyzed by a Carlo-Erba instrument. Previously, 47 mm filters were used by cutting out 1 cm circles. Replicate subsampling from the same filter showed a considerable patchiness and this method was abandoned.

Both total carbon and total organic carbon, where inorganic carbon is removed by HCl fumes, are determined. Both precision and accuracy are adequate at the levels present in coastal water. The detection limit is $1 \mu g$.

Total phosphorus collected on 20 mm GF/F glass fiber filters is digested by persulfate and analyzed by autoanalyzer.

9. Particulate Matter in Anoxic Seawater

Analytical work on anoxic water creates special problems with respect to handling of the samples. The problem is increased with the level of sulfide present. Some fjords in Norway have stagnant basins where the levels of sulfide are in the mM range [Skei, 1988].

Important geochemical processes take place at the redox interface involving particulate phases (i.e. Fe, Mn, etc.). Within the anoxic zone precipitation of minerals may take place due to supersaturation. Consequently, studies of the chemistry of particulate matter in anoxic environments are of interest.

It is essential to avoid any exposure of oxygen to the anoxic water during handling and filtration. The sample

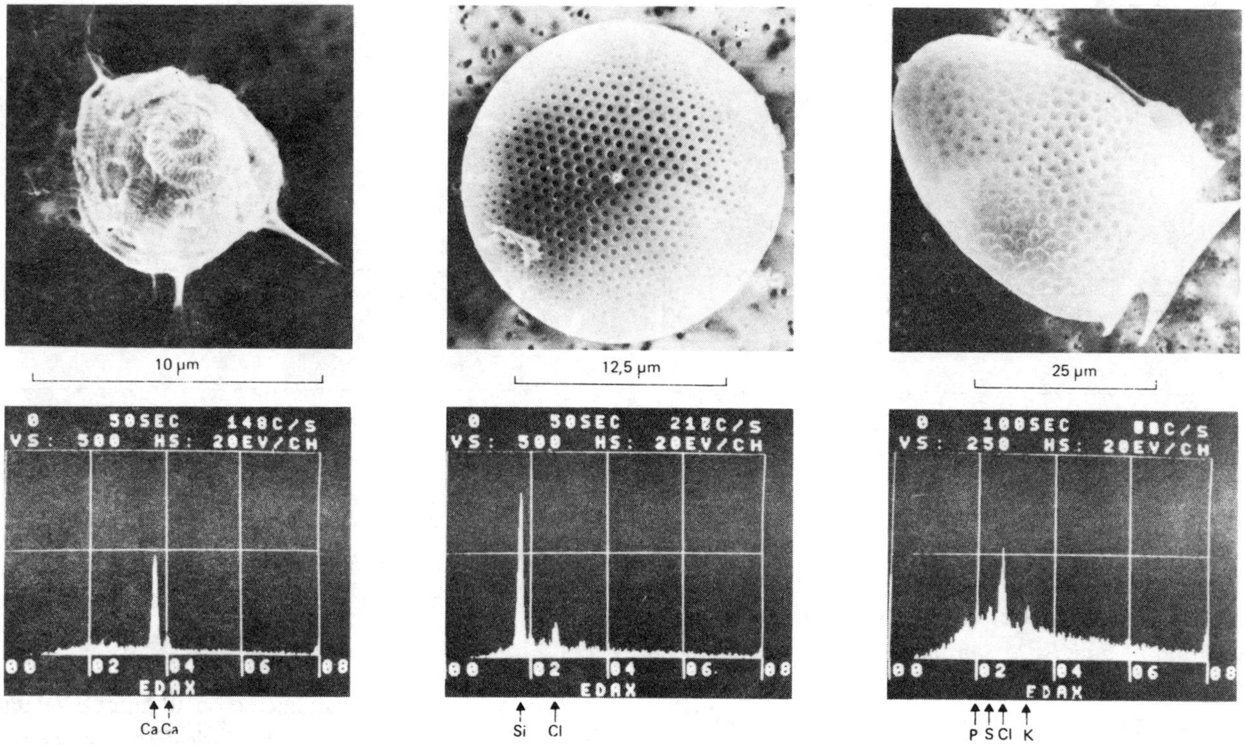

Fig. 5-33. (Left) Dinoflagellate with cellulose skeleton. (Middle) Diatom with silica skeleton. (Right) Coccolithophorid with calcite skeleton.

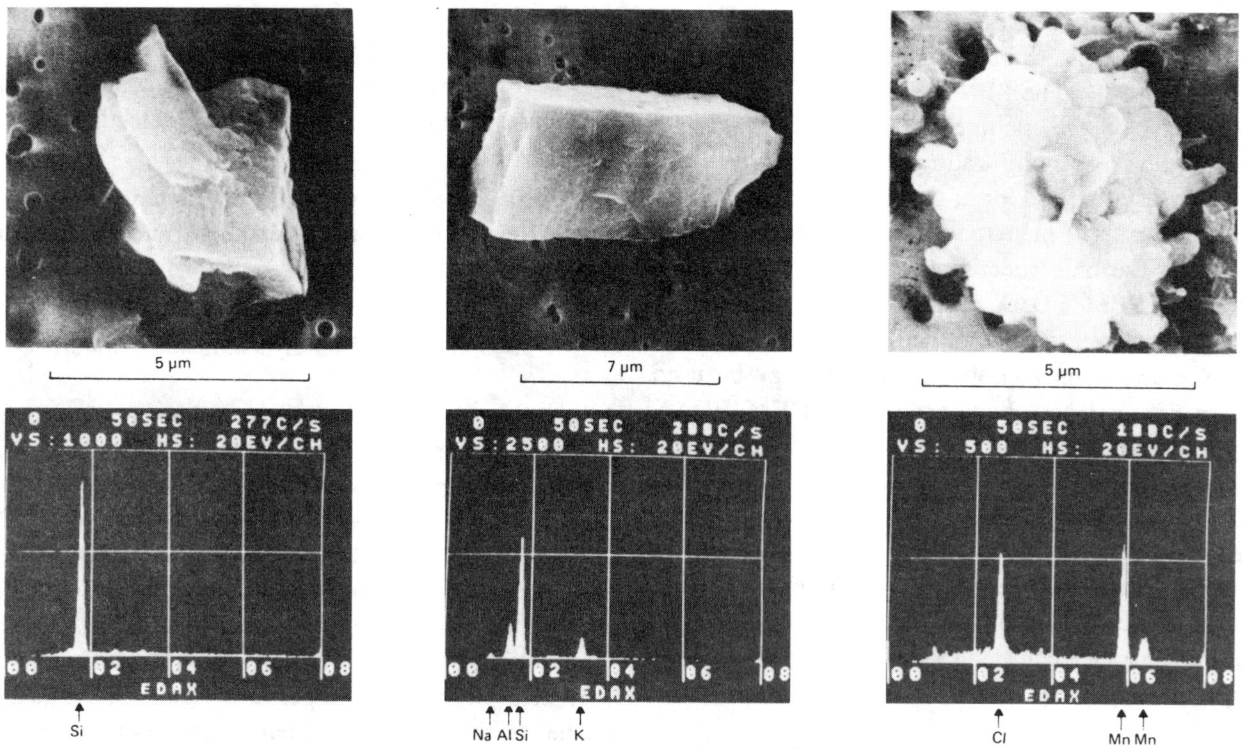

Fig. 5-34. (Left) Precipitate of manganese at a redox interphase. (Middle) Feldspar particle. (Right) Quartz particle.

is filtered immediately under nitrogen pressure either directly in the sampling bottle or in a filtration device which is continually flushed with nitrogen. Under normal conditions one should maintain anoxic conditions in the sample during filtration. The Nuclepore filter may be dried under nitrogen atmosphere, but from then on the filter will be exposed to air (during XRF and SEM/EDAX analyses). It is assumed that no mineral changes will take place on the filter after drying but this assumption has not been tested.

Filtration of anoxic water will normally involve small particle sizes such as finely dispersed Fe-mono sulfides. The effect of the particle size on the retention on 0.4 μm Nuclepore filters was tested by comparing with 0.2 μm Nuclepore. Analyses of particulate Fe and S retained on 0.2 μm filters showed 50 to 70% more Fe and S compared with the 0.4 μm filters. The water contained between 2-6 mmol H_2S [Skei, 1988]. The levels of particulate Al showed the same on the two types of filters. It is assumed that Fe-mono sulfides are bleeding through the 0.4 μm filters.

Analyses of metal sulfide phases in anoxic environments are of great importance to understanding fundamental geochemical processes. The use of SEM/EDAX helps in identifying these phases as long as both S and the relevant metals can be measured.

Studies of framboidal pyrite in the anoxic water of Framvaren, South Norway are based on SEM/EDAX [Skei, 1983, 1988]. By supplementing with electron microprobe analyses of 22 framboids it was ruled out that the stoichiometry was $FeS_{1.97}$ or close to FeS_2. [Skei, 1988]. It could also be stated that the framboids contained small quantities (<1%) of Mn, Cu, Zn, Ni and Co [Jacobs et al., 1985]. In the super-anoxic water of Framvaren (greater than 5 mmol S) individual precipitates of ZnS and Cu-sulfides were also identified.

References

Anderson, L. G., D. Dyrssen, and J. M. Skei, Formation of chemogenic calcite in super-anoxic seawater, Framvaren, South Norway, Mar. Chem., 20, 361-376, 1987.

Hovind, H., and J. M. Skei, Intercomparison exercise on the trace metals in suspended particulate matter (SPM); Phase 1: The capabilities of the laboratories to analyze small samples accurately and reproducibly, Preliminary report presented at WGMS meeting in Lisbon, Portugal, April 2 to 5, 1990.

Jacobs, L., S. Emerson, and J. M. Skei, Partitioning and transport of metals across the O_2/H_2S interface in a permanently anoxic basin, Framvaren fjord, Norway, Geochim. Cosmochim. Acta, 49, 1433-1444, 1985.

Loring, D. H., and R. T. T. Rantala, Sediments and suspended particulate matter: total and partial methods of digestion, ICES Techniques in Marine Environmental Sciences, No. 9, 14 pp., 1990.

Price, N. B., and S. E. Calvert, A study of the geochemistry of suspended particulate matter in coastal waters, Mar. Chem., 1, 169-189, 1973.

Price, N. B., and J. M. Skei, Areal and seasonal variations in the chemistry of suspended particulate matter in a deep water fjord, Estuarine Coastal Mar. Sci., 3, 349-369, 1975.

Skei, J. M., Geochemical and sedimentological considerations of a permanently anoxic fjord, Framvaren, South Norway, Sediment. Geol., 36, 131-145, 1983.

Skei, J. M., Formation of framboidal iron sulfide in the water of a permanent anoxic fjord, Framvaren, South Norway, Mar. Chem., 23, 345-353, 1988.

Skei, J. M., and S. Melsom, Seasonal and vertical variations in the chemical composition of suspended particulate matter in an oxygen deficient fjord, Estuarine Coastal Shelf Sci., 14, 61-78, 1982.

Skei, J. M., N. B. Price, and S. E. Calvert, Particulate metals in waters of Sørfjord, West Norway, Ambio, 2, 122-124, 1973.

COLLECTION AND ANALYSIS OF MARINE PARTICLES FOR TRACE ELEMENTS

John H. Trefry and Robert P. Trocine

Department of Oceanography and Ocean Engineering, Florida Institute of Technology, Melbourne, Florida 32901

1. Introduction

Field and laboratory practices for the collection and analysis of suspended particles from the marine environment have evolved considerably during the past two decades. However, many of these practices have not been discussed or disseminated in great detail throughout the literature. The timely organization of a workshop on "The Analysis and Characterization of Marine Particles" encourages reevaluation and sharing of collective experiences.

We have been fortunate to collect particles from a wide variety of environments including the Mississippi River and Delta [Trefry and Presley, 1976; Trefry et al., 1985, 1986] , the hypersaline Orca Basin [Trefry et al., 1984] , and hydrothermal vent plumes in the Atlantic and Pacific Oceans [Trefry et al., 1985; Nelsen et al., 1986/1987; Trocine and Trefry, 1988]. Some of our practices and thoughts follow, focusing primarily on sampling and analysis for trace elements. Some additional details for sampling hydrothermal plumes are also provided.

2. Pre-Cruise Preparations

The quality of trace element studies of marine particles is highly dependent on precautions taken during filter preparation and sampling. For trace elements, polycarbonate filters produced by the Nuclepore and Poretics Corporations are favored for their strength and versatility. We have used a variety of filter diameters including 25, 47, 90, 142 and 297 mm and selected pore sizes from 0.2-1.0 μm. Each filter size introduces some differences in handling as described below.

Use of 47 mm diameter filters with 0.4 μm pores has been the most frequent in our research. Initial preparation of the filters is carried out in a clean room in our laboratory. The filters are placed in a 4-5 N solution of redistilled nitric acid for 24 hours to maximize the removal of metal contaminants from the filters. Potential problems from Cr contamination by spacers used to separate the filters may require an even more rigorous treatment in some cases, especially if Cr is an analyte of interest. The filters are then rinsed with distilled-deionized water three times. Each filter is placed in an individual, acid-washed plastic petri dish and allowed to dry at 21 to 23° C and 50% relative humidity in a separate drying rack in our clean room. Filters should not be weighed prior to some pre-cleaning treatment because a 5 to 10% loss in filter weight occurs when surfactants are washed away during acid-washing. Such a weight loss could introduce a significant error in calculation of suspended matter concentrations.

After 48 hours, the washed filters are weighed to the nearest μg using a balance calibrated during every use with standard weights. Humidity and temperature control are maintained as above for all drying and weighing operations. A ^{210}Po anti-static device (Staticmaster, Nuclear Products Co.) is placed inside the balance. One anti-static device is usually sufficient for a 47 mm diameter filter under the climatic conditions described. Each filter is weighed three times in a random order.

A series of control and blank filters are also prepared at this time. Control filters are used to monitor filter masses. Some controls are kept in our clean room to monitor static and climate parameters over time. Others are brought to sea as a check on possible shipboard contamination. The same procedure is followed for blank filters that will be used for determining procedural blanks for the elements of interest. All filters are stored in petri dishes that are double-wrapped in Ziploc plastic bags.

Filters with diameters of 90 and 142 mm are weighed to ± 10 μg by placing the pre-cleaned filters on the balance pan in an acid-washed, plastic petri dish of the appropriate size. Several ^{210}Po sources are positioned within 3 cm of the filter because static problems increase with increasing filter size. Sometimes several minutes pass before the initial static charge in the balance is neutralized. Because the half-life of the ^{210}Po alpha source is short (138 days), one must stay cognizant of the manufacture date stamped on the back and replace the source within two half-lives or the effectiveness will decrease significantly. When ordering sources, specifically request the most recent manufacturing run. Other than concerns during weighing, these larger filters may be treated in the same manner as the 47 mm

Marine Particles: Analysis and Characterization
Geophysical Monograph 63

diameter membranes.

We have restricted use of 297 mm diameter filters to areas with suspended matter loadings in excess of 100 mg L^{-1} such as in the Mississippi River. These large filters are more difficult to prepare. For our application, they are washed in 1 N HNO_3 and rinsed with distilled, deionized water in a laminar flow hood at sea. After filtration of 30 L (greater than 3 g of particles), the filter is rinsed as described below and folded into a 90° wedge and stored in an acid-washed Ziploc bag. After normal drying, the suspended matter cake on the filter is easily broken up and removed, thus avoiding weighing these cumbersome filters.

A key aspect of pre-cruise planning is also adequate preparation of the sample bottles and filtration apparatus. We have used a variety of water samplers and recently in situ pumps. The general use of in situ pumps is described by Sherrell (this volume, page 285), and thus use of Go-Flo bottles produced by General Oceanics will be described. The general principles can be applied to any other sampler used. The Go-Flo bottles are completely disassembled before each cruise and cleaned with 10% hydrochloric acid and rinsed with distilled-deionized water. All o-rings and seals are checked and cleaned. The Teflon coating inside the bottle is inspected. The bottles are reassembled and placed in large plastic bags in wooden crates for sea duty. At sea, where possible, we carry out at least one flushing cast of the bottles to a depth of about 1000 m. The bottles are tripped during the upcast at 200 to 500 m and the water is discarded once the bottle is returned to the clean area. Experience has shown that this practice minimizes anomalous data points and pre-conditions the sample bottles. Additionally, we often plan our cruise track to move from areas of lowest to highest suspended matter loadings to minimize the need for rigorous cleaning of the bottle between samples.

3. Sampling

Setting up at sea is simplified if a clean van is available. We have used a clean van in collaboration with the Atlantic Oceanographic and Meteorological Laboratories of NOAA on many occasions. The double-chambered unit consists of an inner clean room with a vertical laminar-flow clean bench and an outer room for garment changing, sampler preparation and sample transfer. When such facilities have not been available, we isolate a portion of the ship where a large laminar flow bench can be placed and where sample bottles can be placed for the filtration process. This involves construction of a temporary plastic room.

We have deployed Go-Flo bottles on Kevlar line and also on a rosette sampler. The choice of procedures is sometimes limited by the vessel being used or the area/depths sampled. Use of Kevlar line requires special precautions such as plastic coating a special winch with polyethylene sheeting as well as using Teflon messengers and an epoxy-coated bottom weight. Our recent experiences in sampling particles from hydrothermal plumes have been carried out using a rosette sampler. We have compared discrete Go-Flo samples with in situ filtration techniques as described in the introduction by Feely and Trefry in this volume, chapter 1, page 5. Values for total suspended matter concentrations compared very well.

All Go-Flo samplers are pre-set in the clean area and covered with a washed plastic bag over each end. These covers prevent any deck/air particles from becoming lodged in the ball-valve assembly. The samplers are then mounted on a continuously cleaned and well-maintained rosette just before deployment. The plastic bags are removed at the moment that the unit is deployed. We try to ensure that the lower 10 m of wire to the rosette is in clean condition and have taped it on numerous occasions with electrical tape to minimize contamination problems. Decisions of this nature are based on the area being sampled. For the purposes of sampling vent plumes, the clean rosette and clean or protected "leader" of hydrowire have seemed adequate based on data for the trace elements of interest.

The ship continues to serve as a major potential source of contamination and caution is always recommended. Prior to deployment of the samplers, it is important to identify locations of ship discharges such as from laundry, sewage, kitchen and stack wastes and halt these operations during the entire sample cast. Chipping and painting aboard the ship should also be suspended during sampling.

Sampling of vent particles requires precise placement of the sample bottles in the water column. We use real-time nephelometry to identify the presence and shape of the dispersed vent plume [Trefry et al., 1985; Nelsen et al., 1986/87]. Using a SeaMarTec Model 6000AR auto-ranging nephelometer, we can distinguish the onset of a lens with greater than 10-12 μg particles L^{-1} relative to background levels of 6 to 8 $\mu g\ L^{-1}$ (Figure 5-35). The plumes generally are found some 200-500 m above the vent source [Nelsen et al., 1986, 1987; Trocine and Trefry, 1988]. We have tried sampling on the

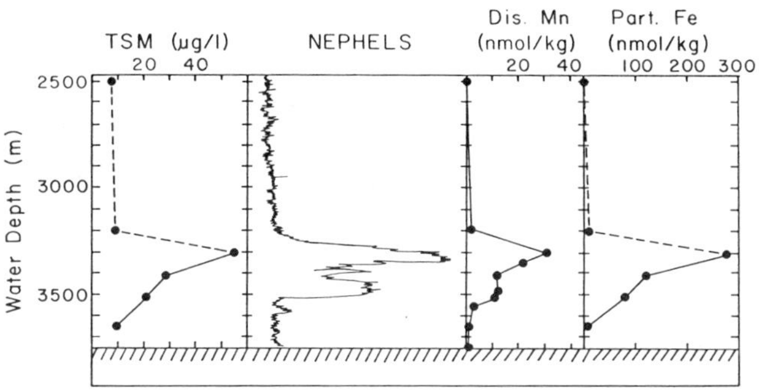

Fig. 5-35. Vertical profiles of total suspended matter (TSM), light scattering as measured by a nephelometer (nephels as arbitrary units), dissolved manganese and particulate iron for a site at 26° N on the Mid-Atlantic Ridge.

upcast and downcast in "background" waters and found no discernible difference in suspended matter concentrations or trace element content for the Mid-Atlantic Ridge [Trocine and Trefry, 1988].

When the rosette sampler is back aboard ship, we place the plastic caps over the ends of the bottles and quickly move the sample bottles to the clean laboratory or clean area. We have used 30 L bottles in our hydrothermal plume work to maximize the mass of material available to us. Salinity and nutrient samples are quickly drawn from the bottle and then the filtration process is initiated immediately. The salinity and nutrient samples are used to ensure that our apparent sample depth is correct. Improperly prepared Go-Flo bottles with loose valves or bad seals can entrain or mix water, generally from shallower depths, into the sampler. Comparison of salinity and nutrient values with CTD data and values from other samplers is essential. We generally have a success rate as indicated by perfect salinity agreement of greater than 90%. In this manner, a problem bottle can be identified at sea and repaired before the next deployment.

4. Filtration

Filtration of water collected from vent plumes with 30 L Go-Flo bottles has been by in-line, pressure filtration. We use acid-cleaned, nylon holders produced by Millipore Corporation. The nylon holders should be cleaned using dilute acid (less than 1 N) to avoid decomposition, cracking or increasing brittleness. A toothbrush works well for cleaning the grooves in the holder. The sample bottle is placed in our clean area and a flow of filtered nitrogen is introduced to the top of the bottle, initially at 20-40 kPa. This pressure is increased as high as 50 kPa during the 2 hour filtration process. The Millipore apparatus is attached at the lower drain valve of the Niskin bottle. The filter is initially wetted by allowing a slow flow of water to cover the filter and force all air out of the system. We filter all of the water in the sample bottle, sometimes shaking the bottle to keep particles in suspension. The filtered water is drained into a large container for later measurement of the total volume. One liter filtered samples for dissolved materials are often collected after the first 2-3 L of water have passed through the membrane.

Many years ago, while collecting our first sample of Mississippi River water [Trefry and Presley, 1976], we learned about particle settling during the filtration process. After filtering 30 L of Mississippi River water (with 200 mg particles L^{-1}) through a 297 mm diameter, 0.4 μm Nuclepore membrane and then removing the endcap of the Niskin bottle, a sizeable amount of Mississippi mud fell onto the floor. We then refitted the sample bottles to drain out the very bottom and forever realized the potential importance of particle settling during filtration.

After all the particles have been collected on the membrane, the filter holder is transferred to a laminar flow hood and the filter is rinsed with pH-adjusted, distilled-deionized water to wash away residual salts. The pH 8 rinse solution is prepared with Ultrex (J.T. Baker Inc.) ammonium hydroxide and is used to prevent dissolution of manganese oxides present on the filter. We prefer an unbuffered solution to minimize addition of contaminants. The pH is checked regularly and a new solution is prepared at least daily. For rinsing in-line filters, we disconnect the filter from the Go-Flo bottle, attach it with Tygon tubing to an acid-washed reservoir (e.g. a buret) and then pass 3-4 separate aliquots of 20-30 mL through the filter by pressure filtration. One important area for salt retention is along the border of the o-ring of the filter holder. Sometimes it is advantageous to remove the top of the filter holder and the o-ring and then carefully use a wash bottle to complete a final rinse of the filter. This process is also carried out in a laminar flow hood, this time by vacuum filtration. After the rinse is completed, the vacuum is maintained a bit longer to draw off any residual water and the filter is returned to its petri dish and Ziploc bags. The sample tends to air dry in less than a day, an approach that is adequate for total element analyses because a final drying is carried out back on shore. In preparation for some leaching techniques, the sample may need to be dried at sea in a vacuum desiccator or stored dry in a nitrogen atmosphere. Such precautions are specific to the final analyses to be carried out.

In all cases, it is advisable to determine the efficiency of the washing process by measuring the concentration of Na or Cl (Cl is easier for XRF users) on the filter. If 5 μL of seawater (salinity = 35 g kg^{-1}) were left on the filter to dry it would deposit about 175 μg of salt or an amount equal to the total amount of suspended matter filtered from a 30 L sample bottle when suspended matter loadings were at deep-sea values of about 6 μg L^{-1}. Thus, checks on the presence of residual salts on the filter are important. One complicating point in the concern for residual salts that we have observed occurs in samples with a large amount of living cells. If the cells have not been ruptured during filtration, the Na or Cl values for the filter may appear higher than normal. Brief inspection of the filters prior to laboratory processing, especially for surface waters, can help identify this distortion.

5. Laboratory Preparation and Elemental Analysis

Upon return to the laboratory, the samples are placed on a drying rack in our clean room (22° C, 50% relative humidity), the lid to the petri dish is cracked open and the filters are left for 48 to 72 hours. Each filter, along with laboratory and at-sea control filters, is then weighed three times.

To determine the total concentration of major elements and trace metals, the filter is placed in an acid-washed Teflon centrifuge tube (15 mL from Berghof/America) fitted with a Teflon stopper. A 47 mm diameter filter can essentially be placed in the tube with Teflon forceps without any overlapping of the membrane around the inner walls of the tube.

Our digestion is modified from that of Eggimann and Betzer [1976] and generally involves addition of about 1 mL of concentrated nitric acid (Ultrex II from J.T. Baker Inc.) and 0.1 mL of hydrofluoric acid (Ultrex II). The tube is heated to 70° C in a water bath and then cooled. Pressure buildup and the unacceptable popped-top is controlled by a lip on the tube that the stopper can be pushed below and a gentle heating process. The heating/cooling procedure is repeated two more times. Between heating cycles the tubes are gently tapped to force acid droplets on the sides of the tube back into the bottom. The acids are then swirled at an angle to wash them across the filter. After heating, the acids are rinsed with about 5 mL of distilled-deionized water and stored in a screw-cap polyethylene bottle. All processing is carried out in a laminar flow hood.

The digestion procedure can be modified to facilitate variations in the amount and composition of the suspended matter. For suspended matter loads on the filter of less than 2 mg, both HNO_3 and HF are added before the first heating cycle. However, when the quantity of material is greater than 2 mg or a large quantity of carbonates or organic matter is expected, only HNO_3 is added before the first heating. Once the tubes have completely cooled, the stoppers are removed and HF is added prior to the second and third heatings. Just prior to the third heating, the progress of the digestion is evaluated by looking through the translucent walls of the tube or by looking inside the tube. If any residue is visible, 0.2 mL of concentrated HCl (Ultrex II) is added. When HCl is present in the tube, it is heated to only 50° C.

We also prepare standard reference samples of sediment or rocks available from the U.S. Geological Survey, the U.S. National Institute of Standards and Technology or the National Research Council of Canada. We typically use 1-10 mg aliquots of these standards and obtain excellent comparison with accepted values thereby giving us confidence in the digestion procedure. Choice of a relevant standard is appropriate where possible so that the trace and major element composition is similar.

Analysis for the elements of interest is followed by atomic absorption spectrophotometry (AAS, flame or Zeeman graphite furnace) or by Inductively-Coupled Plasma-Mass Spectrometry (ICP-MS). Quality assurance is based on processing of replicate samples at selected depths and sites and from analysis of standard reference materials. We use background correction and method of additions checks routinely in AAS analyses. Acid and filter blanks are typically <5% of the particulate metal concentrations. Average precision for metal concentrations is better than 3%.

References

Eggimann, D. W., and P. R. Betzer, Decomposition and analysis of refractory oceanic suspended materials, Anal. Chem., 48, 886-890, 1976.

Nelsen, T. A., G. P. Klinkhammer, J. H. Trefry, and R. P. Trocine, Real-time observation of dispersed hydrothermal plumes using nephelometry: examples from the Mid-Atlantic Ridge, Earth Planet. Sci. Lett., 81, 245-252, 1986/87.

Trefry, J. H., and B. J. Presley, Heavy metal transport from the Mississippi River to the Gulf of Mexico, in Marine Pollutant Transfer, edited by H. L. Windom, and R. A. Duce, pp. 36-79, D. C. Heath & Co., Lexington, Massachusetts, 1976.

Trefry, J. H., B. J. Presley, W. L. Keeney-Kennicutt, and R. P. Trocine, Distribution and chemistry of manganese, iron and suspended particles in the Orca Basin, Geo-Mar. Lett., 4, 125-130, 1984.

Trefry, J. H., R. P. Trocine, G. P. Klinkhammer, and P. A. Rona, Iron and copper enrichment of suspended particles in dispersed plumes along the Mid-Atlantic Ridge, Geophys. Res. Lett., 12, 506-509, 1985.

Trefry, J. H., T. A. Nelsen, R. P. Trocine, S. Metz, and T. W. Vetter, Trace metal fluxes through the Mississippi River Delta system, Rapp. P.-v. Reun. Cons. Int. Explor. Mer., 186, 277-288, 1986.

Trocine, R. P., and J. H. Trefry, Distribution and chemistry of suspended particles from an active hydrothermal vent site on the Mid-Atlantic Ridge at 26° N, Earth Planet. Sci. Lett., 88, 1-15, 1988.

SUSPENDED PARTICULATE SAMPLING AND ANALYSIS FOR TRACE ELEMENTS

Herbert L. Windom and Ralph G. Smith, Jr.
Skidaway Institute of Oceanography, P.O. Box 13687, Savannah, Georgia 31416

1. Introduction

Over the past decade, we have collected and analyzed suspended particles from a variety of aquatic environments including rivers, lakes, estuaries, coastal and shelf waters and the deep sea, although most of our experience has been in estuarine, coastal and shelf waters. Our analyses have included the commonly studied trace metals (i.e., Fe, Mn, Cd, Co, Cu, Ni, Pb, Zn), major elements (i.e., Ca, Mg, Na, K and Al) and particulate organic carbon and nitrogen. More recently we have been able to expand our analyses, through the use of ICP-MS, to include rare earth elements and many others (e.g, Th, U, V, Sc, Ti, Be, Ba, etc.). The techniques we use are described below.

2. Sample Collection

For rivers, lakes and shallow estuaries, a pumping system is used to collect water from depths of up to about 15 m. The system consists of a hand-cranked peristaltic pump through which silicone tubing passes. The inlet end of the tubing is deployed to the desired depth on Kevlar line as shown in Figure 5-36. The system is precleaned by circulating 10% concentrated HCl through the tubing for several minutes, followed by flushing with distilled water. When deployed, several liters of water are flushed through the system before aliquots for filtration are pumped directly into the reservoir of the filtration system shown in Figure 5-37. If samples are collected from small boats, the samples are usually pumped directly into precleaned polyethylene carboys and loaded into the filtration system a few hours later.

For coastal and continental shelf waters, samples are collected using 30 L Go-Flo bottles, deployed on Kevlar line, or Niskin bottles deployed on rosettes using standard stainless steel hydrowire. Immediately after sample collection, Go-Flo bottles are placed on an apparatus that continually rocks them in order to keep particles in suspension during filtration. The bottles are pressurized with filtered nitrogen gas forcing the water through a filtration system similar to that shown in Figure 5-37.

Marine Particles: Analysis and Characterization
Geophysical Monograph 63

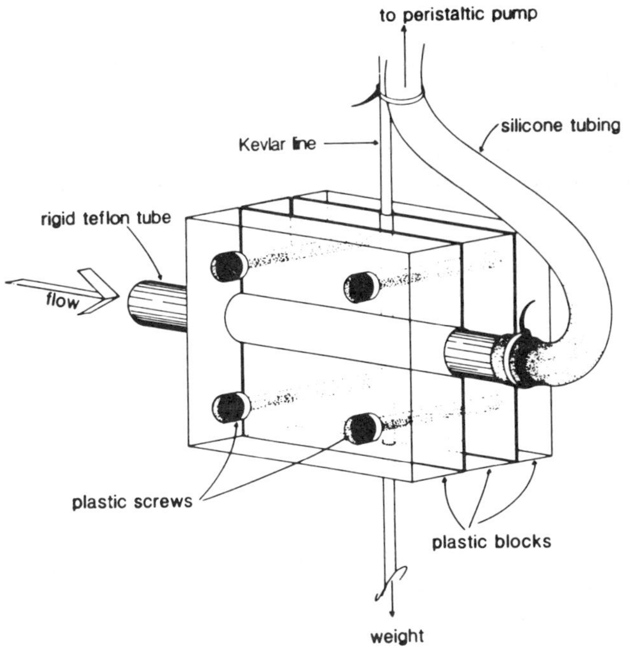

Fig. 5-36. Schematic of pumping system intake.

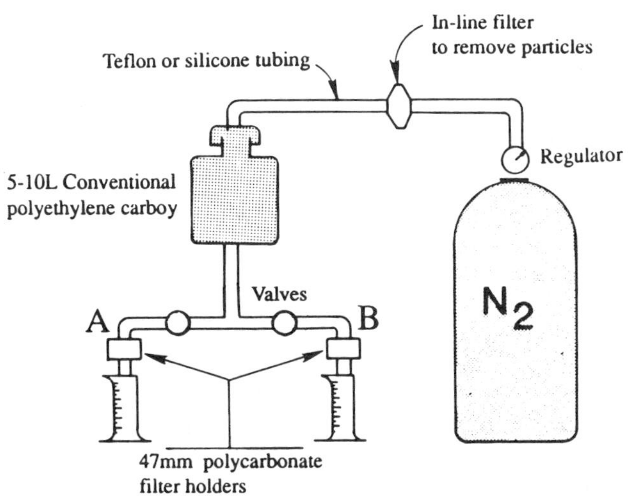

Fig. 5-37. Schematic of pressure filtration system.

Because of the high levels of suspended sediment that are typically encountered in coastal and shelf waters, we have begun to collect our samples from rosette-mounted Niskin bottles. This minimizes station time and has made our data sets more comparable with those of our collaborators and other investigators. Samples are drained directly from the Niskin bottles into polyethylene carboys which are placed into the filtration system (Figure 5-37). No significant contamination has been found to be introduced using Niskin bottles in coastal and shelf environments, but precautions are taken by replacing internal closure springs with silicone tubing and thoroughly cleaning bottles before use.

For the relatively small amount of deep sea work we have done, 30 L Go-Flo bottles have always been used. All rubber used in these bottles is replaced with silicone, metal parts are coated with epoxy paint and drain cocks are replaced with Teflon. The Go-Flo bottles are deployed on a length of Kevlar line shackled to the end of the hydrowire that is already on the winch. A trigger mechanism from a Niskin bottle is used to "bridge" the shackle with messengers.

3. Sample Filtration

Samples collected, as described above, for trace element analyses are filtered using the system schematically shown in Figure 5-37. This system connects, by Teflon tubing, the pressurized polyethylene carboy (or Go-Flo) to duplicate filter holders, containing precleaned filters on a clean bench. A rigid manifold composed of 1/4" Teflon tubing and two ells, two valves and a tee is used for this purpose. An over pressure of less than 1 atmosphere is applied across the filters. The flow rate of the sample through each filter is monitored visually until an obvious slowing occurs indicating that the filter is loaded. The stopcocks are closed and the filter holders are unscrewed from the filter manifold and placed on a vacuum manifold where the remaining seawater is removed and the filters are washed three times with quartz-distilled water. The volume of sample water which passed through each filter is measured and recorded. The filters are folded twice using acid-cleaned, Teflon forceps, placed in precleaned polyethylene vials and stored frozen until digested. All operations are conducted in a clean bench.

The filters used for trace elements are Nuclepore polycarbonate filters with a pore size of 0.4 μm and a diameter of 47 mm. Filters are cleaned by soaking in a 200 mL Teflon jar filled with 10% reagent grade HCl for one week, then rinsed 3 times with quartz-distilled water. The filters are stored wet in quartz-distilled water until loaded onto filter holders. Prior to going to sea, filters are loaded onto modified Nuclepore polycarbonate Swinlok filter holders. The modification involves cutting the Luer connections off the upstream and downstream sides of the holders. This permits the use of 1/4" NPT Teflon adapters which connect to 1/4" tube fittings. The polycarbonate filter holders are cleaned using an ultrasonic bath. The holders are first cleaned in a weak detergent solution by placing them in a 2 liter polyethylene jar filled with deionized water and 1 capful of Micro (International Products Corp.). The holders are sonicated for 1 hour at 180° F then rinsed several times with deionized water. They are sonicated again with 10% HCl for 1 hour at 180° F and rinsed several times with deionized water and sonicated a final time for 15 minutes with quartz-distilled water. The filters are loaded onto the filter holders in a laminar flow clean bench. Teflon forceps are used for loading the filters onto the filter holders which are attached to a vacuum manifold. After loading, the filter holders are rinsed 3 times with quartz-distilled water and the residual water removed by vacuum. The filter holders are then placed in plastic bags until used.

For particulate organic carbon and nitrogen analyses, duplicate aliquots of collected samples are filtered through 25 mm Gelman, precombusted glass fiber filters. Vacuum filtration is carried out using glass filter holders, with 250 mL reservoirs, attached to filter flasks. After filtration, filters are removed from the holders using stainless steel forceps, folded in half, wrapped in aluminum foil, and frozen.

4. Sample Digestion and Analysis

All procedures are carried out in a "clean lab", the access to which is limited. The lab is slightly pressurized by air forced through HEPA filters. Clean benches are also located within the lab, and two class-100 modular clean vans are connected to the area. This clean area is only accessible through a double-doored vestibule where shoe covers are kept to be used by everyone who enters.

Filters for trace element analysis are carefully unfolded using Teflon forceps and placed in 60 mL Teflon digestion vessels (Savillex #561). These containers are rated to withstand 75 psi at room temperature. The filters are treated with 2.5 mL of Ultrex HNO_3 and 200 mL of Ultrex HF and heated in a boiling water bath for 2 hours. Although the filter is not destroyed by this digestion, we assume that metals are quantitatively obtained from the particles on the filter. This assumption is based on the results of analysis of ten replicates of about 25 mg of certified NBS 1646B sediment standard that were treated in a similar manner as sample filters (Table 5-14). After treatment in the water bath, the filter digest is poured into a 10 mL Teflon beaker and the filter carefully rinsed with quartz-distilled

TABLE 5-14.
Results of Analyses of NBS (1643) Estuarine Sediment Standard

	Certified Value	Concentration Found	Yield (%)
Al	6.25 ± .20	6.12 ± 1.08	97.7
Be*	(1.5)	1.73 ± .33	
Cd	0.36 ± .07	0.38 ± .05	105
Ca	0.83 ± .03	0.82 ± .10	98.7
Cr	76 ± 3	70.7 ± 13	93.0
Co	10.5 ± 1.3	9.4 ± .71	89.5
Cu	18 ± 3	17.9 ± 2.1	99.4
Fe	3.35 ± .10	3.35 ± .43	100
Pb	28.2 ± 1.8	27.0 ± 3.5	95.7
Mg	1.09 ± .08	1.08 ± .09	99.1
Mo*	(2.0)	1.8 ± .35	
Mn	375 ± 20	340 ± 21	90.6
Ni	32 ± 3	28.2 ± 2.3	88.1
V	88 ± 10	83.9 ± 7.6	95.4
Zn	138 ± 6	119 ± 7.1	85.6

*Not certified

water and the rinse added to the Teflon beaker. The digest is evaporated to dryness at 120° C under a closed "Pyrex hood" to prevent airborne contamination. The residue is redissolved with 10 mL of 1% HNO_3. The samples are then analyzed by ICP/MS. Filter blanks are determined by digesting unused filters (i.e., that had been loaded in holders but not used) following the procedure described above. Representative trace element concentrations in filter blanks are given in Table 5-15.

A VG-Plasma Quad PQII+ Inductively Coupled Plasma-Mass Spectrometer is used for sample analyses. Sample and skimmer cones are cleaned daily as is the glass sample chamber and torch (see Falkner, this volume, page 243). The instrument is tuned daily with a 100 ng mL^{-1} solution of Li, Co, In, La, Pb and U. The mass calibration and sample resolution are also checked prior to each sample run. The following settings for the ICP-MS are used: coolant gas, 13 L min^{-1}; auxiliary gas flow, 0.5 L min^{-1}; nebulizer gas flow, 0.75 L min^{-1}; RF power, 1350 watts. The data acquisition parameters for all elements except Mg, Al, Ca, Ti, Mn, and Fe are as follows: mass range scanned 6.0 to 250 A.M.U.; skipped scan regions 11.0 to 44.5, 55.5 to 58.5 and 79.5 to 84.5; dwell time 160 μs; number of channels, 4096. Calculations of concentration are made from least squares fit calibration curves.

We use flow injection to introduce samples into the plasma due to their high dissolved solids content. This prevents a buildup of salts on the sample cones which drastically affects sensitivity and stability. Although the sample digests are in 1% HNO_3, interference corrections included in the software are used to remove isobaric overlaps of $^{35}Cl^{16}O$ on ^{51}V and $^{40}Ar^{35}Cl$ on ^{75}As.

The elements noted in the above exceptions such as Mg, Al, Ca, Ti, Mn, Fe are considered major ions and are analyzed using the following data acquisition parameters: mass range scanned, 20.5 to 131 A.M.U.; dwell 320 μs; number of channels 4096; dual mode; skipped scan regions 22.5-23.5; 27.5-42.0, 79.5-80.5. Due to the high concentration of these major ions, a representative number of the samples are diluted 1:100 and a semi-quantitative analysis performed prior to quantitative analysis. This procedure allows for a more appropriate dilution of the samples for quantitative analysis.

Indium is added to all samples and standards as an internal standard to normalize slight changes in the stability of the detector.

Particulate organic carbon and nitrogen analyses are carried out using a Perkin Elmer elemental analyzer. Filters are treated with a drop of 1 N HCl to remove carbonates prior to analysis.

5. Other Comments on Sampling and Analysis

During the filtration of samples, the sample reservoir such as a polyethylene carboy (Figure 5-37) is periodically shaken to keep particles in suspension. A new system has been designed having a reservoir that uses a magnetic stirrer with a Teflon coated stirring bar, to keep particles in suspension.

Our present sampling and analytical techniques allow us to work down to total suspended particle levels of tens of micrograms per liter [Windom and Gross, 1989; Windom et al., 1989]. This limit is controlled by the amount of water we can reasonably collect and process through 47 mm filters. To decrease the analytical uncertainties at suspended particle concentrations of ~ 10 μg L^{-1}, we feel that larger volume samples must be collected. We have, therefore, obtained three Pelagic Electronics Deep Sea Impeller Pumps for in situ filtration. These pumps are powered by Deep Sea Power and Light pressurized batteries, and we plan to use 142 mm Nuclepore polycarbonate filters in filter holders similar to that described by Sherrell [1989].

TABLE 5-15.
Filter Blanks

Element	Filter Blank*		Element	Filter Blank*	
Al	7.5	±3.0	Mo	1.7	±0.47
Ba	2.9	±1.5	Nd	0.48	±0.19
Be	0.53	±0.29	Ni	16.8	±6.3
Ca	11.1	±4.6	Pb	5.3	±2.0
Cd	0.62	±0.23	Pr	0.21	±0.26
Ce	0.72	±0.53	Sc	1.36	±0.39
Co	0.49	±0.15	Sm	0.28	±0 .21
Cr	162	±43	Sn	1.8	± 0.43
Cu	11.1	±4.6	Sr	25.7	±11.3
Dy	0.18	±0.11	Th	0.12	±0.07
Fe	10.5	±3.6	Ti	31.6	±4.1
Ga	0.30	±0.17	U	0.12	±0.07
Gd	0.21	±0.15	V	1.09	±0.29
Hf	0.46	±0.23	Zn	41.7	±15.6
La	0.40	±0.23	Zr	13.9	±3.0
Mn	3.4	±1.6			

*Mean of 10 filter blanks. All concentrations are ng per filter.

All of our results are expressed in concentrations per volume of water filtered. We do not preweigh filters and attempt to quantify total weight of suspended particles collected. We feel that this introduces an additional opportunity to contaminate samples which we prefer to avoid. Regardless, we feel that the total weight of particles suspended in water which we sample can be estimated quite well based on calcium, aluminum and organic carbon concentrations.

References

Sherrell, R. M., The trace metal geochemistry of suspended oceanic particulate matter, Ph.D. dissertation, WHOI/MIT, Woods Hole, MA, 1989.

Windom, H. L., and T. F. Gross, Flux of particulate aluminum across the southeastern U.S. continental shelf, Estuarine Coastal Shelf Sci., 28, 327-338, 1989.

Windom, H. L., R. G. Smith, Jr., and C. Rawlinson, Particulate trace metal composition and flux across the southeastern U.S. continental shelf, Mar. Chem., 27, 283-297, 1989.

COLLECTION AND ANALYSIS OF MARINE PARTICLES

P. A. Yeats

Marine Chemistry Division, Physical and Chemical Sciences, Department of Fisheries and Oceans
Bedford Institute of Oceanography, P.O. Box 1006, Dartmouth, Nova Scotia, Canada B2Y 4A2

1. Introduction

The following is a description of the procedures being used at the present time in the Inorganic Chemistry Section of the Marine Chemistry Division, Department of Fisheries and Oceans, Bedford Institute of Oceanography. These procedures are still being developed so the descriptions represent our present thinking and are subject to change as we gain experience in these areas.

2. Sample Collection Procedures

We have been collecting marine particulate samples in two different ways: first by filtration of water samples collected with Go-Flo bottles, and second by in situ filtration of seawater using a submersible pump.

The Go-Flo sampling procedure differs little in principle from that being used in a large number of laboratories. It is a modification of sampling procedures developed at BIO for particulate matter sampling in coastal seawater by Sundby [1974]. Nuclepore filters in plastic filter holders are attached to the drain spigot of the Go-Flo and water is forced through the filter by applying pressure to the Go-Flo. In our procedure 0.4 μm pore size, 47 mm Nuclepore filters are weighed on a micro-balance, then washed, and stored in numbered plastic petri dishes. On board ship, the filters are transferred to cleaned Sartorius Model SM16511 filter holders in a portable clean bench. Once the unfiltered water samples are drawn from the Go-Flos, the filter holder is attached with a small piece of silicone tubing and filtered N_2 pressure applied through a nylon fitting that replaces the Go-Flo's air vent. All the water in the Go-Flo is filtered, then the filter holder is removed and rinsed by passing two 50 mL aliquots of Milli-Q water through the filter using a filter washing apparatus that consists of a Teflon separatory funnel adapted for pressurization. It is important to wash the filters before they dry out because the salts that precipitate can be very difficult to redissolve. The rinsed filter is then removed from the filter holder in a clean bench, returned to its petri dish and dried in a microwave oven. In the shore lab the filters are reweighed to determine the suspended particulate matter concentrations and the filters retained for particulate metal analysis.

This procedure was developed for sampling of coastal waters and has been found to be adequate for the collection of samples for the determination of suspended particulate matter (SPM) and a number of trace metals. SPM concentrations can be determined with a precision of 10-20% and Mn, Cu, Cd, Fe, Zn with precisions of 20-40% based on the oceanographic consistency of the observed distributions. Individual analyses can be done with greater precision but this analytical precision would give an unreasonably good estimate of the overall capability. These numbers apply for sample loadings of 0.5 to 1 mg per filter that can readily be achieved with 5-10 L water samples given the 50-500 μg L^{-1} SPM concentrations generally observed in coastal waters.

In highly turbid waters this procedure will not work because particles settle to the bottom of the Go-Flo before they can be filtered. For these samples, we take 1-2 L of unfiltered seawater from the Go-Flo immediately after sampling and then filter this sample in a separate filtration apparatus. This system consists of a 1 L Teflon separatory funnel adapted for pressurization that is attached to the filter holder by a piece of silicone tubing. This apparatus has been described and tested in Bewers et al. [1985] except that we have changed from Swinnex to the Sartorius filter holders.

In offshore waters, or low SPM coastal waters, our procedure is limited by the amount of material that can be collected on a filter while processing a reasonable volume of water. If the SPM concentrations are $<10\,\mu$g L^{-1} as is often the case in subsurface waters, filtration of a 30 L sample will yield $<300\,\mu$g of SPM on the filter. While a sample of this size can be weighed for determination of SPM concentrations, the levels of trace metals will be so low (0.1-10 ng) that quantitation of the trace metal concentrations will be very difficult. Atomic absorption sensitivity may be adequate for this sort of analysis but controlling contamination during all the sampling and processing steps at these levels is very difficult.

Filter loadings can be increased by filtering the water from a large number of Go-Flos but the added handling

Marine Particles: Analysis and Characterization
Geophysical Monograph 63

increases the risk of contamination and the added filtration is time consuming. We have filtered up to 100 L through a 47 mm filter but I wouldn't recommend it because of the filtering time involved. Alternatively, SPM samples can be collected with an in situ pumping system.

The in situ pumping system that has been developed at BIO [Beanlands et al., 1987] is a six channel, remotely operated pump that is lowered on a CTD cable and operated from a microcomputer on board ship. Our present depth capability is 5000 m, limited by the length of our CTD cable. The computer is used to set which of the channels is open and to control the pump speed. It also continuously monitors volume filtered, flow rate, and pressure drop across the filter. In the configuration of the pumping system that we have used each channel contains two 47 mm filters in parallel. The filters are at the extreme upstream end of the system and the inlets for the two active filters are the only part of the system that protrudes out of the side of the PVC shell of the system. Flow meter, check valves, pump, etc. are all downstream of the filter.

We have used Nuclepore filters between 0.4 and 12 μm pore size and screens (plankton netting) up to 50 μm. Flow rates for 0.8 μm filters start at about 10 L min^{-1} but decrease over a filtering time of about 1 hour to a few tenths of a L min^{-1}. Total volume filtered will depend on the particle concentrations as our filters generally clog after approximately 1 mg of material has been collected. With the larger pore size filters (12 μm) flow rates start at about 12 L per minute and they don't generally clog so volumes in excess of 500 L can be filtered in less than 1 hour.

Our gravimetric determinations of suspended matter concentrations in the off-shore North Atlantic waters adjacent to eastern Canada based on Go-Flo samples have given concentrations that vary from ~ 10-20 μg L^{-1} for deep waters to 200-300 μg L^{-1} for surface waters, consistent with other investigations in these waters. Comparison of SPM concentrations determined from in situ pump and Go-Flo samples collected at the same time and place has indicated that the pump consistently gives concentrations about 20% lower than those for the Go-Flos. The reason for this discrepancy is not clear to us at present.

With either of these collection techniques contamination control is the biggest impediment to collection of valid samples for trace metal determination particularly for the lightly loaded filters. We have seen marked decreases in the apparent levels of trace metals on particles as we have worked on the problem of reducing contamination levels in the sampling, sample handling and analytical procedures. We have not developed any radically different cleaning procedures; all parts of the various apparatus are washed, soaked in acid and rinsed using clean room techniques in manners similar to those employed in almost all trace metal labs. The actual procedure we use varies somewhat for different parts of the apparatus. Filters are soaked for 24 hrs in 2 N high purity nitric acid then rinsed several times with Milli-Q water. Most of the rest of the apparatus is soaked in 6 N hydrochloric acid for 6 hrs then rinsed with Milli-Q. Gains are made by rigorously and conscientiously applying these techniques rather than by developing "new and better" cleaning procedures.

3. Trace Metal Analysis

The technique described here for trace metal analysis of particulate matter was developed by Dr. D. H. Loring for the analysis of trace metals in marine sediments and adapted for SPM. The analytical procedure, described in Rantala and Loring [1985] consists of a two step fractionation of the sample using 25% acetic acid leach ("non-detrital" metals) and HF/aqua regia digestion ("detrital" metals). The acetic acid procedure involves leaching the SPM on the filter in a Millipore Sterifil vacuum filtration apparatus with Viton A o-rings since the silicone o-rings were found to contaminate the samples with zinc. The samples are leached with 5 mL of 25% acetic acid (Seastar) for 24 hours. After leaching, the sample is vacuum filtered into a 10 mL polypropylene volumetric flask placed within the filtration apparatus and washed with two 2 mL aliquots of Milli-Q water and made up to 10 mL. The filter is then transferred to a Teflon decomposition vessel [Rantala and Loring, 1973] and digested with 1 mL of HF and 1 mL of aqua regia in a boiling water bath for 1 hour. After cooling, the solution is transferred to a 25 mL polypropylene volumetric containing 0.93 g of boric acid and 5 mL of Milli-Q water. The decomposition vessel is rinsed into the volumetric with several small portions of water and the volume made up to 25 mL. Metal concentrations are determined by flame (Si, Cu, Fe, Mn, Ca, Mg, Zn) or graphite furnace (Al, Cd, Pb) atomic absorption spectrophotometry. L'vov platforms are used in the graphite furnace analysis. This procedure was developed for the analysis of estuarine and coastal samples where filter loadings of >2 mg can be achieved. It has, however, been tested for more lightly loaded filters, for example in the ICES intercalibration for metals in suspended matter

[Yeats and Dalziel, 1987].

Most of our work on the trace metal chemistry of marine particles has been focused on estuaries. In the estuaries we have studied, the particles are mostly inorganic and trace metal concentrations are similar to those in nearshore sediments. The exception to this is cadmium, whose concentration in estuarine particles, especially the non-detrital fraction, is elevated by factors of at least two or three. Our experience in the analysis of particles from offshore shelf or deep sea regions is limited. Results for non-detrital metals from two stations in basins on the Scotian Shelf and one slope water station are given in Table 5-16 as examples to indicate the sorts of levels we are detecting.

TABLE 5-16.

Concentrations of Particulate Non-Detrital Metals From Two Stations in Basins on the Scotian Shelf and One Slope Water Station

Depth m	Mn ($\mu g\,g^{-1}$)	Cu ($\mu g\,g^{-1}$)	Cd ($\mu g\,g^{-1}$)
Scotian Slope	Cruise 85-017 stn 10		
20	24	8.7	2.8
100	891	186	5.1
500	368	15	4.3
1000	1010	27	1.3
2000	1270	6.1	4.0
3000	783	15	0.92
4000	851	6.5	0.80
4600	831	5.1	0.37
Scotian Shelf	Cruise 85-017 stn 26		
10	80	12	4.4
25	341	6.7	7.8
50	1590	13	11
100	2660	26	8.4
170	1830	21	7.1
Scotian Shelf	Cruise 85-017 stn 28		
10	475	19	2.1
25	626	17	0.70
50	552	26	3.9
100	2390	9.2	6.9
200	2340	17	4.6
247	1580	14	3.7

An indication of the agreement that can be achieved for trace metals between the in situ pump and Go-Flo samples is illustrated here using manganese results from a station in the Sargasso Sea (Table 5-17).

4. Cross-Flow Filtration

We have collaborated in the past few years with Dr. B. G. Whitehouse of Dalhousie University on the development of cross-flow filtration (CFF) for the separation of colloids from marine waters [Whitehouse et al., 1990]. The emphasis in this work has been on development of a contamination-free system for the ultra-filtration of large volumes of seawater (~ 100 L) with short filtration times. The system consists of a 100 L Teflon reservoir, a self-priming air driven pump (Dominator Maskin AB, Sweden) with all plastic (polypropylene and Teflon) pump head, and a Millipore Pellicon CFF filter housing with a PTGC (polysulfone) membrane filter cassette.

TABLE 5-17.

Concentrations of Particulate Manganese in Samples Collected from Go-Flo Bottles and by Filtration In Situ at a Station in the Sargasso Sea

Depth (m)	Mn ($\mu g\,g^{-1}$)	
	Go-Flo	Pump
200		332
300	372	
400	320	
460	581	
500	1430	2390
800	1210	1400
1000	1500	1860
1200	1600	
1480	1010	
1500	1030	
1900		690
2000	849	940
2010	1130	
3000	1010	
3200	767	

The filters that we use have a nominal molecular weight cut-off of 10,000 and a filter area of 0.46 m^2 and are in an acrylic housing. The system is plumbed together with Teflon tubing (13 mm), tubing connectors and valves. Up to 100 L of water, filtered through 0.4 μm Gelman Versaflow Capsules, is placed in the reservoir and pumped across the filter and returned to the reservoir. The back pressure in the system forces 10-20% of the water through the filter. The flow of most of the water across the filter and the pulsed flow produced by the pump prevents colloid accumulation on the membrane. Filtration rates with a 1/2 inch air-driven pump are about 60 L hr^{-1} with about 300 L hr^{-1} cross-flow. After most of the water has been forced through the filter an aliquot of the permeate or ultrafiltrate is collected and analyzed as a normal dissolved metal sample. Usually about 1 L of the retentate is collected and analyzed using the type of digestion procedure employed for analysis of estuarine water samples [Danielsson et al., 1982].

We had two main concerns when developing this system for trace metal separations:
1) contamination of the sample, and
2) recovery of all the metal in the sample.

We have established in sample contamination tests using Cd, Cu, Zn, and Pb so that the filter cassette in the Pellicon housing can be cleaned by flushing with 50% methanol, followed by 0.1 N hydrochloric acid and then Milli-Q water. Individual parts of the plumbing system can be cleaned by leaching with dilute hydrochloric acid followed by flushing the whole system with Milli-Q water after assembly. Recoveries have been measured by doing mass balances. The only elements for which we have seen significant colloidal concentrations are Cu, Mn, Fe, and Al. Recoveries for these elements have ranged from 71-124%. Recoveries for organic carbon ranged from 76-106% in the same experiments.

In our CFF experiments with marine waters, relatively little of the "dissolved" metal concentration is contained in the colloidal fraction. For iron and manganese the colloidal concentration is only a few percent of the dissolved and for copper perhaps 15%. Only for copper can we consistently detect a difference between filtrate (0.4 μm) and ultrafiltrate (10,000 nominal molecular weight) for off-shore samples (Table 5-18).

TABLE 5-18.
Concentrations of Copper in Filtered (<0.4μm) and Ultrafiltered Sea Water

	Cu (nM)	
depth	filtrate	ultrafiltrate
surface	3.4	2.9
surface	2.4	2.4
500 m	2.0	1.8
500 m	1.9	1.8
2000 m	2.3	1.6

References

Beanlands, B., G. A. Fowler, W. Whiteway, and J. A. Dalziel, The design and development of a particulate in situ sampler (PARISS), Oceans 87 Conf. Proc., 1, 188-193, 1987.

Bewers, J. M., P. A. Yeats, S. Westerlund, B. Magnusson, D. Schmidt, H. Zehle, S. S. Berman, A. Mykytiuk, J. C. Duinker, R. F. Nolting, R. G. Smith, and H. L. Windom, An intercomparison of seawater filtration procedures, Mar. Pollut. Bull., 16, 277-281, 1985.

Danielsson, L.-G., B. Magnusson, S. Westerlund, and K. Zhang, Trace metal determination in estuarine waters by electrothermal atomic absorption spectrophotometry after extraction of dithiocarbamate complexes into freon, Anal. Chim. Acta, 144, 183-188, 1982.

Rantala, R. T. T., and D. H. Loring, New low-cost Teflon decomposition vessel, Atomic Absorp. Newslett., 12, 97-99, 1973.

Rantala, R. T. T., and D. H. Loring, Partition and determination of cadmium, copper, lead and zinc in marine suspended particulate matter, Int. J. Environ. Anal. Chem., 19, 165-173, 1985.

Sundby, B., Distribution and transport of suspended particulate matter in the Gulf of St. Lawrence, Can. J. Earth Sci., 11, 1517-1533, 1974.

Whitehouse, B. G., P. A. Yeats, and P. M. Strain, Cross-flow filtration of colloids from aquatic environments, Limnol. Oceanogr., 35, 1368-1375, 1990.

Yeats, P. A., and J. A. Dalziel, ICES intercalibration for trace metals in suspended particulate matter, J. Cons. Int. Explor. Mer., 43, 272-278, 1987.

CHAPTER 6. MINERALOGY AND SURFACE PROPERTIES

WORKSHOP REPORT

by

Chair: Pierre E. Biscaye, Rapporteur: David J. DeMaster, Assistant Rapporteur: Ursula Ginster
and Members of the Working Group

1. Introduction

Minerals are one of the major types of oceanic particles and include phases of both biogenic and abiogenic origin. The principal biogenic phases are opal, several polymorphs of calcium carbonate and celestite (strontium sulfate). A wide spectrum of abiogenic phases exist including aluminosilicates, oxides, sulfides and sulfates. These phases are formed on the continents, in marine hydrothermal vents and plumes, in the seabed, or in microenvironments within particles in the marine water column. Techniques for sampling, handling, identifying and analyzing these phases are the subject of this section. In addition, discussions of analytical techniques for characterizing surface chemical properties also are provided. This information complements bulk mineralogical measurements because surface phases make up the interface between the mineralogy of a particle and the reactive solutions surrounding it and are the locations of the most chemically reactive sites.

The contributions to this workshop in the area of mineralogy and surface properties represent a range of scientific interests and objectives. At one end of the spectrum are studies that determine the identity and abundance of mineral phases. These studies have the following research objectives:

1. to deduce particle origins, and thereby, the physical, chemical, and biological processes by which they were formed;
2. to determine the degree to which those phases have been modified by the marine environment;
3. to infer where in the continental or marine environments they were formed, that is their provenance; and from that,
4. to infer something about the paths of transport and the associated processes by which these phases got from where they were formed to where they were sampled.

At the other end of the spectrum is research concerning the particle/solution interface, whose objective is to characterize the chemical properties and environment at the particle surface which may be substantially different from the bulk properties of the entire particle. There is some overlap, however, in both the interests of these studies and in the analytical bag of tricks that are available. Both require quantitative analysis of mineralogical or structural characteristics, an endeavor that has lagged considerably behind the quantitative analysis of major and minor elements in marine particulate matter. Many elemental analyses can be done with very low (part per billion or better) detection limits, and with excellent precision and accuracy (of the order of a percent or better). Quantitative analysis of minerals, however, is more difficult, and in some cases, such as clay minerals, still apparently impossible. It is also much less precise and accurate. It is nonetheless productive because of the physico/chemical and oceanographic information inherent in their identification. If bulk elemental analyses of marine particles are made, something about the elemental associations, or speciation, must be inferred from statistical analysis. Minerals represent a degree of elemental organization that inherently contains valuable information. There are, for example, three common polymorphs of calcium carbonate in marine particles, each of which has a different paragenesis, and therefore, the identification of each carries different information relevant to the objectives listed above.

Why is it that minerals are harder to analyze quantitatively in samples of marine particles than are, for example, many major, minor or even trace elements? First, in most oceanic particulate matter (except in bottom-derived nepheloid layers), many of the minerals are present in very low concentration, and the total amount of sample available for analysis may itself be very limited. Second, some of what we call minerals in marine samples represent an entire suite of minerals in the strict definition of the term, viz., a regular crystallographic arrangement of atoms having a regular 3-D repetition, and therefore, a fixed chemical composition. For example, scientists mainly use only a half dozen names for clay minerals in connection with marine particles, but each represents a more-or-less broad range of chemical compositions modifying several basic types of tetrahedral-octahedral layers. Another example, biogenic opal,

Marine Particles: Analysis and Characterization
Geophysical Monograph 63

is an extremely important marine "mineral", but is only weakly crystalline, exhibiting a variably high degree of disorder that can be modified through time by diagenetic reactions. In the study of surface properties, an analogous kind of continuum exists to complicate quantitative analysis; that is, the continuum from the relatively well-defined crystallographic structure of a mineral's interior, to the bonding discontinuities that yield sorption sites at the mineral surface, out to increasingly weak and poorly organized chemical associations in the immediately surrounding water layer.

Having stated the general problems, where do we stand in our ability to identify and quantitatively analyze the minerals found in marine particulate matter?

The group discussion and papers presented in this section represent substantial technical progress in the analysis of mineral particles over the past decade. Some of the analytical techniques used only a decade ago seem primitive by comparison to some of those being used currently. There are, however, also some decades-old techniques still in use because they have not been improved upon or superceded. Our progress in analytical procedures has been driven by oceanographic, geochemical, mineralogical and marine chemical questions seeking new understanding of chemical relationships, reactions and mechanisms.

The papers in this section exhibit some overlap with the section on Inorganic Compounds found in this volume, chapter 5, because of the need for similar analytical data. Some of these papers deal with analytical methods for bulk samples and bulk sample surfaces, some with the analysis of individual particles and particle surfaces, and some with both (see Table 6-1). Many deal with sampling and sample treatment procedures that are peculiar to the kinds of particles they study, or unique to the analytical methods that they subsequently use.

The order of contributions proceeds from the bulk mineral analysis papers towards the single particle and surface properties papers.

- Sjerry Van der Gaast, page 343, provides details of the rationale and procedures of his sophisticated analysis of mineralogy by x-ray diffraction including carbonate, opal and the ever-elusive clay minerals.
- David DeMaster, page 363, reviews the analytical techniques used to measure opal in marine particles with emphasis on the variant procedures for analysis by dissolution and the different objectives and conditions in which each may apply.
- Peter Betzer and Renate Bernstein, page 369, focus on techniques for sampling and analyzing a single class of organisms - Acantharia - which plays a dominant role in the marine geochemistry of a single element - strontium. After describing sampling and preparation techniques designed to prevent the rapid dissolution of the very soluble Acantharia, they discuss the use of computer-controlled scanning electron microscopy and energy-dispersive x-ray spectrometry for analysis of marine particles.
- Andrew Campbell, page 375, reports on the improved sensitivity and resolution in chemically analyzing small particles by SXRF (Synchrotron X-Ray Fluorescence Microprobe Spectroscopy). The other kinds of analyses he is using to study hydrothermal particles include Synchrotron X-Ray Absorption Spectroscopy, Mössbauer Spectroscopy, and ICP-Mass Spectrometry.
- Gerhard Fischer and Gerald Wefer, page 391, report on a whole range of sampling, preparation and analytical techniques used at Bremen to study particles. The analytical techniques include those for certain organics, carbonates, opal, clay minerals, trace elements, and natural radioisotopes as well as carbon and oxygen isotopes.
- Jacques Jedwab, page 399, presents a synopsis of his lifelong study of mineralogy and his two decades of studying opaque minerals in marine particles. He describes the process of using light microscopy combined with Scanning Electron Microscopy-Energy Dispersive X-Ray Spectometry and X-ray diffraction to hunt for and analyze these extremely rare, but information-filled minerals.
- Eric De Carlo and Dirk Koeppenkastrop, page 419, discuss analyses of particles as sorbents from various aquatic systems and sediments, using selective dissolution, x-ray diffraction, and several microscopic techniques. They also report sorption techniques using rare earth element radiotracers on synthetic and natural mineral phases.
- Jim Cowen, page 429, writes about his microanalysis with SEM and STEM (Scanning- and Scanning Transmission Electron Microscope) and EDS (Energy Dispersive fluorescence Spectrometry) examining individual particles in thin sections and on whole particles.
- And finally, Bruce Honeyman, page 437, reviews the approaches that have been taken to understanding chemical complexation of trace elements on particles, and analytical approaches to quantitative characterization of those surface complexation sites by measuring site density and surface area. He reviews

the conceptual models of surface complexation and the evidence for the importance of colloids in understanding trace element behavior.

By comparison with the other Group discussions (e.g., Major Bioelements Group or Inorganic and Radioactive Properties Group), the diversity of the objectives and techniques described in this section is quite large.

TABLE 6-1.
Papers in This Section Were Grouped by Whether They Dealt with Mineralogy or Surface Properties, and by Whether They Described Techniques for the Analysis of Single Particles or of Bulk (Many-Particle) Samples

	MINERALOGY	SURFACE PROPERTIES
BULK ANALYSIS	DeMaster	De Carlo
	Campbell	Honeyman
	Fischer & Wefer	
	Van der Gaast	
SINGLE PARTICLES	Bernstein & Betzer	Cowen
	Campbell	
	Jedwab	

If a greater number of mineral and surface property specialists could have been invited to the workshop, the diversity of topics, objectives and techniques could have been even greater because of the general nature of this type of research (broad, multi-disciplinary, and analytically challenging). The papers in this section, the references therein, and the Mineralogy and Surface Properties Working Group discussions should direct the reader to many, if not most, of the important analytical techniques for the study of this class of oceanic particles.

2. Group Discussions

In this section of the Mineral and Surface Properties Group Report, three aspects of marine particle chemistry are discussed including:

1) sampling and preparative techniques;
2) analytical considerations; and
3) global applications of particulate studies (i.e., shortcomings, recent breakthroughs, and promising areas and techniques of future research).

The section provides information concerning various mineral phases (carbonate, silica, and authigenic minerals) as well as a description of particle surface chemistry (e.g., dissolution, bonding, and micro-morphology). The first part of the Group Report (Sampling and Preparative Techniques) describes several precautions that must be considered when studying the calcium carbonate, biogenic silica, or clay mineralogy content of particle-trap material. In addition, splitting techniques for particle-trap samples are discussed.

2.1. Sampling And Preparative Techniques

2.1.1. Labile phases: calcium carbonate.

Calcium carbonate particles normally comprise an important fraction of the particulate load in the oceans. These particles are formed by a wide variety of organisms including: coccolithophorids, foraminifera, pteropods, heteropods, coralline algae, and bryozoans. In the open ocean these biogenic phases are primarily composed of low-magnesium calcite (coccolithophorids and foraminifera) and aragonite (pteropods and heteropods). The increased pressures and decreased temperatures attendant on increasing depth cause the water to become undersaturated with respect to these phases. Where the dissolution reaction is kinetically favored (at the "lysocline" and below), significant calcium carbonate dissolution can take place. Aragonite dissolution may occur in waters several hundred to several thousand meters shallower than calcite. Betzer et al. [1984] demonstrated that approximately 90% of the aragonite pteropod flux was dissolved in the upper 2.2 kilometers of the North Pacific Ocean. Other sources of reactive calcium carbonate in the oceans are the skeletons of shallow-water benthic organisms. These skeletons are commonly composed of aragonite and high-magnesium calcite with the latter being more soluble than their pelagic counterparts composed of aragonite and low-magnesium calcite [Chave et al., 1962; Mackenzie et al., 1983; Urmos et al., 1986]. In the Pacific Ocean undersaturation with respect to these magnesium calcites occurs at a depth of just a few hundred meters [Sabine and Mackenzie, in press, 1991]. It has been demonstrated that shallow-water sediments can be transported to the open ocean [Mackenzie et al., 1965; Friedman, 1965; Chave and Suess, 1967; Berner et al., 1976; Schlager and James, 1978; Hine et al., 1981; Heath and Mullins, 1984; Boardman and Neumann, 1984]. In fact, magnesium calcite particles have been identified hundreds of kilometers from the nearest source [Chave and Suess, 1967; Berner and Honjo, 1981].

When sampling oceanic particles, care must be taken to prevent the dissolution of carbonates. Long duration sampling devices such as sediment traps may seriously underestimate the mass fluxes of labile carbonates [Betzer et al., 1984]. If the solutions in the sediment-trap sample collection cups are undersaturated and the samples are not processed onto filters soon after collection, dissolution of carbonates can occur. An additional concern in sampling is with the common practice of acid washing equipment. Carbonate particles are very sensitive to the pH changes of the surrounding solution, therefore, it is very important to insure that all equipment is thoroughly rinsed before use, if it has been acid washed. As a general rule, to preserve the integrity of carbonate samples, any solutions that may contact the particles must be kept at or slightly above the saturation state of the most labile carbonate phase in the sample.

2.1.2. Labile phases: biogenic silica.

As outlined in the papers by v. Bodungen (this volume, page 47) and Fischer and Wefer (this volume, page 391), opaline skeletons deposited in sediment trap sampling cups seem to be susceptible to rapid dissolution (in days or weeks). In the supernatant solutions of sampling cups, silicate concentrations between 200 and 1000 μM have been measured, whereas, before deployment, dissolved silicate concentrations were only about 50 μM. The enhanced concentrations are not related to the deployment duration or to water temperature. High silicate concentrations have been measured in particle-trap sample cups from both subtropical regions and the Antarctic (see Fischer and Wefer, this volume, page 391).

Losses of biogenic opal can be estimated by assuming that dissolved silicate did not diffuse out of the sampling cups during the sampling time and after the cups were sealed. When opal fluxes to the traps are high (>50 mg m^{-2} day^{-1}), the estimated losses of biogenic opal are relatively small ($<5\%$) and can probably be neglected. If, however, opal sedimentation is low (on the order of 1 mg m^{-2} day^{-1}), more than half of the trapped opal can be dissolved during and after trapping. Thus, it is recommended to analyze the supernatant solutions for dissolved silicate to estimate opal dissolution and correct the opal flux rates.

Preferential dissolution of fragile opaline skeletons in the trap cups, however, may remove certain opal-forming organisms and leave other, more heavily silicified skeletons intact. Dissolution also may be dependent on the water content and the Al content of the sedimented skeletons [Van Bennekom et al., 1988]. Altered and resuspended biogenic opal is less susceptible to dissolution than are freshly trapped skeletons [Kamatani, 1971].

To minimize these problems, initial nutrient (silicate) concentrations of the trap solutions should be determined by taking a subsample prior to deployment and storing it at the same temperature and light conditions as those in the sampling cups. The estimates of biogenic opal dissolution are conservative because some dissolved silicate may have diffused out when the sampling cup was open. This diffusive loss depends on the salinity of the trap solution, which should be enhanced by a few ‰. Higher salinities in the trap solutions also retard the loss of poisons/preservatives during the sampling period.

For the calculation of silicate losses due to diffusion, internal tracers (e.g., radioisotopes) can be added before deployment, or the changes in sampling cup salinity can be used for this purpose. For certain types of research that require quantitative preservation of easily soluble skeletons, dissolved silicate can be added to the sampling cups prior to deployment at concentrations of about 800 to 1000 μM. This procedure can prevent dissolution of fragile opaline particles. On the other hand, mineral reactions and/or precipitations in the cups due to high dissolved silicate concentrations may occur. This is an area for further experiments and research.

2.2. Selective Sampling by Field Equipment

XRD patterns of particle trap material from SEEP-I traps (North Atlantic, Middle Atlantic Bight) and from a trap in Indonesian waters showed only strong, sharp peaks and an almost complete absence of smectite. This is an indication that small mineral particles were not present in the sediment trap samples. Surface sediments from both areas, however, contain considerable amounts of smectite and display broader peaks in their x-ray patterns, indicating that small particles are transported through the water column to the seabed.

These observations suggest that sediment traps preferentially collect the large mineral particles (greater than 5 or 10 μm) or, in other words, the traps discriminate against small mineral particles. This selective sorting also can be expected to occur with biogenic particles. Alternatively, since the traps in which this effect was noted were either in a shelf environment (Indonesia) or in a dynamically active near-bottom slope environment (SEEP-I, Middle Atlantic Bight), the preferential sorting may be due to strong particle size gradients resulting from resuspension events.

2.3. Pressure and Temperature Effects On Sampling: Carbonate Dissolution

Pressure and temperature are two of the important variables that control the reaction rates of carbonate phases in seawater [Acker et al., 1987; Acker and Byrne, 1989]. In fact, in the deep ocean (depth greater than 1.5 km) water masses are nearly isothermal ($3 \pm 2°$ C), and there are only modest changes in salinity (34 to 36 ppt). As a result, pressure exerts the major influence on the value of saturation concentration for the carbonate ion. The change in reaction rate over a 6000 meter water column can be as much as fourfold [Acker et al., 1987]. Thus, particle-trap sampling strategies for carbonate phases should make note of the relationship between the saturation state and dissolution rates for the three major classes of biogenic carbonate phases produced in the ocean, i.e., low-magnesium calcite, aragonite, and high-magnesium calcite. The important point is that much of the deep ocean is only 50% saturated with respect to these phases with the result that as much as 30% of the mass of a pteropod, foraminifera, or coccolith can be dissolved in a single day. In short, the sampling problem for these phases is much like that for celestite (strontium sulfate) in the sense that the dissolution residence times in much of the ocean are at the very most a few days.

2.4. Splitting of Particle Trap Samples

Large sediment trap samples (≥ several hundreds of mg dry weight) can be split into aliquots prior to analysis by any of several wet splitting techniques. Wet splitting by rotating cups is the procedure probably used by most investigators. At Bremen University a rotary liquid splitter equipped with eight cups is used (see Fischer and Wefer, Figure 6-25, this volume, page 392). If the sample contains a substantial amount of the fraction greater than 1 mm, non-swimmer material, this fraction is separated first from the rest of the sample by sieving. Subsequently, the sieved material is added to a large sample tube of seawater, one end of which is divided into four segments. The sample is agitated into suspension and then allowed to settle into the 4-segment end of the tube.

Recently, a wet-splitting device that uses a peristaltic pump was described by Heussner et al. [1990]. This system takes advantage of the precise volume delivered by each squirt of the pump to divide the sample into as many variable-sized aliquots as desired. Each aliquot consists of many squirts of the pump, which are sequentially added to as many beakers as the number of desired aliquots, while the entire sample is being agitated and recirculated through the pump to maintain a homogeneous suspension. This technique is used on samples from which swimmers have been removed, and physical degradation of the particles being split was not apparent. The precision of this rapid, clean splitting procedure is at least as good as that for other splitting techniques and has averaged less than 5% for trap samples and less than 1% in tests using fine-grained sediments.

3. Analytical Considerations

This section of the Group Report discusses the analytical procedures for several minerals (i.e., clay minerals, biogenic silica, and carbonate) commonly found in the marine environment. When multiple techniques exist in the literature for a particular phase, the advantages and disadvantages of the individual procedures are described. The section concludes by examining filtration as an analytical procedure that may alter certain surface particle properties (e.g., double layer thickness or various adsorption sites).

3.1. Clay Mineralogy (see Working Group paper by Van der Gaast, page 343)

3.2. Biogenic Silica (see Working Group paper by DeMaster, page 363)

3.3. Calcium Carbonate

Carbonates often constitute a significant portion of marine particulate matter and sediments; hence, a variety of different approaches have been developed to meet either the analytical or sample preparation needs of various research projects. The techniques applied to carbonate-rich material generally involve a dissolution or decomposition procedure with the intent of either accurately determining the carbonate content of the material or of eliminating it so that other constituents, often present in lesser quantities, can be studied without interference or dilution effects.

Several papers within this workshop volume present approaches that are individually valid for a specific purpose. These procedures, however, often include treatments that may not be compatible with the particular needs of another project. For example, Fischer and Wefer (this volume, page 391) describe a method that involves the combustion of 2 to 4 mg of particle trap material for the determination of total N and C and treatment of a similar amount of material with 6 N HCl in a silver boat followed by combustion for the determination of organic C. This technique is obviously not amenable to any additional analysis of the particle trap material. These authors also describe the removal of organic matter and decalcification of samples prior to clay mineral analysis by a sequential treatment with 15% H_2O_2 and 20% acetic acid. Van der Gaast (personal communication, 1991) suggested that such a treatment should be carried out with extreme care because the use of H_2O_2 and/or unbuffered acetic acid can attack chlorite (and other clay minerals) and cause problems in subsequent clay analysis.

Honjo [1980] reported the use of 0.5 N HCl for the removal of carbonates from sediments or trap material. Unfortunately, if the sediment material contains coatings of Fe or Mn (oxides or hydroxides), they also will be attacked and any subsequent elemental analysis of the leach solution will include material from other phases. Additionally this treatment also affects some clay minerals, particularly chlorites, thereby precluding their accurate analysis.

Treatment of sediment trap material with vapors of concentrated HCl has also been reported in the literature [Hedges and Stern, 1984]. The effect of this procedure on clay mineral assemblage and surface properties is unknown at present. One difficulty with this technique is that the addition of HCl forms hydrous salts of calcium chloride, which can have variable composition and are difficult to weigh accurately. Recently, Verardo et al. [1990] suggested using an 8% sulfurous acid solution for removal of calcium carbonate prior to organic carbon determination on a Carlo Erba Elemental Analyzer. The sulfurous acid avoids the formation of hydrous salts, and drying the acidified samples avoids loss of acid-soluble organic carbon.

The paper by De Carlo and Koeppenkastrop (this volume, page 419) presents a sequential leaching scheme involving a gentler attack of carbonates using a sodium acetate/acetic acid, pH 5 buffer, leach. While acceptable for major element analysis of the leacheate and subsequent analysis of clay minerals, this procedure is unsatisfactory if the leach is to be analyzed for trace elements. This is the result of trace metal impurities in most reagent grade sodium acetate. Presumably the sodium acetate solution could be purified using Chelex 100 to remove contaminants prior to use. Buffering the system to pH 5 with ammonium acetate, prepared from ultrapure NH_3 and acetic acid, however, would allow reliable trace element analysis, but is unacceptable if clay minerals are to be analyzed in the solid residue.

All the above treatments have a further disadvantage in that they are unable to differentiate between different types of carbonates such as calcite, high-Mg calcite, and aragonite. Additionally, dolomite, found occasionally in marine particles, is known to be more resistant than other carbonates and is not dissolved by the pH 5, buffered leaching procedure or by the acetic acid (pH = 2.8) solution used by Betzer et al. [1977].

3.4. Filtration as an Analytical Technique: Surface Property Considerations

Filtration as a method for separating particles from bulk solution is coming under increasing scrutiny because of the existence of filtration artifacts, [see Buffle, in press, 1991; and references therein]. One potential artifact not usually considered is the displacement of surface reactions from equilibrium as a consequence of particle-particle interactions which result from particle compaction on or near the filter bed. While there is little direct experimental evidence with which to judge the magnitude of this artifact, theoretical arguments by Prieve and Ruckenstein [1978] and others suggest that double-layer overlap and the resulting "regulation" of ion activities within the interfacial region may significantly alter the distribution of ions associated with particle surface sites.

For example, consider two particles falling through the water column. The surface charge of each particle, the result of interactions of particle sites with components of the surrounding solution, is balanced by the accumulation of ions in the diffuse layer. As these particles come into close proximity on the filter surface, inter-particle fluid will be excluded. The loss of electrolyte solution will result in an imbalance of charge, and ions, either sorbed or in the remaining diffuse layer, will adjust their concentrations to restore the charge balance. The result would be a shift in the sorbed concentration of particle-reactive elements. This alteration of surface charge and ion activity could produce element particle/solution distributions (e.g., Kd's) that are different from the corresponding elemental partitioning in bulk solution. The extent of the Kd perturbation, if it occurs, would depend on a variety of factors including the nature of the association of the sorbing material with particle surfaces and the general solution composition.

4. Global Applications of Particulate Studies (GAPS): Shortcomings, Concerns, and New Vistas

This part of the Group Report describes areas of research that are being developed currently or need to be developed in the near future to address major questions of scientific concern. As part of the discussion regarding mineralogy, studies of authigenic and hydrothermal minerals are examined. The later part of the "GAPS" section reviews surface properties and reactions. The chemistry of carbonate, biogenic silica, and metal-oxide surfaces is described, in addition to a discussion of particle complexation sites, ligand chemistry, micromorphology, and the complex interaction of functional groups in solution and on particle surfaces.

4.1. Mineralogy: Authigenic Minerals

Authigenic mineral formation has been discussed in the oceanographic literature for several decades [Mackenzie and Garrels, 1966; see Berner, 1980, for review]. In deep-sea sediments, precipitation of aluminosilicate phases (e.g., sepiolite or clinoptilolite) and barite has been documented by several investigators [e.g., Biscaye, 1965; Hurd, 1973; Church and Wolgemuth, 1972]. Although these authigenic phases generally comprise a small fraction of the deep-sea sediment, they may control to a significant degree the major or minor element chemistry of pore waters and in some cases affect water column chemistry [Willey, 1975]. In the past five to ten years, studies of authigenic mineral formation have examined nearshore environments including estuarine environments [Mackin and Aller, 1984a, b; Mackin, 1986] and salt marshes [Mackin, 1987]. These studies suggest that authigenic mineral formation not only may control the chemistry of certain pore water constituents (e.g., Al, Si, or pH), but also may affect the concentration of certain chemical species up in the water column (i.e., in the river/ocean mixing zone). Additional research is needed to determine the extent of chemical removal by these processes in the ocean and also to evaluate whether or not authigenic mineral formation is typical of most estuarine environments. Most recently, studies of authigenic mineral formation have examined the occurrence of barite in marine waters [Dehairs et al., 1980; Bishop, 1988; Dehairs et al., 1990; Stroobants et al., in press]. The general conclusion from these studies is that barite forms in micro-environments within degrading siliceous aggregates and that the abundance of the mineral in the water column is indicative of past primary production rates.

4.2. Mineralogy: Hydrothermal Minerals and the Application of Mössbauer Spectroscopy and Synchrotron X-ray Diffraction

Mössbauer spectroscopy has been used with considerable success to characterize Fe mineral phases, but generally has not been applied to the study of marine particles. The technique can provide information on the oxidation state, identity, and crystallinity of Fe phases and recently has been used to characterize the mineralogy of hydrothermal vent "black smoker" particles (see Campbell, this volume, page 375). Determining the mineralogy of hydrothermal plume particles is problematic because of their heterogeneous composition and the presence of both well crystallized phases and extremely fine grained, crypto-crystalline phases. Low temperature Mössbauer spectroscopy provides qualitative and quantitative information regardless of the crystallinity of the different phases and allows determination of the bulk mineralogy of the Fe containing phases. This technique may be useful for examining the mineralogy of particulate Fe phases from a variety of marine environments, including anoxic and suboxic basins, estuaries and lagoons, and at the sediment/water interface. The most advantageous application will be in providing a method of unambiguously identifying "amorphous" Fe-oxyhydroxide phases in the marine environment. The main limitation is that several mg of an Fe-containing phase are required.

Bulk characterization by X-ray diffraction (XRD) can reveal a great deal of information about crystalline phases larger than about 30 nanometers (although smaller crystallite sizes may be characterized under certain conditions; Van der Gaast, unpublished data, 1991). Synchrotron XRD (SXRD) has a number of advantages over standard systems. The high photon flux allows the use of much smaller sample amounts for powder techniques and also for single crystal techniques. The use of a monochromator with the "white" light X-rays from the synchrotron allows the selection of specific wavelengths appropriate for different samples and also permits much narrower bandwidths than traditional X-ray sources. The result is that SXRD has a very high resolving power and that closely spaced diffraction lines, which normally overlap, can be completely resolved to baseline. This technique will be especially useful for studying aluminosilicates, opal/amorphous silica, carbonates, and other minerals.

4.3. Surface Properties: Calcium Carbonate Dissolution

There are a host of labile particulate phases present in the ocean system that participate in important chemical cycles. One of the major challenges facing the oceanographic community is a quantitative description of the reactions (sorption/desorption, dissolution, precipitation) that take place on particle surfaces. A recent advance in our ability to accurately measure pH on a microscopic scale has been made by Byrne and coworkers [Byrne, 1987; Byrne et al., 1988; Robert-Baldo et al., 1985; Byrne and Breland, 1989; Acker et al., 1987]. The technique, which utilizes pH-sensitive dyes, is so sensitive that pH can be measured to better than ± 0.0003 units and consequently, has been used to provide continuous dissolution rate data on individual pteropods [Acker, 1988]. The technique also could be applied to the mineralogically and chemically distinct suite of biogenic carbonate phases (low-magnesium calcites, high-magnesium calcites, and aragonite) that are produced in the ocean system. Temperature-controlled, high-pressure vessels, equipped with optical ports, provide straightforward means of establishing the specific response of individual mineral phases to varying pressure and temperature conditions. The important point is that the reaction rates of freshly collected microscopic organisms can be monitored continuously at in situ temperatures and pressures. This represents a major advance in our ability to define the reaction rates of the carbonate phases (foraminifera, coccolithophores, pteropods, heteropods, selected corals, and the suite of high magnesium calcites which are produced by coralline algae and bryozoans).

4.4. Surface Properties: Dissolution Kinetics of Carbonate Fluorapatite and Carbonate Hydroxyapatite

The reactivity of natural carbonate fluorapatites in aqueous solutions is being investigated by Lane and Mackenzie at the University of Hawaii. Of primary interest are the effects that variations of solution and mineral composition have on reaction rates and mechanisms. For example, the carbonate concentration of apatite is known to significantly alter mineral solubility. The experimental approach is to react well-characterized apatite grains in a fluidized bed reactor [Chou and Wollast, 1984]. The system is useful in studying dissolution reactions because the reaction can be carried out far from saturation, which avoids the precipitation of secondary phases.

Preliminary results from the analysis of reacted solutions indicate that apatites dissolve incongruently [Lane and Mackenzie, 1990]. As a consequence of this phenomenon, a surface layer depleted in Ca, F and carbonate, and enriched in P is formed. The existence of an altered surface layer is consistent with systematic changes in solution chemistry. Further decreases in pH rapidly dissolve the outer edge of the depleted surface. Localized etching and pitting are evident when SEM micrographs of unreacted and reacted grains are compared, indicating that the mineral surface is attacked preferentially. Because ions in solution must cross the depleted surface to reach any underlying unreacted mineral, the depleted surface layer controls reaction rates. The composition of the surface layer is currently being investigated using spectroscopic techniques.

4.5. Surface Properties: Dissolution Kinetics of Biogenic Silica

In natural waters the dissolution kinetics of biogenic silica are a function of temperature, pH, particle size, effective surface energy, physical occlusion of the surface by organic or inorganic matter, degree of saturation of solution and salinity [Hurd, 1972, 1973; Hurd and Theyer, 1975; Lawson, et al., 1978; Hurd, 1983a, b; and references therein].

Within the world's oceans, dissolved silica concentrations do not exceed 15-20% saturation and are generally much less; changes in the concentration of dissolved silica in seawater cannot exert more than a 20% effect on dissolution rate. The exception to this condition would be restricted environments such as the sample cups within sediment traps, and perhaps micro-environments within aggregates. It has been calculated that low temperature, low pH, fresh waters may have dissolution rate constants approximately 4 orders of magnitude lower than higher temperature, higher pH, surface seawaters [Hurd, 1983a]. This type of information and understanding is important especially in the calculation of dissolution rates in estuarine environments, where a large range of solution conditions can exist.

Salinity and pH conditions in the world's oceans vary over relatively narrow ranges; laboratory experiments [Hurd, 1973] suggest no more than a factor of two variability in dissolution rate constants resulting from oceanic pH changes of 7.5 to 8.3, and no practical effect of the ocean's narrow salinity range on these constants. Pressure increase with depth in the oceans increases the solubility of biogenic silica [Hurd, 1983a, and references therein], but only by approximately 10%, resulting in another minor effect on the dissolution rate constants. Temperature changes from 2°C to 25°C increase the dissolution rate constants by about a factor of 10; the effect of temperature on the preservation of skeletons is important and will be most noticeable in the upper 1000 meters of the water column [Hurd, 1973].

The preservation of individual siliceous skeletons in seawater will be dependent both on the above solution conditions and the physical (ultra-)structure of the skeleton or frustule. When siliceous skeletons are initially formed, small 3-5 nm spheroids are precipitated and aggregate to form the skeleton. Although many skeletons appear solid, certain groups such as Phaeodarian radiolaria have complex internal structures as well [Hurd, 1983a and references therein]. When dissolution of the skeleton occurs, the surface becomes porous and the effective particle size as measured by gas adsorption becomes smaller [Hurd, 1983b and references therein]. Qualitative estimates of specific surface area can be estimated by dye adsorption [Hurd et al., 1981]. Those skeletons with small effective particle sizes will disappear more rapidly than those with larger ones, and at a certain point when the particles become very small, dissolution rates will begin to increase because of surface energy considerations [Hurd, 1983b].

4.6. Surface Properties: Inhibition of Biogenic Silica Dissolution

Chemical phases forming on or adsorbing to the surface of biogenic silica affect the dissolution of this material in the water column and probably control the ultimate magnitude of pore water silicate values. Although most biota use an organic coating to retard the dissolution of their biogenic silica skeletons, soon after death this material is broken down by microbial or zooplankton activities, leaving the siliceous surfaces directly exposed to seawater. The fact that marine pore waters never attain saturation with respect to biogenic silica and exhibit significant variability in maximum concentration (300 to 900 μM) suggests that a surface phase is retarding dissolution [Schink et al., 1975]. Several investigators [Van Bennekom et al., 1989; Hurd, 1973; Mackenzie et al., 1967; Mackin, 1987; Willey, 1975] have suggested that some type of cation alumino-silicate may be responsible for retarding silica dissolution, whereas Kent and Kastner [1985] indicate that magnesium hydroxysilicate may be the cause of the phenomenon. In the future analytical techniques such as Auger spectroscopy, SEM/EDXA, and Synchrotron XRD may be used to examine the chemical nature of surface phases that exist in the water column and in sediments.

4.7. Surface Properties: Metals and Minerals Associated With Bacteria

Sorption of metals onto bacterial cell surfaces is well documented [Ghiorse and Hirsch, 1979; Nealson and Ford, 1980; Beveridge and Fyfe, 1985; Cowen and Bruland, 1985; Cowen and Li, in press, 1991]. Microbial exudates may also be important in scavenging colloidal materials as demonstrated in the binding of colloidal MnO_2 by bacterial extracellular polysaccharides [Sly et al., 1990]. These mechanisms have important implications with regard to the transformation, distribution and ultimate fate of metals and minerals in aquatic systems [Harvey and Leckie, 1985; Cowen and Bruland, 1985; Cowen and Li, in press, 1991]. Such metal sorption has the apparent effect of stabilization of the polymeric extracellular structures. Low-temperature diagenetic transformations of bacterially produced metallo-organo-polymers have been shown to enhance mineralization of cell structures, resulting in increased preservation of cells in the sedimentary record [Beveridge et al., 1983; Ferris et al., 1986].

Studies of microbe-mineral interactions have included both laboratory and field experiments. Laboratory experiments generally involve the use of cultured isolates. Thus, Hastings and Emerson [1986] employed the catalytic properties of a Bacillus spore to observe the accelerated process of Mn(II) oxidation and mineral maturation from Mn(II) to Mn_3O_4 (hausmanite) to MnO_2. Other bacterial cultures have been used to study the effects of available attachment surfaces [Nealson and Ford, 1980] and a wide range of growth conditions [see Nealson, 1983 for review] on the deposition and oxidation of manganese and iron oxides. Sorption of metals, including Mn, Fe, Cu, Ni, Pb, U and others, onto cell walls has been studied using suspensions of whole cells [Bromfield and David, 1975; Harvey and Leckie, 1985; Nealson and Ford, 1980] or of isolated cell walls [Beveridge and Fyfe, 1985] or biofilms [Ferris et al., 1989]. Beveridge et al. [1983] simulated low-temperature sediment diagenesis using bacterially complexed heavy metals as the sole organic and heavy metal source. Sly et al. [1990] studied the binding of colloidal MnO_2 by the extracellular polysaccharides produced in a culture of fresh water bacteria.

Field studies primarily have involved radiotracer uptake experiments and direct observations of microbe-metal/mineral associations using electron microscopy. Both total and microbially mediated rates of metal binding have been studied using radiotracers in anoxic fjords

[Emerson et al., 1982; Tebo et al., 1984; Tebo and Emerson, 1985], hydrothermal vent plumes [Cowen et al., 1990], and the surface photic zone [Sunda and Huntsman, 1987]. These studies have employed poisons, usually azide, a metabolic inhibitor, and temperature experiments to distinguish metabolic mediation from abiological oxidations.

Direct diagnostic observations of microbe-mineral associations require the use of transmission electron microscopy (TEM) in order to obtain the ultrastructural evidence of the microbial component. A standard method for this approach is discussed by Cowen (this volume, page 429). Modifications of this technique have been used to study 1) bacteria as nucleation sites for the authigenic formation of metal sulfides and silicates in sediments [Ferris et al., 1987, 1989; Beveridge and Fyfe, 1985]; and 2) bacterial metal associations in hydrothermal vent particle plumes [Cowen and Li, in press, 1991], throughout the water column in the Eastern Subtropical Pacific [Cowen and Bruland, 1985], and in marine snow [Cowen and Silver, 1984].

4.8. Surface Properties: Effect of Time on Chemical Equilibrium and Surface Complexation

The accurate determination of element distribution between particle and solution phases at equilibrium is critical for two reasons. First, attainment of chemical equilibrium is a necessary requirement for estimating the intensity of interaction between sorbates and reactive sites on particles. Second, knowledge of the extent to which a system is displaced from equilibrium conditions is needed, in conjunction with required transformation rate constants, for calculating system response time.

Studies of the mechanism of adsorption reactions by chemical kineticists [e.g., Yasunaga and Ikeda, 1986] have shown that metal ions rapidly adsorb, i.e., form co-ordinative complexes, with particle surface sites. The relaxation times for such reactions are on the order of milli-seconds. In contrast to these studies, sorption rate experiments with natural particles (or rate experiments attempting to simulate natural conditions) have produced data with a more complex, and baffling, kinetic behavior. Typically, sorption is characterized by an initial rapid uptake which is followed by a slower, extended removal of the sorbate from solution lasting days to weeks or longer [e.g., Li et al., 1984; Nyffeler et al., 1984; Jannasch et al., 1988; Moran, 1990; Koeppenkastrop and De Carlo, 1990, in press, 1991, Moran, submitted, 1991]. Analysis of such data suggest that sorption likely consists of several trace-element controlling processes coupled in series or acting in parallel [e.g., Jannasch et al., 1988]. These processes include adsorption, absorption and surface precipitation. Discerning the controlling sorption process solely by application of mathematical models to existing data is likely to be inconclusive (Koeppenkastrop and De Carlo, submitted, 1991) because of the mathematical equivalency of process representation (e.g., first order chemical kinetics and diffusion-controlled kinetics [Nkedi-Kizza et al., 1984]).

Sorption rates, determined in situ using radioactive parent/daughter disequilibrium [e.g., McKee et al., 1986; Honeyman et al., 1988], also seem at odds with physico-chemical theory and studies of sorption mechanism because of the relatively small values for sorption constants (sorption times of hours to weeks).

This kinetic behavior has significant ramifications for:

1) modeling of sorption processes;
2) design of laboratory experiments and application of the results to natural systems; and
3) understanding the underlying process(es) contributing to element sorption.

4.9. Surface Properties: Polyfunctional Surface Characteristics

Hochella [1990] recently has summarized the state of knowledge of atomic structure and the reactivity of mineral surfaces. Mineral surface reactivity is dependent on three surface characteristics:

1) chemical composition;
2) atomic structure; and
3) particle microtopography.

Laboratory studies of metal-ion adsorption have suggested that particle surface sites form complexes, of varying intensity, with particle-reactive solutes [e.g., Benjamin and Leckie, 1981]. While a number of methods exist for analyzing mineral surface reactivity, Hochella [1990] states that the role atomic structure plays in affecting the sorption of ions from solution only can be inferred at this time.

An important component of modelling reactive element scavenging by marine particles is the measurement of adsorption constants based on laboratory analyses using specific mineral phases (e.g., Fe, Mn, or Al oxides). Recent work (Hunter and Balistrieri, unpublished, 1990) suggests that laboratory measurements may have been compromised through contamination of the oxide surface by metal ions strongly adsorbed on high energy sites. This effect would mask the surface site properties that are important for reactive element scavenging in the ocean where elemental concentrations are very low. Hunter and Bird (unpublished, 1990) have recently found that 40-50% of naturally occurring ^{234}Th is irreversibly adsorbed to synthetic Mn oxide in seawater electrolytes. These preliminary results suggest that a careful re-examination of ion adsorption studies, which pays attention to the preparation of clean surfaces, is desirable.

4.10. Surface Properties: Adsorption in Complex Mixtures of Sorbents

One of the main objectives of aquatic surface chemistry has been to understand the behavior of elements in natural systems from a knowledge of their basic chemical properties. An expression of this objective, with regard to heterogeneous systems, is the idea that overall the sorptive characteristic of a mixture of sorbents is the "sum" of the sorptive characteristics of the individual sorbents. In other words, the distribution of a sorbate among a suite of particle types is analogous to the partitioning of an element in a homogeneous solution among dissolved ligands [e.g., Luoma and Davis, 1983; Honeyman, 1984].

Experimental work over the past 5 years [e.g., Honeyman, 1984; Anderson and Benjamin, 1990] suggests that the observed sorptive behavior of complex mixtures of adsorbents may reflect interactive effects at odds with the concept of linearly cumulative sorptive properties. Such interactive effects may produce aggregate sorptive behavior markedly enhanced [Honeyman and Santschi, 1988; Schindler, 1990] or reduced [e.g., Anderson and Benjamin, 1990] relative to expectations based on individual component behavior. These interactive effects are particularly likely when a particle consists of an underlying mineral matrix covered by an organic coating (Hunter, this volume, page 251).

The existence of interactive (i.e., "non-linear") effects suggests that an evaluation of approaches to studying sorption of trace elements onto natural particles must be made. For example, will the development of methods to study particle aggregates as a whole ultimately be more fruitful than a more limited approach focusing on discrete particles?

4.11. Surface Properties: Interactive Effects of Particles, Colloids and Solutes

Present mechanistic explanations for reactive element scavenging by marine particles emphasize the importance of colloidal particles as scavengers and their aggregation to form larger, sinkable, particles. However, direct experimental evidence for the presence of reactive elements in this colloidal fraction is relatively scant (see Honeyman, this volume, page 437). Moore and Hunter [1985], for example, used different pore size filters to study marine particles that had been labelled with ^{234}Th by adsorption. They concluded that an insignificant amount of the colloidal ^{234}Th was present in the size range greater than 0.1 μm, implying that any colloidal Th must be present as much smaller particles. This is consistent with more recent studies of colloidal DOM carried out by ultrafiltration which indicates that most of the colloidal DOM is relatively small, i.e. less than 30,000 MW [Moran and Moore, 1989].

Partitioning of particulate, colloidal and dissolved materials by filtration and/or ultrafiltration procedures is potentially confounded by artifacts associated with the filtration. Colloids exist in the form of a self-regulated, particle size spectrum maintained by continuous aggregation to replace larger particles lost from the water column by settling or ingestion. Aggregation theory predicts that at steady state, the shape of the particle size distribution should be relatively constant, but will be considerably disturbed by filtration, especially with filters of smaller pore size. This is expected to initiate aggregation at an increased rate to restore the steady state size spectrum, thus producing further filterable particles. This expectation is consistent with many experimental observations when water samples are re-filtered or when two identical pore size filters are used in tandem. However these effects may also arise because of the way in which filters incompletely retain particles having sizes similar to the nominal pore size. The typical rate of filtration-induced aggregation in seawater systems is not known, but could probably be estimated using digital simulation methods [e.g. Hidy, 1965]. Clearly it has important implications for understanding DOC/POC partitioning, as well as other major components of the DOC pool (N, P, and perhaps S). The significance of this effect to the problem of reactive element scavenging must await the completion of studies using cross-flow ultrafiltration that will be carried out in the near future.

In addition to the role that colloidal material plays in the speciation of particle-reactive materials, the physical behavior of colloids, and the methods used to ascertain the quantities and composition of such material, can have fundamental implications on our understanding of the fate of particle-reactive elements in the oceans. A good example of this effect is trace element scavenging. If colloid/large particle aggregation is the rate-limiting step in trace-element scavenging [Honeyman and Santschi, 1989], then the particle sizes used as an operational definition of "particulate" fraction can directly affect derived scavenging rate constants. In this regard, Moran [1990, submitted, 1991] has demonstrated a qualitative agreement between oceanic scavenging data for both Al and Th and removal rates predicted by the Brownian-pumping model for reactive metal scavenging [Honeyman and Santschi, 1989]. It should be emphasized that while most studies of particle aggregation in the oceans have focused on mechanisms of particle-particle interaction, few have examined the effects of coagulation on sorbed trace elements.

4.12. Surface Properties: Marine Particle Applications of Synchrotron X-ray Absorption Spectroscopy

Synchrotron X-ray spectroscopy has significant applications to marine geochemistry, especially particulate studies. Although it is generally accepted that adsorption and chemisorption onto particle surfaces provide a major removal mechanism for dissolved elements from seawater, little direct evidence exists about the nature of these sites with respect to specific elements or about the chemical transformations that occur. For example, does the organic coating thought to cover marine particles [Balistrieri et al., 1981] affect most uptake sites or are mineral surfaces primarily important (or, perhaps, a synergistic interaction of organic and inorganic species)? Because of the potential complexity of natural systems, this approach will require use of model complexes and experiments.

Analysis of the Extended X-ray Absorption Fine Structure (EXAFS) provides information about the number and type of ligands and bond distances (to about 10A) between an absorbing atom and neighboring atoms that scatter the outgoing fluorescent X-ray photon. This is carried out by a Fourier analysis of the fine structure on the high energy side of the element's K-absorption edge [Stern, 1974; Konigsberger and Prins, 1987; Campbell, this volume, page 375]. Analysis of the X-ray Absorption Near Edge Structure (XANES) provides information about the absorbing atom's oxidation state, coordination symmetry and ligand covalency from the energy position of the absorption edge and the bound state transitions on its low energy side [Bianconi et al., 1983]. Together they provide an extremely powerful, non-destructive technique for elucidating the speciation and coordination environment of metals in different particulate phases.

The use of these techniques has already been established for natural samples with complex matrices including: V in asphaltenes [Goulon et al., 1984], V in nitrogenase [George et al., 1988], S and other elements in coal [Spiro et al., 1984], As in oil shale [Cramer et al., 1988a], Mn in photosynthetic enzymes [George et al., 1989] and Mn-Cu phases from submarine Mn crusts [Stouff and Boulègue, 1989]. The method also has been used to characterize different natural and synthetic Fe and Mn oxides [Manceau and Combes, 1988], to study the formation of ferric gels and their transformation to hematite [Combes et al., 1989, 1990] and most recently to characterize the Fe-oxide phase in submarine hot spring "black smoke" [Campbell et al., 1990; Campbell, this volume, page 375].

The technical features of current EXAFS systems make the approach attractive for particle studies. Since the technique is element specific, it requires no sample pre-treatment to alter the matrix. Hence, artifacts due to sample preparation are minimal. Elements from P to Sr are amenable to analysis with existing systems. The analysis is made in air, or, if desirable, under He; thus wet samples, biological samples, and even fluid samples can be analyzed. Currently the detection limit is about 10 to 100 ppm on a few mg (~ 5) of bulk material run in the transmission mode. In the fluorescence mode [Cramer et al., 1988b; Campbell, this volume page 375] the amount of material needed can be reduced to 50 to 100 μg, and the detection limits are improved (1-10 ppm).

Work is currently in progress to develop single particle (~ 25 μm) analysis and preliminary results are very encouraging (S. Sutton, personal communication, 1990). XANES studies of the M-absorption edge can provide detailed oxidation state and bonding information because it is much more sensitive to valence electronic states than the K edge [George et al., 1989; S. Cramer, unpublished data, 1990]. The on-going construction of more powerful synchrotron sources, such as the Advanced Photon Source at Argonne National Lab, will increase the versatility of this technique by allowing K-edge studies for virtually the entire periodic table, decrease detection limits to the ppb range, and allow single particle studies to be done at the micron-size range.

4.13. Surface Properties: Reaction Rates as a Function of Temperature and Pressure

World ocean surface temperatures vary from near -1.8° C in polar regions to a maximum of 35° C in certain tropical areas. Temperature gradients with depth yield a similar range. Associated with this vertical change is a pressure gradient (1 atm P 10 m^{-1}) leading to pressures exceeding 700 bar in deep trenches. The influence of temperature on the rates of chemical reactions is usually interpreted in terms of the Arrhenius equation [Laidler, 1965; Stumm and Morgan, 1981], whereas pressure effects commonly are treated according to Denbigh [1971]. Other than the recent work by Koeppenkastrop and De Carlo [1990], who studied the kinetics of rare earth element (REE) sorption onto synthetic mineral phases in seawater as a function of temperature in the range 2-40° C and the studies of Tebo and Emerson [1985] who evaluated the kinetics of bacterially mediated metal uptake in restricted fjords, the influence of temperature variations on oceanic sorption processes has not been systematically examined. Similarly, except in the work of Cowen [1989], pressure effects have not been considered in most experimental adsorption studies in the oceanic environment.

Work by Acker et al. [1987] and Acker and Byrne [1989] has revealed that both temperature and pressure are important variables controlling reaction rates of carbonate phases in seawater. This aspect of P and T effects on reaction rates is addressed in the Sampling and Preparative Techniques section of this chapter.

4.14. Surface Properties: Dissolution Rates and Surface Roughness

Surface roughness appears to exert an important influence on the dissolution rate of pteropods in the marine environment [Acker and Byrne, 1989]. These authors found, even at constant saturation state, that increasing the surface roughness of pteropod shells resulted in substantially increased dissolution rates. While there is no data to indicate whether this phenomenon is characteristic of other mineral phases, it would not be unreasonable to expect that the surfaces of other particles, which have been modified by dissolution, also might undergo substantial shifts in reactivity.

Acknowledgements. The authors wish to express their thanks to the other members of the Mineralogy and Surface Properties Working Group for their participation and excellent contributions. This work was undertaken while the authors were supported under Dept. of Energy Grant No. DE FGO2-87-ER60555 (PEB), and National Science Foundation Grant No. OCE-8812975.

References

Acker, J. G., Factors influencing the dissolution of aragonite in the oceanic water column, Ph.D. dissertation, 249 pp., University of South Florida, St. Petersburg, Florida, 1988.

Acker, J. G., and R. H. Byrne, The influence of surface state and saturation state on the dissolution kinetics of biogenic aragonite in seawater, Am. J. Sci., 289, 1098-1116, 1989.

Acker, J. G., R. H. Byrne, S. Ben-Yaakov, R. A. Feely, and P. R. Betzer, The effect of pressure on aragonite dissolution rates in seawater, Geochim. Cosmochim. Acta, 51, 2171-2175, 1987.

Anderson, P., and M. M. Benjamin, Surface and bulk characteristics of binary oxide suspensions, Environ. Sci. Technol., 24, 692-698, 1990.

Balistrieri, L. S., P. G. Brewer, and J. W. Murray, Scavenging residence times of trace metals and surface chemistry of sinking particles in the deep ocean, Deep-Sea Res., 28, 101-121, 1981.

Benjamin, M. M., and J. O. Leckie, Multiple-site adsorption of Cd, Cu, Zn and Pb on amorphous iron oxyhydroxide, J. Colloid Interface Sci., 79, 209-221, 1981.

Berner, R. A., Early Diagenesis: A Theoretical Approach, 242 pp., Princeton University, Princeton, New Jersey, 1980.

Berner, R. A., and S. Honjo, Pelagic sedimentation of aragonite: its geochemical significance, Science, 211, 940-942, 1981.

Berner, R. A., E. K. Berner, and R. S. Keir, Aragonite dissolution on the Bermuda pedestal: its depth and geochemical significance, Earth Planet. Sci. Lett., 30, 169-178, 1976.

Betzer, P. R., D. W. Eggimann, K. L. Carter, D. R. Kester, and S. B. Betzer, Seasonal patterns in suspended calcium carbonate concentration during the dry and wet seasons in the eastern Caribbean, in Fate of Fossil Fuel and CO_2 in the Oceans, edited by N. R. Andersen, and A. Malahoff, pp. 63-79, Plenum Press, New York, 1977.

Betzer, P. R., R. H. Byrne, J. G. Acker, C. S. Lewis, R. R. Jolly, and R. A. Feely, The oceanic carbonate system: a reassessment of biogenic controls, Science, 226, 1074-1077, 1984.

Beveridge, T. J., and W. S. Fyfe, Metal fixation by bacterial cell walls, Can. J. Earth Sci., 22, 1893-1898, 1985.

Beveridge, T. J., J. D. Meloche, W. S. Fyfe, and R. G. E. Murray, Diagenesis of metals chemically complexed to bacteria: laboratory formation of metal phosphates, sulfides and organic condensates in artificial sediments, Appl. Environ. Microbiol., 45, 1094-1108, 1983.

Bianconi, A., L. Incoccia, and S. Stipcich, EXAFS and Near Edge Structure, 420 pp., Springer Verlag, West Berlin, 1983.

Biscaye, P. E., Mineralogy and sedimentation of recent deep-sea clay in the Atlantic Ocean and adjacent seas and oceans, Geol. Soc. Am. Bull., 76, 803-832, 1965.

Bishop, J. K. B., The barite-opal-organic carbon association in oceanic particulate matter, Nature, 332, 341-343, 1988.

Boardman, M. R., and A. C. Neumann, Sources of periplatform carbonates: Northwest Providence Channel, Bahamas, J. Sediment. Petrol., 54, 1110-1123, 1984.

Bromfield, S. M., and D. J. David, Sorption and oxidation of manganous ions and reduction of manganese oxide by cell suspensions of a manganese oxidizing bacterium, Soil Biol. Biochem., 8, 37-43, 1975.

Buffle, J., The use of filtration and ultrafiltration for size fractionation of aquatic particles, colloids and macromolecules, IUPAC Commission on Environmental Chemistry, in press, 1991.

Byrne, R. H., Standardization of standard buffers by visible spectrometry, Anal. Chem., 59, 1479-1481, 1987.

Byrne, R. H., and J. A. Breland, High precision multiwavelength pH determinations in seawater using cresol red, Deep-Sea Res., 36, 803-810, 1989.

Byrne, R. H., G. Robert-Baldo, G. Thompson, and C. T. A. Chen, Seawater pH measurements: an at-sea comparison of spectrophotometric and potentiometric methods, Deep-Sea Res., 35, 1405-1410, 1988.

Campbell, A. C., R. G. Burns, J. M. Edmond, H. Elderfield, and M. Greaves, Geochemistry and mineralogy of bouyant hydrothermal plume particles: TAG and MARK vents Mid-Atlantic Ridge, EOS Trans. Am. Geophys. Union (ABS.), 71, 1651, 1990.

Chave, K. E., and E. Suess, Suspended materials in seawater, Trans. N.Y. Acad. Sci. Ser. II, 29, 991-1000, 1967.

Chave, K. E., K. S. Deffeyes, P. K. Weyl, R. M. Garrels, and M. E. Thompson, Observations on the solubility of skeletal carbonates in aqueous solutions, Science, 137, 33-34, 1962.

Chou, L., and R. Wollast, Study of weathering of albite at room temperatures with a fluid bed reactor, Geochim. Cosmochim. Acta, 48, 2205-2217, 1984.

Church, T. M., and K. Wolgemuth, Marine barite saturation, Earth Planet. Sci. Lett., 15, 35-44, 1972.

Combes, J. M., A. Manceau, G. Calas, and J. Y. Bottero, Formation of ferric oxides from aqueous solutions: a polyhedral approach by X-ray absorption spectroscopy: I. Hydrolysis and formation of ferric gels, Geochim. Cosmochim. Acta, 53, 583-594, 1989.

Combes, J. M., A. Manceau, and G. Calas, Formation of ferric oxides from aqueous solutions: a polyhedral approach by X-ray absorption spectroscopy: II. Hematite formation from ferric gels, Geochim. Cosmochim. Acta, 54, 1083-1091, 1990.

Cowen, J. P., Positive pressure effect on manganese binding by bacteria in deep-sea hydrothermal plumes, Appl. Environ. Microbiol., 55, 764-766, 1989.

Cowen, J. P., and K. W. Bruland, Metal deposits associated with bacteria: implications for Fe and Mn marine biogeochemistry, Deep-Sea Res., 32, 253-272, 1985.

Cowen, J. P., and Y.-H. Li, The influence of a changing bacterial community on trace metal scavenging in a deep-sea particle plume, J. Mar. Res., in press, 1991.

Cowen, J. P., and M. W. Silver, The association of iron and manganese with bacteria on marine macroparticulate material, Science, 224, 1340-1342, 1984.

Cowen, J. P., G. J. Massoth, and R. A. Feely, Manganese in a hydrothermal vent plume, Deep-Sea Res., 37, 1619-1637, 1990.

Cramer, S. P., M. Siskin, L. D. Brown, and G. N. George, Characterization of As in oil shale and oil shale derivatives by X-ray absorption spectroscopy, J. Energy Fuels, 2, 175-180, 1988a.

Cramer, S. P., O. Tench, M. Yocum, and G. N. George, A 13-element Ge detector for fluorescence EXAFS, Nucl. Instrum. Methods Phys. Res., A266, 586-591, 1988b.

Dehairs, F., R. Chesselet, and J. Jedwab, Discrete suspended particles of barite and the barium cycle in the open ocean, Earth Planet. Sci. Lett., 49, 528-550, 1980.

Dehairs, F., L. Goeyens, N. Stroobants, P. Bernard, C. Goyet, A. Poisson, and R. Chesselet, On suspended barite and the oxygen minimum in the Southern Ocean, Global Biogeochem. Cycles, 4, 85-102, 1990.

Denbigh, K. G., The Principles of Chemical Equilibrium: With Applications in Chemistry and Chemical Engineering, third ed., 494 pp., Cambridge University Press, Cambridge, 1971.

Emerson, S., S. Kalhorn, L. Jacobs, B. M. Tebo, K. H. Nealson, and R. A. Rosson, Environmental oxidation rate of Mn (II): bacterial catalysis, Geochim. Cosmochim. Acta, 46, 1073-1079, 1982.

Ferris, F. G., W. S. Fyfe, and T. J. Beveridge, Iron-silica crystallite nucleation by bacteria in a geothermal sediment, Nature (London), 320, 609-611, 1986.

Ferris, F. G., W. S. Fyfe, and T. J. Beveridge, Bacteria as nucleation sites for authigenic minerals in a metal-contaminated lake sediment, Chem. Geol., 63, 225-232, 1987.

Ferris, F. G., S. Schultze, T. C. Witten, W. S. Fyfe, and T. J. Beveridge, Metal interactions with microbial biofilms in acidic and neutral pH environments, Appl. Environ. Microbiol., 55, 1249-1257, 1989.

Friedman, G. M., Occurrence and stability relationships of aragonite, high-magnesian calcite, and low-magnesian calcite under deep-sea conditions, Bull. Geol. Soc. Am., 76, 1191-1196, 1965.

George, G. N., C. L. Coyle, B. J. Hales, and S. P. Cramer, X-ray absorption of *Azotobacter vinelandii* Vanadium nitrogenase, J. Am. Chem. Soc., 110, 4057-4059, 1988.

George, G. N., R. C. Prince, and S. P. Cramer, The Mn site of the photosynthetic water-splitting enzyme, Science, 243, 789-791, 1989.

Ghiorse, W. C., and P. Hirsch, An ultrastructural study of Fe and Mn deposition associated with extracellular polymers of Pedomicrobium-like budding bacteria, Arch. Microbiol., 123, 213-226, 1979.

Goulon, J., A. Retournard, P. Friant, C. Goulon-Ginet, C. Berthe, J. Muller, J. Pobncet, R. Guilard, J. Escalier, and B. Ness, Structural characterization by X-ray absorption spectroscopy EXAFS/XANES of the vanadium chemical environment in borcan asphaltenes, J. Chem. Soc., Dalton Trans. No. 6, 1095-1103, 1984.

Harvey, R. W., and J. O. Leckie, Sorption of lead onto two gram-negative marine bacteria in seawater, Mar. Chem., 15, 333-344, 1985.

Hastings, D., and S. Emerson, Mn(II) oxidation by spores of a marine Bacillus: kinetic and thermodynamic considerations, Geochim. Cosmochim. Acta, 50, 819-826, 1986.

Heath, K. C., and H. T. Mullins, Open-ocean, off-bank transport of fine grained carbonate sediment in the northern Bahamas, in Fine-Grained Sediments: Deep Water Processes and Facies, edited by D. A. V. Stow, and D. J. W. Piper, pp. 199-208, London Blackwell Sci. Publ., Geol. Soc. London Sp. Publ. 11, 1984.

Hedges, J. I., and J. H. Stern, Carbon and nitrogen determinations of carbonate-containing solids, Limnol. Oceanogr., 29, 657-663, 1984.

Heussner, S., C. Ratti, and J. Cabonne, The PPS 3 time-series sediment trap and the trap sample processing techniques used during the ECOMARGE experiment, Continental Shelf Res., 10, 943-958, 1990.

Hidy, G. M., On the theory of the coagulation of non-interacting particles in Brownian motion, J. Colloid Sci., 20, 123-144, 1965.

Hine, A. C., R. J. Wilber, J. M. Bane, A. C. Neumann, and K. R. Lorenson, Offbank transport of carbonate sands along open leeward bank margins: northern Bahamas, Mar. Geol., 42, 327-348, 1981.

Hochella, M. F., Atomic structure, microtopography, composition and reactivity of mineral surfaces, in Mineral-Water Interface Geochemistry, edited by M. F. Hochella, and A. F. White, pp. 87-132, Mineralogical Society of America, Washington, D.C., 1990.

Honeyman, B., Anion and cation adsorption in binary mixtures of adsorbents: an investigation of the concept of adsorptive additivity, Ph.D. thesis, 365 pp., Stanford University, 1984.

Honeyman, B. D., and P. H. Santschi, Critical review: metals in aquatic systems, Environ. Sci. Technol., 22, 862-871, 1988.

Honeyman, B. D., and P. H. Santschi, A Brownian-pumping model for oceanic trace metal scavenging: evidence from Th isotopes, Deep-Sea Res., 47, 951-992, 1989.

Honeyman, B. D., L. S. Balistrieri, and J. W. Murray, Oceanic trace metal scavenging: the importance of particle concentration, Deep-Sea Res., 33, 1137-1152, 1988.

Honjo, S., Material fluxes and modes of sedimentation in the mesopelagic and bathypelagic zones, J. Mar. Res., 38, 53-94, 1980.

Hurd, D. C., Factors affecting the solution rate of biogenic opal in seawater, Earth Planet. Sci. Lett., 15, 411-417, 1972.

Hurd, D. C., Interactions of biogenic opal, sediment and seawater in the Central Equatorial Pacific, Geochim. Cosmochim. Acta, 37, 2257-2282, 1973.

Hurd, D. C., Physical and chemical properties of siliceous skeletons, in Silicon Geochemistry and Biogeochemistry, edited by S. R. Aston, pp. 187-248, Academic Press, New York, 1983a.

Hurd, D. C., and S. Birdwhistell, On producing a more general model for biogenic silica dissolution, Am. J. Sci., 283, 1-28, 1983b.

Hurd, D. C., and F. Theyer, Changes in the physical and chemical properties of biogenic silica from the Central Equatorial Pacific: Part I, Solubility, specific surface area and solution rate constants from acid-cleaned samples, in Analytical Methods in Oceanography, edited by T. R. P Gibb, Jr., American Chemical Soc.. Washington D.C., pp. 833-895, 1975.

Hurd, D. C., H. S. Pankratz, V. Asper, J. Fugate, and H. Morrow, Changes in the physical and chemical properties of biogenic silica from the Central Equatorial Pacific: Part III, Specific pore volume, mean pore size and skeletal ultrastructure of acid-cleaned samples, Am. J. Sci., 281, 833-895, 1981.

Jannasch, H. W., B. D. Honeyman, L. S. Balistrieri, and J. W. Murray, Kinetics of trace element uptake by marine particles, Geochim. Cosmochim. Acta, 52, 567-577, 1988.

Kamatani, A., Physical and chemical characteristics of biogenic silica, Mar. Biol., 8, 84-95, 1971.

Kent, D. B., and M. Kastner, Mg^{+2} removal in the system Mg^{+2}-amorphous silica-water by adsorption and Mg-hydroxysilicate precipitation, Geochim. Cosmochim. Acta, 49, 1123-1136, 1985.

Koeppenkastrop, D., and E. De Carlo, Distribution of rare earth elements beween seawater and synthetic mineral phases, EOS Trans. Am. Geophys. Union, 71, 1417, 1990.

Koeppenkastrop, D., and E. De Carlo, Sorption of REEs from seawater onto synthetic mineral phases: an experimental approach, Chem. Geol., in press, 1991.

Konigsberger, D., and R. Prins, (Eds.), X-ray Absorption Structure, 673 pp., Plenum, New York, 1987.

Laidler, K. W., Chemical Kinetics, second edition, 566 pp., McGraw-Hill, New York, 1965.

Lane, M. E., and F. T. Mackenzie, Mechanisms and rates of natural carbonate fluorapatite dissolution, (ABS.) Geolog. Soc. Am. Annual Meeting, p. 208, 1990.

Lawson, D., D. C. Hurd, and H. S. Pankratz, Silica dissolution rates of decomposing phytoplankton assemblages at various temperatures, Am. J. Sci., 278, 1373-1393, 1978.

Li, Y.-H., L. Burkhardt, M. Buchholtz, P. O'Hara, and P. H. Santschi, Partition of radiotracers between suspended particles and seawater, Geochim. Cosmochim. Acta, 48, 2011-2019, 1984.

Luoma, S., and J. A. Davis, Requirements for modeling trace element partitioning in oxidized estuarine sediments, Mar. Chem., 12, 159-181, 1983.

Mackenzie, F. T., and R. M. Garrels, Chemical mass balance between rivers and oceans, Am. J. Sci., 264, 507-525, 1966.

Mackenzie, F. T., L. D. Kulm, R. L. Cooley, and J. T. Barnhart, *Homotrema Rubrum* (Lamarck), a sediment transport indicator, J. Sediment. Petrol., 35, 265-272, 1965.

Mackenzie, F. T., R. M. Garrels, and O. P. Bricker, Silica in seawater: control by silica minerals, Science, 155, 1404-1405, 1967.

Mackenzie, F. T., W. D. Bischoff, F. C. Bishop, M. Loigens, J. Schoonmaker, and R. Wollast, Magnesian calcites: low-temperature occurrence, solubility, and solid solution behavior, in Carbonates: Mineralogy and Chemistry, edited by R. J. Reeder, BookCrafters, Inc., Chelsea, Michigan, 1983.

Mackin, J. E., The effects of clay mineral reactions on dissolved Al distributions in sediments and waters of the Amazon continental shelf, Cont. Shelf Res., 6, 245-262, 1986.

Mackin, J. E., Boron and silica behavior in salt-marsh sediments: implications for paleo-boron distributions and the early diagenesis of silica, Am. J. Sci., 287, 197-241, 1987.

Mackin, J. E., and R. C. Aller, Dissolved Al in sediments and waters of the East China Sea: implications for authigenic mineral formation, Geochim. Cosmochim. Acta, 48, 281-297, 1984a.

Mackin, J. E., and R. C. Aller, Diagenesis of dissolved aluminum in organic-rich estuarine sediments, Geochim. Cosmochim. Acta, 48, 299-313, 1984b.

Manceau, A., and J. M. Combes, Structure of Mn and Fe oxides and oxyhydroxides: a topological approach by EXAFS, Phys. Chem. Miner., 15, 283-295, 1988.

McKee, B. A., D. J. DeMaster, and C. A. Nittrouer, Temporal variability in the partitioning of thorium between dissolved and particulate phases on the Amazon Shelf: implications for the scavenging of particle-reactive elements, Cont. Shelf Res., 6, 87-106, 1986.

Moore, R. M., and K. A. Hunter, Thorium adsorption in the ocean: reversibility and distribution among particle sizes, Geochim. Cosmochim. Acta, 4, 2253-2257, 1985.

Moran, S. B., The marine geochemistry of aluminum, Ph.D. dissertation, 258 pp., Dept. of Oceanography, Dalhousie University, Halifax, Nova Scotia, Canada, 1990.

Moran, S. B., and R. M. Moore, Distribution of colloidal aluminum and organic carbon in coastal and open-ocean waters off Nova Scotia, Geochim. Cosmochim. Acta, 53, 2519-2527, 1989.

Nealson, K. H., The microbial manganese cycle, in Microbial Geochemistry: Studies in Microbiology, vol. 3, edited by W. E. Krumbein, pp. 191-221, Blackwell Press, Oxford, 1983.

Nealson, K. H., and J. Ford, Surface enhancement of bacterial Mn oxidation: implications for aquatic environments, Geomicrobiol. J., 2, 21-37, 1980.

Nkedi-Kizza, P., J. W. Biggar, H. M. Selim, M. Th. Van Genuchten, P. J. Wierenga, J. M. Davidson, and D. R. Nielsen, On the equivalence of two conceptual models for describing ion-exchange during transport through an aggregated oxisol, Water Resour. Res., 20, 1123-1130, 1984.

Nyffeler, U. P., Y.-H. Li, and P. H. Santschi, A kinetic approach to describe trace-element distribution between particles and solution in natural aquatic systems, Geochim. Cosmochim. Acta, 48, 1513-1522, 1984.

Prieve, D. C., and E. Ruckenstein, The double layer interaction between dissimilar ionizable surfaces and its effect on the rate of deposition, J. Colloid Interface Sci., 63, 317-329, 1978.

Robert-Baldo, G., M. J. Morris, and R. H. Byrne, Spectrophotometric determination of seawater pH using phenol red, Anal. Chem., 57, 2564-2567, 1985.

Sabine, C. L., and F. T. Mackenzie, Oceanic sinks for anthropogenic CO_2, Int. J. Energy Environment Economics, in press, 1991.

Schindler, P., Co-adsorption of metal ions and organic ligands: formation of ternary surface complexes, in Mineral-Water Interface Geochemistry, edited by M. F. Hochella, and A. F. White, pp. 281-307, Mineralogical Society of America, Washington D.C., 603 pp., 1990.

Schink, D. R., N. L. Guinasso, and K. A. Fanning, Processes affecting the concentration of silica at the sediment-water interface of the Atlantic Ocean, J. Geophys. Res., 80, 3013-3031, 1975.

Schlager, W., and N. P. James, Low-magnesian calcite limestones forming at the deep-sea floor, Tongue of the Ocean, Bahamas, Sedimentology, 25, 675-702, 1978.

Sly, L. I., V. Arunpairojana, and D. R. Dixon, Binding of colloidal MnO_2 by extracellular polysaccharides of *Pedomicrobium manganicum*, Appl. Environ. Microbiol., 56, 2791-2794, 1990.

Spiro, C. L., J. Wong, F. W. Lytle, D. H. Maylotte, R. B. Greegor, and S. H. Lamson, X-ray absorption spectroscopic investigation of sulfur sites in coal: organic sulfur identification, Science, 226, 48-50, 1984.

Stern, E. A., Theory of extended x-ray absorption fine structure, Phys. Rev. B., 10, 3027-3037, 1974.

Stouff, P., and J. Boulègue, Geochemistry and crystallinity of oceanic hydrothermal manganese oxyhydroxides showing Mn-Cu association, Geochim. Cosmochim. Acta, 53, 833-843, 1989.

Stroobants, N., F. Dehairs, L. Goeyens, N. Van der Heijden, and R. Van Grieken, Barite formation in the Southern Ocean water column, Mar. Chem., in press, 1991.

Stumm, W., and J. J. Morgan, Aquatic Chemistry, Wiley Interscience, New York, 780 pp., 1981.

Sunda, W. G., and S. A. Huntsman, Microbial oxidation of manganese in a North Carolina estuary, Limnol. Oceanogr., 32, 552-564, 1987.

Tebo, B. M., and S. Emerson, Effect of oxygen tension, Mn(II) concentration, and temperature on microbially catalyzed Mn(II) oxidation rate in a marine fjord, Appl. Environ. Microbiol., 50, 1268-1273, 1985.

Tebo, B. M., K. H. Nealson, L. Emerson, and L. Jacobs, Microbial mediation of Mn(II) and Co(II) precipitation at the O_2/H_2S interfaces in two anoxic fjords, Limnol. Oceanogr., 29, 1247-1258, 1984.

Urmos, J., F. T. Mackenzie, and S. K. Sharma, Characterization of biominerals with Raman spectroscopy, EOS Trans. Am. Geophys. Union, 67, 1721, 1986.

Van Bennekom, A. J., G. W. Berger, S. J. Van der Gaast, and R. T. P. DeVries, Primary productivity and the silica cycle in the Southern Ocean (Atlantic Sector), Paleogeogr. Paleoclimatol. Paleoecol., 67, 19-30, 1988.

Van Bennekom, A. J., J. H. F. Jansen, S. J. Van der Gaast, J. M. Van Iperen, and J. M. Pieters, Aluminum-rich opal: an intermediate in the preservation of biogenic silica in the Zaire (Congo) deep-sea fan, Deep-Sea Res., 36, 173-190, 1989.

Verardo, D. J., P. N. Froelich, and A. McIntyre, Determination of organic carbon and nitrogen in marine sediments using the Carlo Erba NA-1500 Analyzer, Deep-Sea Res., 37, 157-165, 1990.

Willey, J. D., Silica-alumina interactions in seawater, Mar. Chem., 3, 241-251, 1975.

Yasunaga, T., and T. Ikeda, Adsorption-desorption kinetics at the metal-oxide-solution interface studied by relaxation methods, in Geochemical Processes at Mineral Surfaces, edited by J. A. Davis, and K. F. Hayes, Chapter 12, American Chemical Society Symposium Series No. 323, Washington, D.C., 683 pp, 1986.

MINERALOGICAL ANALYSIS OF MARINE PARTICLES BY X-RAY POWDER DIFFRACTION

Sjerry Van der Gaast

Netherlands Institute for Sea Research (NIOZ), POB 59, 1790 AB, Den Burg, Texel, The Netherlands

1. Abstract

X-ray diffraction procedures are discussed including specimen preparation, specimen carriers, diffractometer settings, and raw data treatments in order to obtain optimum results from small amounts (≤5 mg) of sediment trap samples. Low amounts of opal (12%) can be determined if the content of other amorphous materials such as organic matter is low. Relative quantitative mineralogy for lithogenic components of sediment trap samples is possible by using internal standards. An identification technique for small amounts of clay minerals is given which is based on applying different relative humidities to the specimens during scanning.

1. Introduction

Biogenic ocean particles are formed in the upper layers of oceans. They sink to the bottom packed in fecal pellets or they sink as aggregates. Besides biogenic particles the fecal pellets and aggregates also contain lithogenic particles ingested by filter feeders. These lithogenic particles are transported to the oceans by rivers and by the atmosphere. These particles are also supplied by coastal erosion, resuspension of sediment from the shelf and the continental slope and by benthic storms on the ocean floor. Biogenic remains from these sediments are then brought back into suspension. Locally, submarine volcanism may contribute particles. Also low amounts of anthropogenic particles such as fly-ash, exhaust particles from automobiles, and cosmic dust can contribute.

The non-biogenic particles may be a considerable contribution to the downward transported mass, obscuring the biogenic factor. On the other hand the mineral particles may be used to trace the origin of the biogenic particles and possibly their transporting water mass [Honjo, 1980; Wefer et al., 1982].

Biological remains and inorganic particles can change in composition due to ingestion of these particles by filter feeders [Syvitsky and Lewis, 1980] and the particles scavenge chemical elements from the seawater on their way to the bottom, producing new mineral particles (see Fischer and Wefer, this volume, page 391).

The methods normally used with sediment analysis can also be used for mineralogical analysis of trap samples. Marine particles sinking to the bottom are largely of the same composition as the ones that have just arrived at the bottom. At most they are somewhat richer in carbonates, opal (amorphous SiO_2 of biogenic origin) and organic material due to a lesser degree of dissolution and breakdown than is the case with sediment samples. A serious difference, however, lies in the amount of material available. Usually, with sediments, one or more grams of material are used for analysis. With trap samples often only 1 to 10 mgs are available. This restricts analytical procedures severely so that, if fluxes are studied, only main components can be measured. For example, for determination of the carbonate content the sample can be treated with 0.5 N HCl [Honjo, 1980], acetic acid [Honjo et al., 1987; Pilskaln and Honjo, 1987], or orthophosphoric acid [Deuser et al., 1981]. With these chemicals, however, carbonates cannot be differentiated. Besides calcite, high Mg-calcite [Berner and Honjo, 1981] and aragonite [Berner and Honjo, 1981; Betzer et al., 1984] can be present. Dolomite has been found in air and surface water of the Atlantic ocean off the West African coast [Folger, 1970; Chester et al., 1972]. This mineral originated from the Sahara desert.

Opal has been determined by dissolution methods using Na_2CO_3 or NaOH solutions and measuring the dissolved Si. Alumino-silicates, however, will also be attacked and release Si thus giving opal contents of a somewhat too large amount (see D. DeMaster, this volume, page 363).

To determine the lithogenic flux, samples are heated at 500° C. This heating causes loss of absorbed and, partly, hydroxyl water from clay minerals. The amount depends on the clay mineral group and ranges from 5% for illite to about 10% for kaolinite [Grim, 1968; Dankers and Laane, 1983]. Depending on the amount and the types of clay minerals in the sample, the heating method can give too low a lithogenic flux for some %.

To understand more clearly the origin of the particles and the processes which they undergo during their stay in the water column it is necessary to have mineralogical information and preferably also

Marine Particles: Analysis and Characterization
Geophysical Monograph 63

quantitative figures from even the smallest sediment trap samples. In this paper, procedures are described for mineralogical analysis by X-ray powder diffraction (XRD) of sediment trap samples, including the hard parts of biological origin, e.g. opal and carbonates. Limiting factors of the XRD technique for this type of sample are also discussed.

2. Pilot Study

The variability of sediment trap samples, which is discussed in section 3, may be great. This means that for a series of samples an analytical procedure has to be found which gives the best results. It would be a waste, if, after applying some standard procedure which consumes expensive sample material, it turned out that the results did not meet one's expectations. It is preferable to perform pilot studies to determine the most appropriate procedures.

A few mgs of material have been sampled from a large representive sample of a series. With repeated distilled water washings, the chemicals for conservation of the trap sample during its stay in the water column have been removed. From a subsample, a slide for the light microscope (LM) and a mount for the scanning electron microscope (SEM) are prepared. The other part has been exchanged with Ca or Mg and a specimen for XRD analysis is made (see section 4.2.).

From the LM and SEM study, estimates can be made of the following:

a) number of carbonate remains, and from the presence of certain species if, besides calcite, aragonite is present;
b) number of opal organisms, and if they are fresh or aged (resuspended from shelf sediments);
c) number of lithogenic particles, including volcanic glass fragments; and
d) the grain size ranges of each of these groups.

Due to their small sizes, however, the sizes of clay minerals and Fe and Mn (hydr)oxides can often not be distinguished microscopically. The XRD study can give estimates of carbonates, and confirm which carbonate minerals, opal and crystalline lithogenic compounds including clay minerals are present. At this stage it can be decided which internal standards can be used without introducing interference of its peaks with the diagnostic peaks from the components of the sample and which preparation technique should be applied (see section 4.2.).

From these results it can be decided which analytical procedures must be followed to get optimum results with a minimum of damage to the components of the samples.

3. Pretreatment of Samples For XRD Analysis

The composition and particle size ranges of sediment trap samples can be highly variable. Below are listed some pretreatments for application in cases of low concentration of interesting components or if particles must be reduced in size to obtain proper specimens for XRD analysis.

Composition ranges. Depending on the area and period of trap sampling, carbonate, opal, and lithogenic components can be found each in concentrations higher than 50%. Anthropogenic particles and cosmic dust are often present only as traces (see Jedwab, this volume, page 399).

Particle size ranges. The carbonate and the opal hard-parts of the various organisms that produce these phases vary greatly in size. Carbonate organisms such as coccolithophorids, foraminifera, and pteropods can reach sizes of 100, 400, and 12,000 μm, respectively. Opal organisms such as silicoflagellates, diatoms and radiolaria can reach 100, 200, and 2000 μm, respectively. Due to grazing by filter feeders, and in some cases, dissolution, these particles are often fragmented into smaller sizes (E. Ufkes and P. Betzer, personal communication, 1991).

Major components. They can be measured almost always without pretreatments. They mostly consist of a) carbonates; b) opal; and c) some lithogenic compounds which are mostly in coastal areas.

Minor components. In studying minor components, e.g. clay minerals, concentration methods must be applied.

3.1. Concentration by Fractionation

Carbonates and opal can be concentrated by sieving. Clay minerals can be concentrated by collecting the less than 2 μm fraction by sedimentation (centrifugal) methods. With these two methods, only part of the minerals concerned can be collected. For example, with use of a 100 μm sieve, complete coccoliths and foraminifera can be obtained. Smaller pieces and fragments of larger ones are then lost for analysis. And by collecting the less than 2 μm fraction for clay mineral analysis, some of the larger clay particles and the particles that are trapped into the hard-parts of the biogenic particles will not be included. This makes this method only suitable for qualitative analysis. A problem with small samples is that sieving, sedimentation, and weighing their fractions can introduce contamination and serious errors in weight ratio.

3.2. Concentration by Dissolution

Lithogenic particles and opal particles can be concentrated considerably by dissolution of the carbonates with acetic acid buffered at pH 5 with Na-acetate. Treating the sample this way for 1 to 2 hours will not change the minerals measurably.

Carbonates can be concentrated by dissolving the opal with Na_2CO_3 or NaOH. This, however, will harm the clay minerals to a degree that depends on the duration, temperature and the concentration of the chemicals used and of the chemistry, crystalline state, and particle sizes of the mineral components in the clay mineral assemblage. The larger lithogenic particles such as quartz and feldspar are less affected by this treatment.

3.3. Reducing Particle Size

To produce XRD peaks, crystals or crystal fragments must be aligned suitably. In powder diffraction specimens, however, many are not and thus in a specimen with random orientation many particles are necessary to produce diffraction peaks. This is different for clay mineral specimens with oriented particles, which will be mentioned in section 4.2. Unfortunately, trap samples are often small and their particle size ranges are considerable (see above). In XRD analysis, because of particle statistics, the particle size of analyzed substances should be smaller than 5 or 10 μm [Parrish, 1965; DeWolff et al., 1965; Klug and Alexander, 1974]. Moreover, for proper results, smooth specimen surfaces are preferable and this is only possible with small particles. For this reason large carbonate particles from biogenic hard-parts and large lithogenic particles such as quartz and feldspar must be reduced in size, thus providing many crystal fragments. For opal this will be different. Opal particles can be very large in case of complete species, but the crystallite size is very small due to its "amorphous" character [Jones and Segnit, 1971]. From this point of view these particles need no size reduction; however, for producing a smooth specimen surface they do.

Grinding reduces the particle size of sediment trap material, but minerals can structurally be damaged. Intensive grinding of carbonate particles greatly affects the height and the surface area of diffraction peaks [Gavish and Friedman, 1973]. However, mostly, these samples contain soft materials. Lithogenic material, such as quartz and feldspars, can be much harder, but, on the other hand they are often less than 10 μm. Gently grinding sediment trap samples under ethanol or acetone with an agate hand-mortar will often be sufficient. Very small samples can be crushed between microscope glass-slides with addition of a drop of ethanol or acetone.

Below, some pretreatments which are often used in mineralogical analysis of sediments, soils and sometimes trap samples are discussed.

Ultrasonic vibration is applied for disaggregation of loosely consolidated materials. Variations in intensities of XRD lines from a smectite, a kaolinite, and an open illite with this treatment, however, were found by Gâtâ [1964]. Biotite, a clay mineral, that normally shows its (001) line at 10 Å, frequently shifts this line to 13 Å and, less often, to 14 Å, after a 3 minute treatment with a 3 mm microprobe of a 20kHz 150 Watt Ultrasonic Disintegrator apparatus (MSE) at low intensity. No changes were observed after prolonged treatments (Van der Gaast, unpublished data, 1986).

H_2O_2 treatment is often used for removal of organic matter from sediment and soil samples. Douglas and Fiessinger [1971] found a strong degradation of clay minerals by this treatment. By adding a Na-acetate buffer, degradation of the clay minerals decreased but was still noticeable. Langeveld et al. [1978] compared eight methods of removal of organic matter. They found that oxidation with bromine and with H_2O_2 is most effective for organic matter removal. H_2O_2 damages clay minerals the most and bromine the least.

Fe Mn (hydr)oxides which are precipitated on sediment and soil particles can be removed with the citrate-dithionite extraction method [Mehra and Jackson, 1960]. By this method, however, Fe and Al-hydroxides can be extracted from the interlayer of chlorite [Harward and Theisen, 1962; Brewster, 1980] or freshly chloritized smectite, which is then changed back into smectite [Eisma et al., 1978]. Other methods for pretreatment or selective dissolution techniques can be found in the literature [see e.g. Smith and Mitchell, 1987; Moore and Reynolds, 1989b].

3.4. Conclusions

It is impossible to recommend one single analytical procedure for pretreatment of sediment trap samples. For good results, trap samples require a careful, opportunistic approach, in which compromises must often be made, depending on questions posed and the nature of the sample.

It must be kept in mind that most pretreatments can damage, dissolve, or partly dissolve marine particles and thus information carried by the trap samples can be spoiled. If possible, pretreatments must be avoided or at least kept to a minimum. Carbonate removal with buffered acetic acid, however, is a fairly safe method.

4. XRD Techniques

4.1. Principles of XRD Techniques

The principles of XRD are described in several handbooks [see e.g. Cullity, 1967; Klug and Alexander, 1974; Bish and Post, 1989; Moore and Reynolds Jr., 1989a]. From the beginning, XRD patterns were recorded on film with cameras. About 1945 when the diffractometer appeared, Bragg-Brentano para-focussing geometry was most current. It was often equipped with scintillation counters or gas-filled proportional counters. Patterns were displayed directly on recorder-paper. During the last decades, due to an enormous development in electronics, patterns could be stored digitally. Processing of data and automatic pattern recognition became available.

In the last decade the curved position sensitive detector (CPSD) appeared. This detector measures over a wide angle simultaneously as is the case with cameras. Although patterns are obtained in minutes, its sensivity is low compared to the gas proportional detector. Another problem is that bulges appear in patterns due to the diffraction pattern of the glass from the specimen capillary tube. Moreover, due to transmission of X-rays with those types of specimens, diffraction takes place through the whole specimen. With strongly absorbing specimens, however, the thickest part of the capillary absorbs more strongly than the outside of the specimen, thus providing irregular peak profiles of a decreased intensity. When flat specimens have been used with this set-up, intensities of the diffracted X-rays are poor for most angles due to geometry problems. Problems arise also when samples contain some iron, which is often the case with natural samples, and Cu or Co radiation has been applied. Fluorescent radiation increases the background of the pattern drastically. Problems with the glass capillaries and with fluorescent radiation arise for cameras too. For cameras of the transmission type, bulges are recorded from the tape or plastic film that carries the specimen. Moreover, developing camera film delays results and is also time consuming.

With Bragg-Brentano powder diffractometers equipped with a monochromator, the intensity of the detected fluorescent radiation has been brought back to a minimum and most of the time can be neglected. With carefully chosen specimen carrier material, artificial maxima can be brought back to a minimum (see section 4.2.). Recording patterns with this diffractometer takes, however, more time than with the CPSD-equipped diffractometer. On the other hand the effectiveness of it can be increased by using a specimen changer, which makes the diffractometer available 24 hours a day.

If only a few micron-sized grains are available, different XRD techniques can be applied (see Jedwab, this volume, page 399).

In this paper I will stick to the Bragg-Brentano powder diffraction technique because it is a flexible technique since many kinds of specimens can be held in it, and it is the most widespread diffraction technique. There is, however, a serious difficulty: many diffractometers are equipped differently so that results from different laboratories are often not comparable.

In the next section variables which are within reach of the operator, such as specimen preparation and data treatment, are given.

4.2. Specimen Preparation and Specimen Carriers

Usually, powdered specimens are prepared for mineralogical analysis. In these mounts, particles are randomly oriented, at least, that is to be hoped. If random orientation has not been obtained, certain reflections of a mineral will be enhanced and other ones will be depressed. If, moreover, the degree of orientation varies from specimen to specimen in a series it is obvious that the intensity of reflections is not proportional to the minerals present in the specimen. For obvious reasons, then, many techniques have been developed to obtain random orientation [see e.g. Bish and Reynolds, 1989]. With most of these methods, amorphous or crystalline fillers are mixed with the sample in appreciable amounts. Amorphous fillers provide for huge, broad diffraction maxima, which should obscure the presence of opal in the specimen and severely reduce the signal to noise ratio (see section 4.4.). Crystalline fillers provide for peaks that may interfere with the diagnostic peaks of the analyzed material, and, moreover, both types of fillers will strongly decrease peaks from the analyzed material caused by dilution of it. Other techniques like spray-drying would consume irresponsible amounts of trap material. Trap samples often contain appreciable amounts of irregularly shaped biogenic debris, which can be regarded as a filler which provides for unoriented (random) mounts. This is suitable for almost all minerals except for clay minerals (see below).

Care must be taken with this because, with strongly differing amounts of this debris in a series of samples, differences in specimen orientation may occur.

For clay mineral analysis, preferentially oriented specimens are used. In such a specimen the clay mineral particles, which have a platy form, orient strongly, thus providing strong 00*l* series of basal reflections. Patterns of these reflections are very useful for clay mineral identification because each spacing is related to the type of layered structure involved. However, in this case, biogenic debris works against obtaining oriented mounts and the 00*l* reflections will be of decreased intensity. Even fractions less than 2 μm can contain many debris particles.

If the amount of material allows fractionation, centrifugal sedimentation methods can be carried out. Particles of the sample are suspended in distilled water. If flocculation occurs, a stable suspension can be obtained by increasing the pH somewhat, but care must be taken to avoid dissolution of clay minerals. If expandable clay minerals are present in a fraction or can be expected, the fraction should be exchanged with Ca or Mg ions. Then, from a suspension of material, thin specimens can be prepared (see below). Dry material, eventually fractionated, for packed specimens can be obtained by washing the exchanged suspension twice with ethanol using a centrifuge and then by drying it at 60° C. Generally, the type of mount which will be prepared depends on the amount of material and on the minerals of interest.

For large amounts of dry sample material (greater than 2 mg), flat plates can be used in which a depression has been made into which the sample is packed. For this the material can simply be pressed with a microscope slide in the depression of a flat brass plate, which provides for a low background and, unfortunately, one strong peak at 2.87 Å. Figure 6-1A shows an XRD pattern of 2.5 mg trap sample material packed in a brass plate in which a depression has been made of 22 x 4 mm with a depth of 0.3 mm. The sample is packed in the middle of the depression over a length of 5 mm, which makes the irradiated specimen surface area 20 mm^2. The peak from the brass plate has been generated by the irradiated area around the depression (see Figure 6-2). These specimens can be considered as infinitely thick specimens. Thus the intensity of diffraction peaks is proportional to the quantities of minerals present. With these specimens at high angles (e.g. greater than 20° 2θ), a larger volume of the specimen has been irradiated than at low angles resulting in a pattern that

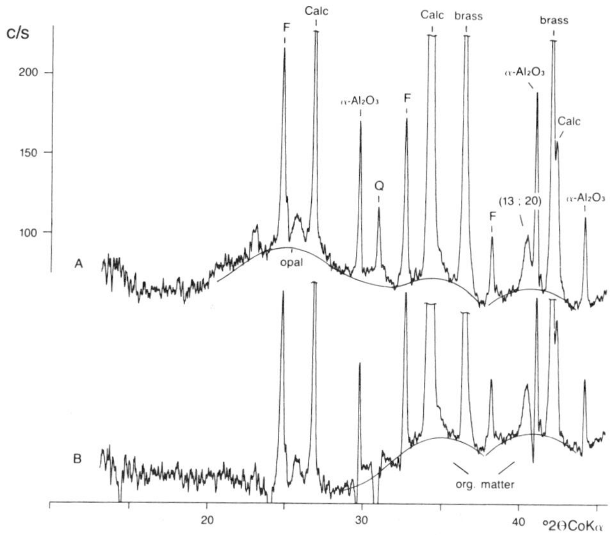

Fig. 6-1 A. XRD pattern from 2.5 mg sediment trap sample D4 (JGOFS). Summation of three scans from unoriented specimens with repacking. Counting time: 2 seconds/0.02° 2θ, variable divergence slit: 5 mm irradiated length, 0.2 mm receiving slit, 1° anti scatter slit. Smoothing over 5 points of equal weight. Bulges from organic matter interfere with that of the opal. B. Subtraction of Figure 6-4B from Figure 6-1A.

provides poor information at low angles, at which clay minerals show their diagnostic 00*l* reflections. Additionally, these reflections are suppressed due to the random orientation of particles in the specimen.

If only small amounts of sample (less than 2 mg) are available, which is often the case if a fraction less than 2 μm has been extracted from a trap sample, only a thin specimen can be made. At increasing angles, these specimens are increasingly transparent to X-rays. How strongly transparent depends on the angle, the amount of material, the absorption of X-rays by the specimen (which depends on its chemical composition) and the applied X-ray wavelength. With such mounts, the intensity of diffraction peaks is not proportional to the minerals present. Consequently, with these transparent specimens, diffraction peaks at low angles are enhanced compared to those at high angles. Because of the presence of debris particles in such a mount, it must be considered as not fully oriented.

Thin specimens may be prepared by sucking the suspension onto

a) a porous ceramic plate, which shows at 5.39Å and at lower values peaks from corundum (JCPDS-10-173), and mullite (JCPDS-15-776), and a broad bulge at

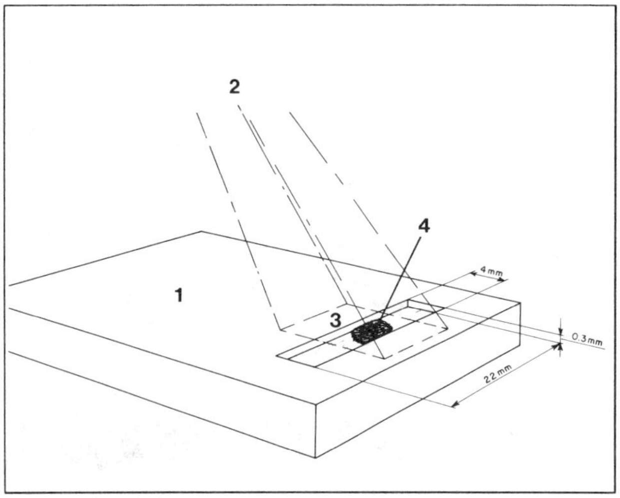

Fig. 6-2. Irradiated specimen and specimen carrier. 1, flat brass plate; 2, incident X-ray beam; 3, irradiated area; 4, specimen in depression.

4.0Å, and thus makes this carrier only suitable for clay mineral analysis at low angles

b) a Ag-filter, which shows peaks at 2.359Å and 2.044Å (JCPDS- 4-783), and drying of a suspension onto a flat, non-porous specimen carrier
c) Si-disc which is a mono crystal cut to the (501) face providing a very low background and absence of diffraction peaks.

The porous plate carrier gives the advantage that the plate acts as a reservoir for water, which makes it suitable for clay mineral analysis (see 5.3.). If a porous plate is placed under the Ag-filter the plate acts also as a reservoir for water. The Si-disc can also be used as carrier for amounts of material far below the mg level.

To prevent differential settling of particles, causing specimen inhomogeneity, preparation time of specimens by suction must be short (minutes). In the literature, specimen carriers such as glass slides and filters of organic material have often been used. Also some types of grease are sometimes used to hold the material onto their carriers. These carriers and greases show huge bulges in XRD patterns [Bish and Reynolds, Jr., 1989]. If opal is to be analyzed, these materials cannot be used because they will obscure or deform the opal bulge, which is measured in opal analysis with XRD (see 5.1.). Bulges from specimen carriers are considered as background thus providing a poor signal to noise ratio (see 4.4.).

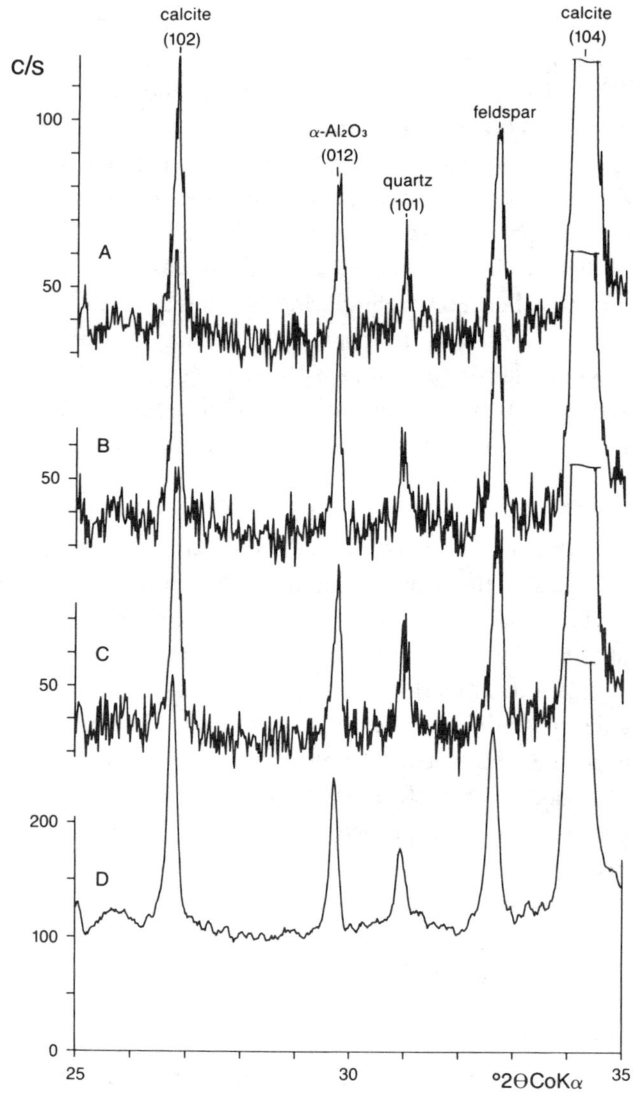

Fig. 6-3. XRD patterns from sediment trap sample D4 (JGOFS) of three scans. The specimen is unoriented and repacked after each scan. Counting time: 2 seconds/0.02° 2θ, variable divergence slit: 8 mm irradiated length, 0.2 mm receiving slit. 1° anti-scatter slit. Patterns A, B, and C are from raw data; D, from summed patterns of A, B, and C, smoothed over 5 points of equal weight.

4.3. Internal Standards

Standard materials are added to samples for several reasons:

a) Improved accuracy in lattice parameter determination. Non-ideal alignment of the diffractometer and aberrations from the instrument are thus ruled out to a great extent.
b) Intensity measurements of peaks in quantitative XRD.

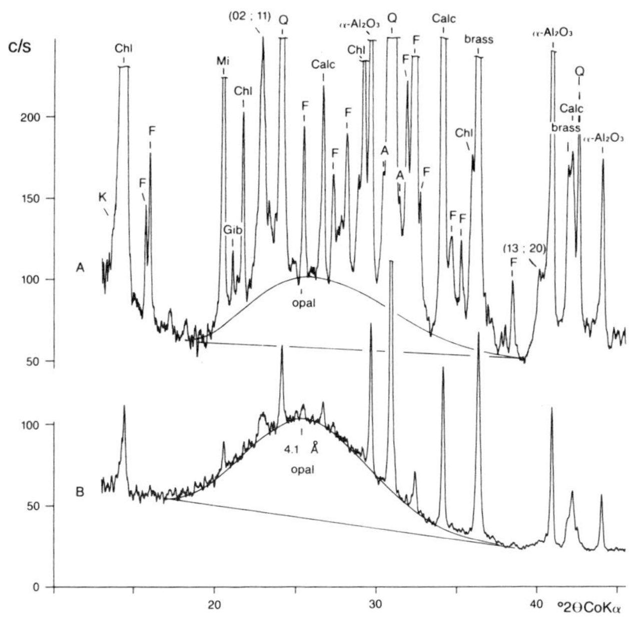

Fig. 6-4. XRD patterns from 5 mg sediment trap sample. Patterns A and B correspond with A and B from Table 6-2. Unoriented specimens in depression in brass plate. Counting time: 2 seconds/0.02° 2θ, variable divergence slit: 8 mm irradiated length, 0.2 mm receiving slit, 1° anti scatter slit. Brass peaks can be seen at 36.8 and 42.0° 2θ. αAl_2O_3 is used as internal standard. Mi=mica/illite; K=kaolinite; Chl = chlorite; F = feldspar; Q = quartz; Gib = gibbsite; A = aragonite; Calc=calcite. (02;11) and (13;20) are reflections from smectite, kaolinite, etc.

c) As orientation indicators in clay mineral analysis.

With the last two applications, a fixed percentage of weight of internal standard is mixed with a sample and the ratio of peak intensities from minerals of the sample to that of a peak from the internal standard is measured. For the orientation indicator a platy material is mixed in a fixed percentage with a sample containing platy minerals. If the orientation of the platy minerals in the specimen changes, the orientation of the orienting internal standard changes to the same degree. Consequently, the ratio of peak intensities, as mentioned above under c), can be found independently of the degree of orientation of particles in the specimen. This method can be applied to clay mineral analysis from oriented and unoriented specimens. In the last case, however, the intensities of the diagnostic 00*l* reflections of these clay minerals will be very weak.

The properties that internal standards must meet are:

a) no interference with diagnostic peaks from the material from the specimen;
b) a strong reflection as close as possible to the diagnostic peak.
c) comparable orientation properties of particles to that of the particles of the specimen.
d) stability under conditions of preparation of specimens and during analysis.
e) in case the sample has to be used for chemical analysis after XRD analysis, the chemistry of the internal standard must be different from the elements that are going to be measured.

For unoriented mounts α-Al_2O_3(corundum), with peaks at 3.479Å, 2.552Å, 2.379Å, and 2.085Å has often been used. Other internal standards are powdered Si, ZnO, Cr_2O_3, CeO_2, As_2O_3, and powdered Cu. Si, CeO_2, and As_2O_3 are less favorable because their largest peak coincides with feldspar reflections in the 3.2Å region. Two of the main peaks of ZnO fall together with a quartz and a clay mineral reflection. From Cr_2O_3 the strongest reflection coincides with Fe-oxide mineral reflections (hematite and goethite) and two other strong reflections from it with a feldspar and with a quartz reflection. Although the 3.479 Å peak of α-Al_2O_3 interferes with the (002) reflection of kaolinite and the (004) reflection of chlorite, for trap samples the α-Al_2O_3 best meets the requirements mentioned above.

For mixing of the internal standard with powdered samples to obtain unoriented mounts, the sample is exchanged with Ca or Mg ions and dried. Then the powder is weighed on a piece of glassine paper and directly after it, on the same piece of paper, close to the sample a certain amount of internal standard is weighed. In case weighing of the desired amount of the internal standard fails, then it can be removed and the sample has not been spoiled. For α-Al_2O_3 as internal standard, I normally use 10% by weight. After weighing, the sample and the internal standard are mixed thoroughly, but carefully, on the paper with a spatula.

For clay mineral analysis with oriented mounts, the platy materials MoS_2, pyrophyllite, talc, and $Zn(OH)_2$ have been used. Pyrophillite, talc, and $Zn(OH)_2$ show strong reflections at 9.20Å, 9.33Å, and 7.89Å, respectively, which is in between some diagnostic 00*l* reflections and often overlaps them somewhat. MoS_2 shows a strong reflection at 6.15Å which is pretty close to some diagnostic peaks, but does not overlap them. Pyrophyllite and talc are clay minerals themselves, and, if these minerals are present in the original sample they cannot be used. $Zn(OH)_2$ is not stable in moist air or at temperatures above 300° C. MoS_2 is not stable above 400° C, but it resists ultrasonic vibration, and is inert to glycerolation [Quakernaat, 1970]. However, temperatures of 500 to 550° C are often applied in clay mineral analysis. It will be shown in section 5.3. that heat treatment can be avoided with such an analysis and that MoS_2 is a suitable orientation-indicator.

For clay mineral analysis, the sample and the standard are mixed in suspension. For this the concentration from a less than 2 μm fraction suspension, for example, is measured by pipetting a small volume of about 100 μL in a preweighed container of a suitable size and weight. Al-combustion boats from Perkin-Elmer used for CN analysis are very suitable for this. After drying the suspension at 105° C, the boat is reweighed and the concentration of the suspension can be calculated. If sufficient material is available the concentration is measured in duplicate. It is obvious that, for this, one needs micro-pipets and a microbalance. From the suspension a certain volume, which holds about 15 mg of material, is pipetted together with a certain amount of internal standard from a suspension of known concentration, and mixed. With MoS_2, 1% by weight is sufficient. Then the specimen can be prepared on a porous plate or a Ag-filter by suction, covering an area of 20 x 15 mm. For Co and Cu radiation, such a specimen is of infinite thickness as far as 15° and 17.5° 2θ, respectively, which is sufficient for measuring the diagnostic 00*l* clay mineral reflections. In case less material is available the specimen surface area, preferably the breadth, must be equally decreased.

If a peak of the internal standard (not the one that is used for measuring ratios of intensities) interferes strongly with an interesting peak from one of the compounds of the specimen, then the peaks of the internal standard can be removed. For this a pattern from the separately measured internal standard has been measured, normalized to the intensities of its lines in the pattern with the sample, and than subtracted (see section 4.6.5.). This is, however, a time-consuming method and, moreover, it will often not be completely successful. Thus this should only be carried out if all the possibilities, which are mentioned above, do not work.

4.4. *Signal and Noise*

Intensity is important in XRD, especially if weakly diffracting minerals and/or small irradiated volumes are analyzed, but even more important is a strong signal to noise ratio. It is also important to know what the noise and what the signal is and which factors contribute to both of them. Then the diffractometer-setting can be changed without making mistakes with the interpretation of the XRD patterns.

A definition of noise is: "every detected radiation that does not belong to diffraction maxima from the analyzed substances". Additionally, this noise is stochastic and is found everywere on the XRD pattern. There are two types of noise:

A) where the intensity is not dependent on the angle of diffraction:

1) proportional and scintillation counters and their electronic circuits generate noise, which cannot be prevented. Recently, a Si(Li) detector (Kevex) using thermo electric (Peltier) cooling has become available which should generate no noise;
2) cosmic rays, which bombard the earth from space, affect counters thus producing noise;
3) radioactivity, which is from materials surrounding us, affects counters as with cosmic rays.

The other type of noise is from the specimen and the specimen area in the diffractometer:

B) where the intensity is dependent on the angle of diffraction:

1) fluorescent radiation of Fe in the specimen, when using Cu or Co radiation. It can be minimized by using a crystal monochromator in the diffracted beam;
2) incoherently scattered radiation from the specimen. This is a diffuse scatter as a consequence of thermal motions of atoms and of the Compton scattering [Klug and Alexander, 1974];
3) incoherently scattered radiation from the specimen carrier;
4) scatter of air above the specimen.

The intensity of this type of noise depends on a) the irradiated volume, and b) the Lorentz-polarization factor (LPF), which decreases with increasing angles. This factor is strong at low angles and weak at higher angles [Klug and Alexander, 1974]. Consequently, below about 10° 2θ, with both types of divergence slits, the noise decreases with increasing angles. With a fixed divergence slit this continues at higher angles due to a slightly decreasing LPF. With a variable divergence slit, however, the noise increases somewhat at angles higher than about 10° 2θ due to a slightly decreasing LPF and an increasing irradiated volume. In practice it means, when using fixed divergence slits, that at low angles narrow slits must be applied to avoid irradiation of the specimen carrier (Table 6-3).

A definition of signal is: "detected radiation that originates from diffraction at the structure of the analyzed substances". Factors affecting the intensity of the signal are:

4.4.1. Structure-sensitive factors.

These factors are properties which are determined by the analyzed substances. They are explained in XRD handbooks [see e.g. Klug and Alexander, 1974; Jenkins, 1989a]:

1) the atomic scattering factor;
2) the structure factor;
3) polarization;
4) multiplicity factor;
5) temperature factor.

TABLE 6-2.
Comparison of Opal Contents of Trap Samples Measured With XRD and With Extraction

Trap sample	Extraction technique Mortlock and Froelich [1989] %	XRD method Eisma and Van der Gaast [1971] %
A	11	12.0
B	68	73.0
C	17	18.5
D	11	13.5
E	23	18.5 (15.5**)
F	23	21.0

** from re-analyzed sample

4.4.2. Instrument-sensitive factors:

A) The intensity of the X-ray source is limited by the capacity of the XRD tubes. Broad Focus (BF) tubes can be used up to 2.7kW and 2.7kW, Normal Focus (NF) tubes at 2kW and 1.8kW, and Long Fine Focus (LFF) tubes at 2.2kW and 1.8kW, for Cu and Co radiation, respectively. BF tubes provide highest intensities due to their large effective focus area. Sharp peaks, however, will be slightly broadened compared with that of NF and LFF tubes. This will, however, often be hardly a problem with the mineralogy of trap samples. With increasing take-off angle of the X-rays the effective focus width will increase and consequently the intensity of the X-ray source. For most diffractometers the take-off angle is fixed. Changing this is beyond the reach of most operators.

B) The angular aperture of the diverged X-ray beam is determined by the width of the divergence slit. The irradiated specimen length (and thus the specimen area) varies with glancing angle θ to approximate relationship:

$$L = \left[\frac{(\gamma)}{(57.3)}\right]\left[\frac{(R)}{(\sin\theta)}\right]$$

where L is the irradiated specimen length in mm, γ is the angular aperture of the divergence slit in degrees, R is the goniometer radius in mm, and θ is the angle between the X-ray beam and the specimen [Wilson, 1987]. In Table 6-3 minimum angles are given at which the various fixed divergence slits should be applied to irradiate 20 mm specimen length. From this Table it can also be seen that at low angles L varies rapidly. With a divergence slit too large, or in other words, with scanning below the minimum angle, the incident beam irradiates the specimen carrier as well as the specimen, causing an increase in noise (see above). From Table 6-3 it can be seen that, for clay mineral analysis, which starts at 2 and 3° 2θ for Cu and Co radiation respectively, a 1/6° slit should be applied. At the end of the scan, at 15° and at 17.5° 2θ, the irradiated specimen length is 3.9 and 3.3 mm, respectively, thus resulting in a pattern of very low intensities. Consequently, the divergence slit should be changed two or three times during a scan to obtain a pattern of a fair intensity. This will be very time consuming, moreover, interpretation of such a sliced pattern will be difficult. In practice operators often compromise by using a 1/2° slit for such measurements and put up with losing some intensity at the high angle side of the pattern and with an increased background at low angles due to noise from the specimen carrier. With trap samples of a few mgs, which are packed in a small depression in a metal plate acting as a specimen carrier (see section 4.2.), large areas of the plate will be irradiated and thus contribute to the noise of the pattern.

The problem with choosing the correct divergence slit has been solved with the variable divergence slit [Jenkins and Paolini, 1974]. The irradiated specimen length remains about 13 mm over the entire range. Intensities, however, do not agree with those obtained with a fixed divergence slit and have to be corrected by multiplying by the sine of the diffraction angle [Jenkins, 1989b]. Namely, with a fixed divergence slit the irradiated specimen volume remains the same over the entire range and with the variable slit the irradiated specimen volume increases with increasing angles. In case of small specimens, with this slit also parts of the specimen carrier will be irradiated and thus contribute to the noise (Figure 6-2). With a variable slit with which the continuous irradiated specimen length can be kept at any length during a scan, noise from the specimen carrier has been reduced (Van der Gaast et al., in preparation). Table 6-4 shows that the noise from the Si-disc is about two times lower than that of the brass plate resulting in an improved detection limit with the Si-disc specimen carrier.

TABLE 6-3.
Minimum Angles at Which Fixed Divergence Slit Widths Irradiate 20 mm Specimen Length

Divergence slit (°)	Minimum angle (°2θ)	Maximum d spacing (Å) for CuKα	Maximum d spacing (Å) for CoKα
1/12	1.4	63	73
1/6	2.9	30.5	35.4
1/4	4.3	20.5	23.9
1/2	8.7	10.2	11.8
1	17.4	5.10	5.92
2	35.2	2.55	2.96
4	74.3	1.28	1.48

TABLE 6-4.
Influence of Two Different Specimen Carriers on the Signal to Noise Ratio of the (012) Reflection of αAl_2O_3 (1.0 μm)

Specimen quantity (mg)	Specimen carrier	
	Brass	Silicon
0.1	3.2	6.7
0.01	0.42	0.85

Notes: Specimens were prepared by pipetting a suspension of $\alpha - Al_2O_3$ particles in water on the carrier and dried at 90° C. Specimens were scanned three times with a counting time of 2 seconds/0.02° 2θ with the variable divergence slit set at 8 mm irradiated specimen length, a receiving slit of 0.2 mm, and a 1° anti-scatter slit. After each scan the specimen was rotated (see section 4.5.1.). The patterns were summed and smoothed over three points of equal weight.

C) With increasing receiving slit widths, peak intensities increase strongly. With a width as wide as 0.4 mm, however, peaks broaden markedly [Parrish, 1965]. Thus, normally a 0.2 mm slit will do.

D) An anti-scatter slit is required at low angles to prevent artificial maxima. With a fixed divergence slit a certain angular aperture of the incident X-ray beam has been determined. At the detector side of the diffractometer the combination of the receiving slit and the anti-scatter slit also determine a certain angular aperture. To prevent intensity losses these apertures should be equal. This means in practice that for example a 1° divergence slit should be used with a 1° anti-scatter slit. The variable divergence slit, however, gives problems here. As mentioned above, its angular aperture varies continuously during a scan. Consequently, the aperture at the detector side should do so as well. As this is not the case, a suitable anti-scatter slit must be applied for every scanning range to prevent intensity loss.

F) For every operator, the detection efficiency is a matter of correct tuning of the detector electronics, which must be checked regularly.

4.4.3. Specimen sensitive factors:

1) *absorption*. With strongly X-ray absorbing specimens, which depends on their chemistry, the irradiated volume is small resulting in decreased intensities.
2) *crystallite size*. One particle can hold many crystal domains, of which many may be oriented differently. These domains must be regarded as the basic diffracting units and thus their size determines the particle size in XRD analysis. Below a certain crystallite size, diffraction peaks will broaden and decrease in intensity.
3) *degree of crystallinity*. With decreasing crystallinity, peaks will also broaden and decrease in intensity. Clay minerals are often poorly crystalline. Opal, a silica glass, is an extreme example of poor crystallinity. It displays only very broad diffraction maxima [Jones and Segnit, 1971].
4) *particle orientation*. In section 4.2. it was already mentioned that orientation of particles in the specimen influences diffraction intensities.

Absorption of X-rays by air along the X-ray path is strong. Van der Gaast and Vaars [1981] applied a vacuum/helium device for the entire X-ray path which resulted in an intensity increase for Cu and Co radiation of 50 and 100%, respectively. With this device He was also flushed in the specimen chamber thus preventing air scatter. The scattering of He is about 10% from that of air thus reducing the scatter above the specimen by a factor of 10.

Summarizing the points that are within reach of the operator:

- use a monochromator to minimize noise from the specimen;
- use the X-ray tube at a high rating;
- the irradiated area, the specimen surface area, and the detected area, must correspond as far as possible, by applying a suitable slit setting;
- for analysis at low angles prepare a specimen with a large surface;
- for analysis at higher angles prepare thick specimens (0.3 mm) but with a surface as large as possible;
- choose specimen carriers made of low-noise material;
- choose a receiving slit as large as possible;
- tune the detector optimally.

4.5. Scanning Conditions

4.5.1. Particle statistics.

Commonly, some mgs of trap sample material are available for XRD analysis. Still, patterns with a reasonable signal to noise ratio can be obtained with these amounts, because the actual volume of a specimen analyzed is also rather small: typically in the range of 2-20 mg, depending on the beam penetration [Jenkins, 1989a].

Another matter of concern is the amount of crystallites in the specimen taking part in diffracting the X-rays. Parrish [1965], DeWolff et al. [1965], and Klug and Alexander [1974] mentioned that reducing sample material to particle sizes below 5 or 10 μm will improve statistics considerably. However, despite the presence of just small particles, statistics can be poor, due to the small amount of sample that is often available. Statistics can be improved by spinning the specimen during scanning. An increased amount of crystallites will then be positioned correctly for and take part in diffracting X-rays. This technique can also be applied with samples on filters. A piece of a filter must then be fixed to a spinning device. If a spinning device is not available and data can be stored digitally, rotation of the specimen can be mimicked by scanning the specimen a few times in which a) for thin specimens on ceramic tiles, filters, etc., the specimen carrier should be rotated for some tens of degrees perpendicular to the specimen surface between the scans taking care to replace the center of the specimen at the same point each scan; b) for unoriented specimens in a depression of a plate, the material should be removed and repacked between the scans. After this the patterns from the same sample are averaged digitally (see section 4.6.5.). In Figure 6-3 an example is shown from a pattern of sediment trap sample D4 (JGOFS) from 25° to 35° 2θ. For this a 2.5 mg sample packed in a depression in a brass plate has been measured three times with repacking of the sample in between. After this the patterns were summed, and the resulting pattern has been smoothed. It can be seen from these measurements that ratios of peak intensities from the successive scans are almost equal thus showing that reproducibility in preparing such a specimen is high.

For specimens containing amounts of material below the mg level and consisting of crystallite sizes around 5 μm, statistics will be poor. Consequently, the intensity of reflections is not proportional to minerals present and will provide only for qualitative information.

4.5.2. Stepsize and counting time.

With specimens displaying just broad bulges, as that from opal or some clay mineral reflections, a 0.1° step size for 2θ combined with a 0.4 mm receiving slit can be applied. With trap samples, however, sharp peaks from Al-Si-minerals, from some clay minerals, and from an eventually admixed internal standard, are often present. Thus, despite the presence of broad peaks, a rather small stepsize of 0.02° in combination with a receiving slit of 0.2 mm must be applied. Larger steps will give problems with broadening of sharp reflections with smoothing (see section 4.6.1.).

For XRD analysis of the bulk mineralogy from an unoriented specimen in a depression in a plate, a scan from 2° to 40° 2θ and 3° to 47° 2θ for CuKα and CoKα radiation, respectively, is often sufficient. For specimens, containing only well crystalline material a counting time of 2 seconds each 0.02° will do and thus will take about 1 hour. If the specimen has to be measured three times, as mentioned in section 4.5.1., the total scanning time for one sample takes about 3.5 hours included repacking time. In case the intensities are weak, increased counting times must be applied.

With clay mineral analysis, applying scans from 2 to 15° 2θ, a counting time of 2 or 4 seconds with steps of 0.02° is normally sufficient and will take 20 or 40 minutes, respectively. If these patterns show poor intensities also, counting times must be increased.

It is obvious that one of the factors which affects the quality of a pattern is the available XRD-apparatus time.

4.5.3. Conditions in the specimen chamber.

Scanning a specimen at room conditions is an often applied step with identification of clay minerals by XRD. A specimen, which is exchanged with Ca^{2+} or Mg^{2+}, is measured at the humidity prevailing then in the X-ray room. Humidities, however, can fluctuate enormously through the year. For example, during rather cool, moist summers in the Netherlands, the relative humidity (RH) in my X-ray room is often 75 to 80%. Smectite, which is a swelling clay mineral, shows then its (001) reflection at about 15Å to 15.2Å. This is an accepted value in the literature. During winter, when the temperature drops below zero, the RH decreases within a few days to values of 12 to 15%. Smectite then loses some water, which is absorbed by the exchanged ions at the interlayer of the mineral, resulting in a decreased c-dimension of the crystal, which is displayed in a shift of the (001) reflection to about 12Å. Obviously, this mineral will be taken mistakenly for some mixed-layer mineral.

If measurements at room conditions are desired, the air in the specimen chamber must be standardized at a RH of 50% or values close to that (see section 5.3.). The effect of RH on basal spacing of some swelling minerals has been described amongst others by MacEwan and Wilson [1980].

4.6. Raw Data Treatment and Pattern Treatment

Raw data from XRD patterns may be statistically of poor quality and/or hold information which is of no consequence and can interfere. To improve patterns and to remove interfering factors the raw data is treated in some sequential steps. For this a variety of software for micro-computers is available. For some of the steps mentioned below a spreadsheet program also will do.

4.6.1. Smoothing.

In case of poor counting statistics, XRD patterns show rather strong fluctuations in intensity. In such cases, smoothing of the pattern is applied. In Figure 6-3D a smoothing of five points of equal weight has been carried out. Peak and background intensities can be determined more accurately, however, sharp peaks decrease about 10% in height with smoothing. The scanning parameters are often a compromise between resolution and available scanning time. Jenkins [1989a] described how step sizes and smoothing of data can influence peak intensities. Different smoothing factors can be applied, such as the running-weighted 5 count average [Hosterman and Dulong, 1989]. In this case the weighted average of the third of five counts is calculated by multiplying the five counts by a factor sequence of 1, 2, 4, 2, and 1 and dividing the total by 10.

4.6.2. Background subtraction.

The background of a pattern is determined by the two factors mentioned in section 4.4. At low angles (less than 10° 2θ) the background is high and sloping due to the LPF. At higher angles broad reflections from poor crystalline components, such as opal, organic matter, Fe and Mn hydroxides, and some clay mineral reflections, are the basis on which crystalline components display their much sharper peaks. Patterns from sediment trap material often contain both types of crystalline components. Thus, broad reflections in patterns from these samples must not be regarded as background and, consequently, must not be removed from the recorded data by subtraction. There is, however, no difficulty in subtracting an average background from the pattern. The total pattern will only shift vertically.

4.6.3. α 2-stripping.

Most XRD patterns are recorded with radiation from the $K\alpha_{1,2}$ doublet [Klug and Alexander, 1974]. Applying a receiving slit of 0.2 mm the α_2 radiation causes a somewhat asymmetric peak profile with peaks at about 30° 2θ. At higher angles this develops as a shoulder at the high-angle side of a peak. With patterns from sediment trap samples α_2-stripping can be omitted unless the chemical composition of carbonate minerals is of special interest (see section 5.2.). Preferably, with mineral identification procedures, α_2-stripping is applied.

With the earlier mentioned Kevex Si(Li) detector, which is also a tunable monochromator, patterns can be recorded with the almost monochromatic $K\beta$ radiation. The α_2-stripping procedure can then be omitted. However, patterns will be much weaker in intensity because the intensity of the $K\beta$ radiation is five times weaker than that of the $K\beta$ doublet.

4.6.4. Peak location and peak intensity.

Mineral identification can be accomplished by a variety of search-match procedures (e.g. from the JCPDS). For good results the diffractometer needs to be aligned well. Aberrations originating from the instrument and the specimen, however, still will have their influence on peak position, peak intensity, and peak profile. This subject is beyond the scope of this paper. The interested reader is referred for this to Wilson [1963], Klug and Alexander [1974], and Jenkins [1989b].

Due to the great variability of clay minerals in each clay mineral group, with respect to crystalline state, chemical composition, and admixed impurities, clay mineral patterns are poorly represented in search-match procedures. It can thus be very useful to make a library of patterns from all kinds of clay minerals. Moreover, they can be measured under identical conditions as that of the samples to be analyzed. A great variety of clay minerals can be obtained from the CMS (Clay Mineral Society, P.O. Box 880, Evergreen, Colorado 80439, U.S.A.) and from WARD'S (Ward's Natural Science Establishment, Inc., 5100 West Henrietta Road, P.O. Box 92912, Rochester, New York, U.S.A.). For small amounts, a clay mineral identification method based on application of different humidities during scanning of the specimen has been given in section 5.3.

Peak intensities are often measured by measuring peak heights after subtraction of a linear background. This method is not reliable since the peak height is strongly influenced by e.g. the crystallinity and the crystallite size (see section 4.4.). The integrated intensity, or peak area, however, will then, because of broadening of the peaks, not change [Klug and Alexander, 1974] and thus can be better used. Moore and Reynolds, Jr. [1989d] obtained increased accuracies by multiplying the height of the peak by the width. This was also satisfying with somewhat overlapping peaks. This procedure, however, must be carried out by hand. With more sophisticated procedures, such as curve fitting and deconvolution methods [Jones, 1989], this can be carried out with a microcomputer. However, a visual inspection of each manipulation must be carried out. At this moment we have to consider this procedure as a future method.

4.6.5. Summing and subtraction of patterns.

As mentioned in section 4.5.1., in case of poor statistics, improved results can be obtained by summing and thus averaging of patterns.

"Hidden" components can be detected by subtraction of patterns. A specimen of a sample is scanned and after a chemical treatment of the sample, with which the material of interest is dissolved, another pattern is recorded from the residue of the sample. From both patterns, the intensities of a certain peak of a mineral from the sample, which is not affected by the treatment, are normalized and patterns are subtracted. If the sample does not contain a suitable mineral, an internal standard, which must also not be affected by the treatment, is then added to the original sample. The difference pattern shows then the component which is removed.

This technique is called Differential-XRD (DXRD) and was first described by Schulze [1981]. He showed with dissolution of iron from soil samples the presence of small amounts of goethite, hematite, and poorly crystalline ferrihydrite. In Figure 6-1B the presence of bulges from organic matter are shown to be clearly present in the pattern of Figure 6-1A by subtracting from it a pattern of an opal-rich sample (Figure 6-4B).

4.7. Quantitative Aspects of XRD Analysis

Quantitative phase analysis by XRD procedures is a difficult matter. Many attempts, with as many methods, have been undertaken to determine the amounts of mineral components in sediment and soil samples.

Wilson [1987] gave a clear review of all kinds of methods for clay minerals and non-clay minerals and tested some of them. His final conclusion was that quantitative phase analysis of unknown materials is a practical impossibility, mainly due to the great variability in diffracting power and, mostly in case of clay minerals, also to specimen preparation techniques. Moore and Reynolds, Jr. [1989d] consider for clay minerals that the sample to sample variations in chemical composition, which influences the structure factor and the atomic scattering factor (see section 4.4.), can be a major source of error in quantitative analysis. However, this is different for opal. In section 5.1. it is shown that the opal content can be determined by XRD. Another factor, which also strongly influences the diffracting power of clay minerals, is the shape of clay mineral crystals. Van der Gaast et al. [1986] found a curled smectite, formed from weathering of volcanic material on land, with such a low diffracting power, that the (001) reflection was completely absent. Not even a trace of a shoulder at the sloping background could be detected. After correction of the pattern for the LPF and the irradiated volume a smectite with an anomalous (001) spacing, due to its shape, appeared. This was in contrast with the strong (001) peaks, well known from standard smectite minerals such as montmorillonite from Wyoming (API standard). Possibly, this "shape" effect is also present with other mineral groups. Halloysite, for example, has often been found as tubular or spheroidal crystals. Actually, such well-crystalline, large-sized clay mineral crystals sometimes provide for non-oriented particles, even in oriented specimens, resulting in poor diffraction peaks from these minerals.

About the still fascinating subject of quantitative phase analysis the interested reader is referred to Klug and Alexander [1974], Brindley [1980], Wilson [1987], Snyder and Bish [1989], Moore and Reynolds, Jr. [1989d], Reynolds [1989], Hosterman and Dulong [1989], Davis and Johnson [1989], Slaughter [1989], and Calvert et al. [1989].

Quantitative phase analysis by XRD is, at least at present, not possible. However, the relative quantity of minerals can be measured by adding a fixed percentage of weight of an internal standard material to the sample. Fluctuations of the amount of a mineral can then be found by measuring the ratio of intensities of a diagnostic peak of a mineral and that of a suitable peak of the internal standard. For clay mineral analysis with oriented specimens an orienting internal standard is used (see sections 4.3., 5.3. and Figure 6-5). For unoriented specimens a non-orienting standard must be added in a fixed percentage of weight. With both methods precision is ≈ 5%. Van der Gaast and Jansen [1984] used unoriented specimens to measure the relative amounts of minerals and opal in sediments of the Angola Basin.

This internal standard method can only be used if the diffracting power of the minerals from sample to sample is identical. If sources change, minerals and their diffracting power may change due to the reasons mentioned above. Van der Gaast and Jansen [1984] found with Angola Basin sediments that a smectite, which was supplied by the Zaire river during interglacial periods, had a different diffracting power than did the smectite which was supplied by the adjacent shelf during glacial periods.

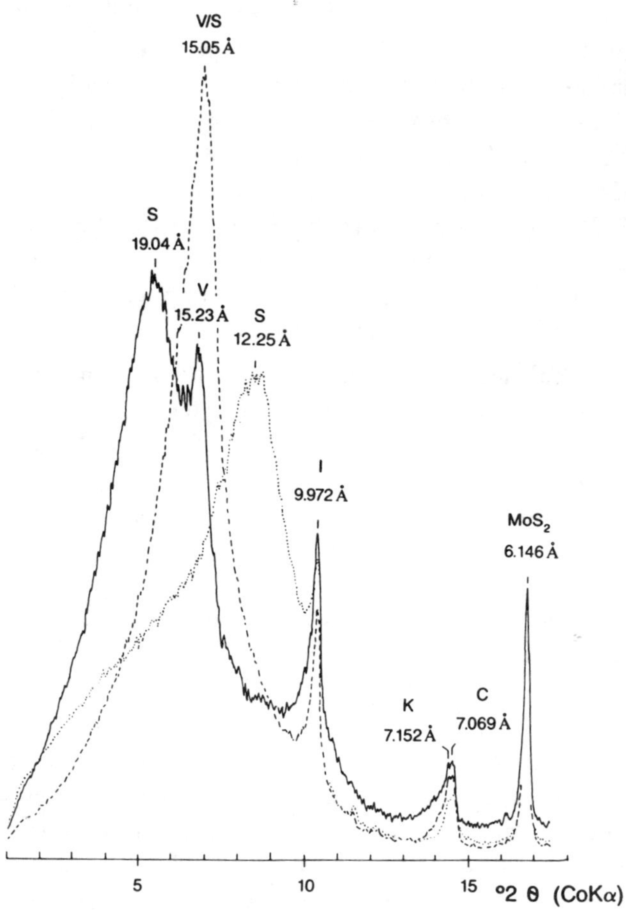

Fig. 6-5. Influence of relative humidity on peak locations of smectite and vermiculite in XRD patterns from an oriented specimen of a Ca-exchanged less than 2 μm fraction of an ODP sample prepared on a ceramic tile with 1% MoS_2 as internal standard. Counting time: 4 seconds/0.02° 2θ, variable divergence slit: 20 mm irradiated length, receiving slit 0.1 mm, anti-scatter slit 0.5°. Patterns are smoothed over 3 points of equal weight and plotted normal to the peak of the internal standard. Full line: 100%RH, broken line: 50%RH, dotted line: 0%RH. S: smectite, V: vermiculite, I: illite, K: kaolinite, C: chlorite. Note that the smectite peak changes in intensity with changing humidity.

5. Mineral Groups

5.1. Opal

A variety of methods exists to determine the opal content of suspended matter and marine sediments (see DeMaster, this volume page 363). In this section it is shown that the XRD method of Eisma and Van der Gaast [1971] can be carried out with a few mgs of trap material. But, just as with other techniques, there are also limitations with this method.

By determining the opal content by XRD the structural properties of the opal are used. Goldberg [1958] measured the height of the 4.04Å (101) peak of α-cristobalite after conversion of the opal to cristobalite by heating the sample at 1000° C. The height of the cristobalite peak, however, is strongly influenced by the presence of clay minerals in the sample [Eisma and Van der Gaast, 1971]. These authors developed a method to determine the opal content by measuring the height of the opal bulge, which was compared with the height of the 2.38Å (110) peak of α-Al_2O_3, which was used as an internal standard. They claim a precision of ± 3% with this method. With this method one is not confronted with the difficulties of quantitative XRD, which are mentioned in section 4.7., because the crystallinity and the chemical composition of fresh opal from sediment trap samples, does not vary to the degree that the diffracting power of the opal is affected. Moreover, due to its small crystallites, which are randomly oriented in the opal remains, there are no particle orientation problems in specimens.

However, problems arise if:

a) aged opal is present. This can occur with

resuspension of sediment from the sea-floor (see section 1). With aging, opal gradually converts to opal-CT, thus changing its diffraction pattern [Jones and Segnit, 1971] and, consequently, its diffracting power.

b) volcanic ash, Fe, Mn-nodules, and organic substances are present. These materials show bulges in diffraction patterns, which will interfere with that of opal [Lapaquellerie, 1987].

To detect the ability of XRD for the small trap samples, the XRD method of Eisma and Van der Gaast [1971] is compared with the extraction method of Mortlock and Froelich [1989]. The trap samples (kindly supplied by P. Biscaye) are from the SEEP-program carried out in the Northern Atlantic Ocean off the coast of the northern U.S.A. For this, 50 mg of Ca-exchanged and dried material was mixed with 10% of weight of α-Al_2O_3(1.0μm, BDH) and from 5 mg of the mixture an unoriented specimen was made in a depression in a brass plate. The specimen was scanned three times from 13 to 46° 2θ at 50%RH with a counting time of 2 seconds for each step of 0.02° 2θ and with a 0.2 mm receiving slit. The results (Van der Gaast, unpublished, 1988) are given in Table 6-2 together with the results of the extraction method carried out on a split of the samples. For samples A to D the XRD method gave somewhat higher amounts and for samples E and F somewhat lower amounts than the extraction method. Sample E, however, gave even a lower amount after re-analysis (value between brackets). This was probably due to inhomogeneity of the sample (P. Biscaye, personal communication, 1989). This is, taking into account the low opal contents for most samples, a quite satisfying result. Because the real amount of opal of these samples is unknown, the most accurate method of the two is unknown. In Figure 6-4 the patterns of samples A and B are presented. Sample A shows many diffraction peaks from lithogenic material, calcite, and some aragonite on a small opal bulge. Sample B shows a strong opal bulge and some peaks of lithogenic material and calcite.

In Figure 6-1A a diffraction pattern is displayed from sample D4 from the JGOFS-program. This sample is from 45.3° N, 25° W at a water depth of 3016 meters. Its carbonate content is 47%, which was measured by difference of weight after dissolution, and the organic C is 14.5%. With a conversion factor of 2 this makes 29% organic matter in the sample. From 2.5 mg material, three scans under identical conditions as those mentioned above were carried out. The three patterns were summed and smoothed over five points of equal weight. At about 4.1Å the opal bulge is visible. In addition broad bulges, probably from organic matter, can be observed at higher angles. However, it might be possible that at the position of the opal bulge another broad reflection from organic matter is present. The degree of interference of it with the opal bulge is not known. In Figure 6-1B the presence of bulges at higher angles than that of opal has been made clearly visible by subtraction of the pattern of Figure 6-4B from that of Figure 6-1A. This is a clear example that the XRD method of Eisma and Van der Gaast [1971] can probably not be used.

The above mentioned examples and the discussion illustrate that the XRD method of Eisma and Van der Gaast [1971] also has limitations, which are met in other amorphous materials than that from fresh opal. In my opinion, a good analytical procedure is to first carry out a pilot study with XRD and microscopical techniques. If volcanic material or organic matter is rather high, then an extraction technique should be used. If the clay mineral content is high, especially that of smectite, then the XRD method should be applied. Namely, Mortlock and Froelich [1989] reported that with their extraction technique, smectite easily can be affected, thus releasing Si, resulting in percentages too high for opal.

5.2. Carbonate

The total carbonate can be determined chemically by dissolution. However, besides calcite, aragonite is formed as hard-parts of pteropods in the surface layers of the ocean and from shallow water benthic organisms, which can be laterally transported [Berner and Honjo, 1981]. The calcite/aragonite ratio can be determined with XRD since these two minerals are structurally different. Berner and Honjo [1981] determined this ratio by preparing mixtures in several ratios from pure planktonic forams (calcite) and pteropods (aragonite). From the peak height of the (104) peak of calcite at 3.035Å and the (111) peak of aragonite at 3.396Å of the mixtures, a calibration curve was prepared. From calcite/aragonite peak ratios the weight percentages were calculated. Droxler [1983] described such a method for carbonate oozes from sediment cores from the Bahamas using integrated peak intensities. He prepared his calibration curve from pure calcite and aragonite samples of comparable crystallinities to that of his samples. It is not known by the writer if an XRD method has been published for the determination of dolomite contents. Possibly, it can be measured using the method described above for calcite and aragonite.

Quantitative analysis is hampered, however, if quartz is present in appreciable amounts, its (101) peak at 3.343Å interferes with the (111) peak of aragonite (see e.g. Figure 6-4a).

Replacement of Ca by Mg in calcite in solid solution occurs frequently resulting in decreasing cell dimensions and a peak shift of the (104) peak towards higher 2θ angles [Goldsmith et al., 1961]. Berner and Honjo [1981] detected in trap samples from the Atlantic Ocean (the Rio Grande Rise) a high-magnesium calcite (HMC) in combination with a low-magnesium calcite (LMC). From the location of the (104) HMC peak its Ca/Mg ratio can be determined [Graf, 1961; Goldsmith et al., 1961]. If the HMC shows its presence as a shoulder on the LMC peak, α_2-stripping or measuring with $K\beta$-radiation may be useful to determine the exact peak location of the HMC component (see section 4.6.3.). Quantitative analysis of HMC by XRD, however, is greatly hampered by a differing crystallinity and diffracting power [Gavish and Friedman, 1973, and references therein].

5.3. Clay Mineral Analysis of Small Amounts of Material

For clay mineral identification, preferentially, oriented specimens are used. The 00*l* basal reflections in patterns are related to the type of layered structure involved. For identification the reader is often referred to Brown and Brindley [1980], Wilson [1987], and Moore and Reynolds, Jr. [1989c]. Their identification procedures are mainly based on treatments of a specimen after which the specimen is scanned. These treatments are carried out for the major part to characterize the swelling minerals such as smectite and for a minor part to detect the presence of e.g. kaolinite. For example a specimen is scanned at room conditions and after glycolation it is scanned again. From another part of the sample, which is K^+ exchanged, a specimen is prepared and scanned at room conditions and scanned after successive heat treatments at 350° C and 500° C. If this analysis is applied to trap samples, the disadvantages are that a) large amounts of material are needed, which, after analysis, only can be used for bulk chemical analysis, and b) the RH conditions for all scans are poorly defined (see section 4.5.3.), except that of glycolation because in that case the RH has no influence on the spacing. Swelling minerals will absorb amounts of water depending on the prevailing RH and thus may show a variety of spacings between 15Å and 12Å. Moreover, the measured spacings of the swelling minerals after the heat treatments will depend on the type of swelling mineral present in the specimen, the prevailing RH, the cooling time, and the scanning time. To prevent the above mentioned aberrations, the RH in the specimen chamber must be controlled during scanning.

If only small amounts of material are available, which is often the case with trap samples, the RH can be stabilized at certain RH's during scanning and manipulated to obtain different stabilized RH values, swelling minerals can be identified by their changing spacings and thus shifting peaks with only one specimen. For this, the windows of the specimen chamber are covered with Mylar foil of 10 to 20 μm thickness (sealed with double-sided tape), thus leaving the windows transparent to the X-rays. The front-side of the chamber must be closed except for a small leak, and a gas (air, He, or N_2) must then be flushed into the chamber via a small tube. The RH of the gas should be tuneable. In the writer's lab a humidity generator was constructed, with which the three mentioned (dry) gases could be moistened from 5 to 92% RH. In case such a generator is not available, a dry gas can be bubbled through saturated salt solutions. A wide range of RH's can be obtained this way. This technique can be found in any good chemistry handbook. A check of the generated RH, however, is necessary, because the obtained RH depends also on the applied flow of the gas through the salt solutions.

An oriented specimen is prepared by suction of a Ca exchanged suspension onto e.g. a ceramic tile. The soaked tile with the wet specimen is directly inserted in the specimen holder of the diffractometer, which is flushed with a gas of greater than 90% RH. During the scan the specimen will be prevented from drying due to the relatively large amount of water in the ceramic tile, which acts as a reservoir feeding the specimen, and also because of the high RH of the flushed gas. Under this condition, the particles are immersed in water and one could say it is 100% RH, smectites expand to 19Å, thus replacing the glycolation method. After this scan the specimen is dried at room conditions and scanned at 50% RH, thus replacing the scan at room condition. The smectite has then lost some of its absorbed water and thus collapses to a spacing of about 15Å.

After flushing dry gas over the specimen for about 30 minutes, little absorbed water is left and the smectite collapses then to a spacing of 11Å to 12Å. How much the smectite collapses probably depends on the presence of some blocking material such as Fe, Al, Mg,-(hydr)oxide or organic matter in the interlayer and on the shape of the clay crystals [Van der Gaast et al., 1986].

In Figure 6-5 the shift of the (001) smectite peak from a specimen of a Ca-exchanged less than 2 μm fraction of a sample from ODP Leg 114, Hole 699 of Miocene age. Recorded intensities are corrected for LPF and the irradiated volume. The three patterns were plotted normal to the peak of the internal standard (MoS_2). This Figure is from Monty et al. [1991]. The patterns show, besides the shifting peaks of smectite, also at 100% RH a peak from vermiculite at 15.2Å. At lower RH the mineral collapses equally with smectite and is thus not visible. Further illite, kaolinite, and chlorite can be identified at 9.97Å, 7.15Å, and 7.07Å, respectively. Peak intensities relative to that of the internal standard (MoS_2) can be measured using the smectite and vermiculite peak of the pattern of 100% RH, the illite of 50% RH, and the combined kaolinite (001) and chlorite (002) peak of any RH. The ratio of kaolinite and chlorite can better be measured with their (002) and (004) reflections, respectively, in the 3.5Å to 3.6Å region and this ratio must then be transferred to their (001) and (002) reflections, respectively.

Unoriented specimens in a depression in a plate can also be measured at the RH values mentioned above. At 0% and 50% RH the specimens are flushed with the gas of the RH. For scans at 100% RH a small amount of water is added from a micro-pipet, so that the material is just moist. During the scan the specimen is flushed with gas of greater than 90% RH. After the scan, the specimen must be checked if it is still moist. For the influence of RH on the spacing of smectite and vermiculite, the reader is referred to MacEwan and Wilson [1980].

6. Conclusions

With a proper choice of pretreatment of samples and preparation techniques; with a correct setting of the XRD apparatus, which must be adapted as much as possible to the dimensions of the specimen, and an adequate treatment of digitized data, the opal content and the contents of carbonate minerals can be determined from about 2.5 mg sediment trap samples. At the same time useful mineralogical information can be obtained from relative quantities of other minerals as well.

References

Berner, R. A., and S. Honjo, Pelagic sedimentation of aragonite, its geochemical significance, Science, 211, 940-942, 1981.

Betzer, P. R., J. G. Byrne, J. G. Acker, C. S. Lewis, R. R. Jolly, and R. A. Feely, The oceanic carbonate system; a reassessment of biogenic controls, Science, 226, 1074-1077, 1984.

Bish, D. L., and J. E. Post, Modern Powder Diffraction, Vol. 20, 369 pp., Reviews in Mineralogy, The Mineralogical Society of America, Washington, D.C., 1989.

Bish, D. L., and R. C. Reynolds, Jr., Sample preparation for X-ray diffraction, in Modern Powder Diffraction, edited by D. L. Bish, and J. E. Post, Vol. 20, Chap. 4, pp. 73-99, Reviews in Mineralogy, The Mineralogical Society of America, Washington, D.C., 1989.

Brewster, G. R., Effect of chemical pretreatment on X-ray powder diffraction characteristics of clay minerals derived from volcanic ash, Clays Clay Miner., 28, 303-310, 1980.

Brindley, G. W., Quantitative X-ray mineral analysis of clays, in Crystal Structures of Clay Minerals and their X-ray Identification, edited by G. W. Brindley, and G. Brown, Chap. 7, pp. 411-438, Mineralogical Society, London (Eng.), 1980.

Brown, G., and G. W. Brindley, X-ray procedures for clay mineral identification, in Crystal Structures of Clay Minerals and their X-ray Identification, edited by G. W. Brindley, and G. Brown, Chap. 5, pp. 305-359, Mineralogical Society, London (Eng.), 1980.

Calvert, C. S., D. A. Palkowski, and D. R. Pevear, A combined X-ray powder diffraction and chemical method for the quantitative mineral analysis of geological samples, in CMS Workshop Lectures, edited by D. R. Pevear, and F. A. Mumpton, Vol. 1, Chap. 6, pp. 154-166, The Clay Minerals Society, Evergreen, Colorado, 1989.

Chester, R., H. Elderfield, J. J. Griffin, L. R. Johnson, and R. C. Padgham, Eolian dust along the eastern margins of the Atlantic Ocean, Mar. Geol., 13, 91-105, 1972.

Cullity, B. D., Elements of X-ray Diffraction, 814 pp., Addison-Wesley, Reading, Massachusetts, 1967.

Dankers, N., and R. Laane, A comparison of wet oxidation and loss on ignition of organic material in suspended matter, Environ. Technol. Lett., 4, 283-290, 1983.

Davis, B. L., and L. R. Johnson, Quantitative mineral analysis by X-ray transmission and X-ray diffraction, in CMS Workshop Lectures, edited by D. R. Pevear, and F. A. Mumpton, Vol. 1, Chap. 4, pp. 104-117, The Clay Minerals Society, Evergreen, CO, 1989.

Deuser, W. G., E. H. Ross, and R. F. Anderson, Seasonality in the supply of sediment to the deep Sargasso Sea and implications for the rapid transfer of matter to the deep ocean, Deep-Sea Res., 28(5), 495-505, 1981.

DeWolff, P. M., J. M. Taylor, and W. Parrish, Experimental study of effect of crystalline size statistics on X-ray diffractometer intensities, in X-ray Analysis Papers, 130-136, Centrex: Eindhoven, 1965.

Douglas, L. A., and F. Fiessinger, Degradation of clay minerals by H_2O_2 treatments to oxidize organic matter, Clays Clay Miner., 19, 67-68, 1971.

Droxler, A., Quaternary aragonite cycles and oxygen isotope record in Bahamian carbonate ooze, Geology, 11, 235-239, 1983.

Eisma, D., and S. J. Van der Gaast, Determination of opal in marine sediments by X-ray diffraction, Neth. J. Sea Res., 5(3), 382-389, 1971.

Eisma, D., S. J. Van der Gaast, J. M. Martin, and A. J. Thomas, Suspended matter and bottom deposits of the Orinoco Delta: turbidity, mineralogy, and elementary composition, Neth. J. Sea Res., 12, 224-251, 1978.

Folger, D. W., Wind transport of land-derived mineral, biogenic, and industrial matter over the North Atlantic, Deep-Sea Res., 17, 337-352, 1970.

Gâtà, Gh., L'effet des champs ultrasonique sur les suspensions de minéraux argileux, Trans. 8th Int. Cong. Soil Sci. Bucharest, 3, 1165-1174, 1964.

Gavish, E., and G. M. Friedman, Quantitative analysis of calcite and Mg-calcite by X-ray diffraction; effect of grinding on peak height and peak area, Sedimentology, 20, 437-444, 1973.

Goldberg, E. D., Determination of opal in marine sediments, J. Mar. Res., 17, 178-182, 1958.

Goldsmith, J. R., D. L. Graf, and H. C. Heard, Lattice constants of the calcium-magnesium carbonates, Am. Mineral., 46, 453-457, 1961.

Graf, D. L., Crystallographic tables for the rhombohedral carbonates, Am Mineral., 46, 1283-1316, 1961.

Grim, R. E., Clay Mineralogy, second ed., 596 pp., McGraw-Hill, New York, 1968.

Harward, M. E., and A. A. Theisen, Problems in clay mineral identification by X-ray diffraction, Soil Sci. Soc. Proc., 26, 335-341, 1962.

Honjo, S., Material fluxes and modes of sedimentation in the mesopelagic and bathypelagic zones, J. Mar. Res., 38, 53-94, 1980.

Honjo, S., B. J. Hay, S. J. Manganini, V. L. Asper, E. T. Degens, S. Kempe, V. Ittekot, E. Izdar, Y. T. Knuk, and H. A. Benli, Seasonal cyclicity of lithogenic particle fluxes at a southern Black Sea sediment trap station, Mitt. Geol.-Paläont. Inst., Univ. Hamburg, 62, 19-39, 1987.

Hosterman, J. W., and F. T. Dulong, A computer program for semiquantitative mineral analysis by X-ray powder diffraction, in CMS Workshop Lectures, edited by D. R. Pevear, and F. A. Mumpton, Vol. 1, Chap. 2, pp. 38-50, The Clay Minerals Society, Evergreen, CO, 1989.

Jenkins, R., Experimental procedures, in Modern Powder Diffraction, edited by D. L. Bish, and J. E. Post, Vol. 20, Chap. 3, pp. 47-71, Reviews in Mineralogy, The Mineralogical Society of America, Washington, D.C., 1989a.

Jenkins, R., Instrumentation, in Modern Powder Diffraction, edited by D. L. Bish, and J. E. Post, Vol. 20, Chap. 2, pp. 19-45, Reviews in Mineralogy, The Mineralogical Society of America, Washington, D.C., 1989b.

Jenkins, R., and F. R. Paolini, An automated divergence slit for the powder diffractometer, Norelco Rep., 1, 9-16, 1974.

Jones, J. B., and E. R. Segnit, The nature of opal; I: Nomenclature and constituent phases, J. Geol. Soc. Aust., 18(1), 57-68, 1971.

Jones, R. C., A computer technique for X-ray diffraction curve fitting/peak decomposition, in CMS Workshop Lectures, Vol. 1, Chap. 3, pp. 52-101, edited by D. R. Pevear, and F. A. Mumpton, The Clay Minerals Society, Evergreen, Colorado, 1989.

Klug, H. P., and L. E. Alexander, X-Ray Diffraction Procedures, second ed., 966 pp., Wiley, New York, 1974.

Langeveld, A. D., S. J. Van der Gaast, and D. Eisma, A comparison of the effectiveness of eight methods for the removal of organic matter from clay, Clays Clay Miner., 26, 361-364, 1978.

Lapaquellerie, Y., Utilisation de la diffractometrie X pour la determination des constituants amorphes dans les sediments marins (silice biogène et cendres volcanique), Clay Miner., 22, 457-463, 1987.

MacEwan, D. M. C., and M. J. Wilson, Interlayer and intercalation complexes of clay minerals, in Crystal Structures of Clay Minerals and their X-ray Identification, edited by G. W. Brindley, and G. Brown, Chap. 3, pp. 197-248, Mineralogical Society, London (Eng.), 1980.

Mehra, O. P., and M. L. Jackson, Iron oxide removal from soils and clays by a dithionite-citrate system buffered by sodium bicarbonate, Clays Clay Miner., Proc., 7th Conf., Nat. Acad. Sci., Natl. Res. Council Publ., pp. 317-327, 1960.

Monty, C. L. V., F. Westall, and S. J. Van der Gaast, The diagenesis of siliceous particles in subantarctic sediments, ODP Leg 114, Hole 699; possible microbial mediation, in Proc. Ocean Drilling Program, Sci. Results, P. F. Ciesielski, et al., 114, pp. 685-710, College Station, Texas, 1991.

Moore, D. M., and R. C. Reynolds, Jr., X-ray diffraction effects, in X-ray Diffraction and the Identification and Analysis of Clay Minerals, Chap. 3, pp. 57-101, Oxford Univ. Press, Oxford, 1989a.

Moore, D. M., and R. C. Reynolds, Jr., Sample preparation techniques for clay minerals, in X-ray Diffraction and the Identification and Analysis of Clay Minerals, Chap. 5, pp. 179-201, Oxford Univ. Press, Oxford, 1989b.

Moore, D. M., and R. C. Reynolds, Jr., Identification of individual clay minerals and associated minerals, in X-ray Diffraction and the Identification and Analysis of Clay Minerals, Chap. 6, pp. 202-240, Oxford Univ. Press, Oxford, 1989c.

Moore, D. M., and R. C. Reynolds, Jr., X-ray diffraction effects, in X-ray Diffraction and the Identification and Analysis of Clay Minerals, Chap. 8, pp. 272-309, Oxford Univ. Press, Oxford, 1989d.

Mortlock, R. D., and P. N. Froelich, A simple method for the rapid determination of biogenic opal in pelagic marine sediments, Deep-Sea Res., 36, 1415-1426, 1989.

Parrish, W., Advances in X-ray diffractometry of clay minerals, in X-ray Analysis Papers, edited by W. Parrish, pp. 105-129, Centrex: Eindhoven, 1965.

Pilskaln, C. H., and S. Honjo, The fecal pellet fraction of biogeochemical particle fluxes to the deep sea, Global Biogeochem. Cycles, 1(1), 31-48, 1987.

Quakernaat, J., Direct diffractometric quantitative analysis of synthetic clay mineral mixtures with molybdenite as orientation-indicator, J. Sediment. Petrol., 40, 506-513, 1970.

Reynolds, R. C., Jr., Principles and techniques of quantitative analysis of clay minerals by X-ray powder diffraction, in CMS Workshop Lectures, edited by D. R. Pevear, and F. A. Mumpton, Vol. 1, Chap. 1, pp. 4-36, The Clay Minerals Society, Evergreen, Colorado, 1989.

Reynolds, R. C., Jr., Principles of powder diffraction, in Modern Powder Diffraction, edited by D. L. Bish, and J. E. Post, Vol. 20, Chap. 1, pp. 1-17, Reviews in Mineralogy, The Mineralogical Society of America, 1989.

Schulze, D., Identification of soil iron oxide minerals by differential x-ray diffraction, Soil Sci. Soc. Am. J., 45, 437-440, 1981.

Slaughter, M., Quantitative determination of clays and other minerals in rocks, in CMS Workshop Lectures, Vol. 1, Chap. 5, pp. 120-151, edited by D. R. Pevear, and F. A. Mumpton, The Clay Minerals Society, Evergreen, Colorado, 1989.

Smith, B. F. L., and B. D. Mitchell, Characterization of poorly ordered minerals by selective chemical methods, in A Handbook of Determinative Methods in Clay Mineralogy, edited by M. J. Wilson, Chap. 8, pp. 275-294, Blackie & Son, Ltd., Glasgow and London, 1987.

Snyder, R. L., and D. L. Bish, Quantitative analysis, in Modern Powder Diffraction, edited by D. L. Bish and J. E. Post, Vol. 20, Chap. 5, pp. 101-144, Reviews in Mineralogy, The Mineralogical Society of America, Washington, D.C., 1989.

Syvitsky, J. P. M., and A. G. Lewis, Sediment ingestion by *Tigriopus Californicus* and other zooplankton. Mineral transformation and sedimentological considerations, J. Sediment. Petrol., 50, 0869-0880, 1980.

Van der Gaast, S. J., and J. H. F. Jansen, Mineralogy, opal, and manganese of Middle and Late Quarternary sediments of the Zaire (Congo) deep-sea fan: origin and climatic variation, Neth. J. Sea Res., 12(304), 313-341, 1984.

Van der Gaast, S. J., and A. J. Vaars, A method to eliminate the background in X-ray diffraction patterns of oriented clay mineral samples, Clay Miner., 16, 383-393, 1981.

Van der Gaast, S. J., C. Mizota, and J. H. F. Jansen, Curved smectite in soils from volcanic ash in Kenya and Tanzania: a low- angle X-ray powder diffraction study, Clays Clay Miner., 34, 665-671, 1986.

Wefer, G., E. Suess, W. Balzer, G. Liebezeit, P. J. Müller, C. A. Ungerer, and W. Zenk, Fluxes of biogenic components from sediment trap deployment in circum-polar waters of the Drake Passage, Nature, 299, 145-147, 1982.

Wilson, A. J. C., Mathematical Theory of X-ray Powder Diffractometry, 128 pp., Gordon & Breach, Science Publ., New York, 1963.

Wilson, M. J., X-ray powder diffraction methods, in A Handbook of Determinative Methods in Clay Mineralogy, edited by M. J. Wilson, Chap. 2, pp. 26-98, Blackie & Son, Ltd., Glasgow and London, 1987.

MEASURING BIOGENIC SILICA IN MARINE SEDIMENTS AND SUSPENDED MATTER

David J. DeMaster

Department of Marine, Earth and Atmospheric Sciences, North Carolina State University, Raleigh, North Carolina 27695

1. Introduction

Measuring the biogenic silica content of marine sediments and suspended matter is essential for a variety of geochemical, biological, and sedimentological studies. Biota forming siliceous skeletal material account for as much as one third of the primary productivity in the ocean [Lisitzin, 1972] and a significant portion (2 to 70% by weight) of open-ocean sediments. Biogenic silica measurements reveal important information concerning the bulk chemistry of suspended material or sediment and are essential in any type of silica flux study in the water column or seabed. Analyses of this biogenic phase in marine plankton are useful in characterizing the basic types of biota present and in comparing the distributions of particulate and dissolved silicate when evaluating nutrient dynamics [Nelson and Smith, 1986]. In the marine environment, diatoms, radiolaria, sponges, and silicoflagellates are the common types of siliceous biota.

Analytical procedures developed to evaluate the biogenic silica content of marine samples include: conversion to cristobalite followed by x-ray diffraction [Goldberg, 1958]; absorption of infrared light [Chester and Elderfield, 1968]; direct x-ray diffraction [Eisma and Van der Gaast, 1971]; bulk chemical analysis coupled with normative calculations [Leinen, 1977]; and differential wet-chemical extraction in alkaline solutions [Bezrukov, 1955; Hurd, 1973; Paasche, 1973; DeMaster, 1979, 1981; Eggimann et al., 1980; Krause et al., 1983; and Mortlock and Froelich, 1989]. Many of these techniques have serious limitations depending on the amount and source (diatoms, radiolaria, sponges, or silicoflagellates) of biogenic silica and the nature of the non-biogenic phase (i.e., crystalline, amorphous, types of clay minerals present). For example, the conversion of biogenic silica to cristobalite is dependent on the types of coexisting clay minerals [Eisma and Van der Gaast, 1971], whereas the absorption/diffraction approaches are sensitive to the crystallinity and water content of the biogenic silica, which often vary depending on the source of the biogenic material. In addition, both of these techniques suffer from significant palagonite interferences [Hurd, 1973]. The normative calculation procedure requires a fairly complete bulk chemical analysis and also is subject to error depending on the accurate characterization and variability of the various end members. The wet-chemical dissolution technique appears to be the most versatile in its application to marine samples of various types and compositions. The remainder of this manuscript examines the most common extraction procedures for biogenic silica existing in the literature and discusses advantages and limitations of the various extraction conditions.

Marine Particles: Analysis and Characterization
Geophysical Monograph 63

2. Discussion

To be fair to the investigators who have proposed the various analytical schemes, the research objectives of the individual studies must be considered during critique and evaluation. For example, Paasche [1973] needed to extract all of the biogenic silica from marine diatoms grown in a chemostat. The investigator used a 10 minute extraction with 0.2 M NaOH in a boiling water bath. Without the matrix problems of coexisting alumino-silicate minerals, this approach works quite well, and in fact is still currently used to evaluate the biogenic silica contents of open-ocean plankton [Nelson and Smith, 1986]. Although this technique was developed primarily for diatoms, the procedure works well with only minor modification for other siliceous biota such as radiolaria, sponges, and silicoflagellates.

The main difficulty in measuring the biogenic silica content of sediments or suspended solids in high turbidity coastal waters is extracting the silica from the biogenic phases without extracting silica from coexisting alumino-silicates. A differential leaching technique commonly is used to maximize biogenic silica dissolution and minimize the silica originating from alumino-silicates. Correcting for the amount of alumino-silicate silica extracted is difficult because Al and Si are not extracted congruently by alkaline leaches [Follett et al., 1965]. Thus, measurement of dissolved Al as an index of alumino-silicate silica is a complicated and somewhat empirical procedure.

Bezrukov [1955] mentions a sodium carbonate leach for the measurement of biogenic silica, but few details of the extraction conditions are available. Hurd [1973] used a 5 hour, 85° C, 5 wt. % sodium carbonate leach to measure biogenic silica and was one of the first to try to compensate for clay mineral contributions to the extraction solution. He combined various amounts of biogenic silica with an artificial sediment standard composed of a "typical" clay mineral assemblage to create a working curve for Equatorial Pacific sediments. The difficulty with this approach is that a clay mineral standard needs to be made for each environment sampled and matching the matrix effects can be difficult.

To resolve the silica extracted from biogenic phases and alumino-silicates, DeMaster [1979, 1981] used a sequential alkaline extraction (85° C, 1% sodium carbonate solution, over a total period of 5 hours). Based on the observation that alumino-silicates release their silica in a linear fashion as a function of time and that most biogenic silica dissolves completely in the first hour or two under these extraction conditions, DeMaster measured the extracted silica after 2, 3, and 5 hours. The silica data as a function of time were extrapolated back to time zero to correct for the silica released from coexisting clay minerals and consequently, the biogenic silica content of the sample was determined (Figure 6-6). The 1% sodium carbonate solution was chosen to maximize the dissolution of biogenic silica in the predominantly diatomaceous sediments examined in the study and minimize the solution of alumino-silicate material (Figure 6-7). DeMaster [1979] recognized that this procedure would not be adequate for sediments or suspended material with significant amounts of robust radiolarian skeletons or sponge spicules and recommended that a stronger alkaline solution (0.1 M NaOH) be used for the extraction of these types of samples. Use of stronger extraction solutions, however, reduces resolution between the biogenic and alumino-silicate sources. The relative precision of the analytical technique is approximately 5%.

Papers by Eggimann et al. [1980], Krause et al. [1983], and Mortlock and Froelich [1989] provide the most recent discussion of wet-chemical methods for measuring biogenic silica. Eggimann et al. [1980] used a 21% (2 M) sodium carbonate solution for a 4 hour leach at 90 to 100° C. The strength of the sodium carbonate leaching solution was chosen based on the concentration necessary to dissolve reagent grade silicic acid, although some dissolution experiments were conducted with cultures of the solution-resistant diatom *Hemidiscus*. If the silica/clay ratio in the sample was greater than 1, no correction for alumino-silicate silica was necessary. When the silica/clay ratio was between 1 and 0.25, dissolved Al measurements were used to estimate the

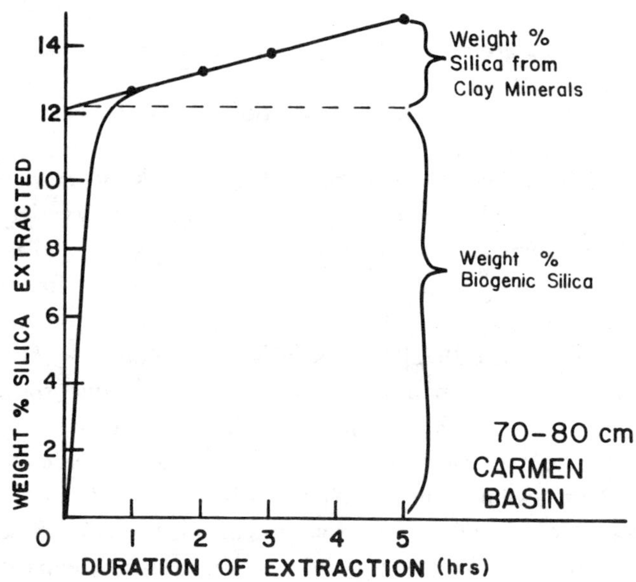

Fig. 6-6. Weight % silica extracted as a function of time for a sediment sample from Carmen Basin, Gulf of California. The weight % silica data from 2, 3, and 5 hours are extrapolated back to time zero to correct for the silica originating from coexisting clay minerals. After 5 hours of leaching, the clay minerals contributed approximately 2 weight % of the total 14 weight % extracted by the alkaline solution.

silica originating from the alumino-silicate phases. The alumino-silicate correction requires knowledge of the sample's mineralogy because initial tests reveal that the Al/Si ratio extracted from four individual clay minerals varied by more than a factor of 2. The investigators acknowledged that in samples containing low amounts of biogenic silica, a sequential extraction was necessary. One disadvantage in using a 21% sodium carbonate leaching solution is that the silica blanks are considerably greater than when using a more dilute solution.

To dissolve biogenic silica, Krause et al. [1983] used a 0.2 M NaOH solution at 100° C for a period of 10 to 15 minutes similar to the Paasche [1973] technique. They report less clay mineral dissolution with these extraction conditions than with a 5-hour, 0.5% Na_2CO_3 leach. DeMaster [1979], however, tried a similar procedure (100° C, 0.1 M NaOH, 12 min. leach) and found that the dissolution of biogenic silica from a diatomaceous Peru sediment was incomplete (~ 50%) and that the levels of clay mineral dissolution were comparable to the Na_2CO_3 technique.

The procedure described by Mortlock and Froelich [1989] is similar to the one by Eggimann et al. [1980] only

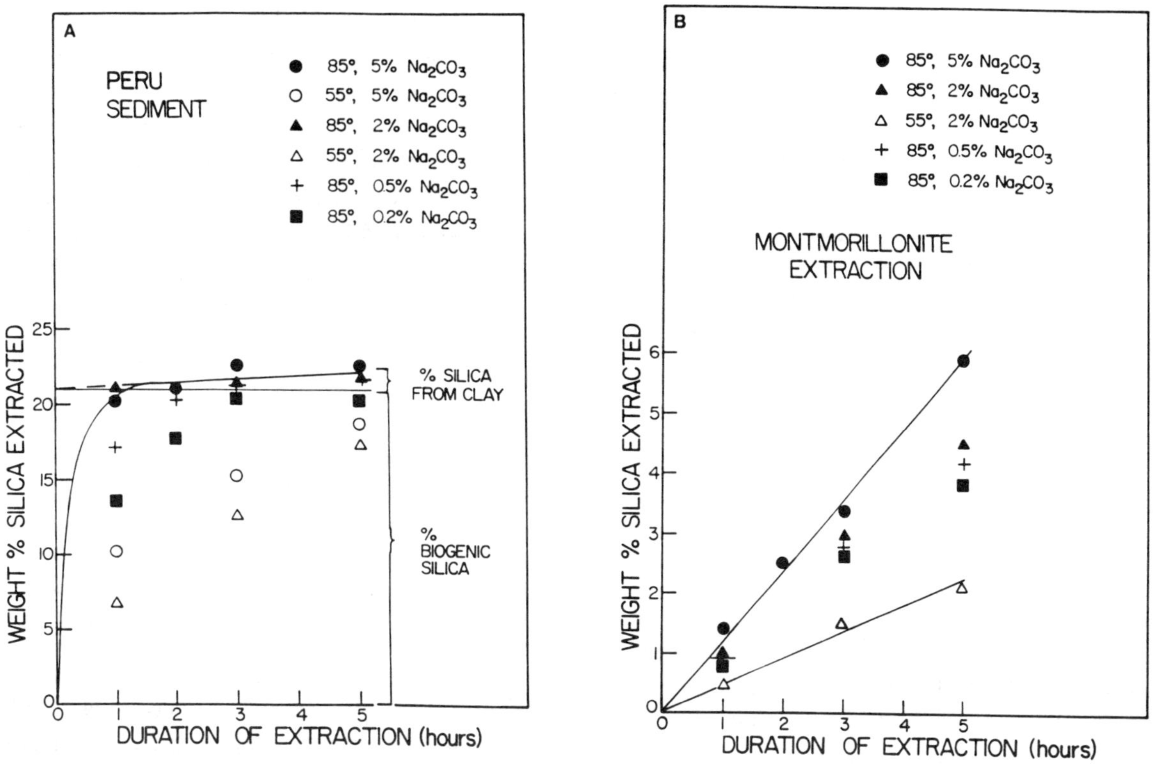

Fig. 6-7. A. Weight % silica extracted as a function of time, sodium carbonate concentration, and temperature from a Peru Shelf sediment sample. The Peru sediment was chosen because it contains minor amounts of clay minerals (detrital fraction composed of mainly feldspars and quartz), which maximizes the break in slope between biogenic silica dissolution (first 1 to 2 hours) and clay mineral dissolution (continuous over 5 hour period). B. Weight % silica extracted as a function of time, sodium carbonate concentration, and temperature from a montmorillonite clay (Wards Scientific, Wyoming montmorillonite). Depending on the abundance and type of clay minerals present and the concentration of the alkaline solution, the silica leached from coexisting alumino-silicates can be as great as 5 weight %.

a pretreatment with hydrogen peroxide and HCl was used, the leach was for a period of 5 hours at 85° C, and no dissolved Al measurements were made. Mortlock and Froelich [1989] chose to lightly crush their samples, because grinding can erroneously increase the silica leached from biogenic silica-poor samples [DeMaster et al., 1983]. The objective of the Mortlock and Froelich [1989] study was to use the biogenic silica measurements to "map" the distribution of biosiliceous components in Recent to Pliocene age deep-sea sediments. Ge/Si ratios were measured in the extraction solution to enable correction for alumino-silicate silica contribution.

This technique has great potential for resolving the source of the extracted silica because of the seven- to tenfold difference for the Ge/Si ratio in biogenic and alumino-silicate silica. Mortlock and Froelich [1989] used an average Ge/Si value of 10×10^{-6} for marine clays to make the correction, but no information is given on how this value might vary with differing alumino-silicate mineral assemblages. Although the corrections for alumino-silicate contribution changed values by as much as one part in six in their study, the investigators do not recommend routine Ge analyses in extraction solutions because the "systematic errors are small compared to the range of opal contents in biogenic sediments".

The use of a pretreatment such as hydrogen peroxide and HCl by Mortlock and Froelich [1989] may enhance the dissolution of biogenic silica by removing certain particle coatings. However, the effect that this pretreatment has on the silica extracted from coexisting alumino-silicates was not discussed by the investigators. Individual clay minerals, both pretreated and not-pretreated, need to be extracted with the alkaline solution to evaluate the variability in Ge/Si ratios as a function of mineral assemblage and determine the overall effect of pretreatment. One disadvantage in using the Ge/Si ratio to correct for clay mineral contributions compared to the sequential leach approach is that additional analytical equipment is necessary and the Ge measurements can be somewhat laborious. The Mortlock and Froelich [1989] paper provides the most recent intercomparison of wet chemical leaching techniques and also discusses standards and blank measurements for the analytical procedure.

More laboratory studies are needed before a general technique for biogenic silica analysis can be identified, and even then, the procedure may vary depending on the specific application. The age of biogenic silica needs to be considered and incorporated into the analytical scheme [Mortlock and Froelich, 1989] in addition to the type of biogenic silica (diatom, radiolarian, silicoflagellate, or sponge). The selection of particular extraction conditions should depend on the objectives of the study, the source of biogenic silica, the nature and amount of the coexisting alumino-silicate material, and the application of the resulting data set.

Two examples should serve to emphasize these points. In the Mortlock and Froelich [1989] study the investigators consider a 2 to 3% clay contamination error to be negligible. In contrast, for a study of global biogenic silica accumulation a 1 or 2% difference in the absolute biogenic silica content can make a dramatic change in the final results. The amount of terrigenous material supplied by rivers to the marine environment is approximately 200×10^{14} gm per year [Holeman, 1968]. If this terrigenous material were deposited along with 1 or 2% biogenic silica, 40 to 80% of the global biogenic silica burial (4 to 5×10^{14} gm silica yr^{-1}) would occur in these terrigenous deposits. The major sinks for biogenic silica in the marine environment appear to be in high latitude siliceous-rich sediments [DeMaster, 1981] rather than in terrigenous deposits with low biogenic silica content, but the calculation does reveal the dependence of global budgets on the sensitivity and accuracy of biogenic silica measurements.

A related example pertains to silica cycling on the Amazon shelf. The Amazon River annually supplies to the ocean 4×10^{13} gm of dissolved silica and 1.1×10^{15} gm of sediment. Several researchers [Milliman and Boyle, 1975; DeMaster et al., 1983, 1986] have determined from dissolved silica versus salinity relationships that as much as 50% of the dissolved silica from the river is utilized by siliceous plankton during river-ocean mixing. An important question is whether or not this biogenic silica is recycled on the shelf or accumulates in the seabed. If all of the dissolved silica extracted from surface waters (50% of the riverine flux) accumulated as biogenic silica in the seabed, the biogenic silica content of the subaqueous delta would be less than 2%. Thus, the accurate measurement of biogenic silica at low levels can by very important in evaluating the extent of silica recycling occurring in major estuarine environments. The technique ultimately selected for measuring biogenic silica must be sufficiently versatile to handle a variety of sample types and also be sufficiently sensitive to characterize silica contents at levels of 1% or less.

References

Bezrukov, P. L., Distribution and rate of deposition of silicate sediments in the Sea of Okhotsk, Dokl. Akad. Nauk, 103, 473-476, 1955.

Chester, R., and H. Elderfield, The infrared determination of opal in siliceous deep-sea sediments, Geochim. Cosmochim. Acta, 32, 1128-1140, 1968.

DeMaster, D. J., The marine budgets of silica and Si-32, Ph.D. dissertation, 308 pp., Yale Univ., New Haven, 1979.

DeMaster, D. J., The supply and accumulation of silica in the marine environment, Geochim. Cosmochim. Acta, 45, 1715-1732, 1981.

DeMaster, D. J., G. B. Knapp, and C. A. Nittrouer, Biological uptake and accumulation of silica on the Amazon continental shelf, Geochim. Cosmochim. Acta, 47, 1713-1723, 1983.

DeMaster, D. J., S. A. Kuehl, and C. A. Nittrouer, Effects of suspended sediments on geochemical processes near the mouth of the Amazon River: examination of biological silica uptake and the fate of particle-reactive elements, Cont. Shelf. Res., 6, 107-125, 1986.

Eggimann, D. W., F. T. Manheim, and P. R. Betzer, Dissolution and analysis of amorphous silica in marine sediments, J. Sediment. Petrol., 50, 215-225, 1980.

Eisma, D., and S. J. Van der Gaast, Determination of opal in marine sediments by x-ray diffraction, Neth. J. Sea Res., 5, 382-389, 1971.

Follett, E. A. C., W. J. McHardy, B. D. Mitchell, and B. F. L. Smith, Chemical dissolution techniques in the study of soil clays. Part I, Clay Miner., 6, 23-34, 1965.

Goldberg, E. D., Determination of opal in marine sediments, J. Mar. Res., 17, 178-182, 1958.

Holeman, J. N., The sediment yield of major rivers of the world, Water Resour. Res., 4, 737-747, 1968.

Hurd, D. C., Interactions of biogenic opal, sediment, and seawater in the Central Equatorial Pacific, Geochim. Cosmochim. Acta, 37, 2257-2282, 1973.

Krause, G. L., C. L. Schelske, and C. O. Davis, Comparison of three wet-alkaline methods of digestion of biogenic silica in water, Freshwater Biol., 13, 73-81, 1983.

Leinen, M., A normative calculation technique for determining opal in deep-sea sediments, Geochim. Cosmochim. Acta, 41, 671-676, 1977.

Lisitzin, A. P., Sedimentation in the World Ocean, edited by K. S. Rodolfo, 218 pp., Soc. Econ. Paleont. Mineral., Sp. Publ. #17, Tulsa, Oklahoma, 1972.

Milliman, J. D., and E. Boyle, Biological uptake of dissolved silica in the Amazon River Estuary, Science, 189, 995-997, 1975.

Mortlock, R. D., and P. N. Froelich, A simple method for the rapid determination of biogenic opal in pelagic marine sediments, Deep-Sea Res., 36, 1415-1426, 1989.

Nelson, D. M., and W. O. Smith, Jr., Phytoplankton bloom dynamics of the western Ross Sea ice edge, II. Mesoscale cycling of nitrogen and silicon, Deep-Sea Res., 33, 1389-1412, 1986.

Paasche, E., Silicon and the ecology of marine plankton diatoms, I. *Thalassiosira pseudonana* (*Cyclotella nana*) grown in a chemostat with silicate as limiting nutrient, Mar. Biol., 19, 117-126, 1973.

LABILE PHASES AND THE OCEAN'S STRONTIUM CYCLE: A METHOD OF SEDIMENT TRAP SAMPLING FOR ACANTHARIANS

Renate E. Bernstein and Peter R. Betzer

Department of Marine Science, University of South Florida, 140 Seventh Street South, St. Petersburg, Florida 33701

1. Introduction and Background

Acantharians are abundant marine planktonic protists and are the only marine organisms that use strontium as a major structural component. These organisms incorporate Sr in the form of celestite ($SrSO_4$) into their skeletons and cysts [Odum, 1951]. Because the ocean is undersaturated with respect to $SrSO_4$ [North, 1974], settling skeletons of dead acantharians and acantharian cysts are readily dissolved. After an initial burst of activity following their discovery in the mid-nineteenth century [Haeckel, 1887; Popofsky, 1909; Schewiakoff, 1926], the study of acantharians lagged during the mid-twentieth century. In part, this hiatus was related to the fact that acantharians are not preserved in the sedimentary record. Another contributing factor was the evolution in sampling and preservation techniques that mitigated against the preservation of these labile phases [Beers and Stewart, 1970; Michaels, 1988]. On the other hand, their susceptibility to dissolution implicates them as important mediators of the ocean's Sr budget [Bernstein et al., 1987]. Our short-term and relatively shallow sediment trap deployments, lack of preservatives, and rapid sample processing permitted the collection of the often elusive acantharian specimens.

In this paper we consider the specific approaches to sample collection and processing that have made it possible to focus on the involvement of these organisms in the ocean's strontium cycle.

2. Sediment Trap Sampling

Free-floating, modified Soutar-type sediment traps [Soutar et al., 1977; Betzer et al., 1984; Figure 6-8] were dispatched for a period of 24 hours. Traps were deployed individually at 400, 1500, and 3200 m depths near the Global Ocean Flux Station in the Atlantic (31° 49.5′N and 64° 08.2′W). Settling particles were captured in the 0.66 m^2 trap and funneled into a 3.8 cm diameter PVC collection cup located at the base of each trap. To prevent the collection of any materials during trap descent, sample cups filled with filtered sea water were sealed in a closed position. Pressure equilibration within the sampling cup was effected with a Millipore filtration device containing a 25 mm diameter, 0.4 μm pore size Nuclepore. The apex of the trap cone was open during descent, thereby allowing the trap to flush. At the desired depth horizon, a preset electronic release allowed the sampling cup to slide into its collecting position at the base of the cone (Figure 6-8). The electronic release also activated a double-seal, ball-valve system that terminated sampling and insulated the sample from any exchange during retrieval. In order to partition the particle flux by size, each sampling cup was equipped with inserts (Figure 6-9). These removable inserts consist of PVC rings covered with Nitex nylon screens that roughly segregate passively settling particles into greater than 500, 500 to 253, 253 to 63 and less than 63 μm size ranges. Our experience with the screens is that they do successfully partition the great majority of collected materials by size. For example, most of our acantharian specimens range from 70 μm to 250 μm and in our sample cups the vast majority of these were isolated on the 63 μm Nitex insert. Size consistent separations were also achieved for other biogenic phases including foraminifera, radiolaria, and pteropods. While the preponderance of coccolithophorids and their plaques were detected in the less than 63 μm size fraction, these particles were also found loosely adhered to larger particles. They are, therefore, distributed among all size categories.

3. Pre-Cruise Laboratory Preparation

Various sizes of Nuclepore polycarbonate filters, 0.4μm pore size were used for all sample filtrations. These filters were most suitable for our purposes which entailed the delineation of very small particles, down to 1 μm, that might have been obscured or lost in the interlocking fibers found in Millipore and Gelman filters.

A considerable portion of the metals associated with Nuclepore membranes can also be leached by an acid wash [Eggimann and Betzer, 1976]. This was accomplished by placing the individual filters into plastic

Marine Particles: Analysis and Characterization
Geophysical Monograph 63

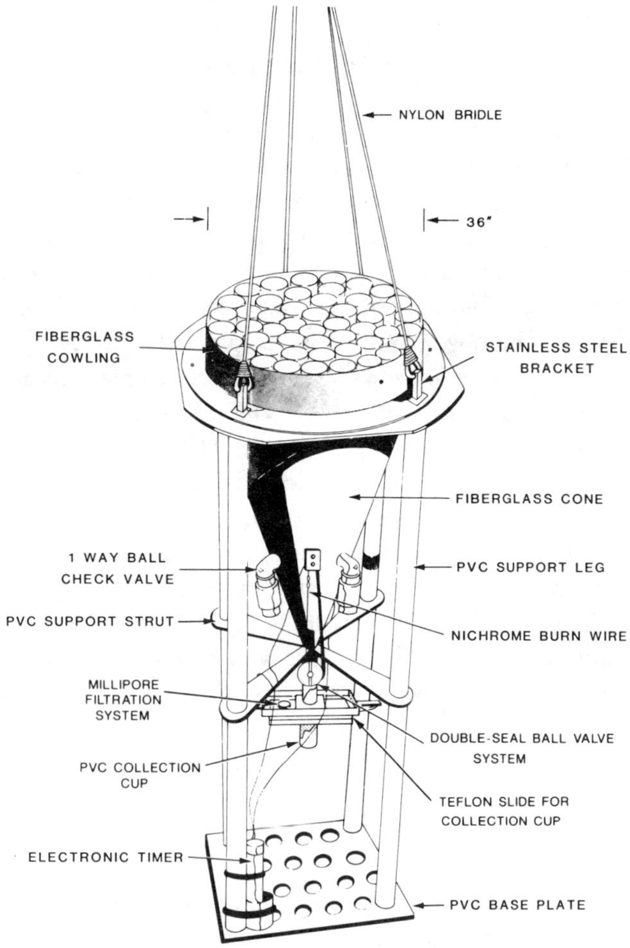

Fig. 6-8. Schematic of sediment trap sampling.

snap-top vials that had already undergone the acid washing procedure that is outlined below:

1. Cover filters with 3 Normal HCl for 24 hours with vial closed.
2. Pour off HCl and rinse three times with distilled-deionized water.
3. Cover filters with distilled-deionized water for 24 hours with vial closed.
4. Pour off water and rinse three times with distilled-deionized water.
5. With filter in the open vial, desiccate under vacuum and over silica gel for at least 48 hours. Silica contamination was obviated by always covering the silica gel in desiccators with Kimwipes.

The acid-washed filters were then preweighed using a Perkin-Elmer auto-balance housed in a temperature- and humidity-controlled room; the conditions were 20° C and 40%, respectively. Prior to use at sea, each filter was transferred and sealed in a numbered petri dish.

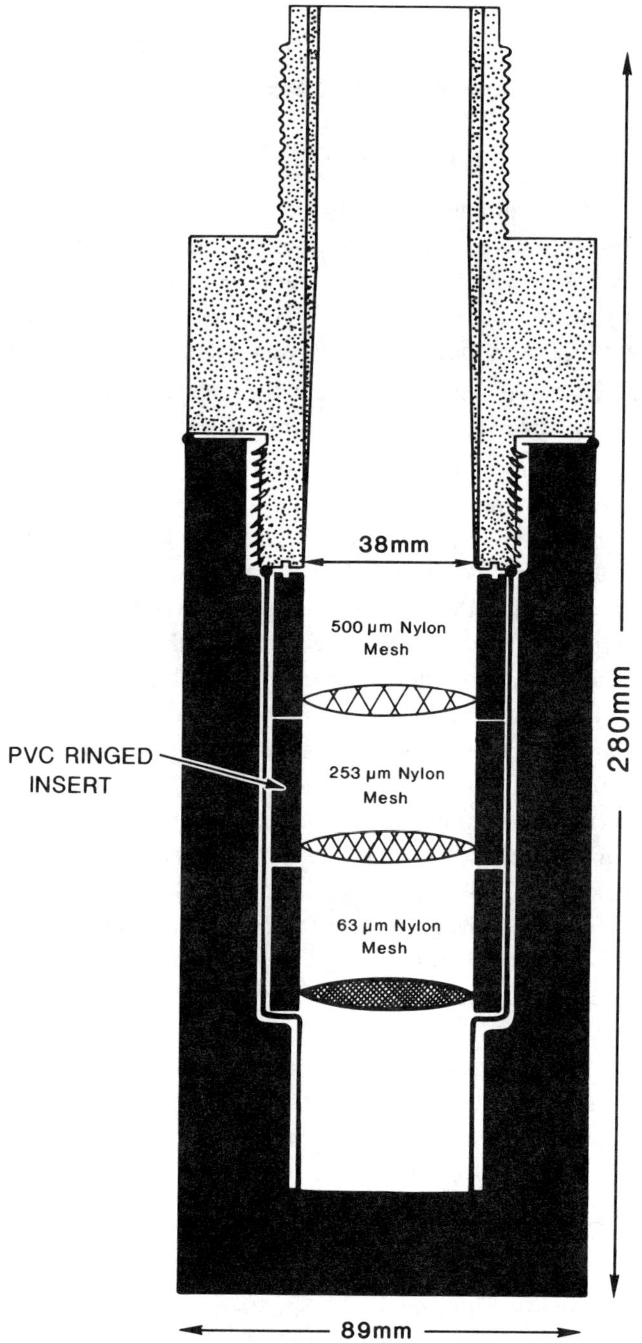

Fig. 6-9. Cross-section of sediment trap sampling cup revealing Nitex-screened inserts.

Weights were recorded accordingly. All other equipment, with the exception of the Nitex screens, that would be in immediate contact with the sample material was also acid washed; since Nitex nylon dissolves in acid it cannot be a part of this cleaning procedure. We must also note, that although the acid washing procedure is not critical to the analyses described in this paper, it is a major factor in subsequent chemical analyses.

4. Sample Processing

During sample processing at sea, we isolated portions of our sample on 47, 25 and 13 mm filters that were used, respectively, to: 1) estimate mass flux; 2) provide a detailed breakdown of the size distribution and chemical composition of the fine-fractioned portion of the sample (less than 63 μm); and 3) examine a small portion of the fine-fractioned sample under high magnification using a scanning electron microscope (SEM).

In all cases, the bulk of the sample (greater than 90%) was filtered on the 47 mm diameter membranes. A much smaller fraction of the less than 63 μm sample (less than 10%) was filtered onto 25 mm diameter filters that were used for individual particle analysis with a computer-controlled scanning electron microscope outfitted with an energy dispersive X-ray system. Finally, we isolated a small amount (less than 5%) of fine-fractioned (< 63 μm) material on 13 mm diameter filters for high resolution scanning electron microscopy.

4.1. Processing Equipment

Millipore Sterifil filtration setups employing 47 mm diameter Nuclepore filters were used to isolate most of the sample. A PVC filtration array containing fritted glass filtration bases and Teflon filtration funnels was devised for the 25 mm diameter Nuclepores. Teflon O-rings placed between the filter and the filtration funnel helped contain sample material during filtration. The array also contained an all-Teflon "spot" sampler (Figure 6-10) equipped with a 13 mm diameter filter that was used to disperse a sample over a 2.5 mm diameter field.

Shipboard sample contamination was minimized with a filtration "theatre" set up. Closed cell, high density polyethylene and high density styrofoam were used to create a five-sided box (top, bottom, and three sides) containing all filtration equipment. Individualized depressions were cut into the polyethylene base to contain and stabilize filtration setups. When not in use, a plastic curtain covered our theatre.

4.2. Filtration Procedure

Immediately upon trap retrieval, samples were filtered under vacuum no greater than 500 mm Hg. Because acantharian specimens dissolve rapidly, we placed a high priority on rapid sample processing so that our samples were always totally processed in less than 1 hour. Using a pipette, approximately 80 mL of material was removed from the sample cup and filtered onto a 47 mm diameter Nuclepore. The top of the cup

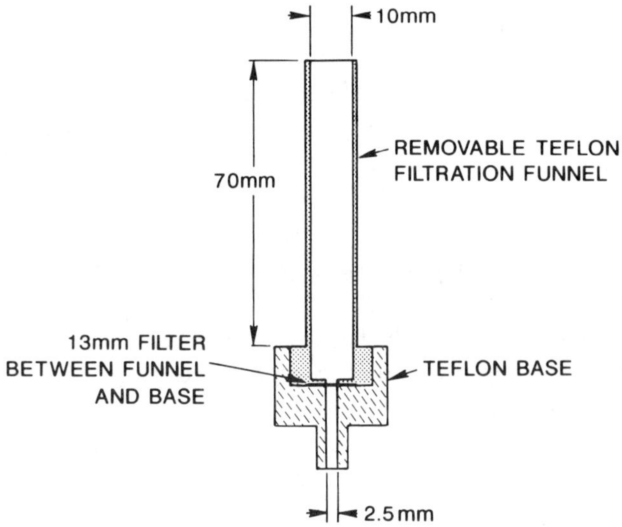

Fig. 6-10. All Teflon "spot" sampler for filtration with 13 mm diameter Nuclepores.

was then removed, thus providing access to the screened inserts (Figure 6-9). Teflon tongs were used to remove individual inserts, the contents of which were filtered onto separate 47 mm diameter Nuclepores, producing initial sample size fractionations approximating >500, 500 to 253, and 253 to 63 μm size ranges.

For the determination of the size and the associated chemical composition of individual particles, a portion (<10%) of the <63 μm size fraction that remained after the above processing was isolated onto 25 mm diameter filters. The analysis of these individual particles is optimized on samples with limited particle densities and good particulate dispersions. To insure that the fine-grained materials were suspended in the sample cup, each sample was stirred. Then, using either eppendorfs or other pipettes, varying amounts of sample material were filtered. To aid in achieving the appropriate particle densities and dispersions, after each filtration the filters were examined under a Wild dissecting microscope at a maximum magnification of 187x. The presence of extremely small and often abundant particles such as coccoliths, which are beyond the resolution capabilities of our dissecting microscope, necessitated bracketing the amounts of filtered material.

Thirteen mm diameter filters were placed into the Teflon "spot" samplers and a small amount of material was filtered for subsequent high resolution photomicroscopic analyses. The remaining less than 63 μm sample was filtered onto another 47 mm Nuclepore.

During each filtration sequence a 47 mm diameter filter was placed into a Millipore setup to produce an atmospheric blank. Periodic filtrations of distilled-deionized water and the filtered sea water used for filling the sample cups were made and used as blanks. To effect the removal of all sample material onto the appropriate filters, all equipment in contact with sample material was thoroughly rinsed with distilled-deionized water. Filters were also rinsed briefly to prevent the formation of salt crystals. All Nuclepores were and should be used with the "shiny" side up. Filters were subsequently transferred to petri dishes which were, in turn, placed, partially open, into vacuum desiccators. Here they were desiccated under vacuum and over silica gel for a minimum of 24 hours. They were then refrigerated while awaiting analysis at a shore-based clean facility.

5. Sample Analyses

To help minimize sample contamination, all shore-based work was conducted under a vertical flow bench or in a clean room where the temperature and humidity were maintained at 20° C and 40%, respectively. Initially, all filters were weighed to estimate mass flux values in mg m^{-2} day^{-1}.

The identification of acantharian specimens in the greater than 63 μm size fractions took place under a Wild reflected light stereomicroscope. These specimens were transferred to a preweighed Nuclepore using the surface tension created after dipping a Japanese calligraphy brush into distilled-deionized water. Depending on the size of acantharian specimens, either a flat Nuclepore membrane or one folded twice into the shape of a cone was used to contain transferred specimens. Specimens in the less than 200 μm size range tend to adhere sufficiently to a flat Nuclepore to allow filter manipulation. Larger specimens don't adhere to Nuclepores as well, and cones therefore, help to contain them. These larger specimens also may more easily be pushed into the apex of the cone. Numbers of specimens dictated the use of either a 25 or a 47 mm diameter membrane. After further desiccation, these isolated acantharians and cysts were weighed using the Perkin-Elmer autobalance. Weights were converted to fluxes of $SrSO_4$ in mg m^{-2} d^{-1}. Static electricity problems posed by desiccated samples were eliminated through the use of ionizing units.

5.1. Computer-Controlled Scanning Electron Microscopic Analysis

Scanning electron microscopic analyses were done on an ISI-DS-130 SEM. The DS refers to a dual stage system whereby the top, or first stage is used for higher magnification. The bottom or second stage on our system contains instrumentation for chemical analysis and image analysis. Chemical analysis is achieved by an energy dispersive X-ray analyzer (EDXA) system (KEVEX 7000). Both the SEM and the KEVEX systems can be computer-controlled by a LeMont image analysis system. In short, our image analysis system can provide both the size and associated chemical signature of the individual particles contained on a sample. Up to 32 elements may be detected. Particles can then be categorized using a chemical classification scheme specified by the operator.

5.2. SEM Sample Preparation

The 25 mm diameter Nuclepores containing the less than 63 μm sample fraction were mounted onto 25 mm diameter carbon SEM stubs. Filters were secured using Ernest F. Fullam Inc. sticky tabs and carbon paste. They were then carbon-coated and their associated particles analyzed by the LeMont system. Our SEM and Kevex systems cannot detect elements below atomic number 11. Because we were interested in specifying the chemical makeup of particles, we were careful to use carbon around our samples [Postek et al., 1980].

Filter pads, of 13 mm diameter, were prepared for scanning electron microscopic top-stage viewing. Because aluminum stubs for the top stage are only 10 mm in diameter, a rectangle containing the sample "spot" was excised from the 13 mm pad with a stainless steel scalpel. The rectangle was then secured on the stub with sticky tabs and silver paste (Figure 6-11). To enhance conductivity and high resolution imaging, samples were coated with Au-Pd [Beaton and Filshie, 1982]. We were now able to obtain high resolution photomicrographs of the less than 63 μm material; the equivalent of which we were analyzing on the bottom stage.

Blanks were prepared for image analysis by excising a portion of the filter and mounting it on a carbon stub using sticky tabs and carbon paste. These samples were carbon-coated and subsequently analyzed on our computer-controlled SEM-EDXA system.

6. Discussion

To our knowledge, our sediment trap samples are the first reported to contain significant numbers of acantharian specimens. Our relatively short, 24 hour, trap deployments, coupled with shallow sampling depths, lack of preservatives, and rapid sample processing help to account for this. However, despite the fact that

Fig. 6-11. Top-stage Al SEM stub with mounted filter rectangle containing sample "spot".

acantharians often outnumber other marine protozoans, the radiolarians and the foraminifera [Bottazzi et al., 1971; Beers and Stewart, 1971; Beers et al., 1975; Bottazzi, 1978; Michaels, 1988], in our samples most of the acantharian specimens are in the form of cysts (Figure 6-12). We, therefore, think that our acantharian specimens represent lower-bound numbers and that dissolution is still a consideration in our traps. It has been suggested that elevating the dissolved Sr content of the sampling cups might help curtail dissolution [Beers and Stewart, 1970; Michaels, 1988]. This process has, however, produced precipitates in certain instances [Michaels, 1988].

Our rapid sample processing in the field was facilitated by the screened inserts contained in the sample cups. Subsequent sample fractionation helped to minimize the potential for filter clogging, thereby promoting rapid filtration. This fractionation also minimized particulate clumping, thus enhancing particulate viewing and "picking".

Furthermore, computer-controlled scanning electron microscopic image analysis of the less than 63 μm size fraction indicated the presence of Sr and S. The top-

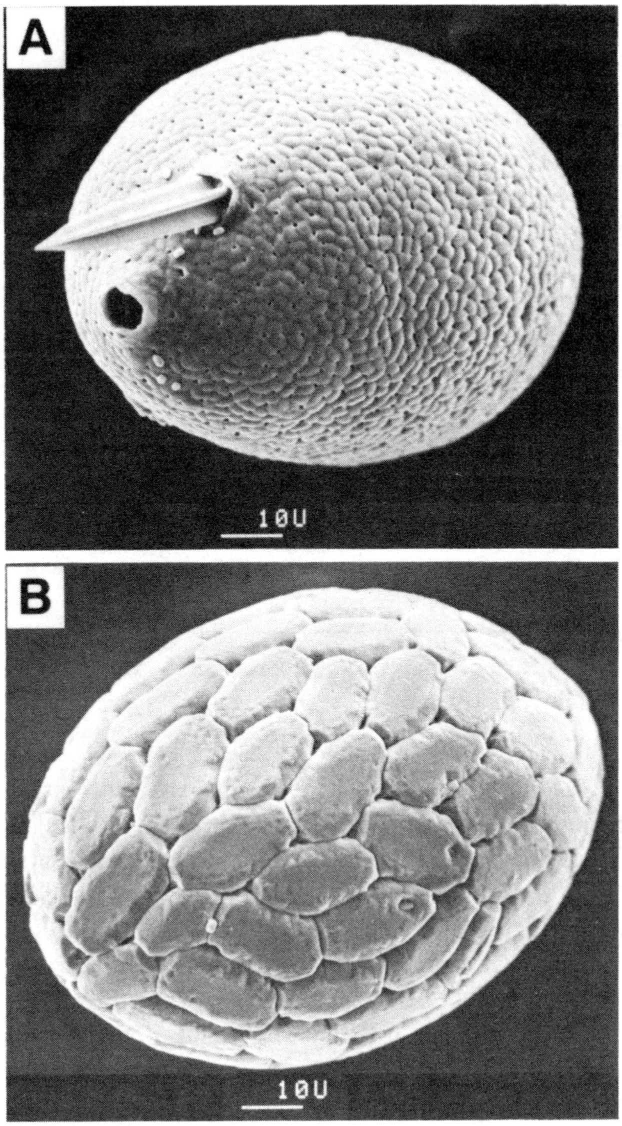

Fig. 6-12. Electron micrographs (A, B) depicting acantharian cysts. Scale bars represent 10 μm.

stage viewing of this same size fraction allowed us to observe heretofore undocumented minute discrete particles of celestite (Figure 6-13). These were also noted in 1500 m samples, where acantharian specimens are rare to nonexistent [Bernstein et al., 1987]. Our sampling technique permitted the disclosure of an extremely labile, minute component of the marine environment. This component is currently under investigation.

In summary, we recognize that the collection and study of such labile phases is fraught with difficulties. While our techniques have improved our understanding of the role acantharians play in the cycling of strontium

Fig. 6-13. Electron micrograph depicting a minute celestite component. Scale bar represents 5 μm.

in the world's oceans, we also appreciate the very marked possibility that improved techniques will provide even greater insights.

Acknowledgements. We wish to thank the captain, Thomas R. J. Tyler, the bosun, Jack E. Buss, and the crew of the R. V. Endeavor for their significant contributions toward achieving a successful cruise. The scientific crew, Walter Bowles, Robert Byrne, Steven Giordano, Giselher Gust, Chris Sabine, and Richard Young worked tirelessly and enthusiastically throughout the cruise. We are particularly grateful to Dr. Robert Byrne for his insightful suggestions during frequent consultations on methodology protocol. This research was funded in part by a National Science Foundation grant, OCE 8813436.

References

Beaton, C., and B. Filshie, Stereoscopic examination of the fine structure of some metal coatings used for scanning electron microscopy, JEOL News, 20E, 23-27, 1982.

Beers, J. R., and G. L. Stewart, The preservation of acantharians in fixed plankton samples, Limnol. Oceanogr., 15, 825-827, 1970.

Beers, J. R., and G. L. Stewart, Micro-zooplankters in the plankton communities of the upper waters of the eastern tropical Pacific, Deep-Sea Res., 18, 861-883, 1971.

Beers, J. R., F. M. H. Reid, and G. L. Stewart, Microplankton of the North Pacific Central Gyre. Population structure and abundance, June 1973, Int. Rev. Gesamten Hydrobiol., 60, 607-638, 1975.

Bernstein, R. E., P. R. Betzer, R. A. Feely, R. H. Byrne, M. F. Lamb, and A. F. Michaels, Acantharian fluxes and strontium to chlorinity ratios in the North Pacific Ocean, Science, 237, 1490-1494, 1987.

Betzer, P. R., W. J. Showers, E. A. Laws, C. D. Winn, G. R. Ditullio, and P. M. Kroopnick, Primary productivity and particle fluxes on a transect of the equator at 153° W in the Pacific Ocean, Deep-Sea Res., 31, 1-11, 1984.

Bottazzi, M. E., Systematic-ecological aspects of Radiolaria with special reference to Acantharia, Boll. Zool., 45, 133-144, 1978.

Bottazzi, M. E., B. Schreiber, and V. T. Bowen, Acantharia in the Atlantic Ocean, their abundance and preservation, Limnol. Oceanogr., 16, 677-684, 1971.

Eggimann, D. W., and P. R. Betzer, Decomposition and analysis of refractory oceanic suspended materials, Anal. Chem., 48, 886-890, 1976.

Haeckel, E., Report on the Radiolaria collected by the H.M.S. Challenger during the years 1873-1876, in Report on the Scientific Results of the Voyage of the H.M.S. Challenger, 1873-1876, Zoology, part I, 18, 888 pp., Johnson Reprint Corp., 1887.

Michaels, A. F., Vertical distribution and abundance of Acantharia and their symbionts, Mar. Biol., 97, 559-569, 1988.

North, N. A., Pressure dependence of $SrSO_4$ solubility, Geochim. Cosmochim. Acta, 38, 1075-1081, 1974.

Odum, H. T., Notes on the strontium content of sea water, celestite radiolaria and strontianite snail shells, Science, 114, 211-213, 1951.

Popofsky, A., Die radiolarian der Antarktis, in Deutsche Suedpolar-Expedition, 1901-03, Vol. 10, pp. 185-305, George Reimer, 1909.

Postek, M. T., K. S. Howard, A. H. Johnson, and K. L. McMichael, Scanning Electron Microscopy. A Student's Handbook, 305 pp., Michael T. Postek, Jr. & Ladd Res. Ind., Inc., 1980.

Schewiakoff, W., Die Acantharia des Golfes Von Neapel, in Fauna e flora del Golfo Di Napoli, Vol. 37, 755 pp., 1926.

Soutar, A., S. A. Kling, P. A. Crill, E. Duffrin, and K. W. Bruland, Monitoring the marine environment through sedimentation, Nature, 266, 136-139, 1977.

MINERALOGY AND CHEMISTRY OF MARINE PARTICLES BY SYNCHROTRON X-RAY SPECTROSCOPY, MÖSSBAUER SPECTROSCOPY, AND PLASMA - MASS SPECTROMETRY

Andrew C. Campbell

Department of Earth, Atmospheric, and Planetary Sciences, Massachusetts Institute of Technology, Cambridge, Massachusetts, 02139
(Current Address: United States Nuclear Regulatory Commission, Mail Stop OWFN-5E2, Washington, D.C. 20555)

1. Introduction

Marine particles have been the subject of intensive geochemical investigations for more than two decades. There is, however, a paucity of mineralogical and speciation information on individual particle phases in comparison to bulk chemical analyses. Hence there is a need to develop and apply new and existing technology to the characterization of marine particulate matter. This is especially important in studies of authigenic hydrothermal plume precipitates, where the exceedingly small grain sizes and amorphous nature of the particles limits the applicability of more standard bulk analysis methods. The focus of the work presented here is the development of an array of materials characterization techniques suitable for determining the mineralogical and trace element composition of cryptocrystalline and amorphous particles in hydrothermal plumes. The long term goal is to apply these tools to particles from a variety of environments including marine, lacustrine and atmospheric particles. One area of special interest is determining the oxidation state and coordination chemistry of adsorbed and coprecipitated metals in Fe and Mn oxide phases, which act as scavenging agents for a host of elements. Several novel analytical techniques are being employed in this endeavor: synchrotron x-ray fluorescence (SXRF) microprobe and x-ray absorption spectroscopy (SXAS), Mössbauer spectroscopy (for Fe phases), and inductively coupled plasma - mass spectrometry (ICP-MS). In addition, sampling and preservation protocols are required in order to obtain representative particle samples from dynamic plumes and to prevent oxidation of reduced chemical species.

2. Sampling and Preservation

2.1. Sampling of Hydrothermal Plumes

Systematic sampling of buoyant hydrothermal plumes is done from the research submersible DSRV Alvin using both conventional and zero-flushing volume samplers, and also in situ filtration systems. Because of the limited diameter of the rising plume it is not possible to systematically sample it from a surface ship using a wire mounted rosette. A typical vent orifice has a thermal power output of more than 1 megawatt and the resulting updraft around the chimney entrains the submersible in the rising plume. This phenomenon can be exploited to sample the plume as the sub is carried upwards at about 10-20 meters per minute from the point of entry into the plume to the neutrally buoyant layer 100-300 meters above bottom. During the January 1990 expedition to the Trans-Atlantic Geotraverse (TAG) and Mid-Atlantic Ridge Kane (MARK) hydrothermal fields, this sampling technique was employed to obtain water and particulate matter samples and measure physical properties with a CTD, nephelometer and light transmissometer [Campbell et al., 1990a]. The hydrothermal fluid component of these samples ranged from less than 0.1 % to about 2 %.

Specific dives were dedicated to plume sampling to avoid contamination of the sampling bottles with fine particles of chimney debris and bottom material. During these dives, the submersible never touched bottom throughout multiple sampling runs. Generally, neutral trim is achieved 50-100 meters above bottom and DSRV Alvin is driven down the water column using the vertical thrusters until bottom is sighted (about 10-15 meters altitude). The submersible is then flown toward the vent site using surface and in-hull navigation, as well as sonar to spot the plume. After entering the plume as close to the chimney tops as safety allows, the boat is controlled primarily by the dynamic upward movement of the rising waters. At the neutrally buoyant layer the upward movement ceases and the procedure is repeated. This routine was dubbed "riding the elevator" by participants.

Conventional 5 L Niskin water samplers were mounted on a small rosette attached to the submarine sample basket and provided samples of fluids and particles. One problem with these devices is that they must be flushed with at least five times their volume to obtain an undiluted sample. Weiss [1971] showed that the vertical movement of a Niskin type bottle by its characteristic flushing length (V/A) dilutes the contents by 1/e.

Marine Particles: Analysis and Characterization
Geophysical Monograph 63

A conventional 5 L Niskin has a flushing length of 1.62 m and requires a vertical movement through ~ 8 meters to be 99% flushed. Hence Niskin samples represent an integral of particles and fluids over several meters rather than an instantaneous signal. On the Mid-Atlantic Ridge the overall scale of the plume from multiple chimney sources was large enough that samples of distinct temperature regions within the plume were obtained. Sampling smaller scale plumes from individual chimneys may be more problematic.

The importance of the flushing volume problem was first recognized by Weiss, who designed a "zero flushing volume" sampler for use on Deep-Tow (Figure 6-14). Upon being tripped, this sampler encapsulates a parcel of water with no entrainment or other artifacts. It was with this device that Weiss et al. [1977], Lupton et al. [1977] and Klinkhammer et al. [1977] first demonstrated that active hydrothermal venting was occurring at a spreading ridge. By mounting a modified version of these samplers on the "forehead" of DSRV Alvin, immediately forward of the sail, the sampling problems discussed above have been eliminated. The sampled water and particles are an instantaneous signal that can be related to the physical properties recorded momentarily before closing of the bottles. In addition these are more appropriate for sampling small scale plumes where flushing volume problems are most important.

To sample particles that initially precipitate as cold, alkaline, seawater is entrained in the hot, acidic and metal rich vent fluids, it is necessary to employ an in situ filtration system. A temperature profile of the initial plume is obtained with high temperature (HT) thermocouple probes up to the reach of the submersible's manipulator arm (about 1 meter). At a specific height above the orifice the entrance nozzle of a simple pumping/filtration system is inserted into the plume and particles and fluid are pumped for about 5-10 minutes.

2.2. Sample Filtration and Preservation

The main factor in preserving the particulate samples from a plume environment is to keep the material separated from atmospheric oxygen as much as possible. Prior to filtration there is little that can be done to prevent reactions in the sampling bottle.

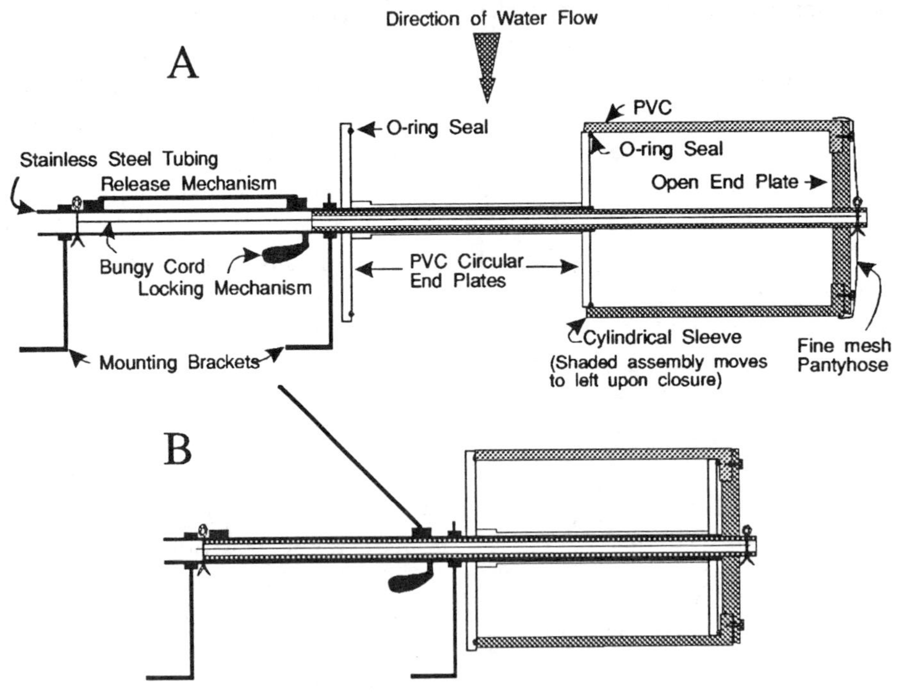

Fig. 6-14. Schematic drawing of Weiss style "zero flushing volume" plume samplers, A) open position, B) closed position. Upon returning to the ship the sampler is removed from the sub. Water is drawn by vacuum through a valve (not shown) directly into a pre-loaded filter cartridge and is collected directly into acid leached plastic bottles. A vent port is connected to a supply of Ar or N_2 gas to inhibit oxidation during filtration.

However, whatever reactions that do occur after closure of the bottle likely would take place in the water column on a similar time scale. The dilution factors for samples collected in the buoyant plume range from about 50 to 10,000 volumes of seawater to 1 volume of vent fluid. Hence reactions that precipitate sulfides, which occur in the one- to tenfold dilution zone [Bowers et al., 1985], are completed before sampling with zero-flushing volume and Niskin bottles. In situ filtration is the only way to obtain particulate samples from the sulfide precipitation zone, which at exit flow rates of a m sec^{-1} is completed in the first meter above the orifice.

Filtration of the samples is best done under an inert Ar (or N_2) atmosphere, either in a glove bag or by pressure/vacuum filtration directly from the sampling bottle. Samples are filtered onto pre-weighed polycarbonate membrane filters (0.4 μm). (For XRF work a low Br variety of filter is now used because the Br K_α line overlaps with the As K_β line in the XRF spectrum. An accurate fitting of the As K_β line is essential to separate the As K_α line overlap with the Pb L_α line in order to obtain accurate data for As and Pb.) After rinsing with ~ 10 mL of filtered and degassed distilled water to remove Cl, residual water is sucked from the filter and it is transferred to a pre-cleaned plastic filter container. The filter is inserted into a PVC food sealable bag, dried under a stream of filtered Ar, and the bag is heat sealed after completely purging with Ar. When done the bag should be quite puffy.

3. Synchrotron X-ray Spectroscopy

3.1. Synchrotron X-ray Fluorescence Microprobe

Synchrotron X-ray fluorescence spectroscopy is a new technique that uses the very intense "white light" synchrotron radiation from an electron storage ring as the primary excitation radiation. Advantages over electron beams or X-ray tubes [Jones and Gordon, 1989] include very high brightness (~ 10^9 times brighter than conventional XRF), intrinsic collimation, low scattering backgrounds due to the polarized nature of synchrotron X-rays, high fluorescent yield relative to electron excitation, high signal to noise ratio (1 x 10^5 better than an electron beam), and low energy loss in target (less radiation damage and target heating). These characteristics make SXRF an ideal analytical tool for determining the trace element composition of small particles. Recently Campbell et al. [1989, 1990b] have used SXRF to determine the trace element distributions in different mineral phases found in "black smoke" precipitates from seafloor hot springs. Many problems in marine geochemistry involve the transfer of elements from dissolved to particulate phases and so the SXRF method has tremendous potential in chemical oceanography and environmental chemistry.

3.1.1. Microprobe system.

The SXRF analyses were carried out on the X-26a beamline microprobe at the National Synchrotron Light Source, Brookhaven National Laboratory (NSLS-BNL). A schematic drawing of the microprobe is shown in Figure 6-15. The beam footprint at 9 meters from the ring (position of the X-26a beamline hutch at NSLS-BNL) is about 15 mm wide and 4 mm high. The brightest, most polarized radiation within this cross section provides the best signal to noise ratio (S/N) and the set-up procedure is designed to find this portion of the beam. The S/N of a well characterized sample is used, typically the Sr K_α line in an Anorthite-100 polished standard with 1400 ppm Sr. Once the position of the 4-jaw slits is centered on the beam they can be varied from ~ 20-300 microns horizontal and vertical width without affecting the signal to noise ratio (the cross section of the optimum part of the beam is ordinarily 1-2 mm vertical by several mm horizontal). The X-rays exit the beamline vacuum (< 10^{-9} Torr) through a water cooled Be window and then enter an ion chamber which measures I_o, the intensity of the incoming radiation.

To further reduce the size of the beam spot on the sample a pinhole that can be varied from 5 μm to 25 μm or larger is used. In the case of single particles the beam spot must irradiate the entire particle. Since the pinhole sub-samples only a small portion of the beam exiting the ion chamber, a second reference counter is necessary. This consists of a scintillation counter that measures the Ar K_α line from the air path to the sample. Its use is essential if the beam intensity must be varied for different samples (to reduce detector dead time) by changing the 4-jaw slits. A typical normalization curve for the scintillation counter relative to Fe in a plume sample is shown in Figure 6-15B.

Up to four samples are held on a 3-axis motorized stage. Particles on polycarbonate membrane filters are mounted in cardboard 35 mm slide holders such that there is no backing for the particles other than the membrane filter. A petrographic microscope, with 5x, 20x, and 40x objectives and a video monitor, provides images of the sample. The detector is a standard Si(Li) or Ge(Li) energy dispersive detector with the option of using various filters (e.g. 170 or 85 μm Al or 3 mil

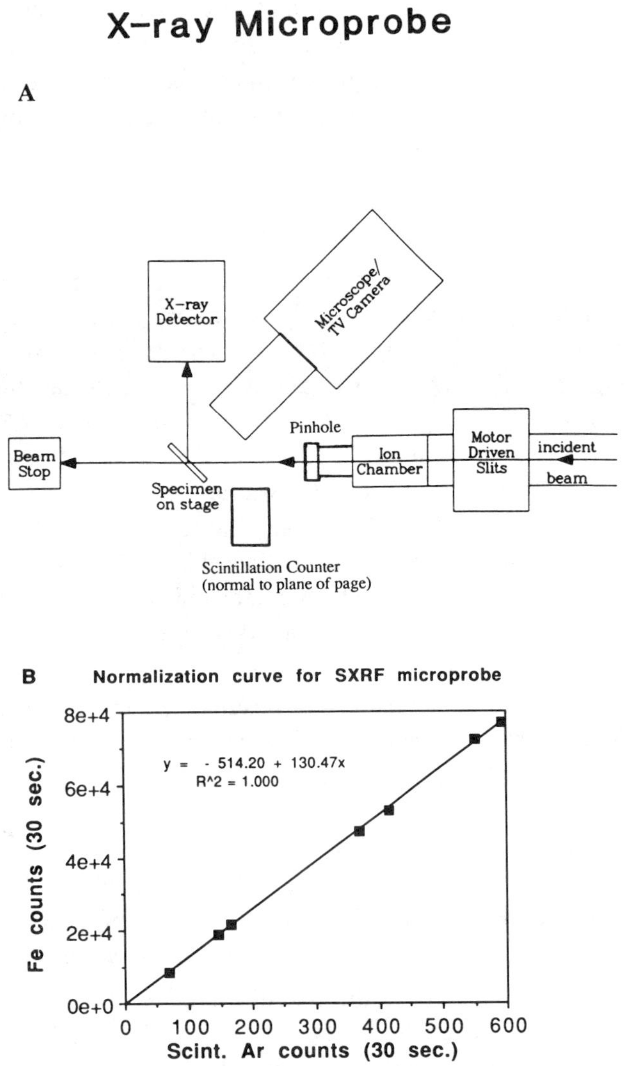

Fig. 6-15. A) Schematic drawing of synchrotron X-ray fluorescence (SXRF) microprobe that uses collimated synchrotron radiation at the X-26a beamline NSLS-BNL. [Figure modified from Lu et al., 1989]. B) Plot of normalization curve for Ar scintillation counts (air path before sample) and Fe counts (thin film of Fe particles (<1 μm) on membrane filter; beam spot ~ 10 μm diameter).

Kapton). Since the X-rays are polarized in the plane of the ring, the detector is placed at 90 degrees relative to the incoming beam to minimize scattering background. In many spectra an 85 μm Al filter was placed over the detector to eliminate elements lighter than Fe and reduce the Fe count rate eightfold. The limit of detection is determined by the total count rate for the detector, hence reducing the Fe counts, the major element in plume particles, significantly reduces detector dead time and improves the detectability for trace elements that have X-rays not significantly attenuated by the Al. The 85 μm Al filter also greatly attenuates or eliminates Fe pileup peaks, which interfere with As and Pb. Typically the system has 1 ppm detection limits on 5-10 μ spots with some variability for different elements. In the present configuration of the SXRF microprobe, soft x-rays are strongly scattered by the air path from the sample to the detector (Figure 6-15). Hence, the fluorescence intensity falls off rapidly at energies below ~ 4keV and little useful data is obtained for elements lighter than sulfur.

3.1.2. Standardization.

The most critical aspect of the technique is converting the background corrected peak intensities into concentrations. In principle, an ab initio procedure can be used to generate concentrations from the known geometric parameters of the SXRF microprobe system, the energy distribution of the synchrotron beam, the sample thickness and density, and the basic physical laws of photon-matter interactions. A computer code developed at the Naval Research Laboratory [Criss, 1977] and modified at NSLS-BNL [Sutton et al., 1986, and personal communication, 1990] follows such an approach. The basic equation for the concentration of element z in the sample for the SXRF system with a 90° angle between the detector and incident beam takes the following form:

$$C_z = (K\, I_z\, [(\mu/\rho)_o + (\mu/\rho)_z]) \,/\, (I_o\, [1 - \exp\{-[(\mu/\rho)_o + (\mu/\rho)_z]\rho T / 0.71\}])$$

where I_z is the fluorescence intensity for element z, I_o is the incident radiation from the synchrotron that can ionize the pertinent shell or subshell for element z, K is a constant composed of the fluorescence cross section for element z at a given I_o and the detector efficiency, geometry, filtration effects, air path, etc., ρT is the sample density and thickness, and $(\mu/\rho)_o + (\mu/\rho)_z$ are the sample mass absorption coefficients for the primary and secondary fluorescent radiation [Jones and Gordon, 1989]. For a "thick target" (one that completely attenuates the incident x-ray beam) the exponential term goes to zero and the equation becomes:

$$C_z = (K\, I_z\, [(\mu/\rho)_o + (\mu/\rho)_z]) \,/\, (I_o)$$

For a "thin film" target (one in which absorption of primary and secondary x-rays is negligible) the equation takes the form:

$$C_z = (K\, I_z)\, (0.71/(I_o \rho T))$$

Recently Lu et al. [1989] estimated that systematic errors arising from K and I_o are on the order of ± 20%.

A major source of uncertainty is due to imperfections in the thickness of the detector filters. However for particles, the greatest uncertainty in C_z originates from imprecise knowledge of the sample thickness and matrix composition.

The use of standards can greatly reduce the sources of error since many of the instrumental parameters will be constant for both sample and standard. Several standardization procedures can be used to calculate trace element compositions. In the thin film approach the mass absorption coefficients for samples and standards are assumed to be negligible. Then the equation for calculating concentrations is as follows:

$$C_{z\,(samp)} = I_{z\,(samp)}(C_{z\,(std)}/I_{z(std)})$$

If a true thin film is being analyzed then a thin film standard can be used. However, particles in the 5 μm to 50 μm thickness range fall between the thick target and thin film approximations and care must be exercised in performing "pseudo thin-film" analyses [Criss, 1976]. Even relatively small particles of less than 5 μm thickness have non-zero mass absorption for lighter elements. Using the thin film approach in this case may be reasonable for heavier elements, but will create systematic artifacts for the lighter ones. If fine grains of standard rock or reference materials are used for calibration of the SXRF microprobe technique, even small differences in the thickness of sample grains and standard grains may cause biases in the calculated concentrations. In the SXRF approach all data are corrected for mass absorption to avoid these types of errors.

In the external standard approach, quantification with polished thick standards or individual particle standards requires a knowledge of the thickness (in μg cm^{-2}) of both the standard and the sample regions irradiated by the X-ray beam. In addition, care must be taken to ensure that the composition of the standard is homogeneous on a fine scale. In this case the NRLXRF program is used to predict the measured x-ray intensity of each element in the standard and in the sample. This approach requires some knowledge of or a guess at the major element composition of the sample. The formula takes the following form:

$$C_{z(samp)} = I_{z(samp)}(C_{z(std)}/I_{z(std)})(\phi z_{(samp)}/\phi z_{(std)})$$

where $C_{z(std)}$ and $C_{z(samp)}$ are the known and calculated concentrations of element z in the standard and sample respectively, $I_{z(samp)}$ and $I_{z(std)}$ are the measured intensities for element z in the sample and standard, and $(\phi z_{(samp)}/\phi z_{(std)})$ is the ratio of the predicted intensities of element z in the sample and standard calculated from NRLXRF. The latter term in the equation essentially represents the relative mass absorption corrections for the sample and standard.

If the concentration of one of the major element components is known or can be assumed on stoichiometric grounds then an internal standard approach can be used. This method also requires a known or estimated particle thickness (in μg cm^{-2}) and an approximate major element composition of the sample matrix. The program NRLXRF is used to calculate the correction for absorption and enhancement by other elements in the particle matrix (ZAF correction) and then predict the intensity ratio (at the detector) of a trace element z to the major element in the sample matrix. The advantage to this approach is that the error in the predicted intensity ratio of two elements from NRLXRF is much smaller than the absolute error in the predicted intensity of a particular element because many of the systematic errors cancel out in normalization. The peak intensities are calculated by fitting a background to each spectrum and then applying a Gaussian peak fitting program to the background corrected spectra. For mid-ocean ridge plume particles the calculated concentrations of the trace elements are calculated relative to the Fe content of the mineral phase, assumed to be stoichiometric in this data set, and any errors in this assumption will affect the value by the percentage error in the assumption. Individual particles are also assumed to be no thicker than their smallest dimension. The concentration of any element z (C_z) in the particle is then calculated from the following equation:

$$C_z = C_{Fe}\,(I_z/I_{Fe})\,(S_z)$$

where C_{Fe} is the assumed (or known) Fe concentration, I_z and I_{Fe} are the background corrected peak intensities of the element of interest and Fe respectively, and S_z is the detector sensitivity for element z relative to Fe [Lu et al., 1989]. One can improve the ZAF correction for the major elements by carrying out this procedure in an iterative fashion. The greatest source of uncertainty in the procedure is the particle mass per unit area. Representative spectra for sulfide particles ranging from 5-100 μm on each side are displayed in Figure 6-16 and the calculated range of trace element concentrations are summarized in Table 6-5. Representative spectra for non-sulfide particles and "thin films" of nano-phase Fe-oxide particles are shown in Figure 6-17 and the calculated range of trace element concentrations is summarized in Table 6-6. These results indicate the basic compositional trends for different particle phases, but are semi-quantitative in nature due to uncertainties in determining the particle thickness.

Filtered Mid-Atlantic Ridge samples consist of a thin layer of Fe-oxide particles (ranging from 0.1 μm to ~4μm) mixed with larger (5-200 μm size) sulfide particles. Since bulk analyses include both types of phases, the XRF microprobe was used to provide an analysis of the nano-phase material independent of the larger particles.

The predominant Fe phase (>95%) in these samples is the nano-phase Fe-oxide (discussed in section 4, below). In this case the bulk mass of particles/cm^2 and the bulk Fe concentration provide reasonable approximations of the sample thickness and Fe content and were used in the equation above to calculate the concentrations. The range and mean values for several elements in the TAG and MARK sub-micron particles are presented in Table 6-7.

More rigorous standardization with thick and thin film standards that were analyzed along with the particle samples is in progress. The data for Fe/V vs. $Si(OH)_4$, which is presented in Figure 6-18, shows the relative increase in V relative to Fe as V is adsorbed onto the Fe phase. Note that the Fe/V ratio of fine grained material in the neutrally buoyant cloud is about 260, very similar to previous work [Trocine and Trefry, 1988].

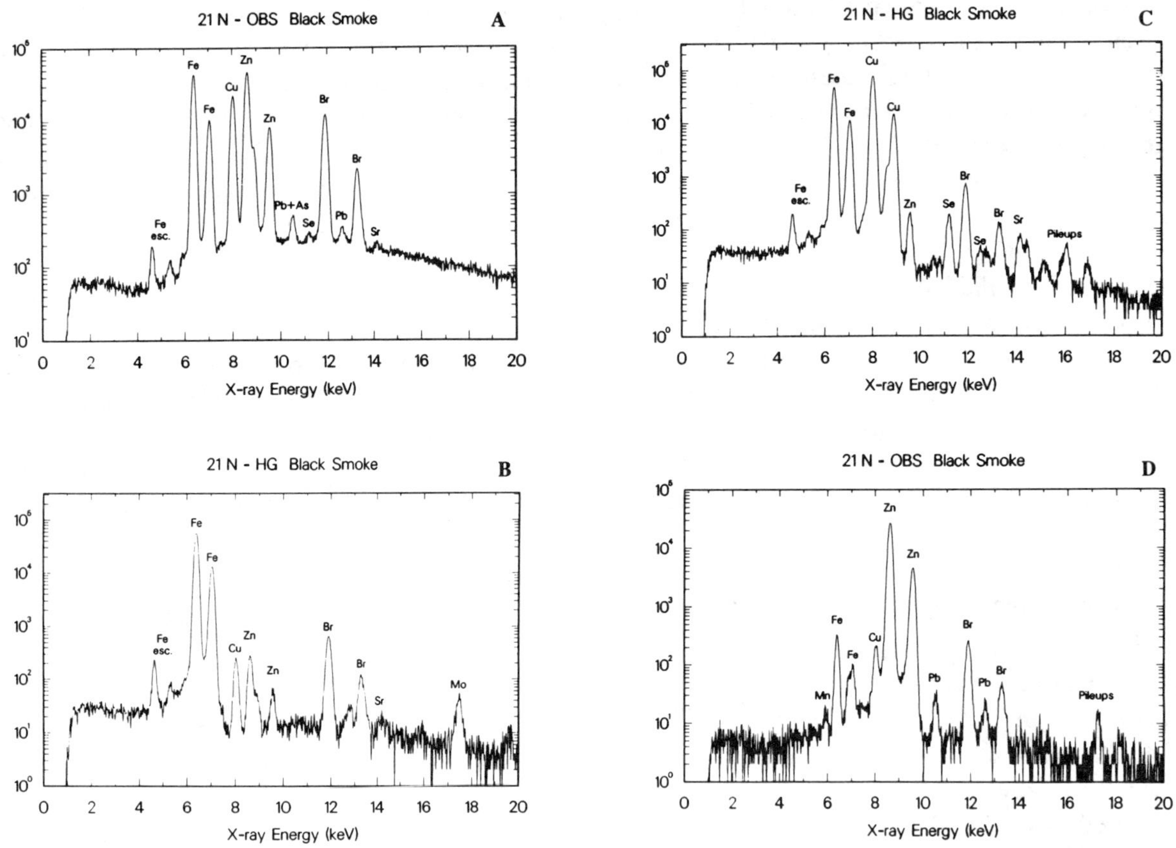

Fig. 6-16. SXRF spectra of individual sulfide particles (5-100μm) from 21° N EPR (note similar types of particles and trends in trace elements seen on TAG, MAR filters). Intensity axis in total counts in a 300 or 600 second collection period. Data collected with Al filter (85 μm) over Si(Li) detector. Particles on Nuclepore membrane filters. Br peaks originate from filter matrix. (Fe escape peak = Fe K_α energy minus Si K_α energy (in detector); energy of pileups = two times most intense peaks). A) Fine grained black smoke particles; B) Fe-S particle; C) Cu-Fe-S particle; D) Zn-S particle.

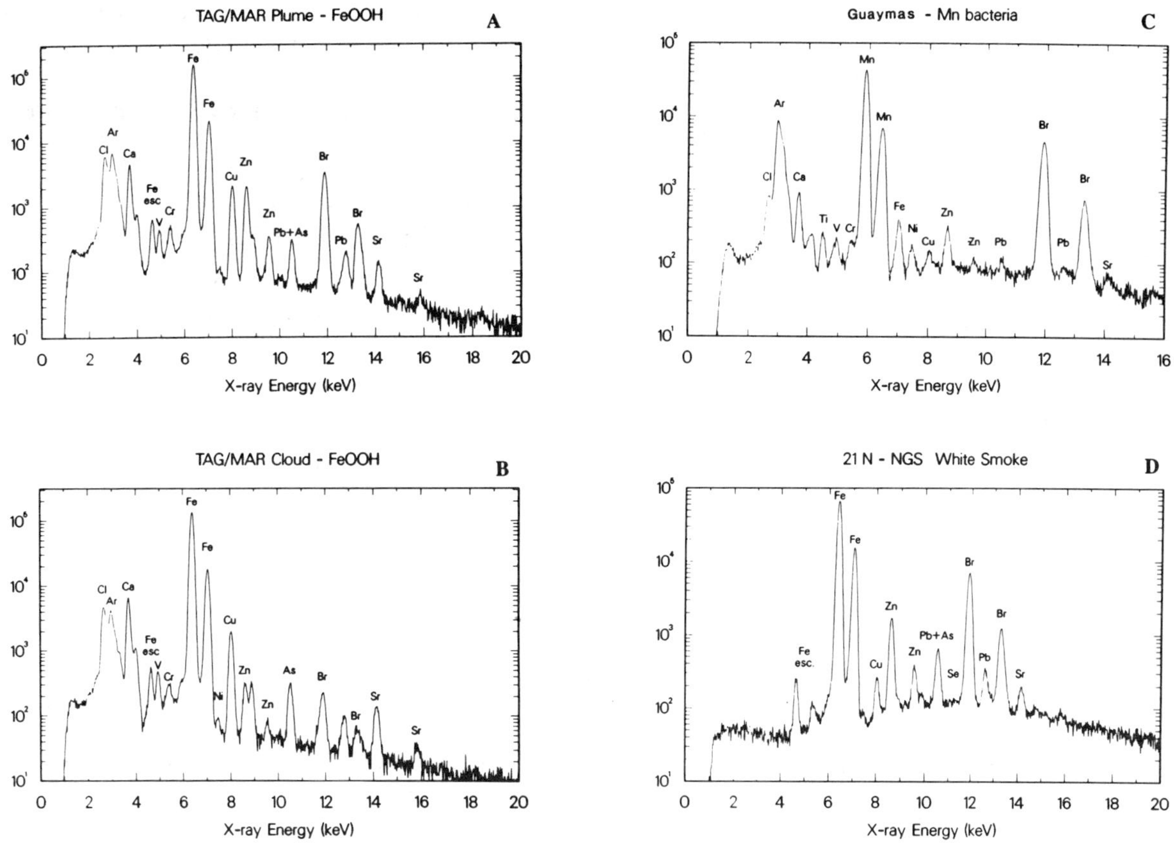

Fig. 6-17. SXRF spectra of non-sulfide phases. Intensity in total counts per spectrum. Data collected without Al filter for A, B, & C to "see" lighter elements. Ar peak from air. A) Fe oxide from plume samplers mounted on ALVIN (mat of sub-micron and micron size particles from buoyant plume); B) Fe oxide from pump samples in cloud (note decrease in Zn/Cu and Pb/As relative to buoyant plume sample in A); C) Mn bacterial particle (<5 microns); D) White smoke Fe-Si filament (<15μm diam.).

3.2. Synchrotron X-ray Absorption Spectroscopy

Synchrotron X-radiation has numerous applications to particulate studies beyond fluorescence analyses. The broad band spectrum allows the selection of specific energies (with a monochromator) for absorption, scattering, diffraction and tomography experiments. X-ray absorption analyses allow the determination of oxidation state, coordination environment and structure to be done on amorphous and poorly crystalline materials by studying the X-ray Absorption Near Edge Structure (XANES) and the Extended X-ray Absorption Fine Structure (EXAFS) [Cramer et al., 1988a, b; George et al., 1988, 1989]. EXAFS provides information about the number and type of ligands and bond distances to the absorbing atom by a Fourier analysis of the fine structure on the high energy side of the element's K-absorption edge. XANES provides information about the absorbing atom's oxidation state, coordination symmetry and ligand covalency from the energy position of the absorption edge and the bound state transitions on its low energy side [Bianconi et al., 1983]. These techniques will be especially useful for determining the speciation and coordination environment of metals in different particle phases. Recently, synchrotron X-ray absorption spectroscopy (SXAS) was used to elucidate structural features of Mn and Fe oxides and oxyhydroxides [Manceau and Combes, 1988] and the speciation and structure of Mn-Cu phases obtained from seafloor Mn crusts [Stouff and Boulègue, 1989].

As it now stands this is a bulk technique and typically requires 1-5 mg of material for transmittance spectra. Details of the technique and various applications can be found in Bianconi et al. [1983]. In the transmission mode, the K edge absorbance of the sample is measured as a function of X-ray energy. The X-ray energy is varied by a beamline monochromator, e.g., Si(100).

TABLE 6-5.
Trace Elements in Sulfide Plume Particles

Semi-quantitative analysis by SXRF microprobe

I. "Black Smoke" particles, ≥ 1 μm - OBS vent 21° N EPR
  Mixture of Fe-, Cu- & Zn Sulfides,
    ~ 10^3 ppm Ca
    10^2-10^3 ppm Co, Pb, As, Cr, Mn, Ga, Sr
    10^1-10^2 ppm Se, Ge, Ni
II. Cu-Fe Sulfides, 5-100 μm ("Black Smoke" only)
  Non-stoichiometric
    10^3-10^4 ppm Zn
    10^2-10^3 ppm Se, Mn, Co, Cr, Sr
    10^1-10^2 ppm As, Pb, Ge, Ni
III. Zn-Fe Sulfides, 5-100 μm ("Black-" & "White Smoke")
  Varying Zn/Fe ratios
    10^3-10^4 ppm As & Pb
    10^2-10^3 ppm Cu, Mn, Cr & Sr
    10^1-10^2 ppm Se, Ni
IV. Fe Sulfides, 5-100 μm ("Black-" & "White Smoke")
    10^3-10^4 ppm Zn > Cu
    10^2-10^3 ppm Co> Mn, Cr, As, Pb
    10^1-10^2 ppm Sr, Ni, Se

Difference in the counts from ion chambers before and after the sample directly measure the absorbance of the X-ray beam. In this approach the absorbance follows Beers law:

$$I(E) = I_0(E)e^{-\mu(E)\rho x}$$

where I(E) is the transmitted radiation at energy E, I_0 is the incident radiation at energy E, μ is the mass absorption coefficient, ρ is the density of the sample, and x is the sample thickness. The total mass absorption, μ(E), originates from both fluorescence and scattering of the X-rays. The experimental setup is similar to any absorption spectroscopy experiment. The sample is usually cut with a low Z material, e.g., sucrose, and ground to a fine consistency (if a mineral phase). The mixture is packed into a simple cell consisting of a rectangular piece of metal or cardboard (about 1-2 mm thick) with a hole about 2-3 cm wide and 3-4 mm high (it must be larger than the dimensions of the X-ray beam). The windows of this cell consist of "Kapton", a low Z, X-ray resistant polymer film. This is placed in the sample holder and the X,Y,Z axes are adjusted so that the beam passes through the sample only.

TABLE 6-6.
Trace Elements in Fe-Oxide, Manganate, and $Fe-SiO_2$ Plume Particles

Semi-quantitative analysis by SXRF microprobe

I. Fe-oxide particles - TAG vent 26° N MAR
  A. buoyant plume (sampled 1986) - DSRV ALVIN
    10^4 ppm Ca, Cu, Zn
    10^3-10^4 ppm Sr, As, V, Cr
    10^2-10^3 ppm Pb, Mn > Ti
    10^1-10^2 ppm Ni
  B. cloud (sampled 1988) - moored in situ pump
    10^4 ppm Ca > Cu
    10^3-10^4 ppm Zn, As, Sr, V,
    10^2-10^3 ppm Mn, Cr, Ti, Pb, Ni
    10^1-10^2 ppm Se, (Ge)
II. Manganese bacteria - Guaymas Basin
  Manganate filaments
    10^4 ppm Fe, Ca
    10^3-10^4 ppm Zn, Ti, Sr, V, Cr, Cu, Pb
    10^2-10^3 ppm Ni > As
III. "White smoke" - NGS vent 21° N EPR
  Fe-silica filaments
    10^3-10^4 ppm Zn
    10^2-10^3 ppm Pb, As, Cr, Sr
    10^1-10^2 ppm Ni, V

TABLE 6-7.
Preliminary Results of SXRF Microprobe Analyses
Thin film of sub-micron plume particles (sampled 1990)

	TAG Particles	mean	MARK Particles	mean
%				
Fe (AA)	33 - 38	35.2	21 - 29	23.9
S	2.6 - 5.1	3.6	26 - 32	29
Ca	1.1 - 2.3	1.7	0.75 - 1.6	1.1
ppm				
V	240- 1120		64 - 530	
Cr	98 - 802	300	15 - 479	174
Ni	2 - 50	11	~ 10	
Cu	3060-12750	7500	1580- 3100	2350
Zn	1107- 8420	2650	6750- 13670	8600

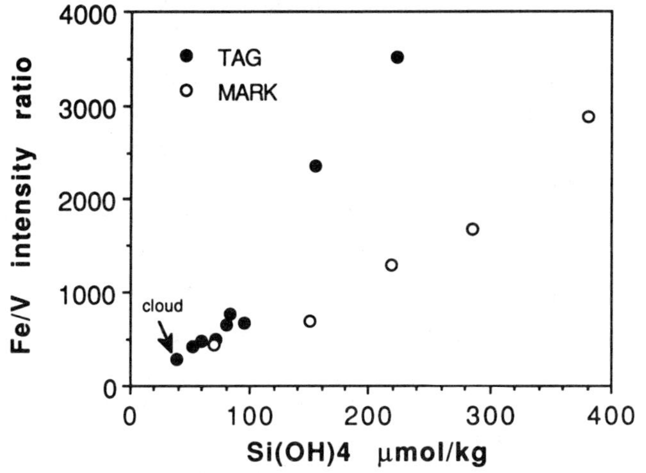

Fig. 6-18. Ratio of Fe/V by SXRF microprobe for thin films of TAG and MARK plume particles plotted vs $Si(OH)_4$ (conservative tracer of hydrothermal component in plume sample). Data uncorrected for absorption effects.

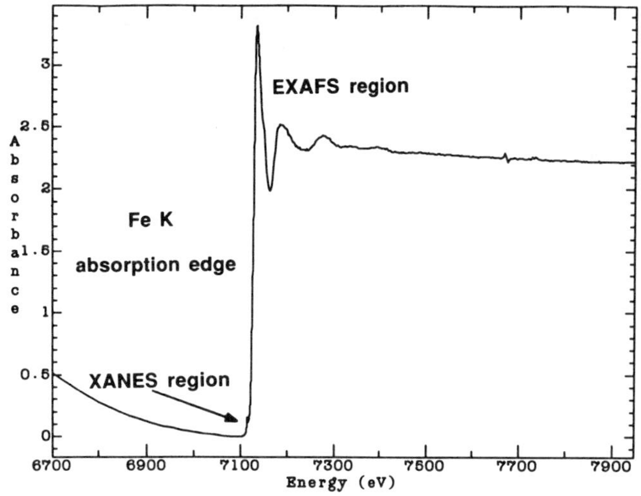

Fig. 6-19. Background corrected Fe K edge absorption spectrum of hydrothermal plume particles. Data collected at 15-20° K.

An alternative approach is to measure the fluorescence of a sample and calculate its absorbance based on predicted yields of fluorescence, scattering and absorption of the X-rays. This approach has the advantage of using much smaller sample sizes (a hundred micrograms or less) and is particularly useful for a mat of particles on a filter. In this case a Ge(Li) detector is used that was developed by S. Cramer and others, Brookhaven National Lab [Cramer et al., 1988b]. This system has 13 independent energy dispersive detectors and the final spectrum is a sum of 13 individual spectra. This results in about a fourfold increase in detection limit per absorbance measurement (usually several absorption runs are made for each sample). An example of such an absorption spectrum for the Fe K edge of Mid-Atlantic Ridge particulate matter is shown in Figure 6-19.

The data workup of the absorption spectrum is relatively straightforward. After establishing E_0 (onset of the K edge), the spectrometer baseline is subtracted and μ and $\mu_{background}$ are established. Subtraction then yields μ_{sample} and division by $\mu_{background}$ yields the EXAFS. The Fourier transform is then performed on the EXAFS to convert to radial distribution space. The resulting plot shows the radial distance (in angstroms) to the different scattering shells around the absorbing atom. The radial distribution function for plume particles from TAG and MARK on the Mid-Atlantic Ridge are shown in Figure 6-20A. These data were collected at 15-20° K using a liquid He cryostat; low temperature damps out vibrational motion of the molecular structure and generally provides better EXAFS data. The radial distribution function for some model Fe-oxide compounds is shown in Figure 6-20B. These data were collected at room temperature. The radial distribution function of the "ferrihydrite" sample clearly shows that it has apparently converted to hematite during storage. Currently work is in progress to do single particle SXAS and preliminary work by S. Sutton (personal communication, 1990) on the SXRF microprobe system at NSLS-BNL clearly differentiates different oxidation states of Cr.

3.3. Future Applications

One area of interest is the distribution of trace elements among different particle phases in the water column and the sediments. Traditionally chemical oceanographers have interpreted the distribution of elements in seawater through bulk analyses of fluids and particles and then hypothesized about the processes that govern partitioning among the different phases. The advent of SXRF and SXAS allows us, for the first time, to begin to see the distribution of trace elements on different particles and the speciation of individual particle phases. Several areas are of special interest:

- The dissolution of aeolian dust and the release of trace elements after deposition to rivers, lakes and the ocean.

- Uptake of trace elements onto biological particles.

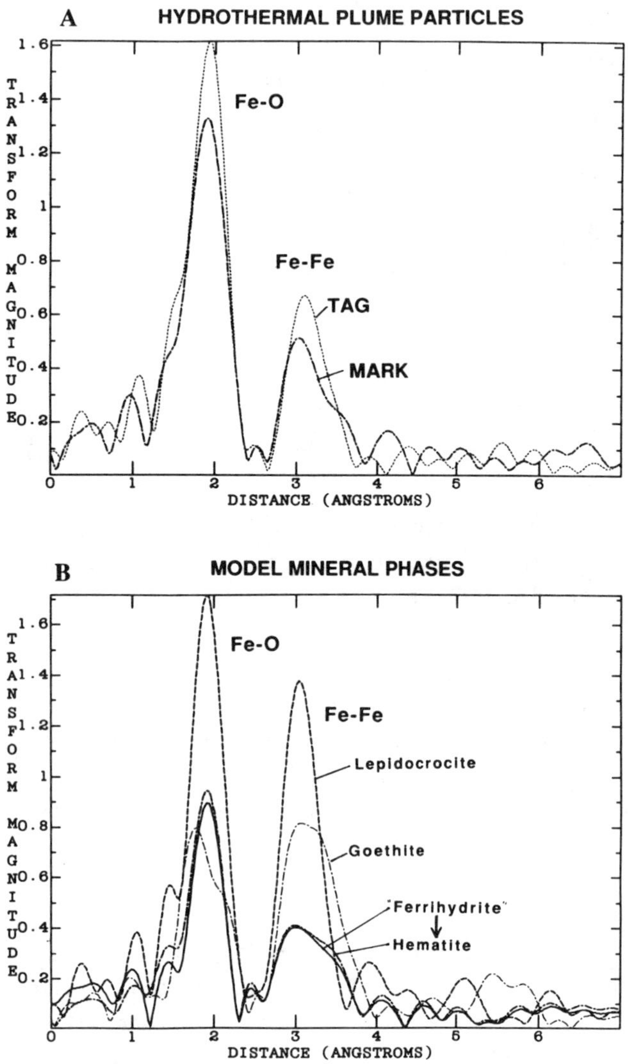

Fig. 6-20. Radial distribution functions from Fe EXAFS data for A) hydrothermal plume particles run at 15-20° K and B) model Fe minerals at room temperature. Scattering atoms at ≤ 2 angstroms are octahedral oxygen, others at about 3 Å are Fe-Fe distances outside inner coordination sphere.

- Desorption, adsorption and coprecipitation processes in estuarine mixing of river water and seawater.

These are but a few of the possible applications of synchrotron X-ray spectroscopy to problems in marine and environmental chemistry. The main features needed are as follows:

- Ability to analyze inorganic particles down to the micron size range at the ppb level.
- Analyses of organic and biological material for trace elements to the ppb level.
- Elemental mapping of individual particles and organic material to determine sites of uptake or release.
- X-ray absorption spectroscopy on individual particles (1-100 microns) to:
  a) determine the speciation and structure of amorphous phases.
  b) determine the oxidation state, ligand covalency, coordination sites, and number and types of ligands for adsorbed and coprecipitated elements.

In the mid-90's synchrotron radiation from the Advanced Photon Source (APS), Argonne National Lab will become available. This next generation synchrotron will use wiggler and undulator systems to greatly enhance the flux of radiation above 20KeV. It will provide an ideal probe for characterizing oceanic mineral and particle phases. The greatly improved spatial resolution and significantly lower detection limits for X-ray fluorescence microprobe spectroscopy at APS compared to current generation bending magnet sources (1-10 ppm on a 5-10 μm spot) will permit non-destructive trace element analysis of a micron size area at the part per billion level (1-10 ppb on 0.5-1.0 μm spot or better). The increased flux of X-rays from 20-100KeV at APS will allow K-line spectroscopy of heavy elements, Pt group elements, rare earths and actinides. The APS system will also allow us to conduct EXAFS/XANES studies on individual particle phases.

4. Mössbauer Spectroscopy

One of the problems in determining the mineralogy of "black smoke" particles is their heterogeneous composition and extremely fine grained, cryptocrystalline nature. Bulk characterization by XRD can reveal crystalline phases only larger than about 30 nanometers and may be misleading quantitatively since relatively crystalline phases will generate more intense diffraction peaks. Ideally one would like to have a bulk technique that can provide qualitative and quantitative information regardless of the crystallinity of the different phases. One such technique is low temperature Mössbauer spectroscopy, which allows one to elucidate the bulk mineralogy of the Fe containing phases. Since the predominant element in plume particles is Fe and nearly all of the important sulfide containing phases contain significant amounts of Fe, this approach has allowed the determination of the relative importance of sulfide and Fe-oxide minerals at different points in the plume.

Mössbauer spectroscopy is a well established method for determining the coordination environments around Fe [Bancroft, 1973]. In the technique the absorption of gamma rays by a sample is measured as a function of small changes in the energy of the γ-rays caused by movement of the sample relative to the source (Doppler broadening). Since the nuclear transitions that give rise to the sample absorption involve a change in spin quantum number, the electron density around the nucleus and the symmetry of the compound affect the observed spectrum. Hence, the technique provides speciation and structural information. It has distinct advantages over XRD in that amorphous compounds can be just as easily analyzed as crystalline materials. The signal is only dependent on the quantity of ^{57}Fe in the sample. The origin of the Mössbauer effect is discussed in detail in Burns [in press, 1991], and Bancroft [1973] and summarized in Figure 6-21. The offset of the energy of absorption of a sample relative to a standard (usually Fe metal) is the isomer shift. An absorber in an octahedral field has two absorption lines corresponding to the transition from the spin $\pm 1/2$ groundstate to the spin $\pm 1/2$ and $\pm 3/2$ excited states. The difference in energy of the two excited states is the quadrupole splitting. In a magnetic field the degeneracies of the ground and excited states is removed resulting in six absorption lines. Thus ferromagnetic, ferrimagnetic, and antiferromagnetic materials display sextets in Mössbauer spectra due to their own intrinsic magnetic fields. Macrocrystalline materials show a sextet of magnetic splitting lines in Mössbauer spectra collected at or above room temperature. However, in superparamagnetic materials, which have extremely small particle or crystallite size, Brownian motion breaks down the intrinsic magnetic field and only a doublet is observed in the Mössbauer spectrum. At very low temperatures (liquid He) this magnetism is restored by decreasing the vibrational motion so that the intrinsic magnetic field of each molecular unit can line up parallel or antiparallel with the other structural units. This removes the spin degeneracy of the nuclear transitions and restores the sextet spectrum. The temperature at which the sextet of magnetic splitting lines converge back to the doublet of the quadrupole splitting lines is a direct function of the particle (or crystal) size.

Mössbauer spectroscopy is a bulk technique that requires about 10 mg or more of Fe containing phases. Typically a sample is transferred (in a glove bag filled with Ar) to a pre-cleaned polyethylene cell that is 1-2 mm deep and 10 mm diam. For samples adhering to a filter, the filter is folded so that the particulate material is more or less evenly distributed over the cross sectional area of the cell. For unlimited sample, the material is cut

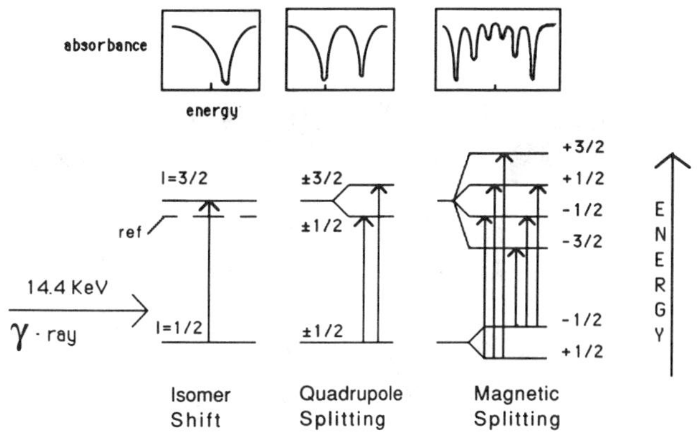

Fig. 6-21. Origin of ^{57}Fe Mössbauer spectra. Absorbance of 14.4 KeV gamma ray by ^{57}Fe in sample promotes $I=1/2$ nuclear spin ground state to $I=3/2$ excited state. Isomer shift originates from difference in sample absorbance relative to reference standard absorbance (usually Fe foil). Quadrupole splitting originates from removal of excited state degeneracy by non-cubic electron or ligand field. Magnetic splitting originates from the complete removal of ground and excited state degeneracies by intrinsic magnetic field of sample. [Figure modified from Burns, in press, 1991].

1:3 with a low Z material like sucrose and this mixture is packed into the cell. The sample cell is placed in the He-cryostat well between the gamma source and a detector. The source is ^{57}Co, which is converted to ^{57}Fe by electron capture, the ^{57}Fe subsequently decays by gamma ray emission. The source (or in some cases the sample) is moved back and forth with respect to the sample holder at varying speeds (typically a few mm sec^{-1}) to obtain Doppler energy shifts of the gamma rays (about 10^{-8} eV for each mm sec^{-1}). The absorption spectrum is measured as a function of the source speed, i.e., its energy. Stripping and fitting programs allow the deconvolution of spectra from multi-phase samples.

Mössbauer spectra have been collected at 293° K and 4.2° K (liq. He) on several plume samples and a pump sample using the facilities at the Francis Bitter National Magnetics Lab, MIT. The Mössbauer spectrum for the pump sample obtained two feet above one of the TAG chimneys is presented in Figure 6-22. Pyrite, chalcopyrite, and an iron oxide phase (identified as ferrihydrite below) occur in proportions of about 7%, 33%, and 60%, respectively, in this sample. The size of the Fe-oxide particles was revealed serendipitously when the liquid He accidently became exhausted during an all night run. The sample cell warmed to about 25° K

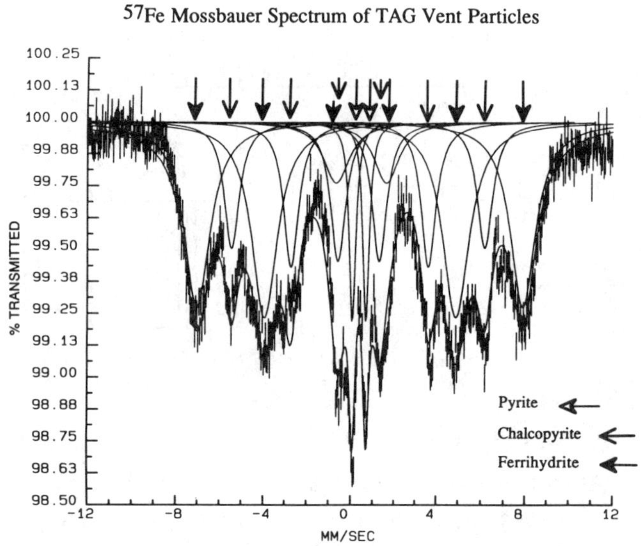

Fig. 6-22. Mössbauer spectra for plume particles from two feet above a vent orifice at TAG. Spectrum obtained at 4.2° K (liquid He). Identity of minerals indicated by arrows (phases constituting less than 5% of total not discernable).

and the magnetic splitting lines for the Fe-oxide began to collapse to a doublet. This indicates very short-range order for the average particle size, on the scale of one nanometer (R. Burns, personal communication, 1990). (Pyrite shows only quadrupole splitting lines in Mössbauer spectra because it is diamagnetic). The buoyant plume samples collected during "elevator rides" only show Fe-oxide (Figure 6-23A). These spectra are virtually identical to a synthetic nano-phase hematite that has a nominal particle size of 1-2 nanometers (Figure 6-23B). However, comparison of the magnetic hyperfine field of this Fe-oxide with an array of model Fe-oxides (Table 6-8) shows that the plume particles are slightly different than "nano-phase" hematite, but both fall within the range of ferrihydrite. The latter is a hydrated ferric-oxide which basically has the hematite structure and, at pH 8, converts to hematite upon aging. The extremely small particle size results in a large free energy of adsorption and explains the rapid uptake of certain elements from seawater by black smoke particles. The dominance of ferrihydrite in these buoyant plume samples and in the pump sample from directly above the vent indicates that it either directly precipitates during mixing of vent fluids and seawater and/or that the initial Fe-sulfide phase(s) rapidly oxidize (in a few hours) to ferrihydrite. Hence, the ubiquitous Fe-oxide phase found in neutrally buoyant hydrothermal clouds is ferrihydrite. The absence of a significant sulfide signal in the Mössbauer spectra of the buoyant plume samples suggests that sulfide particles of 5-100 μm are a relatively minor component of the total Fe present in contrast to the Fe-oxide phase. They may, however, contribute significant amounts of Co, Cu, Zn, Cd, Pb, etc.

5. Inductively Coupled Plasma - Mass Spectrometry

This technique uses an inductively coupled plasma torch as an ion source for a quadrupole mass spectrometer. The advantage of this technique over ICP emission methods is a reduction in the detection limit by several orders of magnitude for most elements. The advantages relative to GFAA are similar if not better detection limits for those elements amenable to atomization in a graphite tube and the ability to analyze the refractory elements that are not (see K. Kenison Falkner, this volume, page 243). Another major advantage is the ability to carry out isotope dilution analyses resulting in increased precision and accuracy. Analyses by ICP-MS are done using both multi-element standard additions and isotope dilution techniques. Dissolved elements in filtered seawater are also done either by standard addition on diluted samples or by using an extraction pre-concentration technique followed by ICP-MS. Isotope dilution methods are particularly useful in the latter case since it eliminates the need for a yield tracer.

Analyses of filtered plume water provide an example of the type of information obtainable by isotope dilution ICP-MS. The Ba and Li plots vs. $Si(OH)_4$ (Figures 6-24A and 6-24B) show linear trends that extrapolate to near the endmember values for these elements in the undiluted hydrothermal fluids. Uranium, however, shows no uptake by the Fe phase (Figure 6-24C) (the mean and standard deviation for all the filtered plume samples is 13.22 ± 0.09 $nMkg^{-1}$) and hence the U anomalies seen in metalliferous sediments must be due to post-depositional processes.

Particle digest techniques used in this lab to recover precipitates from hydrothermal fluid samples and those used to digest suspended refractory material have been adapted to the analysis of hydrothermal particles. The particle digestions are carried out either with a microwave digestion system or by heating a teflon bomb containing the filter and 2 mL HNO_3 (and 50 μL HF if silicates are significant) on a hot plate so that the acid refluxes onto the filter. The acid digest solution is then quantitatively transferred to a pre-weighed bottle and diluted to about 10 mL (g). Analyte solutions for ICP-MS are made by taking 0.1 mL of the diluted digest

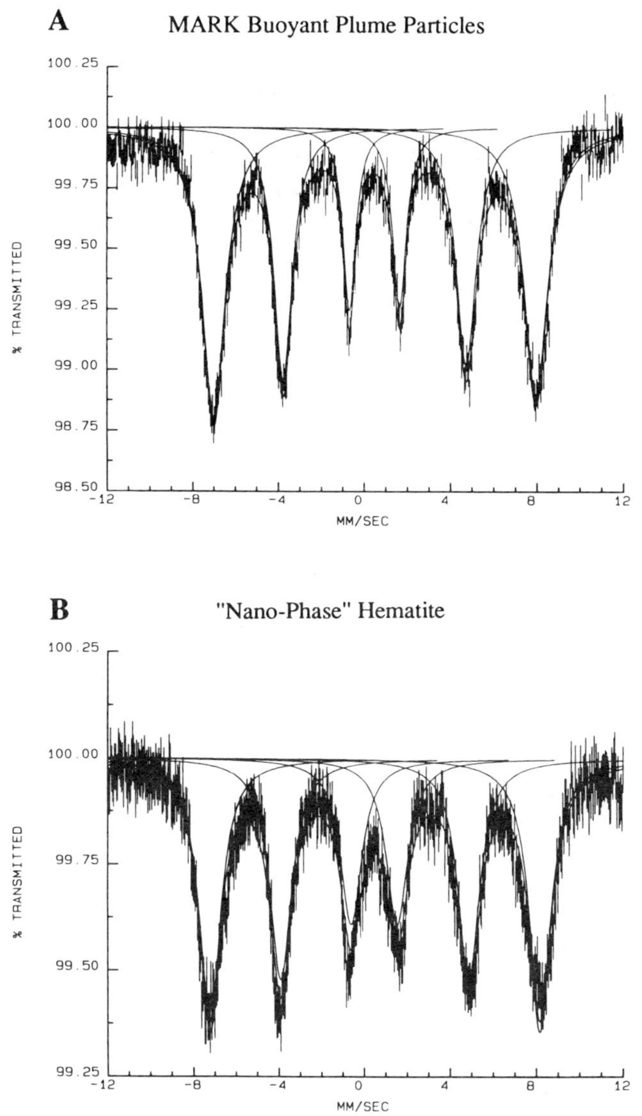

Fig. 6-23. Mössbauer spectra (4.2° K) of Fe-oxide phases in A) buoyant plume at MARK and B) synthetic "nano-phase" hematite.

solution and further dilution to 10 mL in 0.5% HNO_3 containing Be, Sc, In and Bi at the 10 ppb level (as internal standards). Standardization can be done either by standard additions or using a mixed element cocktail standard in filter blank matrix. The range in concentrations of 14 elements in plume particle samples from both the TAG and MARK sites are presented in Table 6-9. Filtered water analyses can be done by adding a stable isotope cocktail spike, pre-concentration by APDC and/or Chelex separation from the major cations, and isotope ratio analyses by ICP-MS. The suite of elements that can be analyzed by ID include V, Cr, Cu, Zn, Ag, Cd, Mo, Tl, and Pb. Elements not amenable to ICP-MS because of molecular ion interferences, e.g., Fe, are done by AA methods.

6. Conclusions

The techniques of synchrotron X-ray spectroscopy, Mössbauer spectroscopy and inductively coupled plasma-mass spectrometry provide new tools for studying the mineralogy and chemistry of marine particles. Synchrotron X-ray fluorescence microprobe analysis allows the determination of the trace elements on individual particles. Current generation systems have detection limits of about 1-10 ppm on a 5-10 μm spot, however, the development of wiggler sources in the near future will allow ppb detection limits on 1 μm or smaller spots. This approach to particle analyses allows us to establish the distribution of trace elements among different particle and mineral phases. Synchrotron X-ray absorption spectroscopy allows the determination of oxidation state, coordination environment and structure to be done on particle phases by studying the X-ray Absorption Near Edge Structure (XANES) and the Extended X-ray Absorption Fine Structure (EXAFS). Mössbauer spectroscopy allows the determination of the mineralogy and oxidation state of Fe containing phases and provides information about the particle (crystallite) size for sub-micron "nano-phase" materials. Inductively coupled plasma - mass spectrometry provides multi-element, bulk analyses of relatively small amounts of material. Together these techniques provide powerful new probes for elucidating the mineralogy and chemistry of particles and allow a better understanding of particle/fluid interactions in the marine environment and the uptake of metals from seawater.

Acknowledgements. The author wishes to acknowledge the efforts of several individuals whose expertise and assistance contributed to this work. Steve Sutton and Mark Rivers - SXRF Microprobe; Steve Cramer - EXAFS; Roger Burns and Duncan Fisher - Mössbauer Spectroscopy. Comments on the manuscript by Gary Klinkhammer greatly helped to improve it. Primary funding for this research was provided to the author by the National Science Foundation Chemical Oceanography Program (Grant Contract #OCE 8918633). Research also carried out (in part) at the National Synchrotron Light Source, Brookhaven National Laboratory, which is supported by the U. S. Department of Energy, Division of Materials Sciences and Division of Chemical Sciences (DOE Contract #DE-AC02-76CH00016).

TABLE 6-8.
Mössbauer Parameters for Iron Oxides (samples run at 4.2° K)

Fe Oxide	Formula	IS	QS	H
plume particles		0.474 to 0.476	0.02 to 0.03	46.4 to 46.5
"nano-phase" hematite		0.463	0.031	47.7
ferrihydrite	$Fe_4O_{12}H_9$	0.47	0.1	46.5 to 50.0
hematite	α-Fe_2O_3	0.487	-0.398	54.56
maghemite	γ-Fe_2O_3	0.442	-0.008	52.97
goethite	α-FeOOH	0.489	0.251	50.86
akaganeite	β-FeOOH	0.499	0.194	48.71
lepedicrocite	γ-FeOOH	0.47	0.04	45.5
feroxyhyte	δ-FeOOH	0.45	0.12	52.9
magnetite	Fe_3O_4	0.67	0	46
faylite	Fe_2SiO_4		3.08	32.3
jarosite	$Fe_2(SO_4)_3$	0.492	0.09	49.72

IS = isomer shift relative to Fe foil
QS = quadrupole splitting
H = magnetic hyperfine field in Teslas
[Table modified from Burns, in press, 1991]

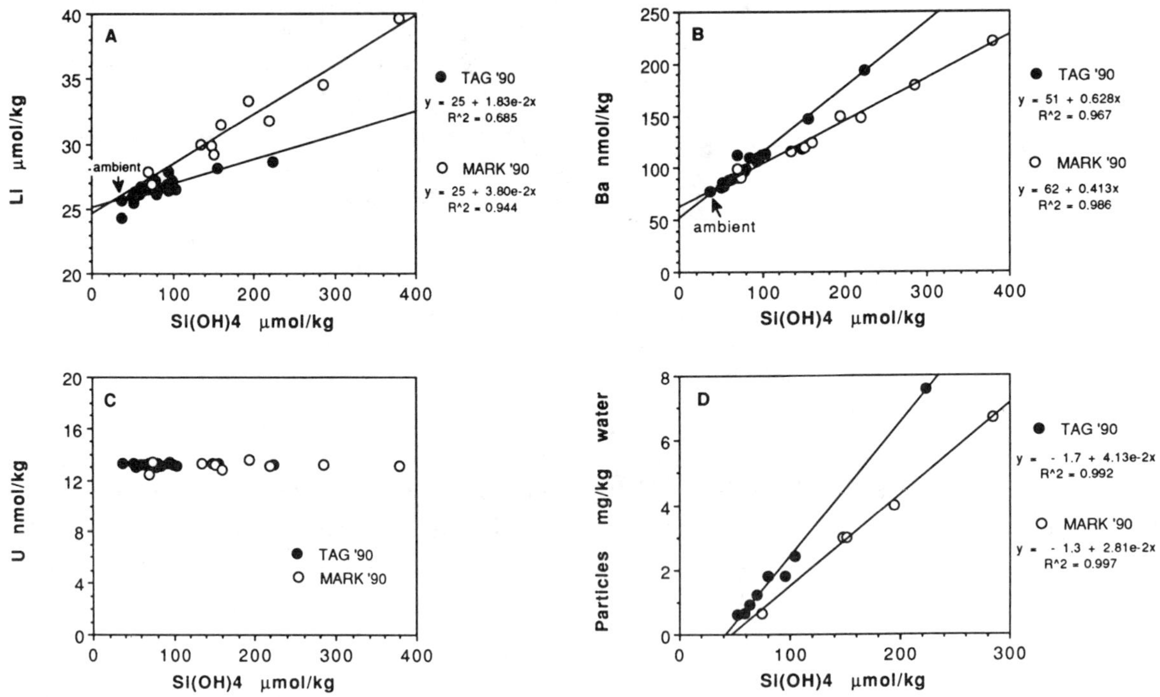

Fig. 6-24. Mid-Atlantic Ridge buoyant plume data. Li, Ba, and U by isotope dilution ICP-MS. Mixing line plots for buoyant plumes at TAG and MARK, MAR. A) Lithium vs. $Si(OH)_4$ (silica from shipboard data), B) Barium vs. $Si(OH)_4$, C) Uranium vs. $Si(OH)_4$, and D) Particulate concentrations in plume from select pre-weighed filters.

TABLE 6-9.
Results of ICP-MS Analyses of Plume Particles
Comparison of bulk particle composition at TAG and MARK sites

	TAG Particles	MARK Particles
Fe % (AA)	35.8 (2.3)	23.9 (3.1)
ppm		
Al	164 - 872	200 - 820
V	157 - 1040	220 - 760
Mn	55 - 101	66 - 92
Co	~ 80	~ 4
Cu	8390 - 12200	2550 - 4180
Zn	2700 - 4750	7110 - 10890
As	150 - 820	330 - 680
Y	~ 8	~ 4
Mo	~ 6	~ 4
Ag	~ 6	~ 11
Cd	~ 25	~ 37
Ba	~ 36	~ 25
Sr	380 - 500	350 - 410
Pb	27 - 320	160 - 205

References

Bancroft, G. M., Mössbauer Spectroscopy - An Introduction for Inorganic Chemists and Geochemists, 251 pp., McGraw Hill Book Co.(UK), London, 1973.

Bianconi, A., L. Incoccia, and S. Stipcich, (Eds.), EXAFS and Near Edge Structure, 420 pp., Springer-Verlag, West Berlin, 1983.

Bowers, T. S., K. L. Von Damm, and J. M. Edmond, Chemical evolution of ridge crest hot springs, Geochim. Cosmochim. Acta, 49, 2239-2252, 1985.

Burns, R. G., Mössbauer spectral characterization of iron in planetary surface materials, in Remote Geochemical Analyses: Elemental and Mineralogical Composition, edited by C. M. Pieters, and A. J. Englert, Lunar Planetary Institute, Houston, Texas, in press, 1990.

Campbell, A. C., M. L. Rivers, and S. R. Sutton, Trace element composition of hydrothermal plume particles by synchrotron X-ray fluorescence spectroscopy, EOS Trans. Am. Geophys. Union, 70, 494, 1989.

Campbell, A. C., R. G. Burns, J. M. Edmond, H. Elderfield, and M. Greaves, Geochemistry and mineralogy of buoyant hydrothermal plume particles: TAG and MARK vents Mid-Atlantic Ridge, EOS Trans. Am. Geophys. Union, 71, 1651, 1990a.

Campbell, A. C., S. Cramer, M. L. Rivers, and S. R. Sutton, Applications of synchrotron X-ray spectroscopy to the geochemistry of hydrothermal plume particles, EOS Trans. Am. Geophys. Union, 71, 1621, 1990b.

Cramer, S. P., M. Siskin, L. D. Brown, and G. N. George, Characterization of As in oil shale and oil shale derivatives by X-ray absorption spectroscopy, J. Energy Fuels, 2, 175-180, 1988a.

Cramer, S. P., O. Tench, M. Yocum, and G. N. George, A 13-element Ge detector for fluorescence EXAFS, Nucl. Instrum. Methods Phys. Res., A266, 586-591, 1988b.

Criss, J. W., Particle size and composition effects in X-ray fluorescence analysis of pollution samples, Anal. Chem., 48, 179-186, 1976.

Criss, J. W., NRLXRF, Naval Research Laboratory Cosmic Program #DOD-00065, Nav. Res. Lab., Washington, D.C., 1977.

George, G. N., C. L. Coyle, B. J. Hales, S. P. Cramer, X-ray absorption of *Azotobacter vinelandii* vanadium nitrogenase, J. Amer. Chem. Soc., 110, 4057-4059, 1988.

George, G. N., R. C. Prince, and S. P. Cramer, The Mn site of the photosynthetic water-splitting enzyme, Science, 243, 789-791, 1989.

Jones, K. W., and B. M. Gordon, Trace element determination with synchrotron-induced X-ray emission, Anal. Chem., 61, 341A-358A, 1989.

Klinkhammer, G. P., M. Bender, and R. F. Weiss, Hydrothermal Mn in the Galapagos rift, Nature, 269, 319-320, 1977.

Lu, F.-Q., J. V. Smith, S. R. Sutton, M. L. Rivers, and A. M. Davis, Synchrotron X-ray fluorescence analysis of rock-forming minerals: 1. Comparison with other techniques, 2. White-beam energy-dispersive procedure for feldspars, Chem. Geol., 75, 123-143, 1989.

Lupton, J. E., R. F. Weiss, and H. Craig, Mantle helium in hydrothermal plumes in the Galapagos rift, Nature, 267, 603-604, 1977.

Manceau, A., and J. M. Combes, Structure of Mn and Fe oxides and oxyhydroxides: a topological approach by EXAFS, Phys. Chem. Min., 15, 283-295, 1988.

Stouff, P., and J. Boulègue, Geochemistry and crystallochemistry of oceanic hydrothermal manganese oxyhydroxides showing Mn-Cu association, Geochim. Cosmochim. Acta, 53, 833-843, 1989.

Sutton, S. R., M. L. Rivers, and J. V. Smith, Applications of synchrotron X-ray fluorescence to extraterrestrial materials, in Proceedings of Ninth Conf. on Appl. of Accelerators in Res. and Ind., edited by J. L. Duggan, and I. L. Morgan, 1138 pp., Denton, Texas, 1987.

Trocine, R. P., and J. H. Trefry, Distribution and chemistry of suspended particles from an active hydrothermal vent site on the Mid-Atlantic Ridge at 26° N, Earth Planet. Sci. Lett., 88, 1-15, 1988.

Weiss, R. F., Flushing characteristics of oceanographic sampling bottles, Deep Sea Res., 18, 653-656, 1971.

Weiss, R. F., P. Lonsdale, J. E. Lupton, A. E. Bainbridge, and H. Craig, Hydrothermal plumes in the Galapagos Rift, Nature, 267, 600-603, 1977.

SAMPLING, PREPARATION AND ANALYSIS OF MARINE PARTICULATE MATTER

G. Fischer and G. Wefer

Geowissenschaften, Universität Bremen, Postfach 330440, 2800 Bremen 33, F.R.G.

1. Introduction

To study the particle transport from the surface waters to the deep oceans, the laboratory of the Sonderforschungsbereich 261 at Bremen University has concentrated on the analysis of marine particles sampled by sediment traps, plankton nets and plankton pumps. This paper will focus on the sampling, splitting and preservation techniques and on the analysis of various inorganic and organic compounds and their isotopes.

2. Sampling and Preservation Techniques

2.1. Particle Sampling With Sediment Traps and Preservation Methods

Generally, bottom-tethered traps are used by our laboratory and deployed in water depths of about 500 m below the surface and 500 m above the seafloor. Current meters, generally equipped with temperature and conductivity sensors and, if possible, transmissometers are deployed in association with the sediment traps. The Kiel traps (model SMT-230/234), with an opening of 0.5 m^2, are cone shaped and fitted with a grid at the top. The traps have computer-controlled devices for collecting 20 samples and were deployed in most cases for 10-15 months.

As poison, $HgCl_2$ is used. Before the traps are deployed, 1 mL of a saturated solution in distilled water is added to 100 mL of seawater. For the next deployments we will probably have to double this concentration to avoid the formation of H_2S in sampling cups with extremely high organic carbon contents. The seawater is taken from a depth at or greater than the planned trap depth and filtrated through a 0.45 μm filter. To increase the salinity in the surrounding trap solution, NaCl is added which increases the salinity by 3‰. One cup of a blank solution is stored at 2 to 4° C and used for the determination of the initial nutrient and element concentrations in the cups (see below). After recovery, 0.5 mL of a concentrated $HgCl_2$ solution is added per 100 mL of trap solution. For shipboard inspection with a light transmission microscope a small amount (0.5 to 1 mL, depending on the mass flux) of the suspended trap material is removed for the preparation of smear slides. The rest of the material is stored at 2 to 4° C for splitting and further processing at a later date in the laboratory.

2.2. Particle Sampling With Plankton Nets

For the collection of phyto- and zooplankton from the upper water column, multi-plankton nets with 63 μm mesh sizes (sometimes 20 μm) are used. For each sediment trap station three to five multi-plankton sets are taken for the analysis of organic compounds and bulk components, planktonic foraminifera and siliceous microfossils. Sampling intervals depend on water column properties and the objectives of the study. For the collection of some planktonic foraminifera and radiolaria, depth intervals reach 1000 to 1500 m with vertical tows. Standard sampling depths for the collection of various planktonic species are: 0-50 m, 50-100 m, 100-150 m, 150-250 m and 250-500 m. For the analysis of organic compounds (e.g. C/N, $\delta^{13}C_{POC}$, alkenones and amino acids), carbonate and biogenic opal, standard depth intervals are: 0-25 m, 25-50 m, 50-100 m, 100-150 m and 150-200 m. All collected material is removed from the nets with seawater, poisoned with $HgCl_2$ as described above and stored at 2 to 4° C.

2.3. Particle Sampling With Membrane Pumps

Particulate matter is collected continuously by pumping large volumes of seawater (1000 to 4000 L) through screens of different mesh sizes (mostly 63 μm). For this purpose the ship's contamination free pumping system with a hull-mounted intake (on RV Meteor 4 m below the sea surface) is used. The material is stored deep-frozen (-25° C) until further processing in the laboratories.

2.4. Large Volume Filtration

A large volume filtration technique is planned by W. Balzer (Bremen) to examine the interaction of dissolved and particulate trace elements in highly productive areas (equatorial Atlantic). For the collection of larger amounts of particulate matter, underwater in situ pumps (Ch. Oceanic SS, England) with nuclepore filters will be used.

Marine Particles: Analysis and Characterization
Geophysical Monograph 63

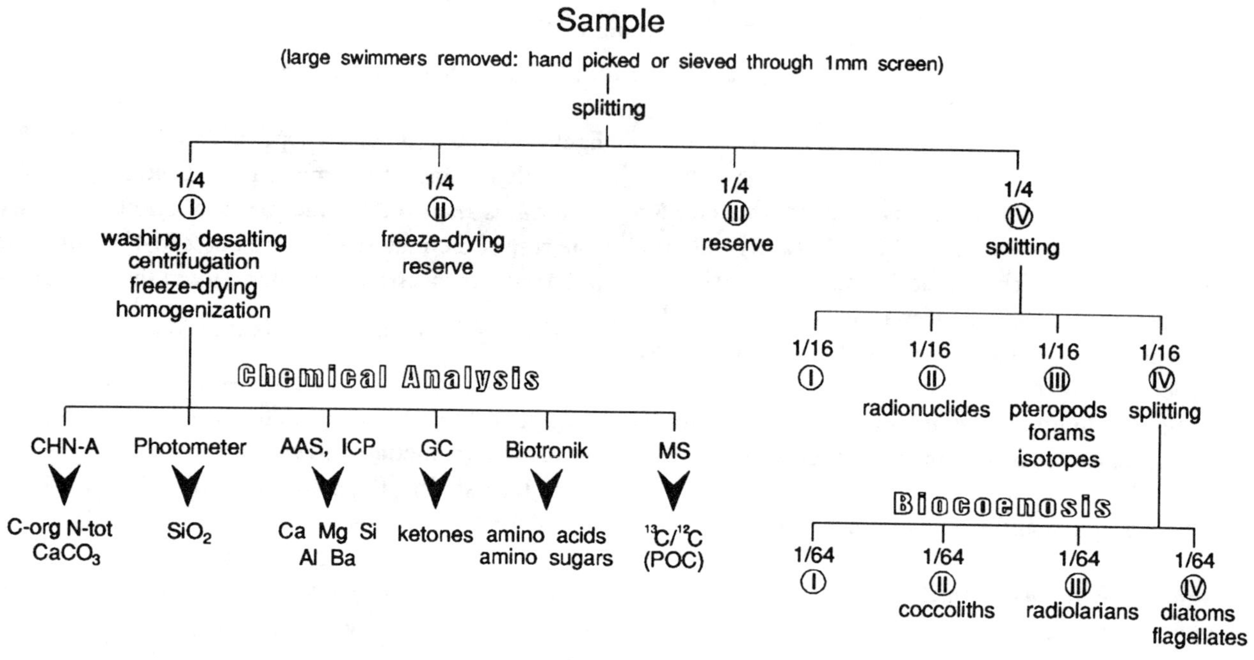

Fig. 6-25. Standard splitting procedure.

3. Sieving and Splitting Procedures (Trap and Plankton Net Material)

3.1. *Trap Material*

First, 50 mL of the surrounding trap solution (total solution about 330 cm^3) is decanted or filtrated through GF/F filters and stored separately at 2 to 4° C for the analysis of dissolved nutrients (e.g. SiO_4^{-4} see below). Depending on the character of the particles sampled, the particulate material is sieved through a 1 mm screen to remove larger particles such as macrozooplankton swimmers such as crustaceans, or the swimmers are hand-picked. Samples with large phaeodarian radiolaria are difficult to handle. These protozoans cannot be hand-picked or separated from the sinking matter by sieving due to their very fragile skeletons. Thus, they are included in the particulate flux. The determination of the size distribution of samples with high amounts of larger fecal pellets such as krill fecal strings is particularly problematic. We suggest that these samples should not be sieved through screens.

After the removal of larger swimmers which are stored in 4% formalin, the particles are split into four aliquots using a rotary liquid splitter with eight cups. To enhance the accuracy of the splitting procedure, the material of two cups standing opposite to one another is added to one split. The standard splitting procedure is described in Figure 6-25. It is used when sufficient material is available. If not, each sample is only split into 1/4 and 1/16 aliquots. We try to do all bulk analysis on one 1/4 split.

3.2. *Net Material*

After the removal of larger swimmers by hand or by a 1 mm screen, this material is split into two aliquots by a rotary liquid splitter.

4. Analytical Methods

4.1. *Organic Compounds*

C/N and $\delta^{13}C$. From the sediment trap material, a 1/4 split is used for chemical analysis (Figure 6-25). Samples from sediment trap, plankton net and plankton pump material are desalted, by three repeated washings with about 50 mL of 1° C cold distilled water to remove hydroscopic NaCl, and centrifuged.

Subsequently, they are freeze-dried at 30° C and homogenized in a mortar. This treatment removes some organic compounds which are easily soluble in water [Müller et al., 1986]. To avoid losses of organic substances, dissolved NH_4Cl is sometimes used for washings by some laboratories, but this procedure biases nitrogen analysis and is not used by us. For C_{org} and N_{tot} measurements, a 2 to 4 mg portion of the freeze-dried material is decalcified with 6 N HCl in a silver boat and dried on a hot plate. No filtration step is involved which could result in losses of various organic compounds. Elemental ratios are determined with a

CHN-Analyzer [Fischer, 1989]. The relative precision is ±2% for organic carbon and ±4% for total nitrogen based on our internal sediment standard (C=2.103; N=0.199). Stable carbon isotope ratios are measured using a Finnigan MAT delta mass spectrometer attached to an HEREAUS-CHN-Analyzer. The samples are combusted at 1050° C in the presence of pure oxygen and He as carrier gas. The total reproducibility for the CO_2 preparation and the measuring procedure is better than 0.15‰.

Alkenones. This study on alkenones is done by P. Müller (Bremen). Total long-chain (C_{37}, C_{38}, C_{39}) alkenones are extracted from sediment or plankton material by soxlet using azeotropic mixtures of toluene or dichloromethane in methanol [e.g. Prahl et al., 1988]. Other useful extraction techniques include ultrasonification or flow-blending [Poynter et al., 1989; Farrimond et al., 1990; ten Haven et al., 1989]. The extracts are analyzed directly by gas chromatography or separated beforehand into lipid classes by medium pressure liquid chromatography. Alkenone abundances are calculated by comparing them with known quantities of internal standards.

Amino acids/amino sugars. These investigations are done by P. Müller (Bremen). Amino acids and amino sugars are released from the freeze-dried sample material by acid hydrolysis, separated by ion exchange chromatography and detected by fluorescence using the o-phthalaldehyde method and post-column derivatization [e.g. Müller et al., 1986]. About 0.5 to 1 mg of the dried and homogenized sample is weighed into glass ampoules, flushed with purified nitrogen/hydrogen (90:10) for 3 minutes after adding 1 mL of 6 N HCl, sealed and then hydrolyzed at 110° C for 22 hr. Appropriate amounts of norleucine or α-aminoadipic acid are added as internal standards prior to hydrolysis. The hydrolysate is evaporated to dryness in a vacuum centrifuge and the residue is taken up in a pH 2 sodium citrate buffer. Aliquots of 50 μL are injected to a Biotronik LC 5001 analyzer. The concentrations of amino sugars (glucosamine and galactosamine) are adjusted by a factor of 1.4 to correct for losses during hydrolysis.

4.2. Mineralogy of Particles

Carbonate. For the small amounts of sediment trap and plankton net material the following method [described in Wefer et al., 1989] is applied: First, about 2 to 4 mg of dried and homogenized material is combusted in an HEREAUS-CHN-Analyzer. This gives total nitrogen and total carbon (organic + inorganic) values. Secondly, about the same amount is treated carefully with 6 N HCL in a silver boat until no CO_2 evolves, dried on a hot plate and measured in the CHN-Analyzer. No filtration step is involved after the treatment with HCl. This gives organic carbon values. Carbonate is calculated as $(C_{total} - C_{org})*8.33$.

Biogenic opal. Opal is determined by sequential leaching [DeMaster, 1981] as modified and automated by Müller and Schneider (Bremen, in preparation). No pretreatment with hydrogen peroxide and HCl, as described by Mortlock and Froelich [1989], was made. A detailed description of various opal determination methods and the problems associated with them is given by DeMaster in this volume, page 363.

In our laboratory the ground sample is extracted under stirring with 100 mL 1 M NaOH at 85° C in a stainless steel vessel. The increase of dissolved silica is simultaneously recorded by continuous flow analysis. For this purpose, a small portion of the leaching solution is continuously passed to an autoanalyzer and analyzed by molybdate-blue spectrometry. A typical extraction curve shows a rapid initial silica increase due to the dissolution of opal, followed by a slower and linear increase representing the dissolution rate of silicate minerals. The intercept value obtained by linear extrapolation to an extraction time of zero gives the biogenic opal content of the sample. The silicon extracted from clay minerals is largely corrected by this evaluation technique. For the elaboration of the method, sponge spicules, recent and Pliocene diatom oozes, radiolarian concentrates, as well as various plankton, sediment trap and sediment samples were measured. The total analysis time depends mainly on the type of the material analyzed and ranges from 10 to 60 minutes. The precision of the overall method is ±4% and ±10% for sediments containing 10% and 2% biogenic SiO_4 respectively.

Another possibility for the control of the leached clay minerals is the parallel measurement of Al (mostly originating from clay particles) according to Eggimann et al. [1980]. For this purpose a sequential wet leach technique (1 M NaOH, 85° C) is applied by Fischer [1989]. Small aliquots (0.5-1 mL) are taken from the leachate after 1, 2, 3, and 5 hours. Si- and Al-determinations are carried out in alkaline solution with an inductively coupled plasma spectrometer (ICP). These concentrations are plotted against the extraction time. As expected, the Al shows a linear increase and the absolute values are much higher for extracted sediments compared with plankton or trap material. Besides the correction of non-amorphous silica with the above described regression-method (y-intercept of SiO_2=biogenic opal) a correction with the Al values using an empirical factor [F=SiO_2/Al_2O_3; Eggimann et al., 1980] was tested. In most cases the Al correction method reveals lower biogenic opal values. Due to the uncertainty of the SiO_2/Al_2O_3 ratios of the clay mineral association we do not recommend this method for an accurate biogenic silica correction. Prior to the application of this correction method, the clay mineral composition of the sample should be determined.

Clay minerals. The method outlined below was developed and tested by H. Berner (Bremen). It is described in more detail by Berner and Wefer [1990] and Berner (in preparation). An excellent review on clay mineral analysis is given by Van der Gaast (this volume, page 343).

For x-ray diffraction analysis in our laboratory, the organic matter of the clay fraction (less than 2 μm) is oxidized with 15% H_2O_2 and decalcified with 20% acetic acid. As outlined in Van der Gaast (this volume) this pretreatment may cause degradation of some clay particles. Saturation of free Mg^{+2} ions is done with 50% $MgCl_2.2H_2O$ and excess $MgCl_2$ is washed out. The treated clay material is suspended in an ultrasonic bath.

The clay fraction, defined as less than 2 μm, of sediments is filtered onto 0.2 μm cellulose nitrate filters (texture). These mudcakes are pressed between wax paper and a porous plate and dried. Subsequently, they are pressed onto double-sided adhesive tape, mounted on sample holders and the filters are peeled off. For automated measurement they are placed into sample collectors of the PHILIPS PW 1700 automatic sample changer.

From the suspended sediment trap material, which is treated as described above, 0.5 mL is carefully dropped onto polished aluminum tiles and dried rapidly to get well-textured layers. Diffraction analysis of very small samples is done by doubling the measuring time.

Clay minerals are expanded with glycol for several hours and recordings are run from 2 to 40° 2θ with 0.2° 2θ increments and a measuring time of 1 second. Peak identifications were done according to Brown and Brindley [1980]. For semiquantitative determinations of smectite, chlorite, kaolinite and illite, a method used by Biscaye [1964, 1965] is applied. Kaolinite was determined by subtracting the chlorite (002) peak which was calculated from the chlorite (003) peak by assuming a chlorite (002)/chlorite (003) peak ratio of two.

Barite. This study on barite in the marine environment is in preparation (G. Bohrmann, AWI, Bremerhaven). According to Bishop [1988] suspended and sedimented barite associated with biogenic opal may indicate the intensity of organic matter regeneration in the water column. The formation of barite will be investigated by scanning electron microscopy after enrichment of barite with a density solution. Combined X-ray diffractometer analysis and atomic absorption spectoscopy should give first information about the Ba contents in opal rich trap samples (Antarctic Polarfront).

4.3. Trace Elements

This study planned by W. Balzer will examine the geochemical behavior of some important trace elements in the dissolved and particulate phase of a highly productive water column (equatorial Atlantic). For the analysis of the dissolved elements from the water column and the supernatant water (trap cups), the flameless AAS-method will be used. Trace metals will be either enriched by liquid-liquid-extraction prior to analysis or measured directly (supernatant in trap cups). The particulate material will be filtrated and digested with $HF/HClO_4$ at 160° C in Teflon beakers. The following elements should be determined with flame or flameless atomic absorption: Al, V, Cr, Mn, Fe, Co, Ni, Cu, Zn, Cd, Ba, Pb.

4.4. Radioisotopes

This study by M. Rutgers van der Loeff and M. Segl investigates the scavenging of Pb and Th in the water column by larger particles. The methods applied are described in Rutgers van der Loeff and Berger [in press]. Water samples are collected with hydrocasts (270 L polyester Gerard bottles) and a centrifugal pump (10 m water depth, pump intake on RV Polarstern). Sixty to 150 L of water are filtrated through 0.45 μm filters. Dissolved $^{210}Pb/^{210}Po$ (20 kg of filtrate) and ^{234}Th (20 kg of filtrate) are coprecipitated in the presence of Pb, ^{208}Po and ^{230}Th spikes. Filter and sediment destruction, purification and platings were done according to Fleer and Bacon [1984] and Anderson and Fleer [1982]. All samples are analyzed and alpha and beta counted on board the ship. ^{210}Pb in the sediment trap material was measured as ^{210}Po and corrected for the decay since deployment. Supported ^{210}Pb was derived from ^{226}Ra measurements and ^{230}Th fluxes were corrected for supported ^{230}Th according to Anderson et

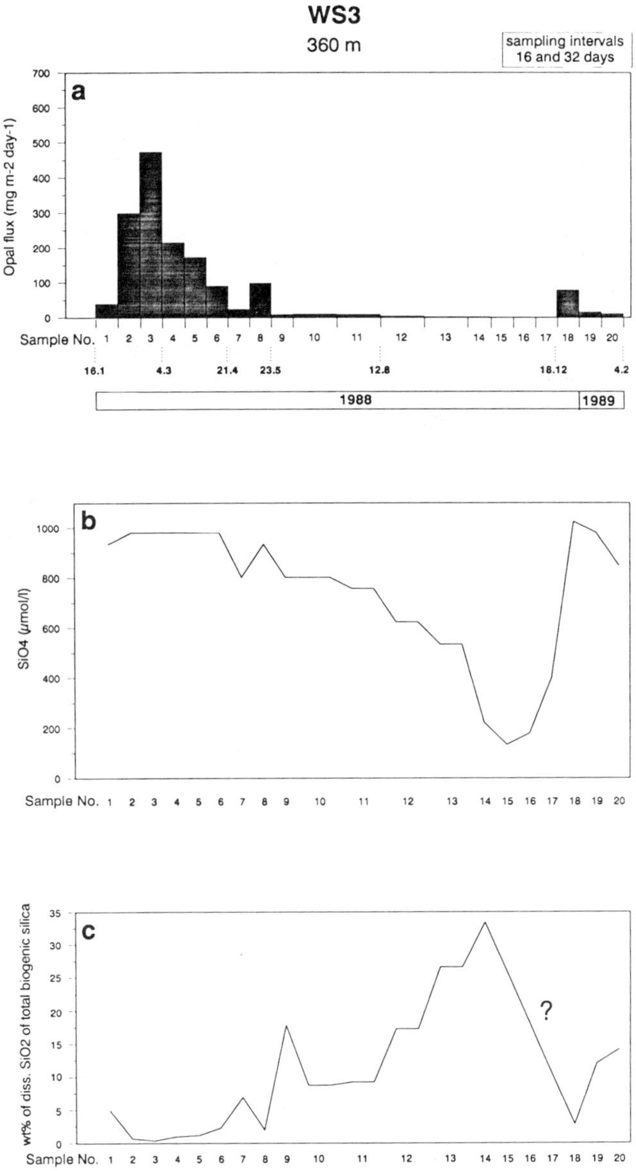

Fig. 6-26. a: Opal flux rates at the trap site WS3 (southeastern Weddell Sea; b: dissolved Si concentrations in the surrounding trap solution (WS3) c: dissolved opal calculated as the percentage fraction of total biogenic silica (WS3).

al. [1983] using the ^{232}Th activities in the sediment trap samples.

4.5. Stable Carbon and Oxygen Isotopes on Planktonic Foraminifera

For this isotope study the foraminifera are handpicked from 1/16 or 1/64 splits (Figure 6-25), counted and dried at 60° C. Isotopic analysis is done using an automated 48 sample KIEL carbonate preparation device. The samples are treated with 100% orthophospheric acid at 75° C and the liberated CO_2 is analyzed in a FINNIGAN MAT 251 mass spectrometer. Based on multiple analysis of laboratory standards, measurement precision is ± 0.1‰ for $\delta^{18}O$ and ± 0.7‰ for $\delta^{13}C$.

4.6. Nutrients in the Supernatant

Nutrients in the surrounding trap solutions are measured according to Grasshoff [1976], except some Si measurements which were done with the method used for biogenic silica determinations (see above).

For the trap site WS1 [northwestern Weddell Sea, Fischer et al., 1989] nutrients show little seasonal variation. The concentrations are in the range of those measured in the water column. In the supernatants of the trap WS3 [Wefer and Fischer, in press] high SiO_4^{-4} concentration are measured ranging from about 200 μM L^{-1} to 1000 μM L^{-1} (Figure 6-26b). High Si concentrations were determined not only in the first cups but also in the cups 18-20 at the end of the deployment time.

These concentrations correlate with highest opal fluxes (Figure 6-26a) and are not related to the deployment time. During times of highest opal fluxes the weight percentages of dissolved SiO_2 with regard to the total biogenic silica (assuming no exchange when the cups are opened) are less than 5% (Figure 6-26c). In contrast, more then 30% of the total biogenic silica can be dissolved when opal sedimentation is low (sample 14, Figure 6-26c). A comparable picture is shown for trap material collected in warmer areas (CB1, Cap Blanc; Figure 6-27a, b, c) although the absolute SiO_4^{-4} values are lower (about 400 to 700 μM L^{-1}). We also calculated the losses of biogenic silica for the whole investigation periods of WS3 and CB1. For the WS3 trap, only about 1.8% of the annual accumulated opal was dissolved due to dissolution in the sampling cups. The annual SiO_2 loss due to dissolution in the CB1 samples amounted to 2.5% of the total biogenic opal flux. These relatively small opal losses may be neglected. Nevertheless, annual losses due to dissolution may be much higher at other trap sites (v. Bodungen, this volume, page 47).

Acknowledgements. The methods and procedures described here have been developed at the University of Bremen in cooperation with the Alfred Wegener Institute for Polar and Marine Research in Bremerhaven. We thank all colleagues and the technical staff for their continuous support. This research was funded by the Deutsche Forschungsgemeinschaft (Sonderforschungsbereich 261 at Bremen University, Contribution No. 34).

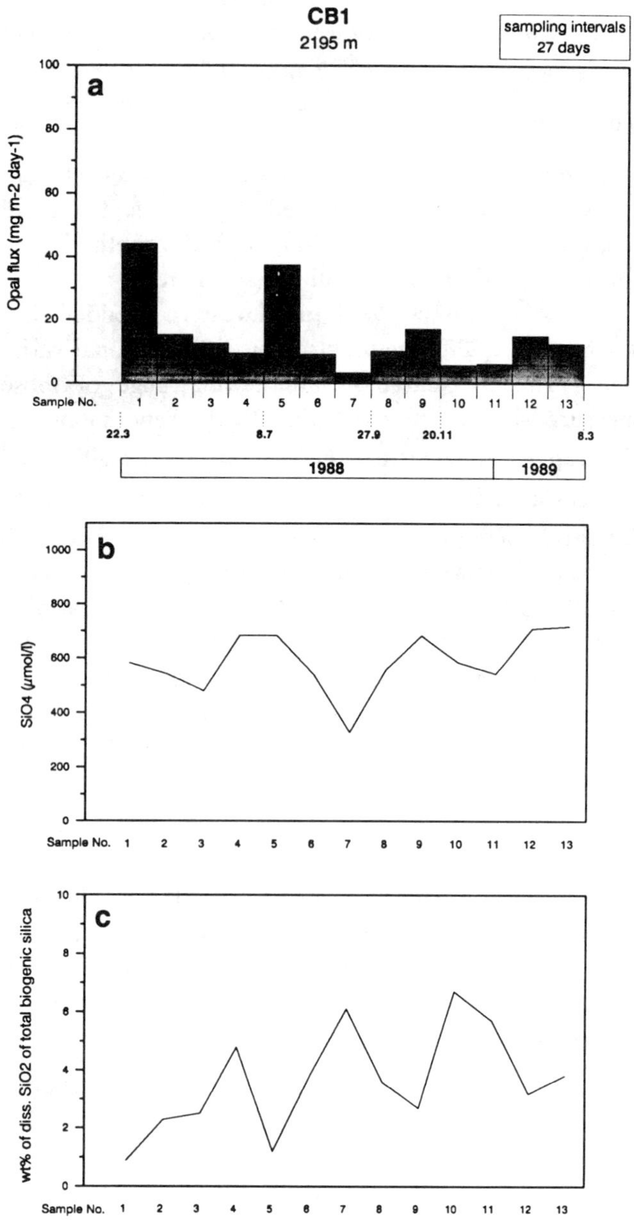

Fig. 6-27. a: Opal flux rates at the trap site CB1 (Cape Blanc, 20° 55.3′ N, 19° 44.5′ W) b: dissolved Si concentrations in the surrounding trap solution (CB1) c: dissolved opal calculated as the percentage fraction of total biogenic silica (CB1).

References

Anderson, R. F., and A. P. Fleer, Determination of natural actinides and plutonium in marine particulate material, Anal. Chem., 54, 1142-1147, 1982.

Anderson, R. F., M. P. Bacon, and P. G. Brewer, Removal of ^{230}Th and ^{231}Pa from the open ocean, Earth Planet. Sci. Lett., 62, 7-23, 1983.

Berner, H., and G. Wefer, Physiographic and biological factors controlling surface sediment distribution in the Fram Strait, in Geological History of the Polar Oceans: Arctic versus Antarctic, edited by U. Bleil, and J. Thiede, pp. 317-335, Kluwer Academic Publ., Dordrecht, Boston, London, 1990.

Biscaye, P. E., Distinction between kaolinite and chlorite in recent sediments by X-ray diffraction, Am. Mineral., 49, 1281-1289, 1964.

Biscaye, P. E., Mineralogy and sedimentation of Recent deep sea clay in the Atlantic Ocean and the adjacent seas and oceans, Geol. Soc. Am. Bull., 76, 803-832, 1965.

Bishop, J. K. B., The barite-opal-organic carbon association in oceanic particulate matter, Nature, 332, 341-343, 1988.

Brown, G., and G. W. Brindley, X-ray procedures for clay mineral identification, in Crystal Structures of Clay Minerals and Their X-Ray Identification, edited by G. W. Brindley, and G. Brown, Chap. 5, pp. 305-359, Mineral. Soc., London, 1980.

DeMaster, D. J., The supply and accumulation of silica in the marine environment, Geochim. Cosmochim. Acta, 45, 1715-1732, 1981.

Eggimann, D. W., F. T. Mannheim, and P. R. Betzer, Dissolution and analysis of amorphous silica in marine sediments, J. Sediment. Petrol., 50, 215-225, 1980.

Farrimond, P., G. Poynte, and G. Eglinton, Molecular stratigraphic study of Peru margin sediments, hole 686B, Leg 112, in Proc. of the Ocean Drill. Program: Sci. Results, edited by E. Suess, and R. von Huene, Vol. 112, pp. 547-553, College Station, Texas, 1990.

Fischer, G., Stabile Kohlenstoff-Isotopen-Verhältnisse in partikulärer organischer Substanz aus dem Südpolarmeer (Atlantischer Sektor), Ph.D. dissertation, 161 pp., Berichte, Fachbereich Geowissenschaften, Universität Bremen, 5, 1989.

Fleer, A. P., and M. P. Bacon, Determination of ^{210}Pb and ^{210}Po in seawater and marine particulate matter, Nucl. Instrum. Methods Phys. Res., 223, 243-249, 1984.

Grasshoff, K., Methods of Seawater Analysis, 317 pp., Verlag Chemie, Weinheim, New York, 1976.

Mortlock, R. D., and P. N. Froelich, A simple method for the rapid determination of biogenic opal in pelagic marine sediments, Deep-Sea Res., 36, 1415-1426, 1989.

Müller, P. J., E. Suess, and C. A. Ungerer, Amino acids and amino sugars of particulate and sediment trap material from waters of the Scotia Sea, Deep-Sea Res., 33, 819-838, 1986.

Poynter, J. G., P. Farrimond, S. C. Brassell, and G. Eglinton, Molecular stratigraphic study of sediments from holes 658A and 660A, Leg 108, in Proc. Ocean Drill. Program: Sci. Results, edited by W. Ruddiman, and M. Sarnthein, Vol. 108, pp. 387-394, College Station, Texas, 1989.

Prahl, G. P., L. A. Muehlhausen, and D. L. Zahnle, Further evaluation of long-chain alkenones as indicators of paleoceanographic conditions, Geochim. Cosmochim. Acta., 52, 2303-2310, 1988.

Rutgers van der Loeff, M. M., and G. W. Berger, Scavenging and particle flux: seasonal and regional variations in the Southern Ocean (Atlantic Sector), Mar. Chem., in press, 1991.

ten Haven, H. L., J. Rullkötter, and R. Stein, Preliminary analysis of extractable lipids in sediments from the eastern North Atlantic (Leg 108): comparison of a coastal upwelling area (Site 658) with a nonupwelling area (Site 659), in Proc. Ocean Drill. Program: Sci. Results, Vol. 108, edited by W. Ruddiman, and M. Sarnthein, pp. 351-360, Ocean Drilling Program, College Station, Texas, 1989.

Wefer, G., and G. Fischer, Annual primary production and export flux in the Southern Ocean from sediment trap data, Mar. Chem., in press, 1991.

Wefer, G., G. Fischer, D. Fütterer, R. Gersonde, S. Honjo, and R. Ostermann, Particle sedimentation and productivity in Antarctic waters of the Atlantic Sector, in Geological History of the Polar Oceans: Arctic versus Antarctic, edited by U. Bleil, and J. Thiede, pp. 363-379, Kluwer Academic Publ., Dordrecht, Boston, London, 1989.

THE SPAS: SELECTED PARTICLE ANALYSIS SYSTEM

Jacques Jedwab

Laboratoire de Géochimie, Université Libre de Bruxelles, 50, Avenue F.D. Roosevelt, B-1050-Brussels, Belgium

1. Introduction

This contribution describes the detailed instruments and operations to find, select, analyse and determine individual particles in the oceanic domains. At the onset of the IDOE-GEOSECS PROGRAM, when I was working on lunar dust and cosmic particles, I was introduced by Edgard Picciotto to Roger Chesselet, who was involved in the particulates branch of that Program. He convinced me to adapt my lunar dust techniques to the study of the GEOSECS filters, which were just being collected. At that time, the bulk composition of the particulate matter was already largely known, but the speciation of the iron and of the other heavy metals remained, for the most part, unknown [Campbell, 1983].

The principal difficulty was in finding the possible metal carriers. The SEM/EMP offered an unprecedented possibility to go directly to the individual particles, instead of extracting data indirectly from total chemical compositions, differential dissolutions or equilibrium calculations. But the direct examination of particles collected on membrane filters is marred by the presence of overwhelmingly numerous Ca, Si and organic particulate species [Broecker and Peng, 1982; Bruland, 1983]. The chances to find one heavy metal mineral (HMM) present in concentrations in the millionths range is prohibitively expensive in terms of instrument time at the scanning electron microscope coupled with the electron microprobe (SEM/EMP). A means of selecting the particles of interest (POIs) must thus be found, in order to go as directly as possible to heavy metal minerals without analysing millions of unwanted ones.

I had already met with a similar problem during the study of meteoritic and lunar dust materials, where some given types of particles are lost in the crowd [Jedwab, 1970; Jedwab et al., 1970; Jedwab, 1971a, b, c, 1973, 1975a, b]. It had then been realized that the human eye-and-brain couple is a very effective instrument in recognizing highly contrasted objects, according to form, color, refractive index, opacity, reflectivity, polarization [Neisser, 1964] but fully ineffective when it comes to chemical and structural determinations, which are irreplaceable when one explores unconventional environments.

In the oceanic realm, a trial experiment showed us that the HMMs, as a group, present positive, distinguishing optical properties, which are not matched by the bulk of particles: they are either opaque, or semi-opaque; when transparent, they are coloured and highly refracting; when transparent and colourless, they display generally a high refractive index and sometimes a very strong light dispersion. They have collectively the property of being highly visible under the light microscope (LM.) On the other hand, the overwhelmingly abundant Ca carbonates, clays, silica and organic debris become inconspicuous, invisible under oil immersion, due to their absence of colour and to their low refractive indices.

The envisioned solution was thus to scan filtered particle collections under the LM, to select or extract noteworthy individual particles, and to complete the analyses under the EMP, where the chemical data would be obtained. One of the most advantageous aspects of this approach would be that a large amount of information could be extracted from very scanty materials, i.e. particles collected only once at the same location and at a given time; but this at great pain and skill, and at very high cost.

The instruments (the LM and the SEM/EMP) were already well established aids, and very effective in their realms, notwithstanding the slowness of the wavelength dispersive spectrometer (WDS). (But who imagined a better and swifter way, like the energy dispersive spectrometer (EDS), which began to be commercially introduced in the late 1960s and early 1970s?)

The most difficult problem was to securely transfer one micron-sized particle, isolated among the thousands or millions of others of trivial interest, from the LM to the SEM/EMP. The first solution which comes to mind is the physical extraction or separation with micromanipulators. But all who tried that way in small particle mineralogy gave up rapidly: electrostatic and capillarity forces are extremely difficult to fight. The solution evolved progressively with the introduction of the membrane filter as the common mediating support: each particle should not be manipulated individually, but should be left on the filter, which would be transferred as a whole from one instrument to the other.

Marine Particles: Analysis and Characterization
Geophysical Monograph 63

The main problem was then the fixation of the particles on the filter, which ought to be immersed in oil for hours for good LM observation, then cleansed and metallized. The transient softening of the ester cellulose filter in an acetate mixture was the solution, which we came to by trial and error [Jedwab et al., 1970].

Another problem was the reliable refinding under the SEM/EMP of a small particle selected under the LM. The worst aspect is that, as long as one uses light illumination and oil immersion, the common minerals are invisible, but reappear massively under the SEM. Here one gets lost and what is more, colour becomes a useless attribute. The development of a mechanical object-marker was part of the answer, suggested by a long microscopic tradition [Johannsen, 1918]. But for the refinding of very small particles, in the range of 1 to 5 μm, supplementary means should be brought into play: the back scattered mode (BSM) of observation under the SEM, where HMMs emerge brightly on the screen [Lloyd, 1987], and the systematic use of cheap Polaroid pictures, as a mediating support between the LM and the SEM/EMP stages.

The SEM/EMP is nevertheless an incomplete means of determining particles: the mineralogy of a certain optical/chemical type of particle cannot always be unambiguously ascertained, and a crystallographic determination is necessary in critical cases. Here again, some established techniques were already available [Bergmann, 1959; Kittrick and Hope, 1967].

The steps of this method evolved progressively: it is possible to see and locate HMMs selectively among the million times more abundant biogenic, Ca and Si compounds, to mark them physically, to bring the filters under the SEM/EMP for chemical analysis. When necessary, the definitive diagnosis of a given HMM could be completed by running an X-ray diagram, obtained by collecting a sufficient number of analysed particles of a given optical/chemical species.

However, severe limitations of the method, which could more appropriately be described as a craft, will become obvious in due course.

1.1. Instrumental Limitations

- the limited number of particles or occurrences which can be practically selected on a given filter surface, due to the physical dimension and entangling of the identification marks, when they are printed in too large numbers.
- the difficulty (or impossibility) of acquiring some optical data which are sometimes critical for a good diagnosis.
- the limitations of the EMP chemical determinations, especially in the case of solid solutions, mechanical mixtures, variously hydrated species, carbonates, oxides and hydroxides.
- the impossibility of obtaining a sufficiently clear XR diagram with only a few particles: at least 30 particles of 10 μm minimum diameter must be collected for each species to be determined.

1.2. Human Limitations

-The researcher should have a good training in the various techniques.

-(S)he should display a personal involvement in the problem, and should personally acquire, follow-up and synthesize the data obtained from the various instruments.

-(S)he should stand the physical strain due to long working sessions at the LM and at the SEM/ EMP.

-(S)he should overcome the unpopularity of visual, qualitative methods in the marine geochemical community.

1.3. Data and Laboratory Management

- Several data will probably be gathered which turn out to be superfluous at the end of the run.
- A detailed documentation, produced by the various observations under the different instruments and by the necessity of precisely localizing each selected particle, must be constituted and managed.
- The instrument times on the LM and SEM/EMP may be considered too long.

It should also be recognized that the SPAS is an elaborate succession of steps, making use of diverse techniques, and integrating a large amount of varied data, some of which are purely qualitative; this would probably not be appealing to the editor of a scientific journal devoted to primary information publication.

The present description is meant for those who need a serious introduction into an effective and proven instrumental method in the search for some of the rarest (but potentially very significant) members of the world of marine particles [Lal, 1977; Sackett, 1978; Campbell, 1983; Bruland, 1983]. Additionally the SPAS is sufficiently flexible that it can be shortened when needed, and some of its most boring aspects, i.e. the detailed book-keeping, can now be alleviated by computer and video techniques (cf. Conclusions).

2. Outline of the SPAS

The SPAS, in the finding and determination of rare, heavy metal minerals (HMMs), interspersed among overwhelmingly abundant, biogenic ones, necessitates the use of three different, but fixed, methods: the light microscope (LM), the scanning electron microscope (SEM), coupled with the electron microprobe (EMP), and X-ray diffraction (XRD).

The method is unidirectional, in the sense that it is impossible (or at least very difficult) to return to the LM, once the SEM/EMP studies have been initiated and completed. The most important reason for this is that after SEM/EMP study, the irradiated filter portions supporting a selected particle become brittle, and the very particles which are the most studied are the most easily lost.

The principal consequence of this unidirectional course is that all LM observations, measurements and documentations must be completed before proceeding to the following steps. This means that several data should be accumulated which could turn out to be of minor or trivial interest.

The whole SPAS, as well as each of its elementary steps, must thus anticipate this unidirectional course. The materials, instruments and actions must thus be well thought out at the start and adapted accordingly, beginning with the particle collection step itself. The particle concentration on the filters must be regulated: heavy loads entail the burial of particles of interest (POIs) under trivial compounds, whereas sparse loads lead to the scanning of larger surfaces. However, it is well known that it is extremely difficult to remobilize particles once settled, and that the danger of contamination increases with repeated manipulation. As a rule of thumb, the filtration of 10 L of deep ocean water through a 47 mm diameter filter ($\sim 0.5\ \mu$m pore-size) yields a reasonable crop.

The six successive steps of the SPAS which will be described in more detail in section 4 can be outlined as follows (excluding the problems of particle collection at sea):

1. Filter mounting
2. LM stage
3. SEM/EMP mounting
4. SEM/EMP stage
5. XRD stage
6. Synthesis of LM, SEM/EMP and XRD data.

But before entering into the description of the method itself, I must spend some time describing the needed materials and instruments, and commenting on their settings and uses.

3. Materials and Instrumentation

3.1. Membrane Filters

We have mainly used two filter types: the cellulose ester mixture and the polycarbonate. The SPAS evolved using Millipore filters (cellulose esters) and Nuclepore filters (polycarbonate). Several other manufacturers and types are now available [Gelman, Schleicher & Schull, etc., cf. Brock, 1983], but since the solvent mixtures have been designed for these specific types and manufacturers, any departure from the original choices should not be attempted without carefully planned trials, and in no case, on valuable, real samples.

The cellulose ester membranes (MF-type, Millipore) may be used directly as particle supports: they are fully transparent and featureless after fixation on the glass slides and under immersion oil. However, their surface appears rough under the SEM, and they melt and crack under the electron beam at 25 kV, the necessary voltage to analyse HMMs for Pb, As, Se, Sr, Br, Y, Zr, and Mo. These membranes are principally used as an intermediate fixation surface for the polycarbonate membranes.

The polycarbonate membranes are not fully transparent under the LM: the etched nuclear tracks, which are the basis of this membrane type, are not filled by the fixation medium or by the immersion oil. They appear as small refracting tubes when inclined, and as black dots when vertical. But one rapidly becomes accustomed to this inconvenience. As they do not stick to the glass slides after softening by the acetate mixture, as the cellulose ester membranes do, a membrane of this latter type is used as an intermediate support. After fixation, the polycarbonate membranes frequently develop a general anisotropy, which may be corrected by adjusting the angular positions of the polarizers.

These membranes have a good resistance to the electron beam, but prolonged observation induces brittleness in the filter and reduces the bonding strength of the irradiated particles to the filters. The latter can be preserved for years.

The polycarbonate membranes have one smooth face and one rough face. Shipboard filtration should thus make sure the smooth face is always "upstream", facing the particles, which is especially important for SEM observation of very small particles. The persons in charge of the on board collections should be seriously warned of this particularity, and trained for quick recognition of the faces.

MF-type Millipore and Nuclepore polycarbonate filters are completely impervious to carbon tetrachloride, which may thus be used as a cleaning agent. A small stock of acetate-resistant filters should be kept for particle removal (SOLVINERT by Millipore).

3.2. Manipulation and Cutting of Filters

The filters are manipulated with flat-tipped, stainless steel tweezers. Large membranes, 47 mm diameter, should be cut to size before fixation on the glass slides. Plastic scissors should be used, as ordinary ones always liberate metallic particles by wear, and contaminate the filters along the cut. Such scissors can be made entirely from polymethacrylate. It is also possible to make a filter trimmer by replacing the steel blade of a stamp collector's trimmer by one made of polymethacrylate. The cutting of the membranes from the slides, after the LM step, is performed with a medical lancet. Membranes are stored in tightly closed petri dishes.

3.3. Fluids

Carbon tetrachloride, p. A.
Amyl acetate, p. A.
Ethyl acetate, p. A.
Immersion oil, silicone base. N=1.515.
(Large capacity bottles (100 mL) generally contain oil of greater fluidity than the small droppers.)

All fluids (except the immersion oil) should be prefiltered to remove possible contaminating particles with compatible membrane filters (~ 0.5 μm porosity). They should be stored in large, tinted glass bottles cleansed with already filtered fluids. Small quantities of carbon tetrachloride and the acetate mixture could be stored in droppers. However, it should be remembered that the acetate mixture shows impaired properties when exposed to air, probably due to differential evaporation.

3.4. Light Microscopy

The operator should be familiar with LM techniques: centering and adjustment operations of transmitted and reflected light, eyepiece adjustment, electric bulb centering, etc. The focusing should be sharp on the slide and on the graduated reticle. The ocular tube separation and focus should be well adjusted to reduce eye strain, headache and nausea. The operator should also have a full command of:

- reproducible and firm slide positioning between the stage clamps
- controlled command of the vertical stage movement
- objective and object-marker interchange
- setting of oil immersion objectives
- introduction and setting of rotating polarizers
- opening, centering and Köhler adjustment of diaphragms
- focusing and stage movement knobs
- reproducible vernier reading.

Microscope. A good research petrographic microscope should be selected from among the best brands. It should be trinocular (two oculars for visual observation and one for photography); double illumination, capable of working at the same time with transmitted and reflected light; adjustable light intensities; polarized light available on light pathways.

Stage. The stage should be rectangular (of the type used in biology): the rectangular movements are actuated by one hand with coaxial knobs in a lowered position, whereas the other hand is used for fine focusing. Verniers should be available along both rectangular directions, for the recording of particle positions. A 0.1 mm precision is sufficient. With such a rectangular stage, one loses the possibilities of the rotating stage (pleochroism, extinction angle, maximum birefringence, conoscopic data, etc.), but hour-long working sessions are not too straining, and particle positions may be recorded with good precision and reproducibility. If necessary, selected and marked particles may later be observed on a rotating stage, which can be fitted on the microscope in place of the rectangular one. The microscope should thus be selected with this provision in mind, or a microscope fitted with a rotating stage should be accessible for short sessions.

Oculars. 8 or 10x magnification, paired. (Unpaired oculars provoke nausea and headache). Spectacle-wearing observers should insist on working with special oculars. A graduated reticle should be fitted in one of the oculars. The graduations should consist of a linear scale with divisons of one hundredths and numbered in tenths. The reticle ocular should be able to revolve freely in its tube as to place the reticle in any angular position. Both observation tubes should allow for separate focusing.

Objective turret. It should be fit to work simultaneously with transmitted and reflected light. In general, turrets for reflected light also work well with transmitted light, provided the objectives are designed to project the images at the infinite. The turret should be provided with at least four sockets: three for the objectives, and one for the object marker. Ideally, each objective should be individually centerable, but such turrets have disappeared from the catalogues.

Objectives.

- Selected for their absence of strain.
- Calculated to project the image to the infinite.
- 4 or 8x objective for synoptic observation: dry.
- 16 to 25x objectives for systematic slide scanning: oil immersion.
- 100x oil immersion objective for close observation of very small particles, and for observation of optical dispersion. (Since these last objectives are very expensive, it could be worthwhile to look around for an unused Zernicke-contrast objective. This also gives good results when used in the SPAS.)

The immersion objectives should have the same tube length, in order to be set at the same working distance without important vertical stage adjustments, when interchanging the objectives.

Object marker. Object markers, which trace a circular mark on the filters with a sharp point are countereffective, since they tear up the filters [Fuess and Reichert types: Johannsen, 1918; Carl Zeiss-Jena]. Punch-type markers, printing a trace in the filters by a vertical movement of the stage, are more adapted to the fragility of the material [Ernest Fullam; Ford, 1974; Sugar, 1963; Williams, 1962]. The Ernest Fullam marker (1976 Catalog, # 1441: EFFA specimen marker) prints a 3 mm diameter circle (too large) around a selected particle, but the marked area may be punched out if the applied pressure is too high. This inconvenience may be avoided by filing down two opposed slots along the circular rim. The marker has no spring, and the glass slide is thus in danger of breaking.

An object marker which has been especially designed for the SPAS, has been developed in Brussels [Jedwab and Delvaux, 1975]. It consists of a main body, which can be screwed into one of the threaded sockets of the revolving turret, as most of the other object-markers (Figure 6-28). The marking blade consists of a vertical, 0.6 mm-wide jeweler's screwdriver's head, in the middle of which a 150 μm slot has been cut (Figures 6-29 and 6-30). When pressed against the membrane filter, this marking piece leaves a straight line interrupted in its middle. When precisely adjusted, the marker leaves a trace, easily recognizable under the SEM, at the center of which lies the selected particle (Figures 6-31 - 6-35).

I describe here the adjustment operation of this object marker. Such adjustment should be conducted before each LM marking session if different people use the LM.

The object marker is tightly screwed into one of the sockets of the turret (always the same socket). First one

Fig. 6-28. General view of the object marker (MP).

imprints a mark anywhere on a blank fixed filter (see "Fixation of the filter on the glass slide", Section 4.1), and one observes its position with the medium power immersion objective. The eyepiece with the reticle is turned until the scale crosses the space amid the two straight branches of the mark. The corresponding graticule division is noted and the ocular is left in this position.

One then chooses any characteristic dust particle on the filter far away from the mark, and brings that particle to the reticle position and division noted previously, by moving the stage (the reticle remains in fixed position). The marker is then set in position, and a mark is imprinted. In principle, the selected particle should now sit just in the middle, between the two straight branches of the mark. The operation must obviously be repeated with any different objective, but it is wise to stay always with the same (medium power).

Practice improves the results. Some experience is needed in applying the marker to the slide by moving the stage vertically. Too much pressure can break the slide, whereas too little leaves an inconspicuous trace. The spring inside the marker can be used to some extent to adjust the pressure. It is recommended to run the initial exercises with a sample of dust from any origin, fixed on a Nuclepore membrane, on which the marks are more

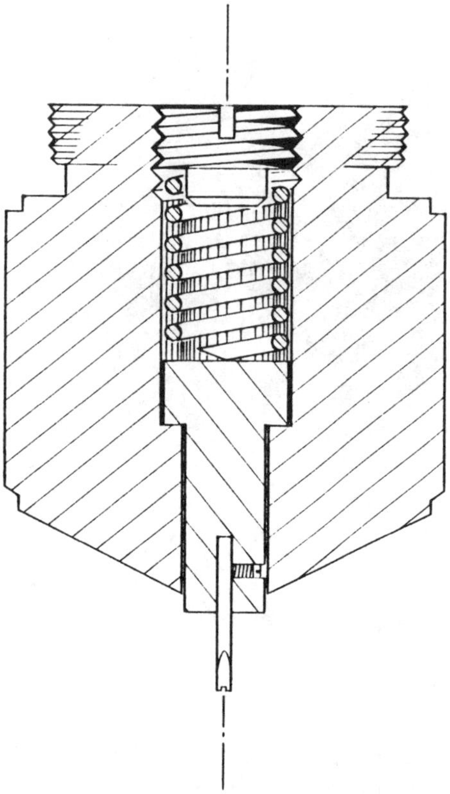

Fig. 6-29. Drawing of the object marker in section.

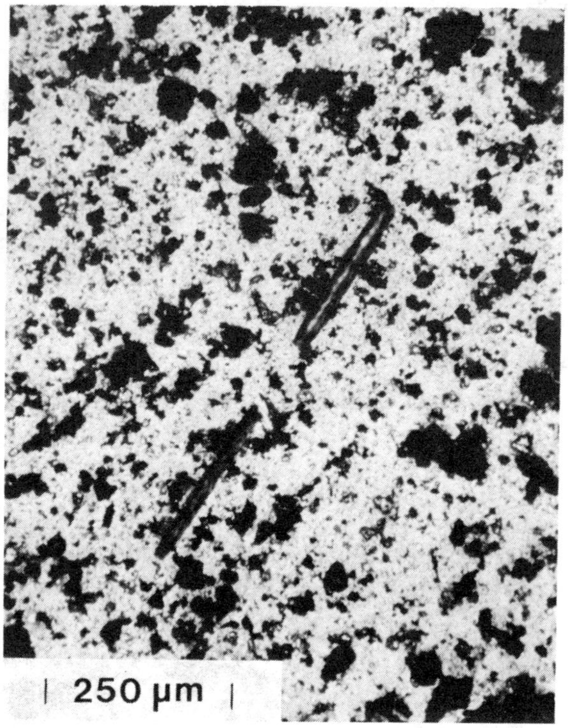

Fig. 6-31. Mark printed in the hand drawn square (LMT-dry obj.).

Fig. 6-30. Tip of the object marker (SEM-SM).

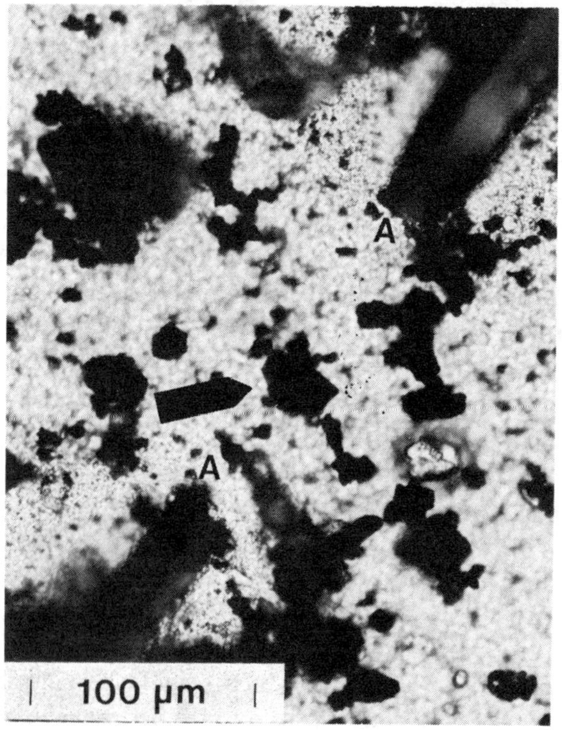

Fig. 6-32. POI (arrow) in line with mark (A). (LMT-dry obj.).

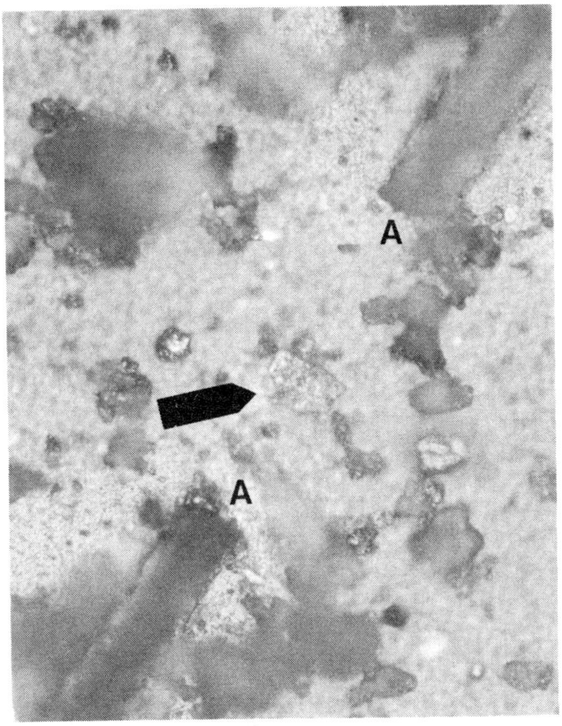

Fig. 6-33. Same under LMR-dry obj.

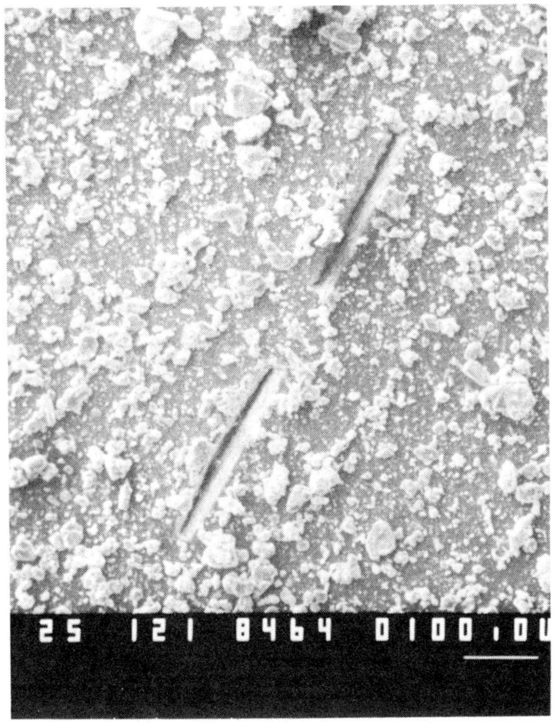

Fig. 6-34. General view of the mark. (Notice that the POI is here still difficult to see) (SEM-SM).

easily displayed with the help of transmitted, polarized light. With clear Millipore filters, it is necessary to close the diaphragm tightly to see the mark.

Causes of trouble are:

- worn out turret stop
- uncontrolled stage movements
- loose fixation of the slide between the stage clamps
- careless initial adjustments
- modifications by different LM users between sessions.

Glass slides. They should be of the dimensions used in biology: 20x70 mm, with one frosted end for labelling, precleaned. Petrographic slides of 25x50 mm are unsuitable, since they are too short, and may be pushed through the stage slot when particles sited on the outer edges of a filter are marked.

Microphotographic equipment. The microscope should be fitted with a complete photographic attachment for 7x10 cm and 10x12 cm Polaroid pictures (8 or 10x projection ocular, focusing eyepiece, cameras, automatic exposure). Polaroid type 667 (B/W, 7x10 cm, 3000 ASA, without negative) is very convenient and relatively inexpensive.

3.5. Binocular Magnifying Glass

Transmitted and reflected light. Magnifications of 6 to 50x. Used for mounting the filters on SEM/EMP mounts.

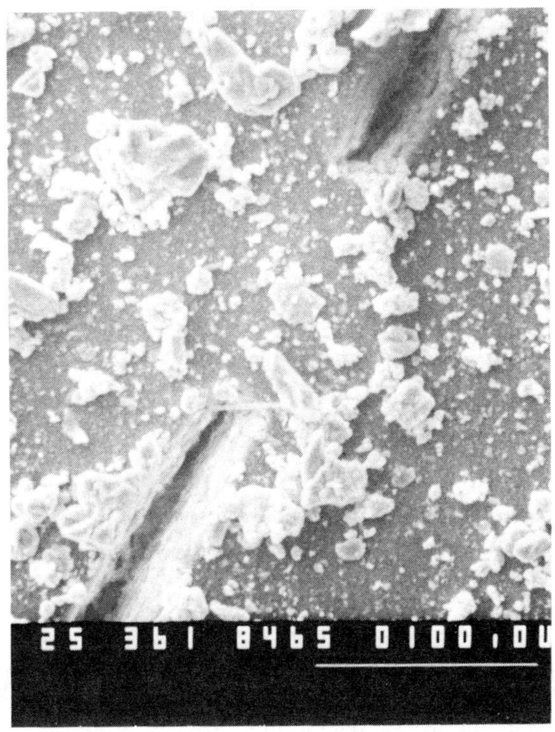

Fig. 6-35. Blow-up showing the POI in the middle of the mark (SEM-SM).

3.6. Scanning Electron Microscope

There are only a few possible instruments, and all are probably quite satisfactory (JEOL, PHILIPS, SIEMENS, CAMBRIDGE, HITACHI, CAMEBAX). One should be certain that the secondary (SM) and back scattered (BSM) modes are effective and easy to switch between. Continuous work with an acceleration voltage of 25kV should be possible. This voltage is necessary to analyze important elements with their best energy peaks situated between 10 and 20kV: Pb, As, Se, Br, Sr, Y, Zr and Mo.

The stage should have two rectangular movements, with legible displacement readings, graduated in mm (readings in tenths of mm are not very reproducible, but verniers are helpful). The sample holders should receive a permanent mark or slot, with which a sample mount can be reproducibly aligned (Figures 6-36 and 6-37).

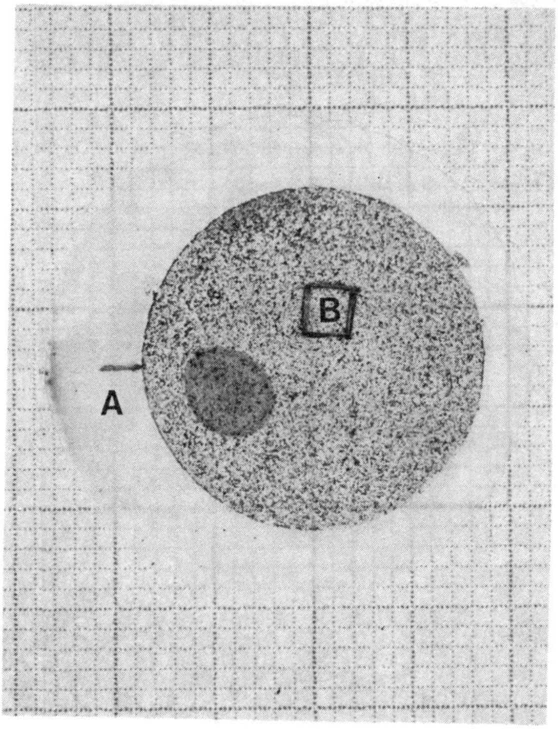

Fig. 6-36. The same filter. A: reference mark. B: hand drawn square region where one mark is printed (here invisible) (MP).

A vacuum metallization apparatus for carbon coating should be available or easily accessible. Gold or gold/palladium coating should not be used, as they strongly absorb the low energy peaks. These metals can be applied at the very end of the process, when one would like to make publication pictures, as they better reveal the fine morphological details of already determined particles.

Aluminum sample mounts, adapted to the available SEM, are prepared by cutting an orientation slot in their edge (Figure 6-37). Double faced tape is used to fix the filters on the mounts. A satisfactory quality is found with the 3M tape used by printers for layout mounts. It keeps its properties over long time spans, and has a low gas pressure.

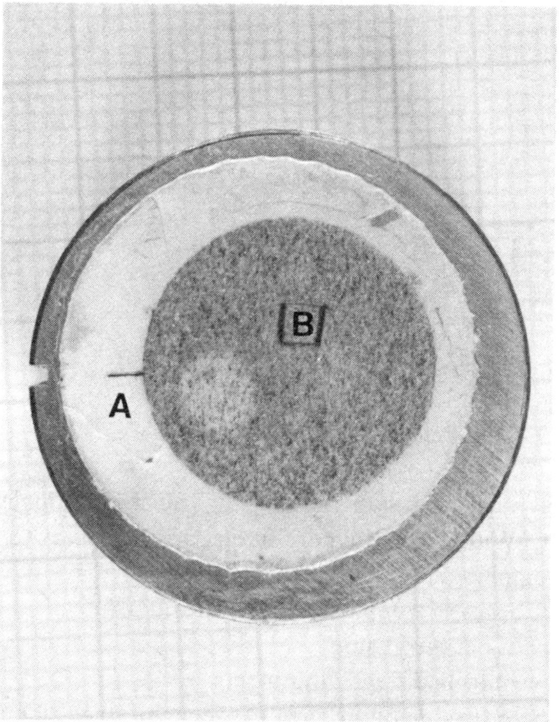

Fig. 6-37. Filter after mounting on SEM stub. (Notice reference mark A in front of orientation slot of the mount) (MP).

3.7. Electron Microprobe (EDS)

There are several available brands, with their pros and cons (TRACOR, LINK, KEEVEX, etc.). A very good knowledge of the elemental spectra is highly recommended, without which painful errors can be made. Not only should the classical interference cases be mastered (Ca/Sb, Se/As/Mg/Al, Ti/Ba, Pb/S/Mo, etc.), but also the relative intensities of the peaks of a single element: even a trained operator could never find Sr, since (s)he would think that the peak at 1.80 KeV is due to the Si $K_{\alpha+\beta}$ peak, and not to the L_α peak of Sr. The same is true for Mo, whose feeble K_α line is sited near the limiting 17 KeV value, and the strong L_α peak which interferes with the S $K_{\alpha+\beta}$ peaks.

The identification marks are relocated with the secondary mode (SM), as well as the selected particles, which should be found between the two straight branches of the marks, if the adjustment of the object marker has been done carefully at the LM stage.

In case of uncertainty, or filter overcrowding, one should make use of the back-scattered mode (BSM). But here, the contrast and brightness levels must be very carefully adjusted, in order to well discriminate the particles. Too high a brightness displays all particles without visible differences. The brightness and contrast could be adjusted on coccoliths and diatoms, which should be barely visible. The operator should have an excellent control of the various switches, allowing him/her to jump quickly and unhesitatingly from one mode to the other.

During the SEM (BSM) observation stage, some attention should be paid to high Z particles, which have not been noticed at the LM stage. Indeed, some HMMs are transparent and colorless, and their refractive indices are relatively low: this is the case with celestite, strontianite, barite, witherite, smithsonite, hydrozincite and the numerous transient zinc hydroxychlorides.

A special case is that of carbonaceous matter (graphite, soot and coal particles), which are easily visible under LM, due to their opacity, but are extremely difficult to find again under the SEM, even under BSM, due to their low Z composition. The projected study of such particles should anticipate this difficulty at the LM stage and high magnification Polaroid pictures should be taken (see Section 5. Applications).

3.8. X-Ray Diffraction

The mineralogic determinations cannot be better completed than with a good XR powder diagram. The oceanic realm is populated with polymorphic and isometallic compounds: there are four polymorphs of Ca carbonate, two of Cu oxide, three of Cu hydroxychloride, etc. Unfortunately, due to the general minuteness of the marine particles, good XR diagrams cannot be obtained with only one particle. The problem is thus not the XR diffraction, but the collection of sufficient particles of identical composition to give a legible diagram. As a rule of thumb, it may be stated that about 30 identical particles of about 10 μm diameter give a good diagram with an exposure time of 1-2 hours.

A mechanical micromanipulator may be used to collect the particles under the LM. To do this, a gelatin capillary (cut from a medicine capsule) is mounted on a rod. The tip is covered with a small drop of drafters' rubber cement. The selected particles are then taken successively, following the noted locations. If large particles are available, an individual particle or group may be taken, and the ball of rubber with the particle(s) gently crushed between two glass slides to obtain a powder. This step may be skipped if the single large particle is already microcrystalline, as observed under the LM or the SEM.

Another method to take selected, analysed, particles is to thoroughly clean the object marker from the immersion oil, and to cover its tip with a droplet of rubber cement. The particles are taken with the object marker, in the same way as they were marked at the onset of the operation, but slightly off center.

The rubber cement ball is then placed in a small diameter (57.4 mm) Debye-Scherrer camera and a diagram is taken. The interpretation of the pattern is generally simplified, as one already has a list of possible candidates, deduced from the chemical compositions. If one obtains a point diagram, instead of lines, it may be read following the principles proposed by Kittrick and Hope [1967].

Good diagrams can also be obtained with a Chesley microdiffraction camera, but in this case, the precision of the readings is very critical [Bergmann, 1959]. An internal standard (colloidal silver: electronic grade) should be added to the sample ball, and the readings of the film should be performed under the light microscope, with a precision better than 10 μm.

3.9. Bench Top Library

General: McCrone and Delly [1973-1980]; Jedwab, [1980a].

Membrane filtration and filters: Brock [1983]. Millipore and Nuclepore current catalogues.

Determinative mineralogy:

1. under EMP: Pierrot [1979]; Pierrot and Cesbron [1988];
2. nomenclature and basic references: Fleisher [1987]; ALEPH ENTERPRISES.
3. LM-transmitted light: Fleisher et al. [1984]
4. LM-reflected light: Picot and Johan [1977]; Ramdohr [1980]; Uytenbogaardt and Burke [1971]
5. XRD: ICDD [1990a, b].

4. Description of Techniques

The flowsheet (Table 6-10) and Figures 6-31 to 6-44 are intended to display the succession of operations of the SPAS. The example described here comes from the sampling of a black smoker hydrothermal vent sited at

EPR 13° N (IFREMER-HYDRONAUT Cruise, Nov. 1987).

4.1. Mounting of the Membrane Filters

Filter preparation. The filters must be thoroughly rinsed of sea water salt immediately after the on board filtration stage, without letting them dry. If this is not done, they will be clogged with halite, calcite and gypsum. After rinsing, they should be dried, possibly aboard; it takes at least 2 hours to dry Millipore filters in the oven set at 60° C, and half an hour for a Nuclepore filter. Filters of large diameter (47 mm) should be cut in two or in four. Filters of 25 mm diameter may be mounted directly on a glass slide, except if a piece needs to be preserved for later use.

Fixation of the filter on the glass slide.

a). *Ester cellulose filters.* An excess of the acetate mixture is poured with a dropper onto a dry slide. The filter, held with tweezers, is slowly dipped into the liquid, beginning with its edge. After thoroughly soaking the filter, which takes about 15 sec, the excess liquid is spilled off by inclining the slide. The latter is set to dry in the oven for 30 min., after which it can be numbered and examined under LM (Figure 6-38).

b). *Polycarbonate filters.* The fixation follows a two-step process: a blank Millipore filter is first deposited on a slide, as described under a), and an excess of acetate mixture is again added on the surface of that blank filter. The Nuclepore filter is then aligned and deposited on the first membrane, the excess acetate is spilled off, and the slide is set to dry as above.

It is possible that the Millipore filter shrivels after the second addition of the acetate mixture. However, the wrinkles will disappear during drying of the filters.

It is highly recommended to acquire a good command of these fixation steps with blank membranes, as it is quite impossible to recover a misplaced filter.

4.2. Light Microscope Scanning of the Slides

Graph paper preparation. Place the slide with fixed filter on the stage, with the frosted end at left. Imprint a reference mark in the middle of the westernmost edge of the filter (Figure 6-36). Read the E-W extreme coordinates of the filter outline and of the reference mark under dry, low power magnification.

Transfer the LM stage coordinates onto graph paper along the respective axes. Adopt a 10x magnification: 1 mm on the stage = 10 mm on the graph. Graduate the axes accordingly. Care should be taken that the reference mark and the filter coordinates appear in their positions as when the slide is seen through the eyepiece (Figure 6-41).

Scanning of the slide. Immersion oil is placed on the slide, and observations may begin with the medium power immersion objective. The slide is scanned, and the coordinates of POIs are noted along with the filter number, date, observer's name, etc. on data sheets. The POIs are given a consecutive number and a short description, along with their coordinates, which should always be registered in the same order, say X/Y. The flow sheet gives a list of the data which should be acquired under the different illumination modes at the LM scanning step.

Ideally, the following observations and measurements could be made at this stage (Figures 6-39, 6-40):

- *Morphology:* habit, crystallinity, fracture, cleavage, dissolutions, inclusions, associations, continuity.
- *Morphometry:* size(s), sizing, elongation, flattening. Particle counts.
- *Optical data (unpolarized light):* transparency, opacity, semi-transparency, color (transmitted and/or reflected light).
- *Optical data (polarized light):* pleochroism, anisotropism, birefringence, extinction, dispersion. The latter is best observed with polarizers exactly crossed, under reflected light, with the high magnification immersion objective and the diaphragm closed at its maximum.

If there is a shortage of material, if the filters carry few particles, or if quantitative data must be acquired, a systematic scanning of the whole filter should be undertaken. The scanning, registration and marking of noteworthy particles on about 3 cm^2 takes about 2-3 hours for a practiced operator.

Marks are not printed at this stage, principally because it is difficult to mark (and find again under the SEM) more than about 30 objects cm^{-2}. Thus it is desirable not to clutter the filter with unnecessary marks.

After the scanning is completed, one selects the best particles to be retained, and they are marked (Figures 6-31, 6-32, 6-33).

Particles which are larger than 100 micrometers receive an outside mark (which should be recorded in the log, as well as any other detail). The selection criteria shoud be the following:

- dimension (very small particles are more difficult to find again);
- representativeness;

TABLE 6-10. SPAS Flowsheet

	DOCUMENTATION	OPERATION	OBSERVATIONS AND MEASUREMENTS
1	Label glass slide.	Desalt filter with distilled water. Dry in oven (2 hrs. at 65° C).	
		Fix filter on glass slide with acetate.	
	Prepare numbered data sheet.	Dry in oven (30 min. at 65° C).	
2	Prepare graph paper with coordinates of reference marks.	Imprint reference marks under LM (dry objective) and read coordinates.	
	Record observations by number on data sheet. Tranfers POIs' coordinates on the graph.	Scan slide under oil immersion. Observe POIs. Read POIs' coordinates.	Morphology: habit, crystallinity, fracture cleavage, inclusions, associations, continuity. Morphometry: size(s), sizing, elongation, flattening. Optical (unpolariz. light): transparency/opacity, color, relief, reflectivity. Optical (polariz. light): pleochroism, anisotropy, birefringence, extinction angle, dispersion.
	Report losses, eccentric POIs, etc.	Select and mark selected POIs.	
		Check numbers on data sheet and graph, marks on slide, coordinates and numbers on graph for omissions and errors.	
	Number photographs and record relevant information.	Make detailed POLAROIDs (oil transmitted and/or reflected light).	
		Remove oil from slide with CCl_4.	
	Number photographs. Report existence of pictures on data sheet.	Shoot localization POLAROID's (dry objective, reflected light).	
3	Prepare SEM stub (double faced tape on one side, label on the other, orientation slot).	Remove filter fron slide. Remove residual oil from filter with CCl_4.	
		Mount filter on stub (match reference mark and orientation slot).	
		Metallize with carbon.	
4		Bring mount under SEM in reproducible position (slot).	
	Register SEM coordinates of marks on graph.	Localize reference marks and POIs.	
	Report SEM observations (SM. and B.S.M. modes). SEM photographs. Chemical spectra and analyses, EMP maps.	Study POIs. Make relevant POLAROIDS. Record EDS spectra.	Morphology: (see LM), sub-phases, inter- & over-growths, homogeneity (high & low Z), organites. Chemical composition: major & minor elements, matrix, mixtures, organics.
		Check data sheet, graph, spectral and photographic documents for omissions.	
5	Record extracted particles on data sheet and on graph.	Extract analysed, monomineralic particles with micromanipulator.	
		Make XR diagram.	Measure reflections, calculate d's.
	Sort out d's of possible candidates.	Match data with possible candidates.	
6		Summarize observations and report conclusions.	
	File documents (data sheets, graphs, photographs, XR diagrams and data, EMP analyses and spectra). Store analysed objects (slide, remaining filters, SEM and XRD mounts.)		

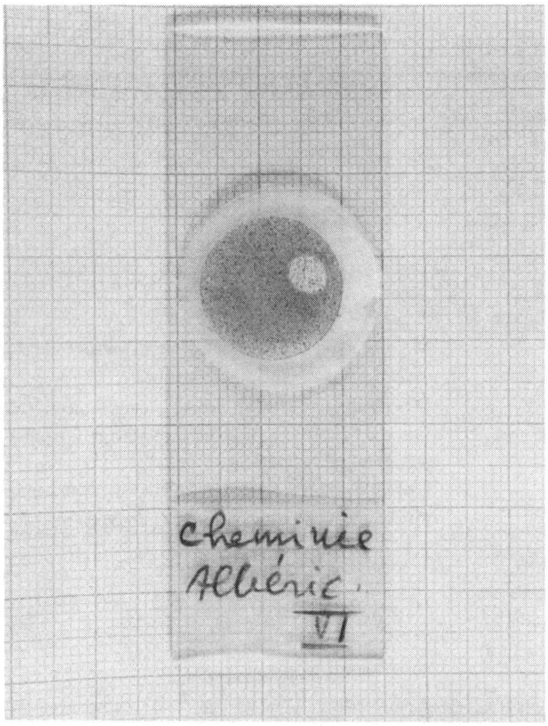

Fig. 6-38. General view of a glass slide with 25 mm diameter filter after fixation and oven drying (MP).

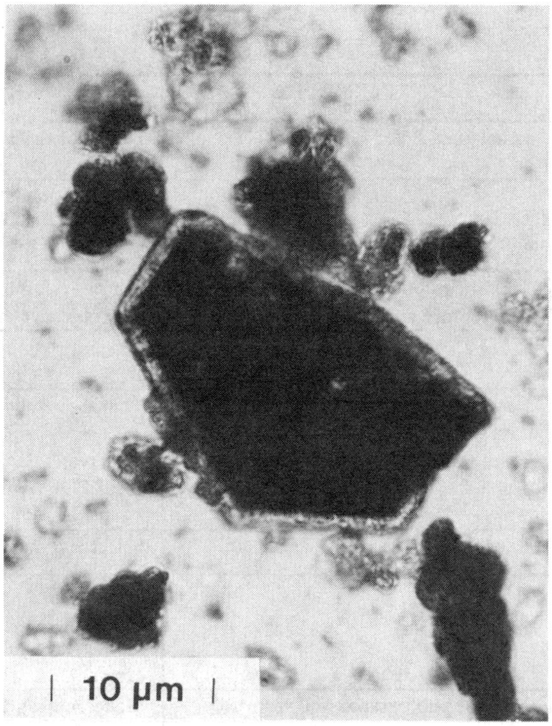

Fig. 6-39. POI under LMT-oil obj. (Notice transparent borders.)

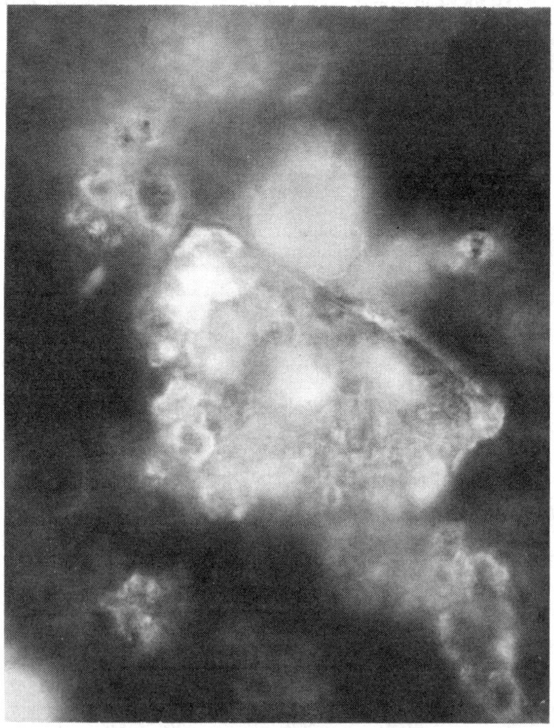

Fig. 6-40. Same, under LMR-oil obj. Notice transparent/dispersive overgrowths.

- isolation from other particles. Especially avoid selecting particles which are evidently part of a lint ball of organics, or which are visibly covered by transparent particles. Attention should also be paid to possible shadowing effects: depending on the relative positions of the particle, a large shadowing particle and the EMP/EDS detector, a particle detected under LM could turn out to be unanalyzable;
- simple mineralogy (if mechanical mixtures are unavoidable, make a sketch of the particle, displaying its different parts).

After marking the selected particles, their coordinates are transferred on to the graph as numbered dots. Check the list, the graph paper and the slide, to verify that each marked particle has a description in the list, and that each number in the list and on the graph has a marked particle on the slide. This seems trivial, but much time is often wasted trying to locate a lost or forgotten particle, when working at the SEM/EMP.

Photographic documents may now be taken. Immersion pictures are made for detailed recording (Figures 6-39, 6-40). For the purpose of precisely locating a particle among several others, the best simulation of what will be observed under the SEM is available with Polaroid photographs made with the low power, dry

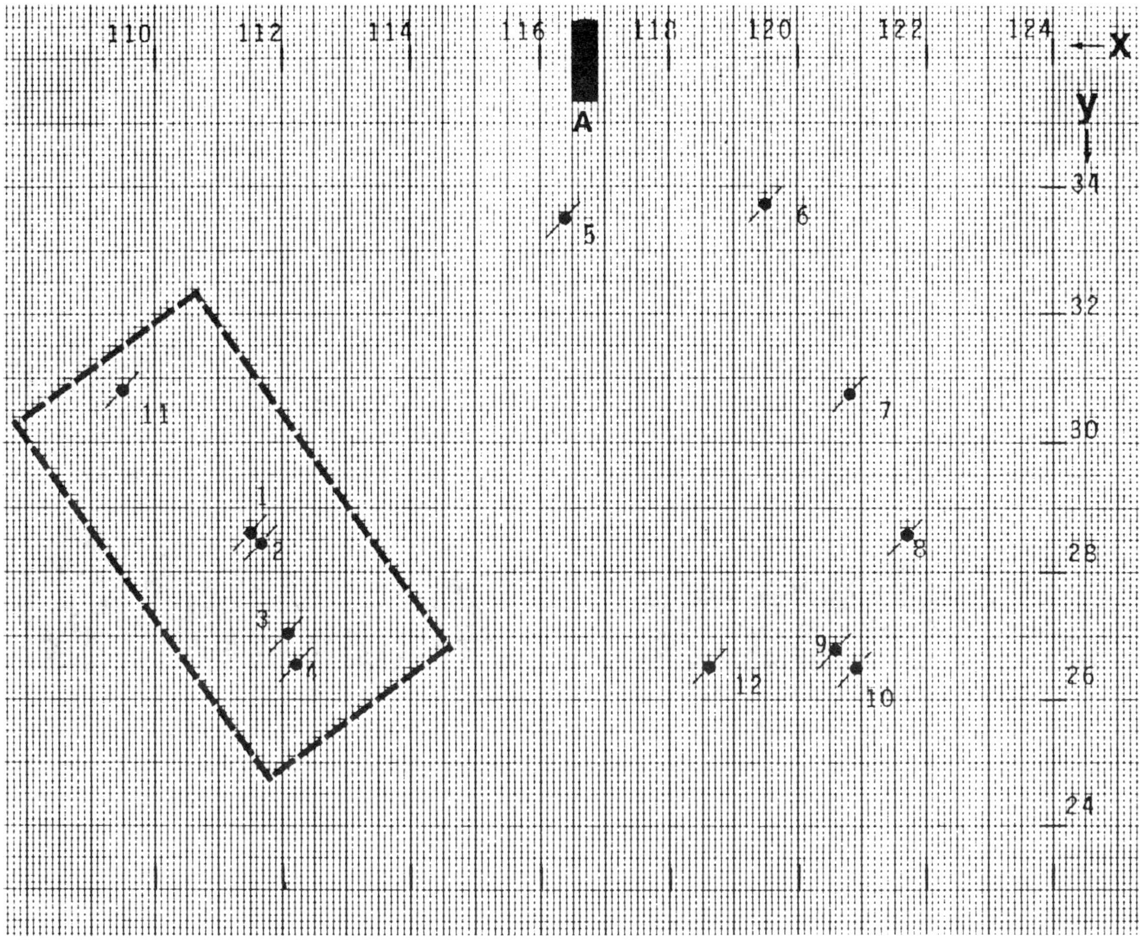

Fig. 6-41. Graph paper showing the X and Y axes, LM stage graduations, the reference mark (A) and the positions of a few POIs marks.

objective and under reflected light (Figure 6-33). The selected particle is marked with an arrow on the picture.

4.3. SEM/EMP Mounting

Cleaning the slide of the immersion oil. To get rid of the immersion oil, the inclined slide is gently flushed with carbon tetrachloride from a squeeze bottle. The removal of the superficial oil is checked after letting the inclined filter dry for 2 minutes. When the slide is cleaned of superficial oil, one may check it once again under the LM, and make dry objective photographs, as described above.

After this is completed, the filter is removed from the glass slide. It is necessary to do this, as oil is generally trapped between the filter and the slide, and it must be completely cleaned before transferring the filter to the SEM/EMP. To remove the filter, it is necessary to raise the border with the lancet, observing under the binocular, and to peel off the filter with the tweezers. If the filter resists, the whole perimeter should be cut with the lancet; taking care, however that the orientation mark is preserved. The filter, still held with the tweezers, is flushed with carbon tetrachloride until all the oil has been washed away.

Preparation of the SEM sample mount. A sample mount of the type used in the available SEM is prepared before receiving the washed and dried filter. Its surface is covered with double faced tape.

Mounting of the filter. The dry filter is deposited on the sample mount with due attention given to aligning of the reference mark on the filter with the slot on the mount (Figure 6-37). This should be done under the binocular magnifying glass. (A large mismatch entails a definitive angular difference between the graph paper positions and the SEM stage readings.) Attention should

Fig. 6-42. Region with mark. SEM-SM.

Fig. 6-44. The same, under SEM-BSM. The overgrowths do not show a very different composition in Z. Since the POI is pyrrhotite, and the overgrowths were transparent under LMR, it is concluded that the whole surface of the POI is coated with the same, optically transparent, material.

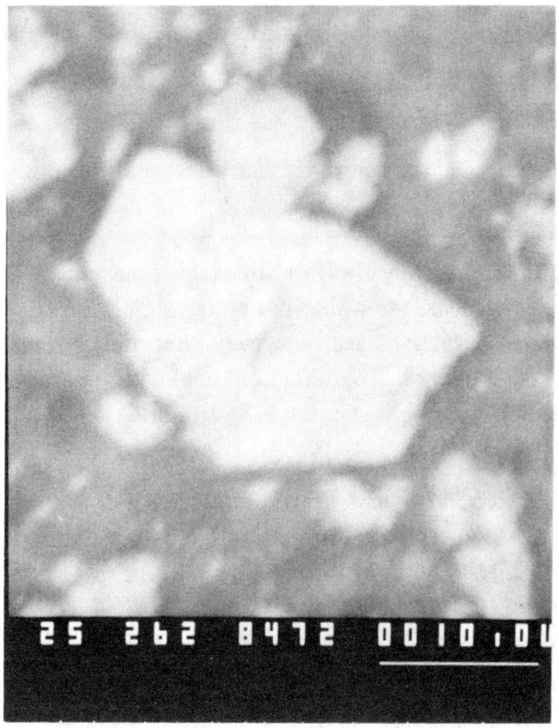

Fig. 6-43. POI under SEM-SM. The details of the overgrown particles are visible.

be paid to electrostatic effects, which could be very detrimental by suddenly attracting the filter onto the mount in an unwanted position. It is advisable to hold the filter with two pairs of tweezers to have full control.

The excess of tape is cut away with the lancet, and removed from the stub with the tweezers or a wooden stick. One should be cautious here of the backlash of stretched tape threads which could spoil the filter (and valuable particles). The mount is finally coated with carbon.

4.4. SEM/EMP Study of the Filter

Relocating the reference and particle marks. The reference mark, located at the northernmost border, is located and its new, SEM coordinates transferred onto the graph paper. From there on, the different particle marks may be found again by counting revolutions of the stage commands (corresponding to millimeters or half of millimeters on the filter). The new SEM coordinates are recorded on the graph for quick reference. It is recommended to relocate all the marks in one continuous run, while confirming chemically only a few particles. This will be very helpful later on, when one will want to go quickly to particles of the same type for systematic comparisons.

Relocating and studying the selected particles. If the marks have been carefully printed, the LM selected particles should lie just in the middle of the two straight lines of each mark (Figures 6-34 and 6-35). As stated above, several LM-invisible, light element particles, could reappear under the SEM. The Polaroid pictures, made under LM-dry objective, will be of great help here to relocate the LM-selected particles. The back-scattered mode (BSM) will also be decisive, if the sought for particle is composed of high Z elements. The EMP chemical compositions are determined, with the use of the BSM, to observe the inhomogeneities of composition in the selected particles.

The following data should ideally be collected during these steps (Figures 6-43 and 6-44):

- Morphology: habit, crystallinity, fragmentation, cleavage, dissolution, inclusions, inter- and overgrowths (especially of diatoms and coccoliths).
- Morphometry: size(s), sizing, elongation, flattening.
- Chemical composition: major and minor elements, matrix, homogeneity, mixtures, inclusions of high/low Z elements, organics. Impregnation of organogenic particles (coccoliths and diatoms) by high Z elements. Organic components, recognizable due to the humped Compton peak.

Two things should be kept in mind: 1) some morphological characteristics, which may seem irrelevant when observed, may turn out to be important when discussing the origin or formation of the particle; 2) some morphological characteristics, like cleavage, are barely visible on carbon coated particles. Thus one should try to complete these observations on gold coated particles.

4.5. X-ray Diffraction

The sample mount from the SEM/EMP stage is securely fixed on a glass slide, and a list of the particles to be extracted is established, taking into consideration the following factors:

- chemical composition, according to EMP analyses
- optical properties
- purity from admixtures
- isolation from other particles.

As stated before, 30 to 50 analysed particles, belonging to the same type, should be collected to obtain a good diffraction. This requirement is obviously in conflict with the advice given before, not to select and mark too many particles. The best way out is to assign a part of the filter to the selection of one same particle category, once the importance of that type has been recognized.

The diagram is made with the collected particles, and after processing, the theta-values of the diffraction peaks are measured, and the corresponding d's calculated. These are compared to those of the possible candidates, extracted from the CHEMICAL TABLE of the ICDD Determinative Volumes.

4.6. Synthesis of Data

The LM, SEM, EMP and XRD data can now be summarized and grouped, in order to get an overall view. The grouping and evaluation of all the data depends, of course, on the aim(s) of the project. It will obviously be a different task to describe and trace the origins of all kinds of particles present, versus to concentrate on one specific kind, for example a given HMM, or graphite, or a bright, opaque component, etc.

One recurrent problem is the recognition of the anthropogenic contributions to the particle load, and the estimation of its intensity. One is faced with the difficult problem of attributing an origin (natural vs. technological) to each kind of particle. Such a discussion for each kind of particle found in the Atlantic Ocean will be found in Jedwab [1980a].

4.7. Output of the SPAS

The output of the SPAS in terms of required time to tackle a given problem or separate filter depends of course very strongly on the skills, commitment, and purposes of the researcher. It will take much more time to make an intensive study of all types of particles, than to aim at the determination of a few types or individuals.

Given a filter surface with a moderate particle load, studied by a skilled operator, it is nevertheless possible to broadly estimate the time needed to complete its study. The filter surface which is here adopted as a practical unit is that of a 25 mm diameter filter, which has a filtration surface of about 3 cm^2. Only stages 2 and 4 (LM and SEM/EMP operations) will be taken into account, since they need to be run individually with some continuity for each sample, whereas the initial stages may be run in batches. The XRD stage should be considered as exceptional, and would take about one working day for one particle type. The synthesis could of course take much more time: that's finally the scientific point.

4.8. LM Stage

LM preparation (reference marks, graph paper, data sheets):	1 hr

LM scanning of the surface, with immediate record of POIs:	2-3 hr
Marking of about 50 selected POIs:	1 hr
Description of properties of representative POIs:	1-2 hr
Polaroid photographs (dry and immersion):	1-2 hr
Checking for errors, filing of documents, etc.:	1 hr
Total:	7-10 hr

4.9. SEM/EMP stage

Mounting and metallization of the filter:	1 hr
Localization of marks:	1/2 hr
Study of POIs:	2 1/2 hr
Production of SEM photographs, recording of EMP spectra:	2 hr
Checking for errors and omissions, filing of documents, etc.:	1 hr
Total:	7 hr.

It is thus clear that, under normal working conditions, it takes at least 2 full working days to complete the most absorbing observational and analytical parts of the study. The utilization of the SEM/EMP would be considered as relatively expensive in terms of present day laboratory management standards. A crucial factor is the operation of the instrument by a specialized technician, or by the researcher him/herself. It is obvious that the better the preliminary localization and documentation work prepared under the LM, the shorter will be the utilization time of the SEM/EMP.

5. Applications

The SPAS has been used with due modifications in a number of research situations for nearly 20 years. The explored fields pertain to:

- Oceanography: [Jedwab, 1979, 1980a; Jedwab and Chesselet, 1979; Dehairs et al., 1980, 1987; Dehairs and Goeyens, 1989; Jedwab and Boulègue, 1984, 1986a, b, 1988a, b, 1989a, b; Boulègue and Jedwab, 1985; Boulègue et al., 1986, 1989; Jedwab et al., 1989].
- Volcanology: [Bernard et al., 1981; Le Guern and Bernard, 1982; Graeber et al., 1983; Bernard, 1985; Bernard and Le Guern, 1986; Bernard et al., 1990].
- Anatomy and medicine: [De Vuyst and Jedwab, 1981; De Vuyst et al., 1981, 1982a, b, 1983, 1986; Dumortier et al., 1986; Fuss et al., 1976; Toussaint et al., 1976; Van de Weyer et al., 1983; Arys et al., 1989].
- Zoology: [Coulon et al., 1988].
- Mineralogy: [Jedwab, 1990].
- Meteorites and lunar dust: [Jedwab 1971c, 1973, 1975b].
- Environment: [Jedwab, 1980c; Jedwab and Dumortier, 1981; Dumortier and Jedwab, 1981].
- Engineering (failure analysis): [Bernard and Rose, 1990].

6. Conclusions

When looking back at the description of the SPAS operation and details, one cannot help thinking that its present state is still transitional. It is obvious that it is complicated, and that it generates a large amount of provisory papers and photographs. It was designed in the early seventies, and has been in use since then without fundamental improvements or updating. Several time-consuming SPAS operations could presently be replaced by electronic methods of recording:

- the written documents can easily be set up by a micro-computer with writing and data base programs.
- the hand-drawn graph paper with the craftsman coordinate reporting can be replaced by a computer cartographic program.
- the LM and SEM Polaroid pictures can be taken on video tape with a video camera on the LM, and from the video signal of SEM's of recent design. Electronic pictures of high quality may be shot economically.
- diagnosis of minerals may be partly achieved with the help of crystallographic and mineralogical data banks (ICDD, 1990a, b; ALEPH ENTERPRISES].
- the synthesis stage can be prepared by integrating the different data generated by the LM, SEM, EMP and XRD techniques with a more elaborate program.

But nothing will replace the observer's skills, and therein lies the main difficulty: I think that with time, dedication and adequate instrumentation (including sufficient time allowed at the LM and SEM/EMP, especially during the training period), a good scientist is able to obtain interesting results and novel observations. The successful operation of the SPAS does not depend on theoretical obstacles, and is based on proven procedures. But the difficulties of interpretation of the novel observations generated by the SPAS, when applied to oceanography, are of another magnitude.

One may ask what are the benefits of the efforts: the suspended oceanic particles are not considered as important tracers, comparable to stable or radioactive isotopes, dissolved gases or elements [Broecker and Peng, 1982].

Only their bulk concentrations have been used for various purposes, which have been dealt with in numerous publications. The selected particles, and especially the HMMs, are even less considerable as tracers. Their study and potential usefulness are marred by several inherent properties and methodological difficulties: multiple, and frequently unclear, origins, patchy distribution, very low relative and absolute concentrations, diagnostic difficulties.

But our error is perhaps to consider that only large scale tracers are worth the efforts. In fact, there are some similarities between the study of selected particles and the planktonic studies in their beginnings. As it appears, the systematic particle studies are still in their infancy, as was the case in marine protistology before the time of Ehrenberg and Haeckel. The best proof is that one still finds unsuspected and unclassifiable categories of particles when changing the observation conditions.

It is thus probably premature to find some "use" for the selected particles. But it is reassuring that presently, the isotope people desperately need a better knowledge of the very nature of the various scavenging particles [JGOFS, 1990]. The particulate iron species are certainly among the most important of these scavengers. But the reader would perhaps have noticed that the iron minerals, which were put in a prominent position in the introduction, have not been dealt with in the text. This obviously results from an intrinsic shortcoming of the SPAS, i.e. the impossibility to distinguish optically and chemically between the different iron hydroxides: lepidocrocite, goethite, akaganeite, feroxyhyte, hydrohematite. The problem is still worsened by the possible presence of jarosite(s), since the so-called iron hydroxides frequently contain considerable amounts of potassium and sulfur. But this problem should be seriously addressed, considering the abundance of the iron species and their strong (and variable) adsorptive properties. A solution could perhaps be found in the recently developed Fourier transform infrared spectrometers coupled with a light microscope. These instruments are capable of recording spectra from particles of 10 μm diameter.

Another interesting aspect emerges from the observation or presence of some particles: one observes several transient species, displaying dissolution or decay phenomena, e.g. the Zn hydroxychloride(s) and hydroxycarbonate, the Fe phosphates(s) and the anhydrite. It should also be stressed that in hydrothermal samples, oxidized Cu species are not observed on the numerous Fe-Cu sulfide particles. They are obviously too near to their formation place and time: one wonders what happens to the oxidized Fe-Cu-Zn sulfides at some distance in the hydrothermal plume.

References

Aleph Enterprises, Mineral/Tracker System, computerized data files for 3900 minerals, Livermore, California, without date.

Arys, A., J. Jedwab, J. J. Pireaux, C. Philippart, and N. Dourov, Brushite in the pulp of primary molars, J. Oral Pathol. Med., 18, 371-376, 1989.

Bergmann, M. E., X-ray micro-diffraction cameras, Norelco Rep., 6, 96-100, 1959.

Bernard, A., Les mécanismes de condensation des gaz volcaniques: Chimie, minéralogie et équilibre des phases condensées majeures et mineures, Ph.D. thesis, 412 pp., Université Libre de Bruxelles, 1985.

Bernard, A., and F. Le Guern, Condensation of volatile elements in high temperature gases of Mount St. Helens, J. Volcanol. Geotherm. Res., 28, 91-105, 1986.

Bernard, A., and W. I. Rose, The injection of sulfuric acid aerosols in the stratosphere by El Chicon volcano and its related hazards to the international air traffic, J. Nat. Hazards, 3, 59-67, 1990.

Bernard, A., J. Jedwab, and F. Le Guern, Discovery of Greenockite CdS in volcanic sublimates from Usu Volcano, Japan, Bull. PIRPSEV, 43, 30-38, 1981.

Bernard, A., R. B. Symonds, and W. I. Rose, Volatile transport and deposition of Mo, W and Re in high temperature magmatic fluids, Appl. Geochem., 5, 317-326, 1990.

Boulègue, J., et J. Jedwab, Conditions profondes de l'hydrothermalisme de la dorsale Pacifique Est à 13° N, C.R. Acad. Sc. Paris Ser. 2, 300, 73-276, 1985.

Boulègue, J., J. L. Charlou, and J. Jedwab, Fluids from subduction zones off Japan, EOS Trans. Am. Geophys. Union, 67, 1204, 1986.

Boulègue, J., A. Mariotti, P. Cros, C. Rangin, and J. Jedwab, Dolomite cement and cementing fluid at 3835 m depth off Japan, IFREMER, Campagnes Océanographiques Françaises, 10, 253-264, 1989.

Brock, T. H., Membrane Filtration; A User's Guide and Reference Manual, 381 pp., Science Tech. Inc., Madison, Wisconsin, 1983.

Broecker, W. S., and T. H. Peng, Tracers in the Sea, 689 pp., Eldigio Press, Palisades, New York, 1982.

Bruland, R. W., Trace elements in sea-water, in Chemical Oceanography, edited by J. P. Riley, and R. Chester, Vol. 8, pp. 157-220, Academic Press, London and New York, 1983.

Campbell, J. A., The geochemical ocean sections study; GEOSECS, in Chemical Oceanography, edited by J. P. Riley, and R. Chester, Vol. 8, pp. 89-155, Academic Press, London and New York, 1983.

Carl Zeiss Jena, Object marker, Cat Nr. 30 58 60 A.

Coulon, P., J. Jedwab, and M. Jangoux, Les cristaux intrakystiques des grégarines (Apicomplexa) parasites du spatangue *Echinocardium Cordatum* (Echinodermata), Ann. Soc. R. Zool. Belg., 118, 177-182, 1988.

Dehairs, F., and L. Goeyens, The biogeochemistry of barium in the Southern Ocean, in Antarctica-Belgian Scientific Research Programme on Antarctica, edited by S. Caschetto, Vol. 2, pp. 1-100, Science Policy Office, Brussels, 1989.

Dehairs, F., R. Chesselet, and J. Jedwab, Discrete suspended particles of barite and the barium cycle in the open ocean, Earth Planet. Sci. Lett., 49, 528-550, 1980.

Dehairs, F., C. E. Lambert, R. Chesselet, and N. Risler, The biological production of marine suspended barite and the barium cycle in the Western Mediterranean Sea, Biogeochem., 4, 119-139, 1987.

De Vuyst, P., et J. Jedwab, L'analyse minéralogique en pathologie respiratoire, Rev. Méd. Brux., 2, 115-119, 1981.

De Vuyst, P., W. Schandevyl, Y. Robience, G. Vandermoten, R. Van de Weyer, J. Jedwab, A. Decoster, and J. C. Yernault, Bronchoalveolar lavage as an aid in the diagnosis of asbestos related diseases, Eur. J. Resp. Dis. Suppl., 113, 54-55, 1981.

De Vuyst, P., J. Jedwab, Y. Robience, and J. C. Yernault, "Oxalate bodies", another reaction of human lung to asbestos inhalation?, Eur. J. Resp. Dis., 63, 543-549, 1982a.

De Vuyst, P., J. Jedwab, P. Dumortier, G. Vandermoten, R. Van de Weyer, and J. C. Yernault, Asbestos bodies in broncho-alveolar lavage, Am. Rev. Respir. Dis., 126, 972-976, 1982b.

De Vuyst, P., M. Mairesse, A. Gaudichet, P. Dumortier, J. Jedwab, and J. C. Yernault, Mineralogical analysis of bronchoalveolar lavage fluid as an aid to diagnosis of "imported" pleural asbestosis, Thorax, 38, 628-629, 1983.

De Vuyst, P., R. Van de Weyer, A. De Coster, F. X. Marchandise, P. Dumortier, P. Ketelbant, J. Jedwab, and J. C. Yernault, Dental technician pneumoconiosis. A report of two cases, Am. Rev. Resp. Dis., 133, 316-320, 1986.

Dumortier, P., et J. Jedwab, La détermination de la minéralogie des métaux lourds dans des retombées, des sols et sur des végétaux. 1981, 13: Caractérisation des particules individuelles, in Progr. Nat. de Rech. & Dévelp. dans le domaine de l'environnement AIR. Rapport scientifique final 1978-1981, edited by A. F. A. Adams, P. Bloch, J. Van Landuyt, G. Van Tendeloo, J. Jedwab, et P. Dumortier, pp. 97-211, 1981.

Dumortier, P., P. De Vuyst, J. C. Yernault, and J. Jedwab, AEM analysis of low concentrations of asbestos in BAL fluids, Proc. XIth Intern. Congr. Electr. Microscopy, Kyoto, 2163-2164, 1986.

Fleisher, M., Glossary of Mineral Species, 227 pp., The Mineralogical Record, Inc., Tucson, Arizona, 1987.

Fleisher, M., R. E. Wilcox, and J. J. Matzko, Microscopic determination of nonopaque minerals, USGS Prof. Pap., B 1627, 453 pp., 1984.

Fullam, E., EFFA specimen or screening marker, cat. #1441, Latham, New York, 1976.

Fuss, M., P. Van Landuyt, and J. Jedwab, A routine method for the integrated study of microcrystals in urine by light and scanning electron microscopy, electron microprobe and micro-X-ray diffraction, in Urolithiasis Research, edited by H. Fleisch, W. G. Robertson, L. H. Smith, and W. Vahlensieck, pp. 269-272, Plenum Press, NY, 1976.

Graeber, E. J., T. M. Gerlach, P. F. Hlava, and A. Bernard, Metal transport and deposition by high temperature gases of the 1983 East Rift eruption, Kilauea volcano, EOS Trans. Am. Geophys. Union, 64, 904, 1983.

ICDD, Mineral Powder Diffraction File-Data Book and Search Manual, Sets 1-35, International Center for Diffraction Data, Swarthmore, Pennsylvania, 1990a.

ICDD, PC-PDF (Powder Diffraction File on CD-ROM), International Center for Diffraction Data, Swarthmore, Pennsylvania, 1990b.

Jedwab, J., Les sphérules cosmiques dans les nodules de manganèse, Geochim. Cosmochim. Acta., 34, 447-457, 1970.

Jedwab, J., Isolement et étude intégrée d'une microparticule minérale par microscopie photonique, microdiffraction X, microscopie et microsonde électroniques, Bull. Soc. Belge. Géol., 80, 177-182, 1971a.

Jedwab, J., Particules de matière carbonée dans des nodules de manganèse des grands fonds océaniques, C. R. Acad. Sci. Paris, 272, 1968-1971, 1971b.

Jedwab, J., La magnétite de la météorite d' Orgueil vue au microscope électronique à balayage, Icarus, 15, 319-340, 1971c.

Jedwab, J., Rare, micron-size minerals in lunar fines, Proc. Lunar Sci. Conf., fourth, 1, 861-874, 1973.

Jedwab, J., A method of extraction and analysis of possible cosmic particles from manganese nodules, Meteoritics, 10, 233-239, 1975a.

Jedwab, J., Cosmic flakes in manganese nodules, Meteoritics, 10, 191-207, 1975b.

Jedwab, J., Copper, zinc and lead minerals suspended in ocean waters, Geochim. Cosmochim. Acta, 43, 101-110, 1979.

Jedwab, J., Rare anthropogenic and natural particles suspended in deep ocean waters, Earth Planet. Sci. Lett., 49, 551-564, 1980a.

Jedwab, J., L'analyse des particules individuelles et ses applications, ULB, Laboratoire de Géochimie, 126 pp., 1980b.

Jedwab, J., Nature minéralogique des particules des retombées en région d'Engis et à Beerse. Journée d'étude branche 2: Transferts. 18 mars 1980, Progr. Nat. de Rech. & Développ. dans le domaine de l'environnement -AIR, 13 pp., 1980c.

Jedwab, J., Native ruthenium in tantalum carbide concentrates from the Ural Mountains, USSR, Mineral. Petrol., 43, 137-146, 1990.

Jedwab, J., and J. Boulègue, Graphite crystals in hydrothermal vents, Nature, 310, 41-43, 1984.

Jedwab, J., and J. Boulègue, Formation of minerals associated with biological colonies in subduction areas off Japan (Nankai trough, Japan trench, Kuril trench), Kaiko Mtg. Abstracts, 2 pp., 1986a.

Jedwab, J., et J. Boulègue, Les spinelles de fer et de fer-zinc trouvés dans des trappes à sédiments amarrées près d'évents hydrothermaux océaniques, Bull. Minér., 109, 635-642, 1986b.

Jedwab, J., et J. Boulègue, Carbone graphite, intercalants du graphite et spinelles Fe-Zn, marqueurs des conditions profondes d'interaction eau-roches dans l'hydrothermalisme océanique (13° N EPR), IFREMER-INSU-ORSTOM-BRGM, Colloque national sur l' hydrothermalisme océanique, Brest, Nov. 22-24, 1988, 100, 1988a.

Jedwab, J., et J. Boulègue, Minéraux rares et objets biologiques non-identifiés récoltés dans les trappes à sédiments CL-I et CL-II sur le site hydrothermal EPR 13° N, IFREMER-INSU-ORSTOM-BRGM, Colloque national sur l'hydrothermalisme océanique, Brest, Nov. 22-24, 1988, 99, 1988b.

Jedwab, J., and J. Boulègue, Formation of minerals associated with biological colonies in subduction areas off Japan (Nankai Trough, Japan Trench, Kuril Trench), IFREMER, Campagnes Océanographiques Françaises, 10, 245-251, 1989a.

Jedwab, J., and J. Boulègue, A vanadium-titanium carbide inclusion in graphite from hydrothermal ejecta at 13° N, East Pacific Rise, Can. Mineral., 27, 617-623, 1989b.

Jedwab, J., and R. Chesselet, Paint pigments; an overlooked group of stable technological tracers, in Biogeochemistry of estuarine sediments, pp. 91-94, Proc. of UNESCO/SCOR workshop, Melreux, Belgium, Nov. 29 to Dec. 3, 1976, UNESCO, Paris, 1979.

Jedwab, J., and G. Delvaux, Patent 830.067, Belgian, 1975.

Jedwab, J., et P. Dumortier, Caractéristiques minéralogiques de la poussière sédimentable, Progr.Nat.de Rech. & Dévelop. dans le domaine de l'environnement AIR, La pollution de l'air par une usine métalurgique: une étude systématique, 27 mars 1981, 11 pp., 1981.

Jedwab, J., M. Defleur-Schenus, and A. Herbosch, Mounting and polishing small quantities of minerals in the micrometer range, Am. Mineral, 55, 1065-1066, 1970.

Jedwab, J., G. Blanc, and J. Boulègue, Vanadiferous minerals from the Nereus Deep, Red Sea, Terra Nova, 1, 188-194, 1989.

JGOFS, The need for process-oriented studies: controls on tracer uptake by particles, in Isotopic tracers, U.S. JGOFS Planning. Rept. Nr. 12, 13, Woods Hole, Massachusetts, 1990.

Johannsen, A., Manual of Petrographic Methods, (p. 611: Section 536. Marking thin sections. Fuess and Reichert section markers.), 649 pp., McGraw-Hill Book Co., Inc., New York and London., 1918.

Kittrick, J. A., and E. W. Hope, A procedure for the identification of small crystals by X-ray diffraction analysis, Am. Mineral., 52, 286-294, 1967.

Lal, D., The oceanic microcosm of particles, Science, 198, 997-1009, 1977.

Le Guern, F., and A. Bernard, A new method for sampling and analysing volcanic sublimates; application to Merapi volcano, Java, J. Volcanol. Geotherm. Res., 12, 133-146, 1982.

Lloyd, G. E., Atomic number and crystallographic contrast images with the SEM; a review of backscattered electron techniques, Mineral. Mag., 51, Part 1, 3-19, 1987.

McCrone, W. C., and J. G Delly, The Particle Atlas, Edition Two. An Encyclopedia of Techniques for Small Particle Identification, Vol. 1-6, 1703 pp., Ann Arbor Sci. Publ., Ann Arbor, Michigan, 1973-1980.

McCrone, W. C., R. G. Draftz, and J. G. Delly, The Particle Atlas. A Photo-Micrographic Reference for the Microscopical Identification of Particulate Substances, 406 pp., Ann Arbor Sci. Publ., Inc., Ann Arbor, Michigan, 1967.

Neisser, U., Visual search, Sci. Am., 210, 194-202, 1964.

Picot, P., et Z. Johan, Atlas of ore minerals, Mém. BRGM, 90, 460 pp., Orléans, 1977.

Pierrot, R. M., Chemical and Determinative Tables of Mineralogy, I: Without the Silicates, 591 pp., Ed. du BRGM, Orléans, 1979.

Pierrot, R. M., and F. P. Cesbron, Chemical and Determinative Tables of Mineralogy, II: Silicates, 356 pp., Ed. du BRGM, Orléans, 1988.

Ramdohr, P., The Ore Minerals and their Intergrowths, Vol. 1, 440 pp., and Vol. 2, 764 pp., second ed., Pergamon Press, Oxford, 1980.

Sackett, W. M., Suspended matter in sea-water, in Chemical Oceanography, edited by J. P. Riley, and R. Chester, vol. 7, pp. 127-172, Academic Press, London and New York, 1978.

Toussaint, D., P. Vereerstraeten, P. Goffin, P. Van Landuyt, J. Jedwab, et J. M. Legrand, Hyperoxalurie primaire, Etude clinique, histologique et cristallographique des lésions oculaires, Arch. Opthalm., 36, 97-112, 1976.

Uytenbogaardt, W., and E. A. J. Burke, Tables for Microscopic Identification of Ore Minerals, 430 pp., Elsevier, Amsterdam, 1971.

Van de Weyer, R., P. De Vuyst, P. Dumortier, et J. Jedwab, Les problèmes diagnostiques des pneumoconioses rares, Rev. Inst. Hyg. Mines, 38, 259-275, 1983.

SAMPLING AND METHODS OF CHARACTERIZATION OF SURFACE REACTIVITY

E. H. De Carlo and Dirk Koeppenkastrop

Department of Oceanography, SOEST, University of Hawaii, Manoa, Honolulu, Hawaii 96822

1. Introduction

Over the past ten to twenty years there has been a strong thrust towards acquiring a better understanding of the chemical and physical properties of particulate matter in the hydrosphere. There is, however, a general lack of knowledge regarding the behaviour of individual phases as most work to date has focused on the bulk properties of particulate matter. Although a knowledge of the reactivity of complex particles is paramount to understanding interactions in nature, it is important to understand the properties of individual simple systems before attempting to integrate such information over bulk material. Mineral assemblages, however, do not necessarily have the same surface properties that one would predict based on the properties of individual phases [Honeyman, 1984; Anderson and Benjamin, 1990]. The separation of natural particulate matter into its individual components for analysis and characterization is, of course, the ideal solution to studying natural systems, but it poses a formidable problem given the current state of the science. Hence, until it is possible to physically separate individual components of particulate matter, there is a need to pursue a parallel approach involving detailed studies of simple systems under controlled laboratory conditions as well as the bulk characterization of natural particulate matter. Results from this approach may then be used to attempt to elucidate the complexity of natural particles.

Research in our laboratory has concentrated on three broadly categorized types of particles. These include: 1) marine sediments and metalliferous deposits, 2) suspended particulate matter in stream, estuarine and nearshore waters, and 3) synthetic mineral particles. Most of the procedures employed in our laboratory with sediments and metalliferous deposits are quite routine and have been reported in the literature [e.g., De Carlo et al., 1983, 1987]. Sampling of deep-sea sediments and ferromanganese deposits has been primarily through shipboard coring and dredging. Subsampling of cores and dredge hauls is then performed to obtain representative samples for further investigations.

The following discussion focuses primarily on techniques used in our laboratory to characterize particles collected from stream, estuarine and coastal waters and on other natural and synthetic particles used in laboratory sorption experiments. Because processes in the ocean involving particles are primarily surface phenomena, our work has focused on reactivity which is influenced by three surface characteristics. These include: 1) the surface composition of the particle; 2) its atomic structure; and 3) the ultra-fine scale morphology. Thus our approach has been designed to acquire data that will allow us to elucidate mechanisms of sorption through a combination of kinetic studies (sorption and desorption) and detailed surface characterization by microscopic techniques.

2. Water Sampling Methods

Near surface water samples are collected manually in acid cleaned 1 L LPE or PTFE bottles. Bottles are thoroughly cleaned in soap and water, rinsed at least five times with distilled deionized water with resistivity >15 Ω cm^{-1}, soaked overnight in 10% HNO_3, and rinsed five times with distilled deionized water. Bottles and caps are inverted and drip dried in a Class 100 laminar flow hood when trace element work is to be performed. Dry bottles are then capped and stored in plastic bags until needed. Vacuum filtration is usually performed either with Gelman or Millipore (1 to 0.22 μm size) cellulose acetate or acid leached Nuclepore polycarbonate membranes when ultra-trace element analysis of the aqueous phase is required.

Manual sampling is carried out by wading into streams or paddling a rubber dinghy to the desired site and collecting the water upstream of the sampler or dinghy position. Near-surface samples are collected by opening acid-cleaned bottles underwater while inverted and righted to fill. The bottles are then raised to the surface, emptied and the process repeated three times. A final sample is collected in this manner except the bottle is capped while still submerged. When collecting saline water samples we have found it useful to pour off a small amount of liquid while protecting the vessel

Marine Particles: Analysis and Characterization
Geophysical Monograph 63

mouth from contamination. This precaution is taken in order to avoid the potential precipitation of dissolved constituents on the threads of the bottle cap. The bottle is then stored upright in a plastic bag until return to the laboratory.

Collection of near-surface water samples can also be carried out with a pole-sampler as described by Boyle et al. [1981]. Such samples are collected in a manner similar to that described above except that the bottle is attached to a Plexiglas holder with Tygon tubing. The holder is itself attached to a PVC pole with Tygon tubing; other pole materials can be used but if a potential for sample contamination is anticipated the end of the pole should be wrapped with a polymeric material to prevent contact of the pole with the collection apparatus. Sampling is then performed from the shoreline or from a vessel underway at slow speed. Pole sampling requires two people; one handles the pole whereas the other handles the sample container. The bottle is removed from the plastic storage bag and uncapped while inverted by a technician wearing plastic gloves and the cap stored in the plastic bag during sampling. Immediately after recovery of the sample, the container is recapped and placed in the plastic bag until return to the laboratory. The advantage of pole sampling is that it allows collection of samples removed from potential contamination such as the boat, bridge, shoreline, etc.

3. Analytical Methods

3.1. Methods Based on Mineral Family

Separation of minerals by group is generally not performed in our laboratory except through the use of selective leaching procedures that solubilize operationally specified fractions. Leaching sediments with a Na Acetate-Acetic Acid acid buffer solubilizes carbonate minerals and a subsequent treatment with hydroxylamine hydrochloride-($NH_2OH{\cdot}HCl$)-dilute HNO_3 (0.01 M) releases easily reducible manganese oxides. Use of $NH_2OH{\cdot}HCl$-Acetic acid however solubilizes both Fe and Mn oxides, whereas a citrate-bicarbonate-dithionate buffer can be employed to reduce Fe(III) oxides. Residues of the final step of the sequential leach procedures are then digested with concentrated $HCl/HNO_3/HF$ for total elemental analysis. The scheme employed in our laboratory is based on modifications of the many leaching schemes that have been described in the literature [Chester and Hughes, 1967; Chao, 1972; Tessier et al., 1979; Chao and Zhou, 1983]. The various methods have been evaluated by Agemian and Chau [1976], Rendell et al. [1980], and more recently by Kheboian and Bauer [1987]. X-ray diffractometry (XRD) is usually performed on residues of each leach step before carrying out the next leaching in order to qualitatively determine the mineralogy. The mass fraction lost by each successive leach is determined by weighing of residues after freeze drying and subsequent oven drying at 110° C. Loss of volatile trace elements such as Hg can occur during drying at 110° C. Mass balance calculations are made from mass loss during the leaching procedure and the elemental analysis of individual leach liquors. The combination of mineralogical and elemental analysis data can often provide insight into processes affecting the studied material in its natural environment. Care must be exercised in the evaluation of the distribution of trace elements between phases when using selective leaching procedures because readsorption of trace elements released by the procedures may occur [e.g., Sholkovitz, 1989].

3.2. Methods Based on Sample Fraction

Clay minerals are operationally defined by size (less than 2μm). These minerals are generally separated by centrifugation to isolate the less than 1 or 2 μm fraction depending on needs or by filtration through successively finer membranes in the 2 to 0.22 μm range. The latter, however, should not be used to acquire size distribution information as it is known that filters can and do trap smaller particles than their nominal size would indicate (A. Campbell, personal communication, 1991). Typical membranes used include Gelman or Millipore cellulose acetate, and Nuclepore porous polycarbonate.

A general characterization of particulate matter is performed by XRD. Qualitative and semi-quantitative mineralogical analyses are made depending on the need. Procedures followed are those commonly used in mineralogical studies (Van der Gaast, this volume, page 343). Powders or slurries of samples are scanned at 1 or 2° 2θ min^{-1} over the range of 2 to 70° 2θ. Identifications are based on peak positions and relative intensities through comparison with the standard powder diffraction file (JCPDS) for crystalline minerals or by methods described by Brindley and Brown [1980] for clay minerals.

Coarser fractions are either separated by centrifugation using the less than 2 μm criterion for separating clays as above or by filtration using 2 μm cutoff size membranes. Separation of the silt fraction (less than 63 μm) from coarser material can be performed either by centrifugation or by sieving through standard SPEX sieves.

3.3. Synthesis and Characterization of Synthetic Mineral Particles

Our interest in the sorptive properties of particulate matter has recently led us to the investigation of surface reactions of the rare earth elements (REE) at the surface of hydrothermal deposits [De Carlo and Koeppenkastrop, 1990]. Because natural marine particles are often complex mixtures whose surface properties are known to be affected by factors such as aging and coating by organic matter [e.g., Balistrieri et al., 1981], it has been difficult to elucidate the specific mechanisms controlling the interaction of dissolved species with the particle surfaces. Much of the work to date involving natural particles has focused on equilibrium studies and recent kinetic studies reveal a very limited understanding of fundamental processes [Jannasch et al., 1988 and references therein]. Recent work addressing the importance of interactions of colloidal matter with dissolved species and truly particulate matter is reviewed by Honeyman (this volume, page 437).

In order to increase our understanding of the fundamental processes occurring at the aqueous solution-mineral particle interface we have carried out laboratory investigations using synthetic mineral particles in seawater [Koeppenkastrop and De Carlo, 1988, 1990a, b, in press, 1991]. Mineral phases used in our laboratory are synthesized by previously described methods [Atkinson et al., 1967; Murray, 1974; Simpson, 1968], then separated from the mother liquor by filtration through 0.22 μm membranes. The particulate matter is equilibrated with seawater prior to the initial characterization of the suspensions and their subsequent use in sorption/desorption experiments.

Aliquots by volume are taken from well-stirred suspensions, vacuum filtered (0.22 μm) from seawater, and subsequently subjected to XRD and infrared spectroscopy (IR) for mineralogical characterization. XRD allows the determination of the basic crystal structure of the synthesized minerals. The information acquired through XRD, although generally not of a quantitative nature, permits an evaluation of the extent of crystallinity of the particles. For example XRD studies of FeOOH synthesized in our laboratory have revealed that both amorphous (Figure 6-45A) and partially crystalline (Figure 6-45B) forms of goethite

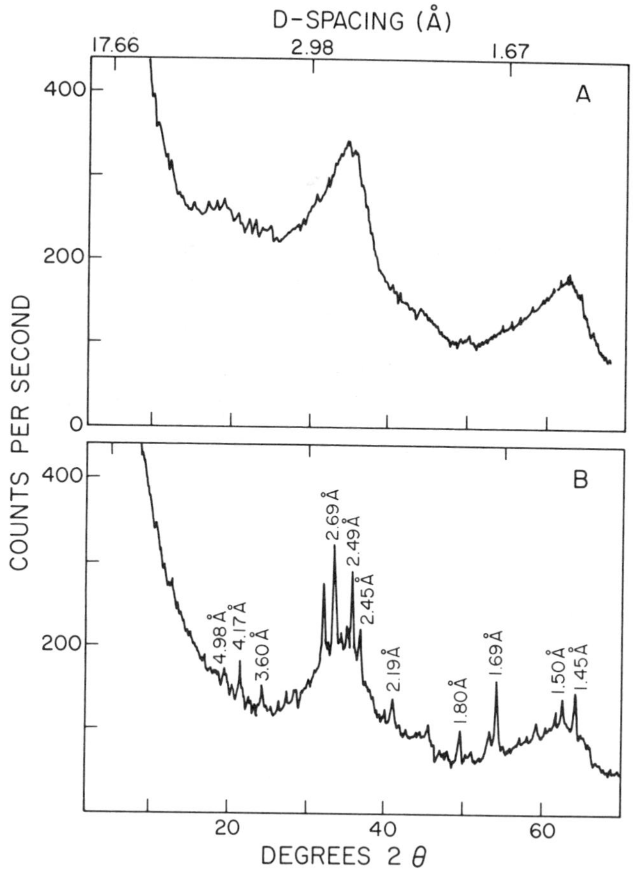

Fig. 6-45. X-ray diffraction patterns of synthetic FeOOH particles obtained on a Scintag PAD-V automated diffractometer equipped with solid state Ge detector. Patterns obtained by irradiation with Cu K_α. A) amorphous FeOOH synthesized by fast reaction procedure. B) mixture of amorphous and crystalline α-FeOOH synthesized by slow reaction procedure. Major d-spacings for goethite are shown and unlabelled peaks are unidentified.

can be prepared from the same reagent mixture if the rate of reaction is controlled. Fast reaction rates favor the formation of amorphous material whereas a slow addition of reagents enhances the crystallinity of the final product. Particles with the same general chemical composition excluding extent of hydration but of differing crystallinities for example amorphous FeOOH versus α-FeOOH appear to exhibit considerably different sorptive properties. These differences are attributable in part to variations in the specific surface area of particles, the lability of lattice ions, and differences in the density of vacant surface octahedral sites that allow exchange with other metal ions [Davis et al., 1989; Koeppenkastrop and De Carlo, in press, 1991].

Infrared spectroscopy has been used in our laboratory primarily to determine the composition of synthetic apatites. Most forms of apatite such as carbonate-, fluor-, hydroxy-, or mixtures thereof display very similar XRD patterns, and elemental analysis by atomic spectroscopy does not provide a sufficiently complete or accurate compositional structure. Shifts in the IR absorption spectrum are influenced by the relative proportions of OH^-, F^-, and CO_3^{-2} in the crystal lattice and allow us to better characterize the form of apatite synthesized in the laboratory. Examples of IR absorption spectra for apatites synthesized in our laboratory are shown in Figure 6-46.

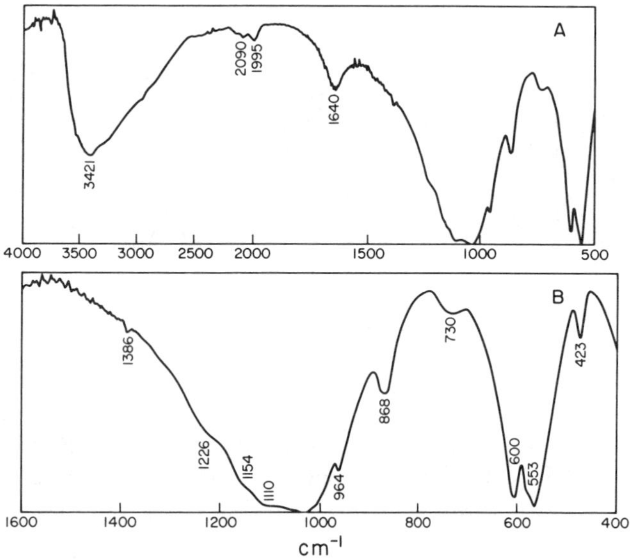

Fig. 6-46. IR spectra of apatite synthesized in our laboratory. A) Scan over range of 4000 to 500 cm^{-1}. B) Scan over range of 1600-400 cm^{-}.

Elemental analysis of synthesized phases is performed by atomic spectroscopy. Filtered aliquots are dried overnight at 110° C, are digested with concentrated acids in sealed Teflon vessels by a microwave method [e.g., De Carlo, 1990], and then subjected to elemental analysis by inductively coupled plasma atomic emission spectroscopy (ICP/AES). Another technique extremely suitable for the characterization of very small quantities of particulate matter is ICP-mass spectrometry (ICP/MS), a technique that combines the multi-element analytical capabilities of ICP/AES with the extremely low detection limits of MS. A general description of the technique is given by Horlick et al. [1987] and Falkner (this volume, page 243), and some details regarding the application of ICP/MS to marine particulate matter are given by Campbell (this volume, page 375).

4. Methods Aimed at Answering Particular Questions

4.1. Sorptive Properties of Particles

This work has been the focus of our research for the past five years. It comprises primarily kinetic and thermodynamic studies of the interaction between dissolved species in seawater and mineral surfaces. A significant portion of this work has involved the use of radiotracers produced by a very simple yet extremely effective technique involving neutron activation of stable isotopes [Koeppenkastrop et al., in press, 1991]. The activated isotopes are then used in sorption studies with synthetic mineral phases generated in our laboratory and with hydrothermal sediments.

A brief description of the procedure used in our sorption experiments with radiotracers follows. Aliquots of carrier-free radiotracer solutions are equilibrated for 24 hrs in previously filtered and UV-oxidized seawater; the pH of the solution is monitored and maintained within 0.05 pH units with NaOH or HCl. Aliquots of the seawater-mineral phase suspensions of synthetic or natural particles are then added to the continuously stirred radiotracer-laden seawater. Experiments are conducted in a refrigerated or heated water bath to maintain constant temperature. At predetermined time intervals 50 mL aliquots are removed with a syringe, and immediately filtered through Nuclepore membranes. The solid and liquid phases are then counted for gamma radiation on a solid-state detector equipped with a multichannel analyzer. Physical limitations of the experimental setup and the operator limit our earliest sampling interval to approximately 1 minute after initiation of the experiment. The temperature dependence of sorption reactions at the surface of mineral particles can be evaluated by conducting sorption studies as a function of time at several temperatures. We have used three temperatures 2, 25, and 40° C in our work in order to determine the pre-exponential factor in the Arrhenius equation [e.g., Laidler, 1987] .

Desorption experiments are carried out in a similar manner to sorption studies except that previously radiotracer-enriched mineral phases are placed in trace metal-free solutions and the release of the metals of interest is monitored as a function of time. It is important to realize that desorption experiments carried out by lowering the pH of the solution to stimulate the release of surface sorbed species are not entirely appropriate. Although conditions of changing pH can occur in the natural environment, this approach suffers from the fact that partial dissolution of the surface of the mineral particles causes the release of an inordinate amount of sorbate, the majority of which was bound to the surface of the particles. The nearly instantaneous release of sorbate that occurs in such experiments is then followed by uptake of the released material with time. Our work has shown this to be more important in experiments with manganese oxides than iron oxides because of the lesser solubility of the latter phase [Koeppenkastrop and De Carlo, 1990b]. Further results of our work and implications thereof for processes occurring in the marine environment will be presented elsewhere (De Carlo and Koeppenkastrop, in preparation). Proper desorption experiments designed to test for the reversibility of the sorption processes should be conducted at the same pH as the sorption experiment. Moore and Hunter [1985] and Moore and Millward [1988] have performed such studies to evaluate the reversibility of Th adsorption onto suspended marine sediments.

Data acquired in sorption experiments with radiotracers are inherently more accurate and precise than those employing higher concentrations of stable isotopes. The difference arises from the extremely high sensitivity of gamma spectrometry relative to typical atomic spectroscopic techniques of analysis. The high affinity of the REE and other particle reactive species for surfaces results in low residual aqueous metal concentrations. These are often near the limit of detection by conventional atomic spectroscopy and lead to a higher degree of analytical uncertainty. The accuracy and precision in gamma spectrometry, however, are only limited by the counting efficiency and by the activity of the measured radionuclide. The latter is a variable that can be controlled by varying the irradiation time during nuclear activation. Statistical evaluations of our data indicate that the reproducibility of the K_d's determined in radiotracer experiments is within 0.05 log units (or approximately 12% relative) as compared to 0.1-0.2 log units (or 26-58%) in the stable isotope experiments. This compares well with the reproducibility of experiments conducted by other researchers (B. Honeyman, personal communication, 1991).

Kinetic data acquired by the above described experiments can be modeled using a variety of computer-assisted techniques. We currently use the MINSQ nonlinear curve fitting and model development program by Micro-Math scientic software (Salt Lake City, Utah). A comprehensive discussion of a variety of sequential and parallel reaction models applied to our REE sorption data is presented elsewhere by Koeppenkastrop and De Carlo (submitted, 1991). This approach is, however, only a parameterization, just as the method applied by Jannasch et al. [1988], and additional experimental techniques are needed to complement such attempts to elucidate actual mechanisms.

4.2. *Effect of Organic Matter on Sorptive Properties*

Sorption studies with organic matter are carried out in a manner similar to that described above except that commercially obtained or natural humic and fulvic substances are added to the mineral suspensions prior to the performance of experiments. This allows us to study the effect of organic matter on the sorptive properties of mineral particles. In such experiments it is necessary to preequilibrate the organic matter with the particle suspension because it is well-known that organic compounds form chelates with metal ions [Davis and Leckie, 1978; McBride, 1985; Choppin and Allard, 1985]. Recent work in our laboratory has shown that a significant portion of apparent distribution coefficients of the REE between aqueous and solid phase is due to the formation of insoluble REE-humates [De Carlo and Koeppenkastrop, 1990]. The distribution of the organic matter between the particles and aqueous phases is determined in blank experiments through total organic carbon (TOC) analysis of the aqueous and solid phases. A recent review by Schindler [1990] describes the co-adsorption of metal ions and organic ligands on mineral surfaces. It should be added that many studies conducted to date, including ours, have used terrestrial or riverine humic/fulvic substances. Such material has often been obtained by chemically harsh processes that are extremely likely to alter the composition and structure of the isolated material; hence, it would be preferable to use true marine organic matter in evaluating the influence of organic coatings on surface reactivity.

4.3. *Surface Structure of Particles*

Filtered and dried particles can be prepared for observation under the scanning electron microscope (SEM) for visual imaging, EDS elemental analysis of individual particles or portions thereof, as well as for size and volume estimation. Use of the Princeton

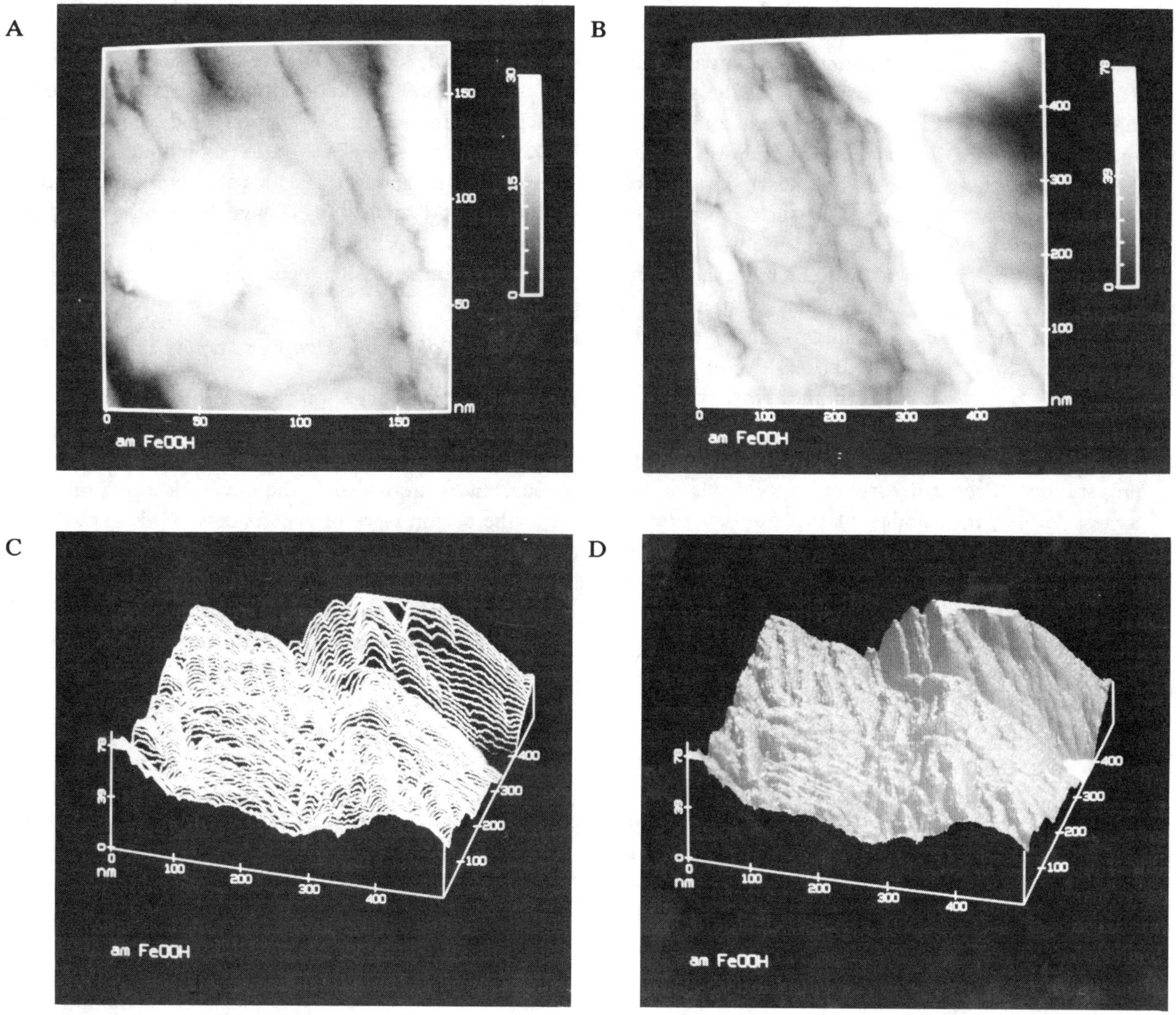

Fig. 6-47. STM images of amorphous FeOOH. A) 500 nm x 500 nm field of view. B) 150 nm x 150 nm field of view of center of image shown in A. C) Computer enhanced topographic image of field of view shown in A. Note the extremely rugged micro-topography which shows features ranging from approximately 30 to 100 nm size and the regular pattern of layers of 30-40 nm thickness in the upper left and bottom right of the field of view. D) Contour topographic map of the region shown in A. Photo credit: Scott Lewis.

Gamma Tech feature analysis statistical package available for SEM allows the evaluation of particle size distribution, a variety of morphological characteristics, and the estimation of surface areas. The reader is referred to the paper by J. P. Cowen in this volume, page 429, for details of electron microscopy. Surface areas of powders are often determined by standard N_2 BET methods [Brunauer et al., 1938]. Unfortunately both SEM and BET methods suffer from the fact that particles are removed from their natural environment, dried, and subjected to a vacuum prior to examination. Hence, their surface characteristics are likely to be altered from those of hydrated particles.

We have recently begun investigating the use of scanning tunneling microscopy (STM) to characterize the surface of our particles on a sub-micrometer scale. This technique can only be employed with immobile particles, thus, particles must be fixed on a substrate such as a membrane or another solid surface prior to imaging. Our first attempts to image the surfaces of wet particles of synthetic amorphous FeOOH and δ-MnO_2 immobilized on filtration media were unsuccessful. The failure has been attributed to the inability to develop a tunneling current due to insufficient conduction of the mineral surface. Results of recent work [e.g., Eggleston et al., 1990; Hochella et al., 1990; Johnsson et al., 1990] and discussions with colleagues (Hochella, personal communication, 1990) indicate that atomic force microscopy (AFM) is the more suitable method for surface studies of non-conducting or semi-conducting surfaces. Nonetheless we have pursued the use of STM to acquire surface micro- or nano-morphological information on FeOOH and MnO_2 particles. In order to image the surface by STM it has been necessary to coat the surface of the particles with carbon or gold to make the surface conducting. Use of the former leads to a surface coating of approximately 20 nm whereas much thinner coatings (a few atomic layers) of gold are possible. Figure 6-47 presents images and computer enhanced topographic contour maps of the surface of an amorphous FeOOH particle we coated with carbon. A thorough review of the relationship between atomic structure, microtopography, composition, and reactivity of mineral surfaces is provided by Hochella [1990].

Several other surface characterization techniques have recently been employed to determine the bonding environment and oxidation states of species at the surface of particulate matter. Techniques with the most potential appear to be X-ray absorption fine structure (EXAFS) and X-ray absorption near edge structure (XANES). Both require synchrotron radiation. EXAFS was recently used by Hayes et al. [1987] to characterize the bonding environment of selenium oxyanions on α-FeOOH. Their work showed that EXAFS can be used to differentiate between outer-sphere and inner-sphere complexation on the metal oxide surface. Further reference to these techniques and others based on synchrotron radiation is provided in this volume by Campbell, page 375.

References

Agemian, H., and A. S. Y. Chau, Evaluation of the extraction techniques for the determination of metals in aquatic sediments, Analyst, 101, 761-767, 1976.

Anderson, P. R., and M. M. Benjamin, Surface and bulk characteristics of binary oxide suspensions, Env. Sci. Tech., 24, 292-298, 1990.

Atkinson, R. J., A. M. Posner, and J. P. Quirk, Adsorption of potential determining ions at the ferric oxide aqueous electrolyte interface, J. Phys. Chem., 71, 550-558, 1967.

Balistrieri, L., P. G. Brewer, and J. W. Murray, Scavenging residence times of trace metals and surface chemistry of sinking particles in the deep ocean, Deep-Sea Res., 28A, 101-121, 1981.

Boyle, E. A., S. S. Huested, and S. P. Jones, On the distribution of copper, nickel, and cadmium in the surface waters of the North Atlantic and North Pacific Ocean, J. Geophys. Res., 86, 8048-8066, 1981.

Brindley, G. W., and G. Brown, Crystal Structures of Clay Minerals and their X-ray Identification, 495 pp., Mineralogical Society, London, 1980.

Brunauer, S., P. H. Emmett, and E. Teller, Adsorption of gases in multimolecular layers, J. Phys. Chem., 60, 309-319, 1938.

Chao, T. T., Selective dissolution of manganese oxides from soils and sediments with acidified hydroxylamine hydrochloride, Soil Sci. Soc. Am. Proc., 36, 764-768, 1972.

Chao, T. T., and L. Zhou, Extraction techniques for selective dissolution of amorphous iron oxides from soils and sediments, Soil Sci. Soc. Am. J., 47, 225-232, 1983.

Chester, R., and M. J. Hughes, A chemical technique for the separation of ferro-manganese minerals, carbonate minerals and adsorbed trace elements from pelagic sediments, Chem. Geol., 2, 249-262, 1967.

Choppin, G. R., and B. Allard, Complexes of actinides with naturally occurring organic compounds, in Handbook on the Physics and Chemistry of the Actinides, edited by A. J. Freeman, and C. Keller, pp. 407-429, Elsevier, New York, 1985.

Davis, J. A, and J. O. Leckie, Effect of adsorbed complexing ligands on trace metal uptake by hydrous oxides, Environ. Sci. Technol., 12, 1309-1315, 1978.

Davis, J. A., B. Rea, R. G. Claypool, and C. Fuller, Surface site density and adsorption capacity of ferrihydrite, Paper #GEOC-40, 198th ACS National Meeting, American Chemical Society, Washington, D.C., 1989.

De Carlo, E. H., Separation of lanthanide series elements in marine Fe-Mn crusts by ion-exchange chromatography and determination by ICP/AES, Sep. Sci. Technol., 25(6), 781-798, 1990.

De Carlo, E. H., and D. Koeppenkastrop, Sorption kinetics and thermodynamics of rare earth elements onto natural and synthetic iron and manganese oxides, EOS Trans. Am. Geophys. Union, 71, 1417, 1990.

De Carlo, E. H., G. M. McMurtry, and H.-Y. Yeh, Geochemistry of hydrothermal deposits from Loihi submarine volcano, Hawaii, Earth Planet. Sci. Lett., 66, 438-449, 1983.

De Carlo, E. H., P. A. Pennywell, and C. M. Fraley, Geochemistry of ferromanganese deposits from the Kiribati and Tuvalu region of the west central Pacific Ocean, Mar. Min., 6, 301-327, 1987.

Eggleston, C. M., M. F. Hochella, and B. M. Bakken, Scanning tunnelling spectroscopy of gold clusters sorbed to galena surfaces, EOS Trans. Am. Geophys. Union, 71, 1654, 1990.

Hayes, K. F., A. L. Roe, G. E. Brown, K. O. Hodgson, J. O. Leckie, and G. Parks, In situ x-ray absorption study of surface complexes; selenium oxyanions on α-FeOOH, Science, 238, 783-786, 1987.

Hochella, J. F., Atomic structure, microtopography, composition, and reactivity of mineral surfaces, in Mineral-Water Interface Geochemistry, edited by M. F. Hochella, and A. F. White, Reviews in Mineralogy, Vol. 23, pp. 87-132, Mineral Society of America, Washington, D.C., 1990.

Hochella, J. F., C. M. Eggleston, V. B. Elings, and M. S. Thompson, Atomic structure and morphology of the albite (010) surface. An atomic force microscope and electron diffraction study, Am. Mineral., 75, 723-730, 1990.

Honeyman, B. D., Cation and anion adsorption in binary mixtures of adsorbents: an investigation of the concept of adsorptive additivity, Ph.D. thesis, 365 pp., Stanford Univ., California, 1984.

Horlick, G., S. H. Tan, M. A. Vaughan, and Y. Shao, Inductively coupled plasma-mass spectrometry, in Inductively Coupled Plasmas in Analytical Atomic Spectrometry, edited by A. Montaser, and D. W. Golightly, pp. 361-398, VCH Publ., New York, 1987.

Jannasch, H. W., B. D. Honeyman, L. S. Balistrieri, and J. W. Murray, Kinetics of trace element uptake by marine particles, Geochim. Cosmochim. Acta, 52, 567-577, 1988.

Johnsson, P. A., C. M. Eggleston, M. F. Hochella, and G. A. Parks, Atomic force microscopy of the hematite (001) surface; applications to studies of surface structure and reactivity, EOS Trans. Am. Geophys. Union, 71, 1654, 1990.

Kheboian, C., and C. F. Bauer, Accuracy of selective extraction procedures for metal speciation in model aquatic sediments, Anal. Chem., 59, 1417-1423, 1987.

Koeppenkastrop, D., and E. H. De Carlo, Adsorption of rare earth elements from seawater onto synthetic mineral phases, EOS Trans. Am. Geophys. Union, 69, 1254, 1988.

Koeppenkastrop, D., and E. H. De Carlo, Interaction of rare earth elements in seawater with iron and manganese oxides. A kinetic approach, EOS Trans. Am. Geophys. Union, 71, 89, 1990a.

Koeppenkastrop, D., and E. H. De Carlo, Distribution of rare earth elements between seawater and synthetic mineral phases, EOS Trans. Am. Geophys. Union, 71, 1417, 1990b.

Koeppenkastrop, D., and E. H. De Carlo, Sorption of rare earth elements from seawater onto synthetic mineral particles. An experimental approach, Chem. Geol., in press, 1991.

Koeppenkastrop, D., M. Roth, and E. H. De Carlo, A method to investigate the interaction of rare earth elements with metal oxides from aqueous solution, J. Radioanal. Nucl. Chem., in press, 1991.

Laidler, K. J., Chemical Kinetics, 531 pp., third ed., Harper and Row, New York, 1987.

McBride, M. C., Influence of glycine on Cu^{2+} adsorption by microcrystalline gibbsite and boehmite, Clays Clay Miner., 33, 397-402, 1985.

Moore, R. M., and K. A. Hunter, Thorium adsorption in the ocean: reversibility and distribution amongst particle sizes, Geochim. Cosmochim. Acta, 49, 2253-2257, 1985.

Moore, R. M., and G. E. Millward, The kinetics of reversible Th reactions with marine particles, Geochim. Cosmochim. Acta, 52, 113-118, 1988.

Murray, J. W., The surface chemistry of hydrous manganese dioxide, J. Colloid Surf. Sci., 46, 357-371, 1974.

Rendell, P. S., G. E. Batley, and A. J. Cameron, Adsorption as a control of metal concentrations in sediment extracts, Environ. Sci. Technol., 14, 314-318, 1980.

Schindler, P. W., Co-adsorption of metal ions and organic ligands: formation of ternary surface complexes, in Mineral-Water Interface Geochemistry, edited by M. F. Hochella, and A. F. White, Reviews in Mineralogy, Vol. 23, pp. 281-308, Mineral Society of America, Washington, D.C., 1990.

Sholkovitz, E. R., Artifacts associated with the chemical leaching of sediments for rare earth elements, Chem. Geol., 77, 47-51, 1989.

Simpson, D. R., Substitutions in apatite: I. Potassium-bearing apatite, Am. Mineral., 53, 432-444, 1968.

Tessier, A., P. G. C. Campbell, and M. Bisson, Sequential extraction procedure for the speciation of particulate trace metals, Anal. Chem., 51, 844-851, 1979.

INDIVIDUAL PARTICLE ANALYSES

James P. Cowen

School of Ocean and Earth Science and Technology, University of Hawaii, Honolulu, Hawaii 96822

1. Individual Particle Analyses

Marine particulate matter consists of a complex, heterogeneous population of reactive and interactive particles. In oligotrophic oceans, particles are produced primarily by phytoplankton within the euphotic zone. Secondary production by bacterioplankton is also an important source of particles and mid-water column chemolithoautotrophy or horizontal advection may provide additional new particles [Sorokin, 1971; Pomeroy, 1974; Sieburth, 1976; Lambert et al., 1981; Karl et al., 1984]. Aeolian input can be an important source of abiogenic particles such as clay minerals [Sutcliffe et al., 1963; Buat-Menard and Chesselet, 1979; Deuser et al., 1981, 1983] while other abiogenic particles may precipitate or flocculate in situ [McCave, 1984; Feely et al., 1990]. In coastal area rivers, runoff, and resuspension can provide additional important sources of particles.

The rate of trace metal sorption by particles is a complex function of the dissolved metal concentration and the concentration and nature of the particles [Neihof and Loeb, 1974; Loeb and Neihof, 1977; Balistrieri et al., 1981; Li et al., 1984; Balistrieri and Murray, 1986; Honeyman et al., 1988a, b; Jannasch et al., 1988; see also Honeyman, this volume, page 437]. Honeyman and Santschi [1988] noted that overall sorption rate constants are "smaller than expected from physicochemical theory or laboratory kinetic experiments". This behavior may be explained by surface site heterogeneity, particle interaction effects and biological activity. However, surface site heterogeneity, or sorption site concentration, is difficult to establish in field studies and particle mass concentration is usually used as a convenient, albeit inadequate, substitute [Honeyman et al., 1988a, b]. Particle-particle interactions are usually lost in the operationally defined terms of filterable "particulate" and filtered "dissolved" fractions [Di Toro et al., 1986; Higgo and Rees, 1986; Morel and Gschwend, 1987; Honeyman and Santschi, 1989]; whereas, biological involvement is often inferred from positive correlation with biogenic materials [Bishop and Fleisher, 1987] or biologic processes such as productivity [Coale and Bruland, 1985]. To distinguish some of these micro-processes it can be advantageous to apply individual micro-analytical techniques.

Marine Particles: Analysis and Characterization
Geophysical Monograph 63

Procedures for processing suspended particle samples for individual particle micro-analysis can be divided into two broad categories: (I) those requiring ultra-structural or electron diffraction information using transmission or scanning transmission electron microscopy, TEM or STEM; and (II) those not requiring the former information but that may or may not require external morphology information using SEM.

1.1. Thin Section Material

Some particle studies require very specific knowledge of the internal structure of the particles, such as ultra-structural or mineralogical considerations or both. Certain bacteria, for example, have mineral deposits associated with extensive fibrillar capsules that are extensions of the cell surfaces (Figures 6-48 and 6-49). While the mineral deposits themselves can be detected and observed using a SEM, the bacterial nature of the association is usually totally obscured by the irregular and coagulated nature of the external morphology of the capsule. Transmission or scanning transmission electron microscopy (STEM) using ultra-thin sectioned samples is necessary to discern the biological nature of the mineral deposit/capsule complex.

A thin section is essentially transparent to the electron beam. Its thickness is very much less than would be the depth of penetration of the electron beam in a bulk sample of the same composition. Since the beam does not have the opportunity to spread, the spatial resolution of the X-ray signal is optimized. Furthermore, the thinness of the section practically negates the effects of absorption and secondary fluorescence. This greatly simplifies the correction process, leaving only the atomic number corrections to be made. Unfortunately, this gain in simplicity must be measured against the greater demands for sample preparation. Counting times of up to 200 seconds are often required due to the very small sample volume being analyzed (thin-sections ~ 85 nm).

Elemental micro-analysis is routine with the appropriately equipped instruments. Major advantages of STEM micro-analysis of particles include superior imaging at high resolutions (to less than 0.2 nm),

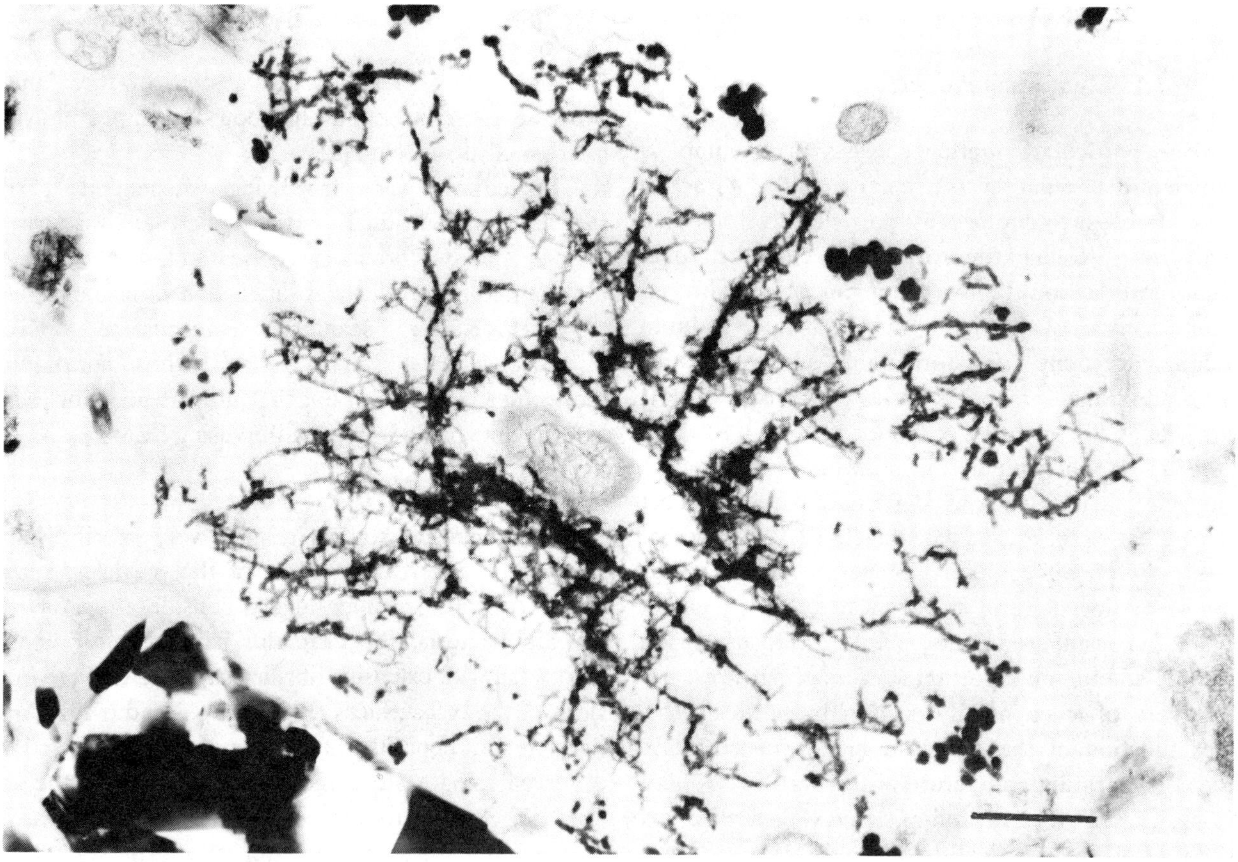

Fig. 6-48. Capsuled bacteria revealing internal cell. TEM, thin sectioned; scale bar, 0.5 μm.

elemental micro-analyses with fine spatial resolution (less than 0.1 μm), and selected area electron diffraction, all on the same particle. The main disadvantage is the time intensive sample preparation. Furthermore, very hard particles like quartz do not section well and can be lost from the sample. Particle loss is readily evidenced by holes in the thin sections. It may be possible to employ alternative methods, such as ion beam thinning, to achieve proper sample thinness. However, this method is most commonly used in conjunction with solid substrates or consolidated sediments and has its own limitations; it has not been extensively applied to suspended particle samples.

1.1.1. Preparation of samples for micro-analysis by STEM.

Water samples intended for individual particle micro-analysis are filtered through 0.2 μm pore size, 47 mm diameter, Nuclepore membrane filters or other membrane filters. Tortuous filters are not appropriate for this procedure as small particles become lodged within the filter matrix and cannot easily be removed. Filters are then carefully "scraped" clean using a soft Teflon spatula under a fine stream of 2% gluteraldehyde-seawater solution. If the filters remain wet throughout the operation the removal is somewhat facilitated. However, fully dried, particle-loaded filters have been successfully cleared of particles using the same procedure after rehydration with the appropriate solution, that is, distilled, deionized water; seawater; glutaraldehyde/seawater; or even "wetted" by non-viscous embedding resin. Even the finest particles are thus quantitatively rinsed from the filter into 1.5 mL centrifuge tubes, as evidenced by examination of representative filters with a scanning electron microscope.

Transfer to the centrifuge tubes is greatly facilitated by the use of a "spouted" petri dish. Hundreds of these can be easily fabricated from standard 50 mm petri dishes by softening a portion of each dish's lip in a flame and bending into a spout with forceps; dishes are then cleaned in the routine manner. The ultimate criteria for the volume of seawater that is required to be filtered is the need to visualize the concentrated particles in

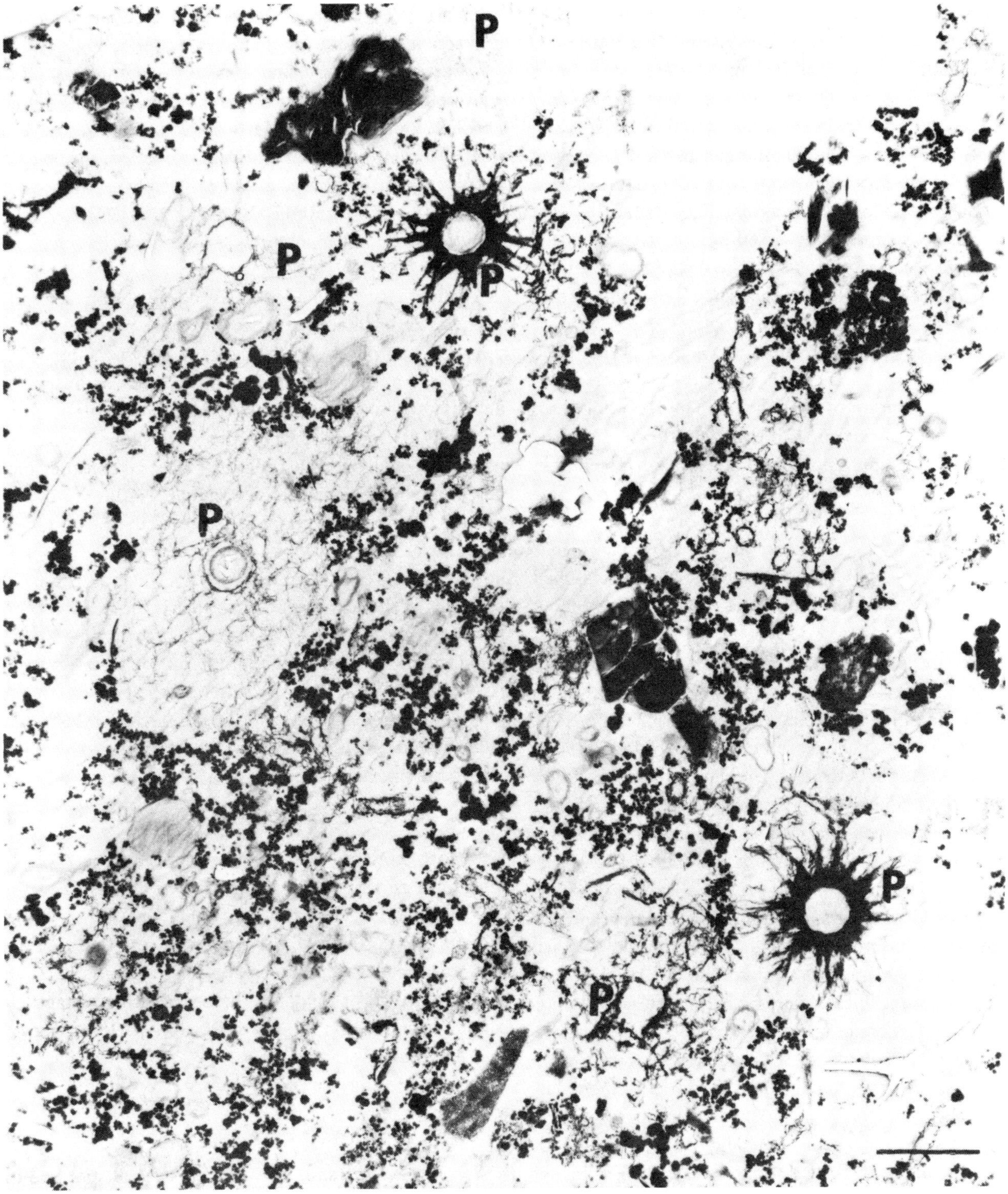

Fig. 6-49. Heterogeneous particle field from hydrothermal vent particle plume. Bacterial capsules, consisting of extracellular polymer material are labeled with a "P". TEM, thin sectioned; scale bar, 1 μm.

subsequent processing steps. Four to 20 L have been routinely filtered in open ocean environments depending on the particle loading. Further sample preparation procedures are described in Cowen and Bruland [1985] as modified from Silver and Bruland [1981].

Basically, the particles are further concentrated into a soft pellet by ultracentrifugation (RCF: 12,800 x G). The entire rinse solution volume can usually be accommodated by 1.5 mL micro-centrifuge tubes and a single centrifugation. However, sequential centrifugations of larger rinse volumes may be done by removing the supernatant of previous centrifugations, adding additional rinse solution to the same tube and recentrifuging.

The pelleted particles are then immobilized in 2% Noble agar. The agar is made molten, then cooled to a constant temperature of 38 to 40° C. The following steps are performed quickly so as to complete them before the agar sets. The supernatant is removed as completely as possible, replaced with molten agar, the sample plug quickly stirred with a clean toothpick-sized rod and re-centrifuged. After cooling in an ice bath the plastic centrifuge tube is cut away and the agar plug is trimmed to isolate the immobilized particles in small (less than 1 mm^2) agar blocks.

Replicate portions of each agar block are treated somewhat differently depending on the type of analysis to be done. Samples for enumeration of organic particles require staining [such as ruthenium red; Luft, 1966], secondary fixation with 1% osmium-tetroxide solution and post-sectioned staining with uranyl acetate and lead citrate as described in Cowen and Bruland [1985]. Samples intended for elemental micro-analysis, however, are directly embedded into resin without osmium, ruthenium red or post-sectioning stains. After embedding in plastic resin [Ringo et al., 1979] the samples are thin-sectioned with an ultra-microtome to silver-yellow sections (~ 85 nm thick). The sections were placed onto formvar-coated copper grids. As noted above, where mobilization of elements by solvents is of concern, air or oven (low heat) dried samples may be directly embedded in resin without stains, liquid dehydration, or sequential infiltration steps. Vacuum embedding with low viscosity resin, for example M-Spurr [Ringo et al., 1979], is highly recommended in this case. All grids are finally coated with an ultra-thin coat of high purity evaporated carbon for stability under the electron beam. Grids are examined under a Scanning Transmission Electron Microscope (STEM) equipped with an energy dispersive X-ray fluorescence spectrometer (EDS). An accelerating voltage of 50kV has been used as a compromise between optimal count rate and spatial resolution. Higher voltages are used when employing selected area electron diffraction techniques.

Particle counts using this thin-sectioning method provide relative numbers of particle types. Estimation of absolute numbers of any particle group requires an independent method of enumerating one of the particle groups, for example bacterial particles by epifluorescent microscopy. A method for converting TEM particle counts to probable numbers is described in Silver and Bruland [1981].

Quantitative micro-analysis of ultra-thin sectioned particles is facilitated relative to whole particle or even polished sample micro-analysis, since specimen relief effects are absent and such matrix effects as absorption and fluorescence are negligible. Micro-analyses are generally performed on portions of unstained, sectioned particles occupying the reduced raster at constant magnification. Semi-quantitative elemental ratios are derived from background continuum subtracted peak integrals. A peak is verified when greater than three standard deviations of the background. The concentration of elements in the analysis volume, that is the portion of the thin sectioned sample analyzed, is determined using standardless or "with-standards" quantitative software. External standards can be made by dissolving an organometallic complex such as tricarbonylmethylcyclopentadienyl-Mn or ferrocene-Fe in the same embedding resin used in sample preparation.

1.2. Electron Diffraction Analysis

Most modern TEMs and STEMs are equipped to image and record electron diffraction patterns. Electron diffraction patterns can be obtained from either sectioned material or whole mounts of very fine particles. The maximum section thickness or diameter of the sample particle depends on the accelerating voltage and sample composition. The diffraction patterns provide the necessary information to obtain "d"-spacings and intensities. The former is derived from the simple formula:

$$R\, d = \lambda\, L,$$

where R is the distance between the central spot and the other diffraction spots, λ is the wavelength of the incident electrons and L is the cameral length. In practice, the product "λ L" is determined using standard calibration materials.

A major advantage of STEM-diffraction of whole or sectioned particles is that individual small particles can be sequentially examined by high resolution SEM-EDS or STEM-EDS and analyzed by electron diffraction without even removing the sample. Thus, information on the morphology or ultrastructure, composition and crystalline or amorphous structure can be obtained nearly simultaneously.

1.3. Whole-Particle Analyses

Scanning electron microscopy (SEM)-based micro-analysis of whole particles provides imaging of external morphology and elemental information usually without regard to internal ultra-structure or compositional heterogeneity. Particles that are identifiable from external morphology or particle fields in which size and elemental composition are the primary concern are amenable to micro-analysis using an SEM.

Distinct advantages of SEM based micro-analysis are the usually minimal sample preparation requirements and amenability to automated analyses. However, quantitative elemental micro-analysis of particles is presently not possible due to the undefined topography of the samples which directly affects the X-ray absorptive path length. Nevertheless, this method can provide qualitative data on particle morphology, size and basic composition, essential to understanding specific particle import/export rates and mechanisms, biogenic and abiogenic particle species identification and other biogeochemical processes. The above methods work equally well for samples collected by sediment traps. Sediment samples may also be processed in this manner although difficulties in the thin sectioning procedure can be encountered. The reader is also referred to the paper by J. Jedwab (this volume, page 399) for detailed discussion of whole particle analyses by light microscopy and SEM.

1.3.1. Preparation of samples for SEM.

The filters are rinsed with several mL of pH adjusted Milli-Q clean deionized water to remove sea salts, dried and weighed. Separate samples may be collected for micro-analysis and bulk analyses. Alternatively, a small disk can be removed using a clean plastic punch. The particles on this disc can be removed following the methods of Tramontano and Church [1984]. The method is quantitative and trace metal clean. Nuclepore filter weight losses are small (1-2%). The proportion of total sample removed is determined both by actual weight measurement mass balance and by consideration of the ratio of disc area to the total filtration area of the original filter. The particles thus removed are refiltered onto a small Nuclepore filter and prepared for analytical micro-analysis. Mineral particles are particularly amenable to this procedure since they are relatively unsusceptible to disruption and can be dispersed appropriately. Samples with interfering levels of organic material can be subjected to plasma etching. This technique selectively removes organics, leaving the more refractory mineral materials exposed for identification, sizing and micro-analysis. The process is very controlled and only molecular layers need be etched away.

The micro-analysis utilizes an energy dispersive X-ray fluorescence spectrometer (EDS) and high resolution digital imaging capability interfaced with a scanning electron microscope (SEM). Several manufacturers offer excellent, complete micro-analysis systems (PGT, LINK, Tracor, Kevex and others). Using dedicated routines to integrate system software for image feature analysis and chemical composition analysis of major elements, the sample particles are individually and automatically analyzed and categorized by size range and predefined chemical class. SEM-EDS analysis of particles is facilitated by the addition of a backscattered electron detector (BSE) which specifically exploits differences in the average atomic weight of the sample components. The signal is strongly influenced by sample composition; the higher the mean atomic number, the greater the positive charge of its nucleus and the more likely an interaction that produces a BSE (higher yield of BSE's). It is particularly advantageous, for example, when specifically interested in the mineral particles present in a mixed particle field of mineral and organic particles.

The remaining filter can be analyzed for total particulate elements (Al, Fe, Mn, Si, Zn, Ca, Pb and Na). Filters can be acid digested using microwave digestion methods [Nadkarni, 1984; Kingston and Jassie, 1986] in special Teflon bombs. This digestion is accomplished by using a mixture of aqua regia (HCl and $HNO3$) and HF. (See De Carlo, this chapter, page 419).

2. Time-Series Sediment Traps (With Respect to Particle Identification and Micro-analysis)

Significant controversy persists as to the optimal trap configuration and even to the relevance of many flux estimates derived from sediment trap data [U.S. JGOFS, 1989]. The problems inherent in present sediment trap technology with respect to collection efficiency, particle leaching and disintegration and, most of all, the swimmer plague are generally recognized. However, it is often felt that the potential informational returns of sediment traps outweigh the unknown biases [U.S. JGOFS, 1989].

The limited studies made on the effect of baffles on collection efficiency on cylinder traps indicate that their effect is variable and could even lead to overestimates of mass flux [Gardner, 1980a, b; Butman, 1986; Martin et al., unpublished data presented in GOFS Report 10, 1989]. The design of the Baker traps appears to be relatively optimal in this regard: high aspect ratio collection cylinder leading to final sample tubes filled with a high density preservative-poison solution. We do not advocate placing screens on top of the trap cylinders because of their potential bias against the collection of certain particle types including aggregate materials; however, attempts should be made to remove obvious swimmers prior to sample processing.

The sample collector tube solution is highly relevant to the differential preservation of specific particle species. In particular, the identification of the relationships between abiogenic components (e.g., abiogenic minerals, metal deposits, scavenged elements) and biogenic structures such as cells, organic polymers, fecal pellets or marine aggregates require a poison and preservative. While variably successful in arresting metabolic activity, most basic poisons or inhibitors like azide, mercuric chloride or antibiotics usually do not preserve organic structures. Formalin, on the other hand, is an effective poison and has the added advantage of being a preservative, maintaining and hardening organic tissues. Thus, the structural integrity of trapped constituents such as fecal pellets and microorganisms is improved, facilitating their identification. Significantly, potential swimmers can also be more easily identified and removed. Collector tubes are filled with a well-buffered formalin and seawater, high density (NaCl) solution, as the preservative-poison, prepared as described by Knauer et al. [1984]. Ca^{++} is added to prevent $CaCO_3$ dissolution [Martin and Knauer, 1984]. Basically the procedure calls for running the solution through a Chelex column followed by filtration (0.2 μm).

Sediment trap design can also be important with respect to sample integrity during trap retrieval and sample manipulation. In the Baker sequentially sampling trap design [Baker and Milburn, 1983] the traps are deployed and recovered in a "blank" position. The sample rosette is easily removed from the trap so that sample collection tubes remain in a closed condition during the entire period of deployment and recovery. The collection tubes can then be removed and processed in a clean van or an on-shore clean room.

Processing of the contents of trap collection tubes includes careful initial siphoning of the part of the supernatant prior to swimmer removal. The latter should be performed while viewing the sample under a dissecting microscope in a laminar flow hood. Samples are visually characterized in terms of their flocculent nature, numbers and sizes of recognizable fecal pellets, large protozoans and large biogenic and abiogenic mineral particles. If abundant, a representative number of fecal pellets can be removed and weighed and analyzed separately. If rare, pellets are usually left with the main sample. The sample is then filtered using tared Nuclepore filters, dried and weighed.

3. Marine Snow Collection for Trace Metal Analyses

Aggregated marine materials, or marine snow, appear to dominate vertical mass flux. In order to differentiate the major mechanisms by which fine mineral and other particles are removed from suspension in the surface waters, the marine snow flux must be clearly separated from the total mass flux using sediment trap collections. Improved knowledge of marine snow composition and specific flux rates should assist in the evaluation of sediment trap collection efficiencies.

Samples of marine snow are collected directly by hand using SCUBA. At each station a clean rubber raft carrying SCUBA divers is deployed and rowed several hundred meters from the mother ship. Divers, wearing plastic gloves, swim into the "current" while collecting aggregates by suction into acid cleaned plastic tubing using a repeating automatic pipette assembly after Hebel [1983]. We have had excellent success with the Fisher brand "Minipet" assemblies. The aggregates have contact only with the tubing. Bevaline tubing is also well suited to the procedure: it is trace metal cleanable because of its Teflon liner, flexible and relatively transparent. Before and after collections, the tube ends are connected forming a continuous loop for transportation shipboard. With experience, approximately 100 to 200 individual aggregates can be rapidly collected into a single 100 mL volume of tubing. The number of aggregates depends on the volume setting on the pipette assembly (0.5-1.0 mL) and its subsequent instantaneous suction. This collection method has had excellent success in oligotrophic gyre waters as well as more productive coastal waters.

Separate tube collections are made for bulk total trace metal analyses, individual particle micro-analysis, particulate organic carbon (POC) and nitrogen (PON) analyses or other analyses. Samples intended for bulk trace metal analyses are filtered onto clean, preweighed Nuclepore filters, rinsed with pH adjusted (pH$\approx$8.0) Milli-Q deionized water, folded and refrigerated until further processing. POC and PON samples are filtered onto preweighed, pre-combusted glass fiber filters and frozen until on-shore analysis on a Hewlett-Packard CHN analyzer. Tubes of ambient seawater only can also be collected and processed as blanks for trace metals, POC and PON. Micro-analysis samples are transferred to a clean test tube and thoroughly agitated in a vortex mixer prior to filtration and further processing. Methods for measuring marine aggregate abundances and flux rates are discussed by Alldredge (this volume, page 43) and Alldredge and Gotschalk [1988].

References

Alldredge, A. L., and C. C. Gotschalk, In situ settling behavior of marine snow, Limnol. Oceanogr., 33, 339-351, 1988.

Baker, E. T., and H. B. Milburn, An instrument system for the investigation of particle fluxes, Cont. Shelf Res., 1, 425-435, 1983.

Balistrieri, L. S., and J. W. Murray, The surface chemistry of sediments from the Panama Basin; the influence of Mn oxides on metal adsorption, Geochim. Cosmochim. Acta, 50, 2235-2243, 1986.

Balistrieri, L., P. G. Brewer, and J. W. Murray, Scavenging residence times of trace metals and surface chemistry of sinking particles in the deep ocean, Deep-Sea Res., 28, 101-121, 1981.

Bishop, J. K. B., and M. Q. Fleisher, Particulate manganese dynamics in Gulf Stream warm-core rings and surrounding waters of the N.W. Atlantic, Geochim. Cosmochim. Acta, 51, 2807-2825, 1987.

Buat-Menard, P., and R. Chesselet, Variable influence of the atmospheric flux on the trace metal chemistry of oceanic-suspended matter, Earth Planet. Sci. Lett., 42, 399-411, 1979.

Butman, C. A., Sediment trap biases in turbulent flows: results from a laboratory flume study, J. Mar. Res., 44, 645-693, 1986.

Coale, K. H., and K. W. Bruland, ^{234}Th: ^{238}U disequilibria within the California Current, Limnol. Oceanogr., 30, 22-33, 1985.

Cowen, J. P., and K. W. Bruland, Metal deposits associated with bacteria, implications for Fe and Mn marine biogeochemistry, Deep-Sea Res., 32, 253-272, 1985.

Cowen, J. P., and M. W. Silver, The association of iron and manganese with bacteria on marine macroparticulate material, Science, 224, 1340-1342, 1984.

Deuser, W. G., E. H. Ross, and R. F. Anderson, Seasonality in the supply of sediment to the deep Sargasso Sea and implications for the rapid transfer of matter to the deep ocean, Deep-Sea Res., 28, 495-505, 1981.

Deuser, W. G., P. G. Brewer, T. D. Jickells, and R. F. Commeau, Biological control of removal of abiogenic particles from the surface ocean, Science, 219, 388-391, 1983.

Di Toro, D. M., J. D. Mahony, P. R. Krichgraber, A. L. O. Byrne, L. R. Pasquale, and D. C. Piccirilli, Effects of non-reversibility, particle concentration and ionic strength on heavy metal adsorption, Environ. Sci. Technol., 20, 50-61, 1986.

Feely, R. A., G. J. Massoth, E. T. Baker, J. P. Cowen, M. F. Lamb, and K. A. Krogslund, The effect of hydrothermal processes on midwater phosphorus distributions in the northeast Pacific, Earth Planet. Sci. Lett., 96, 305-318, 1990.

Gardner, W. D., Sediment trap dynamics and calibration: a laboratory evaluation, J. Mar. Res., 38, 17-39, 1980a.

Gardner, W. D., Field assessment of sediment traps, J. Mar. Res., 38, 41-52, 1980b.

Hebel, D. V. W., Concentration and flux of trace metals, carbon-nitrogen and particulate matter in marine snow, M.A. thesis, 69 pp., San Francisco State Univ., California, 1983.

Higgo, J. J. W., and L. V. C. Rees, Adsorption of actinides by marine sediments; effect of the sediment/seawater ratio on the measured distribution coefficient, Environ. Sci. Technol., 20, 483-490, 1986.

Honeyman, B. D., and P. H. Santschi, Metals in aquatic systems, Environ. Sci. Technol., 22, 862-871, 1988.

Honeyman, B. D., and P. H. Santschi, A brownian-pumping model for oceanic trace metal scavenging, evidence from Th isotopes, J. Mar. Res., 47, 951-991, 1989.

Honeyman, B. D., L. S. Balistrieri, and J. W. Murray, Oceanic trace metal scavenging; the importance of particle concentration, Deep-Sea Res., 35, 227-246, 1988a.

Honeyman, B. D., A. W. Adamson, and J. W. Murray, Panel 2. The nature of reactions on marine particle surfaces, Appl. Geochem., 3, 19-26, 1988b.

Jannasch, H. W., B. D. Honeyman, L. S. Balistrieri, and J. W. Murray, Kinetics of trace element uptake by marine particles, Geochim. Cosmochim. Acta, 52, 567-577, 1988.

Karl, D. M., G. A. Knauer, J. H. Martin, and B. B. Ward, Bacterial chemolithotrophy in the ocean in associated sinking particles, Nature, 309, 54-56, 1984.

Kingston, H. M., and L. B. Jassie, Microwave energy for acid decomposition at elevated temperatures and pressures using biological and botanical samples, Anal. Chem., 58, 2534-2541, 1986.

Knauer, G. A., D. M. Karl, J. H. Martin, and C. N. Hunter, In situ effects of selected preservatives on total carbon, nitrogen and metals collected in sediment traps, J. Mar. Res., 42, 445-462, 1984.

Lambert, C. E., C. Jehanno, N. Silverberg, J. C. Brun-Cottan, and R. Chesselet, Log-normal distributions of suspended particles in the open ocean., J. Mar. Res., 39, 77-98, 1981.

Li, Y.-H., L. Burkhardt, M. Buchholtz, P. O'Hara, and P. Santschi, Partition of radiotracers between suspended particles and seawater, Geochim. Cosmochim. Acta, 48, 2011-2019, 1984.

Loeb, G. I., and R. A. Neihof, Adsorption of an organic film at the platinum-seawater interface, J. Mar. Res., 35, 283-291, 1977.

Luft, J. H., Ruthenium red staining of the striated muscle cell membrane and the myotendinal junction, in Sixth Intern. Congr. Electron Micros. Proc., Vol. II, 794 pp., Kyoto, 1966.

Martin, J. H., and G. A. Knauer, VERTEX: manganese transport with $CaCO_3$, Deep-Sea Res., 30, 411-425, 1984.

McCave, I. N., Size spectra and aggregation of suspended particles in the deep-ocean, Deep-Sea Res., 31, 329-352, 1984.

Morel, F. M. M., and P. M. Gschwend, The role of colloids in the partitioning of solutes in natural waters, in Aquatic Surface Chemistry: Chemical Processes at the Particle-Water Interface, edited by W. Stumm, pp. 405-422, John Wiley, New York, 1987.

Nadkarni, R. A., Applications of microwave oven sample dissolution in analysis, Anal. Chem., 56, 2233-2237, 1984.

Neihof, R. A., and G. Loeb, Dissolved organic matter in seawater and the electric charge of immersed surfaces, J. Mar. Res., 32, 5-12, 1974.

Pomeroy, L. R., The ocean's food web, a changing paradigm, BioScience, 24, 499-504, 1974.

Ringo, D. L., E. H. Cota-Robles, and B. J. Humphrey, Low viscosity embedding resins for TEM, 37th Ann. Proc. Electron Micros. Soc. Am., Vol. 1, edited by G. W. Barley, 707 pp., San Antonio, Texas, 1979.

Sieburth, J. M., Bacterial substrates and productivity in marine exosystems, Ann. Rev. Ecol. Systematics, 7, 259-285, 1976.

Silver, M. W., and K. W. Bruland, Differential feeding and fecal deep-water flora and olive-green "cells", Mar. Biol., 62, 263-273, 1981.

Sorokin, Y. I., On the role of bacteria in the productivity of tropical oceanic waters, Int. Rev. Gesamten Hydrobiol., 56, 1-48, 1971.

Sutcliffe, W. H., E. R. Baylor, and D. W. Menzel, Sea surface chemistry and langmuir circulation, Deep-Sea Res., 10, 233-243, 1963.

Tramontano, J. M., and T. M. Church, A technique for the removal of estuarine seston from Nuclepore filters, Limnol. Oceanogr., 29, 1339-1341, 1984.

U.S. GOFS Working Group, G. A. Knauer and V. Asper, Co-chairs, Sediment trap technology and sampling, Report No. 10, November 1988, 94 pp., Woods Hole, Massachusetts, 1989.

SURFACE CHEMISTRY, COLLOIDS AND TRACE-ELEMENT SCAVENGING

Bruce D. Honeyman

Department of Marine Sciences, Texas A&M University at Galveston, Galveston, Texas 77551
(Address as of January 1992: Environmental Science and Engineering Ecology, Colorado School of Mines, Golden, Colorado, 80401)

1. Introduction

From a marine chemist's point of view, particles represent a class of ligands which are available to form complexes with many elements of interest. For example, marine particles may be represented as containing functional groups of a certain type, for example carboxylic or metal oxide groups, at a specified concentration. Furthermore, stoichiometric and mass-action expressions can be written to describe chemical interactions between particle functional groups and trace elements, and the stability of the resulting complexes can be described in thermodynamic terms. Thus, in one sense, particles, as chemical components of a system, are really no different from many dissolved ligands which are also capable of forming chemical complexes with trace elements.

However, particles pose several problems for marine chemists. First, there is the difficulty of identifying and quantifying particle functional groups. A clear example is the controversy over colloidal material. The second major problem area is the development of a consistent conceptual framework for describing interactions between dissolved components and particle-associated ligands. For example, physico-chemical models for the particle-solution interface are generally based on the notion that particles can be represented as a discrete ligand type, although most environmental particles contain chemical functional groups which exhibit poly-functional characteristics. Another example is the complex kinetic behavior of trace element-particle interactions. Third, the origin, distribution and fate of particles in marine systems is largely unknown. The corollary is that our knowledge of the fate of many elements associated with particles is only as accurate as is our knowledge of the behavior of particles, themselves.

The first and second points mentioned above are directly interrelated. To a large extent, the kind of information needed about particle chemical characteristics, and the methods used to extract that information, are dependant on the assumed conceptual framework governing particle-solution interactions. While the third area of concern is not a problem of particle characterization per se, the roles that particles play in regulating the chemical composition of marine systems, a central theme of marine chemistry, are inextricably connected with the fate of particles.

Marine Particles: Analysis and Characterization
Geophysical Monograph 63

2. The Evolution of Surface Chemistry as Relevant to Marine Chemistry

Until the early seventies, most of the work on trace element-particle interactions was carried out by soil scientists and the emphasis was placed on the interactions of ions with fixed charged surfaces [Thomas, 1977]. The development of surface reactivity models was, in essence, a series of attempts to explain deviations of observations from ideal ion-exchange behavior. Such surface interaction models were physical models in the sense that the primary driving force for trace element-particle associations was considered to be separation of charge, for example, a positively-charged ion interacting with a negatively-charged clay surface.

In the early seventies, however, classically-trained coordination chemists began to study the interaction of metal ions with metal oxide surfaces such as γ-alumina and hydrous ferric oxide. The observation that the tendency of a metal ion to associate with a metal oxide surface correlates with the intensity of solution-phase hydrolysis product formation led to the concept of surface-complex formation [Schindler et al., 1976; Stumm et al., 1976]. Surface complexation models (SCMs) are still "physical" models in the sense that the charges of the sorbing ion and the surface of the sorbent are important factors. However, the possibility of chemical bond formation between surface-active solutes and binding "sites" on mineral surfaces is a new aspect which is a key element in SCMs but was not included in ion-exchange models. While the analogy between solution-phase and surface reactions is exact, the main difference between the various currently-available surface complexation models is the way in which the free energy of sorption is divided into electrostatic and chemical components.

Just as Sillen's model [1961] signaled the beginning of modern chemical oceanography, Schindler [1975] firmly established the link between surface chemistry and

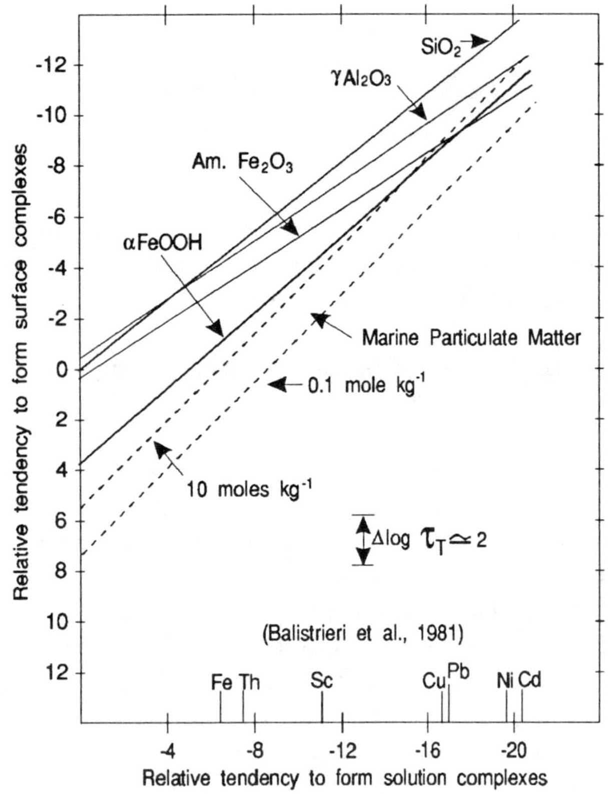

Fig. 6-50. The tendency of a metal ion to form surface complexes is related to the intensity of solution-phase complex formation. This observation led to the establishment of surface complexation models by Schindler, Stumm and co-workers [e.g., Schindler et al., 1976] The total residence time of trace elements (Me) in the water column, τ_T, is related to the fraction of total Me associated with sinking particles, f_p, and the particle residence time, τ_p. An order of magnitude change in Me sorption intensity corresponds to a decade change in residence time, τ_T, of the trace element.

$$\tau_T = \frac{\tau_p}{f_p}$$

Sorption intensity changes with the nature of the trace element and the type and quantity of Me-sorbing particles [from Balistrieri et al., 1981, modified by Honeyman and Santschi, 1988].

the chemical composition of the ocean with his "Zero-Order" model for the removal of trace metals from the ocean waters. In his model, the rate of removal of trace elements from seawater is related to the intensity of interaction between trace elements and particle surface sites, given a constant particle flux. The model correctly predicts the order but not the absolute value of residence times of metals in the ocean, assuming SiO_2 as the model surface. Schindler's model was crucial for two reasons: 1) it provided a formal connection between solution-phase reactions, adsorption and trace metal scavenging; and 2) he established the notion that marine particles are chemical species which compete against dissolved ligands for complexation with trace elements. Following Schindler, Balistrieri et al. [1981] provided the next important step by investigating the extent to which variations in particle surface chemistry would affect trace element scavenging residence times.

3. The Surface Complexation Approach

Figure 6-50 shows several important characteristics of surface chemistry as applied to marine chemistry. First, it illustrates the generally observed relationship between the tendency to form solution-phase hydrolysis products and the intensity of surface complexes. Similar correlations led Stumm, Schindler and co-workers to formulate the analogy between solution and surface complexes:

Aqueous hydrolysis:

$$HOH + Me^{z+} = HOM^{(z-1)+} + H^+$$

Analogous surface complex:

$$SOH + Me^{z+} = SOMe^{(z-1)+} + H^+$$

where SOH represents a protonated particle surface site. Thus, generic stoichiometric expressions for sorption [e.g., Balistrieri and Murray, 1981] usually include the deprotonation of surface sites (or the dissociation of surface-associated water) as part of the net sorption reaction. Second, for a given particle-reactive element, there is significant variation in sorption intensity between particle types. A group of metal oxide surface site types, for example, Al_2O_3, ferrihydrite, MnO_2, etc., can yield a four orders of magnitude range or more in calculated binding intensities for a given metal ion. Third, a certain particle type yields marked variation in sorption intensity throughout the range of particle-reactive elements. Fourth, derived values of sorption intensity are sensitive to the specific site concentration (moles sites-mass particles), measured or assumed, for a selected particle type. A surface complexation model is a framework within which one can postulate surface reactions. In a sense, they are heuristic devices for exploring stoichiometric relationships among system chemical components. For example, Balistrieri et al. [1981], using surface complexation models in this manner, reached the important conclusion that marine particles have the trace metal binding characteristics of organic matter. It is important to recognize, also, that surface chemical models, in their current form, are essentially solution-phase models (such as MINEQL "family" of computer codes) adapted to consider the special case of a charged particle-solution interface. The requirements for the application of surface chemical models are: 1) the representation of particles as ligands and determining their primary chemical characteristics; 2) knowledge of solution-phase chemistry; and 3) selection of a molecular hypothesis (interfacial configuration) for the sorption environment.

3.1. The representation of particles as ligands.

Conceptually, the main difference between dissolved ligands and particle ligands is that the latter are spatially constrained by the particle matrix, itself. The most common representation of particle surface sites is as moles of sites per mass of particles and the two critical parameters whose estimation is required are site density, n_s (sites nm^{-2}), and specific surface area, S_A (m^2 g^{-1}).

3.2. Site density.

One of the direct consequences of the analogy between solution-phase hydrolysis and surface complex formation is that trace element sorption is most frequently represented as an exchange of trace elements for surface-associated protons. Such a representation exists in both semi-generic partitioning equations [Tessier et al., 1989; Honeyman and Santschi, 1988; Balistrieri and Murray, 1983] and in the postulation of surface reactions [HYDRAQL; Papelis et al., 1988]. Thus, the most common methods used to estimate surface site density are those that include the uptake of protons by the sorbent of interest. These methods include tritium exchange [Berube and de Bruyn, 1968] and potentiometric titration [James and Parks, 1982, and references therein]. Such methods are based on the assumption that there is a one to one correspondence between sites accessible to protons and those available to the sorbate. A variety of other methods have also been employed to estimate surface site densities, including crystallographic analysis [Sposito, 1984], reacting the mineral with an OH-labile compound such as CH_3-Mg-I, thermographic analysis and IR spectroscopy [see James and Parks, 1982, and references, therein].

Table 6-11 [compiled by Davis and Kent, 1990] presents site densities for surface functional groups on selected hydrous oxide surfaces, as estimated by the various techniques described above. Site densities range by over an order of magnitude, from 22 sites nm^{-2} for α-FeOOH to less than 2 for kaolinite. While most of the variability in estimated site densities is a direct reflection of differences in mineralogy, some differences can also be ascribed to variability in the results of various techniques for estimating site densities, as shown in Table 6-12. Differences in measured site densities are particularly stark when comparing results for the proton acid-base titrations with "saturation" isotherms for oxyanions which, compared to the proton, are relatively quite large. This indicates that "probe" molecules for surface characteristics should be physically and chemically similar to the sorbate of interest.

Davis and Kent [1990] argue, given the difficulty of accurately measuring site densities, that surface complexation constants for strongly binding sorbates should be derived using a site density of 2.31 sites nm^{-2} (3.84 μM m^{-2}). This suggestion is aimed at developing a self-consistent data set for trace element sorption in complex mixtures of sorbents and is bolstered by recent studies [Hayes et al., in press,1991] that have shown that the ability of surface complexation models to fit adsorption data is relatively insensitive to variations in site density.

3.3. Specific surface area.

The second parameter needed to estimate the specific site concentration (M g^{-1}) of particles is specific surface area, S_A (m^2g^{-1}). An enormous literature exists on the determination of surface area by various methods; Davis and Kent [1990] have recently summarized information about surface area determination of direct relevance to geochemists.

While it is possible to estimate specific surface areas by crystallographic or optical means, surface areas of importance are on a molecular scale and their accurate determination requires the utilization of probe sorbates, either a gas or solute, and then relating the amount adsorbed to the area occupied. The most common gas for surface area determination is N_2, which is sorbed at the boiling temperature of liquid N_2 at atmospheric pressures (ca. 77° K). The most popular procedure for deriving S_A from gas adsorption measurements is the BET method [Brunauer et al., 1938]. Nitrogen adsorption onto materials with relatively low specific surface areas, i.e., less than 1 m^2 g^{-1}, produces erroneous results because of non-ideal N_2 gas behavior and Kr is often employed under those circumstances as an alternative sorbate. However, Kr, itself, is not an ideal sorbent, and the disadvantages of its use have been documented by Gregg and Sing [1982].

TABLE 6-11.

Density of Surface Functional Groups on Oxide and Hydrous Oxide Minerals From Davis and Kent [1990]

Mineral Phase	Range of Site Densities (sites nm^{-2})	Reference	Proton acceptor groups (sites nm^{-2})	Proton donor groups (sites nm^{-2})	Reference
α-FeOOH	2.6-16.8	1,2	4.4	6.7	2
α-Fe_2O_3	5-22	1	-	-	-
Ferrihydrite	0.1-0.9 M per M Fe	3	-	-	-
TiO_2 (rutile)	12.2	1	2.6	4.2	1
TiO_2 (rutile and anatase)	2-12	1	-	-	-
α-$Al(OH)_3$	2-12	4	2.8	5.6	2
α-Al_2O_3	6-9	4	-	-	-
SiO_2	4.5-12	1	0	all	1
kaolinite	1.3-3	5	0.35	1.0	2

References: 1) James and Parks [1982]; 2) Sposito [1984]; 3) Dzombak and Morel [1990]; 4) Davis and Hem [1989]; 5) Fripiat [1964]

Clay minerals pose a special problem because of the porosity of interlayer regions. While N_2 and Kr gas adsorption has been used for clay mineral surface area determination [see Murray and Quirk, 1990], a common alternative is the use of polar organic compounds [glycerol: van Olphen, 1970; ethylene glycol: Dyal and Hendricks, 1950; and ethylene glycol monoethyl ether (EGME): Eltantawy and Arnold, 1973] due to their ability to penetrate clay interregions. Surface area determinations by retention of polar organics are relatively inaccurate for materials with surface areas on the order of 40 to 50 $m^2\ g^{-1}$ and less. Consequently, selection of the appropriate surface area probe requires some knowledge of the mineralogy of the sorptive substrate and the adsorbate of interest.

Organic material, while not "particulate" in the way that mineral species are typically visualized, is, none the less, a major component of many sediments and provides a significant, if not dominant, contribution to the colloidal population of many systems. Organic matter is usually considered a poly-functional material whose complexation capacity (mMg^{-1} carbon) is most often determined by potentiometric titration [see Buffle, 1987; Buffle and Altmann, 1987].

Given the relatively narrow range of measured site densities (generally, 1-7 sites nm^{-2}), the largest contribution to variations in specific site concentrations ($M\ g^{-1}$) is specific surface area, S_A. For example, α-silica (α-SiO_2) has a specific site concentration of ca. 1.5×10^{-5} M sites g^{-1} [Benjamin, 1978] as a consequence of its relatively low specific surface area of 3-4 $m^2\ g^{-1}$. Amorphous silicas have specific surface areas ranging from about 100 to several hundred $m^2\ g^{-1}$, depending on the formation process [Kent and Kastner, 1985], leading to specific site concentrations on the order of a mM g^{-1}.

There is some evidence that site densities are relatively invariant among mineralogical forms of an element [Fetl and Smolkova, 1972, in their study of silicas] and that a metal oxide surface site,

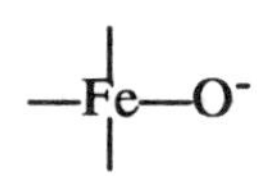

has essentially the same surface chemical characteristics ("acidity") from one mineralogical form to the next e.g., hydrous ferric oxide compared to goethite. This evidence suggests that, in contrast to the enormous variability in mineralogical forms, the surface chemical behavior of chemically-similar mineral species may be relatively easier to classify. For example, iron oxides such as hematite, goethite, ferrihydrite, etc., in spite of their differences in mineralogy, may, on a per site basis, exhibit similar trace element sorptive characteristics.

3.4. Particle Ligands: Discrete or Polyfunctional Character?

Two distinctly-different approaches to representing and determining complexation properties of particle surfaces have emerged over the last fifteen years. Although both approaches explicitly include the notion that particles are ligands with a constrained spatial distribution, they depart in the conceptualization of the chemical heterogeneity of surface sites.

One approach is exemplified by the work of Buffle and co-workers [Buffle and Altmann, 1987] in their application of statistical concepts to trace metal-humic substance interactions. In this and similar approaches, macromolecules and particle surface sites display polyfunctional behavior in which complexation intensity is related, via distribution functions, to the fraction of complexant sites associated with trace elements. Thus, macromolecules or particle sorption sites are characterized by a distribution of site types rather than as a discrete chemical entity. Consequently, the interaction intensity of a trace element with a particle ligand varies with the fraction of particle functional groups which are complexed with the trace element. A distribution of site types may result from such factors as particle surface morphological variations such as edge or pit sites on mineral species or functional group type or location on macromolecules or the aggregate sorption behavior of a suite of competing particle ligands in a particle assemblage.

The second approach, and the one which occupies the largest niche in the geochemical literature, is the direct descendent of the earlier work by Schindler, Stumm and co-workers in which a particle surface is represented as a number concentration of sites of a specific type which form mono-energetic complexes with trace elements.

It makes intuitive sense that particles, especially those found in natural systems, do not have a unique chemical character but are the "sum", in some manner, of the characteristics of the particle components. To a certain extent, the inherent polycharacter of particles is at odds with what has been the major thrust of aquatic surface chemistry, i.e., the direct extension of the solution-phase, metal-discrete-ligand analogy to particle surfaces. Thus, a major question concerning approaches to trace element-particle interactions remains unresolved: What is the best approach to describing particle-trace element interactions?

TABLE 6-12.
Estimates of Site Densities for α-FeOOH by Various Adsorption Techniques From Davis and Kent [1990]

Ion	Adsorption maximum (sites nm^{-2})	Method	Reference
OH^-/H^+	4.0	a	f
OH^-	2.6	a	g
F^-	5.2	b	h
F^-	7.3	c	f
SeO_3^{2-}	1.5	d	i
PO_4^{2-}	0.8	d	i
Oxylate	2.3	b	j
Pb^{2+}	2.6	b	k
Pb^{2+}	7.0	c	k

Methods: a) Acid-base titration; b) Adsorption isotherms; c) Excess F^- at pH 5.5; d) Adsorption isotherm at pH_{pzc} and e) Fit with model.

References: f) Sigg and Stumm [1981]; g) Balistrieri and Murray [1981]; h) Hingston et al. [1968]; i) Hansmann and Anderson [1985]; j) Parfitt et al. [1977]; and k) Hayes [1987].

3.5. Surface Complexation Models of the Particle-Solution Interface

There are four basic tenets of all surface complexation models [Davis and Kent, 1990]:

1) The surface is composed of functional groups which react with solution species to form surface complexes. These surface complexes may be coordinative or ion pairs.
2) The reactions between surface functional groups and dissolved species can be described using mass-action expressions. Corrections are included to account for the work required to bring a charged species from bulk solution to a sorption plane of some net charge.
3) Surface charge and potential are the consequence of chemical reactions of the surface functional groups. They are secondary effects that result from surface coordination reactions.
4) The apparent binding constants determined for the mass-action expressions are empirical parameters related to thermodynamic constants via the rational activity coefficients of surface species [Sposito, 1983].

Hayes [1987] has suggested, since activity coefficients depend on the potentials between the plane of sorption, as well as on the charge of the reacting species, that the ratio of activity coefficients corresponding to particular mass-action expression should be equal to the electrostatic correction term.

A number of different configurations of the particle-solution interface have been developed, including the Constant Capacitance Model [Schindler and Kamber, 1968; Hohl and Stumm, 1976], Diffuse Double-Layer Model [Stumm et al., 1970] and Triple-Layer Model [Davis and Leckie, 1978]. All of these models share: 1) mass law equations for all surface reactions; 2) a mole balance equation for surface sites; 3) equations for computing surface charge; and 4) a set of equations representing the constraints imposed by the particular model (configuration) of the interfacial model. Figure 6-51 presents a schematic drawing of idealized decay of electrical potential with distance from a surface for four different interfacial configurations. The differences are found in the ways in which the free energy of adsorption is resolved into electrostatic and chemical components. More specifically, the contrasts lie in the charge-potential relationships of each interfacial configuration and the planes of assignment for the postulated surface species. For example, the Constant Capacitance Model allows only the association of specifically adsorbing solutes with the particle surfaces (i.e., the formation of inner-sphere complexes) while other ions, Na^+, Cl^-, are considered to be inert with respect to the surface. The Triple-Layer Model, in contrast, has two planes of adsorption, the o-plane for inner-sphere complexes and the $\beta-$plane for outer-sphere, ion-pair complexes. "Inert" electrolyte ions are assigned to the β-plane where they form weak complexes with surface hydroxyls on oxides; other ions, such as Pb^{2+} or SeO_3^{2-} can be assigned to either the o- or β-plane depending on the type of surface complex which is postulated (i.e., inner-sphere or ion-pair, respectively). Because of the charge-potential relationships characteristic of each interfacial configuration, one data set will yield a different surface complexation constant for each model type used.

For example, consider the interaction of a metal ion, Me^{2+}, with a protonated particle surface site. Assuming that $MeOH^+$ is the only species of Me(II) sorbing, there are two possible monodentate surface complexes for consideration:

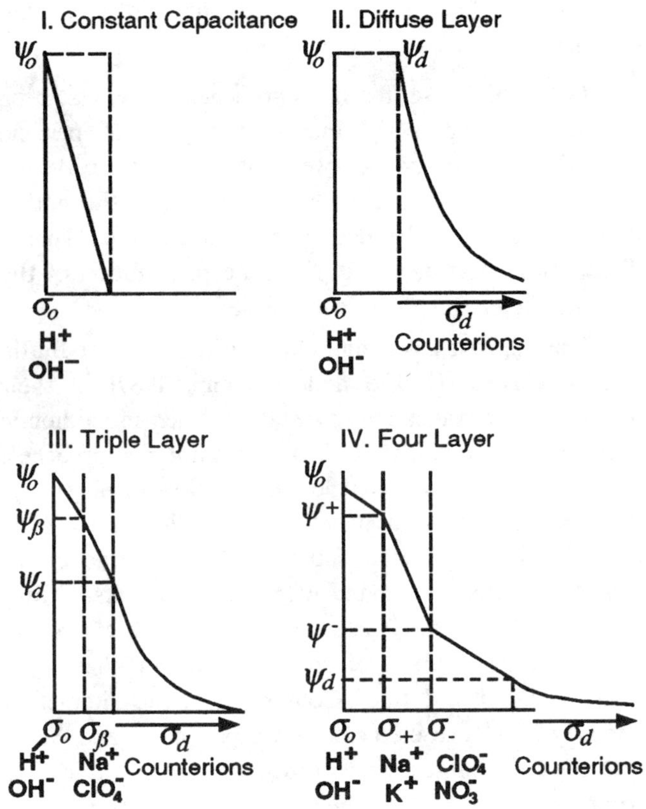

Fig. 6-51. Schematic illustration of interfacial configurations of four surface interaction models [from Davis and Kent, 1990] The decay of potential with distance from the surface is drawn for surfaces immersed in solutions composed of simple electrolytes and at a pH other than the pH_{pzc}. In each case, the interface is conceptualized as consisting of a series of capacitors whose capacitances must be specified as a model parameter. The constant capacitance model contains one capacitor, the triple-layer model has two and the four-layer model, three. Expressions describing the relationship between charge and potential are based on the assumption that potential decays linearly with the distance between charged planes. The charge of the interface taken as a whole must obey conservation of charge. For example, in the case of the triple-layer model: $\sigma_o + \sigma_\beta + \sigma_d = 0$.

Inner sphere:

$$SOH + Me^{2+} + H_2O \overset{K_{Me_1}}{=} SOMeOH^o + 2H^+$$

Ion pair:

$$SOH + Me^{2+} + H_2O \overset{K_{Me_2}}{=} SO^- \text{---} MeOH^+ + 2H^+$$

The equilibrium expressions corresponding to these postulated surface reactions, using the Triple-Layer Model interfacial configuration (Figure 6-51), are:

$$K_{Me_1} = \frac{[SOMeOH^o][H^+]^2}{[SOH][Me^{2+}]}$$

$$K_{Me_2} = \frac{[SO^- \text{---} MeOH^+][H^+]^2}{[SOH][Me^{2+}]} \exp\left(\frac{F(\Psi_\beta - \Psi_o)}{RT}\right)$$

where Ψ_0 and Ψ_β represent potentials at the surface and β planes, respectively, and F is the Faraday constant. In both cases, the empirically-determined concentration ratios are identical but the values of the intrinsic binding constants are highly dependent on the assumed molecular configuration. The exponential term in the mass-action expression for the ion-pair complex can be considered to be equal to the ratio of activity coefficients for the charged surface species. There is no correction for the inner-sphere surface complex, $SOMeOH^o$, because the complex is uncharged.

It should be pointed out, also, that despite differences in self-consistent model parameters among the various models, all of the models are equally capable of reproducing a particular set of titration and sorption data [Westall and Hohl, 1980].

The complexity of natural materials, the variety of techniques used for particle characterization and the refinement of surface complexation models has led to the establishment of a multitude of inconsistent data sets. To counter this, Dzombak and Morel [1990] recently provided a self-consistent, validated data base for ferrihydrate based on a modified form of the Diffuse Double Layer Model. This treatise is an excellent first step in providing aquatic surface chemistry with a much needed benchmark.

4. Surface Chemistry in Application

The power of the surface complexation approach is not in its predictive capability for natural systems, which is thus far minimal, but in its providing a theoretical framework for interpreting observations of trace element behavior. Surface complexation theory, and the body of experimental work supporting its basic tenets, provides a direct linkage between solution-phase reactions and the role of particles in regulating the composition of water. It provides, for example, a rationale for understanding differences in trace element scavenging residence times in oceanic systems [Whitfield and Turner, 1987], the toxicity of trace elements to phytoplankton [Kuwabara et al., 1986] and chemical controls on particle coagulation rates [Liang, 1988]. In its present form, however, surface complexation theory, does not provide a means for understanding many interactive effects. These effects include the competition of ions for particle surface sites [Balistrieri and Chao, 1990], changes to interfacial environments from interacting particles [Prieve and Ruckenstein, 1978], and surface site heterogeneity [Zachara et al., 1989; Benjamin and Leckie, 1981]. Furthermore, it is likely that minerals which are part of mineral assemblages will not have the same surface chemical properties while in the assemblages, that they have in monomineralic systems [Anderson and Benjamin, 1990; Honeyman, 1984]. Finally, surface chemical models provide estimates of the distribution of chemical species at equilibrium. While evidence exists that metal ion sorption on monomineralic surfaces in simple electrolyte solution is rapid, on the order of milliseconds [Yasunaga and Ikeda, 1986], slow coordination reactions [Hering and Morel, 1989] or mass transport limitations [Jannasch et al., 1988; Csanady, 1986] may dramatically hinder the approach of more complicated systems to chemical equilibrium.

5. The Role of Colloidal Material in Marine Chemistry

In very broad terms, the oceanic water column contains three pools of trace-metal binding ligands: 1) macroparticles (that material a few tenths of a micron in size or larger, including marine snow and fecal matter); 2) colloids; and 3) those ligands which are truly dissolved. Not only does the nature of these ligands change spatially and with time but there is an exchange of material between these ligand pools, as well.

As described earlier, one of the basic requirements for the application of surface complexation theory is the accurate identification of chemical nature and distribution of particle-associated ligands. Until quite recently, discussions of the role of particles in regulating the composition of natural waters focussed on particles of relatively large dimension, i.e., a few microns in size and larger [Nozaki et al., 1987; Bacon et al., 1985]. There is accumulating evidence, however, that colloidal material represents a significant pool of trace element-binding ligands and that colloids may have a major impact on the distribution and fate of many particle reactive elements [Moran and Moore, 1989]. It is important to emphasize that the "problem" of colloids is twofold. First, there are questions surrounding the identification (i.e., speciation) of colloidal material. Second, colloidal material appears to control, through its physical behavior, biogeochemical pathways of sorbed elements.

5.1. Evidence For the Importance of Colloids in the Mobility of Trace Elements

Our understanding of the role that colloids play in the geochemical life of trace elements is largely based on circumstantial evidence, although the role of submicron-sized particles in affecting the behavior of particle-reactive solutes has long been recognized. Colloid transport through porous and fractured media has been described by McDowell-Boyer et al. [1986] and Degueldre et al. [1989], respectively, in surface waters by Buffle [1987] and McCave [1984], and the literature on the formation and properties of radiocolloidal dispersions summarized by Kepak [1977]. Evidence for the importance of colloids in oceanic systems is only now emerging.

Buddemeier and Hunt [1988] studied radionuclide transport in groundwaters pumped from a nuclear bomb test cavity and found that the transition and lanthanide radionuclides, which moved considerably further than expected from ^{3}H and other conservative species, were essentially all associated with the less than 0.2 μm particle fraction. Sheppard et al. [1980] concluded that much of the mobility of radionuclides in soil waters is probably due to soil constituents in the 2 to 5 nm range. Recently, Penrose and co-workers [1990] convincingly showed that colloidal material, most likely in the form of colloidal organic carbon, provided for the sub-surface transport of plutonium. Niven and Moore [1988] examined the partitioning of ^{234}Th among dissolved, colloidal and particulate fractions and found that up to 64 percent of the total ^{234}Th was associated with the colloidal (1 nm to 0.2 μm) size fraction. According to Baskaran et al. [1990], 25 to 75% of natural ^{234}Th in Gulf of Mexico water is found in the colloidal fraction. Moran and Moore [1989] estimated that colloidal material exists at about twice the mass concentration of macroparticulate material. Orlandini et al. [1990] found in their study of an oligotrophic lake that submicron colloidal material can dominate the behavior of actinides. Nash and Choppin [1980] estimated that in natural waters at 0.1 mg L^{-1} DOC [humate concentration = (Hu) = 4 x 10^{-7} eq L^{-1}) the ratio of (Th-Hu)/(Th_{diss}) would be about 4 x 10^{10}, suggesting that truly-dissolved thorium is unlikely when macromolecular DOC is present. Sequential extraction and ultrafiltration experiments by Santschi et al. [1983, 1987] have shown that radioisotopes of particle reactive elements are involved in the dynamic cycles of coagulation and disaggregation of macroparticles and colloids in the water column of coastal marine ecosystems; and Buchholtz et al. [1986] demonstrated the aggregation effect of colloidal radionuclides in laboratory batch experiments with marine sediments. Buffle and co-workers [Buffle, 1987, and references therein] have documented trace metal associations with colloidal material.

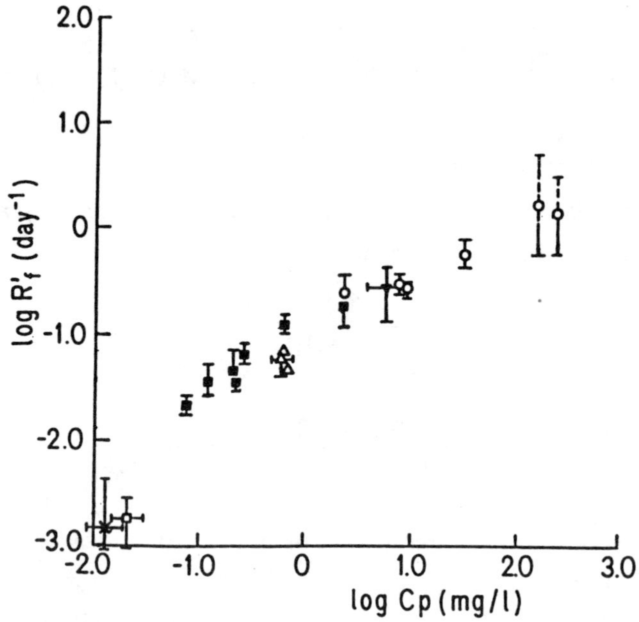

Fig. 6-52. Log of the scavenging rate constant, R_f', versus the log filterable particle concentration, C_p. R_f' was calculated using oceanographic data from the literature [Honeyman et al., 1988, and data sources cited therein] using the following expression

$$R_f' = \log\left[(R_r + \lambda_{234} + \lambda_p) \left(\frac{A_{Th_{part}}}{A_{Th_{diss}}} \right) \right]$$

where R_r is a "desorption" rate constant (assuming 0.007 d^{-1}) for the appearance of Th in the filter-passing fraction of bulk solution and λ_{234} and λ_p are the ^{234}Th decay (= ln2/24.1 days) and particle removal rate constants, respectively, and A represents thorium activity.

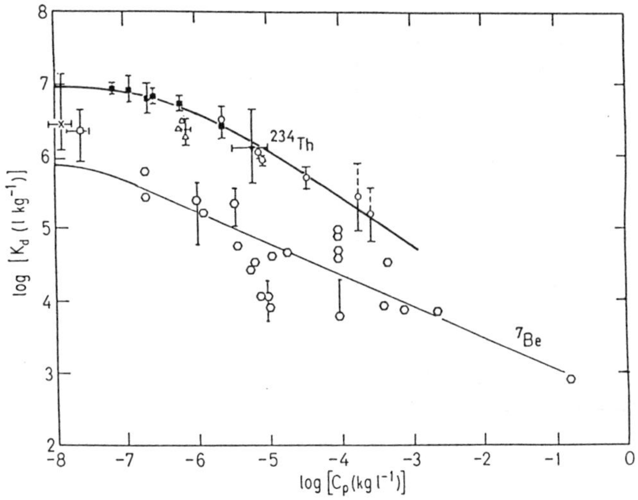

Fig. 6-53. Log K_d versus log C_p [from Honeyman and Santschi, 1988]. Refer to Honeyman et al. [1988] and Honeyman and Santschi [1988] for sources of data. Physico-chemical theory indicates that the intensity of sorption (represented by K_d) should be independent of the ligand concentration (i.e., C_p). This is not observed in many cases. A postulated cause of the "particle concentration effect" is the presence of trace element sorbing colloids in the operationally defined "dissolved" fraction [Morel and Gschwend, 1987]. In other words, it is an artifact of incomplete phase separation.

5.2. Kinetic and Equilibrium Effects of Colloidal Material on Trace Element Behavior

Colloidal material affects our ability to describe trace element behavior for two reasons: 1) it is a large pool of trace element binding ligands which are very difficult to accurately characterize; 2) while often operationally considered as part of the "dissolved" fraction, trace elements associated with colloids will likely have a physico-chemical behavior different from truly dissolved species of the trace element; furthermore, it is likely that there is a significant, albeit slow, transfer of material between the colloid and macroparticle pools.

Thorium associated with colloidal material and operationally considered part of the "dissolved" (i.e., filter-passing) fraction has been implicated [Honeyman and Santschi, 1989] in producing: 1) scavenging rate constants which are relatively small in magnitude and which increase value with particle concentration and 2) the so-called "particle-concentration effect".

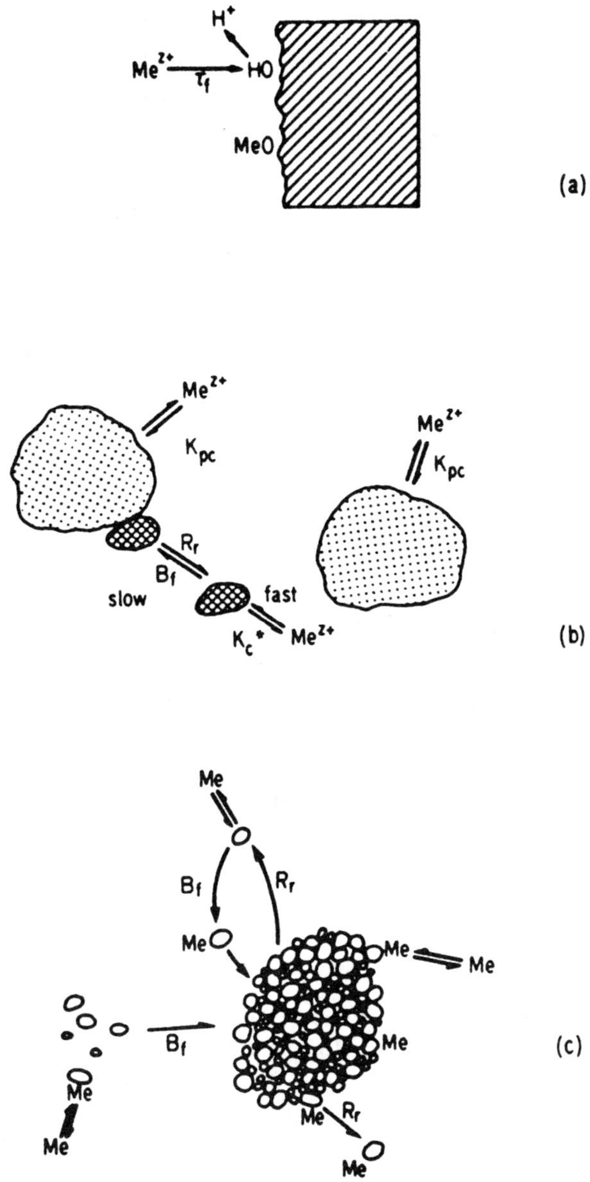

Fig. 6-54. Schematic of colloidal pumping model: coagulation coupled to adsorption [Honeyman and Santschi, 1989]. a) Rapid sorption of trace elements onto particle surfaces. Formation of surface complexes is rapid compared to rates of particle-particle interactions. K_{pc} and K_c* represent conditional sorption constants. b) Slower aggregation, B_f, of colloids with larger particles. Colloidal material includes organic macromolecules, mineral fragments and condensation nuclei. c) Natural particle clusters most likely contain materials of varied size, history and composition.

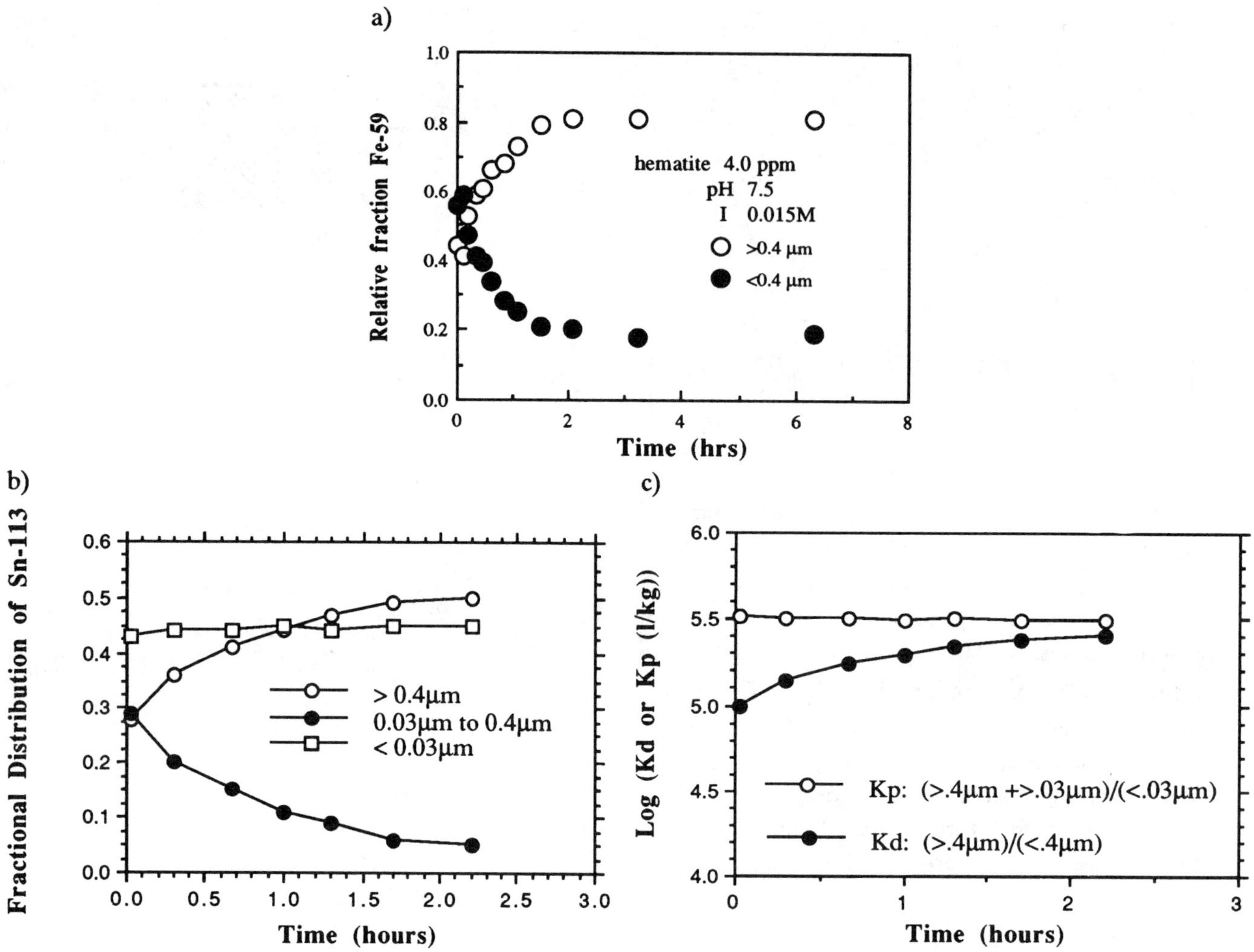

Fig. 6-55. a) Relative amount of 0.25 μm ^{59}Fe-labelled hematite in fractions of bulk solution passing and retained by 0.4 μm Nuclepore filters, as a function of time. b) The effect of particle aggregation on the distribution of ^{113}Sn between the colloidal (0.03 μm to 0.4 μm), particulate (greater than 0.4 μm size fractions) and dissolved (less than 0.03 μm) pools with time. Sorptive equilibrium is rapidly attained (note that the dissolved concentration is constant); Sn is transferred between the particle pools by coagulation. c) K_d is smaller in value than K_p because the 0.4 μm filter-passing fraction, the operationally defined "dissolved" fraction in the K_d determination contains, colloidal (0.03 μm to 0.4 μm) particles. K_p, in contrast, is based on the ratio of sorbed to truly dissolved Sn, its value is time-independent.

Examples of these effects of colloidal material on the behavior of thorium isotopes, and by extension other particle-reactive elements, are shown in Figures 6-52 and 6-53.

Figure 6-54 presents the elements of the colloidal-pumping model described by Honeyman and Santschi [1989]. The conceptual model attempts to couple two characteristics of colloidal material: 1) its participation in chemical reactions; 2) its aggregation with, or transfer to, large particles. When sorption of trace elements onto colloids is rapid (Figure 6-54a) compared to time scales for particle coagulation, the transfer of colloid-sorbed elements to the macroparticle (filter-retained) pool is limited by particle aggregation rates (Figure 6-54b). In this figure large particles are distinguished from colloids by the comparative ease at which they can be separated from bulk solution, for example, with relatively large pore sized filters (0.45 μm). A critical element is the explicit consideration of complexes between trace elements and colloidal (i.e., filter-passing) material. The transport of scavenged trace elements to their final repository on the ocean floor is likely circuitous as the material making up particles periodically cycles through various pools as the consequence of aggregation,

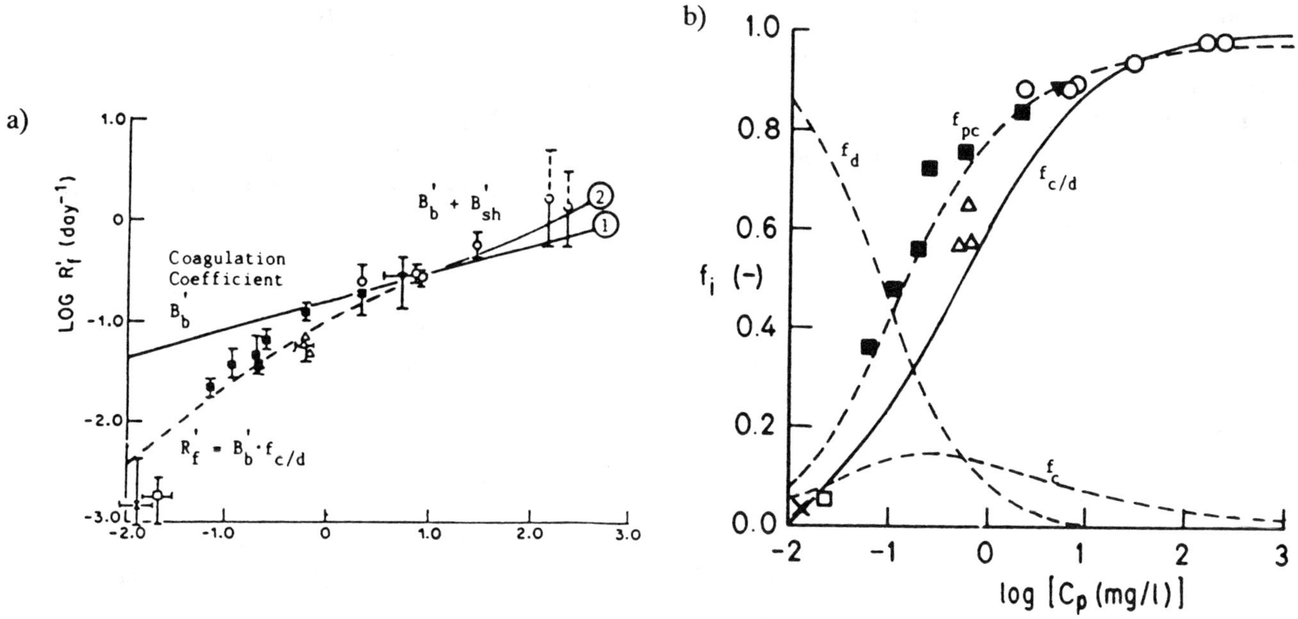

Fig. 6-56. Coupling Farley and Morel's [1986] sedimentation model with sorption and its application to oceanic Th scavenging. [from Honeyman and Santschi, 1989]. a) Data are the same as shown in Figure 6-52. Critical parameter values: Brownian collision efficiency factor, α_b = 0.5, particle density, ρ_p = 2.5 gcm^{-3} and mean depth, h = 6 m (the average depth of a water column that a sorbed ^{234}Th atom will be in contact with before decay). Solid lines represent the scavenging rate constant when a trace element or radionuclide in the filter-passing fraction is completely associated with colloidal particles (i.e., K_c* is "infinite"): 1) Brownian coagulation; 2) Brownian + shear coagulation. In other words, only particle-particle interactions are considered. The dashed line includes the effect of Th solution/colloid partitioning on scavenging rate with log K_c = 10^7. b) Calculated distribution of Th between dissolved, f_d, colloidal, f_c, and filter-retained, f_{pc}, fractions. $f_{c/d}$ represents the portion of the bulk solution passing the filter but which is associated with colloids. Note that as $f_{c/d}$ approaches 1, that is, when little truly dissolved Th exists, that the scavenging rate constant approaches the maximum rate of coagulation + sedimentation. Actual values of f_{pc} correspond to the data in Figure 6-52.

disaggregation and remineralization processes.

Laboratory evidence demonstrating the validity of colloidal pumping is shown in Figure 6-55 [Honeyman and Santschi, in press]. Figure 6-55a shows the transfer of 0.25 μm ^{59}Fe-labeled hematite particles between filter-passing and filter-retained fractions of bulk solution. This transfer occurs because singlet particles (0.25 μm) pass the 0.4 μm filter while aggregates of doublet size and larger are retained. As the particle size distribution of the initially homodisperse system evolves, total particle mass shifts from the pool of singlets to larger and larger aggregates.

The addition of a sorbing element (Sn as ^{113}Sn) shows the effect of particle aggregation (followed simultaneously by ^{59}Fe activity) on Sn transfer to the filter-retained particle pool (Figure 6-55b). Note that truly-dissolved Sn, represented by the less than 0.03 μm portion of bulk solution, remains constant, and that sorptive equilibrium is rapidly attained. Thus, the observed kinetic effect, the transfer of ^{113}Sn between particle pools, is the consequence of particle aggregation. When a clear distinction between sorbed and dissolved is made, the conditional binding coefficient, K_p, is constant in value while the operational coefficient, K_d, increases in value with time, approaching K_p as the fraction of colloidal Sn decreases (Figure 6-55c).

Figure 6-55 illustrates a critical point. While the chemical characterization of colloidal material, and knowledge of the extent of its interaction with trace elements, is important, it is also crucial to understand the physical behavior of colloids. These two aspects of the colloid "problem" are reiterated in Figure 6-56 [Honeyman and Santschi, 1989]. In this case, thorium sorption onto colloidal material and the removal of particle mass (and sorbed thorium) through coagulation-sedimentation are invoked as the primary rate-controlling processes in trace element scavenging. The rate of thorium scavenging depends on 1) the rate of particle aggregation leading to sedimentation and 2) the fraction of thorium associated with colloidal material (i.e., the chemical speciation of thorium and colloidal material).

6. Summary

Surface chemical theory has provided a sound framework for aiding in the interpretation of observations of trace element behavior in many natural systems. Its main contributions have been: 1) systematizing the way in which particles are viewed as chemical species whose surface-bound ligands are available for complexing trace elements and 2) providing a linkage between solution chemistry and the role of particle surfaces. However, several research areas need particular attention. These include: 1) studies of particle assemblages rather than the current focus on monomineralic systems; 2) development of methodologies for quantifying components of the colloid pool; 3) striving for the development of a consensus in configurations for interfacial models and in representations of particles as ligands; and 4) greater effort in understanding the physical behavior of particles.

Acknowledgements. Preparation of this manuscript was funded in part by a grant from the National Science Foundation (OCE-9012103).

References

Anderson, P. R., and M. M. Benjamin, Surface and bulk characteristics of binary oxide suspensions, Environ. Sci. Technol., 24, 292-298, 1990.

Bacon, M. P., C.-H. Huh, A. P. Fleer, and W. G. Deuser, Seasonality in the flux of natural radionuclides and plutonium in the Sargasso Sea, Deep-Sea Res., 32, 273-286, 1985.

Balistrieri, L. S., and T. T. Chao, Adsorption of selenium by iron oxyhydroxide and manganese dioxide, Geochim. Cosmochim. Acta, 54, 739-752, 1990.

Balistrieri, L. S., and J. W. Murray, The surface chemistry of goethite (α-Fe_2O_3) in major ion seawater, Am. J. Sci., 281, 788-806, 1981.

Balistrieri, L. S., and J. W. Murray, Metal-solid interactions in the marine environments: estimating apparent equilibrium binding constants, Geochim. Cosmochim. Acta., 47, 1091-1098, 1983.

Balistrieri, L. S., P. G. Brewer, and J. W. Murray, Scavenging residence times of trace metals and surface chemistry of sinking particles in the deep ocean, Deep-Sea Res., 28, 101-121, 1981.

Baskaran, M., P. H. Santschi, and G. Benoit, Thorium isotope measurements on colloids in surface seawater, EOS Trans. Am. Geophys. Union, 71(43), 1418, 1990.

Benjamin, M. M., Effects of competing metals and complexing ligands on trace metal adsorption at the oxide/solution interface, Ph.D. thesis, 241 pp., Stanford Univ., California, 1978.

Benjamin, M. M., and J. O. Leckie, Multiple-site adsorption of Cd, Cu, Zn and Pb on amorphous iron oxyhydroxide, J. Colloid Interface Sci., 79, 209-221, 1981.

Berube, Y. G., and P. L. de Bruyn, Adsorption at the rutile-solution interface, I: Thermodynamic and experimental study, J. Colloid Interface Sci., 27, 305-318, 1968.

Brunauer, S., P. H. Emmett, and E. Teller, Adsorption of gases in multimolecular layers, J. Phys. Chem., 60, 309-319, 1938.

Buchholtz, M., P. H. Santschi, and W. S. Broecker, Comparison of radiotracer K_d values from batch adsorption experiments with in situ determination in the deep sea using the Manop Lander: the importance of geochemical mechanisms in controlling ion uptake and migration, in Application of Distribution Coefficients to Radiological Assessment Models, edited by T. H. Selby, and C. Myttenaere, pp. 192-206, Elsevier Appl. Sci. Publ., London, 1986.

Buddemeier, R. W., and R. J. Hunt, Transport of colloidal contaminants in groundwater: radionuclide migration at the Nevada Test Site, Appl. Geochem., 3, 535-548, 1988.

Buffle, J., Complexation Reactions in Aquatic Systems: An Analytical Approach, 692 pp., Holsted Press, New York, 1987.

Buffle, J., and R. S. Altmann, Interpretation of metal complexation by heterogeneous complexants, in Aquatic Surface Chemistry: Chemical Processes at the Particle-Water Interface, edited by W. Stumm, pp. 351-384, John Wiley and Sons, New York, 1987.

Csanady, T. S., Mass transfer to and from small particles in the sea, Limnol. Oceanogr., 31, 237-248, 1986.

Davis, J. A., and J. Hem, The surface chemistry of aluminum oxides and hydroxides, in The Environmental Chemistry of Aluminum, edited by G. A. Sposito, pp. 185-219, CRC Press, Baton Rouge, Florida, 1989.

Davis, J. A., and D. B Kent, Surface complexation modeling in aqueous geochemistry, in Mineral-Water Interface Geochemistry, Reviews in Mineralogy, edited by M. F. Hochella, and A. F. White, vol. 23, pp. 177-260, Mineralogical Society of America, Washington, D.C., 1990.

Davis, J. A., and J. O. Leckie, Surface ionization and complexation at the oxide/water interface, II: Surface properties of amorphous iron oxyhydroxide and the sorption of metal ions, J. Colloid Interface Sci., 67, 90-107, 1978.

Degueldre, C., B. Baeyens, W. Goerlich, J. Riga, J. Verbist, and P. Stadelmann, Colloids in water from a subsurface fracture in granitic rock, Grimsel test site, Switzerland, Geochim. Cosmochim. Acta, 53, 603-610, 1989.

Dyal, R. S., and S. B. Hendricks, Total surface of clays in polar liquids as a characteristic index, Soil Sci., 69, 421-432, 1950.

Dzombak, D. A., and F. M. M. Morel, Surface Complexation Modeling: Hydrous Ferric Oxide, 393 pp., John Wiley, New York, 1990.

Eltantawy, I. M., and P. W. Arnold, Reappraisal of ethylene glycol mono-ethyl ether (EGME) method for surface area estimations of clays, J. Soil Sci., 24, 232-238, 1973.

Farley, K. J., and F. M. M. Morel, Role of coagulation in sedimentation kinetics, Environ. Sci. Technol., 20, 187-195, 1986.

Fetl, L., and E. Smolkova, A study of the sorption properties of the "spherosils", J. Chromatogr., 65, 249-254, 1972.

Fripiat, J. J., Surface properties of alumino silicates, Clays Clay Miner., Proc. 12 Natl. Conf. on Clays and Clay Minerals, 327-358, 1964.

Gregg, S. J., and K. S. W. Sing, Adsorption, Surface Area and Porosity, 303 pp., Academic Press, London, 1982.

Hansmann, P. D., and M. A. Anderson, Using electrophoresis in modeling sulphate, selenite and phosphate onto goethite, Environ. Sci. Technol., 19, 544-551, 1985.

Hayes, K. F., Equilibrium, spectroscopic and kinetic studies of ion adsorption at the oxide/aqueous interface, Ph.D. thesis, 260 pp., Stanford Univ., California, 1987.

Hayes, K. F., G. Redden, W. Ela, and J. O. Leckie, Surface complexation models: an evaluation of model parameter estimation using FITEQL and oxide mineral titration data, J. Colloid Interface Sci., in press, 1990.

Hering, J. G., and F. M. M. Morel, Slow coordination reactions in seawater, Geochim. Cosmochim. Acta, 53, 611-618, 1989.

Hingston, E. J., R. J. Atkinson, A. M. Posner, and J. P. Quirk, Specific adsorption of anions on goethite, Nature, 215, 1459-1461, 1968.

Hohl, H., and W. Stumm, Interaction of Pb^{2+} with hydrous α-Al_2O_3, J. Colloid Interface Sci., 55, 281-288, 1976.

Honeyman, B. D., Cation and anion adsorption in binary mixtures of adsorbents: an investigation of the concept of adsorptive additivity, Ph.D. thesis, 365 pp., Stanford Univ., California, 1984.

Honeyman, B. D., and P. H. Santschi, Critical review: metals in aquatic systems, Environ. Sci. Technol., 22, 862-871, 1988.

Honeyman, B. D., and P. H. Santschi, A Brownian-pumping model for oceanic trace metal scavenging evidence from Th isotopes, Deep-Sea Res., 47, 951-992, 1989.

Honeyman, B. D., and P. H. Santschi, Coupling adsorption and particle aggregation: laboratory studies of "colloidal pumping" using ^{59}Fe-labeled hematite, Environ. Sci. Technol., in press, 1991.

Honeyman, B. D., L. S. Balistrieri, and J. W. Murray, Oceanic trace metal scavenging: the importance of particle concentration, Deep-Sea Res., 35, 227-246, 1988.

James, R. O., and G. A. Parks, Characterization of aqueous colloids by their electrical double layer and intrinsic surface chemical properties, Surf. Colloid Sci., 12, 119-216, 1982.

Jannasch, H. W., B. D. Honeyman, L. S. Balistrieri, and J. W. Murray, Kinetics of trace element uptake by marine particles, Geochim. Cosmochim. Acta, 52, 567-577, 1988.

Kent, D. B., and M. Kastner, Mg^{2+} removal in the system Mg^{2+}-amorphous SiO_2-H_2O by adsorption and Mg^{2+}-hydroxysilicate precipitation, Geochim. Cosmochim. Acta, 49, 1123-1136, 1985.

Kepak, F., Behavior of carrier-free radionuclides, in Radionuclide Techniques and Applications, edited by A. Evans, and M. Muramatsu, Chapter 8, pp. 339-374, Marcel-Dekker, Inc., New York, 1977.

Kuwabara, J. S., J. A. Davis, and C. C. Y. Chang, Algal growth response to particle-bound ortho-phosphate and zinc, Limnol. Oceanogr., 31, 503-511, 1986.

Liang, L., The role of surface chemistry in the coagulation of hematite (α-Fe_2O_3) particles, Ph.D. thesis, 159 pp., California Institute of Technology, 1988.

McCave, I. N., Size spectra and aggregation of suspended particles in the deep ocean, Deep Sea Res., 31, 329-352, 1984.

McDowell-Boyer, L. M., J. R. Hunt, and N. Sitar, Particle transport through porous media, Water Resour. Res., 22, 1901-1921, 1986.

Moran, S. B., and R. M. Moore, The distribution of colloidal aluminum and organic carbon in coastal and open ocean waters off Nova Scotia, Geochim. Cosmochim. Acta, 53, 2513-2524, 1989.

Morel, F. M. M., and P. Gschwend, The role of colloids in the partitioning of solutes in natural waters, in Aquatic Surface Chemistry: Chemical Processes at the Particle Water Interface, edited by W. Stumm, pp 405-422, John Wiley, New York, 1987.

Murray, R. S., and J. P. Quirk, Surface area of clays, Langmuir, 6, 122-124, 1990.

Nash, K. L., and G. R. Choppin, Interaction of humic and fulvic acids with Th(IV), J. Inorg. Nucl. Chem., 42, 1045-1050, 1980.

Niven, S. E. H., and R. M. Moore, The effect of natural colloidal matter on the equilibrium sorption of thorium from seawater, in Radionuclides: A Tool for Oceanography, edited by J. Guary, pp. 111-120, Elsevier Applied Science, London and New York, 1988.

Nozaki, Y., H.-S. Yang, and M. Yamada, Scavenging of thorium in the ocean, J. Geophys. Res., 87, 2045-2056, 1987.

Orlandini, K. A., W. R. Penrose, B. R. Harvey, M. B. Lovett, and M. W. Findlay, Colloidal behavior of actinides in an oligotrophic lake, Environ. Sci. Technol., 24, 706-712, 1990.

Papelis, C., K. F. Hayes, and J. O. Leckie, HYDRAQL: A program for the computation of chemical equilibrium composition of aqueous batch systems including surface complexation modeling of ion adsorption at the oxide/solution interface, Tech. Rep. 306, 130 pp., Dept. of Civil Eng., Stanford Univ., California, 1988.

Parfitt, R. L., V. C. Farmer, and J. D. Russell, Adsorption on hydrous oxides: 1. Oxylate and benzoate on goethite, J. Soil Sci., 28, 29-39, 1977.

Penrose, W. R., W. L. Polzer, E. H. Essington, D. M. Nelson, and K. A. Orlandini, Mobility of plutonium and americium through a shallow aquifer in a semiarid region, Environ. Sci. Technol., 24, 228-233, 1990.

Prieve, D. C., and E. Ruckenstein, The double layer interaction between dissimilar ionizable surfaces and its effect of the rate of deposition, J. Colloid Interface Sci., 63, 317-329, 1978.

Santschi, P. H., Y.-H. Li, D. M. Adler, M. Amdurer, J. Bell, and U. P. Nyffeler, The relative mobility of natural (Th, Pb and Po) and fallout (Pu, Am and Cs) radionuclides in the coastal marine environment: results from model ecosystems and Narragansett Bay, Geochim. Cosmochim. Acta, 47, 201-210, 1983.

Santschi, P. H., M. Amdurer, D. Adler, P. O'Hara, Y.-H. Li, and P. Doering, Relative mobility of radioactive trace elements across the sediment-water interface in MERL model ecosystems of Narragansett Bay, J. Mar. Res., 45, 1007-1048, 1987.

Schindler, P. W., Removal of trace metals from the oceans: a zero order model, Thalassia Jugosl., 11, 101-111, 1975.

Schindler, P. W., and H. R. Kamber, Die aciditat von silanolgruppen, Helv. Chem. Acta, 51, 1781-1786, 1968.

Schindler, P. W., B. Furst, R. Dick, and P. U. Wolf, Ligand properties of surface silanol groups. I. Surface complex formation with Fe^{3+}, Cu^{2+} and Pb^{2+}, J. Colloid Interface Sci., 55, 469-475, 1976.

Sheppard, J. C., M. J. Campbell, T. Cheng, and J. A. Kittrick, Retention of radionuclides by mobile humic compounds and soil particles, Environ. Sci. Technol., 11, 1349-1353, 1980.

Sigg, L., and W. Stumm, The interaction of anions and weak acids with the hydrous goethite surface, Colloids Surf., 2, 101-117, 1981.

Sillen, H. G., The physical chemistry of sea water, in Oceanography, edited by M. Sears, pp. 549-581, American Association for the Advancement of Science, Washington, D.C., 1961.

Sposito, G., On the surface complexation model of the oxide-aqueous solution interface, J. Colloid Interface Sci., 91, 329-340, 1983.

Sposito, G., The Surface Chemistry of Soils, 304 pp., Oxford Univ. Press, New York, 1984.

Stumm, W., C. P. Huang, and S. R. Jenkins, Specific chemical interactions affecting the stability of dispersed systems, Croat. Chem. Acta, 42, 223-244, 1970.

Stumm, W., H. Hohl, and F. Dalang, Interaction of metal ions with hydrous oxide surfaces, Croat. Chem. Acta, 48, 491-498, 1976.

Tessier, A., R. Carignan, B. Dubreuil, and F. Rapin, Partitioning of zinc between the water column and the oxic sediments in lakes, Geochim. Cosmochim. Acta, 53, 1511-1522, 1989.

Thomas, G. W., Historical developments in soil chemistry: ion exchange, Soil Sci. Soc. Am. J., 41, 230-238, 1977.

van Olphen, H., Determination of surface areas of clays - evaluation of methods, in Surface Area Determination, edited by D. H. Everett, and R. H. Ottewill, pp. 255-268, Butterworths, London, 1970.

Westall, J. C., and H. Hohl, A comparison of electrostatic models for the oxide/solution interface, Adv. Colloid Interface Sci., 12, 265-294, 1980.

Whitfield, M., and D. R. Turner, The role of particles in regulating the composition of seawater, in Aquatic Surface Chemistry; Chemical Processes at the Particle-Water Interface, edited by W. Stumm, pp. 457-494, Wiley, New York, 1987.

Yasunaga, T., and T. Ikeda, Adsorption-desorption kinetics at the metal-oxide-solution interface studied by relaxation methods, in Geochemical Processes at Mineral Surfaces, edited by J. A. Davis, and K. F. Hayes, American Chemical Society Symposium Series No. 323, Chapter 12, pp. 230-253, Washington, D.C., 1986.

Zachara, J. M., C. C. Ainsworth, C. E. Cowan, and R. L. Schmidt, Adsorption of chromate by subsurface soil horizons, Soil Sci. Soc. Am. J., 53, 418-428, 1989.

Appendix I. SUBJECT INDEX

Appendix II. AUTHOR INDEX

APPENDIX III.

Glossary of Abbreviations used in Text

AA .... atomic absorption
AAS .... atomic absorption spectroscopy, or spectrophotometry
AC .... alternating current
ACS .... American Chemical Society
ADP .... adenosine di-phosphate
AF .... ammonium formate
AFM .... atomic force microscopy
ALOHA .... A Long-term Oligotrophic Habitat Assessment
AMP .... adenosine monophosphate
AMS .... accelerator mass spectrometry
AMU .... atomic mass unit
APDC .... Ammonium pyrrolidine dithio-carbamate
APHA .... American Public Health Association
API .... American Petroleum Institute
APS .... Advanced Photon Source
ATP .... adenosine tri-phosphate
AVS .... acid volatile sulfide
AWI .... Alfred Wegener Institute (for Polar and Marine Research)
BET .... Brunauer, Emmett, Teller
BF .... broad focus
BIO .... Bedford Institute of Oceanography
BNL .... Brookhaven National Laboratory
BSA .... bovine serum albumin
BSE .... backscattered electron detector
BSM .... back scattered mode
BSTFA .... bis(trimethylsilyl)-trifluoroacetamide
CAA .... combined amino acids
CF .... concentration factor
CFF .... cross-flow filtration
CHLA .... chlorophyll a
CHN .... carbon-hydrogen-nitrogen
CI .... chemical ionization
CMS .... Clay Mineral Society
CN .... carbon-nitrogen
CNP .... carbon-nitrogen-phosphorus
CNS .... carbon-nitrogen-sulfur
COC .... concentration of organic colloids, colloidal organic carbon
CP .... carbon phosphorus
CPE .... conventional polyethylene
CPSD .... curved position sensitive detector
CTD .... conductivity-temperature-depth
CV .... coefficient of variation
DAPI 4,6-diamidino-2-phenylindole-2-HCl
DC .... direct current
DCAA .... dissolved combined amino acids
DCI .... desorption chemical ionization
DDW .... distilled, deionized water
DEAE .... diethylaminoethyl
DFAA .... dissolved free amino acids
DIN .... dissolved inorganic nitrogen
DMS .... dimethyl-sulfide
DMSP .... dimethylsulfoniopropionate
DNA .... deoxyribo-nucleic acid
DOC .... dissolved organic carbon
DOM .... dissolved organic matter
DON .... dissolved organic nitrogen
DSRV .... deep submergence research vehicle
DT-DN .... diatoxanthin-diadinoxanthin
DTMS .... deuterated trimethylsilyl
DW .... distilled water
DXRD .... differential x-ray diffraction
ECD .... electron capture detector
EDAX .... energy dispersive x-ray analysis
EDS .... energy dispersive fluorescence spectrometry
EDTA .... ethylene di-amine tetra-acetic acid
EDXA .... energy dispersive x-ray analyzer
EGME .... ethylene glycol monoethyl ether
EI .... electron ionization
EPA .... Environmental Protection Agency
EPM .... extracellular polymer materials
EPR .... East Pacific Rise
ESL .... Environmental Systems laboratory, WHOI
ETS .... electron transport system
EXAFS .... extended x-ray absorption fine structure
FFF .... field flow fractionation
FFFF .... flow field-flow fractionation
FID .... flame ionization detector
FITC .... fluorescein isothiocyanate
FPD .... flame photometric detector
FSU .... Florida State University
FTIR .... Fourier Transform Infrared
GAPS .... Global Applications of Particulate Studies
GC .... gas chromatography
GCMS .... gas chromatography mass spectometry
GC-TCD .... gas chromatography with thermal conductivity detection
GEOSECS .... Geochemical Ocean Sections Study
GFAAS .... graphite furnace atomic absorption spectrophotometry
GPC .... gel permeation chromatography
HAC .... acetic acid
HDPE .... high density polyethylene
HEPA .... high efficiency particulate air filter
HFBBE .... heptafluorobutyryl-n-butyl esters
HFF .... hollow fiber ultrafiltration
HMC .... high magnesium calcite
HMM .... heavy metal mineral
HOT .... Hawaii Ocean Time-series
HPLC .... high performance liquid chromatography
HPTLC .... high performance thin layer chromatography
HT .... high temperature
HTCO .... high temperature catalytic oxidation
HYDRAQL .... a program for the calculation of chemical equilibrium composition of aqueous batch systems
ICDD .... International Center for Diffraction Data
ICES .... International Council for the Exploration of the Sea
ICP .... inductively coupled plasma
ICP-AES .... inductively coupled plasma/ atomic emission spectroscopy
ICPES .... inductively coupled plasma emission spectrometry
ICPMS .... inductively coupled plasma mass spectrometry
ID .... isotope dilution
IDOE .... International Decade of Ocean Exploration.
IEF .... isoelectric focussing
IHSS .... International Humic Substances Society
INAA .... instrumental neutron activation analysis
IR .... infrared, infrared spectroscopy
irm GCMS .... isotope-ratio monitoring gas chromatography-mass spectrometry
IRS .... indented rotating sphere
JCPDS .... Joint Committee for Powder Diffraction Studies

JGOFS-BATS .... Joint Global Ocean Flux Study (at the) Bermuda Atlantic Time-series Station
LC .... liquid chromatogrpahy
LC[FAB]MS .... liquid chromatography [Fast Atom Bombardment] mass spectrometry
LDGO .... Lamont-Doherty Geological Observatory
LFF .... long fine focus
LM .... light microscope
LMC .... low magnesium calcite
LMR .... light microscope, reflected light
LMT .... light microscope, transmitted light
LPE .... linear polyethylene
LPF .... Lorentz polarization factor
LVFS .... large volume filtration system
MARK .... Mid-Atlantic Ridge Kane
MBTH .... the carbohydrate reagent 3-methyl-2-benzothiazolinone hydrazone hydrochloride
MDL .... methanol detection limit, method detection limit, minimum detection limit
MF .... type of Millipore cellulose membrane
MINEQL .... mineral equilibrium computation program
MIT .... Massachusetts Institute of Technology
MP .... macrophotography
MS .... mass spectrometer
MSE .... (ultrasonic disintegrator apparatus?)
MULVFS .... multiple unit large volume in situ filtration system
MW .... molecular weight
NAA .... neutron activation analysis
NAC .... N-acetyl-L-cysteine
NAS .... National Academy of Sciences
NATO .... North Atlantic Treaty Organization
NBS .... National Bureau of Standards
N/C .... nitrogen content
ND .... not detectable
NF .... normal focus
NIOZ .... Netherlands Institute for Sea Research
NIST .... National Institute of Standards and Technology
NIVA .... Norwegian Institute for Water Research
NMR .... nuclear magnetic resonance
NMW .... nominal molecular weight
NOAA .... National Oceanographic and Atmospheric Administration
NPD .... nitrogen-phosphorus detector
NPT .... normal pipe thread
NRLXRF .... Naval Research Laboratory Cosmic Program
NSF .... National Science Foundation
NSLS-BNL .... National Synchrotron Light Source, Brookhaven National Laboratory
O/C .... oxygen content
OCE .... Oceanography Section within National Science Foundation
ODF .... Oceanographic Data Facility (Scripps Institute of Oceanography)
ODP .... Ocean Drilling Project
ODS .... octadecylsilane
OPA .... o-phthaldialdehyde
opal-CT .... cristobalite/tridymite
OSU .... Oregon State University
PAD .... pulsed amperometric detection
PAS .... periodic acid-Schiff (staining)
PC .... particulate carbon
PCAA .... particulate combined amino acids
PCAC .... see PTGC
PCB .... poly-chlorinated biphenyls
PE .... Perkin-Elmer
PEBENOCO .... Pelagic-Benthic Coupling in the Norwegian Coastal Zone
PEEK .... polyetheretherketone
PF .... prefiltered
PFA .... perfluoralkoxy
PIC .... particulate inorganic carbon
PIT .... particle interceptor trap
PMEL .... Pacific Marine Environmental Laboratory
PN .... particulate nitrogen
PNMR .... proton nuclear magnetic resonance
POC .... particulate organic carbon
POI .... particle of interest
POM .... particulate organic matter
PON .... particulate organic nitrogen
POP .... particulate organic phosphorus
POS .... particulate organic sulfur
PP .... particulate phosphorus, polypropylene
PTFE .... polytetrafluoroethylene
PTGC .... Millipore polysulfone cross-flow filter
PTTK .... Millipore polysulfone cross-flow filter
PTHK .... Millipore polysulfone cross-flow filter
PVC .... polyvinyl chloride
PW .... product waters
Py-DCI-MS .... Pyrolysis Desorption Chemical Ionization Mass Spectrometry
Py-GC .... pyrolysis gas chromatography
PY-MS .... pyrolysis mass spectrometry
QA .... quality assurance
RAPPID .... Rotating Automatic Pump for Particulate Inorganics Determination
RCF .... relative centrifugal force
REE .... rare earth elements
REF .... refractory
RG .... reagent grade
RH .... relative humidity
RSD .... relative standard deviation
SCM .... surface complexation model
SCUBA .... self-contained underwater breathing apparatus
SDS .... sodium dodecyl sulfate
SDS-IEF .... SDS isoelectric focusing
SDS-PAGE .... SDS polyacrylamide gel electrophoresis
SEC .... size exclusion chromatography
SEEP .... Shelf Exchange Experiment.
SEM .... scanning electron microscope
SEM-EDS .... scanning electron microscope with an energy dispersive x-ray fluorescence spectrometer
SEM/EMP .... scanning electron microscope coupled with the electron microprobe
SFE .... supercritical fluid extraction
SIMS .... secondary ionization mass spectrometry
SIO-ODF .... Scripps Institution of Oceanography Oceanographic Data Facility
SM .... secondary mode
SOEST .... School of Oceanography and Earth Science and Technology (University of Hawaii)
SPAS .... Selected Particle Analysis System
SPE .... solid phase extraction
SPM .... suspended particulate matter
SRM .... standard reference material
STEM .... scanning transmission electron microscopy
STEM-EDS .... scanning transmission electron microscopy with an energy dispersive x-ray fluorescence spectrometer
STM .... scanning tunneling microscopy
SW .... seawater
SWV .... square wave voltammetry
SXAS .... synchrotron x-ray absorption spectroscopy
SXRD .... synchrotron x-ray diffraction
SXRF .... synchrotron x-ray fluorescence
TAG .... Trans-Atlantic-Geotraverse
TCD .... thermal conductivity detection
TDN .... total dissolved nitrogen
TDP .... total dissolved phosphorus
TEM .... transmission electron microscopy
TEM-EMP .... transmission electron microscope with electron microprobe

TFA .... trifluoroacetic acid
TFF .... tangential flow ultrafiltration
THF .... tetrahydrofuran
TIMS .... thermal ionization mass spectrometry
TLC .... thin layer chromotography
TM .... total mass
TMS .... trimethylsilyl
TOC .... total organic carbon
TPC .... total particulate carbon
TPN .... total particulate nitrogen
TSM .... total suspended matter, thin sheet media
UCM .... unresolved complex mixture
UH/HIMB .... University of Hawaii/Hawaii Institute of Marine Biology
USGS .... United States Geological Survey
UTMSI .... University of Texas Marine Science Institute
UV .... ultraviolet
UV/VIS .... ultraviolet - visible (wavelengths of light)
VENTS .... hydrothermal plume program
VHF .... very high frequency
WCRE .... Warm Core Rings Experiment
WDS .... wavelength dispersive spectrometer
WGMS .... ICES Working Group on Marine Sediments (in relation to pollution)
WHOI .... Woods Hole Oceanographic Institution
WOCE .... World Ocean Circulation Experiment
XANES .... x-ray absorption near edge structure
XRD .... X-ray diffraction
XRF .... X-ray fluorescence
ZAF .... Atomic number absorption fluorescence correction.

APPENDIX IV

List of trade names and manufacturers referenced in this volume

3M
AAII-Technicon
Acrodisc
AESAR (Johnson Mathey)
AKRO
Aldrich Chemical
Aleph Enterprises
American Grease Stick Co.
Amicon
Amicon Centricon
Anopore
Anotec
Antek
Argos
Bakelite
Barnstead Nano-Pure
Beckmann
Belart
Berghof-America
Bev-A-Line (Cole Palmer)
Bevaline
Bio-Gel
Biogel-P
BioRad
Biotroniks
Bodman
Bray
Brownlee
Buna-N
Cahn
Cambridge
Camebax
Canberra
Carl Zeiss-Jena
Carlo Erba
Center for Culture of Marine Phytoplankton
Centricon
Chelex
Chemplast
Chesley
Chirasil-Val
Chopstick
Chromarods
Cole-Parmer
Corvallis Microtechnology
Cuno
Cuno Microwynd
Cyanotech Inc.
Cytofluorograf 2S
Debye-Scherrer
Deep-Sea Power and Light
Dionex
Dominator
Dominator Maskin
Durapore
Dynamax
EG & G
E-Lab
Eldex
Elzone
Environmental Air Control
EPICS
Eppendorf
Ernest F. Fullam Inc.
Europa
Faraday cups
Filtron
Finnigan Incos 50
Finnigan MAT
Fischer
Fisher brand "Minipet"
Fisher Scientific
FIT
Flotec
Fluoromat
Formvar
Fractogel
Freon
Fuess & Reichert
Gelman
General Oceanics
Gerard
Gilson
Gilson Medical Electronics
Go-Flo
Granzow Corp.
Graticules, Ltd.
Hamilton
Heathkit
HEREAUS
Hewlett Packard
Hitachi
Hoffman-LaRoche
HydroBios
Iatron Laboratories
Iatroscan
IBM
IEC
Instruments for Research and Industry
International Products Corp.
Isotech
Jabsco
Jaco
JEOL
Johnson Mathey
J. T. Baker Inc.
Keevex
Kent Meters
Kevex
Kevlar
KIEL
Kimwipe
Kratos
Lab Calc
LaJolla Scientific Co.
LaRoche Regisil
Leeman
Leeman Lab
Lemont
Lexan
Liberty Industries
Link
M-Spurr
Machery-Nagel
Malvern
Mark-7
McLane
McLane Research Laboratories
Meeker
Merck
Mettler
Mg Industries
Micro
Microbio Resources Inc.
Micro-Math
Microsoft
Microwynd
Millipore
Millipore Durapore
Millipore Millipak
Millipore Pellicon
Millipore Sterifil
Milli-Q
Molecular Probes
MSE
Multitrap
Mylar
Nalgene
Niskin
Nitex
Noble
Nonidet
Nova-Pak
NRD, Inc.
NuAire
Nuclear Measurements Corporation
Nuclear Products Co.
Nuclepore
Nylon
Nytex
O. I.
Oceanic Industries
OMS Tech
On-Guard
Osmonics
Oxford
Parafilm
Pasteur
Pecap
Pelagic Electronics
Pelicon
Pellicon
Peltier
Perkin-Elmer
PGT
Philips
Phillips
Pierce
Plexiglas
Polaroid
Porapack
Poretics
Princeton Gamma Tech
PRP (Hamilton)
Pyrex
Raman
Rank
Rank Brothers
Rank Instrument Co.
Regisil
Rubbermaid
Sartorius
Savant
Savillex
Schleicher & Schull
Scientific Products
Scott Paper Co.
Sea Battery
SeaBird
Seal-A-Meal
SeaMar Tec
Seastar
Seastar Instruments
Sephadex
Shimadzu AOC 14
Siemens
Sigma
Sigma Chemical Co.
Silastic
Solvinert by Millipore
Soret
Sovirel
Spectra-Physics
Spherisorb
Staticmaster
Supelco, formerly Fractogel
Super Pellicon
Super-Q
Swagelok
Swin-Lok
Swinnex
Swinney
Technicon
Teflon
Tetko
TFE
Toyopearl TSK
Tracor
TRIGA
Triton
Tygon
U.S. Plastics
Ultrex
Ultrex II
VG
Valco Instrument Co.
Variac
Varian
Viton
VWR
Vycor
Waters Associates
Whatman
Wild
XAD
Zeeman
Zernicke
Zeta Corp.
Ziploc